CHRISTOPHER JON BJERKNES

THE MANUFACTURE
AND SALE OF
SAINT EINSTEIN

IV

Christopher Jon Bjerknes

The Manufacture and Sale of Saint Einstein

Volume IV

Published by
Omnia Veritas Ltd

www.omnia-veritas.com

© Omnia Veritas Ltd - Christopher Jon Bjerknes - 2019.

All Rights Reserved. No part of this publication may be reproduced, distributed, or transmitted in any form or by any means, including photocopying, recording, or other electronic or mechanical methods, without the prior written permission of the publisher, except in the case of brief quotations embodied in critical reviews and certain other noncommercial uses permitted by copyright law.

Table of Contents:

7 Nazism is Zionism --- 7

7.5 All the Best Zionists are Anti-Semites --- 7
7.5.3 Zionists and Communists Delight in Massive Human Sacrifices to the Jewish Messianic Cause --- 57
7.5.4 Einstein Lulls Jews into Complacency—The Zionist Trap --- 60
7.5.4.1 Depressions Make for Fertile Ground for Anti-Semitic Zionist Dictators --- 70
7.5.4.2 Einstein a Subtle Hitler Apologist --- 71
7.5.5 Einstein's Seething Racist Hatred and Rabid Nationalism --- 73
7.5.6 The Final Solution of the Jewish Question is Zionism, but the Final Solution of the German Question is Extermination --- 89
7.6 The Carrot and the Stick --- 116
7.7 British Zionists, in Collaboration with Nazi Zionists, in Collaboration with Palestinian Zionists, Ensured that the Jews of Continental Europe Would Find No Sanctuary Before the War Ended --- 128
7.8 Documented Collaboration Between the Palestinian Zionists and the Zionist Nazis --- 142

8 How the Jews Made the British into Zionists --- 191

8.1 Introduction --- 191
8.2 The Rothschilds and Disraeli Lead the British Down the Garden Path to Palestine --- 191
8.3 Jews Provoke Perpetual War --- 234
8.4 Jewish World Government—A Prophetic Desire --- 246
8.5 Puritans and Protestants Serve Jewish Interests --- 258
8.6 The Planned Apocalypse --- 262
8.7 Cabalistic Jews Calling Themselves Christian Condition the British to Assist in Their Own Demise—Rothschild Makes an Open Bid to Become the Messiah --- 270
8.7.1 The "British-Israel" Deceit --- 304
8.7.2 For Centuries, England is Flooded with Warmongering Zionist Propaganda --- 308
8.7.3 As a Good Cabalist Jew, David Hartley Conditions Christians to Welcome Martyrdom for the Sake of the Jews --- 330
8.7.3.1 Jewish Revolutionaries and Napoleon the Messiah Emancipate the Jews --- 334
8.7.3.2 Hitler Accomplishes for the Zionists What Napoleon Could Not --- 339
8.7.3.3 Zionists Develop a Strategy Which Culminates in the Nazis and the Holocaust as Means to Attain the "Jewish State" --- 381

9 The Priority Myth --- 404

9.1 Introduction --- 404
9.2 Opinions of Einstein and "His" Work --- 417

9.3 THE ÆTHER -- 512
9.4 THE SO-CALLED "*LORENTZ* TRANSFORMATION" -- 522
   9.4.1 Woldemar Voigt's Space-Time Transformation -------------------------- 523
   9.4.2 Length Contraction -- 525
      9.4.2.1 Dynamic Length Contraction -------------------------------------- 529
      9.4.2.2 Kinematic Length Contraction ------------------------------------ 529
   9.4.3 Time Dilatation --- 530
   9.4.4 The Final Form of the Transformation ---------------------------------- 530
   9.4.5 Einstein's Fudge -- 546
   9.4.6 Einstein Begged the Question --- 546
9.5 THE "TWO POSTULATES" --- 567
   9.5.1 The "Principle of Relativity" -- 568
   9.5.2 The "Light Postulate" --- 578

7 NAZISM IS ZIONISM

Zionists have always employed the bogey of "anti-Semitism" to force Jews into segregation. After thousands of years of planning and work, the Jewish bankers had finally accumulated enough wealth to buy Palestine and destroy the Gentile world in fulfillment of Jewish Messianic prophecy. They only lacked one resource needed to become King of the Jews, the Holy Messiah. That one last necessary ingredient for fulfillment of the prophecies of the End Times was the Jewish People, the majority of whom rejected Zionism. The Jewish bankers had an ancient solution for that problem. They manufactured an anti-Semitic dictator who segregated the Jews and filled them with the fear of God. Palestine was for the fearful remnant. Those who would not obey were to have their necks broken and be thrown into the well.

"The way I see it, the fact of the Jews' racial peculiarity will necessarily influence their social relations with non-Jews. The conclusions which—in my opinion—the Jews should draw is to become more aware of their peculiarity in their social way of life and to recognize their own cultural contributions. First of all, they would have to show a certain noble reservedness and not be so eager to mix socially—of which others want little or nothing. On the other hand, anti-Semitism in Germany also has consequences that, from a Jewish point of view, should be welcomed. I believe German Jewry owes its continued existence to anti-Semitism."—ALBERT EINSTEIN[1]

"Hitler will be forgotten in a few years, but he will have a beautiful monument in Palestine. You know the coming of the Nazis was rather a welcome thing. So many of our German Jews were hovering between two coasts; so many of them were riding the treacherous current between the Scylla of assimilation and the Charybdis of a nodding acquaintance with Jewish things. Thousands who seemed to be completely lost to Judaism were brought back to the fold by Hitler, and for that I am personally very grateful to him.'"—EMIL LUDWIG[2]

"[H]ad I been a Jew, I would have been a fanatical Zionist. I could not imagine being anything else. In fact, I would have been the most ardent Zionist imaginable."—ADOLF EICHMANN[3]

7.5 All the Best Zionists are Anti-Semites

The worst enemy of the common Jew has always been the Zionist.

In 1932, Einstein stated, referring to the "deplorably high development of nationalism everywhere"—his own rabid Zionism excepted,

---

[1]. A. Einstein, A. Engel translator, "How I became a Zionist", *The Collected Papers of Albert Einstein*, Volume 7, Document 57, Princeton University Press, (2002), pp. 234-235, at 235.
[2]. M. Steinglass, "Emil Ludwig before the Judge", *American Jewish Times*, (April, 1936), p. 35; *as quoted in:* L. Brenner, *Zionism in the Age of the Dictators*, Chapter 6, Croom Helm, London, L. Hill, Westport, Connecticut, (1983), p. 59.
[3]. A. Eichmann, "Eichmann Tells His Own Damning Story", *Life Magazine*, Volume 49, Number 22, (28 November 1960), pp. 19-25, 101-112; at 22.

"The introduction of compulsory service is therefore, to my mind, the prime cause of the moral collapse of the white race, which seriously threatens not merely the survival of our civilization but our very existence. This curse, along with great social blessings, started with the French Revolution, and before long dragged all the other nations in its train."[4]

Einstein complained to Lorentz on 12 January 1920 that even well-educated persons fell victim to "the illiberal nationalistic standpoint."[5] Einstein called "Nationalism" an "ugly name".[6] Einstein's Zionist hypocrisy did not go unnoticed. He was asked why he stood firmly against Gentile nationalism, while making Zionist nationalism his primary purpose in life. According to Thüring, the *Jüdische Presse* reported on 29 May 1929,

"Man fragte [Einstein], warum er als Verfechter aller internationalen Interessen, als Gegner aller nationalistischen Bestrebungen die jüdische nationale Sache zu seiner eigenen mache. Er erklärte seinen Standpunkt durch ein Gleichnis: Wer einen rechten Arm hat und davon spricht und immer davon spricht, ist ein Narr. Wem aber rechte Arm fehlt, der darf alles tun, um sich das fehlende Glied zu ersetzen. Daher sei er in einer Welt, in der jedes Volk die Bedingungen des nationalen Lebens hat, ein Feind des Nationalismus, als Jude aber ein Anhänger der jüdisch-nationalen Idee, weil den Juden die notwendige und natürliche Voraussetzung ihres nationalen Lebens fehlt."

This clearly elucidates Einstein's nationalistic perspective, which mirrored the Nazis' nationalistic perspective. The Nazis simply pursued the same false reasoning as Einstein and asserted that their right arm was infected with Einstein's self-described foreign and disloyal nationalists. Einstein agreed with the Nazis and saw them as the salvation of the Jews.

Therein lies the potential danger of Einstein's segregationism. Segregationist nationalism is bound to lead to genocidal nationalism. Einstein's tacit premise that citizenry and nationhood be based on ancient territory, ethnicity, race and religion—on *Blut und Boden*, instead of the sovereignty of a group of living persons in a territory, whether homogenous or heterogeneous in its ethnicities and religions, was racist bigotry—commonly held bigotry, but bigotry nonetheless. Einstein's Zionist nationalism, which was no different from Nazi nationalism, would disconnect Jews around the world from the nations in which they were citizens. His racist nationalism definitely did not conform with

---

[4]. A. Einstein, translated by A. Harris, "The Disarmament Conference of 1932. I." *The World As I See It*, Citadel, New York, (1993), pp. 59-60.
[5]. Letter from A. Einstein to H. A. Lorentz of 12 January 1920, English translation by A. Hentschel, *The Collected Papers of Albert Einstein*, Volume 9, Document 256, Princeton University Press, (2004), pp. 214-215, at 214.
[6]. A. Einstein, *The World As I See It*, Citadel, New York, (1993), p. 109.

his internationalist views, which were premised upon a community of nations, which implies a human family. In addition, Einstein voluntarily amputated his right arm, though he pretended that his self-inflicted wound was a congenital defect. Einstein was born a German, not a Palestinian. But Einstein's hypocrisy, his system of double standards, his desire that the Gentiles be consumed in wars and that the Jews reestablish a State and rule the world, were nothing new. They were Judaism.

Einstein was an advocate of world government and a segregated "Jewish State". While this seemed a contradiction to many, including many Jews, especially many secular Jews, Einstein was merely expressing his loyalty to Jewish Messianic myth. Given Einstein's racist Zionism, it is clear that Einstein wished for a day when Jews would rule a world devoid of Gentile government and that they would be segregated from, and reign over, the "Goyim", to use Einstein's term. "Internationalism" was a code word for a world devoid of Gentile government—a Jewish Messianic prophecy. "Zionism" was a code word for Jewish supremacy reigning over the world from Jerusalem in the Jewish Nation. Einstein's "Internationalism" and Einstein's "Zionism" need no reconciliation, they are one in the same objective—Judaism. Rather those who are confused by Einstein's apparent contradictions need only read the Hebrew Bible, where the Jewish prophets tell the Jews to reconstruct the Jewish State and at the same time destroy all the Gentile governments of the world.

After World War II had ended, Einstein's friend Peter A. Bucky also questioned the apparent contradiction in Einstein's political philosophy. Bucky asked Einstein how he reconciled his Zionism with his anti-Nationalism. As a good racist Zionist Jew was wont to do, Einstein exploited modern anti-Semitism to legitimize racist Jewish Nationalism which is at least 2,500 years old,

> "I think that [nationalism] is justified in this special case because the world has forced the Jews to entrench themselves with the continued existence of anti-Semitism."[7]

Einstein felt that Jews owed anti-Semites a great debt of appreciation for forcing Jews to "entrench themselves". He must also have known that the Zionists created the Nazis to force reluctant assimilating Jews to Palestine. Einstein dreaded a world without anti-Semitism, without segregation and without segregated racist Jews like himself. The incentive for Jews to create anti-Semitism is clear. There is abundant evidence that leading Jews have again and again down through history created and sponsored anti-Semitism. In the racist Zionist's view, racist segregationist Judaism and the Jewish tribe cannot continue to exist without manufacturing anti-Semitism to keep them alive.

Given that the vast majority of German Jews during Einstein's lifetime vehemently opposed his bigotry, it is especially odd that Einstein was so

---

[7]. P. A. Bucky, Einstein, and A. G. Weakland, *The Private Albert Einstein*, Andrews and McMeel, Kansas City, (1992), pp. 88-89.

unenlightened and so racist. His own children were assimilated Jews, and he hated them for it.[8] Whereas most German Jews considered the racism of Zionist Eastern Jews primitive and uncivilized, Einstein considered assimilation uncivilized and inhuman, because Einstein believed that European Gentiles were sub-human and incapable of civilization. His Zionist sponsors created wars for, among other things, the purpose of discrediting Gentile government. Einstein owed his fame to Zionists, who used him to publicize their cause. Einstein was more loyal to the Zionists' racism, than he was to his own children. Racism buttered Einstein's bread, his children wanted eat it, though he wouldn't let them—they were sub-human. Fellow Jewish racists kept Einstein in the spotlight and shielded him from criticism.

Einstein, himself, echoed and endorsed the views of the anti-Semites in an interview in which he again revealed himself to be a racist and a segregationist. Zionists intentionally provoked and sought to inspire anti-Semitism, and anti-Semites welcomed the openly racist positions of the Zionists.[9] Einstein went along with the crowd of prominent political Zionists who openly stated that anti-Semitism is welcomed, encouraged and useful to the Zionists. They based their myth on Spinoza's declaration that emancipation leads to assimilation and that the Jews only exist in modern times because glorious anti-Semitism kept them segregated.[10]

Prominent Zionist and author of the *Encyclopaedia Judaica; das Judentum in Geschichte und Gegenwart,* Jakob Klatzkin stated in 1925,

> "The national viewpoint taught us to understand the true nature of antisemitism, and this understanding widens the horizons of our national outlook. [***] In the age of enlightenment antisemitism was included among the phenomena that are likely to disappear along with other forms of prejudice and iniquity. The antisemites, so the rule stated, were the laggard elements in the march of progress. Hence, our fate is dependent on the advance of human culture, and its victory is our victory. [***] In the period of Zionism, we learned that antisemitism was a psychic-social phenomenon that derives from our existence as a nation within a nation. Hence, it cannot change, until we attain our national end. But if Zionism had fully understood its own implications, it would have arrived, not merely as a psycho-sociological explanation of this phenomenon, but also as a justification of it. It is right to protest against its crude expressions, but we are unjust to it and distort its nature so long as we do not recognize that essentially it is a

---

8. P. A. Bucky, Einstein, and A. G. Weakland, *The Private Albert Einstein*, Andrews and McMeel, Kansas City, (1992), pp. 107-108.
9. *Cf.* K. J. Herrmann, "Historical Perspectives on Political Zionism and Antisemitism", *Zionism & Racism: Proceedings of an International Symposium*, International Organization for the Elimination of All Forms of Racial Discrimination, Tripoli, (1977), pp. 197-208.
10. J. Wellhausen, *Sketch of the History of Israel and Judah*, Third Edition, Adam and Charles Black, London, (1891), pp. 201-203.

defense of the integrity of a nation, in whose throat we are stuck, neither to be swallowed nor to be expelled. [***] And when we are unjust to this phenomenon, we are unfair to our own people. If we do not admit the rightfulness of antisemitism, we deny the rightfulness of our own nationalism. If our people is deserving and willing to live its own national life, then it is an alien body thrust into the nations among whom it lives, an alien body that insists on its own distinctive identity, reducing the domain of their life. It is right, therefore, that they should fight against us for their national integrity. [***] Know this, that it is a good sign for us that the nations of the world combat us. It is proof that our national image is not yet utterly blurred, our alienism is still felt. If the war against us should cease or be weakened, it would indicate that our image has become indistinct and our alienism softened. We shall not obtain equality of rights anywhere save at the price of an explicit or implied declaration that we are no longer a national body, but part of the body of the host-nation; or that we are willing to assimilate and become part of it. [***] Instead of establishing societies for defense against the antisemites, who want to reduce our rights, we should establish societies for defense against our friends who desire to defend our rights. [***] When Moses came to redeem the children of Israel, their leaders said to him, 'You have made our odor evil in the eyes of Pharaoh and in the eyes of his servants, giving them a sword with which to kill us.' Nevertheless, Moses persisted in worsening the situation of the people, and he saved them."[11]

Klaus J. Herrmann has collected a great deal of evidence related to Zionist racism in his presentation, "Historical Perspectives on Political Zionism and Antisemitism", *Zionism & Racism: Proceedings of an International Symposium*, International Organization for the Elimination of All Forms of Racial Discrimination, Tripoli, (1977), pp. 197-210. At page 197, Herrmann states, [quoting Constantin Brunner, *Der Judenhass und die Juden*, Berlin, (1918), p. 112; and Ernst Ludwig Pinner, "Meine Abkehr vom Zionismus", *Los vom Zionismus*, J. Kauffmann, Frankfurt, (1928), pp. 32-33; and referencing Houston Stewart Chamberlain, *Die Grundlagen des neunzehnten Jahrhunderts*, F. A. Bruckmann, München, (1899), English translation by John Lees, *Foundations of the Nineteenth Century*, John Lane, New York, (1910)—*see:* F. Kahn, "H. St. Chamberlain (Eine Charakteristik)", *Jüdische Rundschau*, Volume 25, Nummer 63/64, (10 September 1920), pp. 499-500, for a contemporary view of the impact on Jews of Chamberlain's much-read book. His book was popular among Zionists and the English translation of it received a long and favorable review in the *Times Literary Supplement* of 15 December 1910, pp. 500-501.]:

"Jews,' wrote Brunner, 'have been taken in by the racial theories of the Jew-

---

11. J. Klatzkin, *Tehumim: Ma'amarim*, Devir, Berlin, (1925). English translation in J. B. Agus, *The Meaning of Jewish History*, Volume 2, Abelard-Schuman, New York, (1963), pp. 425-426.

haters;' and he accused the Zionists of having taken as their teacher the notorious racist and forger of scholarly documentation Houston Stewart Chamberlain, whose 'confused nonsense revelations' had been 'restammered' in a Zionist book on the subject of race. 'How could Germans of Jewish background begin to talk of a Jewish nation, and to fashion of the worst calumny the dream of their greatest nonsense!'[1]

One of Brunner's disciples, Ernst Ludwig Pinner, who had been a Zionist earlier, bitterly accused the Zionists of having

> taken up Europe's newest nonsense, namely racial theory as the justification of national emotion. Racial arrogance and racial hate poison national emotion, as did previously religious arrogance and religious hatred. Today it is race which is exalted as the banner in whose name everything is justified.

Pinner also designated the Zionists as 'Jews infected by the sickness of racial insanity ... because, similar to the Jew-haters, they drew political consequences out of race-consciousness.'[2] Pinner did absolve Zionists of 'preaching arrogance and hatred;'[3] whether or not he would have done so in later years remains open to conjecture."

At pages 204-205, Klaus J. Herrmann quotes the Zionist ideologist Jakob Klatzkin who stated, among other things, in his book of 1921 *Krisis und Entscheidung im Judentum; der Probleme des modernen Judentums*, Second Enlarged Edition, Jüdischer Verlag, Berlin, pages 61-63, and 118:

> "[I applaud] the contribution of our enemies in the continuance of Jewry in eastern Europe. [***] We ought to be thankful to our oppressors that they closed the gates of assimilation to us and took care that our people were concentrated and not dispersed, segregatedly united and not diffusedly mixed [***] One ought to investigate in the West and note the great share which antisemitism had in the continuance of Jewry and in all the emotions and movements of our national rebirth . [***] Truly our enemies have done much for the strengthening of Judaism in the diaspora . [***] Experience teaches that the liberals have understood better than the antisemites how to destroy us as a nation. [***] We are, in a word, naturally foreigners; we are an alien nation in your midst and we want to remain one."[12]

"Man vergegenwärtige sich, wie groß der Anteil unserer Feinde am

---

12. K. J. Herrmann, "Historical Perspectives on Political Zionism and Antisemitism", *Zionism & Racism: Proceedings of an International Symposium*, International Organization for the Elimination of All Forms of Racial Discrimination, Tripoli, (1977), pp. 197-210, at 204-205. A lengthy quotation from Klatzkin, in English translation, appears in: M. Menuhin, *The Decadence of Judaism in Our Time*, Exposition Press, New York, (1965), pp. 482-483.

Fortbestand des Judentums im Osten ist. [...] Wir müßten beinahe unseren Bedrängern dankbar sein, wenn sie die Tore der Assimilation vor uns schlossen und dafür Sorge trugen, daß unsere Volksmassen konzentriert und nicht zerstreut, abgesondert geeint und nicht zerklüftet vermischt werden[... .] Man untersuche es im Westen, welchen hohen Anteil der Antisemitismus am Fortbestand des Judentums und an all den Regungen und Bewegungen unserer nationalen Wiedergeburt hat. [...] Wahrlich, unsere Feinde haben viel zur Stärkung des Judentums in der Diaspora beigetragen. [...] Und die Erfahrung lehrt, daß die Liberalen es besser als die Antisemiten verstanden haben, uns als Volk zu vernichten. [...] Wir sind schlechthin Wesensfremde, sind — wir müssen es immer wiederholen — ein Fremdvolk in eurer Mitte und wollen es auch bleiben."

Some Jews, and some critics of the Jews, have for thousands of years asserted that Jews always form a separate state within the nations they inhabit. This they attribute to the Jewish religion, with its one God to rule over all—Jews being the chosen people who will one day receive the Messiah who will assist them in ruling the world after all other nations are destroyed, which fatalistic belief system inspires the nationalism many Jews have expressed in the Diaspora.

When Zionists like Herzl, Klatzkin and Rabbi Meir Bar-Ilan, who stated in 1922,

"We have no 'church' that is not also concerned with matters of state, just as we have no state which is not also concerned with 'church' matters—in Jewish life these are not two separate spheres."[13]

confirmed that these ancient religious, nationalistic and political aspirations were current in modern Europe, where Jews had been emancipated, it caused many to view Jews not only with suspicion, but with contempt, most especially so because radical revolutionary organizations were often led by, and populated with, Jews in disproportionate numbers to Gentiles. Many leading figures warned the public that the Bolshevik Jews were seeking world domination. They wanted to end the immigration of Eastern European Jews to Germany, and to expel Eastern European Jews from Germany. The Bolshevik Jews had already conducted successful, though short-lived, revolutions in German territory. Many of the Jews emigrating to Germany from the East were the descendants of the Frankists, who had pledged themselves to destroy the Gentile nations by means of deception and revolution. Frankist Jews were often crypto-Jews who hid their Jewish ethnicity in order to deceive Christians who might not otherwise trust them, to place the blame for their actions on other peoples so as to cause an unjust hatred towards those innocent peoples, and to prevent a backlash against

---

[13]. M. Bar-Ilan, *Kitve Rabi Me'ir Bar-Ilan*, Volume 1, Mosad ha-Rav Kuk, Jerusalem, (1950), pp. 5-16; English translation in A. Hertzberg, *The Zionist Idea*, Harper Torchbooks, New York, (1959), pp. 548-555, at 548.

Jews for the vile actions Jews were taking against other peoples. The Talmud teaches the Jews that they can sin against others with immunity if they hide the fact that they are Jewish such that Jews will not be attacked in retaliation. *Moed Katan* 17a states,

> "R' IL'AI SAYS: [***] IF A PERSON SEES THAT HIS evil INCLINATION IS OVERWHELMING HIM, [***] HE SHOULD GO TO A PLACE WHERE THEY DO NOT RECOGNIZE HIM [***] AND CLOTHE HIMSELF IN BLACK AND WRAP HIMSELF IN BLACK, [***] AND HE SHOULD DO WHAT HIS HEART DESIRES, [***] AND HE SHOULD NOT DESECRATE THE NAME OF HEAVEN OPENLY."[14]

An alternative translation:

> "For R. Il'ai says, If one sees that his [evil] *yezer*[5] is gaining sway over him, let him go away where he is not known; let him put on sordid[6] clothes, don a sordid wrap and do the sordid deed that his heart desires rather than profane the name of Heaven openly.'"[15]

Albert Einstein stated in the *Berliner Tageblatt* on 30 December 1919,

> "It is quite likely that there are Bolshevist agents in Germany, but they undoubtedly hold foreign passports, have at their disposal ample funds and cannot be seized by any administrative measures. The big profiteers among the Eastern European Jews have certainly, long ago, taken precautions to elude arrest by officials. The only [Jews] affected would be *those poor and unfortunate ones*, who in recent months made their way to Germany under inhumane privations, in order to look for work here."[16]

Albert Einstein was himself a racist; and, therefore, a hypocrite when criticizing the racism of others. John Stachel wrote,

> "While he lived in Germany, however, Einstein seems to have accepted the then-prevalent racist mode of thought, often invoking such concepts as

---

[14]. *Moed Katan* 17a, Rabbi Y. S. Schorr, *et al.*, Editors, "Tractate Moed Katan", *Talmud Bavli: the Schottenstein edition: the Gemara: the classic Vilna edition, with an annotated, interpretive elucidation, as an aid to Talmud study / elucidated by a team of Torah scholars under the general editorship of Hersh Goldwurm and Nosson Scherman.*, Volume 21, Mesorah Publications, Ltd., Brooklyn, New York, (1999), p. 17a^2.

[15]. Mo'ed Katan, Rabbi I. Epstein, Editor, "Seder Mo'ed", *The Babylonian Talmud*, Volume 14, The Soncino Press, (1938), p. 107.

[16]. A. Einstein, translated by A. Engel, *The Collected Papers of Albert Einstein*, Volume 7, Document 29, Princeton University Press, (2002), pp. 110-111.

'race' and 'instinct,' and the idea that the Jews form a race."[17]
On 8 July 1901, Einstein wrote to Winteler,

"There is no exaggeration in what you said about the German professors. I have got to know another sad specimen of this kind — one of the foremost physicists of Germany."[18]

Einstein wrote to Besso sometime after 1 January 1914,

"A free, unprejudiced look is not at all characteristic of the (adult) Germans (blinders!)."[19]

After the war Einstein and some of his friends alluded to much earlier conversations with Einstein where he had correctly predicted the eventual outcome of the war. In his diaries, Romain Rolland recorded his conversations with Einstein in Switzerland at their meeting of 16 September 1915,

"What I hear from [Einstein] is not exactly encouraging, for it shows the impossibility of arriving at a lasting peace with Germany without first totally crushing it. Einstein says the situation looks to him far less favorable than a few months back. The victories over Russia have reawakened German arrogance and appetite. The word 'greedy' seems to Einstein best to characterize Germany. [***] Einstein does not expect any renewal of Germany out of itself; it lacks the energy for it, and the boldness for initiative. He hopes for a victory of the Allies, which would smash the power of Prussia and the dynasty... . Einstein and Zangger dream of a divided Germany—on the one side Southern Germany and Austria, on the other side Prussia. [***] We speak of the deliberate blindness and the lack of psychology in the Germans."[20]

Jews often sought to Balkanize nations so as to weaken the power of any faction within a nation and to created perpetual agitation between the nations which could be exploited for profit and other Jewish gains. For example, the

---

17. J. Stachel, "Einstein's Jewish Identity", *Einstein from 'B' to 'Z'*, Birkhäuser, Boston, Basel, Berlin, (2002), pp. 57-83, at 68.
18. A. Einstein to J. Winteler, English translation by A. Beck, *The Collected Papers of Albert Einstein*, Volume 1, Document 115, Princeton University Press, (1987), pp. 176-177, at 177.
19. A. Einstein, English translation by A. Beck, *The Collected Papers of Albert Einstein*, Volume 5, Document 499, Princeton University Press, (1995), pp. 373-374, at 374.
20. R. Romain, *La Conscience de l'Europe*, Volume 1, pp. 696ff. English translation from A. Fölsing, *Albert Einstein: A Biography*, Viking, New York, (1997), pp. 365-367. *See also:* Letter from A. Einstein to R. Romain of 15 September 1915, *The Collected Papers of Albert Einstein*, Volume 8, Document 118, Princeton University Press, (1998); **and** Letter from A. Einstein to R. Romain of 22 August 1917, *The Collected Papers of Albert Einstein*, Volume 8, Document 374, Princeton University Press, (1998).

Rothschilds created the American Civil War and profited from the debts it generated. They hoped to divide America into two nations and to pit these against one another. They were successful. Jews had long been pitting North German Protestants against South German and Austrian Catholics. Jews were the motive force behind the *Kulturkampf*. After creating these divides and promoting perpetual agitations amongst neighbors, Jewry could then fund one side against the other to destroy it whenever Jewry decided to wreck a given nation.

Einstein's dreams during the First World War remind one of the "Carthaginian Peace" of the Henry Morgenthau, Jr. plan for the destruction of Germany following the Second World War. Morgenthau worked with Lord Cherwell (Frederick Alexander Lindemann), Churchill's friend and advisor, who planned to bomb German civilian populations into submission. Lindemann studied under Einstein's friend, Walther Nernst, who worked with Fritz Haber, a Jewish developer of poisonous gas. James Bacque argues that the Allies, under the direction of General Eisenhower, starved hundreds of thousands, if not millions of German prisoners of war to death. Dwight David Eisenhower was called "the terrible Swedish-Jew" in his yearbook for West Point, *The 1915 Howitzer*, West Point, New York, (1915), p. 80. He was also called "Ike", as in... Eisenhower? The Soviets also abused countless German POW's after the Second World War.[21]

Einstein often spoke in genocidal and racist terms against Germany, and for the Jews and England, and he betrayed Germany before, during and after the war. Einstein wrote to Paul Ehrenfest on 22 March 1919,

> "[The Allied Powers] whose victory during the war I had felt would be by far the lesser evil are now proving to be *only slightly* the lesser evil. [***] I get most joy from the emergence of the Jewish state in Palestine. It does seem to me that our kinfolk really are more sympathetic (at least less brutal) than these horrid Europeans. Perhaps things can only improve if only the Chinese are left, who refer to all Europeans with the collective noun 'bandits.'"[22]

While responsible people were trying to preserve some sanity in the turbulent period following World War I, Zionists like Albert Einstein sought to validate and encourage the racism of anti-Semites. The Dreyfus Affair taught them that anti-Semitism had a powerful effect to unite Jews around the world. The Zionists were afraid that the "Jewish race" was disappearing through assimilation. They wanted to use anti-Semitism to force the segregation of Jews from Gentiles and to unite Jews, and thereby preserve the "Jewish race". They

---

21. J. Bacque, *Other Losses: An Investigation into the Mass Deaths of German Prisoners at the Hands of the French and Americans after World War II*, Stoddart,Toronto, (1989).
22. Letter from A. Einstein to Paul Ehrenfest of 22 March 1919, English translation by A. Hentschel, *The Collected Papers of Albert Einstein*, Volume 9, Document 10, Princeton Univsersity Press, (2004), pp. 9-10, at 10.

hoped that if they put a Hitler into power—as Zionists had done in the past, they could use him to herd up the Jews and force the Jews into Palestine against their will. This would also help the Zionists to inspire distrust and contempt for Gentile government, while giving the Zionists the moral high-ground in international affairs, despite the fact that the Zionists were secretly behind the atrocities. Theodor Herzl wrote in his book *The Jewish State*,

> "Oppression and persecution cannot exterminate us. No nation on earth has survived such struggles and sufferings as we have gone through. Jew-baiting has merely stripped off our weaklings; the strong among us were invariably true to their race when persecution broke out against them. This attitude was most clearly apparent in the period immediately following the emancipation of the Jews. Later on, those who rose to a higher degree of intelligence and to a better worldly position lost their communal feeling to a very great extent. Wherever our political well-being has lasted for any length of time, we have assimilated with our surroundings. I think this is not discreditable. Hence, the statesman who would wish to see a Jewish strain in his nation would have to provide for the duration of our political well-being; and even Bismarck could not do that. [***] The Governments of all countries scourged by Anti-Semitism will serve their own interests in assisting us to obtain the sovereignty we want. [***] Great exertions will not be necessary to spur on the movement. Anti-Semites provide the requisite impetus. They need only do what they did before, and then they will create a love of emigration where it did not previously exist, and strengthen it where it existed before. [***] I imagine that Governments will, either voluntarily or under pressure from the Anti-Semites, pay certain attention to this scheme; and they may perhaps actually receive it here and there with a sympathy which they will also show to the Society of Jews."[23]

Albert Einstein wrote to Max Born on 9 November 1919, and encouraged anti-Semitism and advocated segregationism (one must wonder what rôle Albert's increasing racism played in his divorce from Mileva Marić—a Gentile Serb),

> "Antisemitism must be seen as a real thing, based on true hereditary qualities, even if for us Jews it is often unpleasant. I could well imagine that I myself would choose a Jew as my companion, given the choice. On the other hand I would consider it reasonable for the Jews themselves to collect the money to support Jewish research workers outside the universities and to provide them with teaching opportunities."[24]

In 1933, the Zionists publicly declared their allegiance to the Nazis. They

---

[23]. T. Herzl, *A Jewish State: An Attempt at a Modern Solution of the Jewish Question*, The Maccabæan Publishing Co., New York, (1904), pp. 5-6, 25, 68, 93.
[24]. M. Born, *The Born-Einstein Letters*, Walker and Company, New York, (1971), p. 16.

wrote in the *Jüdische Rundshau* on 13 June 1933,

> "Zionism recognizes the existence of the Jewish question and wants to solve it in a generous and constructive manner. For this purpose, it wants to enlist the aid of all peoples; those who are friendly to the Jews as well as those who are hostile to them, since according to its conception, this is not a question of sentimentality, but one dealing with a real problem in whose solution all peoples are interested."[25]

On 21 June 1933, the Zionists issued a declaration of their position with respect to the Nazi régime, in which they expressed a belief in the legitimacy of the Nazis' racist belief system and condemned the anti-Fascist forces.[26]

Michele Besso wrote that it might have been Albert Einstein's racism and bigotry which caused him to separate from his first wife Mileva Marić in 1914. Besso wrote to Einstein on 17 January 1928,

> "[...]perhaps it is due in part to me, with my defense of Judaism and the Jewish family, that your family life took the turn that it did, and that I had to bring Mileva from Berlin to Zurich[.]"[27]

The hypocrisy of racist Zionists often manifested itself in this way. Many had "intermarried". Racist Zionist Moses Hess was married to a Christian Gentile prostitute named Sybille Pritsch.

Einstein may have been affected by his mother's early racist opposition to his relationship with Marić. Another factor in the Einsteins' divorce was, of course, Albert's incestuous relationship with his cousin Else Einstein, and his desire to bed her daughters, as well as his general promiscuity. Albert Einstein opposed his sister Maja's marriage to Gentile Paul Winteler on racist grounds, and Albert thought they should divorce. Albert Einstein wrote to Michele Besso on 12 December 1919, "No mixed marriages are any good (Anna says: oh!)"[28] Besso, himself, was married to a Gentile, Anna Besso-Winteler. Denis Brian

---

25. English translation in: K. Polkehn, "The Secret Contacts: Zionism and Nazi Germany, 1933-1941", *Journal of Palestine Studies*, Volume 5, Number 3/4, (Spring-Summer, 1976), pp. 54-82, at 59.
26. L. S. Dawidowicz, "The Zionist Federation of Germany Addresses the New German State", *A Holocaust Reader*, Behrman House, Inc., West Orange, New Jersey, (1976), pp. 150-155. *See also:* H. Tramer, Editor, S. Moses, *In zwei Welten: Siegfried Moses zum fünfundsiebzigsten Geburtstag*, Verlag Bitaon, Tel-Aviv, (1962), pp. 118.ff; cited in K. Polkehn, "The Secret Contacts: Zionism and Nazi Germany, 1933-1941", *Journal of Palestine Studies*, Volume 5, Number 3/4, (Spring-Summer, 1976), pp. 54-82, at 59.
27. English translation quoted from J. Stachel, "Einstein's Jewish Identity", *Einstein from 'B' to 'Z'*, Birkhäuser, Boston, Basel, Berlin, (2002), pp. 57-83, at 78. Stachel cites M. Besso, A. Einstein, *Correspondance, 1903-1955*, Hermann, Paris, (1972), p. 238.
28. Letter from A. Einstein to M. Besso of 12 December 1919, English translation by A. Hentschel, *The Collected Papers of Albert Einstein*, Volume 9, Document 207, Princeton University Press, (2004), pp. 178-179, at 179.

wrote,

> "When asked what he thought of Jews marrying non-Jews, which, of course, had been the case with him and Mileva, [Albert Einstein] replied with a laugh, 'It's dangerous, but then all marriages are dangerous.'"[29]

On 3 April 1920, Einstein wrote, criticizing assimilationist Jews,

> "And this is precisely what he does *not* want to reveal in his confession. He talks about religious faith instead of tribal affiliation, of 'Mosaic' instead of 'Jewish' because the latter term, which is much more familiar to him, would emphasize affiliation to his tribe."[30]

Albert Einstein often referred to Jews as "tribesmen" and Jewry as the "tribe". Fellow German Jew Fritz Haber was outraged at Albert Einstein's racist treachery and disloyalty. Einstein confirmed that he was disloyal and a racist, and was obligated,

> "[...] to step in for my persecuted and morally depressed fellow tribesmen, as far as this lies within my power[.]"[31]

Einstein bore no such loyalty to Germans, who had feed him and made him famous. In fact, Einstein wanted to exterminate the Germans.

After declaring that Jewish children segregate due to natural forces and that they are "different from other children",[32] not due to religion or tradition, but due to genetic features and "heritage", Einstein continued his 3 April 1920 statement,

> "With adults it is quite similar as with children. Due to race and temperament as well as traditions (which are only to a small extent of religious origin) they form a community more or less separate from non-Jews. [***] It is this basic community of race and tradition that I have in mind when I speak of 'Jewish nationality.' In my opinion, aversion to Jews is simply based upon the fact that Jews and non-Jews are different. [***] Where feelings are sufficiently vivid there is no shortage of reasons; and the feeling of aversion toward people of a foreign race with whom one has, more

---

[29]. D. Brian, *The Unexpected Einstein: The Real Man Behind the Icon*, Wiley, Hoboken, New Jersey, (2005), p. 42.

[30]. A. Einstein, English translation by A. Engel, *The Collected Papers of Albert Einstein*, Volume 7, Document 34, Princeton University Press, (2002), pp. 153-155, at 153.

[31]. A. Einstein quoted in: H. Gutfreund, "Albert Einstein and the Hebrew University", J. Renn, Editor, *Albert Einstein Chief Engineer of the Universe: One Hundred Authors for Einstein*, Wiley-VCH, Berlin, (2005), pp. 314-318, at 316.

[32]. A. Einstein, English translation by A. Engel, *The Collected Papers of Albert Einstein*, Volume 7, Document 34, Princeton University Press, (2002), pp. 153-155, at 153.

or less, to share daily life will emerge by necessity."³³

Einstein made similar comments in a document dated sometime "after 3 April 1920". Einstein was in agreement with Philipp Lenard that a "Jewish heritage" could be seen in intellectual works published by Jews. Einstein stated,

> "The psychological root of anti-Semitism lies in the fact that the Jews are a group of people unto themselves. Their Jewishness is visible in their physical appearance, and one notices their Jewish heritage in their intellectual works, and one can sense that there are among them deep connections in their disposition and numerous possibilities of communicating that are based on the same way of thinking and of feeling. The Jewish child is already aware of these differences as soon as it starts school. Jewish children feel the resentment that grows out of an instinctive suspicion of their strangeness that naturally is often met with a closing of the ranks. [***] [Jews] are the target of instinctive resentment because they are of a different tribe than the majority of the population."³⁴

In a draft letter of 3 April 1920, Einstein wrote that children are conscious of "racial characteristics" and that this alleged "racial" gulf between children results in conflicts, which instill a sense of foreignness in the persecuted child. Einstein wrote,

> "Unter den Kindern war besonders in der Volksschule der Antisemitismus lebendig. Er gründete [s]ich auf die den Kindern merkwürdig bewussten Rassenmerkmale und auf Eindrücke im Religionsunterricht. Thätliche Angriffe und Beschimpfungen auf dem Schulwege waren häufig, aber meist nicht gar zu bösartig. Sie genügten immerhin, um ein lebhaftes Gefühl des Fremdseins schon im Kinde zu befestigen."³⁵

Einstein's racism was perhaps a defense mechanism to depersonalize the attacks he faced as a child and to counter the hurt with a sense of communal love, and communal hatred. Like Adolf Stoecker before him,³⁶ Albert Einstein advocated the segregation of Jewish students. Peter A. Bucky quoted Albert Einstein,

---

33. A. Einstein, English translation by A. Engel, *The Collected Papers of Albert Einstein*, Volume 7, Document 34, Princeton University Press, (2002), pp. 153-155, at 153-154.

34. A. Einstein, English translation by A. Engel, *The Collected Papers of Albert Einstein*, Volume 7, Document 35, Princeton University Press, (2002), pp. 156-157.

35. Letter from A. Einstein to P. Nathan of 3 April 1920, *The Collected Papers of Albert Einstein*, Volume 9, Document 366, Princeton University Press, (2004), p. 492. Also: *The Collected Papers of Albert Einstein*, Volume 1, Princeton University Press, (1987), p. lx, note 44.

36. P. W. Massing, *Rehearsal for Destruction: A Study of Political Anti-Semitism in Imperial Germany*, Howard Fertig, New York, (1967), pp. 278-294.

> "I think that Jewish students should have their own student societies. [***] One way that it won't be solved is for Jewish people to take on Christian fashions and manners. [***] In this way, it is entirely possible to be a civilized person, a good citizen, and at the same time be a faithful Jew who loves his race and honors his fathers."[37]

Einstein stated,

> "We must be conscious of our alien race and draw the logical conclusions from it. [***] We must have our own students' societies and adopt an attitude of courteous but consistent reserve to the Gentiles. [***] It is possible to be [***] a faithful Jew who loves his race and honours his fathers."[38]

On 5 April 1920, Einstein repeated what he had heard from his political Zionist friends, who believed that anti-Semitism was necessary to the preservation of the "Jewish race",

> "Anti-Semitism will be a psychological phenomenon as long as Jews come in contact with non-Jews—what harm can there be in that? Perhaps it is due to anti-Semitism that we survive as a race: at least that is what I believe."[39]

and,

> "I am neither a German citizen, nor is there in me anything that can be described as 'Jewish faith.' But I am happy to belong to the Jewish people, even though I don't regard them as the Chosen People. Why don't we just let the Goy keep his anti-Semitism, while we preserve our love for the likes of us?"[40]

This letter was published in the *Israelitisches Wochenblatt für die Schweiz*, on 24 September 1920, on page 10. It became famous and was widely discussed

---

[37]. P. A. Bucky, Einstein, and A. G. Weakland, *The Private Albert Einstein*, Andrews and McMeel, Kansas City, (1992), p. 88.
[38] A. Einstein, *The World As I See It*, Citadel, New York, (1993), pp. 107-108.
[39]. A. Einstein, English translation by A. Engel, *The Collected Papers of Albert Einstein*, Volume 7, Document 37, Princeton University Press, (2002), p. 159.
[40]. A. Einstein quoted in A. Fölsing, English translation by E. Osers, *Albert Einstein, a Biography*, Viking, New York, (1997), p. 494; which cites speech to the *Central-Verein Deutscher Staatsbürger Jüdischen Glaubens*, in Berlin on 5 April 1920, in D. Reichenstein, *Albert Einstein. Sein Lebensbild und seine Weltanschauung*, Berlin, (1932). This letter from Einstein to the Central Association of German Citizens of the Jewish Faith of 5 April 1920 is reproduced in *The Collected Papers of Albert Einstein*, Volume 9, Document 368, Princeton University Press, (2004).

in newspapers and was used as a political issue. Einstein's racism had already become a weapon for critics of the Jews to wield against German Jews loyal to the Fatherland. Einstein ridiculed the *Central-Verein deutscher Staatsbürger jüdischen Glaubens*, an organization that combated anti-Semitism and vigorously defended and celebrated Jews, because Einstein sought to promote anti-Semitism and because Einstein believed that being "Jewish" was a racial, not a religious, condition. Einstein knew quite well that the letter had been published. The *C. V.* contacted him about it and published a statement regarding it in their periodical *Im deutschen Reich* in March of 1921,

> "So wurde auch in einzelnen Versammlungen der bekannte Brief des Naturforschers Professor Einstein, den dieser an den Central-Verein gerichtet hat, und in welchem er die Bestrebungen des Central-Vereins ablehnt, weil sie zu national-deutsch und zu wenig jüdisch orientiert seien, zum Gegenstand der Erörterungen gemacht. Dieser Brief hat in der öffentlichen Erörterung der jüdischen und judengegnerischen Presse in den letzten Monaten und auch bei den Wahlen eine gewisse Rolle gespielt und Anlaß zu den verschiedenartigsten Betrachtungen je nach der Parteistellung der Versammlungsredner und der verschiedenen Zeitungen gegeben. So hat sich z. B. die jüdisch-nationale „Wiener Morgenzeitung" veranlaßt gesehen, den Central-Verein in wenig vornehmer Weise anzugreifen und ihn wegen seines nationaldeutschen Standpunktes zu verdächtigen. Diese Angriffe würden durch die Auffassung von Professor Einstein nicht gedeckt worden sein, wenn die „Wiener Morgenzeitung" gewußt hätte, daß Professor Einstein ohne nähere Kenntnis der Bestrebungen und der Arbeit des Central-Vereins seinen Brief geschrieben und keineswegs an eine Veröffentlichung, die nur durch eine Indiskretion erfolgt ist, gedacht hat. Erst nach der Veröffentlichung hat er von der Art und Weise der Tätigkeit des Central-Vereins Kenntnis erhalten und hat, wie mit gutem Grund versichert werden kann, infolge dieser Kenntnis eine wesentlich andere Auffassung vom Werte der Arbeit unseres Central-Vereins gewonnen. Auch dieser Vorfall sollte Anlaß geben, Urteile in der Oeffentlichkeit erst dann zu fällen, wenn die Sachlage einigermaßen geklärt ist."[41]

On 24 May 1931, the *Sunday Express* of London published an interview it claimed it had had with Einstein while he was visiting Oxford. The interview contained inflammatory statements similar to those published in the *Israelitisches Wochenblatt für die Schweiz* on 24 September 1920. These statements were repeated in several German language newspapers across Europe together with scathing editorial indictments of Einstein. Einstein claimed that no interview had taken place and the quotations were taken from a letter he had written eleven years prior. Einstein stated in a letter to Michael Traub of 22

---

41. "Zeitschau", *Im deutschen Reich*, Volume 27, Number 3, (March, 1921), pp. 90-97, at 92.

August 1931 that this letter had never been published,[42] though it had been published and Einstein knew quite well that it had been published.

Einstein accused the *Central-Verein deutscher Staatsbürger jüdischen Glaubens e. V.* of instigating the "forgery". The *C.V.* denied that it was behind the publication in the *Sunday Express* and invited Einstein to respond in their official organ the *Central-Verein Zeitung*. Einstein took the opportunity and stated, "Es wurden mir schon wiederholt Auszüge aus einem Artikel der „Sunday Expreß" zugesandt, aus denen ich ersehe, daß es sich um eine glatte Fälschung handelt. Ich habe in Oxford überhaupt kein einziges Zeitungsinterview gegeben. Der Inhalt ist eine böswillige Entstellung eines vor elf Jahren geschriebenen, nicht für die Oeffentlichkeit bestimmten Briefes."[43] He affirmed in 1931 that he had made the statements and did not repudiate them.

In 1932, Einstein stated, referring to the "deplorably high development of nationalism everywhere"—his own rabid Zionism excepted,

"The introduction of compulsory service is therefore, to my mind, the prime cause of the moral collapse of the white race, which seriously threatens not merely the survival of our civilization but our very existence. This curse, along with great social blessings, started with the French Revolution, and before long dragged all the other nations in its train."[44]

Einstein had a reputation as a rabid anti-assimilationist, which is to say that Einstein was a rabid racist segregationist. On 15 March 1921, Kurt Blumenfeld wrote to Chaim Weizmann,

"Einstein [***] is interested in our cause most strongly because of his revulsion from assimilatory Jewry."[45]

Einstein stated in 1921,

"To deny the Jew's nationality in the Diaspora is, indeed, deplorable. If one adopts the point of view of confining Jewish ethnical nationalism to Palestine, then one, to all intents and purposes, denies the existence of a Jewish people. In that case one should have the courage to carry through, in the quickest and most complete manner, entire assimilation. We live in a time of intense and perhaps exaggerated nationalism. But my Zionism does not exclude in me cosmopolitan views. I believe in the actuality of Jewish nationality, and I believe that every Jew has duties towards his coreligionists. [***] [T]he principal point is that Zionism must tend to

---

42. D. K. Buchwald, *et al.*, Editors, *The Collected Papers of Albert Einstein*, Volume 7, Document 37, Princeton University Press, (2002), p. 304, note 8.
43. "Professor Einstein erklärt das „Sunday Expreß"-Interview für gefälscht", *Central-Verein Zeitung*, Volume 10, Number 37, (11 September 1931), p. 443.
44. A. Einstein, translated by A. Harris, "The Disarmament Conference of 1932. I." *The World As I See It*, Citadel, New York, (1993), pp. 59-60.
45. J. Stachel, *Einstein from 'B' to 'Z'*, Birkhäuser, Boston, (2002), p. 79, note 41.

strengthen the dignity and self-respect of the Jews in the Diaspora. I have always been annoyed by the undignified assimilationist cravings and strivings which I have observed in so many of my friends."[46]

In 1921, Einstein declared, referring to Eastern European Jews,

"These men and women retain a healthy national feeling; it has not yet been destroyed by the process of atomisation and dispersion."[47]

Einstein wrote in the *Jüdische Rundschau*, on 21 June 1921, on pages 351-352,

"This phenomenon [*i. e.* Anti-Semitism] in Germany is due to several causes. Partly it originates in the fact that the Jews there exercise an influence over the intellectual life of the German people altogether out of proportion to their number. While, in my opinion, the economic position of the German Jews is very much overrated, the influence of Jews on the Press, in literature, and in science in Germany is very marked, as must be apparent to even the most superficial observer. This accounts for the fact that there are many anti-Semites there who are not really anti-Semitic in the sense of being Jew-haters, and who are honest in their arguments. They regard Jews as of a nationality different from the German, and therefore are alarmed at the increasing Jewish influence on their national entity. [***] But in Germany the judgement of my theory depended on the party politics of the Press[.]"[48]

Einstein also stated,

"The way I see it, the fact of the Jews' racial peculiarity will necessarily influence their social relations with non-Jews. The conclusions which—in my opinion—the Jews should draw is to become more aware of their peculiarity in their social way of life and to recognize their own cultural contributions. First of all, they would have to show a certain noble reservedness and not be so eager to mix socially—of which others want little or nothing. On the other hand, anti-Semitism in Germany also has consequences that, from a Jewish point of view, should be welcomed. I

---

[46]. A. Einstein, "Jewish Nationalism and Anti-Semitism", *The Jewish Chronicle*, (17 June 1921), p. 16.
[47]. J. Stachel, "Einstein's Jewish Identity", *Einstein from 'B' to 'Z'*, Birkhäuser, Boston, (2002), p. 65. Stachel cites, *About Zionism: Speeches and Letters*, Macmillan, New York, (1931), pp. 48-49. For Zionist Ha-Am's use of the image of atomisation and dispersion, *see:* A. Hertzberg, *The Zionist Idea*, Harper Torchbooks, New York, (1959), p. 276.
[48]. A. Einstein, "Jewish Nationalism and Anti-Semitism", *The Jewish Chronicle*, (17 June 1921), p. 16.

believe German Jewry owes its continued existence to anti-Semitism."[49]

Nazi Zionist Joseph Goebbels, sounding very much like political Zionist Albert Einstein, was quoted in *The New York Times*, on 29 September 1933, on page 10,

> "It must be remembered the Jews of Germany were exercising at that time a decisive influence on the whole intellectual life; that they were absolute and unlimited masters of the press, literature, the theatre and the motion pictures, and in large cities such as Berlin, 75 percent of the members of the medical and legal professions were Jews; that they made public opinion, exercised a decisive influence on the Stock Exchange and were the rulers of Parliament and its parties."

On 1 July 1921, Einstein was quoted in the *Jüdische Rundshau* on page 371,

"Let us take brief look at the *development of German Jews* over the last hundred years. With few exceptions, one hundred years ago our forefathers still lived in the Ghetto. They were poor and separated from the Gentiles by a wall of religious tradition, secular lifestyles and statutory confinement and were confined in their spiritual development to their own literature, only relatively weakly influenced by the forceful progress which intellectual life in Europe had undergone in the Renaissance. However, these little noticed, modestly living people had one thing over us: *Every one of them belonged with all his heart to a community*, into which he was incorporated, in which he felt a worthwhile member, in which nothing was asked of him which conflicted with his normal processes of thought. Our forefathers of that era were pretty pathetic both bodily and spiritually, but—in social relations—in an enviable state of mental equilibrium. Then came emancipation. It offered undreamt of opportunities for advancement. The isolated individual quickly found their way into the upper financial and social circles of society. They eagerly absorbed the great achievements of art and science which the Occidentals[50] had created. They contributed to the development with passionate affection, and themselves made contributions of lasting value. They thereby took on the lifestyle of the Gentile world, turning away from their religious and social traditions in growing masses—took on Gentile customs, manners and mentality. It appeared as if they were being completely dissolved into the numerically superior, politically and culturally better organized host peoples, such that no trace of them would be left after a few generations. The complete eradication of the Jewish

---

49. A. Einstein, A. Engel translator, "How I became a Zionist", *The Collected Papers of Albert Einstein*, Volume 7, Document 57, Princeton University Press, (2002), pp. 234-235, at 235.

50. At the time Einstein made his statement, Jews and Gentiles often referred to Jews as "Orientals".

nationality in Middle and Western Europe appeared to be inevitable. However, it didn't turn out that way. It appears that racially distinct nations have instincts which work against interbreeding. The adaptation of the Jews to the European peoples among whom they have lived in language, customs and indeed even partially in religious practices *was unable to eliminate all feelings of foreigness* which exist between Jews and their European host peoples. In short, this spontaneous feeling of foreigness is ultimately due to a loss of energy.[51] For this reason, *not even well-meant arguments can eradicate it*. Nationalities do not want to be mixed together, rather they want to go their own separate ways. A state of peace can only be achieved by mutual tolerance and respect."

Einstein stated that Jews should not participate in the German Government,

"I regretted the fact that [Rathenau] became a Minister. In view of the attitude which large numbers of the educated classes in Germany assume towards the Jews, I have always thought that their natural conduct in public should be one of proud reserve."[52]

Einstein merely parroted the Zionist Party line. Werner E. Mosse wrote,

"While the leaders of the CV saw it as their special duty to represent the interests of the German Jews in the active political struggle, Zionism stood for... systematic Jewish non-participation in German public life. It rejected as a matter of principle any participation in the struggle led by the CV."[53]

In 1925, Einstein wrote in the official Zionist organ *Jüdische Rundschau*,

"By study of their past, by a better understanding of the spirit [Geist] that accords with their race, they must learn to know anew the mission that they are capable of fulfilling. [***] What one must be thankful to Zionism for is the fact that it is the only movement that has given many Jews a justified pride, that it has once again given a despairing race the necessary faith, if I

---

[51]. Einstein repeatedly spoke of the Germans as "greedy" to acquire territory and of the "loss of energy" when different "races" attempted to live together. He have been speaking literally. Georg Friedrich Nicolai wrote of the struggle of life to aquire the energy of the sun and he applied this struggle to humanity. G. Nicolai, *Die Biologie des Krieges, Betrachtungen eines deutschen Naturforschers*, O. Füssli, Zürich, (1917); English translation: *The Biology of War*, Century Co., New York, (1918), pp. 36-39, 44-53.
[52]. R. W. Clarck, *Einstein, the Life and Times*, World Publishing Company, USA, (1971), p. 292. Clarck refers to: *Neue Rundschau*, Volume 33, Part 2, pp. 815-816.
[53]. W. E. Mosse, "Die Niedergang der deutschen Republik und die Juden", *The Crucial Year 1932*, p. 38; English translation in: K. Polkehn, "The Secret Contacts: Zionism and Nazi Germany, 1933-1941", *Journal of Palestine Studies*, Volume 5, Number 3/4, (Spring-Summer, 1976), pp. 54-82, at 56-57.

may so express myself, given new flesh to an exhausted people."⁵⁴

On 12 October 1929, Albert Einstein wrote to the *Manchester Guardian*,

"In the re-establishment of the Jewish nation in the ancient home of the race, where Jewish spiritual values could again be developed in a Jewish atmosphere, the most enlightened representatives of Jewish individuality see the essential preliminary to the regeneration of the race and the setting free of its spiritual creativeness."⁵⁵

Einstein's public racism eventually waned, but he continued to publicly express his segregationist philosophy in the same terms as anti-Semites, as well as his belief that Jews "thrived on" and owed their "continued existence" to anti-Semitism.
Einstein stated in December of 1930 to an American audience,

"There is something indefinable which holds the Jews together. Race does not make much for solidarity. Here in America you have many races, and yet you have the solidarity. Race is not the cause of the Jews' solidarity, nor is their religion. It is something else—which is indefinable."⁵⁶

Einstein's confusing public statement perhaps resulted from his desire to promote multi-culturalism in America, which had the benefit of freeing up Jewish immigration to the United States.⁵⁷ Einstein was also likely parroting, or trying to parrot, a fellow anti-assimilationist political Zionist whose pamphlet was well known in America, Solomon Schechter and his *Zionism: A Statement*, Federation of American Zionists, New York, (1906), in which Schechter states, among other things, "Zionism is an ideal, and as such is indefinable."
Einstein stated in 1938,

"JUST WHAT IS A JEW?"

The formation of groups has an invigorating effect in all spheres of human striving, perhaps mostly due to the struggle between the convictions and aims represented by the different groups. The Jews, too, form such a

---

54. English translation by John Stachel in J. Stachel, "Einstein's Jewish Identity", *Einstein from 'B' to 'Z'*, Birkhäuser, Boston, (2002), p. 67. Stachel cites, "Botschaft", *Jüdische Rundschau*, Volume 30, (1925), p. 129; French translation, *La Revue Juive*, Volume 1, (1925), pp. 14-16.
55. J. Stachel, "Einstein's Jewish Identity", *Einstein from 'B' to 'Z'*, Birkhäuser, Boston, (2002), p. 65. Stachel cites, *About Zionism: Speeches and Letters*, Macmillan, New York, (1931), pp. 78-79.
56. A. Einstein quoted in "Einstein on Arrival Braves Limelight for Only 15 Minutes", *The New York Times*, (12 December 1930), pp. 1, 16, at 16.
57. E. A. Ross, *The Old World in the New: The Significance of past and Present Immigration to the American People*, Century Company, New York, (1914), p. 144.

group with a definite character of its own, and anti-Semitism is nothing but the antagonistic attitude produced in the non-Jews by the Jewish group. This is a normal social reaction. But for the political abuse resulting from it, it might never have been designated by a special name.

What are the characteristics of the Jewish group? What, in the first place, is a Jew? There are no quick answers to this question. The most obvious answer would be the following: A Jew is a person professing the Jewish faith. The superficial character of this answer is easily recognized by means of a simple parallel. Let us ask the question: What is a snail? An answer similar in kind to the one given above might be: A snail is an animal inhabiting a snail shell. This answer is not altogether incorrect; nor, to be sure, is it exhaustive; for the snail shell happens to be but one of the material products of the snail. Similarly, the Jewish faith is but one of the characteristic products of the Jewish community. It is, furthermore, known that a snail can shed its shell without thereby ceasing to be a snail. The Jew who abandons his faith (in the formal sense of the word) is in a similar position. He remains a Jew.

[***]

WHERE OPPRESSION IS A STIMULUS
[***]

Perhaps even more than on its own tradition, the Jewish group has thrived on oppression and on the antagonism it has forever met in the world. Here undoubtedly lies one of the main reasons for its continued existence through so many thousands of years."[58]

Albert Einstein was parroting racist political Zionist leader Theodor Herzl, who wrote in his book *The Jewish State*,

"Oppression and persecution cannot exterminate us. No nation on earth has survived such struggles and sufferings as we have gone through. Jew-baiting has merely stripped off our weaklings; the strong among us were invariably true to their race when persecution broke out against them. This attitude was most clearly apparent in the period immediately following the emancipation of the Jews. Later on, those who rose to a higher degree of intelligence and to a better worldly position lost their communal feeling to a very great extent. Wherever our political well-being has lasted for any length of time, we have assimilated with our surroundings. I think this is not discreditable. Hence, the statesman who would wish to see a Jewish strain in his nation would have to provide for the duration of our political well-being; and even Bismarck could not do that. [***] The Governments of all countries

---

58. A. Einstein, "Why do They Hate the Jews?", *Collier's*, Volume 102, (26 November 1938); reprinted in *Ideas and Opinions*, Crown, New York, (1954), pp. 191-198, at 194, 196. Einstein expressed himself in a similar way to Peter A. Bucky, P. A. Bucky, Einstein, and A. G. Weakland, *The Private Albert Einstein*, Andrews and McMeel, Kansas City, (1992), p. 87.

scourged by Anti-Semitism will serve their own interests in assisting us to obtain the sovereignty we want. [***] Great exertions will not be necessary to spur on the movement. Anti-Semites provide the requisite impetus. They need only do what they did before, and then they will create a love of emigration where it did not previously exist, and strengthen it where it existed before. [***] I imagine that Governments will, either voluntarily or under pressure from the Anti-Semites, pay certain attention to this scheme; and they may perhaps actually receive it here and there with a sympathy which they will also show to the Society of Jews."[59]

Einstein's statements and those of other like-minded racist Zionists threw fuel on the fire and were reflective of the spirit and tone enunciated in *Protocols of the Learned Elders of Zion*, Number 9, which states (no matter who wrote it),

"Nowadays, if any States raise a protest against us, it is only *pro forma* at our discretion, and by our direction, for their anti-Semitism is indispensable to us, for the management of our lesser brethren."[60]

Many Zionist leaders espoused racist nationalism, which made them the darlings of the Nazis, the Nazis they had put into power. Joachim Prinz wrote, among other things, a racist polemic against assimilation in his book published in Germany in the Hitler-era, *Wir Juden* of 1934,

"Die Theorie der Assimilation ist zusammengebrochen. Kein Schlupfwinkel birgt uns mehr. Wir wünschen an die Stelle der Assimilation das Neue gesetzt: *das Bekenntnis zur jüdischen Nation und zur jüdischen Rasse*. Ein Staat, der aufgebaut ist auf dem Prinzip der Reinheit von Nation und Rasse, kann nur vor dem Juden Achtung und Respekt haben, der sich zur eigenen Art bekennt. Nirgendwo kann er in diesem Bekenntnis mangelnde Loyalität dem Staate gegenüber erblicken. Er kann keine anderen Juden wollen, als die Juden des klaren Bekenntnisses zum eigenen Volk. Er kann keine liebedienerischen, kriecherischen Juden wollen. Er muß von uns das Bekenntnis zur eigenen Art fordern. Denn nur jemand, der *eigene* Art und *eigenes* Blut achtet, wird den Respekt vor dem *nationalen Wollen anderer Nationen* haben können.

In dem Bekenntnis des Juden zu seiner eigenen Nation, in seiner Gewißheit, in sich sein eigenes Blut zu tragen, seine eigene Vergangenheit und seine eigene Art — wird er erst beginnen, die Distanz vor den Erlebnissen der anderen Nationen zu wahren, die notwendig ist, um ein ehrliches Miteinander und eine anständige Nachbarschaft zu halten. In dem Augenblick, in dem dieses Bekenntnis zur jüdischen Nationalität die

---

59. T. Herzl, *A Jewish State: An Attempt at a Modern Solution of the Jewish Question*, The Maccabæan Publishing Co., New York, (1904), pp. 5-6, 25, 68, 93.
60. L. Fry, *Waters Flowing Eastward: The War Against the Kingship of Christ*, TBR Books, Washington, D. C., (2000), p. 137.

Mehrheit der Judenheit ergreift, beginnt *die erste ehrliche Aussprache zwischen Juden und Nichtjuden.*"[61]

Prinz wrote of the supposed suicide of the emancipated Jews through assimilation in *liberal* states, and he despised liberalism. His goal was to preserve the alleged purity of the Jewish race in a Jewish nation, *i. e.* the expulsion of the Jews to a new territory which allowed the Zionists to enforce racial segregation. Prinz wrote,

"The brochure of the baptized Jew Karl *Marx* on the Jewish question is an anti-Jewish pamphlet and an autobiographical entry in the chapter of Jewish self-hatred."

"Die Broschüre des getauften Juden Karl *Marx* über die Judenfrage ist ein antijüdisches Pamphlet und ein autobiographischer Beitrag zum Kapitel des jüdischen Selbsthasses."[62]

Prinz was not alone in his condemnation of Karl Marx's anti-Semitism.[63] Hitler and Prinz had much in common. Lenni Brenner documents Prinz' and the Zionists' *kinship* with the Nazis' nationalistic and racial views in his book *Zionism in the Age of the Dictators.*[64]

Dietrich Bronder and Hennecke Kardel[65] state that the top leadership of the Nazi Party and the orchestrators of the "final solution" were of Jewish descent, including Adolf Hitler,[66] Adolf Eichmann, Reinhard Heydrich, Rudolf Hess

---

61. J. Prinz, *Wir Juden*, Erich Reiss, Berlin, (1934), pp. 154-155.
62. J. Prinz, *Wir Juden*, Erich Reiss, Berlin, (1934), p. 44.
63. E. Bernstein, "Jews and German Social Democracy", *Die Tukunft* (New York), Volume 26, (March, 1921), pp. 145ff.; English translation in: P. W. Massing, *Rehearsal for Destruction: A Study of Political Anti-Semitism in Imperial Germany*, Howard Fertig, New York, (1967), pp. 322-330. On Marx's alleged "self-hatred", *see:* H. Hirsch, "The Ugly Marx: Analysis of an 'Outspoken Anti-Semite'", *Philosophical Forum*, Volume 8, (1978), pp.150-162. *See also:* P. L. Rose, *Revolutionary Antisemitism in Germany from Kant to Wagner*, Princeton University Press, (1990), pp. 296-305. *See also:* R. Grooms, "The Racism of Marx and Engels", *The Barnes Review*, Volume 2, Number 10, (October, 1996), pp. 3-8. Communists have always been opportunistic Jew baiters.
64. *See especially* Chapter 5: L. Brenner, *Zionism in the Age of the Dictators*, Croom Helm, London, L. Hill, Westport, Connecticut, (1983).

<http://www.aaargh-international.org/engl/zad/zad.html>

65. D. Bronder, *Bevor Hitler kam: Eine historische Studie*, Hans Pfeiffer Verlag, Hannover, (1964), p. 204 (p. 211 in the 1974 edition). H. Kardel, *Adolf Hitler, Begründer Israels*, Verlag Marva, Genf, (1974); English translation *Adolf Hitler: Founder of Israel*, Modjeskis' Society Dedicated to Preservation of Cultures, San Diego, (1997).
66. H. Koehler, *Inside the Gestapo: Hitler's Shadow Over the World*, Pallas Pub. Co. Ltd., London, (1940). *See aslo:* H. Frank, *Im Angesicht des Galgens; Deutung Hitlers und seiner Zeit auf Grund eigener Erlebnisse und Erkenntnisse. Geschrieben im*

(member of the *Thule-Gesellschaft*, an organization Zionists created to promote anti-Semitism in order to force Jews to accept Zionism), Dietrich Eckart (member of the *Thule-Gesellschaft*), Alfred Rosenberg (member of the *Thule-Gesellschaft*), Julius Streicher (member of the *Thule-Gesellschaft*), Joseph Goebbels, and Hans Frank (member of the *Thule-Gesellschaft*). Dietrich Bronder wrote in 1964,

> "Aus den eigenen Untersuchungen des Verfassers über die führenden Nationalsozialisten sei hier nur mitgeteilt, daß sich unter 4000 Männern der Reichsführung 120 Ausländer von Geburt befanden, viele mit einem oder zwei Elternteilen ausländischer Herkunft und ein Prozent sogar jüdischer Abkunft — also im Sinne der NS-Rassengesetze „untragbar".
> a) So rechnen zu den Auslandsgeborenen:
> Reichsminister und Führerstellvertreter Rudolf Heß (Ägypten); Reichsminister Darré (Argentinien); Gauleiter und Staatssekretär E. W. Bohle und der Reichskommissar Herzog von Sachsen-Coburg (England); Generaloberst Löhr (Jugoslawien); General der Waffen-SS Phleps (Rumänien); Reichsärzteführer und Staatssekretär Dr. Conti und der Berliner Oberbürgermeister Lippert (Schweiz); NSKK-Obergruppenführer G. Wagner (Frankreich); sowie aus Rußland: Reichsminister und Reichsleiter Alfred Rosenberg und die NS-Reichshauptamtsleiter Brockhausen, Dr. von Renteln und Schickedanz, Reichsminister Backe, Präsident Dr. Neubert, Staatsrat Dr. Freiherr von Freytag-Loringhoven und Bischof J. Beermann.
> b) Darüber hinaus stammten von einem oder beiden ausländischen Elternteilen (u. v. a.):
> Der Reichsjugendführer Baldur von Schirach, Generaloberst Rendulic sowie der Generaldirektor Gustav Krupp von Bohlen-Halbach.
> c) Selbst jüdischer Abkunft bzw. mit jüdischen Familien verwandt waren: der Führer und Reichskanzler Adolf Hitler; seine Stellvertreter, die Reichsminister Rudolf Heß und Reichsmarschall Hermann Göring; die Reichsleiter der NSDAP Gregor Strasser, Dr. Josef Goebbels, Alfred Rosenberg, Hans Frank und Heinrich Himmler; die Reichsminister von Ribbentrop (der mit dem berühmten Zionisten Chaim Weizmann, dem 1952 verstorbenen ersten Staatsoberhaupt von Israel, einst Brüderschaft getrunken hatte) und von Keudell; die Gauleiter Globocznik (der Judenvernichter), Jordan und Wilhelm Kube; die hohen SS-Führer und z. T. in der Judenvernichtung tätigen Reinhard Heydrich, Erich von dem Bach-Zelewski und von Keudell II; die Bankiers und alten

---

Nürnberger Justizgefängnis, F. A. Beck, München-Gräfelfing, (1953), pp. 330-331. *See aslo:* D. Bronder, *Bevor Hitler kam: Eine historische Studie*, Hans Pfeiffer Verlag, Hannover, (1964), p. 204 (p. 211 in the 1974 edition). *See aslo:* H. Kardel, *Adolf Hitler, Begründer Israels*, Verlag Marva, Genf, (1974); English translation *Adolf Hitler: Founder of Israel*, Modjeskis' Society Dedicated to Preservation of Cultures, San Diego, (1997).

Förderer Hitlers vor 1933 Ritter von Stauß (Vizepräsident des NS-Reichstages) und von Stein; der Generalfeldmarschall und Staatssekretär Milch, der Unterstaatssekretär Gauß; die Physiker und Alt-Pg.'s Philipp von Lenard und Abraham Esau; die Uralt-Pg.'s Hanffstaengel (NS-Auslandspressechef) und Prof. Haushofer (s. S. 190)."[67]

Inferences can be drawn that these crypto-Jewish Nazi leaders were either motivated by self-hatred, or they were front men under the control of Herzlian political Zionists. Both may have been true of the genocidal Nazi Party leaders. Bryan Mark Rigg estimates the total number of Jewish soldiers and sailors in the Nazi military perhaps ranged upwards to 150,000.[68]

Many Zionists hated themselves and Jews in general and defamed Jews in their literature, especially the relatively impoverished and uneducated Jews of the East, whom the Zionists tried to bribe into migrating to Palestine, though they only largely succeeded in capturing ne'er-do-wells. Herzl considered himself to be a sleazy ultra-Jew in the most pejorative sense of which he could conceive to use the term "Jew". Herzl justified himself by generalizing his character flaws as if they were a racial "Jewish" trait. He hated the masses of poor Jews from the East and the rich Jews of the West, who wanted to assimilate.

In 1845, *The North American Review* wrote of the snobbish class hatred common among Jews, the inter-Jewish racism which has long plagued Jews, and the various dogmatic Jewish sects hatefully at odds with one another (note the misogyny and dogmatic indoctrination of Jews, which continues to this day,[69] and which manifests itself in, among other things, virulent Jewish censorship of others),

"As the Jews were anciently divided into several religious sects,—the Pharisees, the Sadducees, the Essenes,—so we find them distinguished at the present day. Their chief modern denominations, some of which represent the more ancient, are the Caraites, the Zabathaites, the Chasidim, the Rabbinists, or Talmudists, and the Reformed Jews. The Samaritans

[*Footnote:* Mixed descendants of a remnant of the ten tribes left in their own

---

67. D. Bronder, *Bevor Hitler kam: Eine historische Studie*, Hans Pfeiffer Verlag, Hannover, (1964), pp. 203-204 (pp. 210-211 in the 1974 edition).
68. "Who Were Hitler's Jewish Soldiers", *The Jewish Chronicle*, (6 December 1996), p. 1. **See also:** W. Hoge, "Rare Look Uncovers Wartime Anguish of Many Part-Jewish Germans", *The New York Times*, (6 April 1997), p. 16. **See also:** B. M. Rigg, *Hitler's Jewish Soldiers: The Untold Story of Nazi Racial Laws and Men of Jewish Descent in the German Military*, University Press of Kansas, Lawrence, Kansas, (2002); **and** *Rescued from the Reich: How One of Hitler's Soldiers Saved the Lubavitcher Rebbe*, Yale University Press, New Haven, (2004).
69. E. Kaye, *The Hole in the Sheet: A Modern Woman Looks at Orthodox and Hasidic Judaism*, L. Stuart Inc., Secaucus, New Jersey, (1987).

land, and of the Assyrians colonized among them. 2 Kings, xvii. 24, &c. In Christ's time they had a temple on Mount Gerizim, which they held more sacred than Mount Zion and its temple. They receive only the Pentateuch, and perhaps the Books of Joshua and Judges, which are found among them; but confidently wait for the Messiah, and observe the Mosaic laws more strictly than even the Jews. Wolff found fifty families of them at the foot of Gerizim, and they have also been met with in other parts of Palestine and in Egypt.]

are not to be classed among them, though akin to them in many respects. The main point of difference between most of these sects, though not the only one, respects the Talmud. The *Talmud*—a word that means *doctrine*—is a voluminous work of two parts,—the *Mishna*, that is the *second law*, and the *Gemara*, or *completion*. The former, consisting of a divine interpretation of the written law, say the Talmudists, was given to Moses at the same time with that delivered on Mount Sinai, together with rules for its exegesis, all to be orally handed down; and by him it was made known to the whole people, and specially committed to his successors. These traditions were collected in the Mishna, a work ascribed to Judah Hannasi,—the Holy, as he is usually called,—about the middle of the second century. Many glosses upon this text soon accumulated, which the Rabbi Jochanan, about the year 230, threw together in the form of a perpetual commentary upon it, entitled the Gemara; and this, with the Mishna, is called the Jerusalem Talmud; though sometimes the Mishna, and sometimes the Gemara alone, is, by synecdoche, called the Talmud. About a century later, Ashi and Abhina, distinguished Babylonian rabbins, compiled a much larger collection of opinions, which, with the Mishna, is styled the Babylonian Talmud, a work held in much higher esteem than the other, and generally understood when the Talmud, without further specification, is mentioned. It has commonly been published in twelve large folios. The other is comprised in a single folio. The Talmud has been justly described as 'containing things frivolous and superstitious, impieties and blasphemies, absurdities and fables.' As an example of all these in one,—God is represented as having contracted impurity by the burial of Moses, and as washing in fire to cleanse himself. These traditions, many of them the same by which, in Christ's time, the Jews 'made the commandment of God of none effect,' since then, in accumulated instances, have been used to destroy the force of the Old Testament Scriptures; which, indeed, Rabbinists consider of very little importance.

[***]

Rabbinism is the Catholic faith, from which all these sects are, in modern phrase, dissenters. It is the lineal descendant of Pharisaism, and distinguished by its blind adherence to the Talmud. The estimation in which strict Rabbinists hold this book is unbounded. 'He that has learned the Scripture, and not the Mishna,' says the Gemara, 'is a blockhead.' Isaac, a distinguished rabbi, says, 'Do not imagine that the written law is the foundation of our religion, which is really founded on the oral law.' The

Rabbinical doctrine is, ' The Bible is like water, the Mishna like wine, and the Gemara like spiced wine.' [*Soferim* 13*b*] Some even say, that 'to study the Bible is but a waste of time.' [*Baba Mezia* 33*a*] For strict Rabbinism, a melancholy compound of superstition and fanaticism, we must look to Poland, Russia, Hungary, and Palestine, of which we speak, in describing the system. In those countries, the Rabbinists, or Talmudists, discountenance as profane all other study than that of the Bible and Talmud, but are very careful to educate their sons in their religious lore. The Talmud forbids teaching females more than their appropriate domestic arts. 'Whoever instructs his daughter in the Bible is as if he instructed her in abominations.' But it is a disgrace, if boys are not taught to read the Hebrew Bible. The rich provide teachers for their own children, and either permit the poorer to share this provision, or aid them in obtaining masters. So honorable is the office of teacher made, that a bare support is enough generally to secure a competent one. The ordinary method of instruction is very simple. The child, when four years old, is taught the Hebrew letters, and then to pronounce words, the meaning of which he afterwards learns from his tutor; and thus proceeds, without grammar or dictionary, until he can translate the Pentateuch with tolerable ease. Then he begins at Genesis to study exegetically, surrendering his mind, however, entirely to the guidance of some Jewish commentator; and, from first to last, never forming an independent judgment, but implicitly following tradition, and of course never detecting its gross perversions of the Bible. Some stop short of this commentary, with which others conclude their education; while others still, whose parents can afford it, especially if they display quickness in study and fondness for it, pass on to the Talmud,—first the Mishna, then the Gemara, each with its rabbinical commentaries. As an evidence of the ardor sometimes manifested in these studies, and of complete devotion to them, we are told, that a traveller, some years ago, met three young educated rabbins, who 'were born and lived to manhood in the middle of Poland, and yet knew not one word of its language.' A Jewish youth, distinguished for proficiency in Talmudical learning, is anxiously sought in marriage for the daughters of wealthy parents; who look not only at the certain honor of such an alliance, but also at the chance, thus increased, of the Messiah's coming in their line. On the other hand, the Talmud designates by the name of 'people of the land,' equivalent to *peasantry*, those educated in the Bible alone, or not at all; and represents them as an inferior class, fit only for servile labor, with whom others may not intermarry; applying Deut. xxvii. 21,—'Cursed be he that lieth with any manner of beast.' Indeed, the Talmud authorizes every species of oppression towards such, giving them the hope of heaven only if they submit. The Jewish 'peasant' is a servant of servants, ground down by those who have learned, by being oppressed, the art of oppression. In Russia and Poland, where the Jews collect the government taxes among themselves, the rabbins make the peasantry pay nearly the whole. This class, too, where the Jews regulate the conscription, must furnish all the soldiers required.

Some other characteristics of the strict Rabbinists may be briefly noticed. They are the lowest of the Jews in point of morals, and this is sufficiently accounted for by the gross immorality of many Talmudical precepts. On the great yearly *Day of Atonement*, complete absolution from all past sins is pronounced, and from all religious vows, bonds, and oaths taken since the last preceding, and until the next, atonement. This latter absolution, contained in a prayer denominated *col nidre*, being supposed by Christians to extend to all oaths and obligations, civil as well as religious, which the Jews deny, has caused them much trouble in some parts of Europe. The Talmud teaches, moreover, that no respect is due to a Gentile's, or an unlearned Jew's, rights of property; and it accumulates other abominable doctrines, too numerous, and some of them too vile, to mention. Indeed, the modern Rabbinical Jews are generally, in practice, superior to the precepts of the Talmud. They believe in a purgatory, and pray for the souls of the dead; they hold that all Hebrews will rise in the Holy Land, those dying elsewhere rolling painfully under ground until they reach that soil; and that 'all Israel hath part in eternal life.' The dead buried in the Holy Land are expected to be the first to rise in the Messiah's day; [*Kethuboth* 111*a-b*. *Yerushalmi Kilayim* 9, 3. *Ezekiel* 37. *Genesis* 50:25] and so strong has been the desire of burial there, that in the seventeenth century large quantities of Jewish bones were yearly sent thither to be interred. Ship-loads of this melancholy freight might often be seen at Joppa. They believe that a council properly constituted is infallible, and practically, by their implicit confidence in the Talmud, they make the ancient rabbins their 'fathers.' They place a high estimate on the merits of good works, especially those of a ceremonial kind. Thus, though the reading of the Bible is considered hardly a good act, and even as a positive waste of time, the act of taking out the Pentateuch from its depository in the synagogue, and the duty of standing on the left side of the reader, and of closing and removing the roll after service, are considered highly meritorious, and the privilege of performing them is often sold to the highest bidder. A pilgrimage to the Holy Land, much more to pass one's life there, is a superlative merit. They place great confidence in the supererogatory merits of their ancient saints, especially of Abraham, Isaac, and Jacob, for the males, and of Sarah, Rebekah, and Leah, for females. They have daily morning and evening prayer in the synagogue; and all the prayers for public and private devotion are prescribed, and in Hebrew; for the Talmud affirms, that the angels who receive them understand no other language. Women, servants, and children under twelve years of age, are not required to observe the hours of prayer. The Jews of the Holy Land are, perhaps, singular in praying to saints, and honoring and even worshipping relics. They never approach the supposed stones of the temple, some of which are much worn by kissing, without removing their shoes. Every spot where a saint is supposed to be buried is a place of prayer and pilgrimage. The Talmudists do not allow women to attend the synagogue, until they are married; and then, in Poland, Russia, and the East, they occupy a separate apartment.

Public worship among the Talmudical Jews is, for the most part, where the civil power has not interfered, very irreverent and disorderly. A missionary at Beyroot saw comfits thrown among the people in the synagogue, when particular portions of the service were read, *to show the sweetness of the law!* and the audience—some of the adults, and all the boys—tumbling over one another in the scramble for them on the floor. The Talmud declares, that, in observing the feast of Purim, 'Every man must get so drunk, that he cannot distinguish between the phrases, *Blessed be Mordecai*, and *Cursed be Haman*.' While the Talmud imposes many burdensome ceremonies in addition to the Mosaic institutions, it also furnishes multiplied expedients for lightening the latter; and a fertile ingenuity, newly exercised for each emergency, or perpetuated in legendary rules, has extended the dispensation to every desirable point. Stephens, in his travels in the Holy Land, lodged with a Jew, who would not suffer a lamp, lighted the day before, to be extinguished on the Sabbath; but 'described an admirable contrivance he had invented for reconciling appetite with duty;—an oven, heated the night before to such a degree, that the process of cooking was continued during the night, and the dishes were ready when wanted on the Sabbath.' Yet even the Talmudical Jews are generally superior in morals to their Christian neighbours, especially in the point of female purity. No wonder they hate the New Testament, reading it only through the profligate and intolerant conduct of their persecutors.

Hospitality and alms-giving to their brethren are sacred duties among all the Jews. A large majority of those in Palestine are paupers, and, for their support, contributions, averaging fourteen thousand dollars a year, are made in different parts of Europe, deposited at Amsterdam, and thence transmitted to Beyroot. Jerusalem, Hebron, Tiberias, and Saphet are holy cities in Jewish esteem, and in all the Italian synagogues money-boxes are kept, marked, 'For Jerusalem,' 'For Saphet,' &c. The largest collections are in Amsterdam. Leghorn sends about four thousand dollars. But the poor unlearned Jews of Palestine are greatly oppressed by the rabbins, and generally defrauded, wholly or in part, of their share in these charities. When the Hebrew quarter at Smyrna was destroyed by fire, in 1841, Mr. Rothschild, of Vienna, gave 20,000 francs for the sufferers. He and his brothers have lately offered 100,000 francs for founding a Jewish hospital at Jerusalem. Sir Moses Montefiori, during his late visit to Palestine, contributed munificently to the wants of his poor brethren there."[70]

Lenni Brenner cites numerous examples of defamations against Jews by the Jewish Zionists Maurice Samuel, Ben Frommer, Micah Yosef Berdichevsky,

---

[70]. "The Modern Jews", *The North American Review*, Volume 60, Number 127, (April, 1845), pp. 329-368, at 353-354, 357-361.

Yosef Chaim Brenner[71] and Aaron David Gordon.[72] One could add Theodor Herzl's, Berl Katzenelson's[73] and Vladimir Jabotinsky's[74] names to the list. Mussolini called Jabotinsky a "Jewish Fascist" and David Ben-Gurion found Adolf Hitler's writings reminiscent of Jabotinsky's.[75] Lenni Brenner wrote, quoting Chaim Greenberg,

> "In March 1942 Chaim Greenberg, then the editor of New York's Labour Zionist organ, *Jewish Frontier*, painfully admitted that, indeed, there had been:
>
>> a time when it used to be fashionable for Zionist speakers (including the writer) to declare from the platform that 'To be a good Zionist one must be somewhat of an anti-Semite'... To this day Labor Zionist circles are under the influence of the idea that the Return to Zion involved a process of purification from our economic uncleanliness. Whosoever doesn't engage in so-called 'productive' manual labor is believed to be a sinner against Israel and against mankind."[76]

Martin Luther accused the Jews of not sharing the societal burden of manual labor. The ancients also made such accusations against the Jews. Zionists like Theodor Herzl emphasized that Jews must engage in manual labor in their proposed segregated society, so that their gene pool would not be corrupted by foreign laborers, and Herzl stressed his assertions that the poor Jews of Galicia and Eastern Europe were well accustomed to manual labor. Echoing the charges of anti-Semites, Herzl and other Zionists publicly accused the Jews of being "parasites"—to use their term—and the Zionists wanted Jews to take up farming and manual labor allegedly so as to cease to be "parasites". They had other

---

71. J. H. Brenner, "Self-Criticism", in A. Hertzberg, *The Zionist Idea*, Harper Torchbooks, New York, (1959), pp. 307-312. Brenner cites Mendele Moher Sefarim's works (Mendele's real name was Shalom Jacob Abramowitz).
72. A. D. Gordon, *Kitve A. D. Gordon*, In Five Volumes, Tel-Aviv, ha-Va'ad ha-merkazi shel mifleget ha-Po'el ha-tsa'ir, (1927-1930), parts translated to English in A. Hertzberg, *The Zionist Idea*, Harper Torchbooks, New York, (1959), pp. 371-386, *see especially* pp. 372, 376, 377, 379.
73. B. Katzenelson, *Ba-Mivhan*, Tel-Aviv, (1935); parts translated to English in A. Hertzberg, *The Zionist Idea*, Harper Torchbooks, New York, (1959), pp. 390-395, *see especially* pp. 390-391.
74. V. Jabotinsky, *Evidence Submitted to the Palestine Royal Commission*, London, (1937), pp. 10-29; in A. Hertzberg, *The Zionist Idea*, Harper Torchbooks, New York, (1959), pp. 559-570, 1t 560-561.
75. M. Bar-Zohar, *Ben-Gurion: A Biography*, Delacorte Press, New York, (1978), p. 67.
76. L. Brenner, *Zionism in the Age of the Dictators*, Chapter 2, Croom Helm, London, L. Hill, Westport, Connecticut, (1983), p. 24. Brenner cites: C. Greenberg, "The Myth of Jewish Parasitism", *Jewish Frontiers*, (March, 1942), p. 20. Brenner also refers to Yehezkel Kaufman, "Hurban Hanefesh: A Discussion of Zionism and Anti-Semitism", *Issues*, (Winter, 1967), p.106.

ulterior motives. Racists, and there was no one more racist than the Zionists, had long argued that conquered peoples exacted a vengeance of the vanquished by outbreeding, and by overwhelming the cultures of nations which used them as slave labor. The Zionists wanted to avoid any such occurrence by using exclusively Jewish labor in the "Jewish Homeland". They also wanted to strengthen their gene pool, which they believed had been weakened by the ghetto system and urbanization. In addition, in the early 1920's, some, like Lord Sydenham, complained that the Zionists were exporting Jews of poor character from the East to Palestine, people who were not fit for, nor skilled in, the farmwork that was needed in Palestine. Beyond this, Jewish laws forbids non-Jews to live in Jerusalem, even in Greater Israel, and the racist Jews needed a Jewish slave labor force of Eastern European Jews to build them a new nation without violating Jewish law.

Indeed, one of the first objectives of the Jewish Bolshevists was to train Eastern European Jews to farm and perform the trades. A. Borisow wrote in *The Jewish Chronicle* on 22 September 1922 on page 16,

"'Nep and the Jews.

A New Element in Soviet Russia.

By A. BORISOW.

A new persecutor has arisen to plague our long-suffering Russian Jewry in the form of the New Economic Policy, familiarly known in Russia as the 'Nep.'

Most people will look up in surprise when they hear me describe the 'Nep,' the far-famed and much-heralded New Economic Policy of the Soviet Government, as a persecutor. For does not 'Nep' mean the renunciation of the Communist illusions, liberation from the bureaucratic Soviet institutions, the reintroduction of trade into the country, the circulation of money, the right of possession of land and factories? All that is surely a blessing to the Jewish population, mainly an urban and commercial element, and yet I stigmatise it as a persecutor!

Still I repeat that the 'Nep' in Russia is a persecutor of the Jews. During the whole of the last two years the Jews have not suffered economically so much as they have during the few months since the introduction of the 'Nep.' It is not for nothing that the Jews translate the initials of the 'Nep' as the 'Nestchastnaja' ('luckless') Economic Policy.

What is it that the 'Nep' has brought us?

To begin with, it has reduced the number of officials. Many of the Soviet institutions have been closed down. In most of the others, 50 to 60 per cent. of the staff has been dismissed. Viewed on its merits, this is most welcome. It will mean a decrease in the heavy taxation which went to keep all these officials. But for the Jewish population it is a terrible blow. It is no secret that the Soviet institutions, especially in the cities, were staffed almost entirely by Jews. About three-quarters of the total number of officials were

Jews. Tens of thousands of Jewish intellectuals and semi-intellectuals, lawyers, journalists and doctors, managed to earn a crust of bread in the service of the Soviet institutions. They formed the majority of the lettered population. Now they are dismissed, driven out into the streets, condemned to unemployment and to starvation. That is the first blessing which the 'Nep' has brought to the Jews.

Trade in Russia has again become free. People are allowed to exchange commodities, to buy and sell. As that was the usual occupation of the majority of the Jews in pre-war Russia, it should be an excellent thing for the Jews. There is no need now to fear that the 'Cheka' will come down on the traders and have them shot for speculation.

But what is the result?

The reintroduction of trade has meant the annihilation of everything that has been done to foster productive work among the Jews. During the four years of Communism in Russia, the foundations of the old economic order were undermined.

With fire and sword the Communists wiped out every trace of trading in the country. They put a stop to what they called 'speculation.' The 'Cheka' drove our *Luftmenschen* by the fear of death into productive work. No one imagined there would ever be a return to the old conditions. Lest they died of hunger, they were compelled to adapt themselves to the new conditions. They learned some kind of handicraft, or they took to agriculture. Productive co-operatives sprang up in the towns. The younger generation, especially, took to establishing agricultural co-operatives. Thousands of young men and women joined the *Hechaluz*, joined together in a rigid discipline in order to take up agriculture as their life work. The Jewish population, under compulsion, became if not proletarianised, in the sense of becoming a factory population, at least labourised—engaged in direct labour. They did productive work instead of engaging in barter.

And now the 'Nep' has come, and stamped out all these hopeful signs, put a stop to all this new endeavour which has meant so much adaptation and hard work. It has killed the co-operatives, and the *Hechaluz* groups. People have left their handicrafts, their agricultural work, and they have again started their small trading—not only those who were traders in the pre-Revolution period, but also people who had never before in their lives had anything to do with commerce or barter. Men who were intellectuals, lawyers, writers, Government officials, have hailed the 'Nep' as the liberator. People who had grown tired of hungering, who had sold their last garment in order to get a dry crust of bread, who could no longer stand being herded together, ten of them in an unheated cellar, have become drunk with the lust of making money. Hundreds of millions of roubles, they heard, could be made by engaging in trade. So they went into trade. They are 'Nepping.'

It would be ridiculous to blame anyone for that. All we can do is to deplore it. But we must regret that the forced and unwilling, yet nevertheless healthy work of transition of a large part of the Jewish population to productive work has been brought to nothing.

If the 'Nep' at least provided the people with the means of livelihood, if those who have thrown away their handicraft and their agriculture, in order to engage in trade improved their economic position, there would be some sort of justification even for the loss.

Business in Russia to-day is conducted by the million. The slightest transaction involves tens of millions. Where are the people to obtain these huge initial sums with which to start their businesses? Nobody had any money. Most of those who have started in business have sold their effects down to their very last plate or spoon in order to get some sort of a starting capital. They buy up goods for several tens of millions and they sell them again at some hundreds per cent. profit. Splendid, it seems at first sight. But in the interval which the transaction takes to complete, the rate of exchange has generally fallen to such an extent that the total sum realised buys less than the original sum had purchased. Sixty million roubles, for example, to-day buy about as much as 20 to 25 millions bought a short while ago. Nominally, the 'Nep' man has become richer. Actually, he has become poorer.

There is an anecdote in circulation among these 'Nep' people which will serve as an apt illustration.

Somebody bought in the Urals five waggon-loads of nails, brought them to Moscow and sold them at a tremendous profit. He went back again to the Urals, but this time he was unable to get more than three waggon-loads of nails for his money. He came back to Moscow, sold them again at an immense profit, and went back to the Urals. This time he managed to buy no more than one waggon-load of nails. And so it went on and on, until at last he went to the Urals with a simply colossal sum, but all he could get for it was just one nail. So he took that nail and hitched a rope to it and hanged himself.

It is not difficult to earn money, but to become rich or even to make a decent livelihood is impossible, especially with the State shearing the 'Nep' people unmercifully. They are taxed to an enormous extent. And it is not to be wondered at that hundreds and thousands of Jews who at first petitioned for permits to become 'Neppists' are now returning their permits to the Government asking to be released from the honour of being among the builders of the New Economic Policy.

But it is not easy to give it up. When the shopkeepers in Homel, staggering under their heavy taxation, declared a sort of strike, refusing to open their shops and engage in business, they were denounced as counter-revolutionaries, and one of the leaders of the 'Yewsekzie,' the notorious Merejin, published two inflammatory articles in the 'Emess' denouncing the 'first attack by the Jewish bourgeoisie against the Soviet Government,' and demanding that they should be punished as traitors.

Naturally, not all the 'Nep' people go through the same kind of thing. As always, there are exceptions, and there are individuals who have made fortunes, especially in Moscow, which is to-day the greatest, perhaps the only trading centre in Russia. It is to Moscow that the Jews are flocking from every part of the country. Till recently things were not so bad in Minsk, where people managed to do well on contraband trade with Poland. But now there is a Customs office at the railway station in Minsk; all goods are thoroughly examined, and permission to bring goods back into Russia is given only to those who agree to smuggle illegal Communist literature into Poland.

Things are somewhat better for those families who have children over the age of twelve, able to travel round the villages, buy up goods and bring them home to their parents to sell. To have several grown-up children to-day in Russia means to be a rich man. Each child is a bread-winner. So from their earliest days children are being brought up to trade. Speculation is again becoming the forte of the Jews. All education is neglected, in order to train the children to become good business people. Of ideals it is better to say nothing at all.

But the most fortunate under the "Nep' are those families who have been down with typhus. That is an exceptional bit of luck. These people have no fear of again contracting the disease, so they travel about along the railway lines, and bring goods to their homes. There are very few others who will venture to set foot in a train, for the compartments are generally the homes of lice and contagion. They are consequently becoming monopolists. People who want to have things done for them in distant parts usually have to employ these typhus people, who get a good proportion of the profits.

A few individuals become rich, speculating in diamonds and in the exchange rate. The overwhelming majority, however, scuttle about the place like poisoned rats, buying and selling, working sixteen hours in the day, thinking of nothing in the world except their little businesses, and at the end of it all they have gained hardly anything.

Economically, 'Nep' has brought nothing but demoralisation into the life of Russian Jewry.

The moral degeneration is appalling. The mentality of the few new rich is disgusting. Everything is to their view concentrated within their little business transactions. The hunger for profits is stronger than anything else in the world, more potent than social or intellectual interests, for which there is no room left in their minds. A 'Nep' man who has really done well will never give anything away unless he is given a place of honour on some committee or other. The rule of the day in Jewish life in Russia is that 'he who has the money gets the honey.' The few new rich 'Nep' people are the rulers of Jewish life. The old social order has been broken up. The former communal workers have emigrated or have become the poorest of the poor. New people have taken their place.

The story of the 'Nep' is not finished yet. We will not venture to prophesy what it may bring to Russian Jewry in the future. But there is no doubt about what it is to-day. It is a persecution. It is not a New Economic Policy, but, as they say a 'luckless' economic policy."

In an age of Social Darwinism, the Zionists promoted the idea that only young and strong Jews should emigrate to Palestine and that they alone should avoid death at the hands of the Nazis. The infamous stories of the selection process of the *SS*, whereby healthy Jews of childbearing age were selected to survive, while others were selected to die, was, if true, most likely a Zionist directive meant to undue the supposed genetic damage of the ghettoes. The Nazis were also infamous for forcing Jews to perform strenuous manual labor, literally working the old and the weak to death.[77] This practice fulfilled several Zionist objectives—killing off the old and the weak—training Jews to do the dirty work that would be needed to be done in Palestine—fulfilling Jacob's Biblical rôle as an agrarian—and ensuring that the Holy Land would become predominantly Jewish, almost exclusively Jewish, which is also a Biblical goal and one the racist Israelis are still attempting to achieve today. The Nazis devoted a great deal attention to identification of "Jewish racial traits".

While the Zionist Nazis favored Zionist Jews and helped to usher them out of Nazi occupied lands, the Zionist Nazis targeted assimilatory Jewry and Orthodox Jewry, who were largely opposed to Zionism. The Zionists hoped to persuade both assimilated Jews and Orthodox Jews to violate their sensibilities and the Talmud and emigrate to Palestine *en masse* after the Second World War. The Zionists viciously punished these assimilated and Orthodox Jews who had opposed the Zionists after the First World War. The following article appeared in *The Jewish Chronicle* on 11 April 1919 on page 10,

"Jewish Factions in the Polish Parliament.

Co

PENHAGEN [F. O. C.]

When the leaders of the Jewish factions in Poland rose in Parliament to explain the Jewish policy and demands in the new State, there was only one note of agreement struck by them, namely, the loyalty of the Jews to Poland and their goodwill towards the State. Otherwise, a sharp conflict of opinions manifested itself between the Jewish Nationalists (the Zionists and the People's Party) and the Orthodox Group (to which the Assimilation Party also leans on Jewish National questions). The former demanded national minority rights for the Jews, whereas the latter claimed equal rights only. Rabbi Perlmutter, on behalf of the Orthodox Party (speaking in the House without uncovering), declared that he desired to see a great Poland sweeping to the sea. M. Prilutzky, on behalf of the Jewish Nationalists, claimed National rights for the Jews, including special schools and the right to employ Yiddish in Courts of Justice and in State documents. He had a very hostile reception, members shouting at him: 'Let America grant such demands first, and we shall follow.'

---

[77]. See: *The Holocaust Chronicle*, Publications International, Ltd., Lincolnwood, Illinois, (2003), p. 177.

Rabbi Halpern replied to M. Prilutzky that the Orthodox Party, which, as he believed, formed the preponderating group of Jews in Poland, only demanded equal rights. He stated that the Jews loved Poland, and that they believed the declarations of the Polish Party leaders that the Jews would get equal rights. He expressed the fear that the Nationalists would impair the relations between the Poles and the Jews."

Many Jews were aware of the fact that the Zionists were sponsoring anti-Semitism and that Zionists agreed with the precepts of anti-Semitism—were themselves anti-Semites. Some Zionists loudly protested against this truth. On 3 September 1897 on page 12, an article in *The Jewish Chronicle* paraphrased Dr. Birnbaum's statement at the First Zionist Congress,

"Dr. BIRNBAUM mentioned that it had often been contended that Zionism was but a reaction against anti-Semitism. It had not been denied that the growth of Zionism coincided with that of anti-Semitism, and, therefore, the conclusion was arrived at that the former only existed at the mercy of the latter. This was a complete mistake. It should be remembered that every movement had its causes and impetus. Through the former it obtained its pioneers, and through the latter its troops. Zionism could proudly say of itself that all who stood at its head had either long left the anti-Semitic impetus behind them, or that from the beginning their belief originated in the anomaly of the existence of a Jewish people. The want of a land of their own caused this anomaly to be the greater. There was a sentimental feeling in favour of Palestine, but sentiment would not suffice because the land whither they would go did not need special attraction; any country in which their distress would cease would be attractive; what they required was a land which would be able to keep them once they were there, till the grand process of converting them from a mercantile people into a people devoted to all callings, especially agriculture, had been completed, and they would no longer hanker after the flesh pots of Egypt [*Exodus* 16:3]. Palestine was the only country able to accomplish this. The second reason in favour of Palestine was the benefits that would be conferred not on Jews alone, but on mankind in general. A Jewish people in Palestine would not alone be the medium between the social-ethical and political-æsthetical elements of Europeism, but also the long-sought medium between the East and the West. No people is so apt for this as the Jews with their inherited Oriental qualities and their acquired European character. No country is so fitted to be the territorial medium as Palestine, with its proximity to Europe and to the Suez Canal, and as being the inevitable station on the railway to India. Fears had been expressed for the future of the Holy Sepulchre if Jews became the masters of Palestine, but by making the Christian holy places extra-territorial the difficulty would be overcome and all fears would be dissipated."

Much of what Birnbaum stated echoed the sophistry contained in Theodor

Herzl's book *The Jewish State*. If the Zionists were genuinely interested in the best interests of humanity, they would have propped up the Turkish Empire, instead of trying to tear it down. It was the Turkish Empire which had the potential to fulfill the rôles the Zionist European Jews artificially claimed as their own. The Zionists knew that a large Jewish presence in Palestine would have the exact opposite effect of what they claimed. Instead of bringing peace to the region, it would inflame the Moslems and Catholics against the Jews and against one another. The Zionists tossed out the bait that the Suez Canal was of vital interest to European trade, and then falsely asserted that a Jewish presence would secure that interest, when in fact the Jews knew quite well that a Jewish presence would jeopardize European interests by instigating religious conflict. There was nothing that prevented the British and the French from maintaining productive relations with Moslems, other than Jewish Messianic designs.

The Jews did not want to secure the Suez for the sake of the Europeans, rather the Jews wanted the Northern European and British Protestants to secure Palestine for the Jews and protect them from the Catholics and the Muhammadans who would be inflamed by a Jewish colonization of their Holy Lands and shrines. Jewish intolerance of other religions remains a threat today, when Jewish Israelis attack Christians in Bethlehem, violate international law in Jerusalem, and seek to destroy the militaries and societies in Moslem countries, so that the Moslems will have no means with which to fight back when the Jews and Dispensationalist Christians destroy the Dome of the Rock and Al Aqsa Mosque and build a Jewish Temple in their place. The Israelis are also using the military of the United States to take over the territory of Greater Israel, which they know will eventually pass into their hands.

Beginning in the late 1800's, Jewish Zionists heavily promoted anti-Semitism and anti-Semites. Crypto-Jews founded and led anti-Jewish societies, which were financed with Jewish bankers' money. The most prominent Nazis were crypto-Jewish Frankists—Zionist propagandists in anti-Semite's clothing, *agents provocateur*, including Alfred Rosenberg, who took his political ideology from Houston Stewart Chamberlain, Liebenfels and List. Theodor Herzl took his racist political ideology from Eugen Karl Dühring, making Dühring an influence on both Herzlian Zionist racist mythology and Nazi racist mythology.[78] Before Dühring was the Jewish racist Zionist Moses Hess, who created National Socialist racism. Dietrich Eckart proposed that a demagogue should lead the Germans to drive out the Jews—long a Zionist objective.

In 1909, Zionist Max Nordau presented a character profile for the successful revolutionary that fit Hitler, Lenin and Stalin.[79] Nordau, though born and raised in Austro-Hungary, called himself a German, parroted the *Übermensch*

---

[78]. M. Samuel, "Diaries of Theodor Herzl", in: M. W. Weisgal, *Theodor Herzl: A Memorial*, The New Palestine, New York, (1929), pp. 125-180, at 129. T. Herzl, English translation by H. Zohn, R. Patai, Editor, *The Complete Diaries of Theodor Herzl*, Volume 1, Herzl Press, New York, (1960), pp. 4, 111.

[79]. M. Nordau, *Der Sinn der Geschichte*, C. Duncker, Berlin, (1909); English translation: *The Interpretation of History*, Willey Book Co., New York, (1910), pp. 309-315.

philosophy, ridiculed Judaism and Christianity, copied the Germanic Hegelian dialectic, then called the modern world and those philosophers he was copying "degenerate"—a favorite word of Lombroso, who was Jewish,[80] and the Nazis.[81] Disraeli, Nordau and Zollschan promoted the alleged superiority of the "blonde Nordic race" in order to promote the segregation of their own "Jewish race". They asserted that the German and the Jew were superior races to the Slav and the Negro.

---

[80]. I. Zangwill, *The Problem of the Jewish Race*, Judaen Publishing Company, New York, (1914), p. 13; which was first published as an article, "The Jewish Race", *The Independent*, Volume 71, Number 3271, (10 August 1911), pp. 288-295, at 292.
[81]. J. B. Agus, *The Meaning of Jewish History*, Volume 2, Abelard-Schuman, New York, (1963), pp. 410-411, 420.

Cesare Lombroso,[82] who was Jewish,[83] advocated the extermination of

---

[82].C. Lombroso, Italian: *L'Uomo di Genio*, Bocca, Torino, (1888); English: *The Man of Genius*, C. Scribner's Sons, New York,(1891); French: *L'Homme de Génie*, Alcan, Paris, (1889); **and** *Le Crime, Causes et Remèdes*, F. Alcan, Paris, (1907); English: *Crime: Its Causes and Remedies*, W. Heinemann, London, (1911); **and** *L'Uomo di Genio in Rapporto alla Psichiatria, alla Storia ed all'Estetica*, Fratelli Bocca, Torino, (1888); **and** *Applications de l'Anthropologie Criminelle*, Félix Alcan, Paris, (1892); **and** *Les Anarchistes*, E. Flammarion, Paris, (1896); **and** *Le Più Recenti Scoperte ed Applicazioni della Psichiatria ed Antropologia Criminale*, Bocca, Torino, (1893); **and** *Palimsesti del Carcere; Raccolta Unicamente Destinata Agli Uomini di Scienza*, Bocca, Torino, (1888); **and** *L'Anthropologie Criminelle et ses Récents Progrès*, F. Alcan, Paris, (1890); **and** French: *L'Antisemitismo e le Scienze Moderne*, L. Roux e C., Torino, (1894); German: *Der antisemitismus und die Juden im lichte der modernen Wissenschaft*, G.H. Wigand, Leipzig, (1894); **and** French: *L'Antisémitisme*, V. Giard & E. Brière, Paris, (1899); Spanish: *El Antisemitismo*, Viuda de Rodríguez Serra, Madrid, (1900's); **and** *L'Uomo Delinquente, in Rapporto all'Antropologia, alla Giurisprudenza ed alle Discipline Carcerarie: Delinquente-Nato e Pazzo Morale*, Fratelli Bocca, Torino, (1884); **and** *Kerker-Palimpseste; Wandinschriften und Selbstbekenntnisse gefangener Verbrecher*, Verlagsanstalt und Druckerei a.g., Hamburg, (1899); **and** *Criminal Man According to the Classification of Cesare Lombroso*, Patterson Smith, Montclair, New Jersey, (1911); **and** *L'Homme Criminel; Criminel-né, Fou Moral, Épileptique; Étude Anthropologique et Médico-Légale*, Alcan, Paris, (1887) ; **and** *Nuovi Studii sul Genio*, R. Sandrom, Milano, (1902); **and** *Les Applications de l'Anthropologie Criminelle*, Alcan, Paris, (1892); **and** *Crime: Its Causes and Remedies*, Little, Brown, Boston, (1911); **and** *L'Homme Criminel; Criminel né, fou Moral, Épileptique, Criminel fou, Criminel d'Occasion*, F. Alcan, Paris, (1895); **and** *Palimpsestes des Prisons Recueillis*, A. Storck, Lyon, (1894); **and** *Der Verbrecher, in anthropologischer, ärztlicher und juristischer Beziehung*, Richter, Hamburg, (1887-1890); **and** Italian: *Genio e Degenerazione. Nuovi Studi e Nuove Battaglie*, Sandron, Palermo, (1897); German: *Entartung und Genie*, G.H. Wigand, Leipzig, (1894); **and** *Genie und Irrsinn in ihren Beziehungen zum Gesetz, zur Kritik und zur Geschichte*, P. Reclam, Leipzig, (1887); **and** *Neue Fortschritte in den Verbrecherstudien*, Leipzig, (1894); **and** *Neue Verbrecherstudien...* , a.S., Carl Marhold, Halle, (1907); **and** *Die Anarchisten; eine kriminalpsychologische und sociologische Studie*, J.F. Richter, Hamburg, (1895); **and** ; **and** *Die Ursachen und Bekämpfung des Verbrechens*, H. Bermühler, Berlin, (1902); **and** *Genio e Follia; in Rapporto alla Medicina Legale, alla Critica ed alla Storia*, Bocca, Roma, (1882); **and** *Aplicaciones Judiciales y Médicas de la Antropología Criminal*, La España Moderna, Madrid, (1892) ; **and** *Les Palimpsestes des Prisons*, A. Storck, Lyon, G. Masson, Paris, (1894); **and** *Der Selbstmord der Verbrecher insbesondere im Zellengefaengnis*, Berlin, (1901); **and** *L'Homme Criminel: Etude Anthropologique et Psychiatrique*, Félix Alcan Ed., Paris, (1895); **and** *Criminal Man According to the Classification of Cesare Lombroso*, G.P. Putnam's Sons, New York, (1911); **and** *Genie und Entartung: Autorisierte Ubersetzung aus dem Stalienischen*, P. Reclam, Leipzig, (1910); **and** *The Heredity of Acquired Characteristics* ; **and** *L'Uomo Bianco e l'Uomo di Colore Letture su l'Origine e la Varietà delle Razze Umane*, Fratelli Bocca, Torino, Firenze, (1892); **and** *Studien über Genie und Entartung*, P. Reclam, Leipzig, (1910); **and** *Problemes du Jour*, Libr. Universelle, Paris, (1906); **and** *Lombroso und die Criminal-Anthropologie von heute*, Hubertusburg, (1897) ; **and** *L'Amore nel Suicidio e nel Delitto*, Ermanno Loescher, Torino, (1881); **and** *Criminal Anthropology: Its Origin and Application*, Forum Pub. Co.,

alleged criminal phenotypes. His theories later became the model for the Nazis' gassing of political opponents, criminals, the insane and the infirm. Jewish Zionist Max Nordau was in many senses the archetype Nazi. Many Nazis and anti-Semites were of Jewish origin. Both the Zionists and the Nazis loathed the Slavic "race" and brought about its downfall in modern times. They also attempted to wipe out the blonde "race" of Nordics they pretended to admire.

The Old Testament is filled with stories of Jews massacring other Jews and of human sacrifice. The Old Testament and the Talmud instruct pious Jews to kill Jews who abandon Judaism, especially those who sincerely convert to other religions, as well as heathen priests.[84] While addressing the justifications given by religious Zionist terrorists for Yigal Amir's murder of Yitzhak Rabin, Jessica Stern wrote in her book *Terror in the Name of God: Why Religious Militants Kill*,

> "According to the halakah, the rulings of Din Mosser and Din Rodef apply to those Jews who have committed the most despicable crime imaginable— the betrayal of their fellow Jews. The punishment of the Mosser—a person

---

New York, (1895); **and** *The Physiognomy of the Anarchists*, Philadelphia, (1891/1993); **and** *Virchow und die Kriminalanthropologie*, (1896); **and** *The Physiology & Psychology of Crime*, American Institute for Psychological Research, Albuquerque, (1895/1980); **and** *Pazzi ed Anomali; Sàggi*, Lapi, Città di Castello, (1886); **and** *Studj Clinici ed Esperimentali sulla Natura, Causa e Terapia della Pellagra*, G. Bernardoni, Milano, (1870); **and** *Études de sociologie: Les Anarchistes*, E. Flammarion, Paris, (1897). *See also:* C. Lombroso, E. Ferri, R. Garofalo, *Et al.*, *Polemica in Difesa della Scuola Criminale Positiva*, N. Zanichelli, Bologna, (1886). *See also:* C. Lombroso, G. Regnier, and A. Bournet, *L'Homme Criminel; Étude Anthropologique et Médico-Légale*, F. Alcan, Paris,(1887).. *See also:* C. Lombroso, R. Laschi, *Il Delitto Politico e le Rivoluzioni in Rapporto al Diritto, all'Antropologia Criminale ed alla Scienza di Governo. Con 10 Tavole e 21 Figure nel Testo*, Fratelli Bocca, Torino, (1890). *See also:* C. Lombroso, R. Laschi, Rodolfo, and A. Bouchard, *Le Crime Politique et les Révolutions par Rapport au Droit, à l'Anthropologie Criminelle et à la Science du Gouvernement*, F. Alcan, Paris, (1892). *See also:* C. Lombroso, R. Laschi, Rudolfo, H. Kurella, *Der politische Verbrecher und die Revolutionen in anthropologischer, juristischer und staatswissenschaftlicher Beziehung*, Verlagsanstalt und Druckerei, Hamburg, (1891-1892). *See also:* C. Lombroso, G. Ferrero, Italian: *La Donna Delinquente: La Prostituta e la Donna Normale*, L. Roux, Torino, (1893); German: *Das Weib als Verbrecherin und Prostituirte: Anthropologische Studien, gegründet auf eine Darstellung der Biologie und Psychologie des normalen Weibes*,Verlagsanstalt und Druckerei, Hamburg, (1894); French: *La Femme Criminelle et la Prostituée*, F. Alcan, Paris, (1896); English: *The Female Offender*, T.F. Unwin, London, (1895). *See also:* P. Näcke, C. Lombroso, *Ein Willkommengruss von Herrn Lombroso*, Leipzig, (1894). *See also:* L. Fratiny, C. Lombroso, *Une Interview. Criminalité Génialité: C. Lombroso Jugé par Mignozzi-Bianchi*, Firenze, (1909).
83. I. Zangwill, *The Problem of the Jewish Race*, Judaen Publishing Company, New York, (1914), p. 13; which was first published as an article, "The Jewish Race", *The Independent*, Volume 71, Number 3271, (10 August 1911), pp. 288-295, at 292.
84. *Abodah Zarah* 26b.

who hands over sacred Jewish property to the gentile—as well as that of the Rodef—a person who murders or facilitates the murder of Jews—shall be death. Since the execution of the Mosser or the Rodef is aimed at saving the lives of other Jews, there is no need for a trial."[85]

The political Zionists considered non-Zionist Jews to be traitors and they believed assimilation would lead to the death of the mythical "Jewish race". Moses Hess wrote,

"The most touching point about these Hebrew prayers is, that they are really an expression of the collective Jewish spirit; they do not plead for the individual, but for the entire Jewish race. The pious Jew is above all a Jewish patriot. The 'new' Jew, who denies the existence of the Jewish nationality, is not only a deserter in the religious sense, but is also a traitor to his people, his race and even to his family. If it were true that Jewish emancipation in exile is incompatible with Jewish nationality, then it were the duty of the Jews to sacrifice the former for the sake of the latter. This point, however, may need a more elaborate explanation, but that the Jew must be above all a Jewish patriot, needs no proof to those who have received a Jewish education. Jewish patriotism is not a cloudy Germanic abstraction, which dissolves itself in discussions about being and appearance, realism and idealism, but a true, natural feeling, the tangibility and simplicity of which require no demonstration, nor can it be disposed of by a demonstration to the contrary. "[86]

Anti-Semitism was very useful to both the Communists and the Zionists. Politically active anti-Semitic demagogues like Lueger, Ahlwardt, Treitschke and Stoecker had numerous Jewish connections, as did Adolf Hitler—some even had Jewish blood, as did Hitler. Anselm von Rothschild stated that Stoecker was an apostate Jew. The Rothschilds wanted desperately to buy Palestine and establish a Jewish state there, with a Rothschild sitting as king of the world, but the Rothschilds lacked broad Jewish support. The political Zionists later concluded that they could only obtain Jewish support in a climate of advanced anti-Semitism. *The Chicago Tribune*, on 12 December 1881 on page 6, reprinted a letter from Rothschild to Stoecker:

"BARON ROTHSCHILD.

The Letter Written by Him in Defense of the Jews.
Baron Anselm von Rothschild, of Vienna, wrote the annexed letter to

---

85. J. Stern, *Terror in the Name of God: Why Religious Militants Kill*, Ecco, New York, (2003), p. 91.
86. M. Hess, *Rom und Jerusalem: die letzte Nationalitätsfrage*, Eduard Wengler, Leipzig, (1862); English translation, *Rome and Jerusalem: A Study in Jewish Nationalism*, Bloch, New York, (1918/1943), pp. 62-63.

Hof-Prediger Stoecker, of Berlin, the instigator of the anti-Semitic agitation in Germany:

VIENNA, November, 1881.—*To the Court Preacher Stoecker*—SIR: If I am correctly informed, your physician once advised you to take plenty of exercise, and since then you have been almost constantly employed in anti-Semitic movements. This matter really concerns me very little, for, thank God, Austria has not yet advanced so far on the path of intelligence and refinement as to possess a 'Judenhetze,' such as the cultivated city of Berlin can boast of. But still I should like to call the attention of your reverence to certain grave errors which have crept into your speech recently delivered in the German Parliament.

You said in that address, 'Behind me stand the millions.' You are mistaken: the millions stand behind me, and if you doubt this you are respectfully invited to visit my counting-house, where ample proofs shall be given you. You contend that 'the Jewish usurers have ruined all classes of people.' Now, pray tell me, my dear Court-Preacher, who goes to the Jewish usurer? Is it not those whose credit is exhausted? And if their fellow-men will not trust them any longer, are they not already ruined before they seek their last resort—the Jewish usurer? This is only another of many cases where the Jew is made the scapegoat for the offenses of his neighbors. [A Gentile, or Gentile government, could be easily forced to seek loans from a Jewish usurer through the agitation of the Jewish press for war, or an infinite number of other corrupt means—as evinced by the Jewish financiers' destruction of Russia.]

You say further that the Jews, out of all proportion to their numbers, assisted by talent and capital, exercise a mighty influence in the community. I am really surprised that this should surprise you. As if talent had not, from time immemorial, held the sceptre. Would you rather that this world should be ruled by fools than by wise men? And as far as the disproportion of our numbers is concerned we Jews cannot help but feel highly flattered if we possess more talent than our Gentile countrymen. And as for our power as capitalists this is the result of our business genius and our economy. Why do not the Christians imitate us? Do we hinder them from earning money or from saving it? [The answer to this question is obviously yes. Jewish power, wealth and influence in the press result from Jewish racism and Jewish tribalism, and the Cabalist and Talmudic doctrines which encourage Jews to take advantage of Christian integrity in order to exploit Christians. Rich Jews promoted honesty and decentralized power among the Gentiles, while promoting dishonesty and tribalism among their own. This gave the Jews an advantage, which could only be overcome by the Gentiles' sinking to the debased level of the foe, or expelling it. Rothschild's racist arrogance is ample proof of the fact, and if the Gentiles had truly leveled the playing field by sinking to Rothschild's level, they would have quickly crushed the Jews.]

'The Jews should be more modest,' you say. It is true that modesty is a most desirable virtue, suited alike to Jew and Gentile, but as Goethe has it, 'Only scoundrels are modest.' Now, among the Jews there are so few scoundrels, and then really it is much easier for a Court Preacher to be modest than for a Jew. If a Court Preacher displays that commendable virtue, his flock will bow before him and exclaim, 'So mighty and yet so modest.' Let a Jew be modest and he is kicked and spurned, and the mob say, 'Serves him right.' [The reality is that the rich Jews concentrated their wealth and shared it with neither Gentile nor poor Jew. This concentration of wealth gave the Jews enormous power and the resentment this corrupt and undemocratic warmongering power caused was directed at poor Jews, often through the machinations of rich Jews, who sought to keep their poor brethren segregated.]

You aver that the Jew in Lessing's 'Nathan' is no Jew at all, but a Christian. With the same right I might say the Court Preacher Stoecker is no Christian, but an apostate Jew who has banded himself with some barbarous relics of the Middle Ages to prosecute a miserable anti-Semitic agitation. But I will not say this, as I would not desire to so grossly insult my co-religionists. [It is highly interesting that Rothschild called Stoecker an apostate Jew.]

Your friend Bachem is of the opinion that the people are backing him. I will admit that there is a people in his wake, but as a German philosopher once said: 'There are enough wretches in the world to back any bad cause.' Your friend also indulges in the crushing accusation that the Jews are grain speculators. Now, do you know who was the first speculator in grain? None other than Jew, Joseph, in Egypt, although at that time there was no Court Preacher to discover any crime in his action, and the people were grateful to him. [The Egyptian people were not grateful to Joseph, who brought them into slavery (*Genesis* 47).]

In one of your discourses you once exclaimed: 'Look at Herr von Bleichroeder. He has more money than all the evangelical preachers put together.' Now, I am sure that Herr von Bleichroeder has never said: 'Look at Court Preacher Stoecker. He earns more money by a single sermon than a hundred Jewish firms do in a whole year.'

In conclusion, if you will not admit that the Jews have any good qualities, you will at least not envy them for the little money they may possess. If in spite of their wealth they cannot prevent, in the year of 1881, the formation of an agitation against them, what would become of them if they had no money? It is true the Jews put some value on wealth, and I must say that I would rather be a rich Jew than a poor Christian. But then, are there not poor Christians who would rather be rich Jews? Even you, most reverent sir, might, perhaps, be willing to change positions with me (and I flatter myself that you would not make so bad a bargain). For myself, I can only say that if I were not Rothschild, I should still be very far from wishing myself the Court Preacher Stoecker. Very respectfully, A. VON ROTHSCHILD."

The essentially meaningless term "anti-Semitism", a term for anti-Jewish which is employed as a weapon by hypocritical racist Jews with which to smear and threaten, was coined by an "anti-Semitic" crypto-Jew named Wilhelm Marr after Jewish corruption imploded the stock markets of Europe. In its article entitled "ANTI-SEMITISM", *The Universal Jewish Encyclopedia*, Volume 1, The Universal Jewish Encyclopedia, Inc., New York, (1939), pp. 341-409, at 341; states:

> "The word was probably first used by Wilhelm Marr, said to have been a converted Jew, in Der Sieg des Judentums ueber das Germanentum, a pamphlet which he published in 1879, the same year in which he founded the Anti-Semitic League; two years later he began publication of Zwanglose antisemitische Hefte."

Under the heading "Foreign Articles", the following statement appeared in *Niles' Weekly Register*, Volume 17, Number 427, (13 November 1819), p. 169,

> "Mr. Rothschild, the great London banker, indignant at the persecution of his Jewish brethren in Germany, has refused to take bills upon any of the cities in which they are persecuted; and great embarrassments to trade have been experienced in consequence of his determination. ☞ It is intimated that the persecution of the Jews is in part owing to the fact, that Mr. Rothschild and his brethren were among the chief of those who furnished the 'legitimates,' with money to forge chains for the people of Europe."

There would not have been agitations against the Jews, if the Jews in the press had not attacked Christianity and if the Jewish financiers had not attacked Europe with perpetual war[87] throughout the Nineteenth Century. That "little money" in the hands of the Rothschilds alone amounted to some, or one might say "sum" $3,400,000,000.00, acquired through deceitful and inhuman means. It is interesting, though not at all unusual, that Stoecker was a Jew and was behind the anti-Jewish agitation. The same could be said of Goebbels, Streicher, Heydrich, Frank, etc., and, no doubt, Rothschild.

The outspoken Mayor of Vienna Karl Lueger proclaimed that he decided who was, and who was not, a Jew, meaning that he could protect those Jews who helped him—those Jews who put him in power in order to spread anti-Semitism. He had Jewish backers and was an agent for their agenda. Anti-Semite Hermann Ahlwardt advocated the segregation of Jews in the Reichstag in 1895. The segregation of Jews was a Zionist objective. Ahlwardt spoke in anti-assimilationist terms Theodor Herzl would soon use in his book *The Jewish State*. Adolf Stoecker also raised his voice to advocate segregation in the schools, as did racist Zionist Albert Einstein. The dogma of segregation had both Zionist and anti-Semitic origins—for example racist Zionist Moses Hess' *Rom und*

---

[87]. G. E. Griffin, "The Rothschild Formula", *The Creature from Jekyll Island: A Second Look at the Federal Reserve*, Chapter 11, Fourth Edition, American Media, Westlake Village, California, (2002), pp. 217-234.

Jerusalem: die letzte Nationalitätsfrage of 1862 and anti-Semite Eugen Karl Dühring's *Die Judenfrage als Racen-, Sitten- und Culturfrage: mit einer weltgeschichtlichen Antwort* of 1881. Hermann Ahlwardt stated to the Reichstag in 1895,

> "A Jew who was born in Germany does not thereby become a German; he is still a Jew. Therefore it is imperative that we realize that Jewish racial characteristics differ so greatly from ours that a common life of Jews and Germans under the same law is quite impossible because the Germans will perish."[88]

Jewish Zionist Bernard Lazare wrote in 1894,

> "Everything is tending to bring about such a consummation. Such is the irony of things that antisemitism which everywhere is the creed of the conservative class, of those who accuse the Jews of having worked hand in hand with the Jacobins of 1789 and the Liberals and Revolutionists of the nineteenth century, this very antisemitism is acting, in fact, as an ally of the Revolution. Drumont in France, Pattai in Hungary, Stoecker and von Boeckel in Germany are co-operating with the very demagogues and revolutionists whom they believe they are attacking. This antisemitic movement, in its origin reactionary, has become transformed and is acting now for the advantage of the revolutionary cause. Antisemitism stirs up the middle class, the small tradesmen, and sometimes the peasant, against the Jewish capitalist, but in doing so it gently leads them toward Socialism, prepares them for anarchy, infuses in them a hatred for all capitalists, and, more than that, for capital in the abstract. And thus, unconsciously, antisemitism is working its own ruin, for it carries in itself the germ of destruction.
>
> Such, then, is the probable fate of modern antisemitism. I have tried to show how it may be traced back to the ancient hatred against the Jews; how it persisted after the emancipation of the Jews, how it has grown and what are its manifestations. In every way I am led to believe that it must ultimately perish, and that it will perish for the various reasons which I have indicated, because the Jew is undergoing a process of change; because religious, political, social, and economic conditions are likewise changing; but above all, because antisemitism is one of the last, though most long lived, manifestations of that old spirit of reaction and narrow conservatism, which is vainly attempting to arrest the onward movement of the Revolution."[89]

---

[88]. English translation from: P. W. Massing, *Rehearsal for Destruction: A Study of Political Anti-Semitism in Imperial Germany*, Howard Fertig, New York, (1967). p. 304.
[89]. B. Lazare, *Antisemitism: Its History and Causes*, (1894), pp. 182-183. *L'Antisémitisme, son Histoire et ses Causes*, L. Chailley, Paris, (1894).

Dietrich Eckart[90] promoted Adolf Hitler as a viable anti-Semitic demagogue, though many thought that Hitler appeared to be a Jewish actor or comedian spoofing an anti-Semitic demagogue, and they laughed at him. Eckart said,

> "The best would be a worker who knows how to talk... . He doesn't need much brains, politics is the stupidest business in the world, and every marketwoman in Munich knows more than the people in Weimar. I'd rather have a vain monkey who can give the Reds a juicy answer, and doesn't run away when people begin swinging table legs, than a dozen learned professors. He must be a bachelor, then we'll get the women."[91]

Erik Jan Hanussen, a crypto-Jew whose real name was Hermann Steinschneider, coached Hitler on effective public speaking.[92] In 1933, the sick joke, the clown, the Bolshevik Zionist Adolf Hitler rose to power over Germany lifted on the golden purse strings of the Jewish bankers.

In 1934, Jacob R. Marcus incorrectly predicted that Nazis would not carry out the Holocaust, because so many of its prominent leaders were, by Nazism's own standards, "sub-human",

> "The present National Socialist government is too shrewd, in spite of its racial commitments, to lend itself to such extravaganzas. It wants no Brahmanic caste-system in which even the shadow of a low caste Hindu is a pollution. It knows that any attempt toward racial eugenics along purely Nordic lines would disrupt present day Germany with its half-dozen racial strains. Nordicization, if it were literally true to itself, would mean the exclusion from the German state of the following non-Nordic types: the late Paul von Hindenburg; Streicher, the rabid anti-Semitic Nuremberg journalist; Ley, the head of the National Socialist Labor Front; Goebbels, Minister of National Enlightenment and Propaganda; and finally, Hitler, himself. Here is a racial analysis of Hitler made in 1929 by the racial-hygienist, Professor von Gruber, then President of the Bavarian Academy of Sciences and a member of the racially-minded Pan-American Association:
> 'For the first time I saw Hitler at close range. Face and head of poor race, mongrel, low slanting forehead, ugly nose, broad cheek bones, small eyes, dark hair. A short brushlike mustache, no broader than the nose, gives the face a defiant touch. The facial expression is not that of

---

90. A. Hitler and D. Eckart, *Der Bolschewismus von Moses bis Lenin*, Hoheneichen-Verlag, München, (1924); English translation by W. L. Pierce, *Bolshevism from Moses to Lenin: A Dialogue between Adolf Hitler and Me*, World Union of National Socialists, Arlington, Virginia, (1966).
91. D. Eckart, quoted in A. Hitler, translated by R. Manheim with an introduction by A. Foxman, *Mein Kampf*, Houghton Mifflin, Boston, New York, (1999), p. 687.
92. M. Gordon, *Erik Jan Hanussen: Hitler's Jewish Clairvoyant*, Feral House, (2001).

a man who has complete control of himself but of one who is aroused to frenzy. Repeated twitching of the facial muscles. When through, expression of contented self-reliance.' (*Essener Volkswacht*, Nov. 9, 1929.)"[93]

The exposure of the active involvement of Zionists with the Nazi hierarchy—even as instigators of the entire Nazi movement—is shocking, but one is reminded of the willingness of some Jewish religious fanatics to commit suicide and to submit to genocide in order to preserve the integrity of the holy land and of the "race" of the "chosen". Racists like the Jewish Zionist Meir Kahane thrived on conflict. Kahane asked Jews to rejoice at the United Nations Resolution which acknowledged that Zionism is a form of racism. He hoped that it would lead to strife between Gentiles and Jews, because he believed that this would ultimately lead to the destruction of the Gentile world, as Jewish prophecy foretold. Kahane hoped that the entire world would turn against Israel, and falsely tied all Jews to Israel, in the hopes that Jews would be humiliated and then the Gentiles would be destroyed by God, in the form of Zionist subversion. Kahane succinctly wrote, *inter alia*,

"The banding together by the nations of the world against Israel is the guarantee that their time of destruction is near and the final redemption of the Jew at hand."[94]

Jessica Stern, in her book *Terror in the Name of God: Why Religious Militants Kill*, writes of Jews who are willing,

"to risk a world war in pursuit of religious redemption for the Jewish people.30"[95]

Baruch Kimmerling wrote,

"At the center of this culture of death is the remembrance of martyrs—Jews who, in Zionist ideology, had to die so that the state might be born. [***] *A triumphal creed shadowed by death, Zionism transformed the catastrophes of Jewish history into nationalist fables of redemption.*"[96]

---

[93]. J. R. Marcus, *The Rise and Destiny of the German Jew*, The Union of American Hebrew Congregations, Cincinnati, (1934), p. 62.
[94]. M. Kahane, *On Jews and Judaism: Selected Articles 1961-1990*, Volume 1, Institute for the Publication of the Writings of Rabbi Meir Kahane, Jerusalem, (1993), p. 104.
[95]. J. Stern, *Terror in the Name of God: Why Religious Militants Kill*, Ecco, New York, (2003), p. 105. Stern cites E. Sprinzak, "From Messianic Pioneering to Vigilante Terrorism", in D. C. Rapoport, Editor, *Inside Terrorist Organizations*, Frank Cass, London, (1988), pp. 194-216.
[96]. B. Kimmerling, "Israel's Culture of Martyrdom", *The Nation*, (10-17 January 2005), pp. 29-30, 33-34, 36, 38, 40, at 29; which is a review of I. Zertal, *Israel's Holocaust and*

Though Kahane has been rejected by the vast majority of Jews, and by the majority of Israelis, his message is in keeping with Judaism. Kahanism has a romantic allure to some Jews of a promise of community and common enemy. That battles with their better natures and Kahanism threatens to become a broad movement if not checked and exposed again and again as the hateful mythology that it is. This lust for persecution and martyrdom in order to bring death upon the enemies of the Jews, real or imagined enemies, is an ancient tradition for Jews. It is clearly advocated in the writings of Philo the Jew and Josephus, as well as those of Theodor Herzl. Philo the Jew vilified the Egyptians and Caligula with lies—as did Josephus with even more outrageous lies. Josephus fabricated the myth of Masada. These legendary lies are ingrained in the psyches of those who see these lies as their history and who have a "Masada Complex" of imagined persecution and martyrdom.[97] Many modern Jews have created an unhealthy culture of death and persecution around the Holocaust—some say the Holocaust has become a new religion.[98]

In the book of *Numbers*, Chapter 25, Jews were commanded by God to commit genocide against Jews who had assimilated. According to the *Gospel of John* 11:47-53, Caiaphas chose to execute Jesus in order to preserve the nation of the Jews and to gather back its supposedly chosen people:

> "47 Then gathered the chief priests and the Pharisees a council, and said, What do we? for this man doeth many miracles. 48 If we let him thus alone, all men will believe on him: and the Romans shall come and take away both our place and nation. 49 And one of them, named Caiaphas, being the high priest that same year, said unto them, Ye know nothing at all, 50 Nor consider that it is expedient for us, that one man should die for the people,

---

the Politics of Nationhood, Cambridge University Press, (2005); and Y. Grodzinsky, *In the Shadow of the Holocaust: The Struggle Between Jews and Zionists in the Aftermath of World War II*, Common Courage Press, Monroe, Maine, (2004).

97. Philo the Jew, "Flaccus" and "On the Embassy to Gaius", *The Works of Philo*, Hendrick Publishing, U. S. A., (2000), pp. 725-741 and 757-790. Josephus, "Against Apion", *The Works of Josephus*.

98. R. Garaudy, *Les Mythes Fondateurs de la Politique Israélienne*, Samisztat, Paris, (1996); English translations: *The Founding Myths of Israeli Politics*, and *The Mythical Foundations of Israeli Policy*, Studies Forum International, London, (1997) and *The Founding Myths of Modern Israel*, Institute for Historical Review, Newport Beach, California, (2000). **See also:** N. Finkelstein, , *The Holocaust Industry: Reflection on the Exploitation of Jewish Suffering*, Verso, London, New York, (2000); and *Beyond Chutzpah: On the Misuse of Anti-semitism and the Abuse of History*, University of California Press, Berkeley, (2005). **See also:** B. Kimmerling, "Israel's Culture of Martyrdom", *The Nation*, (10-17 January 2005), pp. 29-30, 33-34, 36, 38, 40; which is a review of I. Zertal, *Israel's Holocaust and the Politics of Nationhood*, Cambridge University Press, (2005); and Y. Grodzinsky, *In the Shadow of the Holocaust: The Struggle Between Jews and Zionists in the Aftermath of World War II*, Common Courage Press, Monroe, Maine, (2004).

and that the whole nation perish not. 51 And this spake he not of himself: but being high priest that year, he prophesied that Jesus should die for that nation; 52 And not for that nation only, but that also he should gather together in one the children of God that were scattered abroad. 53 Then from that day forth they took counsel together for to put him to death."

The book of *Matthew* 1:21-23 states that "Jesus"—the Jew—was meant to rescue the Jewish Nation,

"21 And she shall bring forth a son, and thou shalt call his name JESUS: for he shall save his people from their sins. 22 Now all this was done, that it might be fulfilled which was spoken of the Lord by the prophet, saying, 23 Behold, a virgin shall be with child, and shall bring forth a son, and they shall call his name Emmanuel, which being interpreted is, God with us."

If the New Testament is a fiction in part, or in whole, it is a fabrication that fixes blame for the destruction of the Temple and of Jerusalem on Jesus—from the Jewish point of view, instead of on the corruption of some leading Jews against the Roman government and the murder of Caligula for defiling the Temple. It also makes Jesus a means by which to preserve and consolidate the Jewish nation—a human sacrifice. If the New Testament is authentic, then Jesus' murder was a ploy, which enabled Jewish leaders to secure the unity of their people. In either event, the founding of Christianity—the story of the crucifixion of Christ—was a nationalistic attempt to unite the Jews of the world through human sacrifice—an alleged unity that some Jews have since sought and continue to seek at all costs to the themselves and without any regard for the rights and interests of others, both selflessly and selfishly willing to lead the world into an apocalyptic war in order to preserve their nationalistic vision.

At this critical time when humanity faces many important decisions and should be planning for the future of the survival of the human race, the tiny and insignificant country of Israel with a population of only six million receives grossly disproportionate attention on the world stage, draining off resources and time that the other six-and-one-half-billion human beings cannot afford to spare. Humanity would be better served to devote its resources to more important problems and simply impose an equitable solution to the problems in the Middle East with overwhelming force, or overwhelming disinterest. Though it seems the racist Jews will never rest until they have murdered off the Gentiles in way or another.

Early Christians inherited their love of martyrdom from the Jews and mostly were Jews. Ancient writers assert that ancient Jews believed that death by martyrdom was a certain means to immortality. One is further reminded of the countless failed attempts to form a Jewish nation and the desperation of the Zionists to find a means to achieve their ends because they believed the "Jewish race" was on the verge of extinction. The political Zionists embraced anti-Semitism as that meanest of means.

It is a fact that the Nazis in their writings and in their speeches promoted the

Zionists, and that the Zionists in their writings and in their speeches promoted anti-Semitism and the Nazis. It is also a fact that the Zionists advocated the racist position that Jews cannot and should not assimilate and were a foreign, disloyal, and combative nation within Germany.

This common interest between Nazis and Zionists includes financial collusion between Zionists and anti-Semites—the type of financial collusion Herzl advocated in his book *The Jewish State*. Herzl, who exhibited a psychopathic personality, held the majority of the Jews in low regard, and was eager to "sacrifice" them for his cause. His philosophical descendants were even more inhuman. Of course, the guilt of the Zionists who fomented the political climate which precipitated the Holocaust in no way abrogates the guilt of the many Germans and Europeans who participated in murdering millions of innocent men, women and children in the Holocaust and the Second World War. It serves as a warning to us all of the power held by those who mold public opinion and the possibilities for good or evil that control over that force holds. It is presently in the hands of the Zionists and has been for centuries.

7.5.3 Zionists and Communists Delight in Massive Human Sacrifices to the Jewish Messianic Cause

The Second World War ended in 1945 with Albert Einstein's 1915 vision of a divided and destroyed Germany made real. Communism was infinitely stronger than before the war and it looked as if France, Greece, Italy, Germany and even England, in their weakened state, would succumb to it. Zionists used the Nazis' crimes against Jews, which the Zionist Jews intentionally caused, to justify the formation of the State of Israel, and the theft of Palestine, and the perpetual vilification of the Moslems.

Since the ancient Diaspora, all previous attempts to found a State of Israel had failed and the outlook for Jews after the First World War was near total assimilation, and, in the racist minds of political Zionists, the consequent extermination of the "Jewish race". They were, in fact, desperate enough to create the Nazis as a means achieve their ends and they believed Jewish Messianic prophecy fully justified their treachery.

In 1921, political Zionist Jakob Klatzkin stated,

> "[I applaud] the contribution of our enemies in the continuance of Jewry in eastern Europe. [***] We ought to be thankful to our oppressors that they closed the gates of assimilation to us and took care that our people were concentrated and not dispersed, segregatedly united and not diffusedly mixed [***] One ought to investigate in the West and note the great share which antisemitism had in the continuance of Jewry and in all the emotions and movements of our national rebirth . [***] Truly our enemies have done much for the strengthening of Judaism in the diaspora . [***] Experience teaches that the liberals have understood better than the antisemites how to destroy us as a nation. [***] We are, in a word, naturally foreigners; we are

an alien nation in your midst and we want to remain one."[99]

In 1898, Nachman Syrkin wrote,

"Nonetheless, the enemy has *always* considered the Jews a nation, and they have always known themselves as such."[100]

In 1945, after the Zionist Nazi atrocities, Albert Einstein callously reminded the world of the Balfour Declaration and the Palestine Mandate in order to exploit the tragedy of the Holocaust the Zionists had deliberately caused. Einstein used the Holocaust to justify the fulfilment of his pre-Nazi political Zionist agenda. Einstein asserted that the Holocaust proved that the world thought of the Jews as a nation. Genocidal human sacrifice had long been a Judaic tradition, and in more recent times, Friedrich Engels made it clear that the Communists were comfortable with human sacrifices amounting to ten million lives lost in order to prepare the way for revolution and Communist world dominance. In 1887, Frederick Engels knew that the First World War was coming and that it would destroy the Empires of Europe and leave them ripe for revolution,
"No other war is now possible for Prussia-Germany than a world war, and indeed a world war of hitherto unimagined sweep and violence. Eight to ten million soldiers will mutually kill each other off, and in the process devour Europe barer than any swarm of locusts ever did. The desolation of the Thirty Years' War compressed into three or four years and spread over the entire continent: famine, plague, general savagery, taking possession both of the armies and of the masses of the people, as a result of universal want; hopeless demoralization of our complex institutions of trade, industry and credit, ending in universal bankruptcy; collapse of the old states and their traditional statecraft, so that crowns will roll over the pavements by the dozens and no one be found to pick them up; absolute impossibility of foreseeing where this will end, or who will emerge victor from the general struggle. Only *one* result is absolutely sure: general exhaustion and the creation of the conditions for the final victory of the working class."[101]

---

99. K. J. Herrmann, "Historical Perspectives on Political Zionism and Antisemitism", *Zionism & Racism: Proceedings of an International Symposium*, International Organization for the Elimination of All Forms of Racial Discrimination, Tripoli, (1977), pp. 197-210, at 204-205. A lengthy quotation from Klatzkin, in English translation, appears in: M. Menuhin, *The Decadence of Judaism in Our Time*, Exposition Press, New York, (1965), pp. 482-483.
100. N. Syrkin, under the nom de plume "Ben Elieser", *Die Judenfrage und der socialistische Judenstaat*, Steiger, Bern, (1898); English translation in A. Hertzberg, *The Zionist Idea*, Harper Torchbooks, New York, (1959), pp. 333-350, at 343.
101. B. D. Wolfe, *Marxism: One Hundred Years in the Life of a Doctrine*, Dial Press, New York, (1965), p. 67. Wolfe cites: "From Engels's introduction to the reissue of a pamphlet by Sigismund Borkheim. Borkheim's pamphlet, *Zur Errinnerung fuer die deutschen Mordspatrioten 1806-07* [***] The introduction is reproduced in *Werke*, Vol.

In 1945, Einstein wrote, among other things,

"[The Jews'] status as a uniform political group is proved to be a fact by the behavior of their enemies. Hence in striving toward a stabilization of the international situation they should be considered as though they were a nation in the customary sense of the word. [***] In parts of Europe Jewish life will probably be impossible for years to come. In decades of hard work and voluntary financial aid the Jews have restored the soil of Palestine to fertility. All these sacrifices were made because of trust in the officially sanctioned promise given by the governments in question after the last war, namely that the Jewish people were to be given a secure home in their ancient Palestinian country. To put it mildly, the fulfillment of this promise has been but hesitant and partial. Now that the Jews—especially the Jews in Palestine—have in this war too rendered a valuable contribution, the promise must be forcibly called to mind. The demand must be put forward that Palestine, within the limits of its economic capacity, be thrown open to Jewish immigration. If supranational institutions are to win that confidence that must form the most important buttress for their endurance, then it must be shown above all that those who, trusting to these institutions, have made the heaviest sacrifices are not defrauded."[102]

Einstein's statements prove that the human sacrifice of countless Jewish lives in the Zionist Holocaust had not changed the nationalistic racism of the political Zionists at all, but rather had strengthened their hand—in fulfillment of the Zionists' expressed plans. The racist Zionists had no regrets over their mass murder of Jews and they rejoiced at their slaughter of Gentiles. In the 1890's, Bernard Lazare iterated the Zionist mantra:

"It is because the Jews are a nation that anti-Semitism exists. [***] If the cause of anti-Semitism is the existence of the Jews as a nationality, its effect is to make this nationality more tangible for the Jews, to make them more aware of the fact that they are a people."[103]

Albert Einstein told Peter A. Bucky that the Holocaust had the benefit of uniting "all the Jews in the world":

"But the suffering had not been in vain, in Einstein's view. He felt that the Jews who died in Hitler's pogroms had strengthened the bond uniting all of

---

XXI, pp. 350-351."
102. A. Einstein, "Unpublished Preface to a Blackbook", *Out of My Later Years*, Philosophical Library, New York, (1950), pp. 258-259, at 259.
103. B. Lazare, "Jewish Nationalism and Emancipation (1897-1899)", in A. Hertzberg, *The Zionist Idea: A Historical Analysis and Reader*, Garden City, N.Y. Doubleday, (1959), pp. 471-476, at 471.

the Jews in the world."[104]

Einstein was simply repeating the Zionist party line, as expressed by Rabbi Abba Hillel Silver in 1943,

> "Should not, I ask you fellow Jews, ought not, the incalculable and unspeakable suffering of our people and the oceans of blood which we have shed in this war and in all the wars of the centuries; should not the myriad martyrs of our people, as well as the magnificent heroism and the vast sacrifices of our brave soldier sons who are today fighting on all the battle fronts of the world—should not all this be compensated for finally and at long last with the re-establishment of a free Jewish Commonwealth?"[105]

Did it occur to no one that the world, including the Jews, would be far better off if racist Zionism and Jewish tribalism were eradicated, rather than further justified, as a result of yet another massive Jewish tragedy? What, other than Jewish racism, prevented a massive drive for assimilation world-wide after the Holocaust?

7.5.4 Einstein Lulls Jews into Complacency—The Zionist Trap

After the Second World War and the Holocaust were over, few Jews wanted to emigrate to Palestine, despite the racist Zionists' best efforts to destroy their lives and make it impossible for them to live anywhere else. They had had enough of racist segregation. The Zionists then again employed corruption and the manipulation of public opinion to coerce Jews into moving to Palestine against their will and better natures.[106]

---

104. P. A. Bucky, Einstein, and A. G. Weakland, *The Private Albert Einstein*, Andrews and McMeel, Kansas City, (1992), p. 84.
105. A. H. Silver, *Vision and Victory*, Zionist Organization of America, New York, (1949); in A. Hertzberg, *The Zionist Idea*, Harper Torchbooks, New York, (1959), pp. 592-600, at 597.
106. A. M. Lilienthal, *What Price Israel*, Henry Regnery Company, Chicago, (1953), pp. vi-viii, 239. *See also:* "Israel's Flag Is Not Mine", *Reader's Digest*, (September, 1949), pp. 49-53. "The State of Israel and the State of the Jew", *Vital Speeches of the Day*, Volume 16, Number 13, (15 April 1950). *See also:* R. Garaudy, *Les Mythes Fondateurs de la Politique Israélienne*, Samiszdat, Paris, (1996); English translations: *The Founding Myths of Israeli Politics*, and *The Mythical Foundations of Israeli Policy*, Studies Forum International, London, (1997) and *The Founding Myths of Modern Israel*, Institute for Historical Review, Newport Beach, California, (2000). *See also:* B. Kimmerling, "Israel's Culture of Martyrdom", *The Nation*, (10-17 January 2005), pp. 29-30, 33-34, 36, 38, 40; which is a review of I. Zertal, *Israel's Holocaust and the Politics of Nationhood*, Cambridge University Press, (2005); and Y. Grodzinsky, *In the Shadow of the Holocaust: The Struggle Between Jews and Zionists in the Aftermath of World War II*, Common Courage Press, Monroe, Maine, (2004).

Einstein had long known that the Zionists would put a Hitler into power to attack European Jewry. Paul Ehrenfest made an interesting comment in an 8 February 1920 letter to Albert Einstein—a racist political Zionist who believed that anti-Semitism was the salvation of the Jews. Ehrenfest stated that the Zionists had commissioned a Hitler, a Haman, to save them from assimilation,

> "Something quite discontinuous is about to happen in Europe now, isn't that true?—And on this occasion a devil will surely come, on special commission to grab all Jews in Europe *uniformly* and *synchronously* by the scruff of the neck and give them a tremendous shake. Will the great miracle then happen that our prophets foresee, which will awaken and unite us all, orthodox and atheists alike, to a new living faith?—Maybe you have already seen something of it, even just a hint? I can't see it anywhere yet."[107]

The Babylonian Talmud states in *Sanhedrin* folio 97b,

> "R. Joshua said to him, if they do not repent, will they not be redeemed! But the Holy One, blessed be He, will set up a king over them, whose decrees shall be as cruel as Haman's, whereby Israel shall engage in repentance, and he will thus bring them back to the right path."[108]

The Jews put Hitler into power to fulfill this Talmudic commandment, to punish the Jews and thereby win them atonement and the return to Palestine.

Ehrenfest had earlier written to Einstein that an old and very influential Zionist Prof. Oppenheim had warned him that Zionists ought not to mix with secular Jews, who were not, in his view, Jews at all.[109] A sorry fate awaited secular Jews at the hands of the anti-Semites the Zionists had commissioned on special order. After stating that it was not in his nature to lie to the public with the dishonest Zionist propaganda claiming that Einstein was a "Jewish Newton", Ehrenfest expressed doubts about acting immorally and wrote to Einstein on 9 December 1919,

> "But God only knows, this old man may be right: maybe salvation of the masses can only be bought by the *hardest* sacrifice—sacrificing the last remnants of 'purity.' [Please don't read this as elegant empty words!]—Well, maybe that's how it is—but then my powers do not suffice."[110]

---

[107]. Letter from P. Ehrenfest to A. Einstein of 8 February 1920, English translation by A. Hentschel, *The Collected Papers of Albert Einstein*, Volume 9, Document 303, Princeton University Press, (2004), pp. 251-254, at 254.

[108]. I. Epstein, Editor, "Sanhedrin 20b", *The Babylonian Talmud*, Volume 27, The Soncino Press, London, (1935).

[109]. Letter from P. Ehrenfest to A. Einstein of 9 December 1919, English translation by A. Hentschel, *The Collected Papers of Albert Einstein*, Volume 9, Document 203, Princeton University Press, (2004), pp. 173-175, at 174.

[110]. Letter from P. Ehrenfest to A. Einstein of 9 December 1919, English translation by

Disturbed that Jews were perpetually defining themselves by a persecution myth—this many years, decades, centuries, before the appearance of the Holocaust the Zionists themselves created—a myth which made their lives easier in that it gave them unfair advantages in society and unburdened them from an existential quest for *Self;* Ralph Philip Boas identified many of the circumstances in America in 1921, which led to the Holocaust in Europe, including Jewish racism, the Jewish love of manufactured martyrdom, the lack of a genuine *raison d'être* for Judaism in the Twentieth Century, and the need of a common enemy to prevent the Jews from extinction through assimilation—the glorification of the myth that Gentile kindness is the worst enemy of the Jews and that anti-Semitism is the Jews' salvation from integration:

> "DESPITE the fact that we are ceasing to persecute people who disagree with us in religion or politics, we only dimly realize that one of the greatest evils of persecution is the fact that it saves its victims the trouble of justifying themselves. Persecution begets martyrdom, a glory as lacking in reason as its progenitor. Whether Sir Roger Casement was right or not is now only an academic question; his execution, by enshrining him forever in the Pantheon of Irish martyrs, makes the heart rather than the mind his judge. So it is with the Jews. Jews have not troubled themselves to justify, on any rational ground, the tenacious fight of their race against the storms of nineteen centuries of persecution. The fight has been its own justification. Obviously, a race that has endured what theirs has withstood must have some glorious mission to perform; to define that mission would be an element of positive weakness, since their enemies would then have a chance to meet them on the ground of reason, where their peculiar virtues, tenacity, single-mindedness, and pliant heroism, would avail them nothing.
>
> It is, therefore, a happy chance for the American Jew that his age-long persecution has either ended or has degenerated into petty social discrimination. For he must now realize that the day has gone when he could justify himself by recalling his heroic miseries. In other days and other countries he faced only the problems of existence. New ideas and opportunities could not pass the walls of the ghetto; custom made adherence to old ceremonies and beliefs not only easy but imperative. The Sabbath was the one day on which the Jew could be a man instead of a thing; the recurrent holidays gave him his one outlet for the emotions rigidly suppressed in daily life; the study and analysis of the Law and the Talmud furnished the intellectual exercise that his eager mind was denied in the schools and the learned circles of the country which tolerated him. The very fact that he was confined within a pale, therefore, made it easy for him to keep his race a distinct entity.

A. Hentschel, *The Collected Papers of Albert Einstein*, Volume 9, Document 203, Princeton University Press, (2004), pp. 173-175, at 175.

But now, if he is unable to find a rational ground for his religious and racial unity, he will meet a foe more insidious than persecution—the gradual disintegration of race and religious consciousness within the faith. Ironically enough, what pales, pogroms, and ghettos could not accomplish, freedom promises to bring to pass. So the time has come when the Jew in America must decide what he is going to do with and for himself; his enemies can no longer save him the effort of decision.

[***]

What is true of Europe is true also of the United States: the Jew occupies a position the importance of which is out of all proportion to his numbers. Hence the problem of Judaism is of real interest in America, because the influence which the Jew can have upon social life and the current political and financial situation depends almost entirely upon his mode of life and manner of thought. [***] What the Jew is going to do with this self-consciousness may, to Christians, seem of little moment. It is not of that loyal kind which moves men to blow up munition factories, or to plant bombs in steamships. For others, doubtless, its implications are not of great importance. For himself, however, they are everything. His self-consciousness colors his whole point of view. It is not a simple thing. It is compounded of many factors. It is both racial and religious; it makes him both hopeful and despondent; it gives cause both for pride and for a feeling of inferiority; it makes him clannish, and it makes him long for a wider field of acquaintance. [***] Judaism is clannish. Jews undoubtedly hang together. The combination of persecution with its inevitable concomitant, self-justification, acts as a centripetal force in driving Jews upon themselves. Just as Jews have the almost grotesque notion that a man will make his philosophic and religious convictions 'jibe' with his birth, so they have the wholly grotesque notion that a man should choose his friends and his wife from the small group among whom he happens to be born, though later education and environment may move him a thousand miles away. The results of this clannishness are paradoxical. For instance, the average Jew is sure that the chief reason why Anti-Semitism is everywhere ready to show its ugly head, is jealousy of the splendid history and the extraordinary business ability of the race. At the same time he subconsciously assumes the inferiority which has long been attributed to him, covering his feelings, however, by uncalled-for justification and bitter opposition to all criticism. It is torture to him, for example, that *The Merchant of Venice* should be read in the public schools. Who can blame him? For Shylock, although undoubtedly an exaggerated character, nevertheless makes concrete those qualities the portrayal of which hurts because it bears the sting of truth.

The development of committees 'On Purity of the Press' in Jewish societies, and the extraordinary wire-pulling over the Russian treaty and the Immigration bill, show to what lengths this consciousness can go. It is impossible for the Jew to be entirely at ease in the world. He is introspective and suspicious, often unhappy, always sure that, for good or ill, he is a marked man among men.

There are three attitudes which Jews in this country take toward their problem—a few as a result of having thought it through, the majority as a result of the forces of inertia, environment, or chance, forces of which they themselves are perhaps not aware. Some Jews attempt to get rid of their self-consciousness by separating from the group. They deliberately set out to convince themselves that there is no difference between them and other men, and that they can act and live in all respects like other American citizens. A second group find their fellow Jews entirely satisfactory. They are conscious of a difference between themselves and others, but, living as they do in large cities where the Jewish community numbers hundreds of thousands, they feel no need of association with non-Jews other than that which they get in business. They are rich, or at least well-to-do; they have all the comforts that money can buy; they occupy fine streets and build expensive synagogues. They are willing, not only to accept their group-consciousness, but to develop it to the fullest extent by means of societies and fraternal orders. In the third place, there is a small group of Jews keenly conscious of their race, who would like to make Judaism vital as a great religion and a great tradition. They differ from the second group in that they not only accept their individuality but try to justify it. It is not sufficient for them that there should be enough Jewish organizations and undertakings to make a respectable year-book: they are interested in showing why such organizations should exist They not only *are* Jews, but they *want to be* Jews; they want to feel that Judaism really has a mission to fulfill and a message to carry to the questioning world.

The Jew who attempts to solve his problem by separating from his community must leave the great centres of Jewish life and go to some small town where he may make a fresh start. There he will find himself in an anomalous position. He will have neither the support that comes from rubbing elbows with one's own kind, nor the mental and moral stiffening that comes from active opposition. He will be simply an odd fish, and as such will be subject, not to antagonism, but to curiosity. What cordiality he meets with is the cordiality of curiosity. He is a strange creature, similar—on a far lower scale of interest—to a Chinese traveler or a Hindu student. He is engaged in conversation on the 'Jewish problem,' or Jewish customs and history, until he sickens with trading on the race-consciousness that he is striving to forget. With cruel kindliness his friends impress upon him that his Judaism 'makes no difference,' with the result that he finds himself anticipating every imminent friendship by a clear statement of his race, lest the friendship be built upon the sands of prejudice. His social relations must be above reproach. A hasty word, an ill-considered action, in other men to be put down to idiosyncracy, in him is attributed to his birth. Even when there exists the frankest and most open friendship, he is continually seeing difficulties. The fathers have eaten a sour grape and the children's teeth are set on edge. The self-consciousness that he learned in youth reappears in maturity. Whether he will or no, a Jew he remains.

If he finds his situation intolerable he may, of course, utterly and completely deny his Jewish affiliation. He may consort with Christians, join a Christian church, marry a Christian wife, and tread under foot the old associations that will occasionally cast a disagreeable shadow across his life Unfortunately for such a solution, a cloud still hangs about the idea of apostasy. Such a refuge seems to a man of honor despicable. It is a cowardly procedure, surely, to deny one's birth and sail under false colors, the more so since, though it does no harm to others, it gains advantage for one's self. Why ii should it be treason for a Jew to abandon his religion and forget his birth any more than for a Frenchman or a Swede to do so? Probably for the reason that no one cares whether a man was born in France or not, whereas in certain circles it makes a great deal of difference if a man was born in Jewry. Furthermore, Christians feel strongly that the Jew who forsakes the religion into which he was born, does so, not because his eyes have been opened upon the truth, but because he sees in apostasy definite material advantages. The Jew who would take this means of obtaining peace, therefore, would find himself cursed by an irrational idealism which can disturb while it cannot fortify and achieve.

If, however, he returns to some great centre of Jewish life and attempts to affiliate with his own people, he is in a perilous position. He is more than likely to meet with distrust where he seeks sympathy. Jews are so extremely sensitive to criticism and so keenly conscious of the social discrimination which they encounter from Christians, that they can hardly believe that a man who seems to have lived for several years on an equal footing with Christians has not either denied his birth, in which case he has been a traitor, or has not certain qualities of mind which, since they have been palatable to Christians, must be severely critical of Jews.

And, indeed, they have, perhaps, a measure of justice in their position. It is impossible for a Jew to live apart from his race for several years without looking upon his people with a new light. For one thing, distance has enabled him to focus. He has learned to sympathize more than a little with those hotel-keepers whose ban upon Jews is a terrible thorn in the flesh of the man whose money ought to take him anywhere. He has come to see that the clannishness of Jews serves only to intensify what social discrimination may exist, and to make present in the imagination much that does not. He has realized that persecution is not necessarily justification, and that because a Jew was blackballed at a fashionable club does not prove that he was a man of first-rate calibre. And finally, he has perceived that there is an arrogance of endurance as well as an arrogance of persecution, and that for a man to be continually assuming that people are taking the trouble to despise him for his birth, is to postulate an importance that does not exist.

On the other hand, he has, because of his distance, idealized Judaism. In his retirement he studied the history of his people; he thrilled with their martyrdom; he marveled at their tenacity and their fortitude. He built up for himself on the cobweb foundation of boyhood memories, visions of the simple nobility of Jewish ritual and ceremonies, and vague ideals of an inspiring religious faith. He may, perhaps, have met, far more frequently than ill-will, a sentimental and unbalanced adulation of Jews. The cult of the new is with us, and the history, the folk-lore, the literature, and the customs of Judaism have, for many people who pride themselves on their social liberality, the fascination of novelty. It is the easiest thing in the world for a Jew to yield to this sentimental tolerance, and to view his people in a rosy light.

It is, therefore, something of a shock to him when he reënters a great Jewish community, for he finds that the great mass of American Jews have sunk into a comfortable materialism. What persecution could not accomplish, success in business has brought to pass. The innate qualities of the Jew could not save him from the fate of the Christian who has become rich in a hurry—grossness and self-conceit. That Jeshurun waxed fat and kicked is as true now as it ever was, and there is little reason to expect that the race which was hopelessly cankered by national prosperity in the days of Solomon can escape a similar fate in the twentieth century. [***] The sad result is that in prosperity the Jewish self-consciousness ceases to be religious and becomes merely racial.

[***]

The number of immigrants, or children of immigrants, from countries where for centuries they have been trained in an atmosphere of slavish cunning and worship of money, who become rich, is almost incredible. In Russia, Galicia, or Roumania, they cultivated a self-respect by rigid adherence to dignified and beautiful customs; in America the florid exuberance of newly acquired wealth cannot be dignified. Clannishness, exclusion from circles of good taste and good breeding, the infiltration of the parvenu East-European Jews, and imitation of the most obvious aspects of Americanism—its flamboyant and tasteless materialism—all combine to make the thoughtful Jew sadly question what hope lies in the bulk of the Jews who live in the great American cities.

[***]

[Zionism] is actuated by a spirit of helpfulness and by an ideal of racial unity. [***] Aided by persecution and poverty, [American Judaism] furnished admirable discipline to a race naturally stubborn and tenacious. Persecution, poverty, and discipline gone, what is left?—an indistinct monotheism joined to an ethical tradition never formulated into a system, and only vaguely defined. None of the great Jewish philosophers ever succeeded in establishing a Jewish creed; indeed, there was no need of one when common suffering wrought so effectual a bond. [***] At all events it must be remembered that, since the problem of Judaism comes from intense self-consciousness, persecution and sentimental tolerance are both bad for the Jew. The one saves him the trouble of seeking out his reason for existence; the other flatters him into a belief that there is no necessity for the search. If men will treat Jews like other people, instead of nourishing their age-long notions of peculiarity, they will make it easier for time to settle the

Jewish problem as it settles all others."[111]

Kurt G. W. Ludecke wrote in 1937 of Moses Pinkeles, a. k. a. Trebitsch Lincoln, a. k. a. Arthur Trebitsch, a Jew who marched with anti-Semites in the streets of Berlin, scripted their statements, and who funded Adolf Hitler and paid for his Nazi purchase of the newspaper *Münchener Beobachter* from the Thule Society, which became the Nazi Party's official organ the *Völkischer Beobachter*,

> "Another encounter in Vienna lives in my memory as something even more extraordinary. Some one introduced me to Arthur Trebitsch, and I spent a whole evening with him. His name was somewhat known through his books, *Geist und Judentum* and *Deutscher Geist oder Judentum?*, but I for my part had never heard of him; so I found myself quite unprepared for the strange discussion which ensued.
>
> Arthur Trebitsch was a peculiar and pathetic personality, a full-blooded Jew who was an apostate from his people and his religion; who uncompromisingly attacked the Jew and the Jewish spirit in his speeches and writings, yet could not enter into the Gentile world with which he strove to ally himself. Whether the attitude which turned his life into a tragedy sprang from his mind or his emotions, I cannot say. This was the first time I had talked at length with an intellectual and erudite Jew about the German-Jewish problem, and though even among Gentiles I was now discovering a widespread doubt of the Nazi program, I was amazed to find that Trebitsch still passionately endorsed it.
>
> Trebitsch did not consider himself a Jew, either spiritually or physically, in spite of his two Jewish parents. Convinced that he was the result of a phenomenon which biologists call "mutation,' he presented himself as a Gentile. Seriously believing that he looked very much like Houston Stewart Chamberlain, the declared scientific enemy of the Jewish people, he produced as proof one of his pamphlets which showed their pictures facing each other. Looking at his eyes and fair hair, I had to agree that the photographs bore a striking resemblance.
>
> Never before had I considered the Jewish problem from the standpoint of the individual Jew who finds much to condemn in his own people and dares to say so. Trebitsch was an extreme case; yet some of his findings were sound. Discovering that his people were resentful of criticism, he had turned his coat—without finding it any warmer. My mind reverted at once to the two famous apostates, Spinoza and Uriel de Acosta, who were excommunicated from the synagogues, and I reflected that there is no more sorrowful destiny than that which overtakes those who alienate their own people without making friends elsewhere.

---

111. R. P. Boas, "The Problem of American Judaism", *The Atlantic Monthly*, Volume 119, Number 2, (February, 1917), pp. 145-152.

Trebitsch sought to convince me that he could be a valuable ally in the Nazi struggle. Intuition and reason told me to remain reserved. But it was distressing to witness the despair of this exhausted and high-strung man, who beyond question was sincere. Ostracized on one side and rejected on the other, he was indeed an outcast. The tragic overtones of our interview made a deep impression on me, and at the earliest moment I spoke about him at length with Rosenberg. Needless to say, there was no place for him in the Party."[112]

Douglas Reed wrote in 1938 of Moses Pinkeles, a. k. a. Ignatz (Ignatius) Trebitsch-Lincoln, a Jew who financed Hitler, and of Zionists who sponsored Hitler,

"Oblivion for a few years, and then came the Kapp Putsch in Germany, the first of the Nationalist conspiracies to overthrow the democratic liberal regime that was so kind to the Jews, and reinstate the big business men, big landlords, monarchists, militarists, in the seats of the mighty in Germany. Who was a leading figure in this short-lived seizure of power? Trebitsch Lincoln, now a German die-hard. Among the other sympathizers was a relatively unknown man, one Adolf Hitler. Trebitsch Lincoln on the side of the anti-Semites? Of course, he was a Christian. [***] If you doubt me, think of Trebitsch Lincoln leading the anti-Semites down the Wilhelmstrasse to the seat of power. But I can show you the modern counterpart of Trebitsch Lincoln, and I don't mean those pro-Hitler Jews who were said by rumour to have marched round Berlin in the early Nazi days carrying a banner with the legend 'Hinaus mit uns!'—'Chuck us out!'"[113]

Eustace Mullins, Ezra Pound's authorized biographer, stated on Daryl Bradford Smith's radio program *The French Connection*, that the German-Jewish bankers Warburg and Oppenheimer marched with the Nazis carrying signs that read "throw us out".

Prominent Zionist and author of the *Encyclopaedia Judaica; das Judentum in Geschichte und Gegenwart*, Jakob Klatzkin stated in 1925,

"The national viewpoint taught us to understand the true nature of antisemitism, and this understanding widens the horizons of our national outlook. [***] In the age of enlightenment antisemitism was included among the phenomena that are likely to disappear along with other forms of prejudice and iniquity. The antisemites, so the rule stated, were the laggard elements in the march of progress. Hence, our fate is dependent on the advance of human culture, and its victory is our victory. [***] In the period of Zionism, we learned that antisemitism was a psychic-social phenomenon

---

112. K. G. W. Ludecke, *I Knew Hitler: The Story of a Nazi AWho Escaped the Blood Purge*, Charles Scribner's Sons, New York, (1937), pp. 191-218.
113. D. Reed, *Disgrace Abounding*, Jonathan Cape, London, (1939), pp. 249, 251.

that derives from our existence as a nation within a nation. Hence, it cannot change, until we attain our national end. But if Zionism had fully understood its own implications, it would have arrived, not merely as a psycho-sociological explanation of this phenomenon, but also as a justification of it. It is right to protest against its crude expressions, but we are unjust to it and distort its nature so long as we do not recognize that essentially it is a defense of the integrity of a nation, in whose throat we are stuck, neither to be swallowed nor to be expelled. [***] And when we are unjust to this phenomenon, we are unfair to our own people. If we do not admit the rightfulness of antisemitism, we deny the rightfulness of our own nationalism. If our people is deserving and willing to live its own national life, then it is an alien body thrust into the nations among whom it lives, an alien body that insists on its own distinctive identity, reducing the domain of their life. It is right, therefore, that they should fight against us for their national integrity. [***] Know this, that it is a good sign for us that the nations of the world combat us. It is proof that our national image is not yet utterly blurred, our alienism is still felt. If the war against us should cease or be weakened, it would indicate that our image has become indistinct and our alienism softened. We shall not obtain equality of rights anywhere save at the price of an explicit or implied declaration that we are no longer a national body, but part of the body of the host-nation; or that we are willing to assimilate and become part of it. [***] Instead of establishing societies for defense against the antisemites, who want to reduce our rights, we should establish societies for defense against our friends who desire to defend our rights. [***] When Moses came to redeem the children of Israel, their leaders said to him, 'You have made our odor evil in the eyes of Pharaoh and in the eyes of his servants, giving them a sword with which to kill us.' Nevertheless, Moses persisted in worsening the situation of the people, and he saved them."[114]

Who was the "devil" the political Zionists commissioned to shake up the Jews of Europe? When Hitler came to power, some Zionists asked all Jews to let him do as he wished. Some Zionists even hailed him as their savior.

Leon Simon wrote in the introduction to a collection of Einstein's Zionist works, that emancipation posed a greater threat to the Jewish "race" than the problems of the unemancipated. Hitler soon thereafter unemancipated the Jews. Simon wrote in 1930, *inter alia*,

> "THERE are two main ways of approach to Zionism. One starts from those Jews who are made to suffer for being Jews, the other from the smaller number who are not. In the one case Zionism means the transfer of masses of Jews from countries in which they are obviously not wanted to a country

---

114. J. Klatzkin, *Tehumim: Ma'amarim*, Devir, Berlin, (1925). English translation in J. B. Agus, *The Meaning of Jewish History*, Volume 2, Abelard-Schuman, New York, (1963), p. 425-426.

which they might call their own; in the other case it means the re-creation in Palestine of a Hebraic type of life, which will be regarded by all Jews as the embodiment of their own distinctive outlook and ideals, and will thus help to counteract the inevitable tendency of the Jews, when they are not driven back on themselves by external restrictions, to lose their sense of being a separate people.

Of these two conceptions of Zionism, the former has the more direct and obvious appeal. The fact that masses of Jews are made to suffer for the crime of being Jews and wishing to remain Jews is too patent to call for demonstration; and, while it is true that in some countries Jewish disabilities have been removed so far as that can be done by statute, bitter experience engenders a sceptical attitude towards the idea that universal emancipation will provide a panacea for the Jew's troubles. In the first place, the countries with the largest numbers of Jews are not all eager to admit them to full equality; and in the second place, even where equality has been accorded, dislike of the Jew often makes itself felt too strongly for his liking or comfort. Hence, from the point of view of a Jew who wishes to see his people better off in the world than it is to-day, or has been these many centuries, there is much to commend a scheme which sets out to cut at the root of the trouble by removing all the victims of anti-Semitism to a land of their own. By contrast with this perfectly simple and intelligible idea, the other conception of Zionism appears abstruse, almost other-worldly. The problem to which it offers a solution is one of which the existence, let alone the urgency, is not, readily realised by ordinary men and women. It requires no great exercise of thought or imagination to appreciate the unenviable position of the Jewish masses, or the desirability of transporting them to a safe home of refuge. It is less easy to recognise that the emancipated Jew presents, from the point of view of Jewish survival, at least as difficult a problem as the unemancipated; that the very removal of restrictions on the political and economic freedom of the Jews in this or that country creates conditions which are more inimical than persecution to the maintenance of whatever is worthily distinctive of the Jew as such; that the consequent disintegration of an ancient people, involving the disappearance of one of the world's great cultures, is even more tragic than the material ills of the Jewish masses; and that the paramount need of the hour is a safe home of refuge for the Jewish spirit."[115]

7.5.4.1 Depressions Make for Fertile Ground for Anti-Semitic Zionist Dictators

Hitler had little political success until the Great Depression hit the world. The Depression, together with immense funding from Jewish financiers and from

---

[115]. L. Simon, Introduction to A. Einstein, Edited by L. Simon, *About Zionism: Speeches and Letters by Professor Albert Einstein*, Macmillan, New York, (1931), pp. 9-12.

industrialists, propelled Hitler to power in early 1933. Samuel Untermyer called for a boycott of Germany in 1933, and chastised Jewish bankers for financing Adolf Hitler and Nazism,

> "Revolting as it is, it would be an interesting study in psychology to analyze the motives, other than fear and cowardice, that have prompted Jewish bankers to lend money to Germany as they are now doing. It is in part their money that is being used by the Hitler régime in its reckless, wicked campaign of propaganda to make the world anti-Semitic; with that money they have invaded Great Britain, the United States and other countries where they have established newspapers, subsidized agents and otherwise are spending untold millions in spreading their infamous creed.
> The suggestion that they use that money toward paying the honest debts they have repudiated is answered only by contemptuous sneers and silence. Meantime the infamous campaign goes on unabated with ever increasing intensity to the everlasting disgrace of the Jewish bankers who are helping to finance it and of the weaklings who are doing nothing effective to check it."[116]

The political Zionists learned from the financial crisis of 1873, that a financial catastrophe would provide an opportunity to promote political anti-Semitism, which was their goal. At least as early as 1914, Ignatz Zollschan stated,

> "In Germany, in the west European states, and in the United States of America, which enjoy a great economic and political prosperity, and, moreover, have no great percentage of Jewish population, the expropriation of the Jews cannot come into consideration. But should stagnation and depression take the place of prosperity, conditions similar to those in eastern Europe may be expected. In order to verify this statement, we need only cast our glance upon the so-called foundation-years in Germany, and upon the financial crisis in the year 1873. For it was then that birth was given to political anti-Semitism in Germany."

In 1898, Communist Zionist Nachman Syrkin wrote that economic hardships resulted in increased anti-Semitism and the success of criminal anti-Semitic politicians.[117]

7.5.4.2 Einstein a Subtle Hitler Apologist

---

[116]. "Text of Untermyer's Address", *The New York Times*, (7 August 1933), p. 4. *See also:* "Untermyer Back, Greeted in Harbor", *The New York Times*, (7 August 1933), p. 4.
[117]. N. Syrkin, under the nom de plume "Ben Elieser", *Die Judenfrage und der socialistische Judenstaat*, Steiger, Bern, (1898); English translation in A. Hertzberg, *The Zionist Idea*, Harper Torchbooks, New York, (1959), pp. 333-350, at 338-340.

When the "Hitlerites" showed their strength in the elections, political might paid for by Jewish financiers, Einstein and some other Zionist leaders told Jews not worry but to close ranks and unite. Of course, should Hitler lead the country into war, it would benefit bankers, investors, and factory owners. Hitler's anti-Semitism benefitted the political Zionists. An article entitled "Fascists Walk Out of Berlin Council", *The New York Times* on 19 September 1930 on page 9 quoted the *Jewish Telegraphic Agency*, which quoted Albert Einstein,

> "There is no reason for despair,' declared Professor Einstein, 'for the Hitler vote is only a symptom, not necessarily of anti-Jewish hatred but of momentary resentment caused by economic misery and unemployment within the ranks of misguided German youth. I hope that the momentary fever and wave will rapidly fall.
> 'During the more dangerous Dreyfus period almost the entire French nation was to be found in the anti-Semitic camp. I hope that as soon as the situation improves the German people will also find their road to clarity."

Einstein acted as a Nazi apologist and tried to subvert any organized Jewish reaction to Hitler—he effectively promoted Hitler at a critical time in history. Many Jews in Germany failed to respond to Hitler's victory with an organized reaction, in part because treacherous Jews like Albert Einstein led them to believe that Hitler would soon be unseated and that Nazism was an ephemeral malady they need not bother too much about.

At a time when anti-Zionist Jews were desperately trying to organize all Jews to fight against the Fascists, while many Zionists were encouraging the Fascists,[118] Einstein wanted to remove Jews from Germany and was confused by his own racist hypocrisy. Following Hitler's election victory in 1933, Albert Einstein commented, merely parroting the Zionist Party line,[119]

> "For the time being, I see the National Socialist movement as merely a product of the current economic crisis and the teething pains of the Republic. The solidarity of the Jews is for me an eternal commandment, but I feel a specific reaction to the election results would be entirely inappropriate."

> "Ich sehe in der nationalsozialistischen Bewegung einstweilen nur eine Folgeerscheinung der momentanen wirtschaftlichen Notlage und eine Kinder-Krankheit der Republik. Solidarität der Juden halte ich immer für

---

118. *Central-Verein Zeitung*, Volume, 9, Number 28, (11 July 1930); and Volume 9, Number 37, (12 September 1930); and Volume 9, Number 38, (19 September 1930). K. Polkehn, "The Secret Contacts: Zionism and Nazi Germany, 1933-1941", *Journal of Palestine Studies*, Volume 5, Number 3/4, (Spring-Summer, 1976), pp. 54-82.
119. K. Polkehn, "The Secret Contacts: Zionism and Nazi Germany, 1933-1941", *Journal of Palestine Studies*, Volume 5, Number 3/4, (Spring-Summer, 1976), pp. 54-82, at 56-57.

geboten, aber eine besondere Reaktion auf das Wahlergebnis für ganz unzweckmässig."[120]

At the time Einstein made this cavalier statement, he knew that the Nazis were going to annihilate the Jews of Europe—as did Zionist Nazi apologist Ludwig Lewisohn, the dear friend, and the lover, of the famous Hitler-promoter George Sylvester Viereck.[121] Albert Einstein wrote to Gustav Bucky on 15 July 1933,

"I really do believe that any action aimed at keeping Jews in Germany would have the effect of speeding up their annihilation."[122]

In 1933, Einstein told British Prime Minister (1923-1929, 1935-1937) Stanley Baldwin of Hitler's plan for world conquest and that Hitler would perhaps cause a new world war. Baldwin, who was later criticized for not preparing England to face Germany, told Einstein that Great Britain had her allies.[123] Einstein did take a firmer stand against the Nazis and against the Prussian Academy of Sciences in 1933 than many Zionists, and was accused of public anti-Germanism by that Academy. In this exchange, Einstein fought for the rights of Jews to human dignity and the right to equality under the law. What Einstein meant by "annihilation" in 1933 is not necessarily clear. He may have meant the rooting out of Jews from Germany by cutting off their means of earning a living and forcing them to Palestine—as the Nazis and Zionists had planned,[124] or he may have meant mass murder.

7.5.5 Einstein's Seething Racist Hatred and Rabid Nationalism

The smear tactics of Zionists are well known. Einstein's smear tactics gained him and his defenders an international reputation as agitators and reckless defamers. A "Biographical Sketch" issued to U. S. Army Intelligence sometime in 1940 stated,

"The origin of the case is that in Berlin, even in the political free and easy

---

120. P. W. Fabry, *Mutmassungen über Hitler: Urteile von Zeitgenossen*, Droste, Düsseldorf, (1969), p. 130. Fabry cites: *Israelischen Familienblatt*.
121. L. Lewisohn, "A Year of Crisis", in A. Hertzberg, *The Zionist Idea*, Harper Torchbooks, New York, (1959), pp. 488-492, at 489. Hertzberg cites: L. Lewisohn, *Rebirth* (editor), New York, (1935), pp. 290-296.
122. P. A. Bucky, Einstein, and A. G. Weakland, *The Private Albert Einstein*, Andrews and McMeel, Kansas City, (1992), p. 61.
123. P. A. Bucky, Einstein, and A. G. Weakland, *The Private Albert Einstein*, Andrews and McMeel, Kansas City, (1992), p. 62.
124. Letter from A. Einstein to the Prussian Academy of Sciences of 5 April 1933, *The World As I See It*, Citadel, New York, (1993), pp. 82-84, at 83.

period of 1923 to 1929, the Einstein home was known as a Communist center and clearing house. Mrs. and Miss Einstein were always prominent at all extreme radical meetings and demonstrations. When the German police tried to bridge some of the extreme Communist activities, the Einstein villa at Wannsee was found to be the hiding place of Moscow envoys, etc. The Berlin conservative press at the time featured this, but the authorities were hesitant to take any action, as the more radical press immediately accused these reporters as being Anti-Semites."[125]

The historic record bears out the accusation that Einstein and his sponsors had the means and the will to smear innocents in their efforts to redirect public attention away from their own vile actions. It had become a habit for them, and they took every opportunity, no matter how unjustified, to raise the issue of race, paint themselves as victims of racist oppression, and often went so far as to accuse innocent persons of racism. The ridiculous extremes of this political maneuvering were manifest long before the Holocaust, and reached across the English Channel.

In 1919, hypocritical, racist, ethnocentric and insulting Einstein smeared all Germans, all English, and the reporter who had helped to promote him,

> "A final comment. The description of me and my circumstances in *The Times* shows an amusing feat of imagination on the part of the writer. By an application of the theory of relativity to the taste of readers, today in Germany I am called a German man of science, and in England I am represented as a Swiss Jew. If I come to regarded as a *bête noire*, the descriptions will be reversed, and I shall become a Swiss Jew for the Germans and a German man of science for the English!"[126]

Einstein, either directly, or through someone else, took his line from Bernard Lazare's *L'Antisémitisme: Son Histoire et Ses Causes* of 1894,

> "In general the Jews, even the revolutionaries, have kept the Jewish spirit, and if they have given up religion and faith, they have nevertheless been formed, thanks to their ancestry and their education, by the influence of Jewish nationalism. This is true in a very special fashion of the Jewish

---

125. *Reproduced in:* F. Jerome, *The Einstein File*, St. Martin's Press, New York, (2002), second plate following page 170.

126. A. Einstein, "Time, Space and Gravitation", *The Times* (London), (28 November 1919), pp. 13-14; reprinted in *Science* and in E. E. Slossen, *Easy Lessons in Einstein*, Harcourt Brace and Howe, New York, (1920), pp. 109-114. Einstein was perhaps inspired to make this remark by a letter from A. F. Lindemann of 23 November 1919, *The Collected Papers of Albert Einstein*, Volume 9, Document 174, Princeton University Press, (2004). Einstein told Ehrenfest of his joke in a letter of 4 December 1919, *The Collected Papers of Albert Einstein*, Volume 9, Document 189, Princeton University Press, (2004).

revolutionaries who lived in the first half of this century. Heinrich Heine and Karl Marx are two typical examples. Heine is held to be German in France. In Germany he is accused of being French. He was above all a Jew."[127]

A couple of years after Einstein made his comment, in June of 1921, *The Jewish Chronicle* reported,

> "The *Times* of Monday last, by the by, published an interview with Einstein. The interviewer gave minute personal descriptions of the remarkable scientist, and yet did not venture to suggest that he was a Jew. If (the *Jewish World* comments) he had been a Bolshevik or a reprehensible character of any kind, we doubt not the fact would have dawned upon the *Times* correspondent that he was a Jew, and would have found place in what he had to say. Strange how circumstances alter one's point of view!"[128]

Strange, indeed, that no matter what a *Times* correspondent said about Einstein; either Einstein, or some extremist among his supporters, would viciously smear that correspondent as a bigot, without any grounds whatsoever. And for what purpose? This appears to have been a habit for them, a pernicious habit and a divisive habit meant to perpetuate, intensify and generate hatred, fear and conflict—for political Zionist purposes.

Einstein's ardent nationalism became so extreme, that it played into the hands of his political foes, and became an example for their generalizations. Max Nordau described the pernicious habits of racists, with no small measure of hypocrisy, in his address to the First Zionist Congress in 1897,

> "No one has ever tried to justify these terrible accusations by facts. At most, now and then, an individual Jew, the scum of his race and of mankind, is triumphantly cited as an example, and contrary to all laws of logic, the example is made general. This tendency is psychologically correct. It is the practice of human intellect to invent for the prejudices, which sentiment has called forth, a cause seemingly reasonable. Probably wisdom has long been acquainted with this psychological law, and puts it in fairly expressive words: 'If you have to drown a dog,' says the proverb, 'you must first declare him to be mad.' All kinds of vices are falsely attributed to the Jews, because one wishes to convince himself that he has a right to detest them. But the pre-existing sentiment is the detestation of the Jews."[129]

Einstein detested Germans throughout his life. He hated Germans long

---

[127]. D. Fahey, *The Mystical Body of Christ in the Modern World*, Browne and Nolan Limited, London, (1935), p. 77.
[128]. *The Jewish Chronicle*, (17 June 1921), p. 26.
[129]. M. Nordau, "Max Nordau on the General Situation of the Jews", *The Jewish Chronicle*, (3 September 1897), pp. 7-9, at 8.

before the Nazi Party was formed. Einstein's racist nationalism rivaled and perhaps even surpassed Physics in Einstein's self-image, making him the ideological twin of the Nazis—one who wanted to exterminate the Germans—one who wanted to exterminate all Gentile Europeans. He was described in the *Daily Graphic* as,

> "A man of the most simple tastes, he lives in a lofty flat in Berlin. He is an indifferent linguist, and will lecture in German, but he has a passion for music, and beyond this his scientific pursuits and his work for Zionism comprise his sole interests."[130]

While hiding from Arvid Reuterdahl's challenge to a public debate,[131] Einstein announced through his secretary Salomon Ginzberg during his famous stay in America,

> "I came here with one object—the promotion of the establishment of the Hebrew University in Jerusalem. [***] The great purpose of my mission to this country must not be overshadowed by my theory. I will be here a short time, and all of that time must be devoted to the great Palestine reconstruction project."[132]

Einstein stated in an interview following his visit to America,
> "I really went on behalf of the Jewish cause. Yes, I have placed my name and indeed my self in the service of the Zionist movement to make propaganda for Palestine, and the true purpose of the America trip was to collect money for a fund to establish a university in Jerusalem."[133]

Nationalism became so consuming a personal passion for Einstein, that he took advantage of his fraudulently-based fame to promote the political cause. R. S. Shankland stated,

> "About publicity Einstein told me that he had been *given* a publicity value which he did not *earn*. Since he had it he would use it if it would do good; otherwise not. [*Emphasis found in the original*]"[134]

---

130. *Daily Graphic* as quoted in *The Jewish Chronicle*, (17 June 1921), p. 26.
131. *See:* "Challenges Prof. Einstein: St. Paul Professor Asserts Relativity Theory Was Advanced in 1866", *The New York Times*, (10 April 1921), p. 21.
132. "Einstein Refuses to Debate Theory: Dean Reuterdahl's Challenge to Discuss Relativity Declined as Detraction from Mission", *New York American*, (12 April 1921).
133. A. Einstein, "Een interview met Prof. Einstein", *Nieuwe Rotterdamsche Courant*, (4 July 1921). English translation found in, M. Janssen, *et al* Editors, *The Collected Papers of Albert Einstein*, Volume 7, Appendix D, Princeton University Press, (2002), pp. 623.
134. R. S. Shankland, "Conversations with Albert Einstein", *American Journal of Physics*, Volume 31, Number 1, (January, 1963), pp. 47-57, at 56.

His famous trip to America was not made to promote or celebrate the theory of relativity, but to promote his personal vision of nationalism and to raise money for this cause. Though this was absolutely his right, many found Einstein's exploitation of his scientific fame for political purposes distasteful—to the point of being disgraceful.

As early as February of 1914, loyal German Jews publicly protested against anti-German Zionism. Albert Einstein was a virulently racist oddity among German Jews. German Jews knew quite well that the Zionists were planning to deliberately place all Jews in harm's way and ruin Germany. *The New York Times* wrote on 8 February 1914, Section 3, page 3,

"PROTEST AGAINST ZIONISTS.

German-Jewish Organizations Say
They Harm Jews and Fatherland.

Special Cable to THE NEW YORK TIMES.

BERLIN, Feb. 7.—Several Jewish organizations of Germany have joined in a protest against what they call the 'insidious German national Chauvinism,' which is being carried on in the name of German Jews by German Zionists.

It is alleged that the Zionists are resorting to methods that must bring the whole Jewish cause into disrepute at home and abroad and sow seeds of discord between Christians and Jews in Germany itself.

The protest, which has taken the form of a strong public statement, addressed to the press of the country, urges that the mere matter of faith which separates German Jews from their fellow-citizens must not be exploited by overzealous co-religionists to the disadvantage of both Jews and the Fatherland."

In 1930, some German Jews demanded that Albert Einstein stop using his scientific fame to promote racism, disloyalty and "interracial" strife. *The New York Times* reported on 7 December 1930 on page 11,

"The National German-Jewish Union, a small group of extreme nationalist and anti-Zionist Jews, protested against Professor Einstein using his world-fame as a scientist for 'propagating Zionism.'"

After the Second World War, Jews again criticized Einstein for his nationalistic Zionism. Einstein responded,

"In my opinion condemning the Zionist movement as 'nationalistic' is unjustified. [***] Thus already our precarious situation forces us to stand together irrespective of our citizenship."[135]

Einstein parroted the Zionist dogma that ethnic, racial and religious unity among peoples of Jewish descent around the world constituted a sovereignty without physical borders, which should be organized around a community in Palestine, but which sovereign status should be intrinsic to anyone of Jewish descent anywhere in the world—since a Jewish dispersion had allegedly taken place two thousand years ago. Theodor Herzl stated that anti-Semitism was justified and that the only means to end it was segregation. Chaim Weizmann made it very clear that Zionism is not a form of self-defense against prejudice, but is instead an indefensible product of Jewish bigotry. Weizmann proclaimed,

> "The sufferings of Russian Jewry never were the cause of Zionism. The fundamental cause of Zionism was, and is, the ineradicable national striving of Jewry to have a home of its own—a national center, a national home with a national Jewish life."[136]

German Jews around the world had largely assimilated into various nations and cultures and were often quite successful. They were leading and highly productive members of their societies. Eastern European Jews were often living in intolerable conditions and sought to emigrate to the West. They looked to their religious brethren in the West for help, but were often resented and rejected, because they clung to their ancient Jewish racism, and their desire to flee their neighbors and their call to other Jews in other countries was itself a manifestation of their racist tribalism.

Many German Jews feared that these clannish Easterners would inspire anti-Semitism and resented their presence.[137] Weizmann feared that the Russian Revolution would put an end to Zionism, because it achieved the freedom of Russian Jews,

> "At that time the whole world—and the Jews more than anyone else—had been thrilled by the overthrow of the czarist regime in Russia, and the establishment of the liberal Kerensky regime."[138]

Weizmann was a rabid anti-assimilationist.[139] He wasn't simply after social

---

[135]. H. Dukas and B. Hoffmann, *Albert Einstein: The Human Side*, Princeton University Press, (1979), p. 55.

[136]. C. Weizmann, *Trial and Error: The Autobiography of Chaim Weizmann*, Harper & Brothers, New York, (1949), p. 201.

[137]. J. Prinz, *Wir Juden*, Erich Reiss, Berlin, (1934), pp. 50-55. Letter from A. Einstein to M. Born of 22 March 1934, in M. Born, *The Born-Einstein Letters*, Walker and Company, New York, (1971), pp. 121-122.

[138]. C. Weizmann, *Trial and Error: The Autobiography of Chaim Weizmann*, Harper & Brothers, New York, (1949), p. 201. We must be careful not to confuse Kerensky and his "liberalism" with Lenin, Chernyshevsky and "Bolshevism".

[139]. C. Weizmann, *Trial and Error: The Autobiography of Chaim Weizmann*, Harper &

justice for Jews. Weizmann was after self-imposed segregation of the Jews.

The Zionists are the product of an ancient racist and genocidal religious mythology. This religious mythology is largely political and racist, and it affects even secular Jews, some of whom view it as the product of Jewish genes, and therefore of intrinsic value in defining Jews and their actions. The prophets need not have been inspired by God, for they were inspired by a yet more divine source, Jewish blood. The creation myth was turned on its head such that some secular Jews stated that the Jews created a fellow Jew, "God", to express the urges of their "Jewish blood"—their "Jewish soul". Those many secular Jews who rejected this racist viewpoint, also could not have helped but have been somewhat affected by the legacy of centuries of Jewish culture which had evolved in the continuing presence of religious Jewish racism.

The Hebrew Bible contains numerous stories of the segregation, punishment and genocide of assimilationist Jews by anti-assimilationist Jews. For example, *Numbers*, Chapter 25, states:

> "And Israel abode in Shittim, and the people began to commit whoredom with the daughters of Moab. 2 And they called the people unto the sacrifices of their gods: and the people did eat, and bowed down to their gods. 3 And Israel joined himself unto Baal-peor: and the anger of the LORD was kindled against Israel. 4 And the LORD said unto Moses, Take all the heads of the people, and hang them up before the LORD against the sun, that the fierce anger of the LORD may be turned away from Israel. 5 And Moses said unto the judges of Israel, Slay ye every one his men that were joined unto Baal-peor. 6 ¶ And, behold, one of the children of Israel came and brought unto his brethren a Midianitish woman in the sight of Moses, and in the sight of all the congregation of the children of Israel, who *were* weeping *before* the door of the tabernacle of the congregation. 7 And when Phinehas, the son of Eleazar, the son of Aaron the priest, saw *it*, he rose up from among the congregation, and took a javelin in his hand; 8 and he went after the man of Israel into the tent, and thrust both of them through, the man of Israel, and the woman through her belly. So the plague was stayed from the children of Israel. 9 And those that died in the plague were twenty and four thousand. 10 ¶ And the LORD spake unto Moses, saying, 11 Phinehas, the son of Eleazar, the son of Aaron the priest, hath turned my wrath away from the children of Israel, while he was zealous for my sake among them, that I consumed not the children of Israel in my jealousy. 12 Wherefore say, Behold, I give unto him my covenant of peace: 13 and he shall have it, and his seed after him, even the covenant of an everlasting priesthood; because he was zealous for his God, and made an atonement for the children of Israel. 14 Now the name of the Israelite that was slain, *even* that was slain with the Mid'i-anitish woman, *was* Zimri, the son of Salu, a prince of a chief

---

Brothers, New York, (1949), pp. 31-35, 42, 47, 50-53, 65, 82, 156-163, 200-207, 288-289.

> house among the Simeonites. 15 And the name of the Midianitish woman that was slain *was* Cozbi, the daughter of Zur; he *was* head over a people, *and* of a chief house in Midian. 16 And the LORD spake unto Moses, saying, 17 Vex the Midianites, and smite them: 18 for they vex you with their wiles, wherewith they have beguiled you in the matter of Peor, and in the matter of Cozbi, the daughter of a prince of Midian, their sister, which was slain in the day of the plague for Peor's sake."

Since many Zionists were atheists, or pretended to be atheists to assuage Christian and Moslem concerns, as well as secular and religious Jewish fears, and since Herzl and others had made Zionism a political question rather than a religious question, Zionism became strictly a matter of racist segregation.

There was a definite rift between Eastern European Jews and German Jews, who feared that the presence of these Easterners, especially when led by rabidly racist Zionists, would inspire and intensify anti-Semitism. Einstein and Weizmann wanted to force Western European Jews into sponsoring the emigration of Eastern European Jews—who appeared in Western Europe like peoples from another time—and who would make a suitable slave labor force for the Zionists.[140] In turn, these highly racist Eastern European Jews resented the assimilationist Western Jews. Many of the Jews of Palestine also resented the Eastern European Jews for creating conflicts in Palestine, where Jews, Moslems and Christians had been living together in peace.

A racist unity among Jews had long been a goal of the political Zionists despite the resistence they encountered from Jews around the world. Max Nordau wrote, soon after the First Zionist Congress in Basel in August of 1897:

> "Die Voraussetzung des politischen Zionismus ist, dass es ein jüdisches Volk gibt. Das gerade leugnen die Assimilationsjuden und die von ihnen besoldeten geistlosen, salbungsvoll schwatzenden Rabbiner."

and,

> "{*Margin Note:* Die Assimilanten} Viele Juden, namentlich des Westens, haben innerlich vollkommen mit dem Judenthum gebrochen und sie werden es wahrscheinlich bald auch äusserlich thun, und wenn nicht sie, dann ihre Kinder oder Enkel. Diese wünschen ganz unter ihren christlichen Landsleuten aufzugehen. Sie empfinden es als schwere Störung, dass andere Juden neben ihnen ihr besonderes Volksthum laut verkünden und reinlich Scheidung zwischen sich und den anderen Völkern fordern. Ihre grosse Angst ist, in ihrem Geburtslande, dessen freie Bürger sie sind, als Fremde bezeichnet zu werden. Sie fürchten, dass man dies mehr als je vorher thun wird, wenn ein grosser Theil des jüdischen Volkes offen die Rechte eines

---

140. C. Weizmann, *Trial and Error: The Autobiography of Chaim Weizmann*, Harper, New York, (1949), p. 289.

selbständigen Volkes für sich fordert, und nun gar, wenn erst irgendwo in der Welt wirklich ein politisches und culturelles Centrum des Judenthums entsteht, um das sich Millionen national geeinigter Juden gruppieren.

{*Margin Note:* Zwei Millionen gegen zehn} Alle diese Gefühle der Assimilationsjuden sind verständlich. Sie sind auch von ihrem Standpunkt aus berechtigt. Aber sie haben keinen Anspruch darauf, dass der Zionismus ihnen zu Liebe Selbstmord begehe. Die Juden, die in ihrem Geburtslande zufrieden und glücklich sind und die Zumuthung, es aufzugeben, empört zurückweisen, sind etwa ein Sechstel des jüdischen Volkes, sagen wir 2 Millionen von zwölf. Die übrigen fünf Sechstel, zehn Millionen, fühlen sich in ihrem Aufenthaltsorte sehr unglücklich und sie haben auch allen Grund dazu. Diesen zehn Millionen ist nicht zuzumuthen, dass sie sich für immer widerstandslos in ihre Knechtschaft fügen, dass sie jedes Streben nach Erlösung aus ihrer Noth aufgeben, bloss damit das Behagen der zwei Millionen glücklicher und zufriedener Juden nicht gestört werde."[141]

Theodor Herzl wrote of the utility of using Eastern European Jewish peasants as a slave labor force in his book *The Jewish State* and in his diaries. The Zionist Nazis helped the political Zionists to train this slave labor force and to condition them to accept their fate. After the Holocaust, Chaim Weizmann tried to blame assimilatory Jews for the tragic events the Zionists deliberately caused,

"[Rathenau's] attitude was, of course, all too typical of that of many assimilated German Jews; they seemed to have no idea that they were sitting on a volcano; they believed quite sincerely that such difficulties as admittedly existed for German Jews were purely temporary and transitory phenomena, primarily due to the influx of East European Jews, who did not fit into the framework of German life, and thus offered targets for anti-Semitic attacks."[142]

Joachim Prinz explored the issue in his book *Wir Juden*, Erich Reiss, Berlin, (1934), pp. 50-55. Albert Einstein wrote to Max Born on 22 March 1934 that the same impediments Western European Jews had placed against the immigration of Eastern European Jews during their migration to the West were now being instituted against German Jews by the Jews of America, France and England,

"It is particularly unfortunate that the satiated Jews of the countries which have hitherto been spared cling to the foolish hope that they can safeguard themselves by keeping quiet and making patriotic gestures, just as the German Jews used to do. For the same reason they sabotaged the granting

---

141. M. Nordau, *Der Zionismus*, Jüdischen Volksstimme, Brünn, (ca. 1898), pp. 8, 14.
See also: M. Nordau, *Die Tragödie der Assimilation*, Berlin, Wien, R. Löwit, (1920).
142. C. Weizmann, *Trial and Error: The Autobiography of Chaim Weizmann*, Harper & Brothers, New York, (1949), p. 289.

of asylum to German Jews, just as the latter did to Jews from the East. This applies just as much in America as in France and England."[143]

Einstein's personality interfered with his attempts to open up immigration for Eastern European Jews and his bigoted hatred worked against his cause. In the long run, Einstein's racism and provocative statements proved horrifically counter-productive and deliberately aided anti-Semitic racists in their ascent to power in Europe, which might have been his goal all along. Einstein later avowed that the plan for the inhuman carnage of which many Europeans and European governments eventually proved capable under Zionist leadership, appeared in Hitler's *Mein Kampf*, which was written in the 1920's.[144] He knew well what to expect.

Hitler's mentor, Dietrich Eckart, who was a member the Zionists' anti-Semitic propaganda school the *Thule-Gesellschaft*, exploited Jewish racism and anti-Germanism for propaganda purposes. Dietrich Eckart wrote, quoting Hitler, in Eckart's *Der Bolschewismus von Moses bis Lenin: Zwiegespräch zwischen Adolf Hitler und mir*,

> "'Send me a box full of German soil, so that I can at least symbolically defile the accursed country,' wrote the German Jew, Börne; [*Notation:* Ludwig Börne (alias Löb Baruch), *Briefe aus Paris* (Hamburg, 1832), I.] and Heinrich Heine sniffed out Germany's future from a toilet bowl. [*Notation:* Heinrich (alias Chaim) Heine, *Deutschland, ein Wintermärchen* (1844).] The physicist, Einstein, whom the Jewish publicity agents celebrate as a second Kepler, explained he would have nothing to do with German nationalism. He considered 'deceitful' the custom of the Central Association of German Citizens of Jewish Faith [*Notation:* Zentralverein deutscher Staatsbürger jüdischen Glaubens. {Translator}] of concerning themselves only with the religious interests of the Jews and not with their racial community also. A rare bird? No, only one who believed his people already safely in control, and thus considered it no longer necessary to keep up pretenses. In the Central Association itself, the mask has already fallen. A Dr. Brünn frankly admitted there that the Jews could have no German national spirit. [*Notation:* Artur Brünn, *Im Deutschen Reich* (the periodical of the *Zentralverein*) 1913, No. 8.] We always mistake their unprincipled exertions to accommodate themselves to all and everyone for impulses of the heart. Whenever they see an advantage to be gained by adopting a certain pose, they never hesitate, and certainly wouldn't let ethical considerations stand in their way. How many Galician Jews have first become Germans, then Englishmen, and finally Americans! And every time in the twinkling

---

143. Letter from A. Einstein to M. Born of 22 March 1934, in M. Born, *The Born-Einstein Letters*, Walker and Company, New York, (1971), pp. 121-122.

144. A. Einstein, *Ideas and Opinions*, Crown, New York, (1954), p. 213. **See also:** P. A. Bucky, Einstein, and A. G. Weakland, *The Private Albert Einstein*, Andrews and McMeel, Kansas City, (1992), p. 63.

of an eye. With startling rapidity they change their nationality back and forth, and wherever their feet touch, there resounds either the 'Watch on the Rhine,' or the 'Marsellaise,' or 'Yankee Doodle.' Dr. Heim does not once question the fact that our Warburgs, our Bleichroders, or our Mendelssohns are able to transfer their patriotism as well as their residence of today to London or to New York on the morrow. 'On the sands of Brandenburg an Asiatic horde!' Walther Rathenau once blurted out about the Berlin Jews. [*Notation:* Walther Rathenau, *Berliner Kulturzentren*, 1913. Rathenau was a Jewish war profiteer in World War I and later a minister in the Weimar government. He was executed by German patriots in 1922. {Translator}] He forgot to add that the same horde is on the Isar, the Elbe, the Main, the Thames, the Seine, the Hudson, the Neva, and the Volga. And all of them with the same deceit toward their neighbors."[145]

Should Albert Einstein be forgiven as an ethnocentric and racist victim of his time and political affiliations, who defended "his people" from what appeared to him to be a threat to their very existence—the dangers of assimilation and philo-Semitism? Early on, Jews with far more sense than Einstein organized to defend themselves from the fanatical and racist Zionists, knowing that the political games of the racists on both sides of the "Jewish question" would result in tragedy and trauma for the world's Jews. Klaus J. Herrmann wrote,

"To counter the coalition of antisemites and Zionists, in 1912, within the Association for Liberal Judaism, a number of distinguished leaders of Germany's Jewish communities decided to form an *Anti-Zionist Committee*. This Committee [***] took on the task of 'enlightening the German Jews on and combating Zionism.'"[146]

Paul Ehrenfest saw the harm racist and segregationist Zionist Jews were doing to his fellow Jews.[147] Since all reasonable Jews knew the destruction that would inevitably follow from Einstein's ideology, Einstein should have known it, and indeed he did know it. One outgrowth of these anti-Zionist organizations, which formed to protect themselves, is Neturei Karta. Rabbi Moshe Shonfeld documented the collaboration of the Zionists with the Nazis and the deliberate

---

145. D. Eckart and A. Hitler, *Der Bolschewismus von Moses bis Lenin: Zwiegespräch zwischen Adolf Hitler und mir*, Hoheneichen-Verlag, München, (1924); English translation by W. L. Pierce, "Bolshevism from Moses to Lenin", *National Socialist World*, (1966). URL: <http://www.jrbooksonline.com/DOCs/Eckart.doc> p. 7.
146. K. J. Herrmann, "Historical Perspectives on Political Zionism and Antisemitism", *Zionism & Racism: Proceedings of an International Symposium*, International Organization for the Elimination of All Forms of Racial Discrimination, Tripoli, (1977), pp. 197-210, at page 208.
147. Letter from P. Ehrenfest to A. Einstein of 9 December 1919, *The Collected Papers of Albert Einstein*, Volume 9, Document 203, Princeton University Press, (2004).

human sacrifice of innocent Jews in order to establish the "Jewish State".[148] Numerous other Jewish authors have chastised Zionist Jews for their behavior towards other Jews during the Holocaust.[149] Rabbi E. Schwartz published a statement on behalf of the American Neturei Karta in *The New York Times* on 18 May 1993,

> "To achieve the goal of statehood the Zionists have always deliberately provoked anti-Semitism. [***] Their interest was not to save Jews, on the contrary, more spilling of Jewish blood would strengthen their demand of the nations for the creation of their state."[150]

Albert Einstein, the "Person of the Century" who sought to promote and foment anti-Semitism wherever he went, stated in 1921,

> "On the other hand, anti-Semitism in Germany also has consequences that, from a Jewish point of view, should be welcomed. I believe German Jewry owes its continued existence to anti-Semitism."[151]

Contrast this with Nobel Peace Prize laureate Elie Wiesel's statement in 1968,

> "Every Jew, somewhere in his being, should set apart a zone of hate—healthy, virile hate—for what the German personifies and for what persists in the German. To do otherwise would be a betrayal of the dead."[152]

Lieutenant General Rafael Eytan, outgoing Chief of Staff of the Israeli Army, stated on 12 April 1983,

> "When we have settled the land, all the Arabs will be able to do about it will

---

148. M. Shonfeld, *The Holocaust Victims Accuse: Documents and Testimony on Jewish War Criminals*, Neturei Karta of U.S.A., Brooklyn, (1977).
149. Rabbi M. D. Weissmandel, *Min ha-metsar: zikhronot mi-shenot 702-705*, Hotsa'at Emunah, New York, (1960); **and** *Ten Questions to the Zionists*, (1974). *See also:* L. Brenner, *Zionism in the Age of the Dictators*, Croom Helm, London, L. Hill, Westport, Connecticut, (1983); **and** *51 Documents: Zionist Collaboration with the Nazis*, Barricade Books Inc., Fort Lee, New Jersey, (2002). *See also:* M. J. Nurenberger, *The Scared and the Doomed: The Jewish Establishment Vs. The Six Million*, Mosaic Press, Oakville, New York, (1985). *See also:* W. R. Perl, *The Holocaust Conspiracy: An International Policy of Genocide*, Shapolsky Publishers, New York, (1989). *See also:* T. Segev, *The Seventh Million: The Israelis and the Holocaust*, Hill and Wang, New York, (1993).
150. Rabbi E. Schwartz, "WHY DO YOU VIOLATE G'D'S ORDER? IT WILL NOT SUCCEED", *The New York Times*, (18 May 1993), p. A16.
151. A. Einstein, A. Engel translator, "How I became a Zionist", *The Collected Papers of Albert Einstein*, Volume 7, Document 57, Princeton University Press, (2002), pp. 234-235, at 235.
152. E. Wiesel, *Legends of Our Time*, Schocken Books, New York, (1982), p. 142.

be to scurry around like drugged roaches in a bottle."[153]

In an article entitled, "An Israeli Mayor Is Under Scrutiny", *The New York Times* reported on 6 June 1989, on page 5,

> "Rabbi Yitzhak Ginsburg had offered biblical justification for the view that the spilling of non-Jewish blood was a lesser offense than the spilling of Jewish blood. 'Any trial based on the assumption that Jews and goyim are equal is a total travesty of justice,' he said."

Rabbi Yaacov Perrin was quoted by Clyde Haberman, in an article entitled, "Arafat Dismisses Rabin's Moves as 'Hollow'", *The New York Times*, (28 February 1994), p. 1. Rabbi Perrin stated,

> "One million Arabs are not worth a Jewish fingernail[.]"

Wiesel has stressed his view that the Holocaust should be seen as a uniquely tragic event in History. However, this exclusivist view of Jewish History predates the Holocaust by at least a century, for example in a statement from 1845,

> "The sufferings of the Jews—whether the 'wringing out of the dregs of a cup of trembling' from Jehovah, or not—have far exceeded all other experience, and the common measure of human endurance."[154]

After the First World War Einstein and some of his friends alluded to much earlier conversations with Einstein where he had correctly predicted the eventual outcome of the war. In his diaries, Romain Rolland recorded his conversations with Einstein in Switzerland at their meeting of 16 September 1915,

> "What I hear from [Einstein] is not exactly encouraging, for it shows the impossibility of arriving at a lasting peace with Germany without first totally crushing it. Einstein says the situation looks to him far less favorable

---

153. D. K. Shipler, "Most West Bank Arabs Blaming U. S. for Impasse", *The New York Times*, (14 April 1983), p. A3; **and** "Israel's Military Chief Retires and Is Replaced by His Deputy", *The New York Times*, (20 April 1983), p. A8; **and** "The Israeli Army Signs a Political Truce", *The New York Times*, Section 4, (15 May 1983), p. 3. *See also:* A. Lewis, "Hope Against Hope", *The New York Times*, Section 4, (17 April 1983), p. 19; and "The New Israel; Away from the Early Zionist Dream", *The New York Times*, (30 July 1984), p. A21. *See also:* J. Kuttab, "West Bank Arabs Foresee Expulsion", *The New York Times*, (1 August 1983), p. A15. *See also:* A. Cowell, "Israel Frees More Prisoners, But Arabs Are Not Mollified", *The New York Times*, (4 March 1994), p. A10. *See also:* Y. M. Ibrahim, "Palestinians See a People's Hatred in a Killer's Deed", *The New York Times*, (6 March 1994), p. E16.
154. "The Modern Jews", *The North American Review*, Volume 60, Number 127, (April, 1845), pp. 329-368, at 330.

than a few months back. The victories over Russia have reawakened German arrogance and appetite. The word 'greedy' seems to Einstein best to characterize Germany. [***] Einstein does not expect any renewal of Germany out of itself; it lacks the energy for it, and the boldness for initiative. He hopes for a victory of the Allies, which would smash the power of Prussia and the dynasty... . Einstein and Zangger dream of a divided Germany—on the one side Southern Germany and Austria, on the other side Prussia. [***] We speak of the deliberate blindness and the lack of psychology in the Germans."[155]

Jews often sought to Balkanize nations so as to weaken the power of any faction within a nation and to created perpetual agitation between the nations which could be exploited for profit and other Jewish gains. For example, the Rothschilds created the American Civil War and profited from the debts it generated. They hoped to divide America into two nations and to pit these against one another. They were successful. Jews had long been pitting North German Protestants against South German and Austrian Catholics. Jews were the motive force behind the *Kulturkampf*. After creating these divides and promoting perpetual agitations amongst neighbors, Jewry could then fund one side against the other to destroy it whenever Jewry decided to wreck a given nation.

Einstein's dreams during the First World War remind one of the "Carthaginian Peace" of the Henry Morgenthau, Jr. plan for the destruction of Germany following the Second World War. Morgenthau worked with Lord Cherwell (Frederick Alexander Lindemann), Churchill's friend and advisor, who planned to bomb German civilian populations into submission. Lindemann studied under Einstein's friend, Walther Nernst, who worked with Fritz Haber, a Jewish developer of poisonous gas. James Bacque argues that the Allies, under the direction of General Eisenhower, starved hundreds of thousands, if not millions of German prisoners of war to death. Dwight David Eisenhower was called "the terrible Swedish-Jew" in his yearbook for West Point, *The 1915 Howitzer*, West Point, New York, (1915), p. 80. He was also called "Ike", as in... Eisenhower? The Soviets also abused countless German POW's after the Second World War.[156]

Einstein often spoke in genocidal and racist terms against Germany, and for the Jews and England, and he betrayed Germany before, during and after the First World War. Einstein wrote to Paul Ehrenfest on 22 March 1919,

---

155. R. Romain, *La Conscience de l'Europe*, Volume 1, pp. 696ff. English translation from A. Fölsing, *Albert Einstein: A Biography*, Viking, New York, (1997), pp. 365-367. *See also:* Letter from A. Einstein to R. Romain of 15 September 1915, *The Collected Papers of Albert Einstein*, Volume 8, Document 118, Princeton University Press, (1998); **and** Letter from A. Einstein to R. Romain of 22 August 1917, *The Collected Papers of Albert Einstein*, Volume 8, Document 374, Princeton University Press, (1998).
156. J. Bacque, *Other Losses: An Investigation into the Mass Deaths of German Prisoners at the Hands of the French and Americans after World War II*, Stoddart,Toronto, (1989).

"[The Allied Powers] whose victory during the war I had felt would be by far the lesser evil are now proving to be *only slightly* the lesser evil. [***] I get most joy from the emergence of the Jewish state in Palestine. It does seem to me that our kinfolk really are more sympathetic (at least less brutal) than these horrid Europeans. Perhaps things can only improve if only the Chinese are left, who refer to all Europeans with the collective noun 'bandits.'"[157]

Einstein avowed *circa* 3 April 1920, that,

"If what anti-Semites claim were true, then indeed there would be nothing weaker, more wretched, and unfit for life, than the German people".[158]

Einstein avowed that the anti-Semites' beliefs were true. Therefore, Einstein must have believed at least as early as 1920 that the Germans ought to be exterminated. When discussing the meaning of life, Einstein spoke to Peter A. Bucky about persons and creatures who "[do] not deserve to be in our world" and are "hardly fit for life."[159] Einstein's language is quite similar to the language of Hitler's "T4" "*Euthanasia-Programme*".

After siding with Germany's enemies in the First World War—while living in Germany, and after intentionally provoking Germans into increased anti-Semitism, which he thought was good for Jews, and after defaming German Nobel Prize laureates in the international press to the point where they felt obliged to join Hitler's cause, which cause eventually resulted in the genocide of Europe's Jews; Einstein sponsored the production of genocidal weapons to mass murder Germans, whom he had hated all of his life, in the famous letter to President Roosevelt that Einstein signed urging Roosevelt to begin the development of atomic bombs. Einstein signed this letter before the alleged mass murder of Jews had begun.[160]

Genocidal Einstein callously asserted that the use of atomic bombs on civilian populations was "morally justified". I quote Einstein without delving into the question of who first bombed civilian centers,

"It should not be forgotten that the atomic bomb was made in this country as a preventive measure; it was to head off its use by the Germans, if they

---

[157]. Letter from A. Einstein to Paul Ehrenfest of 22 March 1919, English translation by A. Hentschel, *The Collected Papers of Albert Einstein*, Volume 9, Document 10, Princeton Univsersity Press, (2004), pp. 9-10, at 10.
[158]. A. Einstein, English translation by A. Engel, *The Collected Papers of Albert Einstein*, Volume 7, Document 35, Princeton University Press, (2002), pp. 156-157.
[159]. P. A. Bucky, Einstein, and A. G. Weakland, *The Private Albert Einstein*, Andrews and McMeel, Kansas City, (1992), p. 111.
[160]. A. Unsöld, "Albert Einstein — Ein Jahr danach", *Physikalische Blätter*, Volume 36, (1980), pp.337-339; **and** Volume 37, Number 7, (1981), p. 229.

discovered it. The bombing of civilian centers was initiated by the Germans and adopted by the Japanese. To it the Allies responded in kind—as it turned out, with greater effectiveness—and they were morally justified in doing so."[161]

Einstein advocated genocidal collective punishment,

"The Germans as an entire people are responsible for these mass murders and must be punished as a people if there is justice in the world and if the consciousness of collective responsibility in the nations is not to perish from the earth entirely."[162]

and,

"It is possible either to destroy the German people or keep them suppressed; it is not possible to educate them to think and act along democratic lines in the foreseeable future."[163]

Albrecht Fölsing has assembled a compilation of post-WW II quotations by Albert Einstein, which evince Einstein's lifelong habit of stereotyping people based on their ethnicity. Einstein again expressed his hatred after the war—a temptation Max Born had resisted,

"With the Germans having murdered my Jewish brethren in Europe, I do not wish to have anything more to do with Germans, not even with a relatively harmless Academy. [***] The crimes of the Germans are really the most hideous that the history of the so-called civilized nations has to show. [***] [It was] evident that a proud Jew no longer wishes to be connected with any kind of German official event or institution. [***] After the mass murder committed by the Germans against my Jewish brethren I do not wish any publications of mine to appear in Germany."[164]

Einstein wrote to Born on 15 September 1950 that his pathological hatred towards Germans predated the Nazi period,

---

161. A. Einstein, "Atomic War or Peace", *Atlantic Monthly*, (November, 1945, and November 1947); *as reprinted in:* A. Einstein, *Ideas and Opinions*, Crown, New York, (1954), p. 125.
162. A. Einstein, "To the Heroes of the Battle of the Warsaw Ghetto", *Bulletin of the Society of Polish Jews*, New York, (1944), reprinted in *Ideas and Opinions*, Crown, New York, (1954), pp. 212-213.
163. A. Einstein, quoted in O. Nathan and H. Norton, *Einstein on Peace*, Avenel Books, New York, (1981), p. 331.
164. A. Einstein quoted in A. Fölsing, *Albert Einstein: A Biography*, Viking, New York, (1997), pp. 727-728.

"I have not changed my attitude to the Germans, which, by the way, dates not just from the Nazi period. All human beings are more or less the same from birth. The Germans, however, have a far more dangerous tradition than any of the other so-called civilized nations. The present behavior of these other nations towards the Germans merely proves to me how little human beings learn even from their most painful experiences."[165]

and on learning that Born would return to Germany, Einstein wrote on 12 October 1953,

"If anyone can be held responsible for the fact that you are migrating back to the land of the mass-murderers of our kinsmen, it is certainly your adopted fatherland — universally notorious for its parsimony."[166]

Einstein wanted to carry out the extermination of the Germans he had been planning for many decades before the Holocaust. Einstein could not forgive the fact that other nations forgave the Germans and did not take the opportunity the Zionists had created for the complete extermination of the German People, the extermination of Amalek.

7.5.6 The Final Solution of the Jewish Question is Zionism, but the Final Solution of the German Question is Extermination

The generally accepted history of the Wannsee-Konferenz of 20 January 1942 holds that the Nazis first proposed the party policy of the genocidal extermination of Jews on this date. Lesser known today is the fact that a Jewish American named Theodor Newman Kaufman advocated the genocidal sterilization of all Germans as a "final solution" in 1941 in his book *Germany Must Perish!*, Argyle Press, Newark, New Jersey, (1941), for which an ad was posted in *The New York Times* on 1 March 1941 on page 13. Kaufman had called for the sterilization of all Americans in 1939.

Kaufman promoted his book by sending out small black cardboard coffins with a note inside which read, "Read GERMANY MUST PERISH! Tomorrow you will receive your copy," to leading figures and persons in the media. This was followed by a copy of the book the next day. This book was briefly noted in "Latest Books Received", *The New York Times*, (16 March 1941), Book Reviews Section, pp. 28-30, at 29; which simply states, "A plan for permanent peace among civilized nations." *Time Magazine*, under the heading "A Modest Proposal", described the odd book, the strange method by which Kaufman had

---

[165]. M. Born, *The Born-Einstein Letters*, Walker and Company, New York, (1971), p. 189.
[166]. M. Born, *The Born-Einstein Letters*, Walker and Company, New York, (1971), p. 199.

promoted it, and the peculiar history of Theodor Newman Kaufman, who claimed to have known members of Winston Churchill's family.[167] In an interesting aside, Albert Einstein's personal physician, Professor Janos Plesch, became Winston Churchill's personal physician.[168]

Kaufman's book advocating the genocide of Germans was known to most Germans. *Germany Must Perish!* was condemned in German publications, which alleged that President Roosevelt had sponsored it and had even written passages in it. The book, which proposed the genocide of the Germans, provoked attacks on Jews in Germany.[169] To the Germans, *Germany Must Perish!* represented the climax of the generalized vilification of all Germans propagandized by enemies of Germany in the First World War, like Émile Durkheim.[170] At least as early as the 1860's, recalling the myth of Esau and Amalek, Zionist racist and National Socialist Moses Hess[171] argued that the "German race" had a genetically programmed antagonism towards the "Jewish race"—the implication being that one must destroy the other in order to survive. Hess cushions his blows by mentioning enlightened Germans who have supposedly overcome their alleged genetic compulsions to destroy Jews, but his genocidal hatred of Germans is clear.

Hess was an interesting figure. He married a Christian prostitute. He wrote together with Marx, then criticized him. Hess created many of the elements of National Socialism that would eventually become the National Socialist German Worker's Party, or "Nazi" Party.

---

167. "A Modest Proposal", *Time Magazine*, Volume 37, Number 12, (24 March 1941), pp. 95-96.
168. P. A. Bucky, Einstein, and A. G. Weakland, *The Private Albert Einstein*, Andrews and McMeel, Kansas City, (1992), p. 56.
169. W. Diewerge, *Das Kriegsziel der Weltplutokratie: dokumentarische Veröffentlichung zu dem Buch des Präsidenten der amerikanischen Friedensgesellschaft Theodore Nathan Kaufman "Deutschland muss sterben" ("Germany must perish")*, Zentral Verlag der NSDAP, F. Eher Nachf., Berlin, (1941). *See also: Wenn Du dieses Zeichen siehst...* , NSDAP Propaganda Brochure, (November, 1941). *See also:* H. Goitsch, *Niemals!*, Zentral Verlag der NSDAP, F. Eher Nachf., Berlin, (1944). *See also: Der Angriff*, (23 July 1941). *See also: Das Reich*, (3 August 1941). *See also:* "Nazis Attack Roosevelt", *The New York Times*, (24 July 1941), p. 8. *See also:* "Jews of Hanover Forced from Homes", *The New York Times*, (9 September 1941), p. 4; and Kaufman's response, p. 4.
170. É. Durkheim, *"Germany above All" The German Mental Attitude and the War*, Librairie Armand Colin, Paris, (1915). *See also:* "By a German", *I Accuse! (J'Accuse!)*, Grosset & Dunlap, New York, (1915). *See also:* W. F. Barry, *The World's Debate: An Historical Defence of the Allies*, George H. Doran, New York, (1917). *See also:* W. T. Hornaday, *A Searchlight on Germany: Germany's Blunders, Crimes and Punishment*, American Defense Society, New York, (1917). *See also:* D. W. Johnson, *Plain Words from America: A Letter to a German Professor*, London, New York, Toronto, Hodder & Stoughton, (1917).
171. M. Hess, *Rom und Jerusalem: die letzte Nationalitätsfrage*, Eduard Wengler, Leipzig, (1862); English: *Rome and Jerusalem: A Study in Jewish Nationalism*, Bloch, New York, (1918).

With Kaufman's *Germany Must Perish!* as evidence, the Nazis told the German public that the Americans, under the direction of Jews, planned to exterminate the "German race" if the Allies won the war. This life and death struggle between the "German race" and the "Jewish race" was foretold in Hess' book of 1862, *Rom und Jerusalem: die letzte Nationalitätsfrage*, Eduard Wengler, Leipzig, (1862); English: *Rome and Jerusalem: A Study in Jewish Nationalism*, Bloch, New York, (1918).

Goebbels proclaimed that the inhumane crimes Germans had committed against Jews compelled Germany to fight to the very end, thereby maximizing German and Allied and European casualties and the destruction of Europe. At the end of the war, Hitler called for Germans to kill themselves, because they had proven themselves unworthy to live in the fight for survival. Some have alleged that Hitler was sent to destroy Germans, who many Jews had alleged were genetic or cultural enemies of Jews predisposed to destroy them. Hitler destroyed Europe with perpetual war and he destroyed "Red Assimilationist" Jews in order to punish them and to shock American Jews into embracing Zionism.

Einstein's genocidal statements hint at the proposed measures advocated in Kaufman's book of 1941. Among other things, Kaufman wrote,

"A final solution: Let Germany be policed forever by an international armed force? *Even if such a huge undertaking were feasible life itself would not have it so. As war begets war, suppression begets rebellion. Undreamed horrors would unfold.* Thus we find that there is no middle course; no act of mediation, no compromise to be compounded, no political or economic sharing to be considered. There is, in fine, no other solution except one: That Germany must perish forever from this earth! [***] There remains then but one mode of ridding the world forever of Germanism — and that is to stem the source from which issue those war-lusted souls, by preventing the people of Germany from ever again reproducing their kind. This modern method, known to science as Eugenic Sterilization, is at once practical, humane and thorough. Sterilization has become a byword of science as the best means of ridding the human race of its misfits: the degenerate, the insane, the hereditary criminal. [***] The population of Germany, excluding conquered and annexed territories, is about 70,000,000, almost equally divided between male and female. To achieve the purpose of German extinction it would be necessary to only sterilize some 48,000,000—a figure which excludes, because of their limited power to procreate, males over 60 years of age, and females over 45. [***] Reviewing the foregoing case of sterilization we find that several factors resulting from it firmly establish its advocacy. Firstly, no physical pain will be imposed upon the inhabitants of Germany through its application, a decidedly more humane treatment than they will have deserved. As a matter of fact it is not inconceivable that after Germany's defeat, the long-suffering peoples of Europe may demand a far less humane revenge than that of mere sterilization. Secondly, execution of the plan would in no way disorganize the present population nor would it cause any sudden mass upheavals and dislocations The consequent gradual disappearance of the Germans from Europe will leave no more negative effect upon that continent than did the gradual disappearance of Indians upon this."[172]

Perhaps inspired by the accusations against Jews of poising the wells in the 1300's, some Jews unsuccessfully attempted revenge against the Germans for the Holocaust after the Second World War by poisoning the water supply of Germany. Tom Segev wrote in his book *The Seventh Million: The Israelis and the Holocaust*,

"Kovner therefore set six million German citizens as his goal. He thought in apocalyptic terms: revenge was a holy obligation that would redeem and purify the Jewish people. The group divided into cells, each with a commander. Their primary goal, Plan A, was 'to poison as many Germans as possible.' Plan B was to poison several thousand former SS men in the American army's POW camps. Reichman succeeded in infiltrating some members of the group into the Hamburg and Nuremberg water companies. Kovner went to Palestine to bring the poison—and, he hoped, to receive the blessing of the Haganah."[173]

It is often alleged that a group of high ranking Nazi officials met at a conference in Wannsee and settled on a plan to exterminate the Jews of Europe in concentration camps. There is a purported transcript of this meeting. Some have disputed the authenticity of the minutes of the Wannsee-Konferenz. At any rate, the minutes of the Wannsee Conference do not contain any statements plotting the deliberate murder of the Jews or the extermination of all Jews. The "final solution of the Jewish question" proposed in the purported minutes of the Wannsee Conference was not murder or complete extermination; but was instead the deportation of Jews to the East in conformity with the wishes of the Zionist Jews.[174]

Zionist Nazi propagandist Julius Streicher affirmed at the Nuremberg Trials that the Nuremberg Laws of 1935 were patterned after Jewish Law,

"Yes, I believe I had a part in it insofar as for years I have written that any further mixture of German blood with Jewish blood must be avoided. I have written such articles again and again; and in my articles I have repeatedly emphasized the fact that the Jews should serve as an example to every race,

---

172. T. N. Kaufman, *Germany Must Perish!*, Argyle Press, Newark, New Jersey, (1941), pp. 88-89, 93, 94, 96.
173. T. Segev, *The Seventh Million: The Israelis and the Holocaust*, Hill and Wang, New York, (1993), p. 142.
174. An English translation of the minutes appears in: R. S. Levy, "Wannsee Conference on the Final Solution of the Jewish Question", *Antisemitism in the Modern World: An Anthology of Texts*, D.C. Heath, Toronto, (1991), pp. 252-258; *see also:* pp. 250-252.

for they created a racial law for themselves—the law of Moses, which says, 'If you come into a foreign land you shall not take unto yourself foreign women.' And that, Gentlemen, is of tremendous importance in judging the Nuremberg Laws. These laws of the Jews were taken as a model for these laws. When, after centuries, the Jewish lawgiver Ezra discovered that notwithstanding many Jews had married non-Jewish women, these marriages were dissolved. That was the beginning of Jewry which, because it introduced these racial laws, has survived throughout the centuries, while all other races and civilizations have perished."[175]

Dr. Marx asked Julius Streicher,

"Were you of the opinion that the 1935 legislation represented the final solution of the Jewish question by the State?"[176]

Streicher responded that Zionism was the final solution of the Jewish question,

"With reservations, yes. I was convinced that if the Party program was carried out, the Jewish question would be solved. The Jews became German citizens in 1848. Their rights as citizens were taken from them by these laws. Sexual intercourse was prohibited. For me, this represented the solution of the Jewish problem in Germany. But I believed that another international solution would still be found, and that some day discussions would take place between the various states with regard to the demands made by Zionism. These demands aimed at a Jewish state."[177]

Nazi Secretary of State in the Interior Ministry Wilhelm Stuckart, who attended the Wannsee-Konferenz, was questioned by Robert M. W. Kempner at his Nuremberg trial and denied that the extermination of the Jews was discussed,

> "No, I don't believe that I am wrong in saying that there was no discussion of the final solution of the Jewish question, in the sense in which it is now understood.
>
> KEMPNER: Heydrich related clearly, in your presence, what it was about?

---

175. *Trial of the Major War Criminals Before the International Military Tribunal, Nuremberg, 14 November 1945 — 1 October 1946*, Volume 12, Secretariat of the Tribunal, Nuremberg, Germany, p. 315.
176. *Trial of the Major War Criminals Before the International Military Tribunal, Nuremberg, 14 November 1945 — 1 October 1946*, Volume 12, Secretariat of the Tribunal, Nuremberg, Germany, p. 316.
177. *Trial of the Major War Criminals Before the International Military Tribunal, Nuremberg, 14 November 1945 — 1 October 1946*, Volume 12, Secretariat of the Tribunal, Nuremberg, Germany, p. 316.

> STUCKART: That is absolutely out of the question—otherwise I would have known what it meant."[178]

Refer to the Nuremberg trial transcripts of 22 November 1945, where Stuckart was quoted as referring to the "final solution" in the late 1930's, as a political solution, some years before the Wannsee-Konferenz occurred, meaning the formation of a "Jewish State". This quotation was cited prior to the first appearance of the purported "Protocols of the Wannsee Conference". Again, some have called into questions the authenticity of these "Protocols".

Though Eichmann stated that the "final solution" had always meant a Zionist political solution to him, Eichmann alleged many years after the war that he had heard from third party sources that Hitler changed course in the middle of the war and planned to exterminate the Jews.[179] David Irving has argued that Hitler never had any such plan.

Accusations that Hitler was out to exterminate the Jews predated the Holocaust by many years, and served the interests of the Zionists, just as the Holocaust served and serves the interests of the Zionists. *The New York Times* reported on 8 February 1923, on page 3, in an article entitled,"SAYS FORD AIDS ROYALISTS. Auer Charges Financial Help to Bavarian Anti-Semites.":

> "Henry Ford was accused of financing a Bavarian monarchist revolution by Herr Auer, Vice President of the Bavarian Diet, who came to Berlin today to report to President Ebert on the situation. Herr Auer informed The Tribune that Henry Ford's financial as well as moral backing had been given to Bavarian revolution-makers during the past year because a part of the program of Herr Hitler, leader of the Monarchists, is the extermination of the Jews in Germany."

It would be interesting to determine the exact German word Auer used, which had been translated as "exterminate". Was it *Ausrottung*, or perhaps *Vernichtung*? There has been a dispute over the meaning of Hitler's many statements against the Jews in the original German, which hinges on whether or not he meant to simply rid Germany of Jews by deporting them, or whether he was out to exterminate all Jews. At the time, Hitler was calling for the expulsion of Jews from Bavaria and from all German lands. The money scandal drew attention in the newspapers in Germany, but most attention was paid to the French connection. Hitler's agent Kurt G. W. Ludecke failed in his attempts to

---

178. M. Roseman, *The Wannsee Conference and the Final Solution: A Reconsideration*, Henry Holt, New York, (2002), p. 105. Roseman cites: R. M. W. Kempner, *Eichmann und Komplizen*, Europa Verlag, Zürich, (1961), pp. 152-153.

179. Refer to Eichmann's testimony at trial, and: A. Eichmann, "Eichmann Tells His Own Damning Story", *Life Magazine*, Volume 49, Number 22, (28 November 1960), pp. 19-25, 101-112; and "Eichmann's Own Story: Part II", *Life Magazine*, (5 December 1960), pp. 146-161.

solicit monies for the Nazis from Henry Ford.[180]

The "Hamburg Resolutions of the German Social Reform Party" proclaimed in 1899,

"The strivings of Zionism are a fruit of the antisemitic movement. [***] Unfortunately [any hope that all Jews will emigrate to Palestine] appears to be infeasible. [***] As such, [the Jewish question] should be solved in common with other nations and result finally in full separation, and—if self-defense demands—in final annihilation [*Vernichtung*] of the Jewish race."[181]

Adolf Hitler wrote in an article entitled "Staatsmänner oder Nationalverbrecher" in the *Völkischer Beobachter*, Volume 35, Number 22, (15 March 1921), p. 1-2, that the fight against Bolshevism in Russia entailed the rooting out (*Ausrottung*) of the Jews. On 30 January 1939 Hitler famously threatened before the Reichstag that if Jewish finance again led the world into war, it would not mean a victory for "world Jewry", but "the annihilation [*Vernichtung*] of the Jewish race in Europe",

"[I want to be a prophet again today:] If international finance Jewry in and outside Europe succeeds in plunging the peoples into another world war, then the end result will not be the Bolshevization of the earth and the consequent victory of Jewry but the annihilation of the Jewish race in Europe."[182]

"Ich will heute wieder ein Prophet sein: Wenn es dem internationalen Finanzjudentum in und außerhalb Europas gelingen sollte, die Völker noch einmal in einen Weltkrieg zu stürzen, dann wird das Ergebnis nicht die Bolschewisierung der Erde und damit der Sieg des Judentums sein, sondern die Vernichtung der jüdischen Rasse in Europa!"[183]

The Jewish Zionist Nazi tyrant of Poland, Dr. Hans Frank, stated at a Cabinet Session on 16 December 1941,

"As far as the Jews are concerned, I want to tell you quite frankly, that they

---

180. K. G. W. Ludecke, *I Knew Hitler: The Story of a Nazi AWho Escaped the Blood Purge*, Charles Scribner's Sons, New York, (1937), pp. 191-218.
181. English translation in: R. S. Levy, *Antisemitism in the Modern World: An Anthology of Texts*, D. C. Heath and Company, Toronto, (1991), pp. 127-128, at 128.
182. English translation in: R. S. Levy, *Antisemitism in the Modern World: An Anthology of Texts*, D. C. Heath and Company, Toronto, (1991), pp. 222-223, at 223. An alternative translation appears in: "Holocaust", *Encyclopaedia Judaica*, Volume 8, Macmillan, Jerusalem, (1971), col. 852.
183. A. Hitler in M. Domarus, Editor, *Hitler: Reden und Proklamationen, 1932-1945: Kommentiert von einem deutschen Zeitgenossen*, Süddeutscher Verlag, München, (1965), pp. 1057-1058.

must be done away with in one way or another. The Fuehrer said once: should united Jewry again succeed in provoking a world war, the blood of not only the nations which have been forced into the war by them, will be shed, but the Jew will have found his end in Europe"[184]

Did the crypto-Jewish Zionists Adolf Hitler and Hans Frank mean that they would exterminate the Jews of Europe in death camps, or did they mean that they would deport the Jews of Europe to Palestine as a final solution to the Jewish question? Frank was a long-term Zionist who wanted to segregate the Jews in Polish concentration camps and then ship them to Palestine—not to say that he did not intend to kill off a large percentage of his brethren in the process. In the fall of 1933 in Nuremberg on *Reichsparteitag*, Frank stated that his goal was to secure a "Jewish State",

"Unbeschadet unseres Willens, uns mit den Juden auseinanderzusetzen, ist die Sicherheit und das Leben der Juden in Deutschland staatlich, reichsamtlich und juristisch nicht gefährdet. Die Judenfrage ist rechtlich nur dadurch zu lösen, dass man an die Frage eines jüdischen Staates herangeht."[185]

The expression "final solution of the Jewish question (*or:* "problem")" was a commonplace in the parlance of the Zionists long before the Wannsee Konferenz.[186] Jewish Zionist Nahum Sokolow wrote in the introduction of his *History of Zionism* of 1919,

"The progress of modern civilization has come to be regarded as a sort of 'Messiah' for the final solution of the Jewish problem."[187]

Sokolow spoke in reference to the "Jewish mission" of reformed Jews under the influence of Moses Mendelssohn. The Zionists believed this "final solution of the Jewish problem" resulted in fatal assimilation, whereas the Zionists were pitching Palestine as the "final solution of the Jewish problem". Many others believed that assimilation was the only viable "final solution to the Jewish

---

184. H. Frank, (16 December 1941), quoted in: *Nazi Conspiracy and Aggression*, Volume 2, United States, Office of Chief of Counsel for the Prosecution of Axis Criminality, Washington, D. C., United States Government Printing Office, (1946), p. 634. ***See also:*** Y. Arad, Yitzhak, I. Gutman, A. Margaliot, Abraham,Editors, *Documents on the Holocaust: Selected Sources on the Destruction of the Jews of Germany and Austria, Poland, and the Soviet Union*, Yad Vashem in cooperation with the Anti-Defamation League and Ktav Pub. House, Jerusalem, (1981).
185. H. Frank quoted in H. Kardel, *Adolf Hitler, Begründer Israels*, Verlag Marva, Genf, (1974).
186. The exact phrasing depends upon translation, but one finds such phrases in: A. Ha-Am, "The Negation of the Diaspora", in A. Hertzberg, *The Zionist Idea*, Harper Torchbooks, New York, (1959), pp. 270-277, at 272-273, 277.
187. N. Sokolow, *History of Zionism 1600-1918*, Volume 1, Longmans, Green and Co., New York, (1919), p. xvii.

question".[188]

Boris Brasol wrote in 1921,

> "When the Zionist claim was first established, and Theodore Hertzl, in 1897, came out with his specific program of a Jewish State, the world at large gave a sigh of relief as it was trusted that henceforth the Jews would have a country of their own where they would be able to develop freely and unhampered their racial peculiarities, their cultural traditions and their religious thought. Christian countries have been so accustomed to innumerable complaints made by the Jews of their oppression, of anti-Semitism breeding throughout the world, of pogroms ravaging the Jewish masses, that there was every reason to hope that the Jews would dash to Palestine, leaving those cruel Christians to their own destinies. What better scheme for a fair solution of the Jewish problem could be hoped for by both Gentiles and Jews?"[189]

The Zionists wrote in the official organ of the German Zionist Organization, *Jüdische Rundschau*, on 13 June 1933, shortly after Hitler assumed power,

> "Zionism recognizes the existence of the Jewish question and wants to solve it in a generous and constructive manner. For this purpose, it wants to enlist the aid of all peoples; those who are friendly to the Jews as well as those who are hostile to them, since according to its conception, this is not a question of sentimentality, but one dealing with a real problem in whose solution all peoples are interested."[190]

Jewish Zionist Joachim Prinz stated in 1937,

> "Everyone in Germany knew that only the Zionists could responsibly represent the Jews in dealings with the Nazi government. We all felt sure that one day the government would arrange a round table conference with the Jews, at which—after the riots and atrocities of the revolution had passed—the new status of German Jewry could be considered. The government announced very solemnly that there was no country in the world which tried to solve the Jewish problem as seriously as did Germany. Solution of the Jewish question? It was our Zionist dream! We never denied the existence of the Jewish question! Dissimilation? It was our own appeal! ... In a statement notable for its pride and dignity, we called for a

---

188. P. S. Mowrer, "The Assimilation of Israel", *The Atlantic Monthly*, Volume 128, Number 1, (July, 1921), pp. 101-110.
189. B. L. Brasol, *The World at the Cross Roads*, Small, Mayhard & Co., Boston, (1921), pp. 371-379.
190. English translation in: K. Polkehn, "The Secret Contacts: Zionism and Nazi Germany, 1933-1941", *Journal of Palestine Studies*, Volume 5, Number 3/4, (Spring-Summer, 1976), pp. 54-82, at 59.

conference."¹⁹¹

In 1917, Jewish Zionist Elisha M. Friedman made several references to the "solution of the Jewish question",

> "Recent events have served to accentuate Zionism as an attempt at the solution of the Jewish question. [***] And only yesterday, as it were, Adolph Lewinsohn, whose activities transcend creed, has likewise joined those that see in Zionism a solution to the Jewish question. [***] Insofar as it affords no relief to the assimilationist and intensifies the loyalty of the great mass of a dispersed people, the policy of partial assimilation defeats its own ends. It is purposeless. It has been tested out, as a solution of the Jewish question, and has proven an eloquent failure."¹⁹²

In 1911, Jewish Zionist Israel Zangwill made reference to the "solution of the Jewish Question",

> "But if the prospect of a territorial solution of the Jewish Question, whether in Palestine or in the New World appears remote, it must be admitted that the Jewish race, in abandoning before the legions of Rome the struggle for independent political existence, in favor of spiritual isolation and economic symbiosis, discovered the secret of immortality, if also of perpetual motion."¹⁹³

In 1898, an American Jewish Zionist, Richard Gottheil, proposed a Zionist "final solution of the Jewish question". Gottheil feared the "extermination" of the Jewish race, not through violent genocide, but by "a final solution of the Jewish question" of "assimilation". Gottheil proposed that Jews form a nation in Palestine in order to maintain the Jewish race. Note that Gottheil mentions "those Jews who are forced to go" to Palestine. Gottheil's speech appeared in *The World's Best Orations*, Volume 6, F. P. Kaiser, St. Louis, (1899), pp. 2294-2298:

"THE JEWS AS A RACE AND AS A NATION

(Peroration of the Address, ‹The Aims of Zionism,› Delivered in New York City, November 1st, 1898)

---

191. J. Prinz, "Zionism under the Nazi Government", *Young Zionist* (London), (November, 1937), p. 18; *as quoted in:* L. Brenner, *Zionism in the Age of the Dictators*, Chapter 5, Croom Helm, London, L. Hill, Westport, Connecticut, (1983), p. 47.

192. E. M. Friedman, "Zionism and the American Spirit", *Forum*, Volume 58, (July, 1917), pp. 67-80; *reprinted as: Zionism and the American Spirit: A New Perspective*, University Zionist Society, New York, (1917).

193. I. Zangwill, *The Problem of the Jewish Race*, Judaen Publishing Company, New York, (1914), p. 18; which was first published as an article, "The Jewish Race", *The Independent*, Volume 71, Number 3271, (10 August 1911), pp. 288-295, at 294.

I KNOW that there are a great many of our people who look for a final solution of the Jewish question in what they call «assimilation.» The more the Jews assimilate themselves to their surroundings, they think, the more completely will the causes for anti-Jewish feeling cease to exist. But have you ever for a moment stopped to consider what assimilation means? It has very pertinently been pointed out that the use of the word is borrowed from the dictionary of physiology. But in physiology it is not the food which assimilates itself into the body. It is the body which assimilates the food. The Jew may wish to be assimilated; he may do all he will towards this end. But if the great mass in which he lives does not wish to assimilate him — what then? If demands are made upon the Jew which practically mean extermination, which practically mean his total effacement from among the nations of the globe and from among the religious forces of the world, — what answer will you give? And the demands made are practically of that nature.

I can imagine it possible for a people who are possessed of an active and aggressive charity which it expresses, not only in words, but also in deeds, to contain and live at peace with men of the most varied habits. But, unfortunately, such people do not exist; nations are swayed by feelings which are dictated solely by their own self-interests; and the Zionists in meeting this state of things, are the most practical as well as the most ideal of the Jews.

It is quite useless to tell the English workingman that his Jewish fellow-laborer from Russia has actually increased the riches of the United Kingdom; that he has created quite a new industry, — that of making ladies' cloaks, for which formerly England sent £2,000,000 to the continent every year. He sees in him some one who is different to himself, and unfortunately successful, though different. And until that difference entirely ceases, whether of habit, of way, or of religious observance, he will look upon him and treat him as an enemy.

For the Jew has this especial disadvantage. There is no place where that which is distinctively Jewish in his manner or in his way of life is *à la mode*. We may well laugh at the Irishman's brogue; but in Ireland, he knows, his brogue is at home. We may poke fun at the Frenchman as he shrugs his shoulders and speaks with every member of his body. The Frenchman feels that in France it is the proper thing so to do. Even the Turk will wear his fez, and feel little the worse for the occasional jibes with which the street boy may greet it. But this consciousness, this ennobling consciousness, is all denied to the Jew. What he does is nowhere *à la mode;* no, not even his features; and if he can disguise these by parting his hair in the middle or cutting his beard to a point, he feels he is on the road towards assimilation. He is even ready to use the term «Jewish» for what he considers uncouth and low.

For such as these amongst us, Zionism also has its message. It wishes to give back to the Jew that nobleness of spirit, that confidence in himself, that belief in his own powers which only perfect freedom can give. With a home of his own, he will no longer feel himself a pariah among the nations, he will nowhere hide his own peculiarities, — peculiarities to which he has a right as much as any one, — but will see that those peculiarities carry with them a message which will force for them the admiration of the world. He will feel that he belongs somewhere and not everywhere. He will try to be something and not everything. The great word which Zionism preaches is conciliation of conflicting aims, of conflicting lines of action; conciliation of Jew to Jew. It means conciliation of the non-Jewish world to the Jew as well. It wishes to heal old wounds; and by frankly confessing differences which do exist, however much we try to explain them away, to work out its own salvation upon its own ground, and from these to send forth its spiritual message to a conciliated world.

But, you will ask, if Zionism is able to find a permanent home in Palestine for those Jews who are forced to go there as well as those who wish to go, what is to become of us who have entered, to such a degree, into the life around us, and who feel able to continue as we have begun? What is to be our relation to the new Jewish polity? I can only answer: Exactly the same as is the relation of people of other nationalities all the world over to their parent home. What becomes of the Englishman in every corner of the globe? What becomes of the German? Does the fact that the great mass of their people live in their own land prevent them from doing their whole duty towards the land in which they happen to live? Is the German-American considered less of an American because he cultivates the German language and is interested in the fate of his fellow-Germans at home? Is the Irish-American less of an American because he gathers money to help his struggling brethren in the Green Isle? Or are the Scandinavian-Americans less worthy of the title Americans, because they consider precious the bonds which bind them to the land of their birth, as well as those which bind them to the land of their adoption?

Nay! it would seem to me that just those who are so afraid that our action will be misinterpreted should be among the greatest helpers in the Zionist cause. For those who feel no racial and national communion with the life from which they have sprung should greet with joy the turning of Jewish immigration to some place other than the land in which they dwell. They must feel, for example, that a continual influx of Jews who are not Americans is a continual menace to the more or less complete absorption for which they are striving.

But I must not detain you much longer. Will you permit me to sum up for you the position which we Zionists take in the following statements: —

We believe that the Jews are something more than a purely religious body; that they are not only a race, but also a nation; though a nation without as yet two important requisites — a common home and a common language.

We believe that if an end is to be made to Jewish misery and to the exceptional position which the Jews occupy, — which is the primary cause of Jewish misery, — the Jewish nation must be placed once again in a home of its own.

We believe that such a national regeneration is the fulfillment of the hope which has been present to the Jew throughout his long and painful history.

We believe that only by means of such a national regeneration can the religious regeneration of the Jews take place, and they be put in a position to do that work in the religious world which Providence has appointed for them.

We believe that such a home can only naturally, and without violence to their whole past, be found in the land of their fathers — in Palestine.

We believe that such a return must have the guarantee of the great powers of the world in order to secure for the Jews a stable future.

And we hold that this does not mean that all Jews must return to Palestine.

This, ladies and gentlemen, is the Zionist program. Shall we be able to carry it through? I cannot believe that the Jewish people have been preserved throughout these centuries either for eternal misery or for total absorption at this stage of the world's history. I cannot think that our people have so far misunderstood their own purpose in life, as now to give the lie to their own past and to every hope which has animated their suffering body.

Bear with me but a few moments longer while I read the words which a Christian writer puts into the mouth of a Jew. «The effect of our separateness will not be completed and have its highest transformation, unless our race takes on again the character of a nationality. That is the fulfillment of the religious trust that molded them into a people, whose life has made half the inspiration of the world... . Revive the organic centre; let the unity of Israel which has made the growth and form of its religion be an outward reality. Looking toward a land and a polity, our dispersed people in all the ends of the earth may share the dignity of a national life which has a voice among the peoples of the East and the West — which will plant the wisdom and skill of our race so that it may be, as of old, a medium of transmission and understanding. Let that come to pass, and the living warmth will spread to the weak extremities of Israel. Let the central fire be kindled again, and the light will reach afar. The degraded and scorned of the race will learn to think of their sacred land, not as a place for saintly beggary to await death in loathsome idleness, but as a republic, where the Jewish spirit manifests itself in a new order founded on the old, purified, enriched by the experiences which our greatest sons have gathered from the life of the ages. A new Judea, poised between East and West — a covenant of reconciliation. The sons of Judah have to choose, that God may again choose them. The Messianic time is the time when Israel shall will the planting of the national ensign. The divine principle of our race is action, choice, resolved memory. Let us help to will our own better future of the world — not renounce our higher gift and say: ‹Let us be as if we were not among the populations,› but choose our full heritage, claim the brotherhood of our nation, and carry into it a new brotherhood with the nations of the Gentiles. The vision is there; it will be fulfilled.»

These are the words of the non-Jewish Zionist, George Eliot. We take hope, for has not that Jewish Zionist said: «We belong to a race that can do everything but fail.»"

On 22 August 1897, on page 12, in an article entitled, "Jews Against Zionism", *The New York Times* wrote,

> "Many of them thought that a purely philanthropic movement would always be but a palliative, and would never lead to a solution of the Jewish question."

Like countless other Jewish Zionists, Theodor Herzl spoke of Zionism as the "solution of the Jewish question". In fact the very title of Herzl's seminal book makes the reference, *Der Judenstaat; Versuch einer modernen Lösung der Judenfrage*, M. Breitenstein, Leipzig, Wien, (1896). English translation: *A Jewish State: An Attempt at a Modern Solution of the Jewish Question*, The Maccabæan Publishing Co., New York, (1904). Herzl stated in this book,

> "This guard of honour would be the great symbol of the solution of the Jewish Question after eighteen centuries of Jewish suffering."[194]

In an article entitled "Zionist Congress in Basel", *The New York Times* quoted Theodor Herzl, on 31 August 1897, on page 7,

> "I think we shall find Palestine at our disposal sooner than we expected. Last year I went to Constantinople and had two long conferences with the Grand Vizier, to whom I pointed out that the key to the preservation of Turkey lay in the solution of the Jewish question."

In his opening address to the First Zionist Congress, Herzl stated,

> "Wir Zionisten wünschen zur Lösung der Judenfrage nicht etwa einen internationalen Verein, sondern die internationale Diskussion."[195]

Herzl's statements were recorded in, "The Zionist Congress: Full Report of the Proceedings", *The Jewish Chronicle*, (3 September 1897), pp. 10-15, at 11, 12 and 15,

> "We Zionists desire for the solution of the Jewish Question. [***] But it is not the solution of the Jewish Question, and cannot be so in its present form. [***] The financial help which the Jews are able to offer to Turkey is not

---

194. T. Herzl, *A Jewish State: An Attempt at a Modern Solution of the Jewish Question*, The Maccabæan Publishing Co., New York, (1904), p. 29.
195. L. Kellner, "Eröffnungsrede zum ersten Kongress", *Theodor Herzls Zionistische Schriften*, Jüdischer Verlag, Berlin, (1920), p. 139-144, at 140.

small, and would serve to put an end to many an evil from which the country is suffering. If a part of the Oriental question can be solved, together with the Jewish question, this surely is in the interest of all nations. [***] In this way we understand, we expect the solution of the Jewish Question. [***] On the day when the Jews again held the plough in Palestine, on that day would the Jewish Question be solved."

In examining the history of expressed threats of genocide, it should be mentioned that long before Kaufman's genocidal book *Germany Must Perish!* advocated the extermination of Gentile Germans, anti-Semite Eugen Karl Dühring implicitly advocated the genocide of Jews in the 1901 edition of his *Die Judenfrage*, Chapter 5, Sections 4-9, which concludes with the statement:

"Precisely this situation must however urge the determined component of better humanity only so much more to act in order to create communities and communal life whose principles extend over the earth and thereby also, obviously, do not leave any room for Hebrew life."[196]

Jörg Lanz-Liebenfels advocated the deportation and sterilization of "inferior races" in his book *Theozoologie, oder Die Kunde von den Sodomsäfflingen und dem Götter-Elektron eine Einführung in die älteste und neueste Weltanschauung und eine Rechtfertigung des Fürstentums und des Adels...* , Moderner Verlag, Wien, (1905).[197] Hitler's racial views came in part from Lanz-Liebenfels, who promoted the procreation of blond-haired people and the sterilization of the "ape-men" of the "inferior races"—he was also a Zionist who encouraged the formation of a Jewish State, and his mythologies may have been derived from the Jewish myth that angels bred with humans to produce a unique race. One example of the political Zionists' equivalent of Liebenfels prescriptions for the ideal "Aryan", was Elias Auerbach's article "Rassenkunde" in Zionist Martin Buber's journal *Der Jude*, Volume 5, Number 1, (1920-1921), pp. 49-57, which discusses eye and hair color, skeletal proportions, etc. of the average Jew. In 1909, Buber himself romanticized that a Jew awakening to his heritage undergoes many stages of racial self-awareness,

"He perceives then what commingling of individuals, what confluence of blood, has produced him, what round of begettings and births has called him forth. He senses in this immortality of the generations a community of

---

196. E. K. Dühring, *Die Judenfrage als Racen-, Sitten- und Culturfrage: mit einer weltgeschichtlichen Antwort*, H. Reuther, Karlsruhe, (1881); English translation by A. Jacob, *Eugen Dühring on the Jews*, Nineteen Eighty Four Press, Brighton, England, (1997), pp. 211-212. *See also:* E. K. Dühring, *Der Werth des Lebens: Eine Denkerbetrachtung im Sinne heroischer Lebensauffassung*, Fifth Edition, Reisland, Leipzig, (1894), p. 9.
197. *Confer:* W. Daim, *Der Mann, der Hitler die Ideen gab: Jörg Lanz von Liebenfels*, Third Improved Edition, Ueberreuter, Wien, (1994).

blood, which he feels to be the antecedents of his I, its perseverance in the infinite past. To that is added the discovery, promoted by this awareness, that blood is a deep rooted nurturing force within individual man; that the deepest layers of our being are determined by blood ; that our innermost thinking and our will are colored by it. Now he finds that the world around him is the world of imprints and influences, whereas blood is the realm of a substance capable of being imprinted and influenced, a substance absorbing and assimilating all into its own form. And he therefore senses that he belongs no longer to the community of those whose constant elements of experience he shares, but to the deeper-reaching community of those whose substance he shares. [***] Whoever, faced with the choice between environment and substance, decides for substance will henceforth have to be a Jew truly from within, to live as a Jew with all the contradiction, all the tragedy, and all the future promise of his blood."[198]

Josef Ludwig Reimer published *Ein pangermanisches Deutschland. Versuch über die Konsequenzen der gegenwärtigen wissenschaftlichen Rassenbetrachtung für unsere politischen und religiösen Probleme*, F. Luckhardt, Berlin, Leipzig, (1905); which advocated dividing human beings into three categories with the rulers being blond-haired, blue-eyed supermen, who ruled the "mixed-race" and middle class, and the lowest grouping, the non-Germanics.[199] The non-Germanics would be sterilized or prevented by law from bearing children. Extremist and violent Social Darwinism appeared in Germany in the Nineteenth Century in the writings of Friedrich von Hellwald, and Ernst Haeckel advocated Eugenics.[200]

The "Eugenics" of Sir Fancis Galton[201] has a long and complex history dating back to the Greeks and includes such famous persons as Charles Darwin and Alexander Graham Bell. Prior to the Nazi regime, Eugenics was most enthusiastically promoted in the United States, where there was active governmental interest in the field, and where Eugenics influenced legislation. It

---

198. M. Buber, "Das Judentum und die Juden", *Drei Reden über das Judentum*, Rütten & Loening, Frankfurt a. M., (1911); English translation:"Judaism and the Jews", *On Judaism*, Schocken Books, New York, (1967), pp. 11-21, at 15, 19.
199. J. R. Marcus, *The Rise and Destiny of the German Jew*, The Union of American Hebrew Congregations, Cincinnati, (1934), pp. 61-62.
200. F. v. Hellwald, *Culturgeschichte in ihrer natürlichen Entwicklung bis zur Gegenwart*, Lampart & Comp., Augsburg, (1875); **and** "Der Kampf ums Dasein im Menschen- und Völkerleben", *Das Ausland*, Volume 45, (1872), pp. 105ff., *see also: Das Ausland*, (1872), 901ff., 957ff. *See also:* R. Weikart, *The Human Life Review*, Volume 30, Number 2, (Spring 2004), pp. 29-37; **and** *From Darwin to Hitler: Evolutionary Ethics, Eugenics, and Racism in Germany*, Palgrave Macmillan, New York, (2004).
201. F. Galton, *Hereditary Genius: An Inquiry into its Laws and Consequences*, Macmillan, London, (1869); **and** *Inquiries into Human Faculty and its Development*, Macmillan and Co., London, (1883); **and** *The Possible Improvement of the Human Breed under the Existing Conditions of Law and Sentiment*, Washington, D. C., (1902). *See also:* The journal *Biometrika*.

was also welcomed in England. The colonial powers sought scientific justification for their un-Christian treatment of their fellow human beings, as if inferior. America sought to limit the immigration and political power of the so-called "inferior races". The Nazis instituted their "T4" "*Euthanasie-Programme*" in 1939.

German Jews had endured increasingly hostile agitations since the end of the First World War, and the Hitler regime enacted discriminatory laws against the Jews long before Kaufman's book found its way into print, which segregationist laws had an ancient history in Europe and were endorsed by Heinrich Class under the *nom de plume* Daniel Frymann, *Wenn ich der Kaiser wär': politische Wahrheiten und Notwendigkeiten*, Dieterich, Leipzig, (1912); even before the First World War.

In naming the important historical incidents of genocidal propaganda and acts, it must also be mentioned that Biblical passages in the Old Testament and the New, as well as Talmudic writings, prophesied the genocide and enslavement of Gentiles and the ascent of a master race of Jews. Writing on Thomas Jefferson's religious views, William D. Gould wrote,

"Jefferson praised the philosophers of antiquity for their insistence on the necessity of governing the passions, but found that they did not deal adequately with social duties. They taught well the obligation of being just in dealing with one's neighbor or countryman, but felt under no constraint to cultivate a love for all mankind. Even the Jews in Jesus' day, he believed, entertained many erroneous ideas concerning religion and morality. In addition to the fact that he felt that a number of their conceptions of God were incorrect, their ethics, in respect to other nations, were, he thought, decidedly antisocial."[202]

Jefferson criticized ancient philosophers and the ancient Jews in his *Syllabus*. He wrote, *inter alia*, in a letter to Dr. Benjamin Rush of 21 April 1803 responding to rumors that he was not a Christian,

"*Syllabus of an Estimate of the Merit of the Doctrines of Jesus, Compared with Those of Others.*

In a comparative view of the Ethics of the enlightened nations of antiquity, of the Jews and of Jesus, no notice should be taken of the corruptions of reason among the ancients, to wit, the idolatry and superstition of the vulgar, nor of the corruptions of Christianity by the learned among its professors.

Let a just view be taken of the moral principles inculcated by the most esteemed of the sects of ancient philosophy, or of their individuals; particularly Pythagoras, Socrates, Epicurus, Cicero, Epictetus, Seneca,

---

202. W. D. Gould, "The Religious Opinions of Thomas Jefferson", *The Mississippi Valley Historical Review*, Volume 20, Number 2, (September, 1933), pp. 191-208, at 202.

Antoninus.

I. Philosophers. 1. Their precepts related chiefly to ourselves, and the government of those passions which, unrestrained, would disturb our tranquillity of mind.[*Footnote:* To explain, I will exhibit the heads of Seneca's and Cicero's philosophical works, the most extensive of any we have received from the ancients. Of ten heads in Seneca, seven relate to ourselves, viz. *de ira, consolatio, de tranquilitate, de constantia sapientis, de otio sapientis, de vita beata, de brevitate vitae*; two relate to others, *de clementia, de beneficiis*; and one relates to the government of the world, *de providentia*. Of eleven tracts of Cicero, five respect ourselves, viz. *de finibus, Tusculana, academica, paradoxa, de Senectute*; one, *de officiis*, relates partly to ourselves, partly to others; one, *de amicitia*, relates to others; and four are on different subjects, to wit, *de natura deorum, de divinatione, de fato*, and *sommium Scipionis*.] In this branch of philosophy they were really great.

2. In developing our duties to others, they were short and defective. They embraced, indeed, the circles of kindred and friends, and inculcated patriotism, or the love of our country in the aggregate, as a primary obligation: towards our neighbors and countrymen they taught justice, but scarcely viewed them as within the circle of benevolence. Still less have they inculcated peace, charity and love to our fellow men, or embraced with benevolence the whole family of mankind.

II. Jews. 1. Their system was Deism; that is, the belief in one only God. But their ideas of him and of his attributes were degrading and injurious.

2. Their Ethics were not only imperfect, but often irreconcilable with the sound dictates of reason and morality, as they respect intercourse with those around us; and repulsive and anti-social, as respecting other nations. They needed reformation, therefore, in an eminent degree."[203]

Ancient Jewish myths enunciate a nationalistic and destructive racism by a master nation of Israel on a holy mission to mercilessly subjugate the other nations of the world, supposedly pursuant to God's will. For example, *Deuteronomy*, Chapter 7, states,

"When the LORD thy God shall bring thee into the land whither thou goest to possess it, and hath cast out many nations before thee, the Hittites, and the Girgashites, and the Amorites, and the Canaanites, and the Perizzites, and the Hivites, and the Jebusites, seven nations greater and mightier than thou; 2 And when the LORD thy God shall deliver them before thee; thou shalt smite them, *and* utterly destroy them; thou shalt make no covenant

---

[203]. Letter from T. Jefferson to B. Rush of 21 April 1803, *The Writings of Thomas Jefferson*, Volume 10, Issued Under the Auspices of the Thomas Jefferson Memorial Association of the United States, Washington, D.C., (1905), pp. 379-385, at 381-382.

with them, nor show mercy unto them: 3 Neither shalt thou make marriages with them; thy daughter thou shalt not give unto his son, nor his daughter shalt thou take unto thy son. 4 For they will turn away thy son from following me, that they may serve other gods: so will the anger of the LORD be kindled against you, and destroy thee suddenly. 5 But thus shall ye deal with them; ye shall destroy their altars, and break down their images, and cut down their groves, and burn their graven images with fire. 6 For thou *art* an holy people unto the LORD thy God: the LORD thy God hath chosen thee to be a special people unto himself, above all people that *are* upon the face of the earth. 7 The LORD did not set his love upon you, nor choose you, because ye were more in number than any people; for ye *were* the fewest of all people: 8 But because the LORD loved you, and because he would keep the oath which he had sworn unto your fathers, hath the LORD brought you out with a mighty hand, and redeemed you out of the house of bondmen, from the hand of Pharaoh king of Egypt. 9 Know therefore that the LORD thy God, he *is* God, the faithful God, which keepeth covenant and mercy with them that love him and keep his commandments to a thousand generations; 10 And repayeth them that hate him to their face, to destroy them: he will not be slack to him that hateth him, he will repay him to his face. 11 Thou shalt therefore keep the commandments, and the statutes, and the judgments, which I command thee this day, to do them. 12 Wherefore it shall come to pass, if ye hearken to these judgments, and keep, and do them, that the LORD thy God shall keep unto thee the covenant and the mercy which he sware unto thy fathers: 13 And he will love thee, and bless thee, and multiply thee: he will also bless the fruit of thy womb, and the fruit of thy land, thy corn, and thy wine, and thine oil, the increase of thy kine, and the flocks of thy sheep, in the land which he sware unto thy fathers to give thee. 14 Thou shalt be blessed above all people: there shall not be male or female barren among you, or among your cattle. 15 And the LORD will take away from thee all sickness, and will put none of the evil diseases of Egypt, which thou knowest, upon thee; but will lay them upon all *them* that hate thee. 16 And thou shalt consume all the people which the LORD thy God shall deliver thee; thine eye shall have no pity upon them: neither shalt thou serve their gods; for that *will be* a snare unto thee. 17 If thou shalt say in thine heart, These nations *are* more than I; how can I dispossess them? 18 Thou shalt not be afraid of them: *but* shalt well remember what the LORD thy God did unto Pharaoh, and unto all Egypt; 19 The great temptations which thine eyes saw, and the signs, and the wonders, and the mighty hand, and the stretched out arm, whereby the LORD thy God brought thee out: so shall the LORD thy God do unto all the people of whom thou art afraid. 20 Moreover the LORD thy God will send the hornet among them, until they that are left, and hide themselves from thee, be destroyed. 21 Thou shalt not be affrighted at them: for the LORD thy God *is* among you, a mighty God and terrible. 22 And the LORD thy God will put out those nations before thee by little and little: thou mayest not consume them at once, lest the beasts of the field increase upon thee. 23 But the LORD thy

God shall deliver them unto thee, and shall destroy them with a mighty destruction, until they be destroyed. 24 And he shall deliver their kings into thine hand, and thou shalt destroy their name from under heaven: there shall no man be able to stand before thee, until thou have destroyed them. 25 The graven images of their gods shall ye burn with fire: thou shalt not desire the silver or gold *that is* on them, nor take *it* unto thee, lest thou be snared therein: for it *is* an abomination to the LORD thy God. 26 Neither shalt thou bring an abomination into thine house, lest thou be a cursed thing like it: *but* thou shalt utterly detest it, and thou shalt utterly abhor it; for it *is* a cursed thing."

Deuteronomy, Chapter 28, proclaims the punishment to befall the assimilated,

"And it shall come to pass, if thou shalt hearken diligently unto the voice of the LORD thy God, to observe *and* to do all his commandments which I command thee this day, that the LORD thy God will set thee on high above all nations of the earth: 2 And all these blessings shall come on thee, and overtake thee, if thou shalt hearken unto the voice of the LORD thy God. 3 Blessed *shalt* thou *be* in the city, and blessed *shalt* thou *be* in the field. 4 Blessed *shall be* the fruit of thy body, and the fruit of thy ground, and the fruit of thy cattle, the increase of thy kine, and the flocks of thy sheep. 5 Blessed *shall be* thy basket and thy store. 6 Blessed *shalt* thou *be* when thou comest in, and blessed *shalt* thou *be* when thou goest out. 7 The LORD shall cause thine enemies that rise up against thee to be smitten before thy face: they shall come out against thee one way, and flee before thee seven ways. 8 The LORD shall command the blessing upon thee in thy storehouses, and in all that thou settest thine hand unto; and he shall bless thee in the land which the LORD thy God giveth thee. 9 The LORD shall establish thee an holy people unto himself, as he hath sworn unto thee, if thou shalt keep the commandments of the LORD thy God, and walk in his ways. 10 And all people of the earth shall see that thou art called by the name of the LORD; and they shall be afraid of thee. 11 And the LORD shall make thee plenteous in goods, in the fruit of thy body, and in the fruit of thy cattle, and in the fruit of thy ground, in the land which the LORD sware unto thy fathers to give thee. 12 The LORD shall open unto thee his good treasure, the heaven to give the rain unto thy land in his season, and to bless all the work of thine hand: and thou shalt lend unto many nations, and thou shalt not borrow. 13 And the LORD shall make thee the head, and not the tail; and thou shalt be above only, and thou shalt not be beneath; if that thou hearken unto the commandments of the LORD thy God, which I command thee this day, to observe and to do *them*: 14 And thou shalt not go aside from any of the words which I command thee this day, *to* the right hand, or *to* the left, to go after other gods to serve them. 15 But it shall come to pass, if thou wilt not hearken unto the voice of the LORD thy God, to observe to do all his commandments and his statutes which I command thee this day; that all these curses shall come upon thee, and overtake thee:16 Cursed *shalt* thou *be* in the city, and cursed *shalt* thou *be* in the field. 17 Cursed *shall be* thy basket and thy store. 18 Cursed *shall be* the fruit of thy body, and the fruit of thy land, the increase of thy kine, and the flocks of thy sheep. 19 Cursed *shalt* thou *be* when thou comest in, and cursed *shalt* thou *be* when thou goest out. 20 The LORD shall send upon thee cursing, vexation, and rebuke, in all that thou settest thine hand unto for to do, until thou be destroyed, and until thou perish quickly; because of the wickedness of thy doings, whereby thou hast forsaken me. 21 The LORD shall make the pestilence cleave unto thee, until he have consumed thee from off the land, whither thou goest to possess it. 22 The LORD shall smite thee with a consumption, and with a fever, and with an inflammation, and with an extreme burning, and with the sword, and with blasting, and with mildew; and they shall pursue thee until thou perish. 23 And thy heaven that *is* over thy head shall be brass, and the earth that is under thee *shall be* iron. 24 The LORD shall make the rain of thy land powder and dust: from heaven shall it come down upon thee, until thou be destroyed. 25 The LORD shall cause thee to be smitten before thine enemies: thou shalt go out one way against them, and flee seven ways before them: and shalt be removed into all the kingdoms of the earth. 26 And thy carcase shall be meat unto all fowls of the air, and unto the beasts of the earth, and no man shall fray *them* away. 27 The LORD will smite thee with the botch of Egypt, and with the emerods, and with the scab, and with the itch, whereof thou canst not be healed. 28 The LORD shall smite thee with madness, and blindness, and astonishment of heart: 29 And thou shalt grope at noonday, as the blind gropeth in darkness, and thou shalt not prosper in thy ways: and thou shalt be only oppressed and spoiled evermore, and no man shall save *thee*. 30 Thou shalt betroth a wife, and another man shall lie with her: thou shalt build an house, and thou shalt not dwell therein: thou shalt plant a vineyard, and shalt not gather the grapes thereof. 31 Thine ox *shall be* slain before thine eyes, and thou shalt not eat thereof: thine ass *shall be* violently taken away from before thy face, and shall not be restored to thee: thy sheep *shall be* given unto thine enemies, and thou shalt have none to rescue *them*. 32 Thy sons and thy daughters *shall be* given unto another people, and thine eyes shall look, and fail *with longing* for them all the day long: and *there shall be* no might in thine hand. 33 The fruit of thy land, and all thy labours, shall a nation which thou knowest not eat up; and thou shalt be only oppressed and crushed alway: 34 So that thou shalt be mad for the sight of thine eyes which thou shalt see. 35 The LORD shall smite thee in the knees, and in the legs, with a sore botch that cannot be healed, from the sole of thy foot unto the top of thy head. 36 The LORD shall bring thee, and thy king which thou shalt set over thee, unto a nation which neither thou nor thy fathers have known; and there shalt thou serve other gods, wood and stone. 37 And thou shalt become an astonishment, a proverb, and a byword, among all nations whither the LORD shall lead thee. 38 Thou shalt carry much seed out into the field, and shalt gather *but* little in; for the locust shall consume it. 39 Thou shalt plant vineyards, and dress *them*, but shalt neither drink *of* the wine, nor gather *the grapes*; for the worms shall eat them. 40

Thou shalt have olive trees throughout all thy coasts, but thou shalt not anoint *thyself* with the oil; for thine olive shall cast *his fruit*. 41 Thou shalt beget sons and daughters, but thou shalt not enjoy them; for they shall go into captivity. 42 All thy trees and fruit of thy land shall the locust consume. 43 The stranger that *is* within thee shall get up above thee very high; and thou shalt come down very low. 44 He shall lend to thee, and thou shalt not lend to him: he shall be the head, and thou shalt be the tail. 45 Moreover all these curses shall come upon thee, and shall pursue thee, and overtake thee, till thou be destroyed; because thou hearkenedst not unto the voice of the LORD thy God, to keep his commandments and his statutes which he commanded thee: 46 And they shall be upon thee for a sign and for a wonder, and upon thy seed for ever. 47 Because thou servedst not the LORD thy God with joyfulness, and with gladness of heart, for the abundance of all *things*; 48 Therefore shalt thou serve thine enemies which the LORD shall send against thee, in hunger, and in thirst, and in nakedness, and in want of all *things*: and he shall put a yoke of iron upon thy neck, until he have destroyed thee. 49 The LORD shall bring a nation against thee from far, from the end of the earth, *as swift* as the eagle flieth; a nation whose tongue thou shalt not understand; 50 A nation of fierce countenance, which shall not regard the person of the old, nor show favour to the young: 51 And he shall eat the fruit of thy cattle, and the fruit of thy land, until thou be destroyed: which *also* shall not leave thee *either* corn, wine, or oil, *or* the increase of thy kine, or flocks of thy sheep, until he have destroyed thee. 52 And he shall besiege thee in all thy gates, until thy high and fenced walls come down, wherein thou trustedst, throughout all thy land: and he shall besiege thee in all thy gates throughout all thy land, which the LORD thy God hath given thee. 53 And thou shalt eat the fruit of thine own body, the flesh of thy sons and of thy daughters, which the LORD thy God hath given thee, in the siege, and in the straitness, wherewith thine enemies shall distress thee: 54 *So that* the man *that is* tender among you, and very delicate, his eye shall be evil toward his brother, and toward the wife of his bosom, and toward the remnant of his children which he shall leave: 55 So that he will not give to any of them of the flesh of his children whom he shall eat: because he hath nothing left him in the siege, and in the straitness, wherewith thine enemies shall distress thee in all thy gates. 56 The tender and delicate woman among you, which would not adventure to set the sole of her foot upon the ground for delicateness and tenderness, her eye shall be evil toward the husband of her bosom, and toward her son, and toward her daughter, 57 And toward her young one that cometh out from between her feet, and toward her children which she shall bear: for she shall eat them for want of all *things* secretly in the siege and straitness, wherewith thine enemy shall distress thee in thy gates. 58 If thou wilt not observe to do all the words of this law that are written in this book, that thou mayest fear this glorious and fearful name, THE LORD THY GOD; 59 Then the LORD will make thy plagues wonderful, and the plagues of thy seed, *even* great plagues, and of long continuance, and sore sicknesses, and of long continuance. 60

Moreover he will bring upon thee all the diseases of Egypt, which thou wast afraid of; and they shall cleave unto thee. 61 Also every sickness, and every plague, which *is* not written in the book of this law, them will the LORD bring upon thee, until thou be destroyed. 62 And ye shall be left few in number, whereas ye were as the stars of heaven for multitude; because thou wouldest not obey the voice of the LORD thy God. 63 And it shall come to pass, *that* as the LORD rejoiced over you to do you good, and to multiply you; so the LORD will rejoice over you to destroy you, and to bring you to nought; and ye shall be plucked from off the land whither thou goest to possess it. 64 And the LORD shall scatter thee among all people, from the one end of the earth even unto the other; and there thou shalt serve other gods, which neither thou nor thy fathers have known, *even* wood and stone. 65 And among these nations shalt thou find no ease, neither shall the sole of thy foot have rest: but the LORD shall give thee there a trembling heart, and failing of eyes, and sorrow of mind: 66 And thy life shall hang in doubt before thee; and thou shalt fear day and night, and shalt have none assurance of thy life: 67 In the morning thou shalt say, Would God it were even! and at even thou shalt say, Would God it were morning! for the fear of thine heart wherewith thou shalt fear, and for the sight of thine eyes which thou shalt see. 68 And the LORD shall bring thee into Egypt again with ships, by the way whereof I spake unto thee, Thou shalt see it no more again: and there ye shall be sold unto your enemies for bondmen and bondwomen, and no man shall buy *you*."

Isaiah, Chapter 34,

"Come near, ye nations, to hear; and hearken, ye people: let the earth hear, and all that is therein; the world, and all things that come forth of it. 2 For the indignation of the LORD *is* upon all nations, and *his* fury upon all their armies: he hath utterly destroyed them, he hath delivered them to the slaughter. 3 Their slain also shall be cast out, and their stink shall come up out of their carcases, and the mountains shall be melted with their blood. 4 And all the host of heaven shall be dissolved, and the heavens shall be rolled together as a scroll: and all their host shall fall down, as the leaf falleth off from the vine, and as a falling *fig* from the fig tree. 5 For my sword shall be bathed in heaven: behold, it shall come down upon Idumea, and upon the people of my curse, to judgment. 6 The sword of the LORD is filled with blood, it is made fat with fatness, *and* with the blood of lambs and goats, with the fat of the kidneys of rams: for the LORD hath a sacrifice in Bozrah, and a great slaughter in the land of Idumea. 7 And the unicorns shall come down with them, and the bullocks with the bulls; and their land shall be soaked with blood, and their dust made fat with fatness. 8 For *it is* the day of the LORD's vengeance, *and* the year of recompenses for the controversy of Zion. 9 And the streams thereof shall be turned into pitch, and the dust thereof into brimstone, and the land thereof shall become burning pitch. 10 It shall not be quenched night nor day; the smoke thereof shall go up for ever: from generation to generation it shall lie waste; none shall pass through it for ever and ever. 11 But the cormorant and the bittern shall possess it; the owl also and the raven shall dwell in it: and he shall stretch out upon it the line of confusion, and the stones of emptiness. 12 They shall call the nobles thereof to the kingdom, but none *shall be* there, and all her princes shall be nothing. 13 And thorns shall come up in her palaces, nettles and brambles in the fortresses thereof: and it shall be an habitation of dragons, *and* a court for owls. 14 The wild beasts of the desert shall also meet with the wild beasts of the island, and the satyr shall cry to his fellow; the screech owl also shall rest there, and find for herself a place of rest. 15 There shall the great owl make her nest, and lay, and hatch, and gather under her shadow: there shall the vultures also be gathered, every one with her mate. 16 Seek ye out of the book of the LORD, and read: no one of these shall fail, none shall want her mate: for my mouth it hath commanded, and his spirit it hath gathered them. 17 And he hath cast the lot for them, and his hand hath divided it unto them by line: they shall possess it for ever, from generation to generation shall they dwell therein."

Isaiah, Chapter 60:12, 16,

"For the nation and kingdom that will not serve thee shall perish; yea, *those* nations shall be utterly wasted. [***] Thou shalt also suck the milk of the Gentiles, and shalt suck the breast of kings: and thou shalt know that I the LORD *am* thy Saviour and thy Redeemer, the mighty One of Jacob."

Isaiah, Chapter 61,

"The spirit of the Lord GOD *is* upon me; because the LORD hath anointed me to preach good tidings unto the meek; he hath sent me to bind up the brokenhearted, to proclaim liberty to the captives, and the opening of the prison to *them that are* bound; 2 To proclaim the acceptable year of the LORD, and the day of vengeance of our God; to comfort all that mourn; 3 To appoint unto them that mourn in Zion, to give unto them beauty for ashes, the oil of joy for mourning, the garment of praise for the spirit of heaviness; that they might be called trees of righteousness, the planting of the LORD, that he might be glorified. 4 And they shall build the old wastes, they shall raise up the former desolations, and they shall repair the waste cities, the desolations of many generations. 5 And strangers shall stand and feed your flocks, and the sons of the alien *shall be* your plowmen and your vinedressers. 6 But ye shall be named the Priests of the LORD: *men* shall call you the Ministers of our God: ye shall eat the riches of the Gentiles, and in their glory shall ye boast yourselves. 7 For your shame *ye shall have* double; and *for* confusion they shall rejoice in their portion: therefore in their land they shall possess the double: everlasting joy shall be unto them. 8 For I the LORD love judgment, I hate robbery for burnt offering; and I will direct their work in truth, and I will make an everlasting covenant with them. 9 And their seed shall be known among the Gentiles, and their offspring among the people: all that see them shall acknowledge them, that they *are* the seed *which* the LORD hath blessed. 10 I will greatly rejoice in the LORD, my soul shall be joyful in my God; for he hath clothed me with the garments of salvation, he hath covered me with the robe of righteousness, as a bridegroom decketh *himself* with ornaments, and as a bride adorneth *herself* with her jewels. 11 For as the earth bringeth forth her bud, and as the garden causeth the things that are sown in it to spring forth; so the Lord GOD will cause righteousness and praise to spring forth before all the nations."

The Nazis' infamous *Lebensborn* program, the program to breed "Aryan" children for the Reich, was perhaps instead a means for racist Jews to interject Jewish blood into the German "race" so as to dilute the blood of Esau. This is pure speculation, but it is based upon the fact that the Jews viewed Germans as Esau, wanted to destroy or weaken Esau, had control over the Third Reich, and had numerous Jewish members in the *SS* who could have fathered these children.

After the war, many people began to notice that a large number of children in Israel were tall, blond and blue eyed. They could have passed for Swedes. The entire Holocaust may well have been a eugenics program for racist Jews to clean up their blood, which they believed had been damaged by the Ghetto system of Europe. Jewish prophecy and lore teaches that in the Messianic Era Jews will be tall, fair-skinned (radiant: *Isaiah* 60:5) and handsome.

The especially interesting thing about these tall, blond, blue-eyed children in Israel, is that many were allegedly orphans—orphans who believed that they were Gentiles and who were shocked when told that their parents had been Jewish. This has led some to conclude that Jews kidnaped Gentile children and brought them to Israel.[204] This leads to speculation that after anointing their Messiah, racist Jews will use Gentile slaves to breed them children, so that they can populate the world with the children of breeding slaves and completely kill off Gentiles born and raised by Gentiles. They may plan to steal the children fathered and mothered by Gentiles, and they may plan to use Gentile woman as surrogate mothers to bear children of Jewish parents on a massive scale. This speculation is based on many Jewish writings, including, but not limited to, *Isaiah*, Chapter 49,

"Listen, O isles, unto me; and hearken, ye people, from far; The LORD hath called me from the womb; from the bowels of my mother hath he made mention of my name. 2 And he hath made my mouth like a sharp sword; in the shadow of his hand hath he hid me, and made me a polished shaft; in his quiver hath he hid me; 3 And said unto me, Thou *art* my servant, O Israel, in whom I will be glorified. 4 Then I said, I have laboured in vain, I have spent my strength for nought, and in vain: *yet* surely my judgment *is* with

---

204. R. H. Williams, *The Ultimate World Order—As Pictured in "The Jewish Utopia"*, CPA Book Publisher, Boring, Oregon, (1957?), pp. 43-47.

the LORD, and my work with my God. 5 And now, saith the LORD that formed me from the womb *to be* his servant, to bring Jacob again to him, Though Israel be not gathered, yet shall I be glorious in the eyes of the LORD, and my God shall be my strength. 6 And he said, It is a light thing that thou shouldest be my servant to raise up the tribes of Jacob, and to restore the preserved of Israel: I will also give thee for a light to the Gentiles, that *thou* mayest be my salvation unto the end of the earth. 7 Thus saith the LORD, the Redeemer of Israel, *and* his Holy One, to him whom man despiseth, to him whom the nation abhorreth, to a servant of rulers, Kings shall see and arise, princes also shall worship, because of the LORD that *is* faithful, *and* the Holy One of Israel, and he shall choose thee. 8 Thus saith the LORD, In an acceptable time have I heard thee, and in a day of salvation have I helped thee: and I will preserve thee, and give thee for a covenant of the people, to establish the earth, to cause to inherit the desolate heritages; 9 That *thou* mayest say to the prisoners, Go forth; to *them* that *are* in darkness, Shew yourselves. They shall feed in the ways, and their pastures *shall be* in all high places. 10 They shall not hunger nor thirst; neither shall the heat nor sun smite them: for he that hath mercy on them shall lead them, even by the springs of water shall he guide them. 11 And I will make all my mountains a way, and my highways shall be exalted. 12 Behold, these shall come from far: and, lo, these from the north and from the west; and these from the land of Sinim. 13 Sing, O heavens; and be joyful, O earth; and break forth *into* singing, O mountains: for the LORD hath comforted his people, and will have mercy upon his afflicted. 14 But Zion said, The LORD hath forsaken me, and my Lord hath forgotten me. 15 Can a woman forget her sucking child, that *she* should not have compassion on the son of her womb? yea, they may forget, yet will I not forget thee. 16 Behold, I have graven thee upon the palms of *my* hands; thy walls *are* continually before me. 17 Thy children shall make haste; thy destroyers and they that made thee waste shall go forth of thee. 18 Lift up thine eyes round about, and behold: all these gather themselves together, *and* come to thee. *As* I live, saith the LORD, thou shalt surely clothe thee with them all, as with an ornament, and bind them *on thee*, as a bride *doth*. 19 For thy waste and thy desolate places, and the land of thy destruction, shall even now be too narrow by reason of the inhabitants, and they that swallowed thee up shall be far away. 20 The children which thou shalt have, after thou hast lost the other, shall say again in thine ears, The place *is* too strait for me: give place to me that I may dwell. 21 Then shalt thou say in thine heart, Who hath begotten me these, seeing I have lost my children, and *am* desolate, a captive, and removing to and fro? and who hath brought up these? Behold, I was left alone; these, where *had* they *been?* 22 Thus saith the Lord GOD, Behold, I will lift up mine hand to the Gentiles, and set up my standard to the people: and they shall bring thy sons in *their* arms, and thy daughters shall be carried upon *their* shoulders. 23 And kings shall be thy nursing fathers, and their queens thy nursing mothers: they shall bow down to thee *with their* face *toward* the earth, and lick up the dust of thy feet; and thou shalt know that I *am* the LORD: for they shall not be ashamed that wait for me. 24 Shall the prey be taken from the mighty, or the lawful captive delivered? 25 But thus saith the LORD, Even the captives of the mighty shall be taken away, and the prey of the terrible shall be delivered: for I will contend with him that contendeth with thee, and I will save thy children. 26 And I will feed them that oppress thee with their own flesh; and they shall be drunken with their own blood, as with sweet wine: and all flesh shall know that I the LORD *am* thy Saviour and thy Redeemer, the mighty One of Jacob."

Isaiah, Chapter 60:12, 16,

"For the nation and kingdom that will not serve thee shall perish; yea, *those* nations shall be utterly wasted. [***] Thou shalt also suck the milk of the Gentiles, and shalt suck the breast of kings: and thou shalt know that I the LORD *am* thy Saviour and thy Redeemer, the mighty One of Jacob."

There are many instances in the Old Testament of the use of slaves taken from other nations to bear the ancestors of the Jews young, for example Abraham and Hagar. *Isaiah* 66 states, note that the "Lord" who is speaking is the voice of genocidal Jewish racism and absolute Jewish religious intolerance,

"1 Thus saith the LORD, The heaven *is* my throne, and the earth *is* my footstool: where *is* the house that ye build unto me? and where *is* the place of my rest? 2 For all those *things* hath mine hand made, and all those *things* have been, saith the LORD: but to this *man* will I look, *even* to *him that is* poor and of a contrite spirit, and trembleth at my word. 3 He that killeth an ox *is as if* he slew a man; he that sacrificeth a lamb, *as if* he cut off a dog's neck; he that offereth an oblation, *as if he offered* swine's blood; he that burneth incense, *as if* he blessed an idol. Yea, they have chosen their own ways, and their soul delighteth in their abominations. 4 I also will choose their delusions, and will bring their fears upon them; because when I called, none did answer; when I spake, they did not hear: but they did evil before mine eyes, and chose *that* in which I delighted not. 5¶ Hear the word of the LORD, ye that tremble at his word; Your brethren that hated you, that cast you out for my name's sake, said, Let the LORD be glorified: but he *shall* appear to your joy, and they shall be ashamed. 6 A voice of noise from the city, a voice from the temple, a voice of the LORD that rendereth recompence to his enemies. 7 Before she travailed, she brought forth; before her pain came, she was delivered of a man child. 8 Who hath heard such *a thing?* who hath seen such *things?* Shall the earth be made to bring forth in one day? *or* shall a nation be born at once? for as soon as Zion travailed, she brought forth her children. 9 Shall I bring to the birth, and not cause to bring forth? saith the LORD: shall I cause to bring forth, and shut *the womb?* saith thy God. 10 Rejoice ye with Jerusalem, and be glad with her, all ye that love her: rejoice for joy with her, all ye that mourn for her: 11 That ye may suck, and be satisfied with the breasts of her consolations; that ye may milk out, and be delighted with the abundance of her glory. 12 For thus saith the LORD, Behold, I will extend peace to her like a river, and the glory of the Gentiles like a flowing stream: then shall ye suck, ye shall be borne upon *her* sides, and be dandled upon *her* knees. 13 As one whom his mother comforteth, so will I comfort you; and ye shall be comforted in Jerusalem. 14 And when ye see *this,* your heart shall rejoice, and your bones shall flourish like an herb: and the hand of the LORD shall be known toward his servants, and *his* indignation toward his enemies. 15 For, behold, the LORD will come with fire, and with his chariots like a whirlwind, to render his anger with fury, and his rebuke with flames of fire. 16 For by fire and by his sword will the LORD plead with all flesh: and the slain of the LORD shall be many. 17 They that sanctify themselves, and purify themselves in the gardens behind one *tree* in the midst, eating swine's flesh, and the abomination, and the mouse, shall be consumed together, saith the LORD. 18 For I *know* their works and their thoughts: it shall come, that I will gather all nations and tongues; and they shall come, and see my glory. 19 And I will set a sign among them, and I will send those that escape of them unto the nations, *to* Tarshish, Pul, and Lud, that draw the bow, *to* Tubal, and Javan, *to* the isles afar off, that have not heard my fame, neither have seen my glory; and they shall declare my glory among the Gentiles. 20 And they shall bring all your brethren *for* an offering unto the LORD out of all nations upon horses, and in chariots, and in litters, and upon mules, and upon swift beasts, to my holy mountain Jerusalem, saith the LORD, as the children of Israel bring an offering in a clean vessel *into* the house of the LORD. 21 And I will also take of them for priests *and* for Levites, saith the LORD. 22 For as the new heavens and the new earth, which I *will* make, *shall* remain before me, saith the LORD, so shall your seed and your name remain. 23 And it shall come to pass, *that* from one new moon to another, and from one sabbath to another, shall all flesh come to worship before me, saith the LORD. 24 And they shall go forth, and look upon the carcases of the men that have transgressed against me: for their worm shall not die, neither shall their fire be quenched; and they shall be an abhorring unto all flesh."

7.6 The Carrot and the Stick

The violent Bolshevik revolution inspired great trepidation in the West. *The Morning Post* of London published numerous essays attacking Bolshevism and the *Protocols of the Learned Elders of Zion*, and asserting that Bolshevism signified proof of the genuineness of the *Protocols* and of the alleged alliance of the Freemasons and a vast Jewish conspiracy to rule the world.[205] In England, Lord Northcliffe drew attention to the *Protocols* in his newspaper *The London*

---

205. *The Cause of World Unrest*, G. Richards, ltd., London, G.P. Putnam, New York, (1920); which reproduces articles which first appeared in *The Morning Post* of London.

Times and called for an investigation of the Zionists.[206] Henry Ford's THE DEARBORN INDEPENDENT did much the same thing in America and personally attacked many Jewish leaders in America.[207] In Germany, Alfred Rosenberg and Adolf Hitler, among many others, focused public attention on the *Protocols*.

In *Genesis* 12:1-3 the Jews offer the world a carrot and a stick:

> "Now the LORD had said unto Abram, Get thee out of thy country, and from thy kindred, and from thy father's house, unto a land that I will shew thee: 2 And I will make of thee a great nation, and I will bless thee, and make thy name great; and thou shalt be a blessing: 3 And I will bless them that bless thee, and curse him that curseth thee: and in thee shall all families of the earth be blessed."

"Mentor" intimated in *The Jewish Chronicle* on 4 April 1919 on page 7, that Bolshevism was revenge on the Gentile nations for not allowing the Jews their own nation and for not willfully succumbing to the Jewish Messianic prophecy of a world government run by Jews,

"PEACE, WAR— AND BOLSHEVISM.

By MENTOR.

WHAT is written here is pendent to what appeared in this column last week. As I intimated, I propose to revert to the subject then referred to.

BOLSHEVISM is at once the most serious menace to, and the best hope of, Civilisation. Paradoxical as this may sound, but a little thought will show it to be abundantly true. The menace of Bolshevism is manifest. It pulls down what, until now, it has shown itself unable efficiently to replace. In the name of freedom, it imposes galling slavery. In the name of humanity, it inflicts the direst evil upon the men, women, and children who come under its sway. It protests against class domination and itself imposes the domination of class wherever it can obtain power. It knows no bounds either in justice or in liberty. It murders, imprisons and tortures with the ruthlessness of an autocracy drunk with new-found authority. It is ruthless, relentless, all-engulfing. It falls upon the country it infects like a dire pestilence which casts people prone. It is a political disease, an economic infliction, a social disaster.

---

206. "The Jewish Peril, a Disturbing Pamphlet: Call for Inquiry", *The London Times*, (8 May 1920).

207. **English:** *The International Jew: The World's Foremost Problem*, In Four Volumes, (1920-1922); which reproduces articles which first appeared in THE DEARBORN INDEPENDENT. **German:** *Der internationale Jude*, Hammer-Verlag, (1922). **Russian:** Mezhdunarodnoe evreistvo: perevod s angliiskago, (1925). **Italian:** *L'Internazionale Ebraica. Protocolli dei "Savi Anziani" di Sion*, La Vita Italiana, Rassegna Mensile di Politica, Roma, (1921). **Spanish:** B. Wenzel, *El Judío Internacional: Un Problema del Mundo*, Hammer-Verlag, Leipzig, (1930). **Portuguese:** S. E. Castan and H. Ford, *O Judeu Internacional*, Revisão, Porto Alegre, RS, Brasil, (1989).

YET, none the less, in Bolshevism there lies, to-day, the hope of Humanity. For in essence, it is the revolt of peoples against the social state, against the evil; the iniquities—and the inequalities—that were crowned by the cataclysm of the War under which the world groaned for upwards of four years. It is a revolution against a social state which suffered Tsarism to exist in Russia and militarism in Prussia and which still allows, alas, so many a crying wrong in countries that plume themselves on their freedom and boast of their liberty. Bolshevism is the signal to mankind to halt in its social, political, and economic ways of old; to stay and examine them in the light of the sacrifice of the millions of youth who have gone down to darkness eternal, of the millions of treasure which war has wasted, and to ponder them in the light of the incalculable, ineffable burden which the years of struggle have placed upon Society, and, heaviest of all, upon the poor—in light of the war which was proof in all surety that the old order was doomed if civilisation was to survive. That Bolshevism broke out first in the country most oppressed is nothing for wonder; it is merely natural. For centuries Russia had been the forcing ground of every infamy imposed by power and every wickedness done in the name of Government. That the creed has spread to a country whose national aspirations were for generations crushed, and where autocracy ruled, is nothing for wonder. Nor is the protest of Bolshevism merely a matter for Russia and Hungary, or a menace only to bayonet-ridden Germany. It is a challenge to the world—not least to the nations of freedom and liberty. It is a challenge to all the nations including the peoples who nourish liberty and freedom as precious principles, but who have passively allowed a state of affairs to grow and putrefy into the infamies of Russian Tsarism, the iniquity of Hungary, and the wickedness of German militarism; to the world that has suffered Society to fester into these and to break out into the prurient, gaping, sloughing, agonising tumour of such a war as that which is not ended, though it is suspended. And the fact that this protest has been made is the world's best hope. It is a demand for another order of things, for a social state which will render humanity immune from the wickedness and such evil as resulted in the greatest war mankind has ever known. It asks for some guarantee against a system which dragged peoples innocent of any intention of killing, slaying, and slaughtering into the vortex of War—peacefully intentioned peoples who loathed and hated War (such as was England before that fateful day in August, 1914)—from which even the most innocent of belligerents, and even those who stood aside from the contest are suffering to-day; though none were wholly guiltless of it, because for generations all passively concurred in the system. If the world, as a result of the War, had received no such warning as Bolshevism, the evil would, in all probability have gone on, deepening in its wrong, becoming ever blacker. Bolshevism is a social fever which indicates a high blood temperature. It gives the warning of mischief that may be fatal. A wise doctor takes note of the fever and seeks to remove the cause. He does not call the fever ugly names or denounce it, nor is he so stupid as to confuse the patient's consequent delirium with his normal condition, as so many are confusing the delirium of Bolshevism with the normal state of the countries in which it is finding vogue.

ALL such indications on the part of the body politic that there is a disease that must be removed, else the patient must go under, are as unpleasant, as inimical, as is the delirium of the fever-stricken patient distressing. The French Revolution drowned Paris in blood. Its excesses were far greater than anything that even the most malicious has attributed to Bolshevism. It instituted a Reign of Terror. It massacred Royalty. It condemned men and women day by day to the tumbril; so commonly indeed, that the men and women walking in the streets of Paris hardly looked round when some victim of the Jacobins was being taken to the Guillotine. Nothing and nobody was safe from the raging, tearing fever of the Revolution. For years it inflicted upon France a series of infamies, of torture, of horror, of bloodshed almost unparalleled in history. Yet, at the end of it all, and notwithstanding its reaction in Napoleonism, a great English writer declared that there had been nothing greater and more glorious in all history than the French Revolution. By common consent what liberty, equality, and fraternity—liberty, equality, and fraternity which the French Revolution never gained, and which in seeking after it demeaned and disgraced—the rest of the world possesses to-day, it draws in large measure from the days in which France was bathed in the anarchy of revolt. That is because the motive-spring which set the French Revolution into being was an ideal for the betterment of mankind, a protest against the social, political and economic infamies which will for ever be associated with the *régime* of the Bourbons, a striving for a social state that would not allow unbridled luxury, lascivious prodigality, selfish extravagence, inhuman carelessness, to thrive in the Court and to go on side by side with poverty, hunger, a life of groaning and moaning in the alleys hard by. And, even now, while the terror of Bolshevism is in full swing, a writer in an English Daily paper is brought to declare, as one did the other day, that at root Bolshevism in ideal has nothing comparable to it since the teachings which Jesus of Nazareth gave to the world. The writer had, there is little doubt, recollected the parable of the rich man, torn with suffering in Hell, pleading to Lazarus, the beggar whose sores the dogs licked, resting in the bosom of Abraham in Heaven. It is the parable of the ideals of Bolshevism.

IT is not difficult to see why a people which has managed to subsist through Tsardom, because of the religious ideals and ideas which it nourished throughout all its classes, and not least among its peasantry, has been attacked by the ideals of Bolshevism, and why, released from Tsardom, it has, pendulum-like, swung into the arms of Lenin, looking to the ideals of his creed, and not to its wickedness or its excesses. The same reason obtains for the number of Jews who are to be found in the Bolshevist ranks. The Jew is an idealist. He will give much for an ideal. He thirst for idealism as a goal of life. This may seem strange to those who associate the Jew with materialism. But the capacity of the Jew for idealism is such that he notoriously idealises even the material. The fact that there are so many of our people who have associated themselves with the ideals of Bolshevism, even although as Jews its excesses must be repugnant to them, has to be placed in conjunction with another fact. These men will be found for the most part unassociated with or dissociated from the Synagogue. In the ordinary way of speaking they are not observing Jews. Is it not patent that the

Synagogue, having failed to attract them by its idealism, and no other ideal, not even a material ideal, having been provided for them—for they are not men of wealth and substance, such as are usually to be found among the *bourgeoisie*—they have ranged themselves on the side of Bolshevism, because here was no Jewish ideal to which these Jews could devote their sentiments and their energies? I cannot understand how people who for generations have, unprotesting, allowed the Jew, particularly in Eastern Europe, in Russia, to suffer pogroms, to be massacred and ill-treated, and tortured and murdered, and for two thousand years have kept our people outside the ambit of the most potent source of idealism that can appeal to men—that associated with National being—now have the hypocrisy, the soulless impertinence, to complain that so many of our people are Bolshevists! That Jews have been chosen to the extent they have to take a leading part in the movement in Russia and in Hungary, is merely because they are heavily endowed with intellectualism and capacity, as compared with the rest of the population. But the world must not surprised that the Jew, who is an idealist or nothing, has turned to the idealism of Bolshevism, which a British writer has declared to be comparable to the idealism preached by the founder of Christianity. It were surprising, really, were it otherwise. You cannot keep a people out of their rightful place amid the nations of the world, and then complain because they take the leading part which their abilities entitle them to in the nations among whom you have scattered them. The fact that a timorous millionaire afraid, and doubtless with good cause, of Bolshevism, which he probably has never taken the trouble, or perhaps has not the capacity to appreciate in full measure, places a ban of religious excommunication upon those Jews who are Bolshevists, is a thing for the gods to laugh at!

THERE is much in the fact of Bolshevism itself, in the fact that so many Jews are Bolshevists, in the fact that the ideals of Bolshevism at many points are consonant with the finest ideals of Judaism, some of which went to form the basis of the best teachings of the founder of Christianity—these are things which the thoughtful Jew will examine carefully. It is the thoughtless one who looks upon Bolshevism only in the ugly repulsive aspects which all social revolutions assume and which make it so hateful to the freedom-loving Jew—when allowed to be free. It is the thoughtless one that thus partially examines the greatest problem the modern world has been set, and as his contribution to the solution dismisses it with some exclamation made in obedient deference to his own social position, and to what for the moment happens to be conventionally popular."

    Sir Winston Churchill offered the world a carrot and a stick on behalf of the Zionist Jews in a statement published in the *Illustrated Sunday Herald*, on 8 February 1920, on page 5. Churchill threatens God's wrath, in the form of Bolshevism, on any nation that does not commit itself to the Zionist cause, and promises God's gifts to any nation which sponsors Zionism. It was an ancient Zionist appeal to superstitious fear.

    Churchill traveled to Palestine and was an outspoken champion of Zionism in the British Government. Some argue that Churchill was also crypto-Jew, who had a Jewish mother. Churchill aggressively spoke out on behalf of Zionism in

June of 1921 before the House of Commons in an effort to justify the unfair appropriation of the nation by a minority population of Jews.[208]

In 1948, when Israel became a nation-state, Churchill wrote to Chaim Weizmann, "[what a fine moment it was] for an old 'Zionist' like me!"[209] Christopher Sykes details much of Churchill's Zionist activities in his book *Crossroads to Israel*, where he states, "Mr Churchill had always been a Zionist, albeit of a very Gentile and unorthodox kind, since his days as Colonel Secretary."[210] Though Churchill's newspaper article is today seen by many as anti-Semitic, it was written on behalf of the Zionists, who quickly seized upon the opportunity of its publication to point out that in their opinion the only option for all Jews is Zionism.[211]

Where Churchill paints Jews in a bad light, it is done as a threat to Gentiles, not as an attack on Jews, and his arguments were planted in his head by his Zionist cohorts, as evinced by Chaim Weizmann's speech in Jerusalem in January of 1920, as captured in an article, "Eine große Rede Weizmanns in Jerusalem Vor der Abreise aus Palästina", *Jüdische Rundschau*, Volume 25, Number 4, (16 January 1920), p. 4, which stated, *inter alia*,

> "Professor Weizmann emphatically declared that the *beauty of the ideals of the Jewish renaissance* was critical for the English Declaration. It is a misconception, that England made the proposal to us only out of self-interest. *Lloyd George* once said: I know the Palestinian front far better than the French, because I am well acquainted with every borough and every brook from the Bible. For the English, Palestine is above all else a Biblical issue. The English still believe in the Bible more than many classes of Jewry. Therefore, the *idealistic reasons* came first, and afterwards the material reasons were added. It was we who made it clear to the English political leaders that it was in England's interest to unite with us to spread the wings of the British eagle out over Palestine. We did not achieve the Declaration by way of miracles, but rather through persistent propaganda, through constant demonstration of the vigor of our people. We told the people in charge: We are taking over Palestine whether you like it, or not.

---

208. *See:* "The Government and Palestine. Mr. Churchill's Statement", *The Jewish Chronicle*, (17 June 1921), pp. 17-19. *See also:* "Mr. Churchill's Vision", *The London Times*, (15 June 1921), p. 10.

209. W. Churchill quoted in B. Avishai, *The Tragedy of Zionism: Revolution and Democracy in the Land of Israel*, Farrar Straus Giroux, New York, (1985), p. 349. *See also:* J. B. Agus, *The Meaning of Jewish History*, Volume 2, Abelard-Schuman, New York, (1963), p.431. *See also:* A. Hertzberg, *The Zionist Idea*, Harper Torchbooks, New York, (1959), p. 594.

210. C. Sykes, *Crossroads to Israel*, Indiana University Press, (1973), p. 207.

211. N. Goldman, "Zionismus und nationale Bewegung", *Der Jude*, Volume 5, Number 1, (1920-1921), pp. 45-47, at 46. Goldman's articles continued with: "Zionismus und nationale Bewegung", *Der Jude*, Volume 5, Number 4, (1920-1921), pp. 237-242; and "Zionismus und nationale Bewegung", *Der Jude*, Volume 5, Number 7, (1920-1921), pp. 423-425.

You can accelerate or delay our arrival, but it is better for you to help us, because if you don't our constructive power will turn destructive and overthrow the entire world."

"Professor Weizmann betonte, daß die Schönheit des Ideals der jüdischen Renaissance das Entscheidende für die englische Deklaration war. Es sei eine irrtümliche Auffassung, daß England uns den Vorschlag nur aus eigenem Interesse heraus machte. Lloyd George sagte einmal: Ich kenne die Palästinafront viel genauer als die französische, denn jeder Flecken und jeder Bach ist mir aus der Bibel vertraut. Palästina ist für England vor allen Dingen ein Gegenstand der Bibel. Die Engländer glauben an die Bibel noch mehr als manche Schichten im Judentum. Zuerst kamen also die ideellen Gründe, nachher kamen die materiellen hinzu. Wir sind es, die den englischen politischen Führern klargemacht haben, daß es im Interesse Englands ist, sich mit uns zu vermählen, die Fittiche des britischen Adlers über Palästina auszubreiten. Wir erreichten die Deklaration nicht durch Wundertaten, sondern durch beharrliche Propaganda, durch unaufhörliche Beweise von der Lebenskraft unseres Volkes. Wir sagten den maßgebenden Persönlichkeiten: Wir werden in Palästina sein, ob Ihr es wollt oder es nicht wollt. Ihr könnt unser Kommen beschleunigen oder verzögern, es ist aber für Euch besser, uns mitzuhelfen, denn sonst wird sich unsere aufbauende Kraft in eine zerstörende verwandeln, die die ganze Welt in Gärung bringen wird."

Though the Zionists dominated the proceedings of the Treaty of Versaille and later dominated the proceedings of the League of Nations and the Palestine Mandate, the masses of Jews did not want to go to Palestine. Since the Jewish masses failed to heed the Zionists' threats, Weizmann and his fellow Zionists brought world-wide tumult upon the Gentiles, as well as the Jews, just as Chaim Weizmann had promised.

Weizmann took his cue from Theodor Herzl, who strongly believed that anti-Semitism was the best means to achieve a "Jewish State". Herzl unwisely believed that he could threaten the governments of the world with absolute impunity,

"The governments will give us their friendly assistance because we relieve them of the danger of a revolution which would start with the Jews—and stop who knows where!"[212]

Herzl wrote in his book *The Jewish State*,

"When we sink, we become a revolutionary proletariat, the subordinate

---

[212]. T. Herzl, English translation by H. Zohn, R. Patai, Editor, *The Complete Diaries of Theodor Herzl*, Volume 1, Herzl Press, New York, (1960), p. 183.

officers of the revolutionary party; when we rise, there rises also our terrible power of the purse. [***] Again, people will say that I am furnishing the Anti-Semites with weapons. Why so? Because I admit the truth? Because I do not maintain that there are none but excellent men amongst us? Again, people will say that I am showing our enemies the way to injure us. This I absolutely dispute. My proposal could only be carried out with the free consent of a majority of Jews. Individuals or even powerful bodies of Jews might be attacked, but Governments will take no action against the collective nation. The equal rights of Jews before the law cannot be withdrawn where they have once been conceded; for the first attempt at withdrawal would immediately drive all Jews rich and poor alike, into the ranks of the revolutionary party. The first official violation of Jewish liberties invariably brings about economic crisis. Therefore no weapons can be effectually used against us, because these cut the hands that wield them."[213]

Churchill's Weizmannesque and Herzlian article of 8 February 1920, originally published in the *Illustrated Sunday Herald*, on page 5, again issued the ancient threat of *Genesis* 12:3, "And I will bless them that bless thee, and curse him that curseth thee: and in thee shall all families of the earth be blessed[,]" and offered the Goyim the carrot and the stick:

"ZIONISM versus BOLSHEVISM.

A STRUGGLE FOR THE SOUL OF THE JEWISH PEOPLE.

By the Rt. Hon. WINSTON S. CHURCHILL

Some people like Jews and some do not; but no thoughtful man can doubt the fact that they are beyond all question the most formidable and the most remarkable race which has ever appeared in the world.

Disraeli, the Jew Prime Minister of England, and Leader of the Conservative Party, who was always true to his race and proud of his origin, said on a well-known occasion: 'The Lord deals with the nations as the nations deal with the Jews.' Certainly when we look at the miserable state of Russia, where of all countries in the world the Jews were the most cruelly treated, and contrast it with the fortunes of our own country, which seems to have been so providentially preserved amid the awful perils of these times, we must admit that nothing that has since happened in the history of the world has falsified the truth of Disraeli's confident assertion.

Good and Bad Jews.

The conflict between good and evil which proceeds unceasingly in the

---

[213]. T. Herzl, *A Jewish State: An Attempt at a Modern Solution of the Jewish Question*, The Maccabæan Publishing Co., New York, (1904), pp. 23, 99.

breast of man nowhere reaches such an intensity as in the Jewish race. The dual nature of mankind is nowhere more strongly or more terribly exemplified. We owe to the Jews in the Christian revelation a system of ethics which, even if it were entirely separated from the supernatural, would be incomparably the most precious possession of mankind, worth in fact the fruits of all other wisdom and learning put together. On that system and by that faith there has been built out of the wreck of the Roman Empire the whole of our existing civilisation.

And it may well be that this same astounding race may at the present time be in the actual process of producing another system of morals and philosophy, as malevolent as Christianity was benevolent, which, if not arrested, would shatter irretrievably all that Christianity has rendered possible. It would almost seem as if the gospel of Christ and the gospel of Antichrist were destined to originate among the same people; and that this mystic and mysterious race had been chosen for the supreme manifestations, both of the divine and the diabolical.

'National' Jews.

There can be no greater mistake than to attribute to each individual a recognisable share in the qualities which make up the national character. There are all sorts of men—good, bad and, for the most part, indifferent—in every country, and in every race. Nothing is more wrong than to deny to an individual, on account of race or origin, his right to be judged on his personal merits and conduct. In a people of peculiar genius like the Jews, contrasts are more vivid, the extremes are more widely separated, the resulting consequences are more decisive.

At the present fateful period there are three main lines of political conception among the Jews, two of which are helpful and hopeful in a very high degree to humanity, and the third absolutely destructive.

First there are the Jews who, dwelling in every country throughout the world, identify themselves with that country, enter into its national life, and, while adhering faithfully to their own religion, regard themselves as citizens in the fullest sense of the State which has received them. Such a Jew living in England would say, 'I am an Englishman practising the Jewish faith.' This is a worthy conception, and useful in the highest degree. We in Great Britain well know that during the great struggle the influence of what may be called the 'National Jews' in many lands was cast preponderatingly on the side of the Allies; and in our own Army Jewish soldiers have played a most distinguished part, some rising to the command of armies, others winning the Victoria Cross for valour.

The National Russian Jews, in spite of the disabilities under which they have suffered, have managed to play an honourable and useful part in the national life even of Russia. As bankers and industrialists they have strenuously promoted the development of Russia's economic resources, and they were foremost in the creation of those remarkable organisations, the Russian Co-operative Societies. In politics their support has been given, for the most part, to liberal and progressive movements, and they have been

among the staunchest upholders of friendship with France and Great Britain.
International Jews.

In violent opposition to all this sphere of Jewish effort rise the schemes of the International Jews. The adherents of this sinister confederacy are mostly men reared up among the unhappy populations of countries where Jews are persecuted on account of their race. Most, if not all, of them have forsaken the faith of their forefathers, and divorced from their minds all spiritual hopes of the next world. This movement among the Jews is not new. From the days of Spartacus-Weishaupt to those of Karl Marx, and down to Trotsky (Russia), Bela Kun (Hungary), Rosa Luxembourg (Germany), and Emma Goldman (United States), this world-wide conspiracy for the overthrow of civilization and for the reconstitution of society on the basis of arrested development, of envious malevolence, and impossible equality, has been steadily growing. It played, as a modern writer, Mrs. Webster, has so ably shown, a definitely recognizable part in the tragedy of the French Revolution. It has been the mainspring of every subversive movement during the Nineteenth Century; and now at last this band of extraordinary personalities from the underworld of the great cities of Europe and America have gripped the Russian people by the hair of their heads and have become practically the undisputed masters of that enormous empire.

Terrorist Jews.

There is no need to exaggerate the part played in the creation of Bolshevism and in the actual bringing about of the Russian Revolution by these international and for the most part atheistical Jews. It is certainly a very great one; it probably outweighs all others. With the notable exception of Lenin, the majority of the leading figures are Jews. Moreover, the principal inspiration and driving power comes from the Jewish leaders. Thus, Tchitcherin, a pure Russian, is eclipsed by his nominal subordinate Litvinoff, and the influence of Russians like Bukharin or Lunacharski cannot be compared with the power of Trotsky, or of Zinovieff, the Dictator of the Red Citadel (Petrograd), or of Krassin or Radek—all Jews. In the Soviet institutions the predominance of Jews is even more astonishing. And the prominent, if not indeed the principal, part in the system of terrorism applied by the Extraordinary Commissions for Combating Counter-Revolution has been taken by Jews, and in some notable cases by Jewesses. The same evil prominence was obtained by Jews in the brief period of terror during which Bela Kun ruled in Hungary. The same phenomenon has been presented in Germany (especially in Bavaria), so far as this madness has been allowed to prey upon the temporary prostration of the German people. Although in all these countries there are many non-Jews every whit as bad as the worst of the Jewish revolutionaries, the part played by the latter in proportion to their numbers in the population is astonishing.

'Protector of the Jews.'

Needless to say, the most intense passions of revenge have been excited in the breasts of the Russian people. Wherever General Denikin's authority could reach, protection was always accorded to the Jewish population, and strenuous efforts were made by his officers to prevent reprisals and to punish those guilty of them. So much was this the case that the Petlurist propaganda against General Denikin denounced him as the Protector of the Jews. The Misses Healy, nieces of Mr. Tim Healy, in relating their personal experiences in Kieff, have declared that to their knowledge on more than one occasion officers who committed offences against Jews were reduced to the ranks and sent out of the city to the front. But the hordes of brigands by whom the whole vast expanse of the Russian Empire is becoming infested do not hesitate to gratify their lust for blood and for revenge at the expense of the innocent Jewish population whenever an opportunity occurs. The brigand Makhno, the hordes of Petlura and of Gregorieff, who signalised their every success by the most brutal massacres, everywhere found among the half-stupefied, half-infuriated population an eager response to anti-Semitism in its worst and foulest forms.

The fact that in many cases Jewish interests and Jewish places of worship are excepted by the Bolsheviks from their universal hostility has tended more and more to associate the Jewish race in Russia with the villainies which are now being perpetrated. This is an injustice on millions of helpless people, most of whom are themselves sufferers from the revolutionary regime. It becomes, therefore, specially important to foster and develop any strongly-marked Jewish movement which leads directly away from these fatal associations. And it is here that Zionism has such a deep significance for the whole world at the present time.

A Home for the Jews.

Zionism offers the third sphere to the political conceptions of the Jewish race. In violent contrast to international communism, it presents to the Jew a national idea of a commanding character. It has fallen to the British Government, as the result of the conquest of Palestine, to have the opportunity and the responsibility of securing for the Jewish race all over the world a home and a centre of national life. The statesmanship and historic sense of Mr. Balfour were prompt to seize this opportunity. Declarations have been made which have irrevocably decided the policy of Great Britain. The fiery energies of Dr. Weissmann, the leader, for practical purposes, of the Zionist project, backed by many of the most prominent British Jews, and supported by the full authority of Lord Allenby, are all directed to achieving the success of this inspiring movement.

Of course, Palestine is far too small to accommodate more than a fraction of the Jewish race, nor do the majority of national Jews wish to go there. But if, as may well happen, there should be created in our own lifetime by the banks of the Jordan a Jewish State under the protection of the British Crown, which might comprise three or four millions of Jews, an event would have occurred in the history of the world which would, from every point of view, be beneficial, and would be especially in harmony with the truest interests of the British Empire.

Zionism has already become a factor in the political convulsions of

Russia, as a powerful competing influence in Bolshevik circles with the international communistic system. Nothing could be more significant than the fury with which Trotsky has attacked the Zionists generally, and Dr. Weissmann in particular. The cruel penetration of his mind leaves him in no doubt that his schemes of a world-wide communistic State under Jewish domination are directly thwarted and hindered by this new ideal, which directs the energies and the hopes of Jews in every land towards a simpler, a truer, and a far more attainable goal. The struggle which is now beginning between the Zionist and Bolshevik Jews is little less than a struggle for the soul of the Jewish people.

Duty of Loyal Jews.

It is particularly important in these circumstances that the national Jews in every country who are loyal to the land of their adoption should come forward on every occasion, as many of them in England have already done, and take a prominent part in every measure for combating the Bolshevik conspiracy. In this way they will be able to vindicate the honour of the Jewish name and make it clear to all the world that the Bolshevik movement is not a Jewish movement, but is repudiated vehemently by the great mass of the Jewish race.

But a negative resistance to Bolshevism in any field is not enough. Positive alternatives are needed in the moral as well as in the social sphere; and in building up with the utmost possible rapidity a Jewish national centre in Palestine which may become not only a refuge to the oppressed from the unhappy lands of Central Europe, but which will also be a symbol of Jewish unity and the temple of Jewish glory, a task is presented on which many blessings rest."

The Zionists were playing a very dangerous game with the lives of millions of innocent Jews. Israel Cohen saw the dangers of the false association of Bolsheviks with all Jews and wrote, *inter alia*, in *The Jewish Chronicle* of London, on 12 December 1919, on page 17,

"THE 'JEW-BOLSHEVIST' LIE.

By ISRAEL COHEN.

The systematic attempts that are now being made to identify the Jews of Russia with the Bolshevists, to represent Bolshevism as a Jewish movement, and thus to hold up the entire Jewish people to obloquy and attack, are based solely upon the fact that a certain number of prominent Bolshevist Commissaries are of Jewish birth. Upon this fact, which has never been denied, have been built up all sorts of fantastic accusations, such as that the Jews wish to wreak revenge upon Russia for the persecutions under Tsardom, and that they aim at sweeping away Christian civilization so as to enthrone Judaism as the dominant faith throughout the world. How utterly absurd these calumnies are, and how grotesquely exaggerated are most of the stories of Jewish participation in Bolshevism can be proved by an unimpassioned examination of the ascertainable facts and figures."

See also: Fritz Rodeck, "Judentum und Bolschewismus", *Jüdische Zeitung*, Volume 14, Number 25, (3 September 1920), pp. 5-6.

While it was true that most Jews were not Bolsheviks—even in Russia, it is also true that many Jews in lands "liberated" by Bolsheviks welcomed and embraced the mass murderers and aggressively participated in the destruction of their Gentile neighbors' lives. It is further true that the rise and spread of Bolshevism primarily occurred through Jewish communities around the world. It is yet further true that Bolshevism fulfilled Jewish Messianic prophecy, which was no coincidence. However, this does not mean that a majority of Jews have ever been Bolshevists, or even that a majority of Bolshevists have ever been Jews, but there is no doubt that Bolshevism was a Jewish movement meant to accomplish Jewish Messianic prophecies.

While leading Jews in the West decried Bolshevism when the connection to Jews became obvious, they did little to undo the damage Jewish financiers had done through Bolshevism to Russia and other nations. In fact, Jewish leadership instead continued to covertly perpetuate Bolshevism. Had Jewish leaders genuinely opposed Bolshevism, they would have organized and funded massive campaigns to stamp it out and to repair the damage done, and given their wealth and influence, they would have succeeded.

7.7 British Zionists, in Collaboration with Nazi Zionists, in Collaboration with Palestinian Zionists, Ensured that the Jews of Continental Europe Would Find No Sanctuary Before the War Ended

After the First World War, the Zionists had their Peace Conference and their League of Nations and their Palestine Mandate, but they lacked the broad support of the Jewish People. They decided to bring on a Second World War, which would result in another Peace Conference; and, the second time around, they would torture the Jewish People into embracing Zionism.

Lenni Brenner wrote in his exposé *Zionism in the Age of the Dictators*, "The Wartime Failure to Rescue", Chapter 24, Lawrence Hill Books, Chicago, (1983), pp. 235-238 [Brenner cites in his notes: "22. Michael Dov-Ber Weissmandel, *Min HaMaitzer* (unpublished English translation). 23. Ibid. 24. Ibid. (Hebrew edn), p. 92. 25. Ibid., p. 93."],

"'For only with Blood Shall We Get the land'

The Nazis began taking the Jews of Slovakia captive in March 1942. Rabbi Michael Dov-Ber Weissmandel, an Agudist, thought to employ the traditional weapon against anti-Semitism: bribes. He contacted Dieter Wisliceny, Eichmann's representative, and told him that he was in touch with the leaders of world Jewry. Would Wisliceny take their money for the lives of Slovakian Jewry? Wisliceny agreed for 50,000 in dollars so long as it came from outside the country. The money was paid, but it was actually raised locally, and the surviving 30,000 Jews were spared until 1944 when they were captured in the aftermath of the furious but unsuccessful Slovak partisan revolt.

Weissmandel, who was a philosophy student at Oxford University, had Volunteered on 1 September 1939 to return to Slovakia as the agent of the world Aguda. He became one of the outstanding Jewish figures during the Holocaust, for it was he who was the first to demand that the Allies bomb Auschwitz. Eventually he was captured, but he managed to saw his way out of a moving train with an emery wire; he jumped, broke his leg, survived and continued his work of rescuing Jews. Weissmandel's powerful post-war book, *Min HaMaitzer* (From the Depths), written in Talmudic Hebrew, has unfortunately not been translated into English as yet. It is one of the most powerful indictments of Zionism and the Jewish establishment. It helps put Gruenbaum's unwillingness to send money into occupied Europe into its proper perspective. Weissmandel realised: 'the money is needed here – by us and not by them. For with money here, new ideas can be formulated.'[22] Weissmandel was thinking beyond just bribery. He realised immediately that with money it was possible to mobilise the Slovak partisans. However, the key question for him was whether any of the senior ranks in the SS or the Nazi regime could be bribed. Only if they were willing to deal with either Western Jewry or the Allies, could bribery have any serious impact. He saw the balance of the war shifting, with some Nazis still thinking they could win and hoping to use the Jews to put pressure on the Allies, but others beginning to fear future Allied retribution. His concern was simply that the Nazis should start to appreciate that live Jews were more useful than dead ones. His thinking is not to be confused with that of the Judenrat collaborators. He was not trying to save some Jews. He thought strictly in terms of negotiations on a Europe-wide basis for all the Jews. He warned Hungarian Jewry in its turn: do not let them ghettoise you! Rebel, hide, make them drag the survivors there in chains! You go peacefully into a ghetto and you will go to Auschwitz! Weissmandel was careful never to allow himself to be manoeuvred by the Germans into demanding concessions from the Allies. Money from world Jewry was the only bait he dangled before them.

In November 1942, Wisliceny was approached again. How much money would be needed for all the European Jews to be saved? He went to Berlin, and in early 1943 word came down to Bratislava. For $2 million they could have all the Jews in Western Europe and the Balkans. Weissmandel sent a courier to Switzerland to try to get the money from the Jewish charities. Saly Mayer, a Zionist industrialist and the Joint Distribution Committee representative in Zurich, refused to give the Bratislavan 'working group' any money, even as an initial payment to test the proposition, because the 'Joint' would not break the American laws which prohibited sending money into enemy countries. Instead Mayer sent

Weissmandel a calculated insult: 'the letters that you have gathered from the Slovakian refugees in Poland are exaggerated tales for this is the way of the '*Ost-Juden*' who are always demanding money'.[23]

The courier who brought Mayer's reply had another letter with him from Nathan Schwalb, the HeChalutz representative in Switzerland Weissmandel described the document:

> There was another letter in the envelope, written in a strange foreign language and at first I could not decipher at all which language it was until I realised that this was Hebrew written in Roman letters, and written to Schwalb's friends in Pressburg [Bratislava] ... It is still before my eyes, as if I had reviewed it a hundred and one times. This was the content of the letter:
>
> 'Since we have the opportunity of this courier, we are writing to the group that they must constantly have before them that in the end the Allies will win. After their victory they will divide the world again between the nations, as they did at the end of the first world war. Then they unveiled the plan for the first step and now, at the war's end, we must do everything so that Eretz Yisroel will become the state of Israel, and important steps have already been taken in this direction. About the cries coming from your country, we should know that all the Allied nations are spilling much of their blood, and if we do not sacrifice any blood, by what right shall we merit coming before the bargaining table when they divide nations and lands at the war's end? Therefore it is silly, even impudent, on our part to ask these nations who are spilling their blood to permit their money into enemy countries in order to protect our blood—for only with blood shall we get the land. But in respect to you, my friends, *atem taylu*, and for this purpose I am sending you money illegally with this messenger.'[24]

Rabbi Weissmandel pondered over the startling letter:

> After I had accustomed myself to this strange writing, I trembled, understanding the meaning of the first words which were 'only with blood shall we attain land'. But days and weeks went by, and I did not know the meaning of the last two words. Until I saw from something that happened that the words '*atem taylu*' were from '*tiyul*' [to walk] which was their special term for 'rescue'. In other words: you, my fellow members, my 19 or 20 close friends, get out of Slovakia and save your lives and with the blood of the remainder—the blood of all the men, women, old and young and the sucklings—the land will belong to us. Therefore, in order to save their lives it is a crime to allow money into enemy territory—but to save you beloved friends, here is money obtained illegally.
>
> It is understood that I do not have these letters, for they remained there and were destroyed with everything else that was lost.[25]

Weissmandel assures us that Gisi Fleischman and the other dedicated Zionist rescue workers inside the working group were appalled by Schwalb's letter, but it expressed the morbid thoughts of the worst elements of the WZO leadership. Zionism had come full turn: instead of Zionism being the hope of the Jews, their blood was to be the political salvation of Zionism."

Zionist Anglican Chaplain to the British Embassy in Vienna, William Henry Hechler,[214] published *The Restoration of the Jews to Palestine According to the Prophets* in 1884. He contacted racist Zionist Theodor Herzl as soon as he learned of Herzl's book *Der Judenstaat*, in 1896. Hechler knew the Jewish Zionist Leon Pinsker. Queen Victoria requested that Hechler transmit a letter from her to the Sultan of Turkey asking him to allow Russian Jews to take asylum in Palestine, but the British embassy would not transmit the message.[215]

Theodor Herzl changed paths from converting Jews to Christianity in order to end anti-Semitism, to converting anti-Semites to Zionism in order to end anti-Semitism. Herzl's ungodly betrayal of the Jewish People ultimately led the Zionists to create and install the Nazi régime.

Like the prophet Isaiah, Hechler and Herzl envisioned Jerusalem as the new capital of the world. Herzl's vision is revealed in his book *Altneustadt*.

Like many Zionists, Hechler relished the fact that anti-Semitism encouraged Jews to embrace Zionism out of fear for their lives. Isaiah Friedman wrote,

"On 26 March 1896, Hechler wrote to [Frederick the Grand Duke of Baden] about Herzl's project, noting with satisfaction that the antisemitic movement had made the Jews see that they were 'Jews first and [only] secondly Germans, Englishmen, etc.' It reawakened in them a longing to return 'as a nation to the Land of Promise... Palestine belongs to them by right.' Should Germany and England give their support and take the Jewish State, declared neutral, under their protection, the Return of the Jews would

---

214. B. W. Tuchman, *Bible and Sword: England and Palestine from the Bronze Age to Balfour*, New York University Press, New York, (1956). *See also:* C. Duvernoy, *Le prince et le prophète*, Le département des publications de l'Agence juive, Jérusalem, (1966); English translation: *The Prince and the Prophet*, Land of Promise Productions, Paradise, California, (1973); Christian Action for Israel, (1979). *See also:* "Hechler, William Henry", *Encyclopaedia Judaica*, Volume 8 He-Ir, Encyclopaedia Judaica, Jerusalem, (1971), cols. 237-239. *See also:* E. Newman, "Non-Jewish Pioneers of Zionism", in A. W. Kac, Editor, *The Messiahship of Jesus: What Jews and Jewish Christians Say*, Moody Press, Chicago, (1980), pp. 291-297. *See also:* M. Ould-Mey, "The Non-Jewish Origin of Zionism", *The Arab World Geographer / Le Géographe du monde arabe*, Volume 5, Number 1, (2002), pp. 34-52:
<http://mama.indstate.edu/users/mouldmey/The%20Non-Jewish%20Origin%20of%20Zionism.PDF>
215. "Hechler, William Henry", *Encyclopaedia Judaica*, Volume 8 He-Ir, Encyclopaedia Judaica, Jerusalem, (1971), cols. 237-239, at 237.

be a great blessing and would put an end to antisemitism, which was detrimental to the welfare of European nations. He also suggested that the issue be laid before the kaiser, the duke's nephew."[216]

Hechler knew beforehand that the First World War would occur. He also knew beforehand that the Holocaust would occur. He took comfort in his knowledge of these events. Elias Newman wrote,

"To the German-Jewish philosopher Martin Buber, Hechler said in 1913: 'Your fatherland will soon be given back to you. For a serious crisis will occur, whose deep meaning is the liberation of your Messianic Jerusalem from the yoke of the — nations... We are moving towards a world war...' Shortly before his death [in 1931], he said this to the family of the Zionist leader Nahum Sokolov: 'Part of European Jewry is going to be sacrificed for the resurrection of your biblical fatherland.'"[217]

In the context Hechler's foreknowledge of the Holocaust, Claude Duvernoy writes,

"Political, rational and 'scientific' anti-Semitism, born in Austria, spread all over Europe where the ground had been well prepared by centuries of bad Christian catechism. With Moscow opposing Zionism as a heretical movement and London already pursuing its policy of suffocation, one really could not see how this ferocious Nazi could fail in his plan for Jewish genocide. In closing Palestine to Jewish immigrants (which was done in 1939) London delivered up millions of European Jews to the ovens of the crematoriums soon to come—without wishing this, of course. [***] As there was need of a first world war, to force the liberation of Jerusalem from the pagan yoke of the Turk, undoubtedly a second world conflict was inevitable to form a Jewish homeland through much suffering and blood, but he did not dare to think of it."[218]

---

[216]. I Friedman, "The Political Activity of Theodor Herzl", in G. Shimoni and R. S. Wistrich, Editors, *Theodor Herzl: Visionary of the Jewish State*, Herzl Press, New York, (1999). Friedman cites: H. Ellern and B. Ellern, *Herzl, Hechler, the Grand Duke of Baden and the German Emperor, 1896-1904. Documents Found by Hermann and Bessi Ellern Reproduced in Facsimile*, Tel Aviv, (1961), pp. 1-8.

[217]. E. Newman, "Non-Jewish Pioneers of Zionism", in A. W. Kac, Editor, *The Messiahship of Jesus: What Jews and Jewish Christians Say*, Moody Press, Chicago, (1980), pp. 291-297, at 297.

[218]. C. Duvernoy, *Le prince et le prophète*, Le département des publications de l'Agence juive, Jérusalem, (1966); English translation: *The Prince and the Prophet*, Land of Promise Productions, Paradise, California, (1973); Christian Action for Israel, (1979). Here quoted from the English translation by Jack Joffe as found at:
<http://www.israelinprophecy.org/live_site/english/pdf_documents/CLAUDE%20DUVERNOY_P&P_090903_w-gen-index_reduced.pdf>, (1979/2003), pp. 119-120.

After Hechler came David Ben-Gurion, who stated,

"The First World War brought us the Balfour Declaration. The Second ought to bring us the Jewish State."[219]

Before Hechler was Benjamin Disraeli. Soon after Queen Victoria ascended the throne in 1837, the campaign to "restore the Jews to Palestine" gained political support, which was driven by the Rothschild family in hopes that a Rothschild would ascend the throne in Jerusalem to become King of the Jews, a. k. a. the Messiah, and, therefore, King of the World. Victoria believed that she was descended from King David and during her reign Lionel Nathan Rothschild and Disraeli were able to secure the Suez Canal for England as a means to expedite shipping to and from India, and as a means to tie England's fate to the goal of forming a Jewish State in Palestine.[220] Disraeli wrote in 1852,

> "We have shown that the theological prejudice against the Jews has no foundation, historical or doctrinal; we have shown that the social prejudice, originating in the theological but sustained by superficial observations irrespective of religious prejudice, is still more unjust, and that no existing race is so much entitled to the esteem and gratitude of society as the Hebrew. It remains for us to notice the injurious consequences to European society of the course pursued by the communities of this race, and this view of the subject leads us to considerations which it would become existing statesmen to ponder.
> The world has by this time discovered that it is impossible to destroy the Jews. The attempt to extirpate them has been made under the most favourable auspices and on the largest scale; the most considerable means that man could command have been pertinaciously applied to this object for the longest period of recorded time. Egyptian pharaohs, Assyrian kings, Roman emperors, Scandinavian crusaders, Gothic princes, and holy inquisitors, have alike devoted their energies to the fulfilment of this common purpose. Expatriation, exile, captivity, confiscation, torture on the most ingenious and massacre on the most extensive scale, a curious system of degrading customs and debasing laws which would have broken the heart of any other people, have been tried, and in vain. The Jews, after all this havoc, are probably more numerous at this date than they were during the reign of Solomon the wise, are found in all lands, and unfortunately prospering in most. All which proves, that it is in vain for man to attempt to baffle the inexorable law of nature which has decreed that a superior race shall never be destroyed or absorbed by an inferior.
> But the influence of a great race will be felt; its greatness does not

---

[219]. M. Bar-Zohar, *Ben-Gurion: The Armed Prophet*, Prentice-Hall, Englewood Cliffs, New Jersey, (1967), p. 69.
[220]. B. W. Tuchman, *Bible and Sword: England and Palestine from the Bronze Age to Balfour*, New York University Press, New York, (1956), pp. 122, 126, 165-166, 169.

depend upon its numbers, otherwise the English would not have vanquished the Chinese, nor would the Aztecs have been overthrown by Cortez and a handful of Goths. That greatness results from its organisation, the consequences of which are shown in its energy and enterprise, in the strength of its will and the fertility of its brain. Let us observe what should be the influence of the Jews, and then ascertain how it is exercised. The Jewish race connects the modern populations with the early ages of the world, when the relations of the Creator with the created were more intimate than in these days, when angels visited the earth, and God himself even spoke with man. The Jews represent the Semitic principle; all that is spiritual in our nature. They are the trustees of tradition, and the conservators of the religious element. They are a living and the most striking evidence of the falsity of that pernicious doctrine of modern times, the natural equality of man. The particular equality of a particular race is a matter of municipal arrangement, and depends entirely on political considerations and circumstances; but the natural equality of man now in vogue, and taking the form of cosmopolitan fraternity, is a principle which, were it possible to act on it, would deteriorate the great races and destroy all the genius of the world. What would be the consequences on the great Anglo-Saxon republic, for example, were its citizens to secede from their sound principle of reserve, and mingle with their negro and coloured populations? In the course of time they would become so deteriorated that their states would probably be reconquered and regained by the aborigines whom they have expelled, and who would then be their superiors. But though nature will never ultimately permit this theory of natural equality to be practised, the preaching of this dogma has already caused much mischief, and may occasion much more. The native tendency of the Jewish race, who are justly proud of their blood, is against the doctrine of the equality of man. They have also another characteristic, the faculty of acquisition. Although the European laws have endeavoured to prevent their obtaining property, they have nevertheless become remarkable for their accumulated wealth. Thus it will be seen that all the tendencies of the Jewish race are conservative. Their bias is to religion, property, and natural aristocracy; and it should be the interest of statesmen that this bias of a great race should be encouraged, and their energies and creative powers enlisted in the cause of existing society.

But existing society has chosen to persecute this race which should furnish its choice allies, and what have been the consequences?

They may be traced in the last outbreak of the destructive principle in Europe. An insurrection takes place against tradition and aristocracy, against religion and property. Destruction of the Semitic principle, extirpation of the Jewish religion, whether in the Mosaic or in the Christian form, the natural equality of man and the abrogation of property, are proclaimed by the secret societies who form provisional governments, and men of Jewish race are found at the head of every one of them. The people of God co-operate with atheists; the most skilful accumulators of property ally themselves with communists; the peculiar and chosen race touch the hand of all the scum and low castes of Europe! And all this because they wish to destroy that ungrateful Christendom which owes to them even its name, and whose tyranny they can no longer endure.

When the secret societies, in February 1848, surprised Europe, they were themselves surprised by the unexpected opportunity, and so little capable were they of seizing the occasion, that had it not been for the Jews, who of late years unfortunately have been connecting themselves with these unhallowed associations, imbecile as were the governments the uncalled for outbreak would not have ravaged Europe. But the fiery energy and the teeming resources of the children of Israel maintained for a long time the unnecessary and useless struggle. If the reader throws his eye over the provisional governments of Germany, and Italy, and even of France, formed at that period, he will recognise everywhere the Jewish element. Even the insurrection, and defence, and administration of Venice, which, from the resource of statesmanlike moderation displayed, commanded almost the respect and sympathy of Europe, were accomplished by a Jew—Manini, who by the bye is a Jew who professes the whole of the Jewish religion, and believes in Calvary as well as Sinai, 'a converted Jew', as the Lombards styled him, quite forgetting, in the confusion of their ideas, that it is the Lombards who are the converts—not Manini.

Thus it will be seen that the persecution of the Jewish race has deprived European society of an important conservative element and added to the destructive party an influential ally. Prince Metternich, the most enlightened of modern statesmen, not to say the most intellectual of men, was, though himself a victim of the secret societies, fully aware of these premises. It was always his custom, great as were the difficulties which in so doing he had to encounter, to employ as much as possible the Hebrew race in the public service. He could never forget that Napoleon in his noontide hour had been checked by the pen of the greatest of political writers; he had found that illustrious author as great in the cabinet as in the study; he knew that no one had more contributed to the deliverance of Europe. It was not as a patron, but as an appreciating and devoted friend, that the high chancellor of Austria appointed Frederick Gentz secretary of the congress of Vienna—and Frederick Gentz was a child of Israel."[221]

In her autobiography *My Life*, Golda Meir, like Adolf Hitler,[222] could not understand why the British refused to allow Jews to emigrate to Palestine during the Nazi régime—one should also note in this context that British and American Jews prevented the exodus of Jews from the Continent to England and the United States. Meir tries very hard to blame the Palestinians, the Germans and the British for all the horrors that befell the Jews of Europe, but her pangs of guilt

---

221. B. Disraeli, *Lord George Bentinck: A Political Biography*, Chapter 24, Third Revised Edition, Colburn, (1852), pp. 494-499.
222. A. Hitler, "My Political Testament", English translation in: R. Payne, *The Life and Death of Adolf Hitler*, Praeger Publishers, New York, (1973), pp. 589-591.

for her own actions reveal themselves in her constant need to justify herself and to try to explain away the shared and greater guilt of the Zionists. She even justifies the Holocaust as the only means that would accomplish a "Jewish State". However, the only real obstacle to the formation of a lasting "Jewish State" was the reality that the vast majority of Jews did not want to live in such a racist State. Many of those Jews under Nazi persecution who emigrated to Palestine returned to Europe in disgust.

Meir headed the Jewish Agency and missed an opportunity to save Jewish lives at the Evian Conference in July of 1938. Some believe that Zionists sabotaged this effort, because they believed there was no gain for them to be had from shuffling Jews who had been assimilating in Germany to a new destination like America or Britain, where they would again have the opportunity to assimilate. Many Zionists had no concern for Jewish lives or Jewish deaths unless they Jews were sent directly to Palestine, thereby furthering Zionist ambitions. Jewish tradition held that Jews who assimilated ought to be killed and if a Holocaust would restore Jewish fear in God, so much the better—in the view of many Zionists. Meir wrote,

> "I suppose I must have tried a thousand times since 1939 to explain to myself—let alone to others—just how and why it happened that during the very years that the British stood with so much courage and determination against the Nazis, they were also able (and willing) to find the time, energy and resources to fight so long and as bitter a war against the admittance of Jewish refugees from the Nazis to Palestine. But I have still not found any rational explanation—and perhaps there is none. All I know is that the State of Israel might not have come into being for many years if that British war within a war had not been waged so ferociously and with such insane persistence.
>
> As a matter of fact, it was only when the British government decided—in the face of all reason or humanity—to place itself like an iron wall between us and whatever chance we had of rescuing Jews from the hands of the Nazis that we realized that political independence was not something that we could go on regarding as a distant aim: The need to control immigration because human lives depended on such control was the one thing that pushed us into making the sort of decision which might otherwise have waited for much better (if not ideal) conditions. But the 1939 White Paper—those rules and regulations laid down for us by strangers to whom the lives of Jews were obviously of secondary importance—turned the entire subject of the right of the *yishuv* to govern itself into the most pressing and immediate need that any of us had ever known. And it was out of the depth of this need, essentially, that the State of Israel was founded, only three years after the end of the war."[223]

---

223. G. Meir, *My Life*, G. P. Putnam's Sons, New York, (1975), p. 162.

Zionists were by no means as innocent as Golda Meir would have us believe, nor were British Gentiles, as a group, responsible for what happened. British Jews sabotaged the efforts of German Jews to flee to safety in England, because the Zionists wanted the assimilationist Jews to suffer and die so that the Jewish remnant that eventually moved to Palestine would remain in the country out of fear. They also insisted that any Jewish exodus that took place must force the fleeing Jews to Palestine and nowhere else. The anti-Zionist Jews feared the emigration of large numbers of Jews to the West would provoke anti-Semitism in their home countries, and so the Jews obstructed the emigration of other Jews seeking sanctuary. British Jews, not a small number of them the former Sephardic Jews of Spain, had little love for Russian, or even German, Ashkenazi Jews.

Zionists had long been committing acts of terrorism against the British and the British had no legitimate reason to believe that a "Jewish State" in Palestine was in their, or anyone else's, best interests. That said, many British Gentiles were duped, or bought, into embracing the Zionist cause. Most British Gentiles and British Jews, who tended to be anti-Zionists, believed that a Jewish nation in the Middle East would inflame Moslem passions against England and jeopardize British interests in the region and her trade route to her Asian colonies. The Zionists determined that both British and French interests in the region had to be destroyed by the Nazis and the Imperial Japanese, before the Jews could take Palestine (even Greater Israel) from the Palestinians—and the British—and the French.

The Second World War accomplished many things for the Zionists. It frightened Jews into accepting Zionism. It segregated Jews and taught them the skills needed to live in an agrarian country. It killed off weak Jews and anti-Zionist Jews. It largely destroyed British and French Imperial interests in Asia, thereby lessening their incentive to remain in Palestine and Greater Syria, which they had wanted to keep for themselves as an allegedly important trade route to their colonies.

It took the Second World War, the Holocaust, the independence of India from Great Britain and the creation of Pakistan, as well as pervasive corruption both clerical and profane to overcome the political and religious obstacles to Zionism which remained after the First World War. The Jews used the French under Napoleon, and then the British in the First World War, to chase the Turks out of Palestine and Greater Syria. The Jews lured the French and the British into the region by leading them to believe that a route to their East Asian colonies was vitally important to their national interests.

The Jews created the illusion that only Jews could be the Europeans' friends in the Middle East to secure this route, while Moslems could not. The opposite was true as both the French and the British soon learned after the First World War. When the Turks were finally forced out of Palestine and Greater Syria, the French and British almost went to war over who would control this region, into which they had been led by the Jews. The Jews then felt a need to destroy the French and the British Imperial interests in Asia. The Jews accomplished this goal in the Second World War with their Zionist National Socialists, with the

Nazis; and with their old friends, the Imperial Japanese. Zionist Jews murdered one hundred million people in two world wars in order to create a racist "Jewish State" in Palestine, which would house one to five million Jews in a place where they did not want to live. They murdered another hundred million through Communism.[224]

In 1921, Boris Brasol told of the Zionists' plan in 1920 to create a Socialist German army that would crush British Imperialism and secure Palestine for the Jews, and note that this army became the Nazi army, an army Walther Rathenau began to build in cooperation with the Bolsheviks in 1922 with the Rappallo Treaty (Poale-Zion were Russian Jewish Communist Zionists),

> "Mr. Eberlin, a Jew himself, and one of the foremost leaders of the Poale-Zionist movement, in a book recently published in Berlin, entitled 'On the Eve of Regeneration,' stated:
>
> 'The foreign policy of England in Asia Minor is determined by its interests in India. There was a saying about Prussia that she represents the army with an admixture of the people. About England it could be said that she represents a colonial empire with a supplement of the metropolis... . It is obvious that England desires to use Palestine as a shield against India. This is the reason why she is feverishly engaged in the construction of strategic railroad lines, uniting Egypt to Palestine, Cairo to Haifa, where work is started for the construction of a huge port. In the near future Palestine will be in a position to compete with the Isthmus of Suez, which is the main artery of the great sea route from the Mediterranean to the Indian Ocean.'[*Footnote:* Translation from Russian, 'On the Eve of Regeneration,' by I. Eberlin, pp. 129, 130, Berlin, 1920.]
>
> But this Poale-Zionist goes a step farther when he asserts that:
>
> 'It is only Socialism attainted in Europe which will prove capable of giving honestly and without hypocrisy Palestine to the Jews, thus assuring them unhampered development... The Jewish people will have Palestine only when British Imperialism is broken.'"[225]

The Second World War unhitched England from the East and largely destroyed British Imperialism. The Zionists deliberately caused those events and created those circumstances. The lost lives and misery were a deliberate human sacrifice the Zionists made to their Jewish God.

One group of Zionists openly fought against the British and called for an alliance of the Zionists with the Nazis. Francis R. Nicosia has demonstrated that

---

224. S. Courtois, *et al.*, *The Black Book of Communism : Crimes, Terror*, Repression, Harvard University Press, Cambridge, Massachusetts, (1999).
225. B. L. Brasol, *The World at the Cross Roads*, Small, Mayhard & Co., Boston, (1921), pp. 371-379.

the Nazis were not only anti-assimilationists, but were also very pro-Zionist.[226] Michael Bar-Zohar wrote in his book *Ben-Gurion: The Armed Prophet*,

> "The danger soon became a reality. Many were unable to distinguish between the British Government and the British people, and when war broke out, the extremists adopted radical methods. Supporters of Abraham Stern, who dreamed of a Kingdom of Israel extending from the Nile to the Euphrates, fired the first shots against the British. They even committed the unpardonable crime of recommending an alliance with Nazi Germany, against Britain. When the British shot Stern, his gang avenged him by bomb attacks. These men were few in number and represented a very small part of the *Yishuv*, but their terrorist activities began a new, violent phase in the struggle against the British, a phase which was to lead to open warfare between various factions and groups in Palestine, when Jew fought against Jew and disaster almost came to the Zionist cause."[227]

David Ben-Gurion showed his utter disregard for the value of Jewish life,

> "If I knew that it would be possible to save all the children in Germany by bringing them over to England, and only half of them by transporting them to *Eretz Yisrael*, then I would opt for the second alternative. For we must weigh not only the life of these children, but also the history of the People of Israel."[228]

Zionists obstructed efforts to rescue Jews by not allowing them into countries like England and America. The Zionists wanted to ensure that the Jews felt that the only country that would receive them was Palestine and that the only community that would welcome them was the "Yishuv"—but only after the undesirable (in the minds of the Zionists) "7 million Jews" had been murdered with the approval, if not the active planning, of the Zionist Jews.[229]

According to Johannes Buxtorf in 1603, Jewish authors had long ago planned the decimation of their own people and had planned that the rest of the world should turn a blind eye to the injustice and murder. Buxtorf recounts that 14th Century Jewish author Machir of Toledo's (this is perhaps a pseudonym and the work may have been fabricated by Turkish Jews) *Avkat Rokhel*, Constantinople/Istanbul, (1516), states:

---

226. F. Nicosia, *The Third Reich and the Palestine Question*, University of Texas Press, Austin, (1985), pp. 52-60.
227. M. Bar-Zohar, *Ben-Gurion: The Armed Prophet*, Prentice-Hall, Englewood Cliffs, New Jersey, (1967), p. 68.
228. D. Ben-Gurion, quoted in: Y. Gelber, "Zionist Policy and the Fate of European Jewry (1939-1942), *Yad Vashem Studies*, Volume 13, Martyrs' and Heroes Remembrance Authority, Jerusalem, (1979), pp. 169-210, at 199.
229. T. Segev, *The Seventh Million: The Israelis and the Holocaust*, Hill and Wang, New York, (1993), pp. 99-101.

"The sixth miracle, God shall permit the kingdom of *Edom* (to whit that of the Romans) to bear rule over the whole world. One of whose Emperours shall reign over the whole earth nine moneths, who shall bring many great kingdoms to desolation, whose anger shall flame towards the people of Israel, exacting a great tribute from them, and so bringing them into much misery and calamity. Then shall Israel after a strange manner be brought low and perish, neither shall they have any helper: of this time *Esay* [Isaiah] prophesied, {Esa. 59.16.} *And he saw that there was no man, and wondered that there was no intercessor: therefore his arm brought salvation unto him.* After the expiration of these nine moneths, God shall send the *Messias son of Joseph*, who shall come of the stock of *Joseph*, whose name shall be *Nehemiah*, the son of *Husiel*. He shall come with the stem of Ephraim, *Benjamin* and *Manasses*; and with one part of the sons of *Gad*. As soon as the Israelites shall hear of it, they shall gather unto him out of every City and nation, as it is written: {Jer. 3.14.} *Turn ye backsliding children saith the Lord, for I will reign over you, I will take you one of a City, and two of a tribe, and bring you to Sion.*

Then shall *Messias* the son of *Joseph*, make great war against the king of *Edom*, or the *Pope of Rome*, and being conqueror shall kill a great part of his army, and also cut the throat of the king of *Edom*, make desolate the Roman Monarchie, bring back some of the holy vessels to Jerusalem, which are treasured up in the house of *Ælianus*. Moreover the king of Egypt shall enter into league with Israel, and shall kill all the men inhabiting about Jerusalem, Damascus, and Ascalon: which thing once noised over the whole earth, a horrid dread and astonishment shall overwhelm the inhabitants thereof."[230]

Nothing could put greater fear into apostate and anti-Zionist Orthodox Jews than the fact that almost no one interceded to rescue the assimilating Jews of Europe from the Nazi régime. The horrific indifference of the world to the mistreatment, degradation, humiliation and murder of Jews was a key factor in establishing the will of formerly non-Zionist, or anti-Zionist, Jews around the world to found the "Jewish State" in Palestine. Without the Nazis, and without the indifference of heads of state to the plight of Europe's Jews, there would have been no Israel. Christopher Sykes wrote,

"[...]Zionist leaders were determined at the very outset of the Nazi disaster

---

230. J. Buxtorf, *Synagoga Judaica: Das ist Jüden Schul ; Darinnen der gantz Jüdische Glaub und Glaubensubung... grundlich erkläret*, Basel, (1603); as translated in the 1657 English edition, *The Jewish Synagogue: Or An Historical Narration of the State of the Jewes, At this Day Dispersed over the Face of the Whole Earth*, Printed by T. Roycroft for H. R. and Thomas Young at the Three Pidgeons in Pauls Church-Yard, London, (1657), pp. 319-320.

to reap political advantage from the tragedy."²³¹

Zionist leaders were planning these events thousands of years before the Nazi disaster. After the Nazi disaster, Zionist Matin Buber wrote in 1958,

"Effects of Hitlerism

This organic phase of the settlement in Palestine went on till the days of Hitler. It was Hitler who brought Jewish masses to Palestine, not selected people who felt that here they must fulfill their lives and prepare the future. So, selective organic development was replaced by mass immigration and the indispensable necessity to find political force for its security. This was the hour when my great friend, the late Judah Leib Magnes, and I, and other friends felt that we must state clearly our own proposals. *But the majority of the Jewish people preferred to learn from Hitler rather than from us.* Hitler showed them that history does not go the way of the spirit but the way of power, and if a people is powerful enough, it can kill with impunity as many millions of another people as it wants to kill. This was the situation that we had to fight."²³²

Hannah Arendt wrote in her book *Eichmann in Jerusalem: A Report on the Banality of Evil*,

"Of greater importance for Eichmann were the emissaries from Palestine, who would approach the Gestapo and the S.S. on their own initiative, without taking orders from either the German Zionists or the Jewish Agency for Palestine. They came in order to enlist help for the illegal immigration of Jews into British-ruled Palestine, and both the Gestapo and the S.S. were helpful. They negotiated with Eichmann in Vienna, and they reported that he was 'polite,' 'not the shouting type,' and that he even provided them with farms and facilities for setting up vocational training camps for prospective immigrants. ('On one occasion, he expelled a group of nuns from a convent to provide a training farm for young Jews,' and on another 'a special train [was made available] and Nazi officials accompanied' a group of emigrants, ostensibly headed for Zionist training farms in Yugoslavia, to see them safely across the border.) According to the story told by Jon and David Kimche, with 'the full and generous cooperation of all the chief actors' (*The Secret Roads: The 'Illegal' Migration of a People, 1938-1948*, London, 1954), these Jews from Palestine spoke a language not totally different from that of Eichmann. They had been sent to Europe by the communal

---

231. K. Polkehn, "The Secret Contacts: Zionism and Nazi Germany, 1933-1941", *Journal of Palestine Studies*, Volume 5, Number 3/4, (Spring-Summer, 1976), pp. 54-82, at 58; citing C. Sykes, *Crossroads to Israel*, London, (1965); *Kreuzwege nach Israel; die Vorgeschichte des jüdischen Staates*, C. H. Beck, München, (1967), p. 151.
232. M. Buber, "Old Zionism and Modern Israel", *Jewish Newsletter*, Volume 14, Number 11, (2 June 1958), front page.

settlements in Palestine, and they were not interested in rescue operations: 'That was not their job.' They wanted to select 'suitable material,' and their chief enemy, prior to the extermination program, was not those who made life impossible for Jews in the old countries, Germany or Austria, but those who barred access to the new homeland; that enemy was definitely Britain, not Germany. Indeed, they were in a position to deal with the Nazi authorities on a footing a mounting to equality, which native Jews were not, since they enjoyed the protection of the mandatory power; they were probably among the first Jews to talk openly about mutual interests and were certainly the first to be given permission 'to pick young Jewish pioneers' from among the Jews in the concentration camps. Of course, they were unaware of the sinister implications of this deal, which still lay in the future; but they too somehow believed that if it was a question of selecting Jews for survival, the Jews should do the selecting themselves. It was this fundamental error in judgment that eventually led to a situation in which the non-selected majority of Jews inevitably found themselves confronted with two enemies—the Nazi authorities and the Jewish authorities. As far as the Viennese episode is concerned, Eichmann's preposterous claim to have saved hundreds of thousands of Jewish lives, which was laughed out of court, finds strange support in the considered judgment of the Jewish historians, the Kimches: 'Thus what must have been one of the most paradoxical episodes of the entire period of the Nazi regime began: the man who was to go down in history as one of the arch-murderers of the Jewish people entered the lists as an active worker in the rescue of Jews from Europe.'"[233]

7.8 Documented Collaboration Between the Palestinian Zionists and the Zionist Nazis

History records numerous well-documented instances where Zionist leaders, like Rudolf Kastner who assisted in the deportation of one-half-million Jews to concentration camps,[234] collaborated with Zionist Nazi leaders, including Adolf Eichmann, to help them control mass Jewish populations allegedly destined for extermination, in order to save comparatively scant numbers of prominent Zionist Jews—an act which some allege was pardoned by the Israeli Government after the war so as to prevent an investigation into the broader collaboration between Zionists and Nazis in the persecution of Jews.[235] Indeed, Adolf Eichmann—who was of Jewish descent—called himself a Zionist in 1939 in a conversation with Anny Stern,

---

233. H. Arendt, *Eichmann in Jerusalem: A Report on the Banality of Evil*, Viking, New York, (1963), pp. 55-56; in the revised 1964 edition at pp. 60-61.
234. B. Kimmerling, "Israel's Culture of Martyrdom", *The Nation*, (10 January 2005).
235. *See, for example:* B. Hecht, *Perfidy*, Messner, New York, (1961).

"'Are you a Zionist?' Adolph Eichmann, Hitler's specialist on Jewish affairs, asker her. 'Jawohl,' she replied. 'Good,' he said, 'I am a Zionist, too. I want every Jew to leave for Palestine.'"[236]

There were many Zionists in Palestine who placed the acquisition of land from Palestinians above saving European Jewish lives during the Holocaust.[237] Herzl actively conspired with the Sultan of Turkey to cover up the atrocities committed against Armenian Christians in Herzl's efforts to acquire Palestine and force the expulsion of Jews from other nations of the world and drive them into Palestine. Many prominent and highly respected Jews have, over the course of many years, expressed concern and outrage over the alliance of Zionists and anti-Semites during the Hitler régime.

Though certainly not an endorsement of racism, anti-Semitism or Nazism, Samuel Landman's statements in 1936 evince that some Zionists saw an opportunity to forward their agenda as Hitler's persecution of Jews escalated. This disincentive to fight Hitler directly and with all available means, due to a wish to promote Zionism among reluctant, especially assimilationist, Jews before the Holocaust began, is highly troubling; especially so when it comes from English sources. England did little to combat Nazism and prepare for war. It is also disturbing to note that Jews were among those who most strongly opposed the emigration of German Jews out of Germany. Landman, and every sensible person in the world, should have been calling for England to take immediate action against Nazism and to absorb Jews who wished to leave Germany or were forced out. Landman wrote,

"The rise of Hitler to power in Germany, with its ruthless forms of anti-Semitism, has driven home the Zionism of Herzl and given a tremendous impetus to Jewish national feeling all over the world. A few years ago, the view, adopted by Sir Herbert Samuel in 1921, that a smallish Jewish model settlement in Palestine living on healthy national lines would provide spiritual sustenance for the vast majority of Jewry outside Palestine still had a good few adherents, but to-day, German anti-Semitism and its repercussions in other lands, has all but given this doctrine its *coup de grâce*. Every Jew now sees clearly that without a physical and political as well as a spiritual centre, Jewry stands very little chance of survival. This conviction has spread much more rapidly than certain Zionist leaders, who have lost touch with the masses, realise. The Jewish land hunger has grown immeasurably and the Jewish masses feel that Palestine without Transjordan is far too small for the urgent and imperative need of Jewish emigration. Transjordan was originally part of the mandated territory of Palestine to which the Jewish National Home applied. Hence one of the other main

---

236. L. Dickstein, "Hell's Own Cookbook", *The New York Times*, Book Review Section, (17 November 1996), p. 7.
237. M. Shonfeld, *The Holocaust Victims Accuse: Documents and Testimony on Jewish War Criminals*, Neturei Karta of U.S.A., Brooklyn, (1977).

points in the platform of the new Zionist Organisation is the opening of Transjordan to Jewish immigration. [***] The British Empire can afford to wait or hasten slowly; but it will be conceded that in their tragic plight the choice before Jewry is either speedily to rebuild Palestine or slowly to perish in the Diaspora. The words of the traditional Jewish toast—"Next year in Jerusalem" (Leshana Habaa Birushalayim)—are therefore no longer conventional words, but inspiriting and instinct with meaning and action and must assuredly appeal to the sense of humanity and fair play of the British Government and people."[238]

Political Zionists wanted *Ostjuden* to emigrate to a "Jewish homeland", not to England and America. Some assimilated American Jews had long opposed the immigration of more Jews to America for fear it would cause anti-Semitism. Richard Gottheil stated in 1898,

"They must feel, for example, that a continual influx of Jews who are not Americans is a continual menace to the more or less complete absorption for which they are striving."[239]

The New York Times reported on 20 September 1920 on page 16,

"F. Warburg Seeks to Check Exodus Here of Jews in Europe

PARIS, Sept. 18.—Felix M. Warburg of New York, Chairman of the Joint Distribution Committee for American Jewish Relief Funds, who is here, is endeavoring to impress Jewish leaders in Europe with the necessity of discouraging European Jews from flocking to the United States, in order to keep Jewish emigration within reasonable limits.

In this connection Mr. Warburg has conferred with a number of leading Jews in Paris, including Nahum Sokolow, head of the Jewish Delegations Committee."

Albert Einstein wrote to Max Born on 22 March 1934 that the same impediments Western European Jews had put in place against the emigration of Eastern European Jews during the Pogroms were now being instituted against German Jews by the Jews of America, France and England,

"It is particularly unfortunate that the satiated Jews of the countries which have hitherto been spared cling to the foolish hope that they can safeguard themselves by keeping quiet and making patriotic gestures, just as the

---

238. S. Landman, *Great Britain, the Jews and Palestine*, New Zionist Press (New Zionist Publication Number 1), London, (1936), pp. 12-13, 15.
239. R. Gottheil, "The Jews as a Race and as a Nation", *The World's Best Orations*, Volume 6, F. P. Kaiser, St. Louis, (1899), pp. 2294-2298, at 2296.

German Jews used to do. For the same reason they sabotaged the granting of asylum to German Jews, just as the latter did to Jews from the East. This applies just as much in America as in France and England."[240]

The Zionists obstructed the migration of Jews away from Hitler to any sanctuary other than Palestine, allegedly in the belief that this meant certain death and eventually in the knowledge that Jews could not emigrate to Palestine. They wanted the Jews to feel that no other country would allow Jews in their borders and that no other people would want them than Zionist Palestinian Jews. The Zionists used their strong and powerful influence to bring this fate upon the helpless Jews of Europe.[241]

The Nazis were eager to expel Jews and the only reason Jews could not escape Continental Europe was because other Jews stood in their way. Hitler was interviewed in the *Staatszeitung* of New York and stated,

"Why does the world shed crocodile's tears over the richly merited fate of a small Jewish minority? But what happened to the conscience of the world when millions in Germany were suffering from hunger and misery? I ask Roosevelt, I ask the American people: Are you prepared to receive in your midst these well-poisoners [*Brunnenvergifter*] of the German people and the universal spirit of Christianity? We would willingly give everyone of them a free steamer-ticket and a thousand-mark note for travelling expenses, if we could get rid of them. Am I to allow thousands of pure-blooded Germans to perish so that all Jews may work, live, and be merry in security while a nation of millions is a prey to starvation, despair, and Bolshevism?"[242]

Hitler's 7 April 1933 speech to the "Doctors' Union" was paraphrased in *The Speeches of Adolf Hitler April 1922-August 1939*, Volume 1, Howard Fertig, New York, (1969), page 728, as follows:

"*He said that* America of all countries had the least ground to object to these measures. America's own Immigration Laws had excluded from admission those belonging to races of which America disapproved, while America was by no means prepared to open the gates to Jewish 'fugitives' from Germany. 'As a matter of fact the Jews in Germany had not a hair of their heads rumpled.'"

The New York Times published an interview with Hitler by Anne O'Hare

---

240. Letter from A. Einstein to M. Born of 22 March 1934, in M. Born, *The Born-Einstein Letters*, Walker and Company, New York, (1971), pp. 121-122.
241. T. Segev, *The Seventh Million: The Israelis and the Holocaust*, Hill and Wang, New York, (1993), pp. 99-101.
242. A. Hitler, *The Speeches of Adolf Hitler April 1922-August 1939*, Volume 1, Howard Fertig, New York, (1969), pp. 727-728. "Cited from Fritz Seidler, *The Bloodless Pogrom*. London, 1934."

McCormick on 10 July 1933 on the front page extending to page 6. Hitler disclosed his revolutionary and Marxist ideals—Hitler admired Henry Ford for removing class distinctions with his Model T and hinted at the *Volkswagen*. He also stated,

> "'As to the 'persecuted' Jews, whom you see peacefully walking in the streets and dining in all the best cafés in Berlin,' he continued, 'I would be only too glad if the nations which take such an enormous interest in Jews would open their gates to them.
>
> 'It is true we have made discriminatory laws, but they are directed not so much against the Jews as for the German people, to give equal economic opportunity to the majority.
>
> 'You say the Jews suffer, but so do millions of others. Why should not the Jews share the privations which burden the entire nation?
>
> You must remember our fight is not primarily against the Jews as such but against the Communists and all elements that demoralize and destroy us. When I proceed against a Communist, I do not ask if he is a Saxon or a Prussian. What I mean is that I cannot spare a Communist because he is a Jew."

On 24 October 1933, Hitler delivered a speech in the *Sportpalast* in Berlin,

> "In England people assert that their arms are open to welcome all the oppressed, especially the Jews who have left Germany. England can do this! England is big, England possesses vast territories. England is rich. We are small and overpopulated, we are poor and without any possibility for living. But it would be still finer if England did not make her great gesture dependent on the possession of £1,000—if England should say: Anyone can enter—as we unfortunately have done for thirty or forty years. If we too had declared that no one could enter Germany save under the condition of bringing with him £1,000 or paying more, then to-day we should have no Jewish question at all. So we wild folk have once more proved ourselves to be better humans—less perhaps in external protestations, but at least in our actions! And now we are still as generous and give to the Jewish people a far higher percentage as their share in possibility for living than we ourselves possess."[243]

American Zionists, who sponsored the emigration of Eastern European Jews through the 1920's, had come to despise Russian ("red assimilationist") Jews in the 1930's. At this time in Russia, the man behind Stalin's genocide and anti-Semitism was an alleged "self-hating Jew",[244] Lazar Moiseyevich Kaganovich.

---

[243]. A. Hitler, *The Speeches of Adolf Hitler April 1922-August 1939*, Volume 1, Howard Fertig, New York, (1969), pp. 729-730.
[244]. S. Kahan, "Preface", *The Wolf of the Kremlin*, William Morrow and Company, Inc., New York, (1987).

American Communists, many, if not most, of whom were ethnic Jews, largely turned a blind eye to these atrocities, which cost tens of millions of Gentile lives. Kaganovich may well have been a Zionist who wanted to both punish assimilatory Jews ("red assimilationist") and develop in them a keen interest in Palestine. Kaganovich, perhaps the bloodiest mass murderer in history, was the power behind the throne of the Stalinist Regime. Kaganovich directed the genocide of the Ukrainians, as well as "Stalin's purges" and anti-Semitic campaigns.

Perhaps the most compelling evidence that Soviet anti-Semitism was a ploy meant to force reluctant, assimilating Jews into Zionism against their will, was the fact that the most virulent anti-Semitic purges began after the failed attempt to create a "Jewish State" in the far Eastern regions of the Soviet Union, the Jewish Autonomous Oblast in Khabarovsk Krai in the districts of Birobidzhansky, Leninsky, Obluchensky, Oktyabrsky and Smidovichsky.[245] This plan failed, in part, due to the interference of some Zionist Socialists, who insisted that Palestine was the Jews' national home. An even earlier attempt to found a "Jewish State" in Russia in the districts of Homel, Witebsk and Minsk,[246] also failed, largely due to a lack of Jewish interest. The Zionists insisted that anti-Semitism alone could force the Jews to segregate. When the Zionists put Hitler in power, they had the needed impetus to force Jews to flee Europe and the Zionists attempted to steal Chinese territory for a "Jewish homeland" with the help of the Imperial Japanese under the "Fugu Plan".[247] Zionist Jews sought to establish a "Jewish State" in China, which had been taken over by the Imperial Japanese whom the Jews had been financing since the days when Jacob Schiff loaned them $200,000,000.00 in the Russo-Japanese War. The Zionists used the Imperial Japanese to destroy the Chinese government in preparation for the formation of a Jewish nation in China under the "Fugu Plan" in Manchuria or Shanghai. The Jews even promoted the *Protocols of the Learned Elders of Zion* to the Japanese as evidence as to how powerful they were. The "Fugu Plan" failed to attract enough Jews, even under Nazi pressure, and die hard Zionists wanted Palestine. The Zionists then arranged for war between the United States and Japan. When America declared war on Japan, Hitler, seemingly inexplicably, declared war on the United States ensuring the ultimate defeat of Germany. Hitler also went to war with the Soviets, which gave him access to large numbers of Jews the Zionists could then segregate and ready for deportation to Palestine.

American Zionists took Hitler's rise to power as an opportunity to promote

---

245. "Jews", *Great Soviet Encyclopedia: A Translation of the Third Edition*, Volume 2, Macmillan, New York, (1973), pp. 292-293, at 293.
246. I. Zangwill, "Is Political Zionism Dead? Yes", *The Nation*, Volume 118, Number 3062, (12 March 1924), pp. 276-278, at 276.
247. M. Tokayer and M. Swartz, *The Fugu Plan: The Untold Story of the Japanese and the Jews During World War II*, Paddington Press, New York, (1979). D. Goodman and M. Miyazawa, *Jews in the Japanese Mind: The History and Uses of a Cultural Stereotype*, Free Press, New York, (1995).

Jewish nationalism, force Jewish ethnocentricism, and consolidate Zionist-Jewish power in the United States to the detriment of Europe's Jews—those Jews the prominent and well read European Zionist Jakob Klatzkin had said had "seceded" from being Jews,

> "A Jew who no longer wishes to belong to the Jewish people, who betrays the covenant and deserts his fellows in their collective battle for redemption, has thereby abandoned his share in the heritage of the past and seceded from his people. [***] LET US ASSUME that the Galut can survive and that *total* assimilation will not inevitably follow the abandonment of religion. Nonetheless we must assert: The Judaism of the Galut is not worthy of survival. [***] Perhaps it is conceivable that, even after the disintegration of our national existence in foreign lands, there will yet remain for many generations some sort of oddity among the peoples going by the name—Jew. [***] We must increase self-restrictions and prohibitions, for the sake of protecting our identity and apartness, and we must define boundary after boundary between ourselves and the nations among whom we are assimilating. [***] The very culture that engulfs us so transforms our moral and aesthetic sense that we return to our own people, for we have learned to be sensitive to the crime of assimilation and its consequences."[248]

In 1933, prominent American Zionist Ludwig Lewisohn expressed his bitterness towards Jews who dared to disagree with him. Lewisohn issued an ominous warning to Jews who failed to convert to Zionism:

> "[F]or Jews it has become a matter of life and death for each one and for our whole people. A matter of life and death. For the same sparks from which burst forth this year the foul and fatal German conflagration are smoldering, however hid in ashes, however swept out of sight by sincere gentile good will and by unacknowledged Jewish terror, in every land of the dispersion. [***] Hence millions of Jews must be converted, must achieve a *teshuvah* (repentance), each for himself, in order to consent to the saving of their people, in order to consent to the reconstruction of the Jewish communities of the world. Nothing less than a conversion, nothing less than a profound inner change, nothing less than a broken and a contrite Jewish heart, and yet a heart proud in its brokenness and its contrition, will avail. [***] And our books, instead of becoming instruments toward the auto-emancipation of Jewry and the warding off of a catastrophe, were patronized by a few high-brows whose 'ifs' and 'buts' were stamped out in the year 1933 in blood and dirt. [***] The Polish communities, though less catastrophically stricken, are so oppressed and burdened that leadership cannot be expected from them. The Russian Jews are lost to us in this generation by the device

---

[248]. J. Klatzkin, *Tehumim: Ma'amarim*, Devir, Berlin, (1925); English translation by A. Hertzberg in his, *The Zionist Idea*, Harper Torchbooks, New York, (1959), pp. 316-327, at 317, 322, 325.

of Red assimilation, quite analogous to Prussian assimilation and mass baptism during certain decades of the nineteenth century, or to the processes of any polity which, in the period of consolidation, is willing temporarily to admit that assimilation can proceed to the point of paying. Hence the leadership of world-Jewry outside of Palestine devolves upon American Jewry, and American Jewry, the most populous and powerful in the world today, is also the most ignorant and the one in which the crippling sickness of preoccupation without knowledge is most prevalent... It is a necessity and a duty to be brutal today. It is necessary to be brutal even at the risk of being misunderstood. For, given the precise circumstances that confront us from now on, the Jewish ignorance of American Jewry may prove a disaster of incalculable consequences to all Israel."[249]

Ludwig Lewisohn lived with, and had a homosexual relationship with, poet George Sylvester Viereck.[250] Viereck was reputedly the grandson of Kaiser Wilhelm the First and Edwina Viereck, and was the son of the Marxist Louis Viereck. George Sylvester Viereck was one of the chief pro-German propagandists in America during World War I,[251] defended the Kaiser after World War I, was a devoted friend to Sigmund Freud and promoted Albert Einstein—as well as Adolf Hitler. Just as the poet Ezra Pound propagandized for the Fascists in Italy, Viereck propagandized for the Nazis from the 1920's through the 1940's and served time in prison in America for his pro-Nazi activities. Viereck and Lewisohn remained friends after the Second World War—and the Holocaust.[252] Eustace Mullins stated that Viereck was flattered and pleased when Mullins told Viereck that Viereck had cost Germany victory in both world wars.[253]

Morris Raphael Cohen noted the kinship between the Nazis and Zionists like Lewisohn, in Cohen's critique of Lewisohn's *The Answer; the Jew and the World: Past, Present and Future*, Liveright Publishing Company, New York,

---

249. L. Lewisohn, "A Year of Crisis", in A. Hertzberg, *The Zionist Idea*, Harper Torchbooks, New York, (1959), pp. 488-492. Hertzberg cites: L. Lewisohn, *Rebirth* (editor), New York, (1935), pp. 290-296.
250. Refer to the love letters between Lewisohn and Viereck in the "Ludwig Lewisohn papers, 1903-1980's" at the College of Charleston libraries, Special Collections, Third Floor, Mss 28, Box 1, Folders 1, 3 and 5.
251. "Question Dickinson, Agent of Viereck", *The New York Times*, (18 January 1919), p. 4.
252. N. M. Johnson, "George Sylvester Viereck: Poet and Propagandist", *Books at Iowa*, Number 9, (November, 1968), URL:
http://www.lib.uiowa.edu/spec-coll/Bai/johnson2.htm
and *George Sylvester Viereck: Pro-German Publicist in America, 1910-1945*, Dissertation Thesis (Ph. D.), University of Iowa University of Iowa, Iowa City, Iowa, (1971); **and** *George Sylvester Viereck, German-American Propagandist*, University of Illinois Press, Urbana, Illinois, (1972).
253. Daryl Bradford Smith interview of Eustace Mullins of 25 January 2006, "The French Connection", *GCN LIVE*, http://www.iamthewitness.com.

(1939). Cohen stated in his review of Lewisohn's Zionist book of 1939,

> "Yet the answer, which in accordance with the title this book offers us, is clear enough: remove the Arabs from Palestine and Transjordania, over a million of them, (p. 188), and put in their place a majority of the Jewish population of the world (p. 19). [***] Mr. Lewisohn is indeed aware of the fact that not only will a large part of the Jewish population of the world never go to Palestine but that it will take a long time before all those who wish to go can be transported and find room there. [***] Not only are Mr. Lewisohn's ideas hazy, confused, and disdainful of the facts, but his major premises are indistinguishable from the current anti-scientific racial dogmas which threaten to destroy liberal civilization. [***] [...]Mr. Lewisohn resort[s] to downright misrepresentation of Dr. Boas' position when he says (p. 310) that the latter 'hoped that Jewish babies would develop Indian skulls in America.' It is not necessary to refute this absurd and baseless charge; but it is well to call attention to the fact that neither Boas nor Fishberg ever denied the existence of Jews. What they did show by actual measurement is that there are no discoverable hereditary physical traits common to all Jews which distinguish them from other people. Mr. Lewisohn froths at this because it runs counter to the dogma which he shares with Hitler and Mussolini that cultural traits are inherited in the racial blood and cannot be changed (p. 46). [***] It seems cruel to link such an ardent Zionist as Mr. Lewisohn with Hitler and Mussolini, even ideologically. But the fact is that he does agree with them not only in their dogmatic racial fatalism but also in one of the conclusions that they and others draw from it, and that is that the democratic liberal regime of emancipation and toleration has not only failed but cannot and indeed ought not to succeed. From the way Mr. Lewisohn writes, one would suppose that the emancipation of the Jews from the ghetto was a calamity second only to the destruction of the Jewish Commonwealth. By implication he is committed to the view that one born a Jew cannot enter completely into English, French or German culture, not only because he will not be allowed to, but because it is contrary to fate or God's will. [***] The implication that emancipation is responsible for the anti-Semitism in Poland and Rumania belongs to the same class [of misinformation]."[254]

Zionists promised the Jews of the world that if Jews abandoned their calling to nationalism and refused to embrace Zionism, they would face annihilation in Europe. This had long been a common theme of Zionists and anti-Semities. The "Hamburg Resolutions of the German Social Reform Party" proclaimed in 1899,

> "The strivings of Zionism are a fruit of the antisemitic movement. [***]

---

[254]. M. R. Cohen, "Parkes' *The Jew and His Neighbour* and Lewisohn's *The Answer*", *Reflections of a Wondering Jew*, The Beacon Press, Boston, (1950), pp. 116-123, at 119-123.

Unfortunately [any hope that all Jews will emigrate to Palestine] appears to be infeasible. [***] As such, [the Jewish question] should be solved in common with other nations and result finally in full separation, and—if self-defense demands—in final annihilation [*Vernichtung*] of the Jewish race."[255]

Adolf Hitler and Hans Frank later threatened the Jews of the world that if Jewish leadership "forced another world war", then Jews would face annihilation in Europe. Hitler stated before the Reichstag on 30 January 1939,

"[I want to be a prophet again today:] If international finance Jewry in and outside Europe succeeds in plunging the peoples into another world war, then the end result will not be the Bolshevization of the earth and the consequent victory of Jewry but the annihilation of the Jewish race in Europe."[256]

"Ich will heute wieder ein Prophet sein: Wenn es dem internationalen Finanzjudentum in und außerhalb Europas gelingen sollte, die Völker noch einmal in einen Weltkrieg zu stürzen, dann wird das Ergebnis nicht die Bolschewisierung der Erde und damit der Sieg des Judentums sein, sondern die Vernichtung der jüdischen Rasse in Europa!"[257]

The Jewish Nazi tyrant of Poland, Dr. Hans Frank, stated at a Cabinet Session on 16 December 1941,

"As far as the Jews are concerned, I want to tell you quite frankly, that they must be done away with in one way or another. The Fuehrer said once: should united Jewry again succeed in provoking a world war, the blood of not only the nations which have been forced into the war by them, will be shed, but the Jew will have found his end in Europe"[258]

---

255. English translation in: R. S. Levy, *Antisemitism in the Modern World: An Anthology of Texts*, D. C. Heath and Company, Toronto, (1991), pp. 127-128, at 128.
256. English translation in: R. S. Levy, *Antisemitism in the Modern World: An Anthology of Texts*, D. C. Heath and Company, Toronto, (1991), pp. 222-223, at 223. An alternative translation appears in: "Holocaust", *Encyclopaedia Judaica*, Volume 8, Macmillan, Jerusalem, (1971), col. 852.
257. A. Hitler in M. Domarus, Editor, *Hitler: Reden und Proklamationen, 1932-1945: Kommentiert von einem deutschen Zeitgenossen*, Süddeutscher Verlag, München, (1965), pp. 1057-1058.
258. H. Frank, (16 December 1941), quoted in: *Nazi Conspiracy and Aggression*, Volume 2, United States, Office of Chief of Counsel for the Prosecution of Axis Criminality, Washington, D. C., United States Government Printing Office, (1946), p. 634. *See also:* Y. Arad, Yitzhak, I. Gutman, A. Margaliot, Abraham,Editors, *Documents on the Holocaust: Selected Sources on the Destruction of the Jews of Germany and Austria, Poland, and the Soviet Union*, Yad Vashem in cooperation with the Anti-Defamation League and Ktav Pub. House, Jerusalem, (1981).

Did the crypto-Jewish Zionists Adolf Hitler and Hans Frank mean that they would exterminate the Jews of Europe in death camps, or did they mean that they would deport the Jews of Europe to Palestine as a final solution to the Jewish question? Frank was a long-term Zionist who wanted to segregate the Jews in Polish concentration camps and then ship them to Palestine—not to say that he did not intend to kill off a large percentage of his brethren in the process. In the fall of 1933 in Nuremberg on *Reichsparteitag*, Frank stated that his goal was to secure a "Jewish State",

> "Unbeschadet unseres Willens, uns mit den Juden auseinanderzusetzen, ist die Sicherheit und das Leben der Juden in Deutschland staatlich, reichsamtlich und juristisch nicht gefährdet. Die Judenfrage ist rechtlich nur dadurch zu lösen, dass man an die Frage eines jüdischen Staates herangeht."[259]

World War II began soon after, and resulted in the Bolshevization of half of Germany, Eastern Europe, China, North Korea, and ultimately Indochina, with disastrous consequences for a large segment of humanity. Western Europe came very close to falling under the Boshevists' control and endured the Bolshevization of Nazism for many years. The war was also a victory for the Zionists—in their minds.

The Nazis and the Zionists iterated a common message. Even after Germany had initiated war and invaded Poland, where Jews had been forced to gather, the Jews of Europe did not embrace Zionism. They were then annihilated. Lest the Zionists be confused with visionaries who sought to rescue the Jews of Europe, one must bear in mind the Zionists' all too common disdain for "surplus" and assimilationist Jews and the fact that instead of fighting the Nazis, they too often endorsed and encouraged the views of the Nazis—even offered a military alliance with the Nazis.

The unfortunate Jews of Eastern Europe and Germany were caught between Zionists, who hated their "red assimilation", sponsoring racial and nationalistic mythologies in Germany, the Soviet Union, and elsewhere in collusion with the anti-Semites; and powerful and influential assimilated Jews in France, England and America who feared an increase in anti-Semitism should these Eastern Jews be permitted sanctuary in their lands. The Eastern Jews were chased from place to place and often murdered in cowardly cold blood, with the approval of the Zionists.[260] Though many in positions of power around the world could have done much to help the Jews in danger, most did nothing. Immediately after the Second World War ended, the push for Israel became immensely strong among American Jews who had spent the war in relative safety—just as the political

---

259. H. Frank quoted in H. Kardel, *Adolf Hitler, Begründer Israels*, Verlag Marva, Genf, (1974).
260. T. Segev, *The Seventh Million: The Israelis and the Holocaust*, Hill and Wang, New York, (1993), pp. 99-101.

Zionists had always planned would happen. Israel Zangwill stated in 1911,

> "But whether persecution extirpates or brotherhood melts, hate or love can never be simultaneous throughout the diaspora, and so there will probably always be a nucleus from which to restock this eternal type."[261]

The Zionists caused the Holocaust in the twisted belief that American and British Jews would "restock this eternal type" and that "Red assimilationists" and "rich assimilationists" were unworthy of life. Hitler's threat to annihilate the European Jews occurred shortly before England declared war on Germany and Frank's resolution came shortly after Germany declared war on America.

Before Hitler, there was Alfred Rosenberg. Rosenberg, perhaps a Bolshevik agent of what was to become "The Trust", under the tutelage of Houston Stewart Chamberlain,[262] List and Liebenfels, crafted what was to become the party ideology of the NSDAP. The Zionists created the ideology of the Nazi régime through these men. The Russian Jewess Helena Petrovna Blavatsky gave these men their mystic aryanistic dogmas and mythologies. In 1893, Blavatsky created the dogma behind the adoption of the "Aryan" Swastika they and the Nazis adopted—from her.[263]

One of the architects of political Zionism, Max Nordau, wrote extensively[264] on the *Übermensch* and his role in history and politics (continuing the themes of Nietzsche's *Übermensch* in *Thus Spoke Zarathustra* and Dostoevsky's dialectic and Hegelian *Übermensch* turned evil in the form of Raskolnikov in *Crime and Punishment*). Nordau, while formulating a biologically and physiologically

---

261. I. Zangwill, *The Problem of the Jewish Race*, Judaen Publishing Company, New York, (1914), p. 18; which was first published as an article, "The Jewish Race", *The Independent*, Volume 71, Number 3271, (10 August 1911), pp. 288-295, at 294.

262. *Confer:* J. Stolzing, "Houston Stewart Chamberlain. Zu seinem 70. Geburtstage", *Völkischer Beobachter*, Volume 38, Number 137, (9 September 1925), p. 1. *See also:* J. Goebbels, diary entry of 8 May 1926, *Die Tagebücher von Joseph Goebbels, sämtliche Fragmente, Teil I, Aufzeichnungen 1924-1941, Interimsregister*, München , New York, K.G. Saur, (1987), p. 178. *See also:* G. Schott, *Chamberlain, der Seher des Dritten Reiches; das Vermächtnis Houston Stewart Chamberlains an das deutsche Volk, in einer Auslese aus seinen Werken*, F. Bruckmann München, (1934), p.17. *See also:* H.S. Chamberlains letter to Dr. Boepple of 1 January 1924, L. Schmidt, *Houston Stewart Chamberlain. Auswahl aus seinen Werken*, F. Hirt, Breslau, (1934), p.66. *See also:* J. Köhler, *Wagners Hitler: Der Prophet und sein Vollstrecker*, Blessing, München, (1997), p. 385.

263. H. P. Blavatsky, "The Svastika", *The Secret Doctrine: The Synthesis of Science, Religion, and Philosophy*, Volume 2, Part 1, Stanza 4, The Theosophical Publishing Society, (1893), pp. 103-106.

264. M. Nordau, *Entartung*, C. Duncker, Berlin, (1892-1893); English translation: *Degeneration*, D. Appleton, New York, (1895); **and** *Der Sinn der Geschichte*, C. Duncker, Berlin, (1909); English translation: *The Interpretation of History*, Willey Book Co., New York, (1910); **and** *The Drones Must Die*, G.W. Dillingham Co., New York, (1897); **and** with M. A. Lewenz, *Morals and the Evolution of Man*, Lewenz, Funk and Wagnalls Company, New York, (1922).

based political psychology of the superior man—much in agreement with Hitler's later belief system, adopted the ideologies of Hegel, Schopenhauer and their progeny, while viciously criticizing them. He wrote of "Degeneration" in the arts and philosophy by Wagner and Nietzsche and throughout society— political elements which became fundamental in Nazi culture and science through a direct tranference to Jews.

It was something more than common interest and circumstances which drove Rosenberg (in his many rôles as Nazi party leader, Nazi propagandist, and the creator of National Socialist policy) to attempt to fulfill Herzl's goals of a dramatic rise in international anti-Semitism, the distillation of Jews into segregated groups meant for deportation, and the destruction and punishment of upperclass Jews who had opposed Herzl and whom Herzl had repeatedly threatened, and the creation of a "Jewish State". There was common control of the Zionist, Nazi and Communist movements, with the common goals of wreaking havoc on Europe, destroying the genetic and cultural future of Europeans, and herding up the reluctant Jews of Europe for deportation to Palestine, and killing off weak Jews in order to "improve the bloodline" of Israeli Jews, since Palestine could not in any event house the majority of European Jews.

Zionist Jews had no compunctions about killing off a large percentage of assimilatory European Jewry. Bernard Lazare was one of many Zionist Jews who hated wealthy assimilating Jews and wrote in the late 1890's, "It is obvious that the so-called upper class among western Jews, and especially among French Jews, is in an advanced state of decay."[265] Jakob Stern published a rather famous critique of Herzl's *Der Judenstaat*, in which Stern saw Herzl's book as a Utopian Marxist vision. Stern also ridiculed wealthy "Jewish capitalists" who sought sanctuary in "civilized countries", and noted, in his view, "How little racial and tribal kinship and community of religion prevent Jewish capitalists from exploiting Jewish proletarians, could be witnessed again at the International Socialist Congress a short time ago."[266] We learn from Paul Ehrenfest's correspondence that Zionist Jacques Oppenheim believed that secular Jews were not Jews and that all the problems Jews faced were due to educated and influential Jews who had betrayed the "Jewish masses."[267] Einstein joined the chorus,

"The greatest enemies of the national consciousness and honour of the Jews are fatty degeneration—by which I mean the unconscionableness which

---

[265]. B. Lazare, *Job's Dungheap*, Schocken Books, New York, (1948); quoted in A. Hertzberg, *The Zionist Idea*, Harper Torchbooks, New York, (1959), p. 472.

[266]. J. Stern, "Theodor Herzl, *Der Judenstaat. Versuch einer modernen Lösung der Judenfrage*", *Neue Zeit*, Volume 15, Number 1, (1896-1897), p. 186: English translation in: P. W. Massing, *Rehearsal for Destruction: A Study of Political Anti-Semitism in Imperial Germany*, Howard Fertig, New York, (1967), p. 321.

[267]. Letter from P. Ehrenfest to A. Einstein of 9 December 1919, English translation by A. Hentschel, *The Collected Papers of Albert Einstein*, Volume 9, Document 203, Princeton University Press, (2004), pp. 173-175, at 174.

comes from wealth and ease—and a kind of inner dependence on the surrounding Gentile world which comes from the loosening of the fabric of Jewish society."[268]

Anti-Semite Győző Istóczy issued an anti-Semitic Zionistic appeal to rich Jews. In 1878, Istóczy wrote,

"There is only one means of remedying this great international evil: the Jews must be expelled from Europe. [***] In Palestine the Jews will be in position to create a grand state. [***] The innermost, secret wish of most Jews can now become reality if they can overcome those powerful Jews who have acquired power in Europe and for whom it is so very congenial to rule the world from London, Paris, Berlin, Vienna, and Budapest. I appeal to the oft-mentioned patriotism of the Jews; they can now create their own empire; they will surely become a mightier, more influential state. My sincerest and best wishes will accompany the Jews. May the Jews find this acceptable and cease their continuing efforts to exterminate the Christians."[269]

Roman Dmowski iterated a Polish anti-Semite's view of the struggle between wealthy Western Jews and Zionist Jews in his article *The Jews and the War* of 1924,

"Meanwhile there developed a stubborn battle between generally poor idealists, as the Zionists were, and those representing financial power. Englishmen, Americans, Germans, and Frenchmen of Jewish faith were not thinking of leaving the Parises, Londons, Berlins, and New Yorks, with everything they offered. [They] considered Zionism an absurd fantasy. [***] Palestine was never the fatherland of the Jews because they never had a fatherland, but they made Jerusalem their spiritual center; recovering this center along with controlling Palestine, with its non-Jewish population, is the necessary goal of this new current. Yet, at the same time, [this new current] bid them not to forget that they are supposed to 'possess the earth,' that therefore they must be everywhere, and everywhere gain positions and organize their influences."[270]

---

268. A. Einstein, *The World As I See It*, Citadel, New York, (1993), pp. 105-106.
269. G. Istóczy, "Speech to the Hungarian Parliament of 25 June 1878", published in: W. Marr, *Vom jüdischen Kriegsschauplatz eine Streitschrift*, Second Edition, R. Costenoble, Bern, (1879), pp. 41-43; English translation in: R. S. Levy, *Antisemitism in the Modern World: An Anthology of Texts*, D. C. Heath and Company, Toronto, (1991), pp. 100-103, at 102-103.
270. R. Dmowski, "The Jews and the War"; English translation by J. Kulczycki in R. S. Levy, Editor, J. Kulczycki, translator, *Antisemitism in the Modern World: An Anthology of Texts*, D. C. Heath and Company, Lexington, Massachusetts, Toronto, (1991), pp. 182-189, at 184.

Alfred Rosenberg focused attention in Germany on *The Protocols of the Learned Elders of Zion*, which Gottfried zur Beek, under the *nom de plume* Ludwig Müller von Hausen, had translated into German in 1919 and published in early 1920 as *Die Geheimnisse der Weisen von Zion*. Rosenberg published *Die Protokolle der Weisen von Zion und die jüdische Weltpolitik*, Deutsche Volksverlag, München, (1923), in an effort to generate, and promote extant, anti-Semitism.

Alfred Rosenberg, Hitler's mentor, promoted Zionist programs as the state policy of the Nazis, stating in his *Die Spur des Juden im Wandel der Zeiten* of 1920 that,

"The Jews are recognized as a nation living in Germany. [***] Zionism must be powerfully supported, in order to promote yearly a certain number of German Jews to Palestine or, in general, over the borders."[271]

"Die Juden werden als eine in Deutschland lebende Nation anerkannt. [***] Der Zionismus muß tatkräftig unterstützt werden, um jährlich eine zu bestimmende Zahl deutscher Juden nach Palästina oder überhaupt über die Grenze zu befördern."[272]

Though many authentic anti-Semites distrusted Zionism, both Adolf Hitler and Joseph Goebbels encouraged Zionism,[273] as did *SS* officer Baron Leopold Itz von Mildenstein.[274] Adolf Hitler's ethnologist, Hans Günther, embraced and advocated Zionism in 1923, copying verbatim from the amended Zionist Program of 1897,[275]

"Research has shown time and again that the dispersion of the Jews among Gentile Peoples causes endless unrest, and again and again the racial antagonism of necessity escalates into hatred. Having exposed this is one of the most courageous realizations of Zionism. Zionism has clearly shown that the only dignified settlement of relations would be the removal of the Jews from living among the Gentile nations. The creation of a publically, legally secured homeland for the Jewish people in Palestine seems to now

---

271. A. Rosenberg, *Die Spur des Juden im Wandel der Zeiten*, Fourth Edition, Zentral Verlag der NSDAP Franz Eher Nachfolger, (1939), pp. 152-153; English translation by A. Jacob in: *Eugen Dühring on the Jews*, Nineteen Eighty Four Press, Brighton, England, (1997). pp. 44-45.
272. A. Rosenberg, *Die Spur des Juden im Wandel der Zeiten*, Deutscher Volks-Verlag, München, (1920); as quoted in: *Alfred Rosenberg Schriften aus den Jahren 1917-1921 mit einer Einleitung von Alfred Baeumler*, Hoheneichen-Verlag, München, (1944), pp. 320-321.
273. H. Grimm, *Warum, woher, aber wohin?*, Klosterhaus-Verlag, Lippoldsberg, (1954).
274. J. Boas, "A Nazi Travels to Palestine", *History Today*, Volume 30, Number 1, (January, 1980), pp. 33-38.
275. "The Zionist Programme", *The Jewish Chronicle*, (3 September 1897), p. 13.

be politically attainable."

"Immer wieder zeigt die Betrachtung, dass die Zerstreuung der Juden unter nichtjüdischen Völkern eine endlose Unruhe bewirkt, und immer wieder die Artgegensätze bis zum Hass steigern muss. Dies eingesehen zu haben, ist eine der mutigsten Erkenntnisse des Zionismus. Der Zionismus hat es klar eingesehen, dass einzig die Herauslösung der Juden aus dem Zusammenwohnen mit nichtjüdischen Volkstümern eine würdevolle Klärung der Verhältnisse bedeutet. Die Schaffung einer öffentlich rechtlichen gesicherten Heimstätte für das jüdische Volk in Palästina scheint jetzt politisch erreichbar zu sein."[276]

The infamous "Nuremberg Laws" of 1935 forbade Jews from raising the Reich's flag, but Section 4 specifically granted them the right to display the "Jewish colors", which encouraged Zionist nationalism. The Zionists embraced the Nuremberg Laws, which sponsored the racial segregation they desired and which forbade intermarriage or any sexual relations between "Jews" and "Aryans". At least as early as 1914, Zionist racist Ignatz Zollschan iterated the Nazi goals of concentrating and segregating Russian Jews in order to prevent the assimilation of Jews after emancipation. Zollschan asserted that Jews must choose the ghetto or Zionism, if they wished to perpetuate the "Jewish race". Since the vast majority of Jews did not want to segregate and congregate in Palestine, the Zionists and Nazis collaborated to force Jews into ghettoes. Since a very large percent of Jews in Germany were marrying Gentiles, the Zionists and Nazis collaborated to discourage and eventually outlaw such marriages. Political Zionist Zollschan stated at least as early as 1914,

"II.
The Significance of the Mixed Marriage

What can we say with certainty about the purity of the Jewish race? The answer to this question is of vital importance. For if intermarriage with alien races had in former years played a great role among Jews, it is self-evident that we are not justified in speaking of a Jewish race at all. Are the Jews of to-day really the pure descendants of Abraham, Isaac and Jacob?
Nobody assumes to-day that all the Jews are the direct descendants of the three patriarchs; they are derived from the mingling of various stocks which were, however, essentially varieties of one and the same race.
When in the thirteenth century before the current era the Bedouin tribe of the Habiri, that is to say, the Hebrews, took possession of Palestine, they found there a vast native population, the Ganaanites, Hetites, Getites, Aniorites and Phihistines. During the period of the Judges and Kings, the

---

276. H. F. K. Günther, *Rassenkunde des deutschen Volkes*, J. F. Lehmann, München, (1923), p. 430.

Jewish tribes intermarried with all these nations. Their blood was mingled with that of the nations in whose midst they lived. This slow process of intermixture continued till after the first exile, till the time when the powerful word of Ezra severed all existing marriage connections with foreign nations, and henceforth the purity of the race became the dominating principle.

It is quite gratuitous to enter into a controversy about the exact definition and classification of such nations as the Hittites, Amorites, Philistines and others, to which, in a broader sense, the Egyptians, as well as the Babylonians, Assyrians, Phoenicians and Jews belong. Whether we speak of Semites and Hamites in accordance with the inadequate linguistic methods, or of Semites, Hittites, Amorites and Kushites, we regard these nations as related to one another in the racial sense. Ample anthropological evidence exists for this statement, though naturally it cannot be presented in this lecture.

Many historians are of the opinion that the appearance of Ezra did not put an end to the racial intermixture. They think that also in all subsequent centuries the Jews continued to mingle with the nations of the diaspora, just as in the time before the Babylonian exile. They advocate the theory that the Jews of to-day are the descendants of the heathen proselytes during the Hellenistic period, or the offspring of mixed marriages between the Jews and their surrounding nations during the Christian centuries.

We can to-day assert with certainty that the extent of proselytism has been greatly exaggerated. There can indeed be no doubt that Judaism found numerous adherents among the pagan nations during the Roman and Hellenistic and early Christian periods. We have, however, sufficient reason to assume that those proselytes were only the so-called 'proselytes before the gate,' that is to say, converts who practiced the worship of one God, but were never admitted to circumcision or marriage. They were proselytes who later on embraced Christianity.

And in the times that followed immediately, the policy of discriminating between Jew and Gentile was inaugurated. Hadrian's laws forbidding circumcision were, it is true, revoked by his successor, Antoninus Pius, but it was expressly prohibited to make converts to Judaism. In consequence of this, the formal embracing of Judaism became a punishable crime, and it remained such until quite recent times. Even during the periods when the Jews commanded respect to some extent, the Church took good care that the religious boundary-line should be kept intact. In times of persecution and oppression, no appreciable number of adherents of other religions could have gone over to outlawed Judaism. The bars of the Ghetto formed a reliable dividing wall.

But even if we grant that in some cases a few heathens became Jews in every respect prior to the Christian era, they could have been of no significance. As in the Hellenistic period there already existed millions of Jews, the admixture of foreign blood must have been infinitely small. And this foreign blood was, after all, derived from the kindred nations in 'Syria,

Asia Minor and Egypt.

It may be regarded as certain that proselytism almost entirely ceased since the appearance of European Jewish history. Even the invasion of the Khazars in the eighth century does not alter the fact that during the Middle Ages not much of foreign blood was added to the Jews. For already in the tenth century the empire of the Khazars was confined to a small territory, something like Crimea of to-day, and in the eleventh century it was entirely wiped out. A small remnant of Khazarite Jews are still living in Crimea to-day, and belong to the Karaitic sect. But even if we assume that the entire nation of the Khazars embraced Judaism, and professed that religion for a long time, this admixture would still be a *quantite negligeable* and would not alter the ethnical character of the Jewish race. Moreover, it is doubtful whether this conversion was not confined to the rulers and the ruling classes of the Khazars. We would be losing sight of historical proportion, if we were to infer from the conversion of the Khazars that the Jews have any remarkable admixture of foreign blood.

As far as legal mixed marriages are concerned we know that they actually existed in the times of high material culture, namely, in Egypt during the Hellenistic period and in Spain during the twelfth and thirteenth centuries. But, as is the case now in Europe, where there is a strong leaning towards intermarriage, the offspring of those marriages preponderantly went over to Christianity. Besides this, those early periods quickly passed away owing to the changed political conditions, the reaction of orthodoxy and the decisions of the councils of the Christian Church. Moreover, this movement at that time, in contradistinction to the general spread of intermarriage of to-day, was only confined to one country. Intermarriage with northern nations never took place in former years to any considerable extent.

The Jewish nation accordingly has propagated itself in an essentially pure manner from the time of Ezra until to-day, and for more than two thousand years represents an ethnically peculiar race, which was not diluted by foreign blood. It is self-evident that a few drops of foreign blood must have found their way among the Jews during the long time in the diaspora. But these admixtures were too insignificant to have any essential influence upon the ethnical character of the nation. Thus the Cohanim, who were absolutely excluded from mixed marriages, are typically the same as the other Jews. The state of affairs can best be described in one sentence: A great deal of blood was exported from Jewry, but little indeed was imported from outside. And, consequently, we can assume with certainty, that the blood which flows to-day in the veins of the Jews, is the same as that of two thousand years ago.

That Ezra's commandments, among which is also the one about purity of blood, have been kept for thousands of years, is due to the fact that they claimed to be religious ordinances coming from God. It is the case with all nations that social institutions which are interwoven with, and supported by religion are kept most tenaciously. In addition to this, Ezra's prescriptions owe their strength to the circumstance that they consisted in the practical laws of the cult, and not in theoretical doctrines; and that the Jews, after being scattered among other nations, were forced to social and economic isolation.

The true consideration of this circumstance, indicates the great significance of the solution of the problem of intermarriage in our own times. Economic and social isolation and the power of religious legislation, account for the fact that up till to-day this people did not fall a victim to intermarriage, despite its wanderings among strange nations for the last 2,000 years.

As long as ceremonial religion was a great power in the civilized life of all nations, this influence of religion was easily explained. But nowadays, for reasons which will presently become apparent, this influence upon the great masses is confined to the Ghetto environment. As soon as the Jew leaves this Ghetto environment, and participates in the national industry of his country as a factor of equal rights, and adapts himself to the speech and culture of his native land, he begins to free himself from the power of ceremonial religion. A century of free activity in the world of capital; combined with a secular education, entirely estranged the Jews, in all countries where the system of capital is developed, from their former mode of life. The pressure of changed economic conditions and the scientific materialistic conception of our age, sap the vitality of orthodox Judaism and undermine its foundation.

Now since ceremonial religion on the one hand and economic and social isolation on the other, together with the prohibition of Church and State, were the only reasons why intermarriage with foreign nations did not take place on a larger scale, it necessarily follows that affairs to-day have reached a critical stage. Free legislation in countries where the system of capital is developed, has done away with the economic and political isolation; rationalism has shaken ceremonial religion, and no State nowadays prohibits mixed marriages.

In countries where one or another of these important conditions is not fulfilled, as, for instance, Galicia, Russia, and the Orient, Judaism is still kept alive, though the lot of the masses residing there is by no means to be envied. But in the Occident, and in all countries where the Jews are allowed to develop themselves freely, their lot is the same as that of other nations in a similar situation.

Without exception, all the nations who were compelled to leave their native soil and who never formed a compact majority in any part of the world, but were scattered in small communities, have vanished through intermarriage. And the Jews likewise would be swept away by the immense tide of the human race in the five continents, if all obstacles were removed. As can be easily shown, Jews have always married outside the fold whenever conditions were favorable. But never were conditions which make for the disintegration of Judaism as powerful as to-day. Nations who dwell together always mingle, unless intermarriage is made impossible by outside pressure of law or religion. The Jews nowadays come into contact with other nations, the civil law permits intermarriage, and the authority of religion is beginning to wane. The laws of love and material interests are mightier than all religious barriers, especially when the latter are weakened and enfeebled as they are to-day. The result of these considerations is, that to-day more than ever, Judaism is in danger of being dismembered.

The facts derived from statistics confirm this conclusion in all its details.

The first impulse to abrogate the laws forbidding marriage between Christians and Jews went forth from the French Revolution, and gradually spread from country to country—to Holland, Belgium, Denmark and Scandinavia; to England and the United States; to Germany, Italy and Hungary. It is even permitted in the Balkan States. On the other hand, it is still prohibited to-day in Austria, Russia, Spain and Portugal, and in Mohammedan countries. The most favorable places for mixed marriages are naturally those countries in which Jews have been domiciled for a considerable time and where they have attained prosperity. This is especially the case in the States of Western Europe.

The losses to Judaism in these western countries cannot be numerically ascertained, as there are no statistics in Italy, France and England relating to mixed marriages. Among the high-class Jewish families in Italy, for instance, it has almost become a rule to marry their children to Christians. All observers are unanimous in declaring that mixed marriages are extremely frequent in that country. As early as 1881, the mixed marriages in the province of Rovigo formed 34 per cent. of the pure Jewish marriages. Mixed marriages are also very common in Sweden, Denmark, Australia and France. In the last-named country, the highest aristocracy has often intermarried with Jewish heiresses. The Jews who had been domiciled in England for several generations, have occasionally allied themselves to the aristocracy, during the nineteenth century. On the other hand, the Jewish population that immigrated to that country in the last few decades from Russia, Galicia and Roumania, is averse to intermarriage. The same holds good of France. In Sweden, the number of mixed marriages is actually greater than that of pure Jewish marriages.

Three-fourths of the Denmark Jews reside in Copenhagen. In that city, the average percentage of mixed marriages from 1880 to 1905, amounted to 69 per cent. of the pure Jewish marriages. The mixed marriages showed a tendency to increase, whereas pure Jewish marriages gradually decreased, as may be seen from the following table:

1880-1889 ... ... 55.8%
1890-1899 ... ... 68.7%
1900-1905 ... ... 93.1%

According to the latest statements it is 96 per cent. It also appears that the Jewish population of Denmark did not increase from 1840 to 1901, but rather relatively decreased. In 1840, 0.30 per cent. of the general population were Jewish, while in 1901 there were only 0.14 per cent. The Jewish percentage, accordingly, was reduced to less than a half. The chief reason for this phenomenon is to be sought, along with the fact that fewer children were born to each family in the mixed marriages, through which the Christian population has gradually encroached upon the Jewish. In the other Scandinavian countries, as has already been remarked, the number of mixed marriages is actually greater than pure Jewish marriages.

In the United States, where no confessional statistics exist, conditions resemble those of England. The few Jews who had settled there for some time and who mostly belong to the wealthy classes, as, for instance, those of the Portuguese congregations now in process of disintegration, incline towards intermarriage, while the great masses of Jews who immigrated there since 1881, keep away from mixed marriages. But even here, at least in the congested districts of New York, marriages with the surrounding elements, such as the Irish and particularly the Italian, occur with growing frequency.

In Prussia, the number of couples who intermarried rose from 2,100 in the year 1885, to 5,100 in the year 1905. The marriage of a Jew to a Christian woman is, as a rule, more frequent than the opposite case. Along with the growth of mixed marriages, the number of children resulting from such marriages has naturally increased. Where the husband is Jewish about a fourth only of the offspring remained Jews; while where the woman is Jewish, only one-fifth—four-fifths falling to the lot of Christianity. In Germany, the mixed marriages in 1905 amounted to 21 per cent. and in 1910, to 26 per cent. of the pure Jewish marriages. This average was greatly exceeded in the large cities. Thus the number of mixed marriages amounted to 45 per cent. in Berlin, and to 60 per cent. in Hamburg. And even in Frankfort on the Main, which has the reputation of being orthodox, there were about 30 per cent. of mixed marriages in the year 1908.

In Austria, intermarriage between Christians and Jews is forbidden, while intermarriage between Jews and nonconformists is permitted. Marriage is, accordingly, only possible when one of the parties embraces the religion of the other, or belongs to no denomination. It is obvious, for this reason, that the number of mixed marriages is much smaller in that country. The greater number of such marriages are contracted in Vienna. In the year 1906, they amounted to 13 per cent. While in Austria, as a rule, intermarriages between Jews and nonconformists are pretty rare, they are rather frequent in Triest. The following is a table of the average percentage of mixed marriages in the last few decades:

$$\begin{aligned}
&1877\text{-}1890 \text{ about} \ldots \ldots \ldots \ldots \ldots . 33\% \\
&1891\text{-}1895 \text{ " } \ldots \ldots \ldots \ldots \ldots . 38\% \\
&1896\text{-}1899 \text{ " } \ldots \ldots \ldots \ldots \ldots . 41\% \\
&1900\text{-}1903 \text{ " } \ldots \ldots \ldots \ldots \ldots . 62\%
\end{aligned}$$

This is to be accounted for by the fact that Triest is on the border of Italy where, as is the case also in Denmark and Australia, the increasing frequency of mixed marriages actually threaten the existence of the Jewish population.

In Hungary, mixed marriages have been permitted since 1895, and they have become very numerous since that time. The capital towns of all countries offer the best opportunities for mixed marriages. In Hungary, the greater part of such marriages are contracted at Budapest. They amount in that town to 20 per cent.

The majority of Holland Jews reside in Amsterdam. Here also, mixed marriages between Jews and Christians show a constant increase. In 1903, they formed a fifth part of all the pure Jewish marriages.

Statistical figures recently obtained show a steady progress in the same direction. The language of these statistics is so eloquent and forceful, that it almost renders all discussion superfluous.

If we wish to draw up a summary of the above data, we can divide the countries, where mixed marriages are contracted, into four classes, according to Ruppin's scheme.

The first place must be accorded to the great mass of Jews whom modern culture has not reached as yet, and who remain in the same stage of civilization as they were during the Middle Ages. To this class belong the vast lower masses of the Jews in Russia, Roumania and Galicia, the native Jews of Morocco, Asia and European Turkey. They have their own vernaculars, the so-called Yiddish and Ladino, respectively. They dwell in their national exclusiveness, wear their peculiar garb and live for the greater part according to the old Jewish laws. The greater bulk are poor workmen or artisans and store keepers of precarious existence. It is in those countries that we still find the home of religious fervor and talmudic study. At the utmost, two mixed marriages out of a hundred pure marriages are contracted there.

The second class has been somewhat influenced by European culture, and speak the language of the country either exclusively or along with their jargon. They have abandoned their peculiar garb, and are dressed like their Christian neighbors. They still practice Jewish observances, but the intolerance towards the non-Jewish is abated, and the imitation of Christian manners and the occupation with non-Jewish literature, are no longer regarded as reprehensible. The members of this class mostly live in conditions free from care, and some of them have even attained decided prosperity. To this class belong chiefly the Russian and Galician Jews who immigrated to America, the Jews of Hungary and of the small towns of Austria and Germany. Their number amounts to three millions. Mixed marriages occur there from two to ten per cent. The third class have renounced all Jewish ceremonial practices, especially the Sabbath, speak exclusively the language of the country and no longer occupy themselves with Jewish literature. The fact that the people of this class belong to Judaism is only proved by their contracting marriages with Jews, by circumcising their sons, and by attending synagogue during the High Festivals. To this class which, as a rule, lives in good material conditions, belongs the wealthy Jewish class of the large cities in Europe and America. Their number amounts to about two millions. In this class mixed marriages take place from ten to thirty per cent.

The fourth and last class has severed all connections with Judaism and religion. It still remains Jewish, because a sense of honor, family and social ties prevent it from going over to Christianity. To this class belong the Jews in the capital towns, and those who possess an academic education. Their number may be computed as something like a million. Mixed marriages are very frequent in this class—from thirty to fifty per cent.

These four classes, however, which I have attempted to portray with a few bold strokes, are not fixed groups, but cross-cuts at at different positions, of a constantly flowing stream whose source to-day is in orthodox Judaism of eastern Europe, and which wends its way into the sea of Christianity. The process of infiltration of modern culture into Judaism goes on incessantly, and in the same manner, orthodox Judaism constantly yields to the members of the second tolerant class. The latter gradually yields to the class of reformers and freethinkers, and finally baptism, and especially intermarriage, leads the Jews to Christianity. These four classes can also be represented as four consecutive generations. Four or five generations intervene between our own age and the time of Mendelssohn. It is a melancholy reflection, that hardly one of the Jews who lived at that time in Berlin has any Jewish descendants.

This process would also assume equally large dimensions in Russia, if the Jews were granted equal rights and if the Pale of Settlement were removed. The amelioration of the material conditions would remove the Ghetto environment which is one of the factors in preserving orthodox Judaism. But still more important would be the elimination of the second factor, namely, the keeping together of the Jews in one compact mass. If it were possible for the Russian Jews to spread themselves over the immense Russian Empire, the Jewish population in that country would not be denser than in western Europe. Thereby the progressive changes which exercise their destructive influences upon the western Jews would also apply to their Russian brethren. For the country that is more developed, serves as a picture of the future of the one that is less developed. Accordingly, eastern Jews will after some time apparently find themselves in the same position as the western Jews are to-day.

We may epitomise our conclusions from the processes described above, as follows: When the Jews in the diaspora became prosperous, assimilation which appears on the scene takes them away more or less from Judaism. It is mainly when they are oppressed, when they are in economically unfavorable conditions, that the Ghetto environment, in its old sense, is still retained. And although conditions to-day are not favorable in all countries, the beginning of this development can he recognized everywhere. Under favorable material conditions, and through the prevalence of secular education, Judaisrn, on account of its being scattered among nations of an alien race, is in danger of being disintegrated and destroyed, since the influence of ceremonial religion is waning.

It is not for the first time that we notice this process of disintegration. There were similar phases in all countries and throughout all ages. In accordance with the laws of historical evolution ever since the exile, this process has appeared in every country where a high culture brought about freedom from political pressure, from care for a livelihood, and from superstition. These phenomena appeared in those cases where Judaism actually imported foreign cultures, as for instance the Greek culture in the second century before the present era up till the first century of the common era, and afterwards the Arabian culture from the eighth till the twelfth century. Greek culture, from whose combination with Judaism, Christianity sprang, brought Judaism to the brink of ruin, and deprived it of a great part of its adherents. The million of Jews who, during the first century after Christ lived in Egypt, which was then the center of Hellenistic culture, appear to have gone over to Christianity. And the intimate and friendly intercourse which prevailed later on between Jews and Mohammedans in Babylon and Spain, caused the frequent recurrence of mixed marriages and conversions to Islam. The fact that in the empire of Castile, from the year 1290 till the year 1474, the number of Jews was reduced from 850,000 to 150,000, may serve as a proof for this assertion.

It is impossible to deny the resemblance of these two periods with the process of disintegration of our own times. Only, nowadays, the beginning of this process exists in all countries, and it has the tendency of becoming ttniversal. Formerly, these processes were only partial, confined to certain domains of culture. Modern culture, however, has broken all boundaries, and has become a world culture.

In China, the Jews who in former centuries were quite numerous, have almost entirely disappeared without leaving a trace, through intermarriage with the Chinese. And finally, we have to take into consideration the ten tribes who disappeared among the foreign nations, because at that time religion had not yet become ceremonial in the same sense as it existed after the Reformation of Ezra and Nehemiah. And also to-day it is possible, that wherever religion ceases to be ceremonial, the greater part of Jewry in the diaspora will, in the near future, become absorbed among the nations in whose midst it exists.

From all these considerations it becomes clearly manifest, how significant the problem of intermarriage is to-day. An inexorable process of disintegration is in progress. Although this process of breaking up Judaism is only gradual, from individual to individual, from family to family, it is of significance on account of the principle and inevitable result that it involves. The future of the Jews is seriously menaced by economic impoverishment in the East, and by baptism and mixed marriages in the west of Europe. In addition to this, there is a decrease in the birth-rate of the latter.

The Jewish people which existed almost from the time when the history of the world began, which flourished in antiquity, which defied fire and sword in the Middle Ages, which is the only one of the nations that survived from the earliest times until to-day, whose representatives even to-day have brilliant achievements to their credit—it is just to this people that culture and the development of civilization have brought nothing but misfortune; they have estranged many of its best sons, and through political and economic anti-Semitism have slowly but surely taken away the ground from under the feet of the great masses.

It is therefore not impossible that Judaism may be disbanded in the near future—to be more precise, when the amelioration of the lot of the Jews will enable them to spread themselves still more. Are we justified in hindering these historical processes, which may mean the termination of thousandfold tribulations? Can the continued existence of a nation which is externally persecuted by fate, be of any value to us? What our sentiment says is quite clear; but what answer do we get from positive Science? Would it not be perhaps of great benefit to the development of civilization if the Jews were to assimilate with other races of high standing?

These are questions and problems which cannot be solved from our subjective point of view, but we must seek for an answer in Sociology, History and Natural Science. Which is better when considered from the general point of view, race-mixture or race-purity? The point of view which modern Science adopts towards the important questions of race-mixture and in-breeding is totally different from that which prevailed up till the last quarter of the preceding century.

Whereas it was formerly believed that in the intermarriage of two different races, the qualities of both component parts would appear in the offspring, we know now that the question of race-mixture is by no means so simple. It is possible, but not certain, that only kindred elements could improve through crossing. On the other hand, the interbreeding of totally different nations produces a bastard type whose character is far below the level of either parent.

The observations made in countries which have a population of half-breeds, have pointed to the unfavorable effect of crossing. In India, the progress of race-crossing caused civilization to retrograde. We also know very well the wretched conditions of Central and South Amertea, which are inhabited by half-breeds, whose cultural stagnation stands in striking contrast to the rapid and ambitious development of the United States and Canada. It is certain that the conditions in Central and South America must, to some extent, be considered as the result of race-crossing. It is true that also in North America the population arose from a blending of various nationalities. But here it was chiefly Englishmen, Frenchmen, Spaniards, Dutchmen and Germans; that is to say, nations which were closely related to one another, who were amalgamated; whereas in Soiiith America it was Spaniards, Indians, Negroes and Mongolians who formed affinities.

Colonization in newly discovered countries has always succeeded in those places where, like in North America, the conquering nations have avoided crossing. In Brazil, on the other hand, there rules an indescribable mixed type whose bodily, intellectual and moral energy is exceedingly enfeebled. The natives of South Africa have a proverb: 'God created the white man, God created the black man, but the devil created the mulatto.'

According to the laws of Nature, the general instinctive abilities, from which depth of talent and character emanate, dwindle among half-breeds; while individual abilities often become more pronounced. Almost all observers are unanimous that through cross-breeding, bodily shapeliness, facility of talent could be gained, but resistibility of body and strength of character are impaired. Furthermore, the ability to achieve anything great and extraordinary, as well as nobility of mind are, as a rule, unknown to half-breeds. The latter characteristics form the constitutional ability, and the former the individual. The constitutional type becomes enfeebled through crossing, and the more distant the two races are, the more pronounced is this weakening.

Let us take a few examples. On the coast of Labrador there are a great number of half-breeds which are the offspring of Eskimos and Scotch immigrants. The old Scotch settlers were able to brave the adverse surroundings more easily than the new generation. At present, tuberculosis is raging there. Also the other polar nations, who have for thousands of years defied the most dreadful influences of their surroundings, are now retrogressing, after crossing found its way among them. The only exceptions are the Tunguses, who with their own culture, withstood the European settlers. The same recurrence is repeated elsewhere. Wherever the intermixture is limitless, as in Hawaii, that type which is numerically weaker, gradually dies out without even increasing the number of half-breeds. The crossing of the Hawaiians chiefly takes place with the Chinese. Besides, those islands were exempt from war, pestilence and starvation, which are otherwise the causes of the destruction of uncivilized peoples. The Tasmanians and Australian negroes have vanished through crossing. The Eurasians at Java, who are the offspring of Europeans and Indian natives, are weaklings who are rapidly perishing. The Spanish mulattoes in the Philippines are a bastard race, doomed to destruction. The bushmen have for hundreds of years waged the battle for existence under the hardest conditions, for Hottentots and Bantus were their superior enemies. And yet it is only now, after general intermixture stepped in, that they are about to disappear.

These examples will suffice to prove that crossing is one of the principal causes of the destruction of nations, and that the interbreeding of widely different types leads to the reduction of fertility and vitality. The difference of race and character leads as also animal breeders assert, to the formation of discordant, irresolute characters. It is for this reason that all half-breeds who are the offspring of widely different races have a had reptttation in respect to character.

In history there are many examples of the impossibility for half-breeds, even when their parents did not belong to races very far from each other, to reach a state capable of developing a living culture. This impossibility is also observed in cases where each nationality in itself possessed very great ability. All investigations thus point to the ennobling influences of racial purity, and to the destructive effects of racial chaos.

One calls to mind the flourishing nations of the ancient Orient: the Indians, Persians, Egyptians and Greeks. One also compares their former creativeness and influence with those of the time when the tide of foreign nations began to overwhelm them. How brave were the old Romans, and how capable did the Germanic race that mingled with them prove to be later on; and yet how wretched was the product of this crossing! After the barbarism of the Middle Ages, it took about a millenium before men of firmly rooted greatness arose once more, and before the national character strongly and harmoniously asserted itself! How changed were the inhabitants of Greece after they absorbed the Slavonic tribes! What became of the Indians after the Arabs and Mongolians broke into their country? Each of these racial components proved itself capable of high culture, and yet the result was always a change for the worse for both parent-races. That these results were not due to historical and social conditions alone, can be seen from the case of smaller nations like the Armenians and Jews who have retained their racial purity, and have consequently preserved and increased their cultural ability despite their unhappy lot. One calls to mind the high cultural ability of the Moors and the Goths, and one considers the result of the mixture in Spain, when the Gothic population absorbed the former after the destruction of the Moorish rule. One also thinks of the racial medley of Germans, Slavs and Tartars in Russia. It becomes evident from these examples, not speaking of the single individuals, but of the greater majority, what a bad effect the mixture of races has. The normal historical development does not tend towards the effacement of race, but rather towards making the racial features more pronounced, and is thus combatting political influences. The quintessence of race is the hero, the genius.

From experiments on and observations of our domestic animals, we also learn that thoroughbred animals which possess superior characteristics, become deteriorated with respect to these very characteristics, through intermixture. The same holds good of the human races. It is now regarded as certain, that virtues and superior qualities are mostly to be found among races which have kept themselves pure, while mixed breeds usually develop the defects and vices of their parents, but none of their good sides.

There is thus no doubt that the power of heredity is more powerful among pure races. Potential cultural energy will always predominate in pure untainted races. It is only among such races that ingenious creative power as well as artistic and moral genius find a favorable home.

These are, accordingly, the answers which Science gives to-day to the above questions. Even from the cosmopolitan point of view, therefore, it would be no advantage if Judaism were to disappear through assimilation with those Slavonic nations, in whose midst the bulk of the Jews reside to-day. Such an event would be detrimental to both sides. We have to strive after race-purity, not after racial chaos. Greatness of intellect, and character in the highest degree, and genius, can only emanate from the rich source of instincts which are to be found in pure races. In order to get an exact idea of the power of instincts and the effect of heredity, we ought to bear in mind that every man, in twenty generations, is the product of more than a million forefathers, and in thirty generations he is the product of a thousand million forefathers. If all these forefathers descend from one race, this enormous sum of similar instinctive talent, and with it the strength of constitutional capacity, becomes manifest. For it is this constitutional type, as above indiacted, which produces bodily resistance, depth of intellect and strength of character. And this constitutional type becomes enfeebled through crossing. Accordingly, if a nation wishes to achieve something great and powerful for itself and mankind, its policy with reference to the future must have only one aim: to force its way from racial chaos to racial purity.

We have proved by our investigations that the Jews have racial purity and that an extraordinary high racial value falls to their share. Their disappearance would not only be a national loss, but also an irretrievable loss for the general culture. But unfortunately, even at this present moment, this race is in danger of being destroyed. The conservation and further development of the distinguished possibilities that are found in this ancient race owing to its long-standing purity, are just now being questioned. For there is not so much danger to the Jew from baptism, as is usually maintained, as there is from intermarriage. In the first place, because baptism only finds its way among Jews of ignoble character, while intermarriage is found among all classes; and secondly, because intermarriage is practiced even in countries where baptism, for one reason or another, is of rare occurrence.

In conclusion, I wish to repeat the following sentence which contains the social law appertaining to mixed marriages, and for which we have previously cited statistical proofs: Tribes which live together always intermarry when such marriages are not forbidden by law or religion. Since they have been scattered all over the globe, the Jews have mingled with other nations. Civil law to-day permits mixed marriages, and religion has actually begun to lose its authority.

In order to preserve the Jews for Judaism two remedies are possible: to preserve the Ghetto with its external and social influences, or to abolish the diaspora. The first alternative can only mean a continued morbid existence.

This is the Jewish question in a quite different sense from that in which it is usually conceived, namely, the question about the future lot of the Jewish race, which, after thousands of years of splendid development and stubborn resistance, now presents the sad picture of the body of a people which is partly perishing in misery and partly in course of decomposition."[277]

---

[277]. I. Zollschan, "The Significance of the Mixed Marriage", *Jewish Questions: Three Lectures*, New York, Bloch Pub. Co., (1914), pp. 20-42.

Both the Nuremberg Laws and Zollschan's racist Zionist tracts are derivative of Theodor Fritsch's *Antisemiten-Katechismus: eine Zusammenstellung des wichtigsten Materials zum Verständniss der Judenfrage*, H. Beyer, Leipzig, (1893), pp. 358ff., the first edition of which was published in 1887 under the *nom de plume* Thomas Frey. An English translation of Fritsch's "Ten German Commandments of Lawful Self-Defense" is found in P. W. Massing, *Rehearsal for Destruction: A Study of Political Anti-Semitism in Imperial Germany*, Howard Fertig, New York, (1967), pp. 306-307, which book also contains translations of other early political anti-Semitic works, as does R. S. Levy's *Antisemitism in the Modern World: An Anthology of Texts*, D. C. Heath and Company, Toronto, (1991). Fritsch went on to publish numerous anti-Jewish works in collaboration with Adolf Hitler, including a German translation of *The International Jew: The World's Foremost Problem*. Earlier racist proscriptions against intermarriage are found throughout the Old Testament, including, among other places:

"26:34 And Esau was forty years old when he took to wife Judith the daughter of Beeri the Hittite, and Bashemath the daughter of Elon the Hittite: 26:35 Which were a grief of mind unto Isaac and to Rebekah. [***] 28:1 And Isaac called Jacob, and blessed him, and charged him, and said unto him, Thou shalt not take a wife of the daughters of Canaan. 28:2 Arise, go to Padan-aram, to the house of Bethuel thy mother's father; and take thee a wife from thence of the daughters of Laban thy mother's brother. 28:3 And God Almighty bless thee, and make thee fruitful, and multiply thee, that thou mayest be a multitude of people; 28:4 And give thee the blessing of Abraham, to thee, and to thy seed with thee; that thou mayest inherit the land wherein thou art a stranger, which God gave unto Abraham. 28:5 And Isaac sent away Jacob: and he went to Padan-aram unto Laban, son of Bethuel the Syrian, the brother of Rebekah, Jacob's and Esau's mother. 28:6¶ When Esau saw that Isaac had blessed Jacob, and sent him away to Padan-aram, to take him a wife from thence; and that as he blessed him he gave him a charge, saying, Thou shalt not take a wife of the daughters of Canaan; 28:7 And that Jacob obeyed his father and his mother, and was gone to Padan-aram; 28:8 And Esau seeing that the daughters of Canaan pleased not Isaac his father; 28:9 Then went Esau unto Ishmael, and took unto the wives which he had Mahalath the daughter of Ishmael Abraham's son, the sister of Nebajoth, to be his wife."—*Genesis* 26:34-35; 28:1-9

"14 For thou shalt worship no other god: for the LORD, whose name *is* Jealous, *is* a jealous God: 15 Lest thou make a covenant with the inhabitants of the land, and they go a whoring after their gods, and do sacrifice unto their gods, and *one* call thee, and thou eat of his sacrifice; 16 And thou take of their daughters unto thy sons, and their daughters go a whoring after their gods, and make thy sons go a whoring after their gods."—*Exodus* 34:14-16

"20:24 But I have said unto you, Ye shall inherit their land, and I will give it unto you to possess it, a land that floweth with milk and honey: I *am* the LORD your God, which have separated you from *other* people. [***] 20:26 And ye shall be holy unto me: for I the LORD *am* holy, and have severed you from *other* people, that ye should be mine. [***] 21:14 A widow, or a divorced *woman*, or profane, *or* an harlot, these shall he not take: but he shall take a virgin of his own people to wife."—*Leviticus* 20:24, 26; 21:14

"This *is* the thing which the LORD doth command concerning the daughters of Zelophehad, saying, Let them marry to whom they think best; only to the family of the tribe of their father shall they marry. So shall not the inheritance of the children of Israel remove from tribe to tribe: for every one of the children of Israel shall keep himself to the inheritance of the tribe of his fathers."—*Numbers* 36:6-7

"When the LORD thy God shall bring thee into the land whither thou goest to possess it, and hath cast out many nations before thee, the Hittites, and the Girgashites, and the Amorites, and the Canaanites, and the Perizzites, and the Hivites, and the Jebusites, seven nations greater and mightier than thou; 2 And when the LORD thy God shall deliver them before thee; thou shalt smite them, *and* utterly destroy them; thou shalt make no covenant with them, nor show mercy unto them: 3 Neither shalt thou make marriages with them; thy daughter thou shalt not give unto his son, nor his daughter shalt thou take unto thy son. 4 For they will turn away thy son from following me, that they may serve other gods: so will the anger of the LORD be kindled against you, and destroy thee suddenly."—*Deuteronomy* 7:1-4

"12 Else if ye do in any wise go back, and cleave unto the remnant of these nations, *even* these that remain among you, and shall make marriages with them, and go in unto them, and they to you: 13 Know for a certainty that the LORD your God will no more drive out *any* of these nations from before you; but they shall be snares and traps unto you, and scourges in your sides, and thorns in your eyes, until ye perish from off this good land which the LORD your God hath given you."—*Joshua* 23:12-13

"3:5 And the children of Israel dwelt among the Canaanites, Hittites, and Amorites, and Perizzites, and Hivites, and Jebusites: 3:6 And they took their daughters to be their wives, and gave their daughters to their sons, and served their gods. 3:7 And the children of Israel did evil in the sight of the LORD, and forgat the LORD their God, and served Baalim and the groves. 3:8 Therefore the anger of the LORD was hot against Israel, and he sold them into the hand of Chushan-rishathaim king of Mesopotamia: and the children of Israel served Chushan-rishathaim eight years. [***] 14:1 And Samson went down to Timnath, and saw a woman in Timnath of the daughters of the Philistines. 14:2 And he came up, and told his father and his mother, and said, I have seen a woman in Timnath of the daughters of the Philistines: now therefore get her for me to wife. 14:3 Then his father and his mother said unto him, *Is there* never a woman among the daughters of thy brethren, or among all my people, that thou goest to take a wife of the uncircumcised Philistines? And Samson said unto his father, Get her for me; for she pleaseth me well. 14:4 But his father and his mother knew not that it *was* of the LORD, that he sought an occasion against the Philistines: for at that time the Philistines had dominion over Israel."—*Judges* 3:5-8; 14:1-4

"But king Solomon loved many strange women, together with the daughter of Pharaoh, women of the Moabites, Ammonites, Edomites, Zidonians, *and* Hittites: 2 Of the nations *concerning* which the LORD said unto the children of Israel, Ye shall not go in to them, neither shall they come in unto you: *for* surely they will turn away your heart after their gods: Solomon clave unto these in love. 3 And he had seven hundred wives, princesses, and three hundred concubines: and his wives turned away his heart. 4 For it came to pass, when Solomon was old, *that* his wives turned away his heart after other gods: and his heart was not perfect with the LORD his God, as *was* the heart of David his father. 5 For Solomon went after Ashtoreth the goddess of the Zidonians, and after Milcom the abomination of the Ammonites. 6 And Solomon did evil in the sight of the LORD, and went not fully after the LORD, as *did* David his father. 7 Then did Solomon build an high place for Chemosh, the abomination of Moab, in the hill that *is* before Jerusalem, and for Molech, the abomination of the children of Ammon. 8 And likewise did he for all his strange wives, which burnt incense and sacrificed unto their gods."—I *Kings* 11:1-8

"9:1 Now when these *things* were done, the princes came to me, saying, The people of Israel, and the priests, and the Levites, have not separated themselves from the people of the lands, *doing* according to their abominations, *even* of the Canaanites, the Hittites, the Perizzites, the Jebusites, the Ammonites, the Moabites, the Egyptians, and the Amorites. 9:2 For they have taken of their daughters for themselves, and for their sons: so that the holy seed have mingled themselves with the people of *those* lands: yea, the hand of the princes and rulers hath been chief in this trespass. 9:3 And when I heard this thing, I rent my garment and my mantle, and plucked off the hair of my head and of my beard, and sat down astonied. 9:4 Then were assembled unto me every one that trembled at the words of the God of Israel, because of the transgression of those that had been carried away; and I sat astonied until the evening sacrifice. 9:5 And at the evening sacrifice I arose up from my heaviness; and having rent my garment and my mantle, I fell upon my knees, and spread out my hands unto the LORD my God. 9:6 And said, O my God, I am ashamed and blush to lift up my face to thee, my God: for our iniquities are increased over *our* head, and our trespass is grown up unto the heavens. 9:7 Since the days of our fathers *have* we *been* in a great trespass unto this day; and for our iniquities have we, our kings, *and* our priests, been delivered into the hand of the kings of the lands, to the sword, to captivity, and to a spoil, and to confusion of face, as *it is* this day. 9:8 And now for a little space grace hath been *shewed* from the LORD our God, to leave us a remnant to escape, and to give us a nail in his holy place, that our God may lighten our eyes, and give us a little reviving in our bondage. 9:9 For we *were* bondmen; yet our God hath not forsaken us in our bondage, but hath extended mercy unto us in the sight of the kings of Persia, to give us a reviving, to set up the house of our God, and to repair the desolations thereof, and to give us a wall in Judah and in Jerusalem. 9:10 And now, O our God, what shall we say after this? for we have forsaken thy commandments, 9:11 Which thou hast commanded by thy servants the prophets, saying, The land, *unto* which ye go to possess it, *is* an unclean land with the filthiness of the people of the lands, with their abominations, which have filled it from one end to another with their uncleanness. 9:12 Now therefore give not your daughters unto their sons, neither take their daughters unto your sons, nor seek their peace or their wealth for ever: that ye may be strong, and eat the good of the land, and leave *it* for an inheritance to your children for ever. 9:13 And after all that is come upon us for our evil deeds, and for our great trespass, seeing that thou our God hast punished us less than our iniquities *deserve*, and hast given us *such* deliverance as this; 9:14 Should we again break thy commandments, and join in affinity with the people of these abominations? wouldest not thou be angry with us till *thou* hadst consumed us, so that *there should be* no remnant nor escaping? 9:15 O LORD God of Israel, thou *art* righteous: for we remain *yet* escaped, as *it is* this day: behold, we *are* before thee in our trespasses: for *we* cannot stand before thee because of this. [***] 10:17 And they made an end with all the men that had taken strange wives by the first day of the first month. 10:18¶ And among the sons of the priests there were found that had taken strange wives: *namely*, of the sons of Jeshua the son of Jozadak, and his brethren; Maaseiah, and Eliezer, and Jarib, and Gedaliah. 10:19 And they gave their hands that they would put away their wives; and *being* guilty, *they offered* a ram of the flock for their trespass. [***] 10:44 All these had taken strange wives: and *some* of them had wives by whom they had children."—*Ezra* 9; 10:17-19, 44

"9:2 And the seed of Israel separated themselves from all strangers, and stood and confessed their sins, and the iniquities of their fathers. [***] 13:3 Now it came to pass, when they had heard the law, that they separated from Israel all the mixed multitude. [***] 13:23¶ In those days also saw I Jews *that* had married wives of Ashdod, of Ammon, *and* of Moab: 13:24 And their children spake half in the speech of Ashdod, and could not speak in the Jews' language, but according to the language of each people. 13:25 And I contended with them, and cursed them, and smote certain of them, and plucked off their hair, and made them swear by God, *saying*, Ye shall not give your daughters unto their sons, nor take their daughters unto your sons, or for yourselves. 13:26 Did not Solomon king of Israel sin by these *things?* yet among many nations was there no king like him, who was beloved of his God, and God made him king over all Israel: *nevertheless* even him did outlandish women cause to sin. 13:27 Shall we then hearken unto you to do all this great evil, to transgress against our God in marrying strange wives? 13:28 And *one* of the sons of Joiada, the son of Eliashib the high priest, *was* son in law to Sanballat the Horonite: therefore I chased him from me. 13:29 Remember them, O my God, because they have defiled the priesthood, and the covenant of the priesthood, and of the Levites. 13:30 Thus cleansed I them from all strangers, and appointed the wards of the priests and the Levites, every one in his business;"—*Nehemiah* 9:2; 13:3, 23-30

In September of 1935, the Nazis passed the anti-miscegenation Nuremberg Laws, which proscribed intermarriage and sexual contact between "Jews" and "Aryans". Many Zionists were delighted. Despite the fact that these laws needlessly caused many Jews great pain and suffering, the Zionists, many of them hypocrites, rejoiced in the fact that the "race" of Jews had been saved from the death of assimilation. Their religion taught them to oppose "intermarriage" and to consider Jews who intermarried as traitors against God who must be killed.

The Old Testament is filled with proscriptions against "intermarriage". Those who fabricated the Old Testament riddled it with racist messages to frighten anyone who would marry outside of the fold, and to provide the community with a justification to murder those who elected to "intermarry". The ills of the Jews were often blamed on "intermarriage", which allegedly brought down God's wrath upon them. Even Solomon the wise is said to have been ruined by "intermarriage". While tied to religion, the real motivation behind the myths is racism. The Biblical stories tell the Jews to keep the seed of Abraham pure so that there will be a pure race of God's chosen to rule the world and subjugate the allegedly inferior Gentiles. In the Old Testament, God punished the Jews for "intermarriage" with death—God often instructed the Jews to murder their own people who "intermarried". [For example, *Malachi* 2:12—to "cut off" means to kill.]

On 26 April 1946, Nazi propagandist Julius Streicher affirmed at the Nuremberg Trials that the Nuremberg Laws of 1935 were patterned after Jewish Law,[278]

> "Yes, I believe I had a part in it insofar as for years I have written that any further mixture of German blood with Jewish blood must be avoided. I have written such articles again and again; and in my articles I have repeatedly emphasized the fact that the Jews should serve as an example to every race, for they created a racial law for themselves—the law of Moses, which says, 'If you come into a foreign land you shall not take unto yourself foreign women.' And that, Gentlemen, is of tremendous importance in judging the

---

278. *Genesis* 28:1, 6. *Exodus* 34:16. *Leviticus* 20:26. *Numbers* 23:9. *Deuteronomy* 7:1-6. *Ezra* 9. *Nehemiah* 9:2; 13:3, 23-30.

Nuremberg Laws. These laws of the Jews were taken as a model for these laws. When, after centuries, the Jewish lawgiver Ezra discovered that notwithstanding many Jews had married non-Jewish women, these marriages were dissolved. That was the beginning of Jewry which, because it introduced these racial laws, has survived throughout the centuries, while all other races and civilizations have perished."[279]

Dr. Marx asked Julius Streicher, and note that the "1935 legislation" called for the segregation of Jews, not the extermination of the Jews, and was lauded by Zionists like Georg Kareski,[280]

"Were you of the opinion that the 1935 legislation represented the final solution of the Jewish question by the State?"[281]

Streicher responded that Zionism was the final solution of the Jewish question,

"With reservations, yes. I was convinced that if the Party program was carried out, the Jewish question would be solved. The Jews became German citizens in 1848. Their rights as citizens were taken from them by these laws. Sexual intercourse was prohibited. For me, this represented the solution of the Jewish problem in Germany. But I believed that another international solution would still be found, and that some day discussions would take place between the various states with regard to the demands made by Zionism. These demands aimed at a Jewish state."[282]

Bernhard Lösener found common ground with the Zionists in the new Nuremberg Laws. He stated in November of 1935,

"If the Jews already had their own state in which the greater part of their people were settled, then the Jewish question could be considered completely resolved today, also for the Jews themselves. The least amount of opposition to the underlying ideas of the Nürnberg Laws has been raised by Zionists, because they know at once that these laws represent the only

---

[279]. *Trial of the Major War Criminals Before the International Military Tribunal, Nuremberg, 14 November 1945 — 1 October 1946*, Volume 12, Secretariat of the Tribunal, Nuremberg, Germany, p. 315.

[280]. "Georg Kareski Approves of Ghetto Laws. Interview in Dr. Goebbels' 'Angriff'", *The Jewish Chronicle*, (3 January 1936), p. 16.

[281]. *Trial of the Major War Criminals Before the International Military Tribunal, Nuremberg, 14 November 1945 — 1 October 1946*, Volume 12, Secretariat of the Tribunal, Nuremberg, Germany, p. 316.

[282]. *Trial of the Major War Criminals Before the International Military Tribunal, Nuremberg, 14 November 1945 — 1 October 1946*, Volume 12, Secretariat of the Tribunal, Nuremberg, Germany, p. 316.

correct solution for the Jewish people as well. For each nation must have its own state as the outward form of appearance of its particular nationhood."[283]

The new laws did indeed meet with much applause from the political Zionists, who had for years been vocal advocates of such a policy.[284] The political Zionists even went so far as to take credit for the Nuremberg Laws unto themselves, as if it were an honor. A. I. Berndt, an editor, published a statement of solidarity with the Nazi restrictions in the *Jüdische Rundshau* on 17 September 1935, stating, *inter alia*,

"Germany has merely drawn the practical consequences from this and is meeting the demands of the International Zionist Congress when it declares the Jews now living in Germany to be a national minority. *Once the Jews have been stamped a national minority* it is again possible to establish normal relations between the German Nation and Jewry. The new Laws give the Jewish minority in Germany their own cultural life, their own national life. In future they will be able to shape their own schools, their own theater, their own sports associations; in short, they can create their own future in all aspects of national life. On the other hand, it is evident that from now on and for the *future* there can be *no interference* in questions connected with the Government of the German people, that there can be no interference in the national affairs of the German Nation."[285]

Georg Kareski, "the Jew who has accepted office under the Nazi Government as Reich Commissioner for Jewish Cultural Affairs",[286] whom Lenni Brenner called a "Hitler's Zionist Quisling before Quisling",[287] stated in an interview in the Nazi Party's *Der Angriff* in late 1935, as quoted in "Georg Kareski Approves of Ghetto Laws. Interview in Dr Goebbels' 'Angriff '", *The Jewish Chronicle* on 3 January 1936 on page 16,

"I have for many years regarded a complete separation between the cultural activities of the two peoples as a condition for a peaceful collaboration and I have always been in favour of such a separation, provided it is founded on

---

[283]. F. Nicosia, *The Third Reich and the Palestine Question*, University of Texas Press, Austin, (1985), p. 53.

[284]. I. Zollschan, *Das Rassenproblem unter besonderer Berücksichtigung der theoretischen Grundlagen der jüdischen Rassenfrage*, W. Braumüller, Wien, (1910); **and** *Jewish Questions: Three Lectures*, New York, Bloch Pub. Co., (1914).

[285]. A. I. Berndt, "Comment in the German News Agency on the Nuremberg Laws", *Jüdische Rundschau*, Number 75, (17 september 1935); English translation from Y. Arad, I. Gutman and A. Margaliot, Editors, *Documents on the Holocaust*, Eighth Edition, University of Nebraska Press, Lincoln, Nebraska, London, (1999), pp. 82-83.

[286]. "Georg Kareski Approves of Ghetto Laws. Interview in Dr Goebbels' 'Angriff '", *The Jewish Chronicle*, (3 January 1936), p. 16.

[287]. L. Brenner, *Zionism in the Age of the Dictators*, Chapter 12, Croom Helm, London, L. Hill, Westport, Connecticut, (1983), pp. 135-141.

the respect for the alien nationality. [***] The Nuremberg Laws of September 15th, 1935, seem to me, apart from their legal provisions, entirely to conform with this desire for a separate life based on mutual respect."[288]

The racist legacy of political Zionism, and of Judaism, lingers. Israeli Supreme Court Justice Haim Cohn was quoted in *The London Times* on 25 July 1963 on page 8:

"It is one of the bitterest ironies of fate that the same biological or racist approach which was propagated by the Nazis and characterized the infamous Nuremberg laws should, because of an allegedly sacrosanct Jewish tradition, become the basis for the official determination or rejection of Jewishness in the state of Israel."[289]

Years later, Zionist Meir Kahane sought to establish the Nuremberg-style laws in Israel.[290] Kahane wrote on 11 May 1979,

"We will also act to end the relationships between Arab men and Jewish women that is now growing and that so desecrates the Name of G-d."[291]

After leading Jews announced that "Judea Declares War on Germany"[292] in March of 1933, and instituted a boycott of German goods following Hitler's election, there was a very short-lived boycott of Jewish businesses in Germany,

---

288. Reprinted in: L. Brenner, Editor, *51 Documents: Zionist Collaboration with the Nazis*, Barricade Books Inc., Fort Lee, New Jersey, (2002), pp. 155-156. An alternative English translation appears in: F. Nicosia, *The Third Reich and the Palestine Question*, University of Texas Press, Austin, (1985), p. 56.
289. *See also:* J. Badi, *Fundamental Laws of the State of Israel*, Twayne Publishers, New York, (1961), p. 156.
290. Y. Kotler, *Heil Kahane*, Adama Books, New York, (1986), pp. 153, 195, 198-212. L. Brenner, *Jews in America Today*, L. Stuart, Secaucus, New Jersey, (1986), pp. 298, 301. G. Cromer, "Negotiating the Meaning of the Holocaust: An Observation on the Debate About Kahanism in Israeli Society", *Holocaust and Genocide Studies*, Volume 2, Number 2, (1987), pp. 289-297, at 292-294.
291. M. Kahane, *On Jews and Judaism: Selected Articles 1961-1990*, Volume 1, Institute for the Publication of the Writings of Rabbi Meir Kahane, Jerusalem, (1993), p. 81.
292. "JUDEA DECLARES WAR ON GERMANY", *Daily Express*, (24 March 1933), front page, banner headline. *See also:* Chaim Weizmann's letter of 29 August 1939 to Prime Minister Chamberlain, that the Jews had declared war on Germany, "Jews Fight for Democracies" *The London Times*, (6 September 1939), p. 8. *See also: The Jewish Chronicle*, (8 September 1939). *For an extensive analysis of Jewish declarations of war against Germany, see:* H. Stern, *Jüdische Kriegserklärungen an Deutschland: Wortlaut, Vorgeschichte, Folgen*, FZ-Verlag, München, Second Edition, (2000), ISBN: 3924309507; **and** *KZ-Lügen: Antwort auf Goldhagen*, FZ-Verlag, München, Second Edition, (1998), ISBN: 3924309361.

on 1 April 1933. Nazis placed yellow and black emblems in the storefronts of Jewish owned shops, despite the fact that most German Jews were loyal to the Fatherland. Strangely, Robert Weltsch published an editorial in the *Jüdische Rundshau*, which was the official party organ of the Zionist Federation of Germany, in which he blamed assimilationist Jews for Nazism. He called on Jews to bear the medieval stigmata with pride. The resentment Weltsch expressed towards assimilationist Jews leads one to wonder if the Nazis were created in order to sponsor Zionism and eventually to punish those who would not embrace the cause after being warned of the consequences of a failure to do so—should the Jews of Europe continue to resist emigrating away from their homes after being warned. Weltsch wrote, *inter alia*,

> "What should be recommended at this time is that the work which witnessed the infancy of Zionism, Theodor Herzl's The Jewish State, be disseminated among Jews and non-Jews in hundreds of thousands of copies. If there is still left any feeling for greatness and nobility, gallantry and justice, then every National Socialist who looks into this book is bound to shudder at his own blind actions. Every Jew who reads it would also begin to understand and would be consoled and uplifted by it. Page after page of this booklet, which first appeared in 1896, would have to be copied to show that Theodor Herzl was the first Jew dispassionate enough to examine anti-Semitism in connection with the Jewish question. And he recognized that an improvement cannot be effected by ostrich-like behavior, but only by dealing with facts frankly and in full view of the world...
>
> We Jews who have been raised in Theodor Herzl's spirit want to ask ourselves what our own guilt is, what sins we have committed. At times of crisis throughout its history, the Jewish people has faced the question of its own guilt. Our most important prayer says, 'We were expelled from our country because of our sins.' Only if we are critical toward ourselves shall we be just toward others.
>
> Jewry bears a great guilt because it failed to heed Theodor Herzl's call and even mocked it in some instances. The Jews refused to acknowledge that 'the Jewish question still exists.' They thought the only important thing was not to be recognized as Jews. Today we are being reproached with having betrayed the German people; the National Socialist press calls us the 'enemies of the nation,' and there is nothing we can do about it. It is not true that the Jews have betrayed Germany. If they have betrayed anything, they have betrayed themselves and Judaism.
>
> Because the Jews did not display their Jewishness with pride, because they wanted to shirk the Jewish question, they must share the blame for the degradation of Jewry... "[293]

---

[293]. R. Weltsch, English translation by L. S. Dawidowicz, "Wear the Yellow Badge with Pride!", *A Holocaust Reader*, Behrman House, Inc., West Orange, New Jersey, (1976), pp. 147-150, at 147-148.

Though many Jews and philo-Semites organized an international boycott of German goods in hopes of defeating the Hitler régime, there was one place where German products and services were not only welcomed, they were commissioned. In Palestine, Zionists worked in collusion with the Nazis to extort monies from Jews emigrating from Germany to Palestine and to use those funds to buy German products, thus annulling the effect of the boycotts and stimulating Hitler's economy with investment capital. This conspiracy to take the wealth of German Jews and use it to further persecute European Jews in the interest of forcing Jews into Palestine was called the "Ha'avara Agreement",[294] which fulfilled Herzl's plan for both Zionists and anti-Semites to profiteer from the suffering and expulsion of assimilationist Jews. Hitler, the *SS* and the Gestapo, being staunch Zionists, supported Ha'avara over the objections of the more authentically anti-Semitic, pro-German, members of the Nazi Party.[295] A good deal of evidence of the collaboration of Nazis and Zionists is presented in Roger Garaudy's book *Les Mythes Fondateurs de la Politique Israélienne*, Samiszdat, Paris, (1996); English translations: *The Founding Myths of Israeli Politics*, and *The Mythical Foundations of Israeli Policy*, Studies Forum International, London, (1997) and *The Founding Myths of Modern Israel*, Institute for Historical Review, Newport Beach, California, (2000).

The Zionist Federation of Germany (*Zionistische Vereinigung für Deutschland*, or ZVfD) welcomed Hitler and the Nazi Party as their best hope for forcing Jews into Zionism.[296] The Federation celebrated the emergence of governmental and nationalistic racism in Germany. On 21 June 1933, soon after Hitler was elected, they sent a memorandum to the Nazi Government embracing and encouraging Nazism. This memorandum iterated many prevalent Zionist myths, such as the belief that the emancipation of Jews by the French Revolution caused assimilation which was destructive to the "Jewish race" and to "Gentile races". The memorandum also anticipated the segregationist spirit of the Nuremberg Laws. The Zionist Federation of Germany's memoranda stated, among other things,

> "The emancipation of the Jews, begun at the end of the 18th, beginning of the 19th century, was based on the idea that the Jewish question could be solved by having the nation-state absorb the Jews living in its midst. This view, deriving from the ideas of the French Revolution, discerned only the individual, the single human being freely suspended in space, without regarding the ties of blood and history or spiritual distinctiveness.

---

294. E. Black, *The Transfer Agreement: The Dramatic Story of the Pact Between the Third Reich and Jewish Palestine*, Brookline Books, Cambridge, Massachusetts, (1984/1999). T. Segev, *The Seventh Million: The Israelis and the Holocaust*, Hill and Wang, New York, (1993), pp. 19-22, 24-29, 33-34.
295. K. Polkehn, "The Secret Contacts: Zionism and Nazi Germany, 1933-1941", *Journal of Palestine Studies*, Volume 5, Number 3/4, (Spring-Summer, 1976), pp. 54-82.
296. K. Polkehn, "The Secret Contacts: Zionism and Nazi Germany, 1933-1941", *Journal of Palestine Studies*, Volume 5, Number 3/4, (Spring-Summer, 1976), pp. 54-82.

Accordingly, the liberal state demanded of the Jews assimilation into the non-Jewish environment. Baptism and mixed marriage were encouraged in political and economic life. Thus it happened that innumerable persons of Jewish origin had the chance to occupy important positions and to come forward as representatives of German culture and German life, without having their belonging to Jewry become visible. [***] On the foundation of the new state, which has established the principle of race, we wish so to fit our community into the total structure so that for us too, in the sphere assigned to us, fruitful activity for the Fatherland is possible. We believe it is precisely the new Germany that can, through bold resoluteness in the handling of the Jewish question, take a decisive step toward overcoming a problem which, in truth, will have to be dealt with by most European peoples—including those whose foreign-policy statements today deny the existence of any such problem in their own midst. [***] Thus, a self-conscious Jewry here described, in whose name we speak, can find a place in the structure of the German state, because it is inwardly unembarrassed, free from the resentment which assimilated Jews must feel at the determination that they belong to Jewry, to the Jewish race and past. We believe in the possibility of an honest relationship of loyalty between a group-conscious Jewry and the German state. [***] We are not blind to the fact that a Jewish question exists and will continue to exist. From the abnormal situation of the Jews severe disadvantages result for them, but also scarcely tolerable conditions for other peoples. Our observations, presented herewith, rest on the conviction that, in solving the Jewish problem according to its own lights, the German Government will have full understanding for a candid and clear Jewish posture that harmonizes with the interests of the state."[297]

In 1937, evidently referring to the above cited memoranda of the Zionist Federation of Germany, Zionist Joachim Prinz recalled,

"Everyone in Germany knew that only the Zionists could responsibly represent the Jews in dealings with the Nazi government. We all felt sure that one day the government would arrange a round table conference with the Jews, at which—after the riots and atrocities of the revolution had passed—the new status of German Jewry could be considered. The government announced very solemnly that there was no country in the world which tried to solve the Jewish problem as seriously as did Germany. Solution of the Jewish question? It was our Zionist dream! We never denied the existence of the Jewish question! Dissimilation? It was our own appeal! ... In a statement notable for its pride and dignity, we called for a

---

[297]. English translation by L. S. Dawidowicz, "The Zionist Federation of Germany Addresses the New German State", *A Holocaust Reader*, Behrman House, Inc., West Orange, New Jersey, (1976), pp. 150-155.

conference."²⁹⁸

On 4 August 1933, the *Jüdische Rundschau*, the official party organ of the Zionist Federation of Germany, published an article entitled "Rasse als Kulturfaktor" on page 392, which stated,

"We who live here as a 'foreign race' have to respect racial consciousness and the racial interest of the German people absolutely. This however does not preclude a peaceful living together of people of different racial membership. The smaller the possibility of an undesirable mixture, so much less is there need for 'racial protection'... There are differentiations that in the last analysis have their root in ancestry. Only rationalist newspapers who have lost feeling for the deeper reasons and profundities of the soul, and for the origins of communal consciousness, could put aside ancestry as simply in the realm of 'natural history'."²⁹⁹

Heinz Höhne wrote,

"Alongside this majority [of patriotic German-Jews who did not wish to leave Germany], however, a small group of Zionist spokesmen was at work, and their object was to turn the minds of German Jewry away from their traditional German patriotism and direct them towards Palestine. Initially therefore they regarded the advent of National-Socialism as by no means a catastrophe; in their eyes it presented Zionism with a unique opportunity to fulfil its object, the return to a Jewish State and Jewish national consciousness. The rise of anti-Semitism in Germany exerted a curious fascination over the Zionists, for in it they saw the defeat of westernised Jewry which, they considered, was striving to identify itself with the non-Jewish industrialised peoples. After the Nazi seizure of power the Zionist newspaper *Jüdische Rundschau* proclaimed on a note of triumph: 'An ideology has collapsed; we will not lament it but will think of the future.'³⁸

Many were tempted to regard 30 January 1933 as a favourable turning-point in Jewish history—'Jewry for the Jews' could become the watchword once more. This remark was to be found in an article entitled 'We Jews' written by a young Rabbi, Dr Joachim Prinz. (Hans Lamm, the historian of German Jewry under the Third Reich, described it as 'a curious, almost apologetic, interpretation of the anti-Semitic phenomenon.') Prinz

---

298. J. Prinz, "Zionism under the Nazi Government", *Young Zionist* (London), (November, 1937), p. 18; *as quoted in:* L. Brenner, *Zionism in the Age of the Dictators*, Chapter 5, Croom Helm, London, L. Hill, Westport, Connecticut, (1983), p. 47.

299. "Rasse als Kulturfaktor", *Jüdische Rundschau*, Volume 38, Number 62, (4 August 1933), pp. 391 (front page) -392, at 392; *as quoted and translated in:* L. Brenner, *Zionism in the Age of the Dictators*, Chapter 5, Croom Helm, London, L. Hill, Westport, Connecticut, (1983), p. 51.

considered that 'there can be no further evasion of this Jewish problem; emancipation has forced the Jew to accept anonymity and deny his Jewish nationality.' But this, he continued, had not profited the Jews at all. 'Among those who nevertheless realised that a man was a Jew, this anonymity gave rise to the tensions generated by mistrust and the sense of contact with a foreigner.' What solution could there be to the Jewish tragedy other than to take the road to Palestine? Prinz continued: 'No subterfuge can save us now. In place of assimilation we desire to establish a new concept— recognition of the Jewish nation and Jewish race.'[39]

For the Jewish nationalists the prospect was tempting; under the pressure of German racialism and with its assistance the Zionist ideal might win that victory denied it in the humanitarian and democratic atmosphere of the Weimar republic. If both the Zionists and National-Socialists regarded race and nationhood as universally valid criteria, some common ground must be discoverable between the two. As early as 13 June 1933, the *Jüdische Rundschau* had come out into the open: 'Zionism recognises the existence of a Jewish problem and desires a far-reaching and constructive solution. For this purpose Zionism wishes to obtain the assistance of all peoples, whether pro- or anti-Jewish, because in its view, we are dealing here with a concrete rather than a sentimental problem, in the solution of which all peoples are interested.'[40]

At this point von Mildenstein stepped in. The task of the SD, he argued, was to turn the German-assimilated Jews back into 'conscious Jews, to promote 'dissimilation' in order to awaken in the breasts of the largest possible number of Jews the urge to go to Palestine, the only country open at the time to large-scale Jewish immigration. Himmler seized on Mildenstein's plan and set him to work. Within the SD Hauptamt Mildenstein set up a Jewish desk (entitled II 112); a period of SS Jewish policy began in which, according to Hans Lamm, 'the adoption or affectation of a pro-Zionist attitude' was in order.[41]

The new SS policy made its first appearance in the columns of the *Schwarze Korps*; in place of the paper's anti-Jewish tirades references began to appear to the 'sensible, totally unsentimental Jew' of the Zionist movement. The paper forecast: 'The time cannot be far distant when Palestine will again be able to accept its sons who have been lost to it for over a thousand years. Our good wishes together with our official goodwill go with them.'[42],[300]

---

300. H. Höhne, *The Order of the Death's Head: The Story of Hitler's S. S.*, English translation by R. Barry, Coward-McCann, Inc., New York, (1970), pp. 331-333. Höhne cites: [38] *Jüdische Rundschau*, (28 April 1933). [39] H. Lamm, *Über die innere und äussere Entwicklung des deutschen Judentums im Dritten Reich*, p. 94. [40] H. Lamm, *Über die innere und äussere Entwicklung des deutschen Judentums im Dritten Reich*, p. 156. [41] H. Lamm, *Über die innere und äussere Entwicklung des deutschen Judentums im Dritten Reich*, p. 149. A. Eichmann, *Record of Interrogation*, Volume 1, Column 67. [42] *Das Schwarze Korps*, (15 May 1935).

An *SS* officer, Baron Leopold Itz von Mildenstein traveled to Palestine and reported on his impressions in the official Nazi Party organ *Der Angriff* in a series of twelve articles under the heading "Ein Nazi fährt nach Palästina" from 26 September 1934 to 9 October 1934. As director of the central office of the "Jewish desk" in the intelligence branch of the *SS*, Mildenstein promoted the Zionist cause in Nazi Germany. His primary goal was to convert reluctant Jews to Zionism. Jacob Boas wrote,

> "The gist of that policy was to assist the expansion of Zionist influence among Germany's Jews who, despite the oppressive conditions under which they lived, still showed no great desire to emigrate to Palestine. By making a distinction between race-minded, emigration-conscious Zionists and 'assimilationists' out to destroy National Socialism, the S. S. strove to strengthen the Zionist position in the Jewish community. Accordingly, S. S. officials were instructed to encourage the activities of Zionists and to discourage those of non-Zionists. Zionists were given privileges denied to other groups. A police decree of March, 1935, for example, ordered officers to favour Zionist youth groups over non-Zionist ones; the former were to be allowed to don uniforms but not the latter. The S. S. also looked with favour on the Zionist vocational and agricultural training centres which groomed young Jews for a life of toil in Palestine, and access to Nazi functionaries generally proved easier for Zionists than for assimilationists. Even the Nuremberg Laws (September 15th, 1935), which deprived Jews of their German citizenship and condemned them to pariah status, contained a special 'Zionist' provision: forbidden to fly the German colours, Jews were given the right to hoist their own flag, i. e. the Zionist emblem, the blue Star of David between stripes, also blue, against a white background."[301]

The *SS* issued a report in the summer of 1934 which recommended that Jewish youth be trained for the laborious task of improving Palestine for massive settlement. The report recommended that the German Government sponsor Zionism and persuade German-Jewish leadership to promote the Zionist cause. Should this fail, other measures would have to be taken.[302] On 26 September 1935, *Das Schwarze Korps*, the official organ of the *Schutzstaffeln* (*SS*), reported,

> "In the context of its *Weltanschauung*, National Socialism has no intention of attacking the Jewish people in any way. On the contrary, the recognition of Jewry as a racial community based on blood, and not as a religious one, leads the German government to guarantee the racial separateness of this

---

301. J. Boas, "A Nazi Travels to Palestine", *History Today*, Volume 30, Number 1, (January, 1980), pp. 33-38, at 38.
302. K. A. Schleunes, *The Twisted Road to Auschwitz: Nazi Policy Toward German Jews 1933-1939*, University of Illinois Press, (1970), pp. 178-182.

community without any limitations. The government finds itself in complete agreement with the great spiritual movement within Jewry itself, the so-called Zionism, with its recognition of the solidarity of Jewry throughout the world and the rejection of all assimilationist ideas. On this basis, Germany undertakes measures that will surely play a significant role in the future in the handling of the Jewish problem around the world."[303]

This statement relates to the fact that the Zionists had reacted negatively to Moses Mendelssohn's reforms of Judaism to make it a universal spiritual religion, as opposed to the racist and nationalistic religion found in the Old Testament. Zionists like Moses Hess asserted in consort with anti-Semites, that Judaism is not a religion, but a race and a nation, and that Jews produced their religion as a product of their unique racial characteristics. In 1862, racist Zionist Moses Hess called the "new Jew" a traitor to the "Jewish race",

"The most touching point about these Hebrew prayers is, that they are really an expression of the collective Jewish spirit; they do not plead for the individual, but for the entire Jewish race. The pious Jew is above all a Jewish patriot. The 'new' Jew, who denies the existence of the Jewish nationality, is not only a deserter in the religious sense, but is also a traitor to his people, his race and even to his family. If it were true that Jewish emancipation in exile is incompatible with Jewish nationality, then it were the duty of the Jews to sacrifice the former for the sake of the latter. This point, however, may need a more elaborate explanation, but that the Jew must be above all a Jewish patriot, needs no proof to those who have received a Jewish education. Jewish patriotism is not a cloudy Germanic abstraction, which dissolves itself in discussions about being and appearance, realism and idealism, but a true, natural feeling, the tangibility and simplicity of which require no demonstration, nor can it be disposed of by a demonstration to the contrary. "[304]

In Hess' view, better the Ghetto and persecution than emancipation, if emancipation meant assimilation. Hess asserted that a "race war" was needed to subjugate the German People to submit to Hess' racist Zionist ideology,

"The democrats of 1848 undoubtedly fully demonstrated their superiority over the demagogues of the 'War of Liberation,' the Romantic lads of the Jahn and Arndt type, whom they left far behind on the road of progress. And yet, on the basis of my long experience, I feel inclined to assert that Germany as a whole, in spite of its collective intellectuality, is in its practical

---

[303]. F. Nicosia, *The Third Reich and the Palestine Question*, University of Texas Press, Austin, (1985), p. 57.
[304]. M. Hess, *Rom und Jerusalem: die letzte Nationalitätsfrage*, Eduard Wengler, Leipzig, (1862); English translation, *Rome and Jerusalem: A Study in Jewish Nationalism*, Bloch, New York, (1918/1943), pp. 62-63.

social life far behind the rest of the civilized nations of Europe. The race war must first be fought out and definitely settled before social and humane ideas become part and parcel of the German people, as was the case with the Romance peoples which, after a long historical process, finally defeated race antagonism."[305]

Hess described Judaism as a national cult and argued that the essence of Judaism is national, and that pure Judaism, which balances spiritualism with materialism, would supplant the spiritual extremism of the Christian Judaic cult, which Hess alleged was out of balance and therefore unstable. Hess believed that things evolve in three stages and that the modern age is the Messianic Age, begun by Spinoza and the French Revolution. Hess adopted the racism of Judaism and of German Folkish mythology and expressed his beliefs that there are various races which each serve their function in the human organism led by Jews, allegedly the true People of God. He wanted to kill off the "German race"—eliminate Esau—with "Jewish love" in this third era of human history, so that the Jews can lead the world into a Utopia dominated by the "Jewish race", as prophesied in the Hebrew Bible,

"The laws of universal history, I mean the history of the universe, namely, those of the cosmic, organic and social life, are as yet little known. We have particular sciences, but not a science of the universe; we still do not know the unity of all life. One thing, however, is certain, that a fusion of cults, an ideal to which so many aspire, and which was realized, at least in part, for thousands of years by Catholic Rome, will as little establish a lasting peace in human society as the philanthropic but unscientific belief in the absolute equality of men. In their attempt to base the granting of equal rights to all men on the primitive uniformity of all races and types, the humanitarians confound the organization of social life on the basis of solidarity, which is the result of a long and painful process of historical development, with a ready-made, inorganic equality and uniformity, which becomes rarer and rarer the farther back we go in history. The reconciliation of races follows its own natural laws, which we can neither arbitrarily create nor change. As to the fusion of cults, it is really a past stage in the development of social life. It was the watchword of that religion which, owes its existence to the death of the nations of antiquity, i.e., Christianity. To-day the real problem is how free the various oppressed races and folk-types and allow them to develop in their own way. The dangerous possibility that the various nationalities will separate themselves entirely from each other or ignore each other is to be feared as little as the danger that they will fight among themselves and enslave one another.

The present-day national movement not only does not exclude

---

305. M. Hess, *Rom und Jerusalem: die letzte Nationalitätsfrage*, Eduard Wengler, Leipzig, (1862); English translation, *Rome and Jerusalem: A Study in Jewish Nationalism*, Bloch, New York, (1918/1943), p. 80.

humanitarianism, but strongly asserts it; for this movement is a wholesome reaction, not against humanism, but against the things that would encroach upon it and cause its degeneration, against the leveling tendencies of modern industry and civilization which threaten to deaden every original organic life-force, by introducing a uniform inorganic mechanism. As long as these tendencies were directed against the antiquated institutions of a long-passed historical period, their existence was justified. Nor can this nationalistic reaction object to them, insofar as they endeavor to establish closer relations between the various nations of the world. But, fortunately, people have gone so far in life, as well as in science, as to deny the typical and the creative; and as a result the vapor of idealism, on the one hand, and the dust of atomism on the other, rest like mildew on the red corn, and stifle the germinating life in the bud. It is against these encroachments on the most sacred principles of creative life that the national tendencies of our time react, and it is against these destructive forces that I appeal to the original national power of Judaism.

Like the general universal cosmic life which finds its termination in it, and the individual microcosmic life in which all the buds and fruits of the spirit finally ripen. Humanity is a living organism, of which races and peoples are the members. In every organism changes are continually going on. Some, quite prominent in the embryonic stage, disappear in the later development. There are organs, on the other hand, hardly noticeable in the earlier existence of the organism, which become important only when the organism reaches the end of its development.

To the latter class of members of organic humanity (which class is really the creative one) belongs the Jewish people. This people was hardly noticeable in the world, where it was greatly oppressed by its powerful, conquering neighbors. Twice it came near being destroyed; namely, in the Egyptian and Babylonian captivities; and twice it rose to new spiritual life and fought long and successfully against the mightiest as well as the most civilized peoples of antiquity—the Greeks and the Romans. Finally, in the last struggle of the ancient world, it was this people which fertilized the genius of humanity with its own spirit, so as to rejuvenate itself, along with the regeneration of humanity. To-day, when the process of rejuvenation of the historical peoples is ended and each nation has its special function in the organism of humanity, we are for the first time beginning to conceive the special significance of the various organs of humanity.

England, with its industrial organization, represents the nerve-force of humanity which directs and regulates the alimentary system of mankind; France, that of general motion, namely, the social; Germany discharges the function of thinking; and America represents the general regenerating power by means of which all elements if the historical peoples will be assimilated into one. We observe that every modern people, every part of modern society, displays in its activity as an organ of humanity a special calling, then he must also determine the importance and function of the only ancient people which still exists to-day, as strong and vigorous as it was in days of old, namely, the people of Israel.

In the organism of humanity there are no two peoples which attract and repel each other more than the Germans and the Jews; just as there are no two mental attitudes which are simultaneously akin to each other and still diametrically opposed, as the scientific-philosophical and the religious-moral. Religion, in its higher form, is the spiritual tie which binds the creature to the Creator, the infinite thread, the end of which returns to its source, the bridge which leads from one creation to the other, from life to death, and from death back to life. It not only brings man to know the Absolute more intimately, but it inspires and sanctifies his whole life with the divine spirit. In religion, as in love, especially in a religion like Judaism, which is neither one-sidedly materialistic nor one-sidedly spiritualistic, body and spirit merge into one another. The greatest and most dangerous enemy of the Jewish religion in antiquity was the religion of gross sensualist, the material love of the Semites, namely, Baal worship. In mediæval ages, the enemy was represented by the embodiment of spiritualistic love—Christianity. The Jewish people which, thanks to its prophets of antiquity and rabbis of the Middle Ages, kept its religion from both extremes of degeneration, was, and is still to-day that organ of humanity which expresses the living, creative force in universal history, namely, the organ of unifying and sanctifying love. This organ is akin to the organ of thought, but is, at the same time, opposed to it. Both draw their force from the inexhaustible well of life. But, while the religious genius individualizes the infinite, philosophic, scientific thought abstracts from life all its individual, subjective forms and generalizes it. Objective philosophy and science have no direct connection with life; religious teaching is intimately united with it, for either religion is identical with the national, social and moral life, or it is mere hypocrisy.

I have wandered from my trend of thought. I merely wanted to explain to you why I do not ally myself with the humanitarian aspirations which endeavor to obliterate all differentiation in the organism of humanity and in the name of such catch words as 'Liberty' and 'Progress,' build altars to arbitrariness and ignorance, on which our light-minded youth offers its best energies and sacrifices."[306]

Die Geheime Staatspolizei (the Gestapo) also assisted the Zionists, as Zionist leader Hans Friedenthal noted,

"The Gestapo did everything in those days to promote emigration, particularly to Palestine. We often received their help when we required anything from other authorities regarding preparations for emigration. This position remained constant and uniform the entire time, until the year

---

306. M. Hess, *Rom und Jerusalem: die letzte Nationalitätsfrage*, Eduard Wengler, Leipzig, (1862); English translation, *Rome and Jerusalem: A Study in Jewish Nationalism*, Bloch, New York, (1918/1943), pp. 121-126.

1938."³⁰⁷

In April of 1936, Zionist Meyer Steinglass quoted Zionist Emil Ludwig in the *American Jewish Times*,

> "'Hitler will be forgotten in a few years, but he will have a beautiful monument in Palestine. You know', and here the biographer-historian seemed to assume the role of a patriarchal Jew—'the coming of the Nazis was rather a welcome thing. So many of our German Jews were hovering between two coasts; so many of them were riding the treacherous current between the Scylla of assimilation and the Charybdis of a nodding acquaintance with Jewish things. Thousands who seemed to be completely lost to Judaism were brought back to the fold by Hitler, and for that I am personally very grateful to him.'"³⁰⁸

In 1937, it was becoming increasingly clear to both the Nazis and the Zionists that the mere existence of the Nazi regime was not enough to drive Jews into Zionism, and that even if it were, Great Britain and other nations had placed too many obstacles in the way of a massive migration to Palestine for Zionism to succeed. The Ha'avara Agreement was a failure. The British had long wanted Palestine for a route to India and later to oil, ironically thoughts which were implanted into the British mind by opportunistic Jewish Zionists. Many of the German Jews who had fled to Palestine quickly became disenchanted with the desert and returned to Germany. The Nazis soon began to target Jews, especially healthy rich assimilated male Jews, for arrest and imprisonment in concentration camps. It was inexplicable act of self-destruction for the Germans headed by two Jews, Reinhard Heydrich and Adolf Eichmann.

Theodor Herzl had long ago warned rich assimilated Jews that if they did not follow the political Zionists, there would be dire consequences for them. Herzl wrote in his book *The Jewish State*,

> "The Governments of all countries scourged by Anti-Semitism will serve their own interests in assisting us to obtain the sovereignty we want. [***] Great exertions will not be necessary to spur on the movement. Anti-Semites provide the requisite impetus. They need only do what they did before, and then they will create a love of emigration where it did not previously exist, and strengthen it where it existed before. [***] I imagine that Governments will, either voluntarily or under pressure from the Anti-Semites, pay certain attention to this scheme; and they may perhaps actually receive it here and

---

307. F. Nicosia, *The Third Reich and the Palestine Question*, University of Texas Press, Austin, (1985), p. 57.
308. M. Steinglass, "Emil Ludwig before the Judge", *American Jewish Times*, (April, 1936), p. 35; *as quoted in:* L. Brenner, *Zionism in the Age of the Dictators*, Chapter 6, Croom Helm, London, L. Hill, Westport, Connecticut, (1983), p. 59.

there with a sympathy which they will also show to the Society of Jews."[309]

In the early 1940's, the Zionists had drawn the inhuman conclusion that since all other avenues had been tried and had failed, the only avenue for success for their tyrannical vision was the literal destruction of assimilatory Jewry. The Zionists had always exhibited an ungodly hubris and believed that they had the God given right to decide for all Jews and for all the world how each individual Jew must act and think. The Zionists' dogma was similar in this respect to the dogmatic insistence of the Marxists that they had a right to ruin the lives of the peoples the world over in order to promote the destruction of Capitalism and set the stage for their Communist world revolution. Marxists, too, believed that they knew better than each individual how that individual must think and how he or she must act. Many Zionists and Marxists believed that those who dared disagree with their "truths" must be rescued from themselves, by death if necessary—or even just convenient—to them life and liberty are cheap and comradeship means blind obedience—ultimately blind obedience to genocidal Jewish bankers seeking to create the "Jewish Utopia" of the "end times" of Jewish Messianic myth. Knowing what was soon to come, knowing the Zionist Nazis were about to turn up the heat on European Jews, some Zionists began to pull away from their public expressions of unity with the Nazis in the late 1930's, while working with Nazi authorities behind the scenes to annihilate the assimilatory and Orthodox anti-Zionist Jewry of Europe.

Zionist leader Feivel Polkes met with several high-ranking Nazi officials in Berlin in 1937, including Adolf Eichmann. The Zionists invited Adolf Eichmann and Herbert Hagen to Palestine to discuss how to purge Europe of Jews and ensure that they ended up in Palestine, so that the Jews could change the demographics of the region and take Palestine from the majority Moslem population. Eichmann and Hagen accepted the invitation and traveled to Palestine under the pretense that they were editors of the *Berliner Tageblatt*. After being refused permission to enter Palestine by the British authorities, they met with Polkes in Egypt, where Zionist Polkes commended the Zionist Nazis for persecuting the Jews. This was recorded in Eichmann and Hagen's reports on the meetings,

"Nationalist Jewish circles expressed their great joy over the radical German policy towards the Jews, as this policy would increase the Jewish population in Palestine, so that one can reckon with a Jewish majority in Palestine over the arabs in the foreseeable future."[310]

In 1938, Albert Einstein stated in his essay "Our Debt to Zionism",

---

[309]. T. Herzl, *A Jewish State: An Attempt at a Modern Solution of the Jewish Question*, The Maccabæan Publishing Co., New York, (1904), pp. 25, 68, 93.

[310]. K. Polkehn, "The Secret Contacts: Zionism and Nazi Germany, 1933-1941", *Journal of Palestine Studies*, Volume 5, Number 3/4, (Spring-Summer, 1976), pp. 54-82, at 74; citing "RFSS film roll 411".

> "Rarely since the conquest of Jerusalem by Titus has the Jewish community experienced a period of greater oppression than prevails at the present time. [***] Yet we shall survive this period too, no matter how much sorrow, no matter how heavy a loss in life it may bring. A community like ours, which is a community purely by reason of tradition, can only be strengthened by pressure from without."[311]

The Zionists proposed a military alliance with the Nazis. The Zionists asked to facilitate the Nazis' "new order in Europe" with a fascistic totalitarian Zionist state in Palestine. Klaus Polkehn wrote,

> "Thus what was on offer was no more and no less than the establishment of a fascist Jewish state in Palestine as an ally of German fascism!"[312]

---

[311]. A. Einstein, "Our Debt to Zionism", *Out of My Later Years*, Carol Publishing Group, New York, (1995), pp. 262-264, at 262.
[312]. K. Polkehn, "The Secret Contacts: Zionism and Nazi Germany, 1933-1941", *Journal of Palestine Studies*, Volume 5, Number 3/4, (Spring-Summer, 1976), pp. 54-82, at 79.

8 HOW THE JEWS MADE THE BRITISH INTO ZIONISTS

The Biblical story of Esau and Jacob teaches the Jews that Gentiles will soldier and slave for the Jews. The Bible also prophesied that the Jews would be dispersed unto the ends of the Earth. Cabalistic Jewish racists believed that Jews must dwell in England in order for the Messianic Era to commence. The Jews gained great power in England and even managed to convince the British that they, the British, were of Jewish descent, and that British Royalty descended from the Messianic line of King David. Zionist Jews used Great Britain to ruin the Turkish Empire, which ruled over Palestine for many centuries. More than a million British have died while killing off millions more Germans, Italians, Turks, Japanese and Iraqis on behalf of the Zionist cause.

"Let their table become a snare before them: *and that which should have been* for *their* welfare, *let it become* a trap."—PSALM 69:22

"¶7 And the children of Israel were fruitful, and increased abundantly, and multiplied, and waxed exceeding mighty; and the land was filled with them. 8 Now there arose a new king over Egypt, which knew not Joseph. 9 And he said unto his people, Behold, the people of the children of Israel *are* more and mightier than we: 10 Come on, let us deal wisely with them; lest they multiply, and it come to pass, that, when there falleth out any war, they join also unto our enemies, and fight against us, and *so* get them up out of the land. 11 Therefore they did set over them taskmasters to afflict them with their burdens. And they built for Pharaoh treasure cities, Pithom and Raamses. 12 But the more they afflicted them, the more they multiplied and grew. And they were grieved because of the children of Israel."—EXODUS 1:7-12

8.1 Introduction

The Old Testament's solution to the Jewish question was two-fold. If the Jews obeyed God and remained segregated, God would give them the land from the Nile to the Euphrates. Note that the Jews were not the original inhabitants of the land and that they promised it to themselves. If the Jews did not obey God and assimilated into the Gentile world, they would be laid to waste in the lands in which they dwelt, and the righteous remnant—the most racist Jews—would steal the Promised Land from its original inhabitants. Note that racist Jews created this religious mythology and only racist Jews feel obliged to fulfill it.

8.2 The Rothschilds and Disraeli Lead the British Down the Garden Path to Palestine

Jewish British Prime Minister Benjamin Disraeli illegally purchased shares in the Suez with a check written on the Bank of Rothschild. In 1875, Lionel Nathan

Rothschild advanced Disraeli £4,000,000.[313] The Rothschilds profited from the purchase with a commission on the huge sum and by its interest—as well as with speculation in the money, stock and commodities markets—Egyptian cotton was quite valuable.

The purchase accomplished little for England, but much for the Zionists. It tied England to the region and gave the Zionists an opportunity to persuade the British that they had an incentive to sponsor a "Jewish State" in Palestine in order to protect the illegal investment to which the Jewish racist Zionist Prime Minister of England Benjamin Disraeli had committed England in 1875. It also provoked hostility between England and Russia, and Zionists had long wished to destroy the Russian Empire. Not coincidently, both the Egyptian Khedive and the Sultan of Turkey were on the verge of bankruptcy when approached by the Zionists for the purchases of the Suez Canal and Palestine—bankruptcy brought on by the Rothschilds, who wanted to secure their loans with Palestine. International finance coupled with bad advice given to a sovereign can easily drive a nation into bankruptcy. What is worse, many a corrupt sovereign were covertly agents of the Jewish financiers.

The Rothschilds profiteered from Disraeli's purchase of shares in the Suez Canal and they were accused of it. Disraeli defended the Rothschilds by arguing that there had been "stock-jobbing" at Waterloo, but that the Rothschilds were honorable and would not do such a thing. In a rather obvious *non sequitur*, Disraeli argued that since the British victory at Waterloo was beneficial to the British Nation and was accompanied by stock-jobbing, stock-jobbing must be good for the British Nation, or at least a necessary consequence of positive events. Everyone knew that the Rothschilds had robbed the British People after the Battle of Waterloo. Disraeli's argument obviously fails, because the British could have won the Battle of Waterloo without the Rothschilds having exploiting the event with lies to steal from their fellow countrymen. However, Disraeli was able to insult the intelligence of the Gentile members of the British Government with impunity, because the Rothschilds had the financial might to shut down the British Empire at any time.

Disraeli purchased the shares without lawful authority and had his friend Lionel Rothschild secure the check, earning the Rothschilds an enormous commission and enabling them to corruptly profit from the purchase on the stock markets with "inside information", as they had earlier done by lying about the outcome of the Battle of Waterloo. Disraeli protested with sophistries, knowing that the Rothschilds could break the Bank of England, if it came to it,

> 'Sir, although, according to the noble Lord, we are going to give a unanimous vote, it cannot be denied that the discussion of this evening at least has proved one result. It has shown, in a manner about which neither the House of Commons nor the country can make any mistake, that had the

---

[313]. J. G. Lockhart, *Cecil Rhodes: The Colossus of Southern Africa*, Macmillan, New York, (1963), pp. 103-104.

right honourable Gentleman the Member for Greenwich been the Prime Minister of this country, the shares in the Suez Canal would not have been purchased... ... The right honourable Gentleman defies me to produce an instance of a Ministry negotiating with a private firm... ... The right honourable Gentleman found great fault with the amount of the commission which has been charged by the Messrs. Rothschild and admitted by the Government; and, indeed, both the right honourable Gentlemen opposite took the pains to calculate what was the amount of interest which it was proposed the Messrs. Rothschild should receive on account of their advance. It is, according to both right honourable Gentlemen, 15 per cent; but I must express my surprise that two right honourable Gentlemen, both of whom have filled the office of Chancellor of the Exchequer, and one of whom has been at the head of the Treasury, should have shown by their observations such a lamentable want of acquaintance with the manner in which large amounts of capital are commanded when the Government of a country may desire to possess them under the circumstances under which we appealed to the House in question. I deny altogether that the commission charged by the Messrs. Rothschild has anything to do with the interest on the advance; nor can I suppose that two right honourable Gentlemen so well acquainted with finance as the Member for Greenwich and the Member for the University of London can really believe that there is in this country anyone who has £4,000,000 lying idle at his bankers. Yet one would suppose, from the argument of the right honourable Gentleman the Member for Greenwich, that such is the assumption on which he has formed his opinion in this matter. In the present instance, I may observe, not only the possibility, but the probability, of our having immediately to advance the whole £4,000,000 was anticipated. And how was this £4,000,000 to be obtained? Only by the rapid conversion of securities to the same amount. Well, I need not tell anyone who is at all acquainted with such affairs that the rapid conversion of securities to the amount of £4,000,000 can never be effected without loss, and sometimes considerable loss; and it is to guard against risk of that kind that a commission is asked for before advances are made to a Government. In this case, too, it was more than probable that, after paying the first £1,000,000 following the signature of the contract, £2,000,000 further might be demanded in gold the next day. Fortunately for the Messrs. Rothschild they were not; but, if they had, there would in all likelihood have been a great disturbance in the Money Market, which must have occasioned a great sacrifice, perhaps the whole of the commission. The Committee, therefore, must not be led away by the observations of the two right honourable Gentlemen, who, of all men in the House, ought to be the last to make them.

But the right honourable Gentleman the Member for Greenwich says we ought to have gone to our constitutional financiers and advisers, the Governor and Deputy Governor of the Bank of England, and, of course, the honourable Member for Galway (Mr. Mitchell Henry), who rose much later in the debate, and who spoke evidently under the influence of strong feeling, also says that we ought to have asked the Governor of the Bank of England to advance the £4,000,000. But they forget that it is against the law of this country for the Bank to advance a sum of money to the Ministry.

But then it may be said—'Though the Bank could not have advanced the £4,000,000, you might have asked them to purchase the shares.' But how could they have purchased the shares? They must have first consulted their legal adviser, who probably would have told them that they had not power to do it; but, even if that doubtful question had been decided in the affirmative, they must have then called a public Court in order to see whether they could be authorized to purchase those shares to assist the Government. Now, I ask the Committee to consider for a moment what chance would we have had of effecting the purchase which we made under the circumstances, and with the competitors we had to encounter, and the objects we had to attain, if we had pursued the course which the right honourable Gentleman opposite has suggested? 'But,' says the Member for the University of London—and this also has been echoed by his late right honourable Colleague—'you would have avoided all this, if you adopted the course which we indicate, and which I have just reminded the Committee is illegal, if you had only taken the illegal course we recommend, you would have got rid of this discreditable gambling, because although the Messrs. Rothschild, some of whom have been Members of this House, are men of honour, yet they have a great number of clerks who are all gambling on the Stock Exchange.' Now, my belief is that the Messrs. Rothschild kept the secret as well as Her Majesty's Government, for I do not think a single human being connected with them knew anything about it. And, indeed, it was quite unnecessary for the Messrs. Rothschild to have violated the confidence which we reposed in them, and quite unnecessary even for the Members of Her Majesty's Government to hold their tongues, for no sooner was the proposal accepted than a telegram from Grand Cairo transmitted the news to the Stock Exchange, and it was that telegram which was the cause of all the speculation and gambling to which the right honourable Gentleman has referred. It is a fact that while the matter was a dead secret in England, the news was transmitted from Cairo. That was the intelligence on which the operations occurred. But I wish to say one word respecting the moral observations which have been made. As to gambling on the Stock Exchange, are we really to refrain from doing that which we think is proper and advantageous to the country because it may lead to speculation? Why, not a remark was made by the noble Lord, who has just addressed the House, the other night, or by me in reply, that would not affect the funds. On the one side people would say—'The Government are in great difficulty, and probably a Vote of Censure will arise out of this Suez Canal speculation,' while other persons would observe—'There is evidently something coming about Egypt, and he is not going to let it all out.' Ought we to refrain from doing what is necessary for the public welfare because it leads to stock-jobbing? Why, there is not an incident in the history of the world that led to so much stock-jobbing as the battle of Waterloo, and are we to regret that that glorious battle was fought and won because it led to stock-jobbing? So much for the operations on the Stock Exchange. I think we have been listening all night to remarks on this transaction that have very little foundation. We have been admonished for conduct which has led to stock-jobbing and we have been admonished because we applied to a private firm when from the state of the law, I have shown that it was absolutely necessary from the character of the circumstances we had to deal with that a private firm should be appealed to."[314]

Disraeli continues in his speech to attempt to justify the purchase of the Suez as if it were England's only hope for securing trade with India and China. Disraeli's hidden plan was to cajole England into the misguided and self-defeating belief that her destiny lay in the hands of the Jews, who Disraeli and his fellow Zionists planned would come to occupy Palestine and regulate trade between the continents. In the Zionists' chimera, England was a helpless child without means, who required the Jews to rescue her. The disingenuous nature of this fallacy is revealed by the fact the Zionists had made precisely the same pitch to the French some ten years prior.

It was far wiser for England, for her own sake, to make alliances with Turkey, Egypt and Russia and improve their economies, than to drive Egypt, Turkey and Russia towards bankruptcy and war with England for the benefit of the Jews, as the Zionists were attempting to do. It was only by manipulating public opinion with lies, that the Zionists were able to vilify the Moslems and drive a wedge between Christians and Islam, despite the fact that Moslems and Jews had lived together for centuries in peace and prosperity. The British would have been far better off allying themselves with the Moslems and suppressing Jewish racism, than alienating and antagonizing the Moslems by creating a racist "Jewish State" in the heart of the Moslem world.

Many crypto-Jewish English Zionists sought to convert Christians to Judaism by asking the Jews to "convert" to Christianity in order to subvert it. They asked Jews to convert to Christianity in order to make the Christians common allies with the Jews against Islam. Zionists feared that if Jewish finance, or a common collection taken from the Jews, were to simply buy Palestine from the Turks, without the appearance of the Jewish Messiah to lead them into Palestine; then Christians would join forces with Islam to crush the Jews, as prophesied in the apocalyptic visions of both the Old and the New Testaments, and in the Koran. Many Gentiles in England realized these facts and sought alliances with Egypt, Turkey and Russia.

A quite similar situation exists today, where it would have been in the interests of England and America to have given Russia greatly more financial aid after the fall of the Soviet Union than they did, and to have joined forces with, and improved the lot of, Turkey, Lebanon, Syria, Jordan, Egypt, Iraq, Iran, etc. against Israel, in order to facilitate international trade through the Middle

---

314. J. H. Park, Editor, *British Prime Ministers of the Nineteenth Century: Policies and Speeches*, New York University Press, New York, (1950), pp. 237-244, at 237-240.

East and Russia. Instead, due in no small part to the corrupting influences of Zionism on public opinion, the Zionists have made Christianity and Judaism the unnatural common enemy of Islam, and Islam the unnatural enemy of an alliance of Judaism and Christianity—to the detriment of Christendom, Islam, Judaism, and the rest of humanity.

The Zionists have successfully blinded Americans and Jews around the world to their own best interests. If the Moslems had played the game by the same rules as the Zionists and sponsored the formation of a political party in America with the agenda of removing Zionists from the Middle East, allying America with the Moslem world to promote trade with India, China and Russia, and working with Russia to flood it with investment capital, while increasing trade with Pakistan, many of the world's problems would be lessened. Instead, the Zionists are leading America into alienation from Russia, China, Pakistan and the Moslem world; which increases world poverty, world-wide instability, and the likelihood of another—though even more disastrous—world war. Six and one half billion people face world war, death and absolute destruction for the sake of about five million obscenely selfish Zionists living in Israel, who stole the Palestinians' land on the racist premise that their religion is a nationalistic religion and that their Jewish God had promised the land to them thousands of years ago. (Note that Jews have long suffered from the superstition that they ought not to count their own, and it is sometimes difficult to know how many Jews have lived at any given time in any given place, *see: Exodus* 30:12. II *Samuel* 24. I *Chronicles* 21. *Hoshea* 2:1. *Yoma* 22b. *Rashi* on *Exodus* 30:11-12. In addition, there are many crypto-Jews throughout the world, who go uncounted as Jews.)

For centuries prior to forming a state, Jewish Zionists incited violence and world war. Subsequent to forming the State of Israel, they have endlessly incited violence and desire another world war.

If the Arabs had invested their oil-monies in advanced education and American media outlets, instead of palaces, limousines and other unproductive ends, they could have helped to form public opinion in America with the facts and turned it against the inhuman Jewish Zionists, who have artificially created a religious war between Christians and Moslems. Jews took Palestine without a Messiah; which in Christianity means that these Jewish Zionists, who reject Christ, are in league with the "anti-Christ" and must be annihilated. Whereas it would be in the mutual best interests of both Christians and Moslems to join forces to defeat racist Jewish Zionism, racist warmongering Jews have turned Christianity against the Christians and made the Christians the artificial enemies of the Moslems. Instead of presenting the American public with a fair analysis of the facts, the media in America is led by tribal racist Jews who defame all Moslems in the American media as if genetically inferior terrorists, who are inherently prone to war, and in consort with the devil. Jews had done the same thing to the Catholics and Protestants, when they fomented the *Kulturkampf*.

The Zionists believed it was in their interests to destroy Catholicism (truly all of Christianity) and the Turkish Empire. They had initially hoped that the French Revolution would accomplish both these ends—as is revealed in the eleventh and twelfth "letters" in Hess' *Rome and Jerusalem* of 1862. Napoleon came close to achieving their ends. Since the Jewish People would not go to Palestine, the Zionists promoted the idea that the purchase of the Suez Canal would benefit France or England, in an attempt to draw the French, or the British, into the region as a means of creating a European commitment to the region that would provide security for the establishment of Jewish colonies. The Jews sold this plan to the French and British public on the false premise that Jews in the region would provide security for French and English interests—the Zionists created a problem where one did not exist, in order to offer themselves as its solution, which they were not.

Only after the Zionist effort to coax the French into purchasing the Suez failed, did the Zionists turn to Disraeli, who deceived England in the 1870's with the same self-defeating mythologies that had been tried upon the French in the 1860's—and yet earlier with Napoleon Bonaparte.

In the 1840's Christian Zionist agents of the Rothschilds had already promoted the myth that a Jewish state in Palestine would benefit England and Christendom.[315] The Zionists' plans eventually resulted the First and Second World Wars, where both England and France were pitted against Germany and Turkey. Racist Jewish Zionist Moses Hess published a book entitled *Rome and Jerusalem* in 1862, which was a direct precursor to Adolf Hitler's *Mein Kampf*. Note the tone of the *Kulturkampf* and the attacks on the Ottoman Turks from the racist Zionists. Note further that Hess discredits Christianity, the alleged divinity of Jesus and claims that Jesus hated Gentiles by quoting extensively from the Jewish historian Graetz in the Epilogue, Part 2, "Christ and Spinoza" [pages 186-211 in the 1943 edition of Hess' *Rome and Jerusalem*] though the later attacks in the *Kulturkampf* were more openly vitriolic, the goal was consistently to tear down Christianity and Islam in order to make way for the Jews in Palestine—a goal often iterated in the Talmudic and Cabalistic writings. Hess wrote,

> "What we have to do at present for the regeneration of the Jewish nation is, first, to keep alive the hope of the political rebirth of our people, and next, to reawaken that hope where it slumbers. When political conditions in the Orient shape themselves so as to permit the organization of a beginning of the restoration of a Jewish State, this beginning will express itself in the founding of Jewish colonies in the land of their ancestors, to which enterprise France will undoubtedly lend a hand. You know how substantial was the share of the Jews in the subscriptions to the fund raised for the benefit of the Syrian war victims. It was Cremieux who took the initiative in the matter, the same Cremieux who twenty years ago traveled with Sir Moses Montefiore to Syria in order to seek protection for the Jews against the persecutions of the Christians. In the *Journal des Debats*, which very seldom accepts poems for publication, there appeared, at the time of the

---

315. R. Sharif, "Christians for Zion, 1600-1919", *Journal for Palestine Studies*, Volume 5, Number 3/4, (Spring-Summer, 1976), pp. 123-141.

Syrian expedition, a poem by Leon Halevi, who at the time, perhaps, thought as little of the rebirth of Israel as Cremieux, yet his beautiful stanzas could not have been produced otherwise than in a spirit of foreseeing this regeneration. When the poet of the *Schwalben* mournfully complains:

Where tarries the hero? Where tarries the wise?
Who will, O my people, revive you anew;
Who will save you, and give you again
A place in the sun?

The French poet answers his query with enthusiastic confidence:

Ye shall be reborn, ye fearsome cities!
A breath of security will always hover
O'er your banks where our colors have fluttered!
Come again a call supreme!
Au revoir is not adieu—
France is all to those she loves,
The future belongs to God.

Alexander Weill sang about the same time:

There is a people stiff of neck,
Dispersed from the Euphrates to the Rhine,
Its whole life centered in a Book
Oft times bent, yet ever straightened;
Braving hatred and contempt,
It only dies to live again
In nobler form.

France, beloved friend, is the savior who will restore our people to its place in universal history.

Allow me to recall to your mind an old legend which you have probably heard in your younger days. It runs as follows:

'A knight [Esau] who went to the Holy Land to assist in the liberation of Jerusalem, left behind him a very dear friend. While the knight fought valiantly on the field of battle, his friend spent his time, as heretofore, in the study of the Talmud, for his friend was none other than a pious rabbi [Jacob].

'Months afterward, when the knight returned home, he appeared suddenly at midnight, in the study room of the rabbi, whom he found, as usual, absorbed in his Talmud. 'God's greetings to you, dear old friend,' he said. 'I have returned from the Holy Land and bring you from there a pledge of our friendship. What I gained by my sword, you are striving to obtain with your spirit our ways lead to the same goal.' While thus speaking, the knight handed the rabbi a rose of Jericho.

'The rabbi took the rose and moistened it with his tears, and immediately the withered rose began to bloom again in its full glory and splendor. And the rabbi said to the knight: 'Do not wonder, my friend that the withered rose bloomed again in my hands. The rose possesses the same characteristics as our people: it comes to life again at the touch of the warm breath of love, in spite of its having been torn from its own soil and left to wither in foreign lands. So will Israel bloom again in youthful splendor; and the spark, at present smoldering under the ashes, will burst once more into a bright flame.''

The routes of the rabbi and the knight dear friend, are meeting to-day. As the rabbi in the story symbolizes our people, so does the knight of the legend signify the French people which in our days, as in the Middle Ages, sent its brave soldiers to Syria and 'prepared in the desert the way of the Lord.'

Have you never read the words of the Prophet Isaiah: 'Comfort ye, comfort ye, my people, saith your God. Speak ye comfortably to the heart of Jerusalem, and cry unto her, that the appointed time has come, that her iniquity is pardoned; for she hath received at the Lord's hand double for all her sins. The voice of one that crieth in the wilderness; prepare ye the way of the Lord, make straight in the desert a highway for our God. Every valley shall be exalted, and every mountain and hill shall be made low, and the crooked shall be made a straight place, and the rough places a plain. And the glory of the Lord shall be revealed, and all flesh shall see it together: for the mouth of the Lord hath spoken it.'[*Footnote:* Isaiah xl, 1-5.]

Do you not believe that in these words, with which the second Isaiah opened his prophecies, as well as in words with which the Prophet Obadiah closed his prophecy,[*Footnote:* 'And saviors shall come up on Mount Zion to judge the mount of Esau; and the kingdom shall be the Lord's.'] the conditions of our own time are graphically pictured? Was not help given to Zion in order to defend and establish the wild mountaineers there? Are not things being prepared there and roads leveled, and is not the road of civilization being built in the desert in the form of the Suez Canal works and the railroad which will connect Asia and Europe? They are not thinking at present of the restoration of our people. But you know the proverb, 'Man proposes and God disposes.' Just as in the West they once searched for a road to India, and incidentally discovered a new world, so will our lost fatherland be rediscovered on the road to India and China that is now being built in the Orient. Do you still doubt that France will help the Jews to found colonies which may extend from Suez to Jerusalem, and from the banks of the Jordan to the Coast of the Mediterranean? Then pray read the work which appeared shortly after the massacres in Syria, by the famous publisher, Dentu, under the title *The New Oriental Problem*. The author hardly wrote it at the request of the French government, but acted in accordance with the spirit of the French nation when he urged our brethren, not on religious grounds, but from purely political and humanitarian motives, to restore their ancient state.[*Footnote:* I have heard that an American writer has discussed this question from a practical point of view, for a number of years. Also representative Englishmen have repeatedly declared themselves in favor of the restoration of the Jewish State.]

I may, therefore, recommend this work, written, not by a Jew, but by a French patriot, to the attention of our modern Jews, who plume themselves on borrowed French humanitarianism. I will quote here, in translation, a few pages of this work, *The New Eastern Question*, by Ernest Laharanne.[*Footnote:* See note IX at end of book.]

'In the discussion of these new Eastern complications, we reserved a special place for Palestine, in order to bring to the attention of the world the important question, whether ancient Judæa can once more acquire its former place under the sun.

'This question is not raised here for the first time. The redemption of Palestine, either by the efforts of international Jewish bankers, or the nobler method, of a general subscription in which all the Jews should participate, has been discussed many times. Why is it that this patriotic project has not as yet been realized? It is certainly not the fault of pious Jews that the plan was frustrated, for their hearts beat fast and their eyes fill with tears at the thought of a return to Jerusalem.[*Footnote:* My friend, Armond L., who traveled for several years through the Danube Principalities, told me that the Jews were moved to tears when he announced to them the end of their suffering, with the words 'The time of the return approaches.' The more fortunate Occidental Jews do not know with what longing the Jewish masses of the East await the final redemption from the two thousand year exile. They know not that the patriotic Jew cannot suppress his cry of anguish at the length of the exile, even in the midst of his festive songs, as, for instance, the patriotic poem which is read on Chanukah, closes with the mournful call:

'For salvation is delayed for us and there is no end to the days of evil.'

'They asked me,' continued my friend, 'what are the indications that the end of the exile is approaching?' 'These,' I answered, 'that the Turkish and the papal powers are on the point of collapse.']

'If the project is still unrealized, the cause is easily cognizable. The Jews dare not think of the possibility of possessing again the land of their fathers. Have we not opposed to their wish our Christian veto? Would we not continually molest the legal proprietor when he will have taken possession of his ancestral land, and in the name of piety make him feel that his ancestors forfeited the title to their land on the day of the Crucifixion?

'Our stupid Ultramontanism has destroyed the possibility of a regeneration of Judæa, by making the present of the Jewish people barren and unproductive. Had the city of Jerusalem been rebuilt by means of Jewish capital, we would have heard preachers prophesying, even in our progressive nineteenth century, that the end of the world is at hand and predictions of the coming of the Anti-Christ. Yes, we have lived to see such a state of affairs, now that Ultramontanism has made its last stand in oratorical eloquence. In the sacred beehive of religion, we still hear a continuous buzzing of those insects who would rather see a mighty sword in the hands of the barbarians, than greet the resurrection of nations and hail the revival of a free and great thought inscribed on their banner. This is undoubtedly the reason why Israel did not make any attempt to become master of his own flocks, why the Jews, after wandering for two thousand years, are not in a position to shake the dust from their weary feet. The continuous, inexorable demands that would be made upon a Jewish settlement, the vexatious insults that would be heaped upon them and which would finally degenerate into persecutions, in which fanatic Christians and pious Mohammedans would unite in brotherly accord—these are the reasons, more potent than the rule of the Turks, that have deterred the Jews from attempting to rebuild the Temple of Solomon, their ancient home, and their State.

'But if this cause explains the lack of courage on the part of patriotic Jews, we cannot refrain from accusing the so-called progressive Jews of indifference to the fate of the Jewish people; for whenever a project for the restoration of the Jewish State is being considered, they display toward it a naïveté that neither does credit to their reasoning power nor to their heart. The explanations offered by them on such occasions are inadmissible both from a moral and from a political point of view. A declaration, composed by the representatives of the progressive Jews at their meeting in Frankfort, contains the following Article:

'We acknowledge as our fatherland only the land where we are born and to which we are inseparably united by the bonds of citizenship.'

'No member of the Jewish race can renounce the incontestable and fundamental right of his people, without at the same time denying the history of the Jews and his own ancestors. Such an act is especially unseemly, at a time when political conditions in Europe will not only not obstruct the restoration of a Jewish State, but will rather facilitate its realization. What European power to-day would oppose the plan that the Jews, united through a Congress, should buy back their ancient fatherland? Who would object if the Jews flung to decrepit old Turkey a few handfuls of gold, and said to her: 'Give me back my home and use this money to consolidate the other parts of your tottering empire?'

'No objections would be raised to the realization of such a plan, and Judæa would be permitted to extend its boundaries from Suez to the harbor of Smyrna, including the entire area of the western Lebanon range. For we will not be eternally engaged in war; the time must come when this wholesale massacre, usually accompanied by the booming of cannon, will be condemned by humanity, so that the nation which desires conquest in addition to commerce, will not dare to carry out its designs. We must therefore prepare and break new ground for the peaceful struggles of industry. European industry has daily to search for new markets as an outlet for its products. We have no time to lose. The time has arrived when it is imperative to call the ancient nations back to life, so as to open new highways and byways for European civilization.'

In another passage, the author speaks with so much enthusiasm, love and reverence for the Jews, that what he says overshadows all that has ever been said by a Jew in praise of his own people.

'There is a mysterious power which rules the destiny of humanity. Once the hand of the Infinite Power has signed the decree of a nation to be banished forever from the fact of the earth, the fate of that nation is irrevocable. But when we see a nation, torn from its cradle in its early childhood, and after having tasted all the bitterness of exile is brought back to its land, only to be tossed again into the wide world; and that nation, during the eighteen centuries of its wandering has displayed such remarkable powers of endurance, suffering age-long martyrdom without extinguishing in its heart the fire of patriotism, then we just admit that we are standing before an infinite mystery, unparalleled in the history of humanity.'

In these few words there is concentrated the whole history of Israel.

What an example! What a race! You, Roman conquerors, led your legions in battle against the already ruined Zion and drove the children of Israel out of their ancestral land. Your European, Asiatic and African barbarians lent your ear to superstition and pronounced your curse upon them. You feudal kings branded the Jews with the mark of shame—the Jews, who, in spite of all your persecutions, supplied you with the necessary gold wherewith to arm your vassals and serfs and who provided your markets with goods. You, grand Inquisitors, searched among the children of the dispersed people of Israel for your richest victims, with whom to fill your prisons and coffers, and in order to feed your auto-da-fe's—and you revoked the edict of Nantes and drove out of the land the remnant that had escaped the destruction of Apostolic fanaticism. And finally, you modern nations have denied these indefatigable workers and industrious merchants civil rights. What persecutions! What tears! What blood you children of Israel have shed in the last eighteen hundred years! But you sons of Judæa, in spite of all suffering are still here. You have overcome the innumerable obstacles which the hatred, contempt, fanaticism and barbarism of the centuries have placed in your way. The hand of the Eternal has surely guided you.

France finally freed you. On the eve of the great world epoch, France, while shattering its own chains, called all nations and also you, into freedom. You became citizens and now you are brothers. The year 1789 was the first step in the process of rehabilitation. Pursuing its mission, liberation, the eye of France searched after all persecuted races, and it found you in your ghetto and shattered its doors forever.[*Footnote:* The old Beneday, who was still alive in 1842, at the time of the publication of the first *Rhenische Zeitung* used to come, from time to time, to the office of that paper to converse with the members of the staff; and on one of these occasions he told us the story, which I had really heard before, how he, at the commission of the first French Republic had laid the ax at the gates of the Bonn Ghetto. Beneday could hardly conceive how his son Jacob could, at one and the same time, be a liberal and yet unfriendly toward the French.

I comforted him by pointing to the progressive German Jews, who in reality have to thank the French for whatever political and civil rights they possess here or elsewhere in Germany, and yet rail, in company with the Germans, against the 'hereditary enemy.'] France invited you to its Chambers. You participated in its triumphs; you shared its happiness and its reverses. You have raised your voice on the day of council, shouted for joy at our victories and wept at our defeats. You are good citizens and devoted brothers. France will perhaps be to you a lighthouse of salvation, a rock against your enemies, who are also the enemies of our modem institutions. It will defend you against the libelers of your nationality, your character and your religion.

You are an elemental force and we bow our heads before you. You were powerful in the early period of your history, strong even after the destruction of Jerusalem, and mighty during the Middle Ages, when there were only two dominant powers—the Inquisition and its Cross, and Piracy with its Crescent. You have escaped destruction in your long dispersion, in spite of the terrible tax you have paid during eighteen centuries of persecution. But what is left of your nation is mighty enough to rebuild the gates of Jerusalem. This is your mission.

Providence would not have prolonged your existence until to-day, had it not reserved for you the holiest of all missions. The hour has struck for the resettlement of the banks of the Jordan. The historical books of the royal prophets can, perhaps, be written again only by you.

A great calling is reserved for you: to be a living channel of communication between three continents. You should be the bearers of civilization to the primitive people of Asia, and the teachers of the European sciences to which your race has contributed so much. You should be the mediators between Europe and far Asia, open the roads that lead to India and China—those unknown regions which must ultimately be thrown open to civilization. You will come to the land of your fathers crowned with the crown of age-long martyrdom, and there, finally, you will be completely healed from all your ills! Your capital will again bring the wide stretches of barren land under cultivation; your labor and industry will once more turn the ancient soil into fruitful valleys, reclaim the flat lands from the encroaching sands of the desert, and the world will again pay its homage to the oldest of peoples.

The time has arrived for you to reclaim, either by way of compensation or by other means, your ancient fatherland from Turkey, which has devastated it for ages. You have contributed enough to the cause of civilization and have helped Europe on the path of progress, to make revolutions and carry them out successfully. You must henceforth think of yourselves, of the valleys of Lebanon and the plains of Gennesareth.

March forward! At the sight of your rejuvenation, our hearts will beat fast, and our armies will stand by you, ready to help.

March forward, Jews of all lands! The ancient fatherland of yours is calling you, and we will be proud to open its gates for you.

March forward, ye sons of the martyrs! The harvest of experience which you have accumulated in your long exile, will help to bring again to Israel the splendor of the Davidic days and rewrite that part of history of which the monoliths of Semiramis are the only witness.

March forward, ye noble hearts! The day on which the Jewish tribes return to their fatherland will be epoch-making in the history of humanity. Oh, how will the East tremble at your coming! How quickly, under the influence of labor and industry, will the enervation of the people vanish, in the land where voluptuousness, idleness and robbery have held sway for thousands of years.

You will become the moral stay of the East. You have written the Book of books. Become, then, the educators of the wild Arabian hordes and the African peoples. Let the ancient wisdom of the East, the revelations of the Zend, the Vedas, as well as the more modern Koran and the Gospels, group themselves around your Bible. They will all become purified from every superstition and all will proclaim alike the principles of freedom, humanity, peace and unity. You are the triumphal arch of the future historical epoch, under which the great covenant of humanity will be written and sealed in your presence as the witnesses of the past and future. The Biblical traditions which you will revive, will also sanctify anew our Occidental society and destroy the weed of materialism together with its roots.

And when you shall have made this wonderful progress, remember, ye sons of Israel, remember Modern France which, from the moment of its rebirth, has loved you continually and has never wearied of defending you.

[***]

If one appreciates fully the infinitely tragic rôle which the Jewish people has thus far played in history, he must also inevitably perceive the only way that will bring salvation to our misery. This solution is at present not as impractical as it may look at first sight. It is in accordance with the sympathies of the French people and with the interests of French politics, that after France's victorious armies shall have overthrown the modern Nebuchadnezzar, France will extend its work of redemption also to the Jewish nation. It is to the interest of France to see that the road leading to India and China should be settled by a people which will be loyal to the cause of France to the end, in order that it may fulfil the historical mission which has fallen to it as a legacy from the great Revolution. But is there any other nation more adapted to carry out this mission than Israel, which was appointed for the same mission from the beginning of its history?

'Frenchmen and Jews!' I hear you exclaim. 'If so, then the Christian German reactionaries were right in their denunciations of the Jews!' Yes, my dear friend, the animal instinct which scents the enemy in the distance is always infallible. Reaction has everywhere recognized its mortal enemy in those who stand midway between reaction and revolution and who act as the midwife of progress, the giant who is to smite reaction over its head. For it is a law of organic and social life history, that the mediate being whose existence is limited to the transition epoch, should pave the way from the imperfect to the more perfect and higher scales of life.

Frenchmen and Jews! It seems that in all things they were created for one another. They resemble one another in their humane and national aspirations, and differ only in such qualities as can only be complemented by another nation, but which are never united in one and the same people. The French people excel in alertness, in the humanistic and sympathetic quality to assimilate all elements; the Jews, on the other hand, possess more ethical seriousness than the French, and in meeting other types, the Jew will rather impress his stamp on his environment than be molded by it. The French can rule the world because they absorbed the best of the entire human race. The Jews can only be masters of their own flock, and with the holy fire which they have kindled in their own midst, they will warm and enlighten a world composed of heterogeneous elements, and thus prevent this world from disintegrating into its elements and relapsing into the chaos out of which it was raised once before by Judaism.

The generous help which France has extended to civilized peoples toward the restoration of their nationality, will be remembered longer by our nation than by any other. How easily will we come to an understanding with this humane French people about our religion and its sacred places in Palestine. But matters have not gone so far yet. The Jewish people must first show itself worthy of the regeneration of its historical cult; it must first feel the necessity of a national restoration if it would reach that point. Until then we need not think about building the Temple; we must win the heart of our brethren for the great work which will finally bring eternal glory to the Jewish nation and salvation to humanity.

For Jewish colonization on the road to India and China, there is no lack, either of Jewish laborers or of Jewish talent and capital. Let only the germ be planted under the protection of the European powers, and the tree of a new life will spring forth by itself and bear excellent fruit."[316]

Just as when the French were unwilling to buy the Suez Canal for the Jews, the Zionists looked to Disraeli in England to accomplish this end; when the English moved toward improving their relations with Russia, Egypt and Turkey, the Zionists looked to Germany as a sword with which to conquer the Turks and the Russians, and with which to manipulate the British and the French, resulting in the First and Second World Wars. When Germany failed them, they turned America against Germany and ruined it. In more modern times in America, when the French, who emancipated the Jews of Continental Europe, opposed war against Islam for Israel's sake, the Zionists stirred up hatred of the French in America, though Hess had long ago tried lure the French into Palestine with the promise that the Jews would forever be loyal to France, the France which had liberated them,

---

[316]. M. Hess, "Eleventh Letter", *Rom und Jerusalem: die letzte Nationalitätsfrage*, Eduard Wengler, Leipzig, (1862); English: *Rome and Jerusalem: A Study in Jewish Nationalism*, Bloch, New York, (1918/1943), pp. 145-159, 167-169.

"It is to the interest of France to see that the road leading to India and China should be settled by a people which will be loyal to the cause of France to the end, in order that it may fulfill the historical mission which has fallen to it as a legacy from the great Revolution. But is there any other nation more adapted to carry out this mission than Israel, which was appointed for the same mission from the beginning of its history?"

Zionists are loyal to none but themselves. When the French failed them, they became eternally loyal to England, and when that failed them, to Germany, and when that failed them, to America. Should America fail to perpetually slave for Israel, they will turn to China.

The Zionists repay the ancient gift (in Jewish myths) of the Persian King Cyrus of the freedom of the Jews from the captivity of Babylon, as well as King Cyrus' restoration of the Jews to Judea and the rebuilding Jerusalem and the Temple, as well as the gift of Persian King Ahasuerus, who assisted Queen Esther and Mordecai to mass murder "the enemies of the Jews"—modern Jews repay these ancient gifts by perpetually destroying Iran and corrupting its leadership to the detriment of the Iranian People. Though the *Book of Esther* is a work of fiction, it provides a model that the Jews have often followed. The Rothschilds often followed the ancient Jewish model of Jacob and Esau, whereby Jacob exploited Esau's deathly hunger to steal Esau's freedom and Esau's land; and the ancient Jewish model of Joseph, whereby Joseph exploited the deathly hunger of the Egyptians to steal the Egyptians' freedom and the Egyptians' land—this in collusion with a corrupt Pharaoh, who helped the Jews destroy the currency—this after the Egyptians had given Jews land in Egypt. *Genesis* 47 tells the Jews to ruin host nations and then leave them taking their wealth,

"1 Then Joseph came and told Pharaoh, and said, My father and my brethren, and their flocks, and their herds, and all that they have, are come out of the land of Canaan; and, behold, they *are* in the land of Goshen. 2 And he took some of his brethren, *even* five men, and presented them unto Pharaoh. 3 And Pharaoh said unto his brethren, What *is* your occupation? And they said unto Pharaoh, Thy servants *are* shepherds, both we, *and* also our fathers. 4 They said moreover unto Pharaoh, For to sojourn in the land are we come; for thy servants have no pasture for their flocks; for the famine *is* sore in the land of Canaan: now therefore, we pray thee, let thy servants dwell in the land of Goshen. 5 And Pharaoh spake unto Joseph, saying, Thy father and thy brethren are come unto thee: 6 The land of Egypt *is* before thee; in the best of the land make thy father and brethren to dwell; in the land of Goshen let them dwell: and if thou knowest *any* men of activity among them, then make them rulers over my cattle. 7 And Joseph brought in Jacob his father, and set him before Pharaoh: and Jacob blessed Pharaoh. 8 And Pharaoh said unto Jacob, How old *art* thou? 9 And Jacob said unto Pharaoh, The days of the years of my pilgrimage *are* an hundred and thirty years: few and evil have the days of the years of my life been, and have not attained unto the days of the years of the life of my fathers in the days of their pilgrimage. 10 And Jacob blessed Pharaoh, and went out from before Pharaoh. 11 And Joseph placed his father and his brethren, and gave them a possession in the land of Egypt, in the best of the land, in the land of Rameses, as Pharaoh had commanded. 12 And Joseph nourished his father, and his brethren, and all his father's household, *with* bread, according to *their* families. 13 ¶ And *there was* no bread in all the land; for the famine *was* very sore, so that the land of Egypt and *all* the land of Canaan fainted by reason of the famine. 14 And Joseph gathered up all the money that was found in the land of Egypt, and in the land of Canaan, for the corn which they bought: and Joseph brought the money into Pharaoh's house. 15 And when money failed in the land of Egypt, and in the land of Canaan, all the Egyptians came unto Joseph, and said, Give us bread: for why should we die in thy presence? for the money faileth. 16 And Joseph said, Give your cattle; and I will give you for your cattle, if money fail. 17 And they brought their cattle unto Joseph: and Joseph gave them bread *in exchange* for horses, and for the flocks, and for the cattle of the herds, and for the asses: and he fed them with bread for all their cattle for that year. 18 When that year was ended, they came unto him the second year, and said unto him, We will not hide *it* from my lord, how that our money is spent; my lord also hath our herds of cattle; there is not ought left in the sight of my lord, but our bodies, and our lands: 19 Wherefore shall we die before thine eyes, both we and our land? buy us and our land for bread, and we and our land will be servants unto Pharaoh: and give *us* seed, that we may live, and not die, that the land be not desolate. 20 And Joseph bought all the land of Egypt for Pharaoh; for the Egyptians sold every man his field, because the famine prevailed over them: so the land became Pharaoh's. 21 And as for the people, he removed them to cities from *one* end of the borders of Egypt even to the *other* end thereof. 22 Only the land of the priests bought he not; for the priests had a portion assigned them of Pharaoh, and did eat their portion which Pharaoh gave them: wherefore they sold not their lands. 23 Then Joseph said unto the people, Behold, I have bought you *this* day and your land for Pharaoh: lo, *here is* seed for you, and ye shall sow the land. 24 And it shall come to pass in the increase, that ye shall give the fifth *part* unto Pharaoh, and four parts shall be your own, for seed of the field, and for your food, and for them of your households, and for food for your little ones. 25 And they said, Thou hast saved our lives: let us find grace in the sight of my lord, and we will be Pharaoh's servants. 26 And Joseph made it a law over the land of Egypt unto this day, *that* Pharaoh should have the fifth *part*; except the land of the priests only, which became not Pharaoh's. 27 ¶ And Israel dwelt in the land of Egypt, in the country of Goshen; and they had possessions therein, and grew, and multiplied exceedingly. 28 And Jacob lived in the land of Egypt seventeen years: so the whole age of Jacob was an hundred forty and seven years. 29 And the time drew nigh that Israel must die: and he called his son Joseph, and said unto him, If now I have found grace in thy sight, put, I pray thee, thy hand under my thigh, and deal kindly and truly with me; bury me not, I pray thee, in Egypt: 30 But I will lie with my fathers, and thou shalt carry me out of Egypt, and bury me in their buryingplace. And he said, I will do as thou hast said. 31 And he said, Swear unto me. And he sware unto him. And Israel bowed himself upon the bed's head."

This story taught the Jews that they could ruin any nation if they could control the nation's money supply. They controlled the money supply by melting down gold and silver and keeping the metals for themselves. Once they had ruined metallic currency, the Jews could then operate on a barter system with the subjugated Gentiles. They learned that with gold reserves, they could loan out more script money than they had gold and silver on reserve, and they could loan it out at interest. They could also buy up debts for foreign goods whether the securities for those supposed goods actually existed, or not. In this manner, the Jews could increase the money supply and earn interest on monies which they never possessed.

Their profits came at the expense of inflation, which again taxed the people for their sake. In Socialist countries taxes gave them complete control over the flow of money. In Capitalist countries, they rigged the system so that the wealthiest paid little or no tax, while benefitting from the infrastructure of the nation and from the protection of their trade and property by the military and courts, which served their interests and their interests alone. Not only did they not pay the taxes, they reaped the profits of the bond markets which also effectively taxed the people. They not only kept the monies which they should have been paying in taxes, they earned interest on the monies which otherwise would have been lost to them in taxes—interest for which the people paid. All of these advantages quickly put virtually all of the wealth of the nation into their hands and prevented others from ever advancing to a point where they could effectively challenge them.

Jews could also contract the money supply by refusing loans, by calling in loans, by selectively issuing different rates for loans in different nations, and by melting down metallic currencies. This is an especially powerful means for garnering international control, because it provides empires with a means of securing protectionism and favoritism, by increasing the costs of production and other costs in colonial and competing nations. In this way, the Jews were able to accumulate much of the world's wealth into a given nation or empire which they effectively ran through corruption, and then take that wealth unto themselves, leaving the nation which otherwise would fight to take back the wealth the Jews had taken from them, in ruins. The Jews would then take the wealth they had stolen to another nation they could build up in order to knock down. The last ruined nation had not the funds with which to attack the new host nation, or host empire, and the Jews obtained security by means of bribery and blackmail. Those who were aware of what the Jews were doing and objected to it were often assassinated. Thomas Jefferson warned Americans, in anticipation of the Great Depression in the Twentieth Century, when he stated in an 1802 letter to Albert Gallatin, Secretary of the Treasury,

"I believe that banking institutions are more dangerous to our liberties than standing armies... . If the American people ever allow private banks to control the issue of their currency, first by inflation, then by deflation, the banks and corporations that will grow up around [the banks]... will deprive the people of all property until their children wake-up homeless on the continent their fathers conquered... . The issuing power should be taken from the banks and restored to the people, to whom it properly belongs."

Even if the issuing power of money is granted to the people, a group acting in collusion can melt down metallic currencies and syphon off the money supply. Fiat money is no guarantee of safety if the money is based on bonds, because the Jews can then tax the people into poverty by instigating wars or government projects which cannot be paid for immediately by direct taxes. Should this fail to give the Jews control over the money supply, as in the case of Russia, the Jews can then instigate a revolution and deliberately cause chaos in a nation. They then spread word that banking reform and a dictatorship are the only means to restore order. Then the Jews install a dictator of their choosing, who funnels off the wealth of the subject nation into Jewish coffers, and who instigates wars of the Jews' choosing, which further profits them.

The Jews again ruined the Egyptians many times in the modern era. They deliberately bankrupted the nation and exploited its cotton markets and water ways. The purchase of the Suez, which was made to draw England into the region to sponsor Zionist ambitions, was then used as an excuse to secure alleged English interests in the region by means of Jewish colonialism. However, had it not been for the corrupt actions of Disraeli and Rothschild which brought England into the region, there would have been no English interests to secure, and placing a Jewish colony in Palestine would have worked against British interests in that it would have destabilized the region.

An article in the *Christian Reader*, Volume 3, Number 67, (19 November 1824), p. 366 evinces that the Jews were not needed by the British to secure British interests in the region, but rather that the British were needed by the Jews to secure Jewish Messianic interests in the region. Note that the Rothschilds and the Jews believed they had an incentive to ruin the Egyptians, in order to promote their own interest in the theft of the land of Palestine. Of course, any Egyptian who reacted to the Jewish attack on their civilization would be called a racist and religiously intolerant, which defamations Jewish racists would employ as an excuse to further ruin the Egyptians.

"CHRISTIAN REGISTER.

BOSTON, FRIDAY, NOVEMBER 19, 1824.

THE JEWS. It is stated with much assurance in the Gazette of Spires, that the Sublime Porte has recently made proposals to the House of Rothschild for the loan of a considerable sum of money, and has offered as a security for payment, the entire country of Palestine. It is stated also that in consequence of this proposal a confidential agent had been dispatched by that House to Constantinople, 'to examine into the validity of the pledge offered by the Turkish Cabinet.'

The editor of the *National Advocate* observes in relation to this report, that he at first supposed it was intended as a satire on the prevailing custom of raising loans for different nations; but on a nearer view of the subject, the proposition might be supposed probable. The Advocate proceeds with some interesting remarks on the subject, tending to show, that if such a proposition had been made it could not be accepted with any prospect, on the part of the Rothschilds, (who are Jews,) of the immediate restoration of their countrymen to Palestine, as it was probably not in the power even of the Turkish government, to guarantee to the Jews the quiet possession of the country against the prejudices and interests of the Egyptians, the Wechabites, the Wandering Arabs, and the Tartar Hordes.

It is also argued that the descrepancy of education, habits, views, and manners, existing between the Jews of different countries, unfit them to amalgamate and become united under one government. They must be prepared for this by the same discipline which their fathers, who went out of Egypt were subjected to under Moses, for forty years in the wilderness, to prepare them for the promised land. 'Our country,' continues the Advocate, 'must be an asylum to the ancient people of God. Here they must reside; here, in calm retirement, study laws, governments, sciences; become familiarly known to their brethren of other religious denominations; cultivate the useful arts; acquire a knowledge of legislation, and become liberal and free. So, that appreciating the blessings of just and salutary laws, they may be prepared to possess permanently their ancient land, and govern righteously.'"

The pretext Disraeli and the racist Zionists used to justify the purchase of the Suez Canal was to persuade England that she had a vital interest in securing a route to India—the same pretext Hess and the racist Zionists had used in their earlier attempts to draw France into the region. The common denominator of this prolonged effort to take land from the Moslems was racist Zionism, not a genuine need for a European presence in the Middle East.

Disraeli flattered the Queen by dubbing her the "Empress of India". Disraeli is perhaps overrated as an intellect and politician. He probably only succeeded because of support from the Rothschilds, who had the ability to shut down the English economy. Disraeli did not create this scheme to draw England into Egypt. Rather, it arose in the mind of an influential American Jew named Mordecai Manuel Noah,[317] who published *Discourse on the Evidences of the*

---

317. M. M. Noah, *Call to America to Build Zion*, Arno Press, New York, (1814/1977); **and** *Discourse Delivered at the Consecration of the Synagogue of [K. K. She`erit Yisra`el] in the City of New-York on Friday, the 10th of Nisan, 5578, Corresponding with the 17th of April, 1818*, Printed by C.S. Van Winkle, New-York, (1818); **and** *Discourse on the Evidences of the American Indians Being the Descendants of the Lost Tribes of*

American Indians Being the Descendants of the Lost Tribes of Israel: Delivered Before the Mercantile Library Association, Clinton Hall: J. Van Norden, New York, (1837); so as to make it appear that the Jews had a greater right to America than the Gentiles. Noah stated in 1837,

> "Firmly as I believe the American Indian to have been descended from the tribes of Israel, and that our continent is full of the most extraordinary vestiges of antiquity, there is one point, a religious as well as a historical point, in which you may possibly continue to doubt, amidst almost convincing evidences.
> If these are the remnants of the nine and a half tribes which were carried into Assyria, and if we are to believe in all the promises of the restoration, and the fulfillment of the prophecies, respecting the final advent of the Jewish nation, what is to become of these our red brethren, whom we are driving before us so rapidly, that a century more will find them lingering on the borders of the Pacific ocean?
> Possibly, the restoration may be near enough to include even a portion of these interesting people. Our learned Rabbis have always deemed it sinful to compute the period of the restoration; they believe that when the sins of the nation were atoned for, the miracle of their redemption would be manifested. My faith does not rest wholly in miracles—Providence disposes of events, human agency must carry them out. That benign and supreme power which the children of Israel had never forsaken, has protected the chosen people amidst the most appalling dangers, has saved them from the uplifted sword of the Egyptians, the Assyrians, the Medes, the Persians, the Greeks and the Romans, and while the most powerful nations of antiquity have crumbled to pieces, we have been preserved, united and unbroken, the same now as we were in the days of the patriarchs—brought from darkness to light, from the early and rude periods of learning to the bright reality of civilization, of arts, of education and of science.
> The Jewish people must now do something for themselves, they must move onward to the accomplishment of that great event long foretold—long promised—long expected; and when they DO move, that mighty power which has for thousands of years rebuked the proscription and intolerance shown to the Jews, by a benign protection of the *whole* nation, will still cover them with his invincible standard.
> My belief is, that Syria will revert to the Jewish nation by *purchase*, and that the facility exhibited in the accumulation of wealth, has been a providential and peculiar gift to enable them, at a proper time, to re-occupy their ancient possessions by the purse-string instead of the sword.

Israel: Delivered Before the Mercantile Library Association, Clinton Hall, J. Van Norden, New York, (1837); **and** *Discourse on the Restoration of the Jews: Delivered at the Tabernacle, Oct. 28 and Dec. 2., 1844*, Harper, New York, (1845); **and** *The Jews, Judea, and Christianity: A Discourse on the Restoration of the Jews*, Hugh Hughes, London, (1849).

We live in a remarkable age, and political events are producing extraordinary changes among the nations of the earth.

Russia with its gigantic power continues to press hard on Turkey. The Pacha of Egypt, taking advantage of the improvements and inventions of men of genius, is extending his territory and influence to the straits of Babelmandel on the Red sea, and to the borders of the Russian empire; and the combined force of Russia, Turkey, Persia and Egypt, seriously threaten the safety of British possessions in the East Indies. An intermediate and balancing power is required to check this thirst of conquest and territorial possession, and to keep in check the advances of Russia in Turkey and Persia, and the ambition and love of conquest of Egypt. This can be done by restoring Syria to its rightful owners, not by revolution or blood, but as I have said, by the purchase of that territory from the Pacha of Egypt, for a sum of money too tempting in its amount for him to refuse, in the present reduced state of his coffers. Twelve or thirteen millions of dollars have been spoken of in reference to the cession of that interesting territory, a sum of no consideration to the Jews, for the good will and peaceable possession of a land, which to them is above all price. Under the cooperation and protection of England and France, this re-occupation of Syria within its old territorial limits, is at once reasonable and practicable.

By opening the ports of Damascus, Tripoli, Joppa, Acre, &c., the whole of the commerce of Turkey, Egypt, and the Mediterranean will be in the hands of those, who even now in part, control the commerce of Europe. From the Danube, the Dneister, the Ukraine, Wallachia and Moldavia, the best of agriculturalists would revive the former fertility of Palestine. Manufacturers from Germany and Holland; an army of experience and bravery from France and Italy; ingenuity, intelligence, activity, energy and enterprise from all parts of the world, would, under a just, a tolerant and a liberal government, present a formidable barrier to the encroachments of surrounding powers, and be a bulwark to the interests of England and France, as well as the rising liberties of Greece.

Once again unfurl the standard of Judah on Mount Zion, the four corners of the earth will give up the chosen people as the sea will give up its dead, at the sound of the last trumpet. Let the cry be Jerusalem, as it was in the days of the Saracen and the lion-hearted Richard of England, and the rags and wretchedness which have for eighteen centuries enveloped the persons of the Jews, crushed as they were by persecution and injustice, will fall to the earth; and they will stand forth, the richest, the most powerful, the most intelligent nation on the face of the globe, with incalculable wealth, and holding in pledge the crowns and sceptres of kings. Placed in possession of their ancient heritage by and with the consent and co-operation of their Christian brethren, establishing a government of peace and good will on earth, it may then be said, behold the fulfillment of prediction and prophecy: behold the chosen and favoured people of Almighty God, who, in defence of his unity and omnipotence, have been the outcast and proscribed of all nations, and who for thousands of years have patiently endured the severest of human sufferings, in the hope of that great advent of which they never have despaired:—and then when taking their rank once more among the nations of the earth, with the good wishes and affectionate regards of the great family of mankind, they may by their tolerance, their good faith, their charity and enlarged liberal views, merit what has been said in their behalf by inspired writers, 'Blessed are they who bless Israel.'"[318]

Noah published *Discourse on the Restoration of the Jews: Delivered at the Tabernacle, Oct. 28 and Dec. 2., 1844*, Harper, New York, (1845); in which he laid out the plan to draw England into the Mideast, which Disraeli and Rothschild fulfilled.

The New York Times reported on 31 December 1897 on page 5 that some Jews—especially those allied with the Puritans, a sect likely created by Cabalist Jews—had long sought America as a new Israel, and told of Judge Noah's plan to draw the British into the region and destroy the Turks:

"America and the Ten Tribes.

Dr. Alder, in reply to Dr. Kohler, contended that Anthony Montecinos originated the idea that America was the abode of the ten tribes.

Dr. Kohler said that the term Arsaveth was never used in Jewish writings. The term these was Eretz Aheret.

Dr. Leo Wiener gave some striking specimens of the folk-lore of the Russian Jews, which, he said, had thus far been virtually ignored in literature. He repeated an amusing story of a little Jewish tailor who set out to discover the lost tribes. He found them at last beyond a great river, and they were giants. One of them put the little tailor in his pocket, and going into the synagogue, forgot all about him. The little tailor made answer of 'Amen,' however, to the prayer that was offered. Then he was taken out of the pocket, was recognized as a Jew, and was greatly honored.

Sarcastic comments upon several of the theories about the lost tribes that have been put forward were made by Dr. H. P. Mendes and others. The Rev. A. H. Neito reported upon some inscriptions upon ancient Jewish tombstones in New York which he had deciphered.

Early Zionist Projects.

A paper by Max J. Kohler on 'Some Early American Zionist Projects' was next read, and engaged the close attention of those present. The most curious part of it, and one which excited both laughter and applause, was an account of the three projects of Mordecai M. Noah, once a distinguished figure in New York, to re-establish the Jewish Kingdom. Mr. Kohler first reviewed the efforts to colonize Jews in this hemisphere, from the establishment of the settlement in Curacao, in 1652, and the scheme of Maurice de Saxe, about 1749, to create a kingdom for himself, peopled by

---

318. M. M. Noah, *Discourse on the Evidences of the American Indians Being the Descendants of the Lost Tribes of Israel: Delivered Before the Mercantile Library Association*, Clinton Hall: J. Van Norden, New York, (1837), pp. 37-40.

the descendants of Abraham, and the projects of Dr. Kayurling and W. D. Robinson in this country, the former in 1783 and the latter in 1819. Judge Noah's first idea, announced in 1818, was that the Jews were to overthrow the Turkish domination in Northern Africa and Western Asia, and to regain possession of Palestine. In 1825 he devised the plan of founding the 'City of Ararat' on Grand Island in the Niagara River. He got some of his friends to constitute him 'Governor and Judge of Israel.' He issued proclamations and decrees, and made appointments which were laughed at and refused. In setting forth his third idea in 1845, in a pamphlet, 'The Restoration of the Jews,' Judge Noah made this remarkable forecast: 'England must possess Egypt, as affording the only secure route to her possessions in India, through the Red Sea.' This, he thought, would lead to the resettlement of the Jews in Palestine, with the consent of the Christian, and for the safety of the neighboring nations. This was to be accomplished by gradual means, the first step being to induce the Sultan to grant to the Jews permission to purchase and hold land in Palestine. Mr. Kohler drew attention to the parallelism of the arguments employed by Noah, from whom he quoted at length, in favor of this scheme, and those of the Zionites of to-day, as represented by the Congress at Basel.

A sketch of the Jewish pioneers of the Ohio Valley by the Rev. Dr. David Philipson of Cincinnati, a paper on 'Ezra Stiles (first President of Yale) and the Jews of Newport,' notes on New York wills by Dr. Herbert Friedenwald, 'A Statement Relative to Manuscripts Belonging to Hyam Solomon,' by Dr. J. H. Hollander of Johns Hopkins University, and a paper entitled 'A Brave Frontiersman,' by the Rev. Henry Cohen of Galveston, Texas, were among the other contributions. Dr. C. D. Spivak of Denver sent an argument in favor of the society making an index of periodical and pamphlet literature and data on Jewish-American history. On motion a committee was appointed consisting of Prof. R. J. H. Gottheil, Dr. Friedenwald, and the Rev. Dr. Mendes, to take charge of the matter.

The selection of the place for holding the next annual meeting was left in the hands of the council. The meeting was then adjourned."

Disraeli and Rothschild artificially created an animosity in England towards Russia. Zionist publications called the Turkish Sultans and the Russian Czars the anti-Christs. The Rothschilds curbed Pan-Slavic interests by regulating Russia's access to funds, in order to promote instead the interests of Pan-Judaism.[319] The Rothschilds, who were already the Kings of the Gentile world, had long been seeking to have one of their own become the official King of the Jews and rule the world from Jerusalem as Messiah, as prophesied in *Isaiah*. On 14 July 1878, *The Chicago Daily Tribune* reported on page 9 that the Rothschilds, and their agents around the world, organized an international Pan-Judaic union, which

---

[319]. "Pan-Judaism", *The Chicago Daily tribune*, (14 July 1878), p. 9. "General Notes", *The Chicago Daily Tribune*, (8 September 1878), p. 9.

would rule the Jews and the world,

"PAN-JUDAISM.
WHAT IS LIKELY TO BE DONE AT THE PARIS CONFERENCE.

An International Jewish Conference will be held this month in Paris for the purpose of discussing measures to improve the political and social condition of the Jews in various parts of the world. Delegates will be sent from Jewish congregations in every quarter of the globe. The veteran Adolphe Crenneix [sic] is expected to preside, and among the delegates will probably be Chief Rabbi Astruc and M. Oppenheim from Belgium, Senator Artom from Italy, Chief Rabbi Cahn and Baron de Rothschild from France, Sir Julian Goldsmid and Baron de Worms from England, Baron de Rothschild and Dr. Jellinck from Austria, Mr. William Seligman from the United States, and a member of the Jewish clergy. Among the matters which occupy the attention of the conference are: The condition of the Jewish residents of the Danubian principalities and of Russia, Morocco, and Persia; the best means for securing industrial and educational advantages for the Jews of Jerusalem; the adoption of measures for the promotion of Hebrew education and for the advancement of Hebrew literature. The most important subject to be considered is a proposition to convene a synod for the purpose of inquiry into the condition of modern Judaism and the authoritative exposition of Jewish ecclesiastical law. Within the past few years two synods have been held, avowedly for this purpose, one at Leipzig, attended chiefly by European Jews, and the other at Philadelphia, attended exclusively by American Jews. The proposition to be considered at the coming conference is to call a synod which shall represent the Jews all over the world.

Since the destruction of the Temple and the dispersion of the Jews there has been no regular priesthood nor any recognized ecclesiastical authority, except such as was assumed by the chief rabbis of the various communities, who frequently differ among themselves. Such changes and modifications as have been made in the Jewish ritual or the Jewish law have been introduced by the various communities on their own responsibility, and are not recognized by the Jews generally. Therefore, if such a synod as it is proposed to call could be convened, it would have a powerful effect upon the condition of the Jews everywhere, and it might result in the establishment of some central recognized ecclesiastical authority which would restore to the synagogue the discipline that it now lacks. Even the most orthodox Jews would pay respectful attention to the opinion of such a body, and, indeed, they are in favor of calling the synod. Mr. M. S. Isaacs, the President of the American Board of Jewish Delegates, says in a recent report:

> There is a choice between an exposition by skillful, learned, competent, authoritative teachers, expounders, and judges of the ecclesiastical law, and the capricious, unreliable, ephemeral decisions of the mere officials in a particular territory, town, or congregation. The latter method is seen in its

full extent in America.... . Such a representative synod, aiming to strengthen Judaism by the recognition of current forces and agencies, by the education and guidance of the general body, without interfering with individual liberty or congregational independence within its spere, would be an intense relief after that groping for a settlement of vexed questions, which has in despair turned in every direction for the counsel and example, and found no resource save in the untrained and deceptive public opinion of a congregation rarely fortunate in a minister at once educated and practical, versed in the law and able to calculate the effect of a novel interpretation, or a conscious departure from an existing ordinance."

When the Czars responded with suspicion towards the Jews of Russia (whom the English Zionists had asked to sponsor attacks on Persia and Turkey and later the Czars, at least since the days of David Alroy, in order to secure Palestine for the Jews), Rothschild feigned indignation and published his "Memorial of the Jews in England to the Czar of Russia" in 1882. *The Chicago Daily Tribune* reported on 19 February 1882 on page 5,

"THE JUDENHETZE.

Text of the Memorial of the Jews in
England to the Czar of Russia.

The following is the full text of the memorial of the Jews of England which was handed to Prince Lobanoff for transmission to the Emperor of Russia, but which the Prince declined to transmit, in accordance with instructions from his Government:

'*To his Imperial Majesty Alexander III., Emperor of All the Russias:* The humble memorial of the Jews of England on behalf of the Jews of Russia. May it please your Imperial Majesty, a grievous cry of suffering has reached us from our brethren in faith in many parts of your Majesty's great empire. For the past nine months large numbers of your Majesty's Jewish subjects, especially those residing in the southern provinces of your Majesty's dominions, have been the victims of serious civil outbreaks. The security of life and property, so many years enjoyed by them, has vanished. Murder, rapine, and pillage have taken its place. The most terrible deed of violence have been perpetrated on helpless women and children. Unarmed and unoffending men have become a prey to the fury of a brutal mob. The survivors, scarcely more fortunate than the slain, live only to find their homes devastated or burned, their fortunes wrecked, and their means of subsistence gone.

'Great, indeed, is our horror at these atrocities, but greater still, we feel certain, must be your gracious Majesty's pain and indignation at the sufferings thus inflicted on thousands of your subjects.

'Until last year Jews and Christians throughout your Majesty's empire lived on terms of amity rarely, if ever, disturbed. No act of the Jews has been committed to warrant the interruption of the friendly attitude of their neighbors or the goodwill of their rulers. Your Jewish subjects love and honor your Majesty, and in their homes and synagogs pray for your welfare. They respect the laws and pay the State its just dues. They serve your Majesty in peace and war, even without hope or chance of promotion, and willingly lay down their lives for the country that has given them birth, and that has hitherto protected them. In truth, they are commanded by our sacred books to promote the welfare of the land which shelters them, to obey its laws, to honor its rulers, and to love as themselves their neighbors, though differing in faith; and the Israelites, acting in conformity with those precepts, are innocent of cause for the oppression that has befallen them.

'We have reason to believe that in most cases it has not been the honest, law-abiding neighbors of the Jews who have originated or perpetrated these lamentable excesses, but professional agitation from a distance, acting upon the turbulent and revolutionary spirits, the enemies of law, loyalty, and order. No better proof of this can be afforded than the fact that the ringleaders have in many localities, with an audacity and shamelessness unparalleled in history, traitorously used the august name of your Majesty as a warrant for their infamous projects, and have published a forged ukase purporting to authorize the general spoliation of the Jews.

'But we fear the cup of affliction of our brethren is not yet full, for the future appears even blacker than the past. For now the enemies of our brethren seek to palliate the atrocities that have been perpetrated, falsely declaring the Jews to have merited their persecution by their own misconduct, by their odious mode of trading, and by their having overreached their neighbors; and these enemies endeavor to induce the Government of your Majesty to impose upon all Israelites such new restrictions as to residence, occupation, and education as will not only prevent their fairly competing with their Christian fellow-subjects, but will practically prevent their becoming useful citizens and servants of the State, and will even debar them from earning their subsistence.

'We have heard with alarm and grief that commissions have been issued with instructions couched in terms of opprobrium and hostility, teeming with charges, assumed, but not true, which would render impossible any result favorable to the Jews. The worst effects are, therefore, apprehended. Even in Poland, where the Israelites have ever dwelt on terms of good fellowship with their neighbors, and where, until the lamentable event of last month, they have always enjoyed immunity from outrage of any kind, like commissions have been issued with similar instructions, so that everywhere throughout your Majesty's dominions the poplace seems to imagine that it has the Imperial sanction for its ill-treatment of our brethren, an idea which we are convinced could never have been, however faintly, conceived by the benignant and humane spirit of your Majesty.

'Already deplorable results have ensued from the terms in which these commissions have been issued. For many local authorities, in anticipation of the reports of the commission, have put in force certain ancient laws of domicile, which had fallen into desuetude, and have forcibly driven the

Jews, still smarting from their recent calamities, away from the towns and villages which they have so long been permitted to inhabit; while others, perhaps a little less inhuman, have allowed them to remain only on condition of their being pent up within the limits of their ancient ghettos.

'With regard to the imputations that have been made upon your Majesty's Jewish subjects, we humbly submit to your Majesty that whatever exceptional social position they may occupy, or whatever failings may be charged to some of them, these are due mainly to the exceptional laws to which they have been so long subjected.

'If, in some places, undue activity has characterized their conduct in certain trades and occupations, we believe it to be because other means of earning a subsistence have been denied them, because they have been too crowded in particular localities, and have, therefore, experienced the greatest difficulty in gaining a livelihood.

'We feel certain that if the special laws affecting the Jews were abolished their exceptional status, social and civil, would come to an end. Complaint would no longer be heard of their undue commercial and economic activity operating to the detriment of others if the Jews were suffered to disperse themselves at will so as to become merged amid their fellow-subjects instead of being concentrated, to the injury of themselves and others, in overcrowded hives of industry.

'Here in England, where perfect civil and religious equality has been granted us, we English Jews can bear testimony to the happy results effected by such complete emancipation. Here all those restrictions—civil, commercial, and educational—which formerly oppressed us have happily been removed, and, as a result, Jew and Christian here live and work side by side on terms of mutual respect and good fellowship, engaged in friendly rivalry, which stimulates public industry and adds to the common weel.

'And so, sire, may it be in the mighty Empire whose destinies you wield with wisdom and enlightenment. For, as the late Emperor, your father, of sainted memory, rendered his name immortal as emancipator of millions of serfs, even so it may be your Mejesty's high destiny to give life and protection to those now trembling on the verge of destruction, to give equal rights to the millions of your loyal Jewish subjects, who in their dread emergency look up to you, sire, Emperor and father of your people, only for leave to live with home and hearth secure from violence.

'Humbly do we present this memorial to your Majesty on behalf of our brethren in the name of humanity—the foundation of all religion; in the name of justice—the heritage of all; in the name of mercy—the prerogative of Imperial power.

'And we shall ever pray that the Supreme King or Kings may bless the efforts of your Majesty for the glory of your mighty Empire and the well-being of your subjects, and that He may grant your Majesty a long, and prosperous, and happy reign.

'Signed, on behalf of the Jews of England, this 19th day of January.

'N. M. DE ROTHSCHILD."

British Jews organized for centuries to destroy Russia and Turkey. They set forth their plans in countless books and articles, which concomitantly called for the "restoration of the Jews to Palestine" and the annihilation of the Russian and/or Turkish "anti-Christs". Jews were behind the revolts in those lands in the Twentieth Century which decimated their empires, cultures and their futures. Jews in general considered Gentiles to be animals, and not their Hebrew "neighbors", and thus Russians were not protected by Jewish law in the sense which Rothschild alleged. In addition, many Jews considered Slavs to be lower than Aryans, and thus beneath the contempt many Jews had for Gentiles in general. Contrary to Rothschild's assertions, Jewish tribalism, racism and corruption did indeed continue after emancipation and became most manifest when Jews were accorded the greatest freedom after the Bolshevik Revolution and took advantage of their liberty as an opportunity to slaughter Gentiles. Most tellingly, when Russian Jews sought to emigrate to England and America, it was English and American Jews who most strongly opposed their emigration, realizing better than anyone else how tribal, racist and corrupt Russian and Galician Jews could be.

The Zionist financiers were so successful in making it appear that Great Britain was acting out of its own best interests by inserting itself into the Turkish Empire, and not acting pursuant to the instigation of the Zionists; that many came to conclude that the Balfour Declaration materialized out of British interests. Ironically, this backfired on the Zionists, and some sectors of the British Government were reluctant to give up Palestine to the Jews, while others were reluctant to incite the French to war by interfering with French interests in the region—all of which frustrated the Zionists' efforts to steal the land from the Palestinians after the First World War. *The London Times* reported on 29 June 1920 on page 15,

"THE POPE AND ZIONISM.

ACRIMONIOUS ITALIAN COMMENT.
(FROM OUR OWN CORRESPONDENT.)

ROME, JUNE 27.

Sir Herbert Samuel, High Commissioner to Palestine, who left Rome last night, visited both the King and the Pope. His visit to the Pope has attracted a certain amount of attention, as it was bound to do.

The *Tempo* comments acrimoniously on British policy in Palestine, saying that England merely supported Zionism in order to find an excuse for establishing herself there, where she had no other excuse to be. But the *Tempo* has never been anything but anti-British. The article finishes by asking whether Sir Herbert Samuel attempted to assure the Pope that fears inspired by Zionism were unfounded, and whether he is likely to have succeeded.

Certainly the Vatican has been nervous about Zionism, and certain utterances have given it cause to be. But there is every reason to believe that

Sir Herbert should be able to still these fears by proving them to be unjustified."

This was, however, a minor obstacle for the Rothschilds when compared with the fact that most Jews did not wish to live in Palestine and did not have the racist mindset of the Zionists. *The London Times* reported on 17 June 1918 on page 5,

"FUTURE OF PALESTINE.

OPPOSITION TO ZIONISM.

The ideals of the League of British Jews in regard to the future of Palestine as distinct from those of the Zionists were expounded by Dr. Israel Abrahams, of Cambridge, at Wigmore Hall yesterday.

What divided the League from the Zionists, he said, was that the former could not assent to the setting up in Palestine of a State composed exclusively of Jews. They maintained that, whatever the government, the State should be absolutely free from any racial or religious test. Citizenship and nationality had nothing to do with religion. As to the Jews outside, the League could not assent to the statement that they constituted a nation. They belonged to many nations, and could neither control Palestinian politics nor be controlled by them. The Jews of the world were not united, but divided by nationality, and now were actually fighting each other. The Palestine of the future was for the Jews who desired to live there, and for those who wished to escape from countries where they had no home.

In a discussion which followed, some opposition to the lecturer's point of view was shown, and one speaker asserted that the League had hindered the colonization of Palestine."

The tribalism of Rothschild is apparent not only his covert designs to destroy Russia and to use English treasure and lives to achieve his ends, not only in the fact that he felt a tribal kinship with the Jews of Russia and rushed to defend them, but in his statement that even after the Jews had been emancipated in England they were at perpetual war with the Christians,

"Here all those restrictions—civil, commercial, and educational—which formerly oppressed us have happily been removed, and, as a result, Jew and Christian here live and work side by side on terms of mutual respect and good fellowship, engaged in friendly rivalry, which stimulates public industry and adds to the common weel."

How did the Rothschilds gain the wealth which fed their arrogance? In part by stealing from the English, who had granted the Jews freedom. This Jewish theft of British treasure took place at a time when England was at war. That was how the Rothschilds repaid English generosity. It was the Rothschilds' method of "friendly rivalry" with the Christians. If the Christians had responded in unkind, the Jews would have been wiped out in a very short while. Perhaps the English example gave the Czar pause.

Concerned that the Rothschilds were moving into America during the Civil War, after having largely ruined the markets of Europe by plundering Europe's wealth, on 2 June 1867 on page 3, *The Chicago Tribune* told part of the story of the Jewish war profiteers and cheats, the Rothschilds of their day. It was one of many stories the *Tribune* ran, which exposed the Rothschilds:

"THE HOUSE OF ROTHSCHILD.

Its Origin and History—The 'Red Shield'—The Power and Wealth of the Rothschilds—Their Operations with American Bonds—The Rothschilds and the Pope.

(Frankfort Correspondence of the Boston Journal.)
THE RED SHIELD.

Come with me to the eastern part of the city—the old town—where you will discover scarcely a sign of modern architecture. The streets are narrow; the houses lean toward each other from opposite sides of the way, as if they were friends about to fall into each other's arms. It is the Jews' quarter. The door-ways are crowded with women and children—all bearing the unmistakable features which, the world over, characterize this historic people—rejected of God, despised of men, scattered everywhere, yet retaining their nationality, endowed with a vitality which has no parallel in the human race.

We turn down the Judengasse, the Jew's alley, from the chief thoroughfare of the modern town. In this street, 124 years ago, lived a dealer in old clothes who had a red shield for a sign, which in German reads *Roth Schild*. It was in 1743 that a child was born to this Israelite. The name given to the boy was Anselm Meyer, who also became a clothes dealer and a pawn broker, succeeding to the business of his father. By degrees he extended his business, lending money at high rates of interest during the wars of the last century, managing his affairs with such skill that Prince William the Landgrave made him his banker. When Napoleon came across the Rhine, in 1806, this clothes dealer was directed to take care of the treasures of the Prince, amounting to twelve million dollars, which he invested so judiciously that it brought large increase to the owner, and especially to the manager.

This banker died in 1812, leaving an estate estimated at $5,000,000—not a very large sum these days—but he left an injunction upon his five sons, which was made binding by an oath given by sons around his death-bed, which has had and still has a powerful influence upon the world. The sons bound themselves by an oath to follow their father's business together, holding his property in partnership, extending the business, that the world might know of but one house of the *red shield!* (Rothschild.)

The sons were true to their oath. Nathan went to Manchester, England, as early as 1797, but afterward moved to London. Anselm remained at

Frankfort, James went to Paris, Solomon to Vienna, and Charles to Naples, the five brothers thus occupying great financial centres. Nathan, in London, amassed money with great rapidity, and the same may be said of all the others, the wars of Napoleon being favorable to the business of the house. Nathan went to the Continent to witness the operations of Wellington in his last campaign against Napoleon, prepared to act with the utmost energy, let the result be as it might. He witnessed the battle of Waterloo, and, when assured of Napoleon's defeat, rode all night, with relays of horses, to Ostend; went across the channel in a fishing smack—for it was before the days of steam—reached London in advance of all other messengers, and spread the rumor that Wellington and Blucher were defeated. The 20th of June in that memorable year was a dismal day in London. The battle was fought on the 18th. Nathan Meyer, of the house of Red Shield, by hard riding, reached London at midnight on the 19th. On the morning of the 20th, the news was over town that the cause of the allies was lost, that Napoleon had swept all before him. England had been the leading spirit in the struggle against Napoleon. The treasury of Great Britain had supplied funds to nearly all of the allied Powers. If their cause was lost what hope was there for the future? Bankers flew from door to door in eager haste to sell their stocks. Funds of every description went down. Anselm Meyer was besieged by men who had funds for sale. He too had stocks for sale. What would they give? But meanwhile he had scores of agents purchasing. Twenty four hours later Wellington's messenger arrived in London; the truth was known. The nation gave vent to its joy; up went the funds, pouring, it is said, five million dollars into the coffers of this one branch of the house of the Red Shield!

Though Frankfort is comparatively a small city, though it has no imperial court, it is still a great money centre, solely because that here is the central house of the Rothschild and other bankers.

The House of the Red Shield is the greatest banking house of the world—the mightiest of all time. Its power is felt the world over—in the Tuileries of Paris, in the ministerial chamber of Berlin, in the imperial palace at St. Petersburg, in the Vatican at Rome, in the Bank of England, in Wall street, State street, and by every New England fireside. The house of the Red Shield, by the exercise of its financial power, can make a difference in the yearly account of every man who reads these words of mine! Though Anselm Meyer has been half a century dead; though several of his sons have gone down to the grave—the house is the same. The grand-children have the spirit of the children. The children of the brothers have intermarried, and it is one family, animated by a common purpose, that the world shall know only one *red shield*.

AMERICAN BONDS.

The house, at an early stage of the American war, took hold of the United States bonds. Germany had confidence in America. England strove for our ruin, but the people of the Rhine believed in the star of American liberty. Fifty years of peace had been long enough to bring wealth to this land, and so with every steamer orders were sent across the Atlantic for investment in American securities. It is supposed that Germany holds, at the present time, about three hundred and fifty millions of United States bonds, and it is said that there have been no less than fifty million dollars profit to the bankers of Frankfort on American securities since 1863!

The great banking houses here make little show. The transactions of the Rothschilds amount to millions a day, and yet the operations are conducted as quietly as the business of a small counting house. You can purchase any stock here. Passing along the street I noticed bonds of the State of California—of several American States—of the United States—bonds in Dutch, Russian, Turkish, Arabic, Spanish, Italian, French—bonds of all lands—of States, cities, towns and companies. The reports of the Frankfort exchange are looked at by European bankers with as much interest as that of London or Paris.

Erlanger, the banker who negotiated the rebel cotton loan, and who fleeced English sympathizers with the South out of fifteen million dollars, has a house here. he has just now taken hold of the new Tunisian loan, but his management of the rebel loan has brought discredit upon his house.

The power of the Red Shield was felt by Prussia last summer. The Prussian Government demanded an indemnity of great amount, twenty-five million dollars, I believe, from the city of Frankfort. The head of the house of the Red Shield informed the Count Bismark that if the attempt was made to enforce that levy he would break every bank in Berlin; that he had the power to do it, and that he should exercise the power. Prussia had won a victory at Konnigratz; but here, in the person of one man, she had met an adversary who had the power to humble her, and she declined the contest. A much lower sum was agreed upon, which was paid by the city.

THE ROTHSCHILDS AND THE POPE.

For fifteen centuries the Jews have been cursed by the Pope, and persecuted by the Roman Church. There is no more revolting chapter of horrors in history than that of the treatment of the Jews at the hands of the Pontiffs. In all lands where the Roman religion is dominant the children of Israel have been treated with barbaric rigor—allowed few privileges, denied all rights, looked upon as a people accursed of God, and set apart by divine ordination to be trampled upon by the church. In Rome, at the present day, the Jews are confined to the Ghetto; they are not allowed to set up a shop in any other part of the city; they cannot leave the city without a permit; they can engage only in certain trades; they are compelled to pay enormous taxes into the Papal treasury; the are subject to a stringent code of laws established by the Pope for their special government; they are imprisoned and fined for the most trivial of offences. They cannot own any real estate in the city; cannot build or tear down or remodel any dwelling or change their place of business, without Papal permission. They are in abject slavery, with no right whatever, and entitled to no privileges, and receive none, except upon the gracious condescension of the Pope. In former times they were unmercifully whipped and compelled to listen once a week to the *Christian* doctrine of the priests. But time is bringing changes. The Pope is in want of money; and the house of the red shield has money to lend on good security. The house is always ready to accommodate Governments. Italy wants money, so she sells her fine system of railroads to the Rothschilds. The Pope wants money, and he sends his Nuncio to the wealthy house of the despised race, offers them security on the property of the church, the Compagna, and receives ten million dollars to maintain his army and Imperial State. That was in 1865. A year passes, and the Pontificial expenditures are five million more than the income, and the deficit is made up by the Rothschilds, who take a second security at a higher rate of interest. Another year has passed and there is a third great annual vacuum in the Papal treasury of six million, which quite likely will be filled by the same house. The firm can do it with as much ease as your readers can pay their yearly subscription to the weekly *Journal*. When will the Pope redeem his loan at the rate he is going? Never. Manifestly the day is not far distant when these representatives of the persecuted race will have all the available property of the Church in their possession. Surely time works wonders."

Russians had many reasons to suspect Russian Jews, who were pledged to retaliate against Russian Gentiles for the persecutions they had faced. *The Chicago Daily Tribune* wrote on 21 July 1878 on page 13,

"BEACONSFIELD'S LUCK.

Bismarck's Hand Disclosed in the
Workings of the Congress
at Berlin.

How the Jew Bankers Revenged
Themselves for Insults to Their
Race.

Correspondence New York Graphic.
LONDON, July 6.—All hail, Beaconsfield!

He is the hero of the hour. He is looked upon by all loyal Englishmen as the pivot on which has turned all the deliberations of the Berlin Congress. But is this the correct view?

Not at all. England's triumphs at Berlin are simply incidents in the 'streak of luck' which has marked the career of this great political adventurer.

I am enabled to furnish the *Graphic* with the first true account of the recent moves on the chess-board of European politics.

The result of the Congress may be briefly stated as the complete humiliation of Russia. True, she receives Batoum, with conditions that render the concession practically valueless. True, she regains her little strip of Bessarabia that had been given to Roumania, and she is permitted to retain Kars. But it is her rivals who have secured the material advantages at the Congress, and, worse than all, it is England, her special rival, who has been made the chief recipient of the fruits of Russia's expenditure of blood and treasure.

It is now certain—it will be published in the journals and confirmed in Parliament ere this letter is 1,000 miles on its way to you—that England is to have Cyprus as her own, and is to acquire a protectorate of the whole of Asiatic Turkey, with practically illimitable possibilities of the extension of trade in the Levant and down the Valley of the Euphrates. Egypt is virtually hers; the Suez Canal is absolutely in her control.

Russia has acquired neither facilities for the extension of her trade nor territory; and she has lost all the prestige acquired by the war.

What does this mean?

The answer to this question involves three names—Rothschild, Bismarck, Andrassy.

First, as to Rothschild. The sympathy of the Hebrews all over the world has been with Turkey and against Russia. Russia, in the nineteenth century, has oppressed and persecuted the Jews with the most bitter and malignant cruelty. The hatred of the Greek Church for the Jews to-day is as intense as was that of some of the bigoted Catholics in the Middle Ages for that long suffering and persecuted race. The success of the Russian arms against Turkey filled the Jews with indignation and alarm. The Turks in their rule in Europe and in Asia have been tolerant alike to Christian and to Jew; it may be said they have been forced to award this tolerance; but it was not in violation of their faith nor of the will of their great Prophet, for to this day there exists the authenticated manuscript of the famous decree of Mohammed, in which he commands the faithful to abstain from persecuting and to treat charity and kindness the Jews and Christians dwelling under their rule. But, against the personal wishes of the Czar, the blind and bitter hatred of the Russians for the Jews continually manifests itself, and their persecution of the chosen people has never ceased.

Russia was forced to make great pecuniary sacrifices to keep her armies in the field; she taxed her monetary resources to the utmost; and when the San Stefano treaty had been negotiated and the question of war or peace hung trembling in the balance, she found to her dismay that if she ventured upon a war with England she must reckon with a potent foe, of whose existence she had hitherto been disdainful, if not ignorant.

This foe was the most powerful element in Continental Europe.

All bankers are not Jews. But the Hebrew element among the money-lenders and money-masters of Europe is so widespread and so powerful that it was easy for it to effect combinations by which Russia was shut out from the privilege of borrowing money to continue to renew her march of conquest.

She tried to borrow in England—no money! She sought to effect a loan in Paris—no money! She intrigued through her most skillful agents in all the minor Bourses of Europe—not a rouble could she obtain. And now, as you will probably learn in a few days, she is in such desperate financial straits that, as a last resort, she is about to call upon her patriotic subjects—if she has any—to put their hands in their pockets and lend her their own money,—if they have any, which is doubtful.

Yes! In the very hour of Russia's military triumph, when, flushed with her dearly-bought victories, and with the Sultan willing to prostrate himself as a vassal at her feet, the despised and persecuted Israelite was able to say to the Czar: 'Thus far and no farther!'

It was not England who forced Russia to appear before the Berlin Congress, and submit to a revision of her extorted treaty with Turkey.

Russia was forced into this humiliation by the Jew bankers of the world.

Once in the Congress, Gortschakoff and Schouvaloff found to their dismay and horror that they were contending single-handed against all Europe.

Bismarck proved to be the arch enemy of Russia in the Congress, the master-spirit who formed the combination to humiliate her by the Treaty of Berlin after her victories more than she had been humiliated by the Treaty of Paris after her defeats.

Now for a State secret, hinted at in various ways, but which has never come to light in any official form, and the details of which cannot be fully known until after Kaiser William and Prince Bismarck are dead.

Bismarck, with true statesmanlike prescience, detests Russia. Russia is a military power of incalculable possibilities, capable, perhaps, in time, of overrunning and conquering all Europe. A war that would increase the military prestige or augment the territorial domain of Russia, Bismarck regarded with alarm and indignation.

Why, then, did he not put an end to the Russian and Turkish war?

The answer is—Kaiser William.

The German Emperor is swayed by his personal affections and his dynastic prejudices. The old gentleman never had much political sense. He supposed his personal honor was pledged to Russia. The Czar had not interfered with Prussia in her wars with Austria and France. He, then, should not interfere in Russia's contest with Turkey. Bismarck had been quite willing to have an amicable understanding with Russia as regarded Austria and France; but he had no intention of permitting Russia to gain a military and territorial predominance that might overshadow Germany.

Thus it was Bismarck who formed the combination that robbed Russia of the fruits of her great victories.

How did he effect this? Here comes in the third name—Andrassy.

The Prime Minister of Hungary, be it remembered, is a Hungarian statesman. Blood with him, also, is thicker than water. He remembers that, when Hungary had German-Austria at her feet in 1848, Russia sent 60,000 troops to the aid of Austria, turned the tide of victory, and crushed out forever the hopes of Hungary for independent neutrality. The hated Slav was thus used to overcome the legitimate and patriotic aspirations of Hungary.

I state upon the best authority that, in the conferences held in the beginning of the late war by Bismarck and Andrassy, the scheme was concocted which culminated in the yet unsigned Treaty of Berlin. It was in these conferences determined that Russia should be despoiled of the fruits of her victories. One of the results is seen in the virtual annexation of Bosnia and Herzegovina by Austria, and the great strengthening of that Power thereby.

Here, then, is the key to the mysteries of the Congress of Berlin. Rothschild, the representative of the Jews, closing the Bourses Europe against Russia; Bismarck, intent on the purpose of curbing and manacling the giant of the North in the interests of Western civilization; Andrassy paving off Russia for the injuries inflicted on Hungary in 1848, and turning her victories into Dead Sea fruit,—pleasant to the sight, but turning to ashes upon the lips.

But how about Disraeli—Beaconsfield? Is he not the real hero of this great dama? Not at all.

True, again, blood with him is thicker than water; and undoubtedly he placed himself in relation with the Jewish money-kings to effect the humiliation of Russia. True, he withdrew the timid and hesitating Lord Derby at the right moment, and put the courageous Marquis of Salisbury in his place. But the cession of Cyprus to England, and investing her with protectorate of Asiatic Turkey, was really the work of Bismarck.

Cyprus should have been given to France. The trade of the Levant properly belongs to her and to Italy more than to England. But Bismarck, in view of the prejudices of his own people,—not that he shares these prejudices, for he is a true statesman, but merely out of deference to these narrow hatreds and dislikes,—was compelled to permit England to take what really belongs to France, and by doing this he has crowned with a new chaplet the brow of that strange personage, the novelist and the political adventurer who is now Premier of England, who will certainly become a Duke, and who is possibly destined—as gossip will have it—to still further honor, to wear the Royal robes of Prince Consort and to occupy the long vacant bed of 'Albert the Good.'"

Bismarck followed the advice of, and was at the mercy of, Jewish bankers. As part of the Bolsheviks controlled opposition, Hitler also argued that Pan-Germany could save Western Civilization from Pan-Slavism and Bolshevism. He expected England's support in this posture. Again and again, from Napoleon onward, Russia was attacked by Western Europe and the central issue was Jews. Whether the pretext was to rescue them or to attack them, the results were to gain control of the Holy Land from Turkey and to use the Jews of Russia to take and to occupy it—then to use the Russian Jews as a slave labor force to construct palatial estates for wealthy Western Jews.

G & C Merriam believed that Bismarck was a Jew, and they expressed this belief, perhaps not coincidently, in the context of Disraeli and Rothschild. *The Chicago Tribune* published the following article on 13 March 1872 on page 3:

"THE DICTIONARY QUESTION.

To Jew, a Verb—Jesuitical—Card from the Merriams.

To the Editor of the Springfield (Mass.) Republican:

Some few days since you commented upon the course of the dictionaries in regard to 'jew' and jesuitical.'

In a recently issued circular of ours, which we hand you herewith, replying to certain strictures upon Webster's definitions of political terms, you will notice the ground the dictionary professes to take in regard to opprobrious and offensive appellations, that of strict impartiality. It is an error of judgment, and not of intention, if that position is not maintained in regard to two words in question. Some few weeks since a respected business acquaintance, Mr. Solomons, of Washington, a Jew, wrote us complaining, in substance, that the use of 'jew, verb, active, to cheat or defraud; to swindle,' in Webster, was unjust and unauthorized;—that is, that it wronged his people, and was unsanctioned by good usage. An examination by us disclosed the fact, after a careful collation, that the word as a verb, in any sense, does not appear in any dictionary ever published in England, so far as we have the means at hand of ascertaining. It is not found in Bailey, Johnson, Richardson, Walker, Reid, Smart, Ogilvie, Knowles, etc. The inference seems fair that the word has no recognized use out of this country. It is found in none of the earlier editions of Webster, and first appears in the present. Our attention is now originally called to it, and how it found its way with us, we know not. We fear it must have been drawn from Worcester, where we first find it. Then, as to popular or recognized usage; we do not recall ever seeing it employed in literary composition,—rarely, if ever, to have heard it used colloquially. In these circumstances it seemed due to truth, to our correspondent, and to literary impartiality, to adopt the course pursued.

You allude to it as a 'Shakspearean word.' Whilst we think the masterly delineation of Shylock the Jew, in the Merchant of Venice, by Shakspeare, thus attaching this offensive characteristic, as a national trait, to the Jewish race, (and a writer of fiction, in a strongly-drawn character, is usually understood as justified in a very considerable exaggeration), if not first, yet most strongly, fastened this feature of a sharp bargainer upon the poor Hebrew, yet we believe you will nowhere find 'jew,' as a verb, employed by him. We speak only from memory, but such is our strong conviction. Sir Walter Scott, in Ivanhoe, more justly and more naturally, because giving a mixed character, presents, in Rebecca the Jewess, one of his loveliest female portraits, and Isaac her father has noble as well as mercenary traits.

Injustice, perhaps, is done to the Jewish race, by not sufficiently considering the past and current conditions of their national, or rather race existence; while the noble traits which characterized them whilst the chosen people of the Lord, and which still exist, are forgotten or overlooked. Who ever heard of one depending upon public charity, or uncared for by his race? Two circumstances seem to have combined to make them a trading people.

The severest civil disabilities, until quite recently enforced against them in nearly all lands, frequent banishments, and the bitterest persecutions, have prevented permanent settlements, and agricultural or mechanical pursuits. They must stand ready to depart at a moment's notice, and a life of traffic seemed their only resource. Men, with beautiful, if misplaced faith (yet eminent Christian scholars, in the light of prophecy, look to their final restoration to Palestine, with something of its pristine glory), they believe they are but strangers and pilgrims in all other lands, and are to find rest only in their own.

The founder of Christianity was himself a Jew, and the race are 'Israelites, to whom pertaineth the adoption, and the glory and the covenants, and the giving of the law and the promises; whose are the fathers, and of whom, as concerning the flesh, Christ came, who is over all, God blessed forever.' Should we not hesitate, on this ground alone, about applying an epithet to the race of somewhat doubtful propriety? So far as our personal observations goes, the Jews are much like other men, neither essentially better, nor worse. Certainly, we have known excellent people among them. One of the most prominent booksellers of Philadelphia a few years since was a Jew, and liberal and equitable in his dealings. Although with Christian partners, the store was invariably and closely closed on Saturdays, (on Sunday's likewise), thus involving much business sacrifice and negativing, certainly, inordinate mercenary views, and so presented a marked aspect on the thronged thoroughfare of Chestnut street. Rothschild, the banker, Disraeli, the statesman (we have the impression Bismarck, the Prussian Premier), all Jews, certainly give evidence of extraordinary intellectual powers, not coupled with unennobling traits. The isolated distinctive existence of the Jewish race, thus secured by Providential causes, as well as by their own religious faith and rites, while yet they mingle without commixing with all people, assures, wonderfully, the fulfillment of prophecies uttered more than twenty centuries ago, and it thus a marked proof of the truth of revelation.

We have but a few words in regard to 'jesuitical.' In preparing for the revision of the dictionary, we applied, through a Roman Catholic friend, to the late Archbishop Hughes of New York, then at the head of the Catholic prelacy in this country, as to the person of highest scholarship in that Church to whom we could intrust the revision and preparation of Roman Catholic terms. He introduced us to Dr. O'Callghan of Albany, by whom that revision was made. These, of course, were subsequently submitted to President Porter, the editor-in-chief, and as left by him now appear in the dictionary. Jesuitical, as now defined, meets the approval of the scholars and dignitaries of the Catholic Church, who accord to it, as employed in popular use, the signification given in the dictionary, which is also accepted by Protestants. This use in neither colloquial nor local, like 'jew,' but is employed by the best writers and speakers, and so has long been. Intelligent men, of whatever faith do not take umbrage at this, and if others do, it is from want of a proper understanding of the province of the lexicographer. Loyola, the founder of the order, as have, presumably, those since connected with it, probably claimed that a 'higher law' in divine and religious obligation, was paramount and superior to civil rule and rulers: and hence justified to themselves measures to thwart the latter, unjustifiable on any other supposition. Hence their practices, and the word growing out of them. As with Jews, there might be some sacred associations with the word Jesus, Jesu-itical, to make undesirable the use of the term in an offensive sense, yet the usage seems too well established to be changed. Do we meet your difficulty?

G. & C. MERRIAM."

How did Disraeli and Lionel Nathan Rothschild, both of whom were Zionists, skirt the laws of England and purchase the Suez with Rothschild's credit? Legend had it that the Rothschilds had demonstrated that they could break the Bank of England at any time. *The Chicago Daily Tribune* published the following article on 21 February 1877 on page 2,

"NATHAN ROTHSCHILD.

His Little Scrimmage with the Bank of England.

Somewhere near a score of years ago, I think, I read the story, then fresh. It has been recalled to my mind by its telling in my presence to an English gentleman, who assured us that he could personally vouch for its truth, he having had business with the old lady of Threadneedle street while the transaction was in progress; and, from this assurance of an eye-witness, I deem the thing worth repeating. I think I remember it as it was told to me.

A bill of exchange, for a large amount, was drawn by Anselm Rothschild, of London. When the gentleman who held it arrived in London, Nathan was away, and he took the bit of paper to the Bank of England, and asked them there to discount it. The managers were very stiff. With haughty assurance they informed the holder that they discounted only their own bills; they wanted nothing to do with the bills of 'private persons.' They did not stop to reflect with whom they had to deal. Those shrewd old fellows in charge of the change of the realm should have known and remembered that that bit of paper bore the sign manual of a man more powerful than they,— more powerful because independent of the thousand-and-one hampers that rested upon them.

'Umph!' exclaimed Nathan Rothschild, when the answer of the bank was repeated to him. 'Private persons! I will give those important gentlemen to know with what sort of private persons they have to deal!'

And then Nathan Rothschild went at work. He had an object in view,— to humble the Bank of England,—and he meant to do it. He sent agents upon the Continent, and through the United Kingdom, and three weeks were spent in gathering up notes of the smaller denominations of the bank's own issue. One morning, bright and early, Nathan Rothschild presented himself at the bank at the opening of the teller's department, and drew from his pocketbook a five pound note, which he desired to have cashed. Five sovereigns were counted out to him, the officers looking with astonishment upon seeing the Baron Rothschild troubling himself personally about so trivial a matter. The Baron examined the coins one by one, and, having satisfied himself of their honesty in quality and weight, he slipped them into a canvas bag, and drew out and presented another five pound note. The same operation was gone through with again, save that this time the Baron took the trouble to take a small pair of scales from his pocket and weigh one of the pieces, for the law gave him that right. Two—three—ten—twenty—a hundred—five hundred pound notes were presented and cashed. When one pocketbook had been emptied another was brought forth; and when a canvas bag had been filled with gold it was passed to a servant who was in waiting. And so he went on until the hour arrived for closing the bank; and at the same time he had nine of the employes of the house engaged in the same work. So it resulted that ten men of the house of Rothschild had kept every teller of the bank busy seven hours, and had exchanged somewhere about £22,000. Not another customer had been able to get his wants attended to.

The English like oddity. Let a man do something original and piquant, and they will applaud even though their own flesh is pricked. So the people contrived to smile at the eccentricity of Baron Rothschild, and when the time came for closing the bank, they were not a tenth part so much annoyed as were the customers from abroad, whose business had not been attended to. The bank officials smiled that evening but—

On the following morning, when the bank opened, Nathan Rothschild appeared again, accompanied by his nine faithful helpers, this time bringing with him as far as the street entrance four heavy two-horse drays, for the purpose of carting away the gold, for to-day the Baron had bills of a larger denomination. Ah, the officers of the bank smiled no more, and a trembling seized them when the banker monarch said, with stern simplicity and directness:

'Ah, these gentlemen refuse to take my bills. Be it so. I am resolved that I will keep not one of theirs. It is the house of Rothschild against the Bank of England!'

The Bank of England opened its eyes very wide. Within a week the house of Rothschild could be demanding gold it did not possess. The gentlemen at the head of affairs saw very plainly that in a determined tilt the bank must go to the wall. There was but way out of the scrape, and they took it. Notice was at once publicly given that thenceforth the Bank of England would cash the bills of Rothschild as well as its own!—*Exchange*."

Under the heading "Foreign Articles", the following statement appeared in *Niles' Weekly Register*, Volume 17, Number 427, (13 November 1819), p. 169,

"Mr. Rothschild, the great London banker, indignant at the persecution of his Jewish brethren in Germany, has refused to take bills upon any of the cities in which they are persecuted; and great embarrassments to trade have been experienced in consequence of his determination. ☞ It is intimated that the persecution of the Jews is in part owing to the fact, that Mr. Rothschild and his brethren were among the chief of those who furnished the 'legitimates,' with money to forge chains for the people of Europe."

Not only could no nation claim to be a democracy while the Rothschilds held so much sway in politics, no nation could claim national sovereignty. Michael Shapiro wrote of the Rothschilds, in Shapiro's book *The Jewish 100: A Ranking of the Most Influential Jews of All Time*,

> "Although their political power would wane after the First World War as more banking houses rose to prominence and competition set in, the Rothschilds helped shape the political fortunes of many of the great figures of the age, including, but certainly not limited to, Napoleon, the Duke of Wellington, Talleyrand, Metternich, Queen Victoria, Disraeli, and Bismarck (and the futures of their countries)."[320]

and of Disraeli,

> "With his sister's fiancé, William Meredith, Disraeli left Britain in 1830 for a 'Grand Tour' of the Mediterranean. The sixteen-month trip made a permanent impression on him. Disraeli was particularly taken with Jerusalem. He began to understand the relationship between his Jewish heritage and Christian assimilation. Indeed, this Middle Eastern journey inspired creation of the protagonist of his novel *Alroy* (1833). Set in an exotic twelfth-century milieu, the character, David Alroy, fails in his attempt to restore the Holy Land to Jewish dominion. Later, in his novel *Tancred*, Disraeli's early Zionism would result in the often quoted line that 'a race that persists in celebrating their vintage although they have no fruits to gather, will regain their vineyards.'"[321]

Rabbi Emil G. Hirsch was quoted in *The Chicago Tribune* on 5 November 1889 on page 10, and capsulized the disparate views of wealthy "assimilated" Jews many of whom were under Rothschild's influence, reformed Jews, and Orthodox Jews,

> "'Many orthodox Jews go to Jerusalem to die. They believe that when the resurrection takes place those who are not buried there will have to go there from their graves. In order to avoid the journey after death they go before. The restoration of the City of Jerusalem was a dream of Disraeli and of

---

320. M. Shapiro, *The Jewish 100: A Ranking of the Most Influential Jews of All Time*, Citadel Press, Secaucas, New Jersey, (1996), p. 113.
321. M. Shapiro, *The Jewish 100: A Ranking of the Most Influential Jews of All Time*, Citadel Press, Secaucas, New Jersey, (1996), p. 87. It perhaps should be noted here that David Alroy was a real person and a false Messiah of the Twelfth Century.

'Daniel Deronda.' The reformed Jews are entirely indifferent to this question, though the orthodox expect the restoration and rebuilding to take place in some miraculous way."

Disraeli admitted that the purchase of the Suez was not made as an investment for England, but a was a purely political maneuver to draw England into Egypt for the benefit of Zionists and to take Palestine from the Turkish Empire and its native population,

> "The noble Lord himself has expressed great dissatisfaction, because I have not told him what the conduct of the Government would be with regard to the Canal in a time of war. I must say that on this subject I wish to retain my reserve. I cannot conceive anything more imprudent than a discussion in this House at the present time as to the conduct of England with regard to the Suez Canal in time of war, and I shall therefore decline to enter upon any discussion on the subject..... What we have to do tonight is to agree to the Vote for the purchase of these shares. I have never recommended, and I do not now recommend this purchase as a financial investment. If it gave us 10 per cent of interest and a security as good as the Consols I do not think an English Minister would be justified in making such an investment; still less if he is obliged to borrow the money for the occasion. I do not recommend it either as a commercial speculation although I believe that many of those who have looked upon it with little favour will probably be surprised with the pecuniary results of the purchase. I have always, and do now recommend it to the country as a political transaction, and one which I believe is calculated to strengthen the Empire. That is the spirit in which it has been accepted by the country, which understands it though the two right honourable critics may not. They are really seasick of the 'Silver Streak.' They want the Empire to be maintained, to be strengthened; they will not be alarmed even it be increased. Because they think we are obtaining a great hold and interest in this important portion of Africa—because they believe that it secures to us a highway to our Indian Empire and our other dependencies, the people of England have from the first recognized the propriety and the wisdom of the step which we shall sanction tonight."[322]

In an allusion to Shakespeare's character Shylock in the play *A Merchant of Venice*, The Chicago Daily Tribune reported on 4 July 1881 on page 7,

"ROTHSCHILD'S POUND OF FLESH

It appears from the report, too, that the foreign bondholders, mostly French and English, still have possession of the country, and are like the leeches of that valley in the days of Moses. There have been some changes

---

[322]. J. H. Park, Editor, *British Prime Ministers of the Nineteenth Century: Policies and Speeches*, New York University Press, New York, (1950), pp. 237-244, at 243-244.

in the physical conditions, and the boundaries of the domain of the security lands have been changed; still, the Government sees to it that the foreign usurers are paid their pound of flesh. Mr. Farman says:

When the decree appeared abolishing the law of the moukabalah, the Rothschilds refused to pay over the balance of the proceeds of the loan then in their hands until other securities were given them. The result was, that, while they consented to the increase of their taxes in an amount of about $500,000, this was not to be paid until their coupons were provided for, and they had also pledged to them, as a further guarantee, the revenues of the Province of Kenah, which contains 283,842 acres of cultivable land, on which the annual tax is $1,478,805. The whole revenues of the province are in excess of this sum.

It will be seen that the interest is amply secured; and that the increase of the taxes caused by the repeal of the law of the moukabalah, so far as relates to lands mortgaged to secure this loan, is only nominal, and cannot injuriously effect the bondholders. In case of a low Nile or bad crops from any other cause, full provision has been made for their coupons. On the occurrence of any such event, it will be the people of Egypt who are to suffer, and not the Parisian or London bankers."

8.3 Jews Provoke Perpetual War

The power and duplicity of Jewish finance again revealed itself in the First World War. In 1920, the Zionist Organization of America, New York, published *A Guide to Zionism*, edited by Jessie E. Sampter, which contained a time-line, which states on pages 238-239, *inter alia*,

"[1914] *Sept.* Whole press in England begins active agitation for Jewish rights in Russia. [***] [1915] *June.* Zionist Organization (in Germany) refuses request of Government that it issue appeal to all Zionists asking for sympathy with Germany, replying that it could not involve the Zionist movement in world politics."

The New York Times reported on 30 December 1917 on page 5,

"JEWS IN GERMANY FIRM.

Won't Support War Loan Until
Palestine Independence Is
Sanctioned.
Special Cable to THE NEW YORK TIMES.

THE HAGUE, Dec. 29.—It is reported here that the leading Jewish financiers of Germany refused to support the German war loan unless the German Government undertook to refrain from all opposition to the establishment of a Jewish State in Palestine, independent of any Turkish suzerainty or control.

By Associated Press.

THE HAGUE, Dec. 29.—The Jewish Correspondence Bureau here has received a telegram from Berlin stating that at a Zionist conference in Germany a resolution was adopted in which satisfaction was expressed that Great Britain had recognized the right of the Jewish people to a national existence in Palestine."

Eduard Bernstein wrote after the war,

"To many Social Democrats the war really seemed to be one for national existence; and to many passionate natures the opposition of so many Jews to the war credits might have seemed to betray un-German or anti-German thinking. How little such feeling had to do with anti-Semitism can be seen from the fact that those Jews who voted for the war loans were more highly esteemed and sought after than ever."[323]

After the war, Kaiser Wilhelm II lived in exile in the Netherlands at Doorn. Many in the Jewish controlled press tried to place the blame for the war on him. Baron Clemens von Radowitz-Nei alleged that he had discussed politics with the former Emperor on May 20th, 21st and 22nd of 1922. The Baron reported on his alleged conversations with the former Kaiser in *The Chicago Daily Tribune* on 3 July 1922 on the front page in an article which continued onto page 4, where the Baron alleged, among other things,

"The former emperor had a very great respect for Dr. Rathenau's ability, but considered him a great danger to Germany. In the first place, Rathenau was a Jew, and the Kaiser has come apparently to the firm conviction that the Jews are at the bottom of most of the troubles in Germany and Europe.

'The much talked of Wiesbaden agreement,' said the former emperor, 'was not an international agreement. It was an understanding between two groups of capitalists, two great trusts—between Rathenau and the interests represented by Loucheur and Giraud.'

And curiously enough, when I saw Dr. Rathenau a few weeks later, he asked me if many people did not think that—in France.
[***]
The Kaiser is convinced that all the evils of the modern world originate with the Jews.

'A Jew cannot be a true patriot,' he exclaimed. 'He is something different—like a bad insect. He must be kept apart, out of a place where he can do mischief—even if by pogroms, if necessary.

---

[323]. P. W. Massing, *Rehearsal for Destruction: A Study of Political Anti-Semitism in Imperial Germany*, Howard Fertig, New York, (1967), p. 325.

'The Jews are responsible for bolshevism in Russia, and Germany, too. I was far too indulgent with them during my reign, and I bitterly regret the favors I showed to prominent Jewish bankers and business men.'

I notice that one of the generals in attendance on him at the time wore the swastika, symbol of an anti-Semetic organization in Germany.

[***]

[The former emperor] was much disturbed by the strong Jewish-Masonic influence in France, and thought that this was at the bottom of much that went wrong.

[***]

The Jewish influence among the Young Turks worries him, and he fears that bolshevist elements are becoming too powerful among them; but he thinks that Turkey and Egypt will form the nucleus, sooner or later, of a Moslem bloc."

The Baron's allegations also appeared in *The New York Times* on 3 July 1922 on the front page continuing onto page 3. However, the following statements, which appeared in *The Chicago Daily Tribune*, were absent in *The New York Times*:

"[...]to the firm conviction that the Jews are at the bottom of most of the troubles in Germany and Europe."

"The Kaiser is convinced that all the evils of the modern world originate with the Jews. 'A Jew cannot be a true patriot,' he exclaimed. 'He is something different—like a bad insect. He must be kept apart, out of a place where he can do mischief—even if by pogroms, if necessary. 'The Jews are responsible for bolshevism in Russia, and Germany, too. I was far too indulgent with them during my reign, and I bitterly regret the favors I showed to prominent Jewish bankers and business men.' I notice that one of the generals in attendance on him at the time wore the swastika, symbol of an anti-Semetic organization in Germany."

"The Jewish influence among the Young Turks worries him, and[...]"

The following statement, which appeared in *The New York Times*, was absent in *The Chicago Daily Tribune*:

"Yet, while the former Emperor disliked Rathenau, on the matter of the treaty with the Russian Bolsheviki signed at Rapallo, he was even more indignant at Baron von Maltzahn, head of the Russian Division of the Foreign Office. That Rathenau should have signed a treaty with the Bolsheviki he thought more or less intelligible, but that a professional diplomat should have thrown in his lot with them was a different and to him far more serious matter."

Kaiser Wilhelm II denied that he had had a political discussion with the Baron, though he admitted that the Baron had visited him. The Kaiser alleged that the visits were limited to non-political small talk about family, and to photo sessions. The Baron reaffirmed that the political discussions took place and *The New York Times* supported the Baron's contention that he had visited the Kaiser over the course of three days.[324]

The publication of these articles soon after Rathenau's assassination tended to place the blame for his murder on the Monarchy and on anti-Semitism. The Kaiser had long ago been under the influence of men like Adolf Stoecker and Heinrich von Treitschke, who, like Rathenau, wanted the Jews to assimilate and give up nationalistic ambitions and disloyalties. They quoted Jewish authors like Heinrich Graetz, who, like Moses Hess, stated that Judaism is more than a mere religion, but represents a racial perspective and national culture.

Nevertheless, this was strange talk coming from Kaiser Wilhelm II, who was the grandson of Queen Victoria, a woman who believed that she was directly descended from King David, making Wilhelm his supposed heir as well. The Messiah was to come from the seed of King David (II *Samuel* 7; 22:44-51; 23:1-5. *Isaiah* 9:6-7. *Jeremiah* 23:5; 33:15, 17). Wilhelm II was the proud owner of the "Spear of Destiny",[325] which had supposedly pierced the side of Jesus and rendered its holder invincible in battle.

General von Ludendorff believed that the Kaiser had betrayed Germany. Ludendorff iterated the common belief that Jews were an enemy of the German People and intimated that they sought to make Germany a Communist state—which in fact did occur in part, in Bavaria, and the Soviets again took over a large part of Germany after the Second World War, creating East Germany out of the Soviet Sector. Ludendorff was quoted in the *Chicago Daily Tribune* on 1 March 1924 on page 3 in article with the header "'I Fought Rule by Red or Jew'—Ludendorff', and his statements were in full agreement with those of Jewish Zionists—it almost appears as if he were scripted by Zionists, like Moses Pinkeles, who discussed such things in his autobiography. *The Tribune* wrote,

> "[...]With this introduction Gen. von Ludendorff launched into a long explanation of the reasons for attempting a coup d'état against the republican government, which he sees undermined by the socialist principles of Marxism and pan-Judaism.
>
> 'There cannot be the slightest doubt of my attitude towards the

---

324. "Ex-Kaiser Denies, Baron Reaffirms", *The New York Times*, (7 July 1922), frontpage. "Text of Ex-Kaiser's Denial", *The New York Times*, (8 July 1922), p. 4.

325. T. Ravenscroft, *The Spear of Destiny: The Occult Power Behind the Spear Which Pierced the Side of Christ*, Spearman, London, (1972). T. Ravenscroft and T. Wallace-Murphy, *The Mark of the Beast: The Continuing Story of the Spear of Destiny*, Sphere, London, (1990). H. A. Buechner and W. Bernhart, *Adolf Hitler and the Secrets of the Holy Lance*, Thunderbird Press, Metairie, Louisianna, (1988). J. E. Smith and G. Piccard, *Secrets of the Holy Lance: The Spear of Destiny in History & Legend*, Adventures Unlimited Press, Kempton, Illinois, (2005).

communists,' he continued. 'Before the war this Marxist world turned against every military power. Philip Scheidemann said to France, 'You are not our enemies, but our friends and allies.'

'In connection with this is the Jewish question. I made its acquaintance during the war. For me it is a question of race. Little as the Englishmen or Frenchmen can be permitted to obtain domination over us, so little can the Jew be permitted. Freedom of the nation cannot be expected from him. Therefore I was against him.

'We want a Germany free of Marxism, semitism, and papal influences.'"

One of the reasons why racist Zionist Jews mass murdered the Armenians, was so that they could justify their planned petition after the First World War, at a planned peace conference, for the break up of the former Empires into small, "racially" segregated nations—so that they could petition for a racist "Jewish State" in Palestine. Political Zionists gave speeches before and during the First World War, which likened the situation of the Zionists, in terms of the war, to the efforts of Mazzini, Garibaldi and Cavour to solve the "Italian Question" through means of a peace conference.

The Sardinians had entered the Crimean War (1854-1856) on the side of the Turks, English and French, against the Russians, solely for the purpose of raising the "Italian Question" at the inevitable peace conference which would of necessity follow the Crimean War.[326] Giuseppe Mazzini, who was Jewish, used Masonic lodges and secret societies to forward the agenda of World Jewry. He joined the "Carbonari". His crypto-Jewish revolutionary "Young Italy" movement served as the model for the crypto-Jewish "Young Turk" movement, which destroyed the Turkish Empire for the Zionists in the Balkan Wars and in the First World War. The crypto-Jewish Zionist Mustafa Kemal "Atatürk" then completely dissolved the Empire and promoted Turkish nationalism and anti-Imperialism.

The Masonic Lodges of Italy spread to Salonika through the Grand Master of the Macedonia Risorta Masonic Lodge, the Jewish Zionist Emmanuel Carasso, who headed the Committee of Union and Progress of the Jewish and crypto-Jewish "Young Turks", which committed genocide against the Armenian Christians, and which destroyed the Turkish Empire on behalf of the Zionists. Carasso worked with the Jewish Zionist Revolutionary terrorists Vladimir Jabotinsky and Israel Lazarevich Helphand, a. k. a. "Alexander Parvus"—who inspired the crypto-Jewish Zionist terrorist Lev Bronstein, a. k. a. "Leon Trotsky", to adopt a policy of "permanent revolution" against the human race—and who brought Lenin from Switzerland to Russia.[327] In collusion with Carasso,

---

326. J. G. Sperling, "CRIMEAN WAR", *Encyclopedia International*, Grolier Incorporated, New York, (1966), pp. 320-321, at 320.
327. D. Fahey, *The Mystical Body of Christ in the Modern World*, Browne and Nolan Limited, London, (1935); **and** *The Rulers of Russia*, Third Revised and Enlarged American Edition, Condon Printing Co., Detroit, (1940). *See also:* J. M. Landau, *Pan-*

Helphand profiteered off of the "Young Turk Revolution" in the grain and arms trade, and worked as a propagandist for World Jewry in the Ottoman Empire, after the Jewish takeover.[328] Jabotinsky also served as a propagandist for World Jewry in the Turkish Empire, after the Jewish takeover.[329]

Following Giuseppe Mazzini's example, the Zionists sought a peace conference. In order to have a peace conference, there must first be war. The Zionists required a world war, in order to destroy the Turkish Empire and free up Palestine for the Jews; and in order to break apart the Russian Empire and free up the Jews for deportation to Palestine. The Zionists promoted a racist vision of small "racially" homogenous nations, and planned to make a plea for a Jewish State after they had ruined the Turkish Empire and brought the World into war.

Racist Zionist Theodor Herzl's ally in the British press,[330] Lucien Wolf wrote in a Letter to the Editor in *The London Times*, on 8 September 1903, on page 5, which was styled, "The Zionist Peril",

> "But I have always urged that, if through any political convulsion in Eastern Europe, it should become possible to secure the Holy Land for a Jewish State under the protection of the Powers, then, no matter what the difficulties might be, the whole Jewish people should strain their utmost endeavour to establish the unassimilated Jewish population of Europe in such a State and to make it a social and political success."

Lucien Wolf was doubtlessly referring, in 1903, to the planned Jewish takeovers which are commonly, erroneously and deceptively referred to as the "Russian Revolution" of 1905 and the "Young Turk Revolution" of 1908. It is also beyond doubt that Wolf had in mind the Zionist orchestrated Balkan Wars and First World War. We know this because the racist Zionist Max Nordau stated in the Zionist Conference of 1903 that the World War was coming and with it a peace conference where the Zionist Jews could make their pitch for their takeover and invasion of Palestine. They already had in the works plans for the "League of Nations", which was a Zionist institution even before it was instituted.

Wolf and the Rothschilds opposed the Uganda scheme of 1903, because they were planning to destroy Russia to free up the Jews of Eastern Europe for

---

Turkism, from Irredentism to Cooperation, Indiana University Press, Bloomington, (1995).
328. "Helphand, Alexander", *The Universal Jewish Encyclopedia*, Volume 5, The Universal Jewish Encyclopedia, Inc., New York, (1941), p. 312.
329. "Jabotinsky, Vladimir", *Encyclopaedia Judaica*, Volume 9 Is-Jer, Macmillan, Jerusalem, (1978), cols. 1178-1186, at 1179. "Jabotinsky, Vladimir", *The Universal Jewish Encyclopedia*, Volume 6, The Universal Jewish Encyclopedia, Inc., New York, (1942), pp. 2-4, at 2.
330. M. R. Buheiry, "Theodor Herzl and the Armenian Question", *Journal of Palestine Studies*, Volume 7, Number 1, (Autumn, 1977), pp. 75-97, at 87.

deportation, to destroy the Turkish Empire to free up Palestine for a European Jewish invasion, and to weaken Europe and pan-Arabia with a war so devastating that all the World would call for World government, a. k. a. the Zionist controlled "League of Nations", on the false premise that it would end war. They planned a war so devastating that it would destroy the Arabs' and Christians' will to fight against the al-Dajjal/anti-Christ Rothschild scheme to artificially fulfill Jewish messianic myth. Though the Jewish bankers succeeded in their geopolitical goals, they were unable to convince European Jews to move to Palestine. The Jewish bankers then placed Adolf Hitler on the throne of Europe in order to chase the European Jews to Palestine.

THE DEARBORN INDEPENDENT published an article alleging that the Zionists knew that the First World War was coming long before it came. The article was titled, "Did the Jews Foresee the World War?", and it appeared on 21 August 1920:

> "Fortunately the clue to the answer is supplied to us by unquestionable Jewish sources. The *American Jewish News* of September 19, 1919, had an advertisement on its front page which read thus:
>
> ### 'WHEN PROPHETS SPEAK
> By Litman Rosenthal
> Many years ago Nordau prophesied the Balfour
> Declaration. Litman Rosenthal, his intimate
> friend, relates this incident in a
> fascinating memoir.'
>
> The article, on page 464, begins: 'It was on Saturday, the day after the closing of the Sixth Congress, when I received a telephone message from Dr. Herzl asking me to call on him.'
>
> This fixes the time. The Sixth Zionist Congress was held at Basle in August, 1903.
>
> The memoir continues: 'On entering the lobby of the hotel I met Herzl's mother who welcomed me with her usual gracious friendliness and asked me whether the feelings of the Russian Zionists were now calmer.
>
> ''Why just the Russian Zionists, Frau Herzl?' I asked. 'Why do you only inquire about these?'
>
> ''Because my son,' she explained, 'is mostly interested in the Russian Zionists. He considers them the quintessence, the most vital part of the Jewish people.''
>
> At this Sixth Congress the British Government ('Herzl and his agents had kept in contact with the English Government'—Jewish Encyclopedia, Vol. 12, page 678) had offered the Jews a colony in Uganda, East Africa. Herzl was in favor of taking it, not as a substitute for Palestine, but as a step toward it. It was this which formed the chief topic of conversation between Herzl and Litman Rosenthal in that Basle hotel. Herzl said to Rosenthal, as reported in this article: 'There is a difference between the final aim and the ways we have to go to achieve this aim.'

Suddenly Max Nordau, who seems at the conference held last month in London to have become Herzl's successor, entered the room, and the Rosenthal interview was ended.

Let the reader now follow attentively the important part of this Rosenthal story:—(the italics are ours)

'About a month later I went on a business trip to France. On my way to Lyons I stopped in Paris, and there I visited, as usual, our Zionist friends. One of them told me that this very same evening Dr. Nordau was scheduled to speak about the Sixth Congress, and I, naturally, interrupted my journey to be present at this meeting and to hear Dr. Nordau's report. When we reached the hall in the evening we found it filled to overflowing and all were waiting impatiently for *the great master*, Nordau, who, on entering, received a tremendous ovation. But Nordau, without paying heed to the applause showered upon him, began his speech immediately, and said:

"You all came here with a question burning in your hearts and trembling on your lips, and the question is, indeed, a great one, and of vital importance. I am willing to answer it. What you want to ask is: How could I—I who was one of those who formulated the Basle program—how could I dare to speak in favor of the English proposition concerning Uganda, how could Herzl as well as I betray our ideal of Palestine, because you surely think that we have betrayed it and forgotten it. Yet listen to what I have to say to you. I spoke in favor of Uganda after long and careful consideration; deliberately I advised the Congress to consider and to accept the proposal of the English Government, a proposal made to the Jewish nation through the Zionist Congress, and my reasons—but instead of my reasons let me tell you a political story as a kind of allegory.

"I want to speak of a time which is now almost forgotten, a time when the European powers had decided to send a fleet against the fortress of Sebastopol. At this time Italy, the United Kingdom of Italy, did not exist. Italy was in reality only a little principality of Sardinia, and the great, free and united Italy was but a dream, a fervent wish, a far ideal of all Italian patriots. The leaders of Sardinia, who were fighting for and planning this free and united Italy, were the three great popular heroes: Garibaldi, Mazzini, and Cavour.

"The European powers invited Sardinia to join in the demonstration at Sebastopol and to send also a fleet to help in the siege of this fortress, and this proposal gave rise to a dissension among the leaders of Sardinia. Garibaldi and Mazzini did not want to send a fleet to the help of England and France and they said: 'Our program, the work to which we are pledged, is a free and united Italy. What have we to do with Sebastopol? Sebastopol is nothing to us, and we should concentrate all our energies on our original program so that we may realize our ideal as soon as possible.'

"But Cavour, who even at this time was the most prominent, the most able, and the most far-sighted statesman of Sardinia, insisted that his country should send a fleet and beleaguer with the other powers Sebastopol, and, at last, he carried his point. *Perhaps it will interest you to know that the right hand of Cavour, his friend and adviser, was his secretary, Hartum, a Jew*, and in those circles, which were in opposition to the government, one spoke fulminantly of Jewish treason. And once at an assembly of Italian patriots one called wildly for Cavour's secretary, Hartum, and demanded of him to defend his dangerous and treasonable political actions. And this is what he said: 'Our dream, our fight, our ideal, an ideal for which we have paid already in blood and tears, in sorrow and despair, with the life of our sons and the anguish of our mothers, our one wish and one aim is a free and united Italy. *All means are sacred if they lead to this great and glorious goal.* Cavour knows full well that after the fight before Sebastopol *sooner or later a peace conference will have to be held*, and at this peace conference *those powers will participate who have joined in the fight*. True, Sardinia has no immediate concern, no direct interest in Sebastopol, but if we will help now with our fleet, *we will sit at the future peace conference, enjoying equal rights with the other powers*, and at this peace conference Cavour, as the representative of Sardinia, will proclaim the free and independent, united Italy. Thus our dream for which we have suffered and died, will become, at last, a wonderful and happy reality. And if you now ask me again, what has Sardinia to do at Sebastopol, then let me tell you the following words, *like the steps of a ladder*: Cavour, Sardinia, the siege of Sebastopol, the future European peace conference, the proclamation of a free and united Italy."

'The whole assembly was under the spell of Nordau's beautiful, truly poetic and exalted diction, and his exquisite, musical French delighted the hearers with an almost sensual pleasure. For a few seconds the speaker paused, and the public, absolutely intoxicated by his splendid oratory, applauded frantically. But soon Nordau asked for silence and continued:

"Now this great progressive world power, England, has after the pogroms of Kishineff, in token of her sympathy with our poor people, offered through the Zionist Congress the autonomous colony of Uganda to the Jewish nation. Of course, Uganda is in Africa, and Africa is not Zion and never will be Zion, to quote Herzl's own words. But Herzl knows full well that *nothing is so valuable to the cause of Zionism as amicable political relations* with such a power as England is, and so much more valuable as England's main interest is concentrated in the Orient. Nowhere else is precedent as powerful as in England, and so it is most important to accept a colony out of the hands of England and create thus a precedent in our favor. Sooner or later the Oriental question will have to be solved, and the Oriental question means, naturally, also the question of Palestine. England, who had addressed a formal, political note to the Zionist Congress—the Zionist Congress which is pledged to the Basle program, England will have the deciding voice in the final solution of the Oriental question, and Herzl has considered it his duty to maintain valuable relations with this great and progressive power. *Herzl knows that we stand before a tremendous upheaval of the whole world. Soon, perhaps, some kind of a world-congress will have to be called*, and England, the great, free and powerful England, will then continue the work it has begun with its generous offer to the Sixth

Congress. And if you ask me now what has Israel to do in Uganda, then let me tell you as the answer the words of the statesmen of Sardinia, only applied to our case and given in our version; let me tell you the following words as if I were showing you *the rungs of a ladder leading upward and upward: Herzl, The Zionist Congress*, the English Uganda proposition, *the future world war, the peace conference* where with the help of England a free and Jewish Palestine will be created.'

'Like a mighty thunder these last words came to us, and we all were trembling and awestruck as if we had seen a vision of old. And in my ears were sounding the words of our great brother Achad Haam, who said of Nordau's address at the First Congress:

"I felt that one of the great old prophets was speaking to us, that his voice came down from the free hills of Judea, and our hearts were burning in us when we heard his words, filled with wonder, wisdom and vision."

The amazing thing is that this article by Litman Rosenthal should ever have been permitted to see print. But it did not see print until the Balfour Declaration about Palestine, and it never would have seen print had not the Jews believed that one part of their program had been accomplished.

The Jew never betrays himself until he believes that what he seeks has been won, then he lets himself go. It was only to Jews that the 1903 'program of the Ladder'—*the future world war—the peace conference—the Jewish program*—was communicated. When the ascent of that ladder seemed to be complete, then came the public talk."

In the English translation of Max Nordau's *The Interpretation of History*, Willey, New York, (1910), p. 293; Nordau employs the image of the ladder,

"The politician uses the parliamentary system as a ladder up which he may climb from being a secretary to a member, parliamentary reporter, or honorary secretary to some political club, to member of a parliamentary committee, member of Parliament itself, party leader, and finally minister."

The London Times reported on 15 August 1914, on page 3,

"JEW AND GERMAN.

A PROTEST AGAINST UNFAIR SUSPICION.

The Editor of the *Jewish Chronicle* and *Jewish World* writes:—

'Instance after instance has come to my knowledge of the ignorant assumption up and down the country that every Jew is necessarily a German and is hence being made an object of hatred as an enemy of this country. In Germany I learn that our Jews are in a somewhat similar case. But there they are not called 'German' Jews, but 'Russian' Jews. The fact is, of course, that Jews are by their tradition and, indeed, by absolute Jewish law, bound in loyalty to the country of which they are citizens. The Jew in Germany is no more German than the German, and the Jew in England is no less English than the English. Even in Russia the Russian Jew, at this hour of Russia's trial, is as Russian as the Russian.

'From end to end of the Empire Jews of all classes have shoulder to shoulder with their fellow-citizens manifested their unswerving loyalty in a hundred directions to this country in the righteous cause for which it has drawn the sword. This attitude of our people is perhaps only natural, seeing what the Jews of all the world owe to England for the example she has set in relation to Jews. But I do think it unwise at this juncture in the nation's affairs that anything should be done or said which it is possible may encourage in the ignorant some doubt about the loyalty of a section of the country's citizens.'"

Karl Lamprecht published an article in the *Berliner Tageblatt*, on 23 August 1914, arguing that the First World War was a racial war.[331] Some Germans were concerned by the success of Serbia against the Young Turks in the Balkan Wars and feared that it would provide Russia, which allegedly sought to unify all Slavs in a Pan-Slavic Russian Empire and to take Constantinople, in order to establish a port and route into the Adriatic and Mediterranean Seas through Albania and Constantinople. This area of the world had long been a source of international conflict throughout the period of the "Eastern Question" with the wars between Turkey, Russia, England, France, etc. These conflicts were fomented by Zionist Jews.

Jews wanted to discredit and ruin the Turkish Empire and the religion of Islam. Jews wanted to remove an ancient enemy from the region—an enemy which would oppose the anointment of a Jewish King in Jerusalem as the crowning of the Anti-Christ. Jews wanted to eliminate a skilled business competitor. Jews wanted to foment a war between Christians and Moslems, which would start in the Balkans and grow into World War I, and which would artificially pit Moslems and Christians, Slavs and Teutons, against one another and leave the Jews standing in Jerusalem. Many were aware of this fact at the time. See, for instance, B. Granville Baker's *The Passing of the Turkish Empire in Europe*, Seeley, London, (1913); *and* Richardson L. Wright's review of this book, "Pan-Slavism: Balkan War a First Step in a Great Racial Conflict", in the Book Review Section of *The New York Times*, (4 May 1913), p. BR6. Political Zionist Moses Hess forecast a "race war" and "last catastrophe" in 1862 in his book *Rom und Jerusalem: die letzte Nationalitätsfrage*, Eduard Wengler, Leipzig, (1862); English translation, *Rome and Jerusalem: A Study in Jewish Nationalism*, Bloch, New York, (1918/1943), p. 80. Some of the corrupt leadership of Germany sought to oblige the Jewish bankers' plan for a race war between the pan-Germans and pan-Slavics. Jewish leaders murdered Christians in the Turkish Empire in order to touch off the Balkan Wars, and the Balkan

---

331.See the forward of the first German edition of Karl Kautsky's "Rasse und Judentum", *Ergänzungshefte zur Neuen Zeit*, Number 20, (1914/1915 Ausgeben am 30. Oktober 1914), pp. 1-5.

Wars to touch off the First World War, which "Battle of Armageddon" they had been planning for centuries, truly for thousands of years.

The political Zionists Theodor Herzl and Max Nordau were both products of the Austro-Hungarian Empire, where the Pan-Germanic and Pan-Slavic forces directly confronted one another.[332] They must have known that this antagonism could provoke a massive conflict. Friedrich August Hayek stated,

> "I think the decisive influence was really World War I, particularly the experience of serving in a multinational army, the Austro-Hungarian army. That's when I saw, more or less, the great empire collapse over the nationalist problem. I served in a baffle in which eleven different languages were spoken. It's bound to draw your attention to the problems of political organization."[333]

At the end of the First World War, the breakup of the Austro-Hungarian Empire into small ethnically segregated nations would provide the precedent and the climate for the Zionists' artificial creation of the nonexistent "small nation" of Israel in Palestine, as if the Jews dispersed among all the nations of the Earth were one small nation among many small nations deserving of recognition and protection by the major powers of Western Europe. Not only would the First World War break up the Austro-Hungarian Empire, it would dissolve Turkish Empire, which owned Palestine. Under the influence of "Colonel" House, the recognition of the rights of minor nations became one of Zionist President Woodrow Wilson's favorite themes. The Zionists knew that a Peace Conference would be held at war's end at which they could petition for the creation of a "Jewish State" in Palestine. The entire war served the interests of the Zionists. They had been planning it and fomenting it for centuries.

Some have argued that the "racial" tribalism of the Pan-Germanic and Pan-Slavic forces was modeled after ancient Judaic tribalism and "racial" nationalism.[334] The political Zionists, many of whom were positivists, were one of many interested parties fanning the fires of "racial", nationalistic and religious discord in Vienna. Some Zionists believed that these Empires harmed Jews by insisting upon assimilation—the case of Czar's proclamation against Zionism being a primary example. Horace Mayer Kallen stated, "Pan-Germanism, Pan-Slavism, and all the other panic movements are assimilationist."[335] The Turkish Empire prevented the formation of a sovereign "Jewish State" in Palestine and encouraged assimilation. Political Zionism preferred smaller democracies where

---

332. J. B. Agus, *The Meaning of Jewish History*, Volume 2, Abelard-Schuman, New York, (1963), pp. 410-411.
333. F. A. Hayek, edited by S. Kresge and L. Wenar, *Hayek on Hayek: An Autobiographical Dialogue*, University of Chicago Press, (1994), p. 48.
334. J. B. Agus, *The Meaning of Jewish History*, Volume 2, Abelard-Schuman, New York, (1963), p. 411.
335. H. M. Kallen quoted in A. Hertzberg, *The Zionist Idea*, Harper Torchbooks, New York, (1959), p. 529.

ethnicities were encouraged to segregate. They had plans to eventually wipe out all of these small nations with the force of Communism and replace all nations with a Jewish world government, after they had formed their "Jewish State" in Palestine. But first they had to break up the Gentile Empires.

8.4 Jewish World Government—A Prophetic Desire

Political Zionist Moses Hess forecast a "race war" and "last catastrophe" in 1862. From the 1870's onward in England, the fabulously wealthy businessman Cecil John Rhodes, who was an agent for the Rothschild family,[336] planned for a world government to be led by the British and Americans; because, so he asserted, the English were a master race which had the moral authority to exploit the inferior races. Rhodes was a "pacifist", who used the liberal sentiment of pacifism to justify tyranny, colonialism and slavery. He was very close to the Rothschild family[337] and Alfred Beit. Rhodes formed a "secret society"—to use his term—of the world's wealthiest persons, which had as its goal the accumulation of the world's wealth for the purpose of world domination.[338] Rhodes advocated the reunification of the "English-speaking race". Rhodes enslaved the blacks he sent to South Africa to work the gold and diamond mines and the British introduced the use of concentration camps to destroy the Boers. Rhodes openly called for a "secret society" patterned after the Jesuits, which he planned would rule the world. *The New York Times* wrote on 9 April 1902,

"MR. RHODES'S IDEAL OF
ANGLO-SAXON GREATNESS

Statement of His Aims, Written

---

336. G. E. Griffin, *The Creature from Jekyll Island: A Second Look at the Federal Reserve*, Fourth Edition, American Media, Westlake Village, California, (2002), p. 208.
337. G. E. Griffin, *The Creature from Jekyll Island: A Second Look at the Federal Reserve*, Fourth Edition, American Media, Westlake Village, California, (2002), p. 208.
338. "Mr. Rhodes's Ideal of Anglo-Saxon Greatness", *The New York Times*, (9 April 1902), p. 1. *See also:* W. T. Stead, "Cecil John Rhodes", *The American Monthly Review of Reviews*, Volume 25, Number 5, (May, 1902), pp. 548-560, at 556-557. *See also:* "The Progress of the World", *The American Monthly Review of Reviews*, Volume 25, Number 5, (May, 1902), pp. 515-598. *See also:* S. G. L. Millin, *Cecil Rhodes*, Harper & Brothers, New York, (1933); **and** *Rhodes*, Chatto & Windus, London, (1952). *See also:* C. Quigley, *Tragedy and Hope: A History of the World in Our Time*, Macmillan, New York, (1966), pp.130-133, 144-153, 950-956, 1247-1278; **and** *The Anglo-American Establishment: From Rhodes to Cliveden*, Books in Focus, New York, (1981), p. ix; **and** "The Round Table Groups in Canada, 1908-38", *Canadian Historical Review*, Volume 43, Number 3, (September, 1962), pp. 204-224. *See also:* J. E. Flint, *Cecil Rhodes*, Little, Brown, Boston, (1974). *See also:* R. I. Rotberg and M. F. Shore, *The Founder: Cecil Rhodes and the Pursuit of Power*, Oxford University Press, New York, (1988). *See also:* A. Thomas, *Rhodes*, St. Martin's Press, New York, (1997).

for W. T. Stead In 1890,

He Believed a Wealthy Secret Society
Should Work to Secure the World's
Peace and a British-American
Federation.

LONDON, April 9.—An article on the Right Hon. Cecil J. Rhodes, by William T. Stead, will appear in the forthcoming number of The American Review of Reviews. The article, excerpts from which follow, consists of a frank, powerful explanation of Mr. Rhodes's views on America and Great Britain, and for the first time sets forth his own inmost aims. It was written mainly by himself for Mr. Stead in 1890. For originality and breadth of thought it eclipses even his now famous will, yet it is merely a collection of disoriented ideas, hurriedly put together by 'The Colossus,' as a summary of a long conversation between himself and Mr. Stead. In those days Mr. Stead was not only one of Mr. Rhodes's most intimate friends, as indeed he was till the last, but also his executor. Mr. Stead's name was only removed from the list of the trustees of Mr. Rhodes's will on account of the Boer war, which forced the two men into such vehement political opposition. Of this, episode Mr. Stead says:

'Mr. Rhodes's action was only natural, and, from an administrative point of view, desirable, and it in no way affected my attitude as political confidant in all that related to Mr. Rhodes's world-wide policy.'

In its three columns of complex sentences the whole of Mr. Rhodes's international and individual philosophy is embraced. Perhaps it can best be summarized as an argument in favor of the organization of a secret society, on the lines of the Jesuit order, for the promotion of the peace and welfare of the world, and the establishment of an American-British federation, with absolute home rule for the component parts.

'I am a bad writer,' says Mr. Rhodes in one part of what might be called his confession, 'but through my ill-connected sentences you can trace the lay of my ideas, and you can give my idea the literary clothing that is necessary.'

RHODES'S ROUGH NOTES UNEDITED.

But Mr. Stead wisely refused to edit or dress it up, saying:

'I think the public will prefer to have these rough, hurried, and sometimes ungrammatical notes exactly as Mr. Rhodes scrawled them off, rather than have them supplied with literary clothing by any one else.'

Mr. Rhodes began by declaring that the 'key' to his idea for the development of the English-speaking race was the foundation of 'a society copied, as to organization, from the Jesuits.' Combined with 'a differential rate and a copy of the United States Constitution,' wrote Mr. Rhodes, 'should be home rule or federation.' An organization formed on these lines In the House of Commons, constantly working for decentralization and not wasting time on trivial questions raised by 'Dr. Tanner, or the important matter of O'Brien's breeches,' would, Mr. Rhodes believed, soon settle the all-important question of the markets for the products of the empire.

'The labor' question,' Mr. Rhodes wrote, 'is important, but that is deeper than labor.'

THE MENACE TO BRITISH TRADE.

America, both in its possibilities of alliance and its attitude of commercial rivalry, was apparently ever present in Mr. Rhodes's mind. 'The world, with America in the forefront,' he wrote, 'is devising tariffs to boycott your manufactures. This is the supreme question. I believe that England, with fair play, should manufacture for the world, and, being a free trader, I believe that, until the world comes to its senses, you should declare war, I mean a commercial war, with those trying to boycott your manufactures. That is my programme. You might finish the war by a union with America and universal peace after a hundred years.' But toward securing this millenium Mr. Rhodes believed the most powerful factor would be 'a secret society, organized like Loyola's, supported by the accumulated wealth of those whose aspiration is a desire to do something,' and who would be spared the 'hideous annoyance' daily created by the thought to which 'of their incompetent relations' they should leave their fortunes. These wealthy people, Mr. Rhodes thought, would thus be greatly relieved and be able to turn 'their ill-gotten or Inherited gains to some advantage.'

Reverting to himself. Mr. Rhodes said:

'It is a fearful thought' to feel you possess a patent, and to doubt whether your life will last you through the circumlocution of the Patent Office. I have that inner conviction that if I can live I have thought out something that is worthy of being registered in the Patent Office. The fear is shall I have time and opportunity? And I believe, with all the enthusiasm bred in the soul of an inventor, that it is not self-glorification that I desire, but the wish to live and register my patent for the benefit of those who I think are the greatest people the world has ever seen, but whose fault is that they do not know their strength, their greatness, or their destiny, but who are wasting their time in minor or local matters; but, being asleep, do not know that through the invention of steam and electricity, and, in view of their own enormous increase, they must now be trained to view the world as a whole, and not only to consider the social questions of the British Isles. Even a Labouchere who possesses no sentiment should be taught that the labor of England is dependent on the outside world, and that, as far as I can see, the outside world, if he does not look, out, will boycott the result of English labor.'

Once again the personal feelings of the man crop out. 'They are calling the new country Rhodesia,' he wrote. 'I find I am human, and should like to be living after my death. Still, perhaps, if that name is coupled with the object of England everywhere it may convey the discovery of an idea which will ultimately lead to the cessation of all wars, and on language throughout the world, the patent being the gradual absorption of wealth and human minds of the higher order to the object.'

Here Mr. Rhodes used the sentence cabled to America, in Mr. Stead's article of April 4:

'What an awful thought it is that if, even now, we could arrange with the present members of the United States Assembly and our House of Commons the peace of the world would be secured for all eternity! We could hold a Federal Parliament, five years in Washington and five in London.'

Mr. Rhodes added: 'The only thing feasible to carry out this idea is a secret society gradually absorbing the wealth of the world, to be devoted to such an object.'

'There is Baron Hirsch,' interpolated Mr. Rhodes, 'with twenty millions, very soon to cross the unknown border and struggling in the dark to know what to do with his money, and so one might go on ad infinitum.'

'Fancy,' Mr. Rhodes goes on to say, the charm to Young America, just coming on, and dissatisfied, for they have filled up their own country and do not know what to tackle next, to share in a scheme to take the government of the whole world. Their present President [Mr. Harrison] is dimly seeing it; but his horizon is limited to the New World, north and south, and so he would intrigue in Canada, Argentina, and Brazil, to the exclusion of England. Such a brain wants but little to see the true solution. He is still groping in the dark, but very near the discovery, for the American has been taught the lesson of home rule and of the success of leaving the management of the local pump to the parish beadle. He does not burden his House of Commons with the responsibility of cleansing the parish drains. The present position of the English House is ridiculous. You might as well expect Napoleon to have found time to have personally counted his dirty linen before he sent it to the wash and to have recounted it upon its return.

'It would have been better for Europe if Napoleon had carried out his idea of a universal monarchy. He might have succeeded if he had hit upon the idea of granting self-government to the component parts.'

COUNTRIES 'FOUND WANTING.'

Dealing with the 'sacred duty of the English-speaking world of taking the responsibility for the still uncivilized world,' and commenting upon the necessary departure from the map of such countries as Portugal, Persia, and Spain, 'who are found wanting.' Mr. Rhodes said:

'What scope! What a horizon of work for the next two centuries for the best energies of the best people in the world!'

In regard to tariffs, Mr. Rhodes was characteristically positive.

'I note,' he wrote, 'with satisfaction that the committee appointed to inquire into the McKinley tariff, reports that in certain articles our trades have fallen off 50 per cent. Yet the fools do not see that if they do not look out they will have England shut out and isolated, with 90,000,000 to feed and capable of internally supporting about 6,000,000. If they had a statesman they would at the present moment be commercially at war with the United States, and would have boycotted the raw products of the United States until she came to her senses; and I say this because I am a free trader.

Your people have not known their greatness. They possess one-fifth of the world and do not know it is slipping away from them. They spend their time in discussing Mr. Parnell and Dr. Tanner, the character of Sir Charles Duke, compensation for beer houses, and omne hoc genus. Your supreme ciuestion at present is the seizure of the labor vote for the next election. Read the Australian bulletins and see where undue pandering to the labor vote may lead you. But, at any rate, the eight-hour question is not possible without a union of the English-speaking world; otherwise you drive your manufactures to Belgium, Holland, and Germany, just as you have placed a great deal of cheap shipping trade the hands of Italy by your stringent shipping regulations.'

Here this 'political will and testament,' as Mr. Stead calls it, abruptly breaks off. Mr. Stead, commenting on this, says:

'It is rough and inchoate and almost as uncouth as one of Cromwell's speeches. but the central idea glows luminous throughout. Its ideal is the promotion of racial unity on the basis of the principles embodied in the American Constitution.'"

Rhodes' statement, sans the literary clothing *The New York Times* supplied, appeared in *The American Monthly Review of Reviews*, Volume 25, Number 5, (May, 1902), pp. 548-560, at 556-557. Stead had founded this journal in order to promote Rhodes' millenniumistic vision. Rhodes' wrote,

"Please remember the key of my idea discussed with you is a Society, copied from the Jesuits as to organization, the practical solution a differential rate and a copy of the United States Constitution, for that is Home Rule or Federation, and an organization to work this out, working in the House of Commons for decentralization, remembering that an Assembly that is responsible for a fifth of the world has no time to discuss the questions raised by Dr. Tanner or the important matter of Mr. O'Brien's breeches, and that the labor question is an important matter, but that deeper than the labor question is the question of the market for the products of labor, and that, as the local consumption (production) of England can only support about six million, the balance depends on the trade of the world.

That the world with America in the forefront is devising tariffs to boycott your manufactures, and that this is the supreme question, for I believe that England with fair play should manufacture for the world, and, being a Free Trader, I believe until the world comes to its senses you should declare war—I mean a commercial war with those who are trying to boycott your manufactures—that is my programme. You might finish the war by union with America and universal peace, I mean after one hundred years, and a secret society organized like Loyola's, supported by the accumulated wealth of those whose aspiration is a desire to do something, and a hideous annoyance created by the difficult question daily placed before their minds as to which of their incompetent relations they should leave their wealth to. You would furnish them with the solution, greatly relieving their minds, and turning their ill-gotten or inherited gains to some advantage.

I am a bad writer, but through my ill-connected sentences you can trace the lay of my ideas, and you can give my idea the literary clothing that is necessary. I write so fully because I am off to Masbonaland, and I can trust you to respect my confidence. It is a fearful thought to feel that you possess a patent, and to doubt whether your life will last you through the circumlocution of the forms of the Patent Office. I have that inner conviction that if I can live I have thought out something that is worthy of being registered at the Patent Office; the fear is, shall I have the time and the opportunity? And I believe with all the enthusiasm bred in the soul of an inventor it is not self-glorification I desire, but the wish to live to register my patent for the benefit of those who, I think, are the greatest people the world has ever seen, but whose fault is that they do not know their strength, their greatness, and their destiny, and who are wasting their time on their minor local matters, but being asleep do not know that through the invention of steam and electricity, and in view of their enormous increase, they must now be trained to view the world as a whole, and not only consider the social questions of the British Isles. Even a Labouchere, who possesses no sentiment, should be taught that the labor of England is dependent on the outside world, and that as far as I can see, the outside world, if it does not look out, will boycott the results of English labor. They are calling the new country Rhodesia, that is from the Transvaal to the southern end of Tanganyika; the other name is Zambesia. I find I am human and should like to be living after my death; still, perhaps, if that name is coupled with the object of England everywhere, and united, the name may convey the discovery of an idea which ultimately led to the cessation of all wars and one language throughout the world [*see: Zephaniah* 3:9—CJB], the patent being the gradual absorption of wealth and human minds of the higher order to the object.

What an awful thought it is that if we had not lost America, or if even now we could arrange with the present members of the United States Assembly and our House of Commons, the peace of the world is secured for all eternity. We could hold your federal parliament five years at Washington and five at London. The only thing feasible to carry this idea out is a secret one (society) gradually absorbing the wealth of the world to be devoted to such an object. There is Hirsch with twenty millions, very soon to cross the unknown border, and struggling in the dark to know what to do with his money; and so one might go on *ad infinitum.*

Fancy the charm to young America, just coming on and dissatisfied—for they have filled tip their own country and do not know what to tackle next—to share in a scheme to take the government of the whole world! Their present President is dimly seeing it, but his horizon is limited to the New World north and south, and so he would intrigue in Canada, Argentina, and Brazil, to the exclusion of England. Such a brain wants but little to see the true solution; he is still groping in the dark, but is very near the discovery. For the American has been taught the lesson of Home Rule and the success of leaving the management of the local pump to the parish beadle. He does not burden his House of Commons with the responsibility of cleansing the parish drains. The present position in the English House is ridiculous. You might as well expect Napoleon to have found time to have personally counted his dirty linen before he sent it to the wash, and recounted it upon its return. It would have been better for Europe if he had carried out his idea of Universal Monarchy; he might have succeeded if lie had hit on the idea of granting self-government to the component parts. Still, I will own tradition, race, and diverse languages acted against his dream all these do not exist as to the present English-speaking world, and apart from this union is the sacred duty of taking the responsibility of the still uncivilized parts of the world. The trial of these countries who have been found wanting—such as Portugal, Persia, even Spain—and the judgment that they must depart, and, of course, the whole of the South American republics. What a scope and what a horizon of work, at any rate, for the next two centuries, the best energies of the best people in the world; perfectly feasible, but needing an organization, for it is impossible for one human atom to complete anything, much less such an idea as this requiring the devotion of the best souls of the next 200 years. There are three essentials (1) The plan duly weighed and agreed to. (2) The first organization. (3) The seizure of the wealth necessary.

I note with satisfaction that the committee appointed to inquire into the McKinley Tariff report that in certain articles our trade has fallen off 50 per cent., and yet the fools do not see that if they do not look out they will have England shut out and isolated with ninety millions to feed and capable internally of supporting about six millions. If they had had statesmen they would at the present moment be commercially at war with the United States, and they would have boycotted the raw products of the United States until she came to her senses. And I say this because I am a Free Trader. But why go on writing? Your people do not know their greatness; they possess a fifth of the world and do not know that it is slipping from them, and they spend their time on discussing Parnell and Dr. Tanner, the character of Sir C. Dilke, the question of compensation for beer-houses, the *omne hoc genus.* Your supreme question at the present moment is the seizure of the labor vote at the next election. Read the *Australian Bulletin* (New South Wales), and see where undue pandering to the labor vote may lead you, but at any rate the eight-hour question is not possible without a union of the English-speaking world, otherwise you drive your manufactures to Belgium, Holland, and Germany, just as you have placed a great deal of cheap shipping trade in the hands of Italy by your stringent shipping regulations which they do not possess, and so carry goods at lower rates."

William Winwood Reade described the origins of the millennium concept, with its one language, nihilistic "last catastrophe" destruction to renew, world government and lasting peace,

"Those Jews of Judea, those Hebrews of the Hebrews, regarded all the

Gentiles as enemies of God; they considered it a sin to live abroad, or to speak a foreign language, or to rub their limbs with foreign oil. Of all the trees, the Lord had chosen but one vine; and of all the flowers but one lily; and of all the birds but one dove; and of all the cattle but one lamb; and of all the builded cities only Sion; and among all the multitude of peoples he had elected the Jews as a peculiar treasure, and had made them a nation of priests and holy men. For their sake God had made the world. On their account alone empires rose and fell. Babylon had triumphed because God was angry with his people; Babylon had fallen because he had forgiven them. It may be imagined that it was not easy to govern such a race. They acknowledged no king but Jehovah, no laws but the precepts of their holy books. In paying tribute they yielded to absolute necessity, but the tax-gatherers were looked upon as unclean creatures; no respectable men would eat with them or pray with them; their evidence was not accepted in the courts of justice.

Their own government consisted of a Sanhedrin or Council of Elders, presided over by the High Priest. They had power to administer their own laws, but could not inflict the punishment of death without the permission of the procurator. All persons of consideration devoted themselves to the study of the law. Hebrew had become a dead language, and some learning was therefore requisite for the exercise of this profession, which was not the prerogative of a single class. It was a rabbinical axiom that the crown of the kingdom was deposited in Judah, and the crown of the priesthood in the seed of Aaron, but that the crown of the law was common to all Israel. Those who gained distinction as expounders of the sacred books were saluted with the title of rabbi, and were called scribes and doctors of the law. The people were ruled by the scribes, but the scribes were recruited from the people. It was not an idle caste—an established Church—but an order which was filled and refilled with the pious, the earnest, and the ambitious members of the nation.

There were two great religious sects which were also political parties, as must always be the case where law and religion are combined. The Sadducees were the rich, the indolent, and the passive aristocrats; they were the descendants of those who had belonged to the Greek party in the reign of Antiochus, and it was said that they themselves were tainted with the Greek philosophy. They professed, however, to belong to the conservative Scripture and original Mosaic school. As the Protestants reject the traditions of the ancient Church, some of which have doubtless descended viva voce from apostolic times, so all traditions, good and bad, were rejected by the Sadduccees. As Protestants always inquire respecting a custom or doctrine, 'Is it in the Bible?' so the Sadduccees would accept nothing that could not be shown them in the law. They did not believe in heaven and hell because there was nothing about heaven and hell in the books of Moses. The morality which their doctors preached was cold and pure, and adapted only for enlightened minds. They taught that men should be virtuous without the fear of punishment and without the hope of reward, and that such virtue alone is of any worth.

The Pharisees were mostly persons of low birth. They were the prominent representatives of the popular belief, zealots in patriotism as well as in religion—the teaching, the preaching, and the proselytising party. Among them were to be found two kinds of men. Those Puritans of the Commonwealth with lank hair and sour visage and upturned eyes, who wore sombre garments, sniffled through their noses, and garnished their discourse with Scripture texts, were an exact reproduction, so far as the difference of place and period would allow, of certain Jerusalem Pharisees who veiled their faces when they went abroad lest they should behold a woman or some unclean thing; who strained the water which they drank for fear they should swallow the forbidden gnat; who gave alms to the sound of trumpet, and uttered long prayers in a loud voice; who wore texts embroidered on their robes and bound upon their brows; who followed minutely the observances of the ceremonial law; who added to it with their traditions; who lengthened the hours and deepened the gloom of the Sabbath day, and increased the taxes which it had been ordered should be paid upon the altar.

On the other hand, there had been among the Puritans many men of pure and gentle lives, and a similar class existed among the Pharisees. The good Pharisee, says the Talmud, is he who obeys the law because he loves the Lord. They addressed their god by the name of 'Father' when they prayed. 'Do unto others as you would be done by' was an adage often on their lips. That is the law, they said; all the rest is mere commentary. To the Pharisees belonged all that was best and all that was worst in the Hebrew religious life.

The traditions of the Pharisees related partly to ceremonial matters which in the written law were already diffuse and intricate enough. But it must also be remembered that without traditions the Hebrew theology was barbarous and incomplete. Before the captivity the doctrine of rewards and punishments in a future state had not been known. The Sheol of the Jews was a land of shades in which there was neither joy nor sorrow, in which all ghosts or souls dwelt promiscuously together. When the Jews came in contact with the Persian priests they were made acquainted with the heaven and hell of the Zend-Avesta. It is probable, indeed, that without foreign assistance they would in time have developed a similar doctrine for themselves. Already in the Psalms and Book of Job are signs that the Hebrew mind was in a transition state. When Ezekiel declared that the son should not be responsible for the iniquity of the father nor the father for the iniquity of the son, that the righteousness of the righteous should be upon him, and that the wickedness of the wicked should be upon him, he was preparing the way for a new system of ideas in regard to retribution. But as it was, the Jews were indebted to the Zend-Avesta for their traditional theory of a future life, and they also adopted the Persian ideas of the resurrection of the body, the rivalry of the evil spirit, and the approaching destruction and renovation of the world.

The Satan of Job is not a rebellious angel, still less a contending god:

he is merely a mischievous and malignant sprite. But the Satan of the restored Jews was a powerful prince who went about like a roaring lion, and to whom this world belonged. He was copied from Ahriman, the God of Darkness, who was ever contending with Ormuzd, the God of Light. The Persians believed that Ormuzd would finally triumph, and that a prophet would be sent to announce the gospel or good tidings of his approaching victory. Terrible calamities would then take place; the stars would fall down from heaven; the earth itself would be destroyed. After which it would come forth new from the hands of the Creator; a kind of Millennium would be established; there would be one law, one language, and one government for men, and universal peace would reign.

This theory became blended in the Jewish minds with certain expectations of their own. In the days of captivity their prophets had predicted that a Messiah or anointed king would be sent, that the kingdom of David would be restored, and that Jerusalem would become the headquarters of God on earth. All the nations would come to Jerusalem to keep the feast of tabernacles and to worship God. Those who did not come should have no rain; and as the Egyptians could do without rain, if they did not come they should have the plague. The Jewish people would become one vast priest-hood, and all nations would pay them tithe. Their seed would inherit the Gentiles. They would suck the milk of the Gentiles. They would eat the riches of the Gentiles. These same unfortunate Gentiles would be their ploughmen and their vine-dressers. Bowing down would come those that afflicted Jerusalem, and would lick the dust off her feet. Strangers would build up her walls, and kings would minister unto her. Many people and strong nations would come to see the Lord of Hosts in Jerusalem. Ten men in that day would lay hold of the skirt of a Jew saying, 'We will go with you, for we have heard that God is with you.' It was an idea worthy of the Jews that they should keep the Creator to themselves in Jerusalem, and make their fortunes out of the monopoly.

In the meantime these prophecies had not been fulfilled, and the Jews were in daily expectation of the Messiah—as they are still, and as they are likely to be for some time to come. It was the belief of the vulgar that this Messiah would be a man belonging to the family of David, who would liberate them from the Romans and become their king; so they were always on the watch, and whenever a remarkable man appeared they concluded that he was the son of David, the Holy One of Israel, and were ready at once to proclaim him king and to burst into rebellion. This illusion gave rise to repeated riots or revolts, and at last brought about the destruction of the city.

But among the higher class of minds the expectation of the Messiah, though not less ardent, was of a more spiritual kind. They believed that the Messiah was that prophet, often called the Son of Man who would be send by God to proclaim the defeat of Satan and the renovation of the world. They interpreted the prophets after a manner of their own: the kingdom foretold was the kingdom of heaven, and the new Jerusalem was not a Jerusalem on earth but a celestial city built of precious stones and watered by the Stream of Life.

Such were the hopes of the Jews. The whole nation trembled with excitement and suspense; the mob of Judea awaiting the Messiah or king who should lead them to the conquest of the world; the more noble-minded Jews of Palestine, and especially the foreign Jews, awaiting the Messiah or Son of Man who should proclaim the approach of the most terrible of all events. There were many pious men and women who withdrew entirely from the cares of ordinary life, and passed their days in watching and in prayer.

The Neo-Jewish or Persian-Hebrew religion, with its sublime theory of a single god, with its clearly defined doctrine of rewards and punishments, with its one grand duty of faith or allegiance to a divine king, was so attractive to the mind on account of its simplicity that it could not fail to conquer the discordant and jarring creeds of the pagan world as soon as it should be propagated in the right manner. There is a kind of natural selection in religion; the creed which is best adapted to the mental world will invariably prevail, and the mental world is being gradually prepared for the reception of higher and higher forms of religious life. At this period Europe was ready for the reception of the one-god species of belief, but it existed only in the Jewish area, and was there confined by artificial checks. The Jews held the doctrine that none but Jews could be saved, and most of them looked forward to the eternal torture of Greek and Roman souls with equanimity, if not with satisfaction. They were not in the least desirous to redeem them; they hoarded up their religion as they did their money, and considered it a heritage, a patrimony, a kind of entailed estate. There were some Jews in foreign parts who esteemed it a work of piety to bring the Gentiles to a knowledge of the true God, and as it was one of the popular amusements of the Romans to attend the service at the synagogue a convert was occasionally made. But such cases were very rare, for in order to embrace the Jewish religion it was necessary to undergo a dangerous operation and to abstain from eating with the pagans—in short, to become a Jew. It was therefore indispensable for the success of the Hebrew religion that it should be divested of its local customs. But however much the Pharisees and Sadducees might differ on matters of tradition, they were perfectly agreed on this point, that the ceremonial laws were necessary for salvation. These laws could never be given up by Jews unless they first became heretics, and this was what eventually occurred. A schism arose among the Jews: the sectarians were defeated and expelled. Foiled in their first object, they cast aside the law of Moses and offered the Hebrew religion without the Hebrew ceremonies to the Greek and Roman world. We shall now sketch the character of the man who prepared the way for this remarkable event.

It was a custom in Israel for the members of each family to meet together once a year that they might celebrate a sacred feast. A lamb roasted whole was placed upon the table, and a cup of wine was filled. Then the eldest son said, 'Father, what is the meaning of this feast?' And the father replied that it was held in memory of the sufferings of their ancestors, and of the mercy of the Lord their God. For while they were weeping and bleeding in the land of Egypt there came his voice unto Moses and said that each father of a family should select a lamb without blemish from his flock, and should kill it on the tenth day of the month Abib, at the time of the setting of the sun; and should put the blood in a basin, and should take a sprig of hyssop and sprinkle the door-posts and lintel with the blood; and should then roast the lamb and eat it with unleavened bread and bitter herbs. They should eat it as if in haste, each one standing with his loins girt, his sandals on his feet, and his staff in his hand. That night the angel of the Lord slew the first born of the Egyptians, and that night Israel was delivered from her bonds.

When the father had thus spoken the lamb was eaten, and four cups of wine were drunk, and the family sang a hymn. At this beautiful and solemn festival all persons of the same kin endeavoured to meet together, and Hebrew pilgrims from all parts of the world journeyed to Jerusalem. When they came within sight of the Holy City and saw the Temple shining in the distance like a mountain of snow, some clamoured with cries of joy, some uttered low and painful sobs. Drawing closer together, they advanced towards the gates singing the Psalms of David, and offering up prayers for the restoration of Israel.

At this time the subscriptions from the various churches abroad were brought to Jerusalem, and were carried to the Temple treasury in solemn state; and at this time also the citizens of Jerusalem witnessed a procession which they did not like so well. A company of Roman soldiers escorted the lieutenant-governor, who came up from Caesarea for the festival that he might give out the vestments of the High Priest, which, being the insignia of government, the Romans kept under lock and key.

It was the nineteenth year of the reign of Tiberius Caesar. Pontius Pilate had taken up his quarters in the city, and the time of the Passover was at hand. Not only Jerusalem, but also the neighbouring villages, were filled with pilgrims, and many were obliged to encamp in tents outside the walls.

It happened one day that a sound of shouting was heard; the men ran up to the roofs of their houses, and the maidens peeped through their latticed windows. A young man mounted on a donkey was riding towards the city. A crowd streamed out to meet him, and a crowd followed him behind. The people cast their mantles on the road before him, and also covered it with green boughs. He rode through the city gates straight to the Temple, dismounted, and entered the holy building. In the outer courts there was a kind of bazaar in connection with the Temple worship. Pure white lambs, pigeons, and other animals of the requisite age and appearance were there sold, and money merchants, sitting at their tables, changed the foreign coin with which the pilgrims were provided. The young man at once proceeded to upset the tables and to drive their astonished owners from the Temple, while the crowd shouted and the little gamins, who were not the least active in the riot, cried out, 'Hurrah for the son of David!' Then people suffering from diseases were brought to him, and he laid his hands upon them and told them to have faith and they would be healed. When strangers inquired the meaning of this disturbance they were told that it was Joshua—or—as the Greek Jews called him, Jesus—the Prophet of Nazareth. It was believed by the common people that he was the Messiah. But the Pharisees did not acknowledge his mission. For Jesus belonged to Galilee, and the natives of that country spoke a vile patois, and their orthodoxy was in bad repute. 'Out of Galilee,' said the Pharisees with scorn, 'out of Galilee there cometh no prophet.'

All persons of imaginative minds know what it is to be startled by a thought; they know how ideas flash into the mind as if from without, and what physical excitement they can at times produce. They also know what it is to be possessed by a presentiment, a deep, overpowering conviction of things to come. They know how often such presentiments are true, and also how often they are false."

Like the firstborn of Egypt, the story of Jesus (the lamb of God) is the story of bloody human sacrifice for the sake of Jewish "restoration to Palestine"—in this instance God sacrifices his firstborn child, just as the Jews had so often sacrificed their own children to Baal.

In Austria, Georg Schönerer, or Georg Ritter von Schönerer, agitated for Pan-Germanism, or an *Alldeutscher Verband*, in which all members of the "German race" or "Aryan Race" would unite to form a unified state with broad borders across Middle Europe. Schönerer advocated the segregation of Jewish children from Christian schools, a goal of the Zionists. He also founded a worker's party, which eventually morphed into the NSDAP. Schönerer was staunchly anti-Catholic and founded the *Los von Rom Bewegung*. In 1892, a thirty page pamphlet appeared entitled *Ein deutsches Weltreich?*, Sammlung deutscher Schriften, Volume 7, Lüstenöder, Berlin. This brochure called for the "German races" to unite and form an empire to rule the world.[339] Between the British Imperialist racists and the German Imperialist racists, between the remnants of the Holy Roman Empire (the Catholics of France and Italy) and the Ottoman Turks, Moses Hess' Pan-Judaic Zionists had the makings of their Biblical race war to end all wars, and they did what they could to provoke it. They planned to eventually replace all the other empires they had pitted against one another with a universal Jewish Empire.

8.5 Puritans and Protestants Serve Jewish Interests

Racist Zionist Moses Hess was one of the founders of the Jewish Communist factions. The Jewish Communists, with their blind and brutal cult following,

---

[339]. "The Pan-Germanic Movement", *The American Monthly Review of Reviews*, Volume 26, Number 1, (July, 1902), p. 93; which cites: Sir Rowland Blennerhassett, "The Pan-Germanic", *National Review* (W. H. Allen, London), (1902?).

looked forward to a devastated Europe, which weakened world would enable them to take over the Earth through violent revolt. The Communists' world of universal "equality", would give every Gentile an equal opportunity to slave for Jewish leadership—as prophesied in *Isaiah*. The Communists justified their dark visions of ultimate destruction with the same false premise as the Jews and Christians, that a new millennium would occur after the devastation, and the Earth would become a Utopia. All their terrible attacks on humanity and their Socialist dictatorships were merely transitional phases working toward the Utopia of Communism, the Jewish "End Times".

It is interesting to note that the Communists in Russia prevented wealth accumulation and the pooling of investment capital for decades, which left Russia, after having shaken off the yoke of Communism, vulnerable to another Jewish takeover led by Jewish financiers. From the beginning, the Communists drew off the wealth of Russia and fed it to the Jewish financiers who had funded and organized the Russian Revolution. Communism always served the interests of Jewish Capitalists.

Like the Communism the Jews gave the Christians, Christianity itself also taught Gentiles that wealth accumulation was immoral. This worked against the interests of the Gentiles, while providing more opportunities for Jews to accumulate the Christians' wealth. Jewish sponsored Christianity led the Romans and Europe into the Dark Ages. It was the more Judaically minded Protestants, with their Judaic concept of the "elect" (*Isaiah* 65. *Enoch*) that justified wealth accumulation, who materialistically prospered under a new form Judaized Christianity—at the expense of the colonial peoples—and resulting in the second destruction of a Roman Empire, the slow decline of the Roman Catholic Church. The Protestants became the parasites of the "Third World" colonies.

One suspects that Cabalists and other Jews may have been the instigators of the Christian Reformation; for they, more than anyone else, were opposed to Catholicism, that second Roman Empire which according to them: worshiped idols, treated the Pope as they would treat a Jewish Messiah, gave the Pope the authority to interpret God's word while taking away that right from Rabbis and individuals, and stood in the way of Jewish desires on Jerusalem. There was also the issue of faith versus works.

Jewish mythology holds that nations which worship idols must be exterminated (*Exodus* 34:11-17. *Isaiah* 65; 66. *Ezekiel*), and that when this divine obligation is accomplished, the Jews will rule the world. The Talmud teaches in one opinion that "heathens" can annul idols and that Jews can use force to make heathens annul their idols (*Abodah Zarah* 43*a*). The annulment of Catholic idol worship was one of the main goals of the Reformation. Frankist Jews became Catholics in order to undermine the religion, in order annul the worship of idols and ruin the authority of the Pope. The Illuminati and Free Masonry sought to destroy "superstitious" religion. The Communists use force to make other religions annul their idols.

Catholicism became the focal point of Jewish genocidal hatred and mythology. They had a model for the Reformation in the lives, writings and practices of Jon Wycliffe and Jan Hus. All they lacked were spokesmen in the Christian community, whom they recruited in the form of their friend Martin Luther, as well as John Calvin (some claim "Calvin" is a corruption via "Cauin" of "Cohen"[340]—the man had a classical Jewish appearance) and the new Enoch—Melchior Hofmann, Ignatius Loyola, the father of the Jesuits, etc.

There are many allegations of a long term plan carried out by Prussian Protestants, French free thinkers, the Illuminati and Freemasonry to convert Catholics to Judaism and eventually atheism. This charge was strongly brought forth after the French Revolution by John Robison[341] and Abbé Barruel.[342] The alleged plan to subjugate the world to a tyranny of hypocrites preaching disingenuous Liberalism took on its ultimate protagonist in Marx's Communism, which failed in its promise of a liberal Utopia, but succeeded quite well in its nihilistic ambitions. More recent accusations include, among many others: George Pitt-Rivers', *World Significance of the Russian Revolution*, B. Blackwell, Oxford, (1920); Nesta Helen Webster's, *Germany and England*, Boswell, London, (1938); and Captain Archibald Henry Maule Ramsay's, *The Nameless War*, Britons Publishing Company, London, (1952).

Martin Luther had direct and indirect connection to Cabalistic Jews, influential Jews and anti-Semitic Jews, who claimed to have converted to Christianity, including: Konrad Mutian (a. k. a. Conradus Mutianus Rufus), Johann Reuchlin, Pico della Mirandola, Jakob Questenberg, Jakob ben Jehiel Loans, Obadja Sforno of Cesena, Johann Pfefferkorn, etc. Note that in the Dualistic and dialectical terms of the Cabalah, both anti-Semites, and the defenders of Judaism as a "racial" and nationalistic sect, serve the same purpose—the beloved hateful segregation of the Jews from the Gentiles, after which the Jews sought.[343]

For Cabalistic Jews, both evil and good are functions of, and serve, God. Contemporary Jews believed that Martin Luther was preparing the way for the arrival of the Jewish Messiah. The *Encyclopaedia Judaica* writes in its article "Messianic Movements":

> "About the same time many Jews pinned their hopes on Martin *Luther as a man who had come to pave the way for the Messiah through gradually

---

[340]. A. H. M. Ramsay, *The Nameless War*, Chapter 1, Britons Publishing Company, London, (1952).

[341]. J. Robison, *Proofs of a Conspiracy Against All the Religions and Governments of Europe: Carried on in the Secret Meetings of Free Masons, Illuminati, and Reading Societies*, Printed for William Creech, and T. Cadell, Junior, and W. Davies, Edinburgh, London, (1797); see especially the fourth edition of 1798, to which Robison added a postscript.

[342]. Abbé Barruel, *Mémoires pour Servir a l'Histoire du Jacobinisme*, De l'Imprimerie Françoise, Chez P. Le Boussonier, Londres, (1797-1798); English translation by R. Clifford: *Memoirs Illustrating the History of Jacobism*, Printed for the translator by T. Burton and Co., London, (1798).

[343]. M. A. Hoffman II, *Judaism's Strange Gods*, Independent History and Research, Coeur d'Alene, Idaho, (2000), pp. 108-109.

educating the Christians away from their idolatrous customs and beliefs."[344]

Luther caused the slaughter of countless Christians, then caused Christian enmity towards Jews—which were Zionist aspirations.

Malachi 3:1 and 4:5 speak of a forerunner of the Messiah who will prepare the way, like John the Baptist (*Matthew* 3; 11:10; 17:10-13). Cabalist Jews considered Martin Luther (1483-1546) to have been this forerunner,

> "Behold, I *will* send my messenger, and he shall prepare the way before me: and the LORD, whom ye seek, shall suddenly come to his temple, even the messenger of the covenant, whom ye delight in: behold, he *shall* come, saith the LORD of hosts. [***] Behold, I *will* send you Elijah the prophet before the coming of the great and dreadful day of the LORD:"

Some Cabalist Jews believed that Isaac ben Solomon Luria (1534-1572) was "the Messiah, son of Joseph" (as opposed to: "the Messiah, son of David").[345] Luria formulated a new Cabalistic dogma, which preached Metempsychosis and emphasized the Messianic prophecies in a way that was forbidden in the Talmud (*Kethuboth* 111*a*). The Lurian Cabalah inspired Shabbatai Zevi and Jacob Frank, both of whom claimed to be the Davidic Messiah—Frank claiming to have received the soul of the Messiah of Shabbatai Zevi through Cabalistic Metempsychosis—the transmigration of souls.

The Messiahship was believed to have been a dynasty of Jewish Kings descended from Joseph and David—in fact many myths alleged that their were two Messiahs, one a sacrificial warrior, and the other a genocidal tyrant to rule over the entire Earth. Many believe that the Lurian Cabalah became the basis for the Hasadic dynasty of the Lubavitchers, whose descendants today claim that the Jewish Messiah is among us. They are eagerly waiting to anoint him King of the Jews.

The followers of this Hasidic dynasty were said to number 3,000,000 strong in 1930,[346] and are, so some claim, the descendants of the Frankists of Poland and Russia. Jacob Frank taught that both good and evil are necessary functions of God, and that Jews should cause rampant evil in the world in order to hasten the coming of the Messianic Era.

Frank taught that since God is hidden in all things and yet controls them, and since Jews are to humanity what God is to the Universe, Jews should act as a hidden force controlling humanity. He taught his followers to feign conversion

---

344. "Messianic Movements", *Encyclopaedia Judaica*, Volume 11 LEK-MIL, Encyclopaedia Judaica, Jerusalem, The Macmillan Company, New York, (1971), cols. 1417-1427, at 1426.

345. "LURIA, ISAAC BEN SOLOMON", *Encyclopaedia Judaica*, Volume 11 LEK-MIL, Encyclopaedia Judaica, Jerusalem, The Macmillan Company, New York, (1971), cols. 572-578, at 574.

346. "Throng Storms Depot to Greet Jewish Leader", *The New York Times*, (10 February 1930), p. 7.

to other religions, as Shabbatai Zevi had, so as to infiltrate other religions and governments and, once in power, destroy them. Frank taught his followers to engage in sexual orgies and practice other forms of depravity—practices allegedly common among some groups of Hasidic Jews as evidenced in their frequents fits of frantic dancing. Many of the descendants of Jacob Frank have come to America and live as crypto-Jews. Hasidic Jews tend to be very secretive.

The Lubavitchers were initially outspoken anti-Zionists. It shocked many when the last of the Lubavitch Dynasty, the Rebbe whose life was to herald the coming of the Messiah, the seventh Rebbe in the line, Rebbe Menachem Mendel Schneerson, declared that he was a Zionist and that the Messiah is here alive among us. One wonders if this influential dynasty had been secretly planning the rise of Israel for centuries—if they had employed Frankist followers to destroy governments and religions and create bloody wars. The Lubavitchers were notoriously racist and considered Gentiles to be something less than human.[347] The Talmud states that Gentiles are subhuman at: *Baba Mezia* 108*b* and 114*b*; *Berakoth* 58*a-b*, *Yebamoth* 60*b*-61*a*, 98*a*; *Kerithoth* 6*b*; *Kiddushin* 68*a*; relying upon *Ezekiel* 23:20 and 34:31.

The Zionist racists who believe that they are the divine leaders of the world, will, if successful, destroy all nations and replace them with a world government led from Jerusalem by the King of the Jews. They will proscribe all religions other than Judaism and will force Gentiles into atheism. Then they will systematically exterminate all Gentiles.

Zionism is based on the Judaic myth of a new Utopian millennium following world-wide nihilistic devastation—which is to say Utopian for "righteous Jews", hellish for non-racist Jews and Gentiles.[348] Judaism and Christianity make the Christians vulnerable to an absolute genocidal Zionist tyranny, in that Judaism asked Gentiles to slave and fight for Jews, and Christianity promotes a slavish mentality and a self-defeating fatalism which sponsors the suicidal belief that the worse one's conditions are, the better off one is in God's eyes.

8.6 The Planned Apocalypse

Since God had not yet brought about the horrific wars prophesied in the end times, some Jews began to intervene on God's behalf. In the 1800's, Baron Edward Bulwer-Lytton wrote of an extraordinary "occult" force called "Vril", which was so destructive that it resulted in peace among those who could control it, because it assured mutual destruction between combatants. His book was

---

[347]. A. Nadler, "Last Exit to Brooklyn: The Lubavitcher's Powerful and Preposterous Messianism", *The New Republic*, (4 May 1992), pp. 27-35. M. A. Hoffman II, *Judaism's Strange Gods*, Independent History and Research, Coeur d'Alene, Idaho, (2000), p. 58.
[348]. *Exodus* 34:11-17. *Psalm* 2; 72. *Isaiah* 1:9; 2:1-4; 6:9-13; 9:6-7; 10:20-22; 11:4, 9-12; 17:6; 37:31-33; 41:9; 42; 43; 44; 61:6. *Jeremiah* 3:17; 33:15-16. *Ezekiel* 20:38; 25:14. *Daniel* 12:1, 10. *Amos* 9:8-10. *Obadiah* 1:18. *Micah* 4:2-3; 5:8. *Zechariah* 8:20-23; 14:9. *Romans* 9:27-28; 11:1-5.

titled *The Coming Race: Or the New Utopia*.[349] Lord Lytton wrote of a race of giants descended from Aryans which lived below the surface of the Earth, flew about on artificial wings, and which would one day again surface to exterminate all who lived on the surface of the Earth.

Lytton's tale recalls the Jewish myth of the race of giants bred from women and angels told in the Old Testament and the book of *Encoch*.[350] It is also derivative of the Hindu myths of the Nagas, a serpent-human cross which lives underground. Lytton had his angelic characters instruct us that humans evolved from tadpoles in a Lamarckian manner and in a process of natural and sexual selection (this before Spencer, Wallace and Darwin, though much after Empedocles), that souls undergo reincarnation pursuant to the principles of Metempsychosis, that the name of God must not be written, that all forces are unified, that form should follow function, that the principle of logical economy made dictatorships more reasonable than democracies, that we should practice vegetarianism, that enemy races must be unemotionally and mercilessly exterminated, etc. Many of these ideas, which stem from various sects of Judaism and Hinudism, found their way into Nazi mythology. In the Bible, the angels, like Lytton's children of the underworld, committed genocide and other atrocities against human beings. In Lytton's book, six children could—like the Lord's angels—destroy thirty million of us, by harnessing the force of "Vril", an æthereal fluid as destructive as nuclear bombs.

This fantasy of alien races and super forces later became a facet of *Thule-Gesellschaft* mythology, which influenced Adolf Hitler and several other prominent Nazis. They taught that Aryans descended from aliens from outer space, which had interbred with humans. Other Earthly sub-human races were akin to the apes.

This mythology has roots in Jewish mythologies centered around the Biblical people called Nefilim, and more broadly around the Gentiles. Jewish myth has it that this people called Nefilim is descended from a mixture of fair humans and angels who fell from heaven to fall in love with the beautiful human women whose beauty had seduced them, much like Adam was tempted to sin by a woman—much like the Sirens who seduced sailors into suicide. For their sin, God banished the angels from the future world as He had banished Adam from the Garden—yet another instance of misogyny in Jewish mythology. Another Jewish myth also holds that the Gentiles are descended from a race begun by the fornication of Eve with the serpent who tempted her—an attack transferred to Jews themselves by the Apostate crypto-Jews who founded the Christian Identity movement in America.[351]

---

349. E. Bulwer-Lytton, *Rienzi: The Pilgrims of the Rhine; The Coming Race*, Brainard, New York, Continental Press, New York, (1848); **and** *Rienzi, Two Volumes in One; The Pilgrims of the Rhine; the Coming Race*, Boston, Dana Estes & Co., 1848; **and** *The Pilgrims of the Rhine. The Coming Race*, Dana Estes & Co., Boston, (1849); **and** *The Coming Race: Or the New Utopia*, Francis B. Felt & Co., New York, (1871).
350. *Genesis* 6:1-5. *Numbers* 13:25-33. I *Enoch*.
351. *Zohar*, I, 25a-25b, 28b-29a; III, 208a. T. R. Weiland, *Eve, Did She or Didn't She:*

There is also a Jewish myth which holds that angels dubbed "Watchers" fell from heaven to Earth, became men, and bred with the fair daughters of the Earth to produce a race of giants, who were evil and destructive. This myth holds that the angels taught humans the secrets of nature and that this is how evil came to the world.[352] After fornicating with women, the angels lost their immortality. This mythology, which again mirrors the story of Adam and Eve, is one of the major themes of the book of *Enoch*, which is probably in large part a plagiarism of the story of Gilgamesh.[353] The apocalyptic book of *Enoch* contains much that later appeared in the Reformation of the Protestants and in the "second Reformation" of the Puritans, with their emphasis on the mythology of the "elect", the destruction of the Earth, damnation, and their hatred and reluctance to look to redeem those who have sinned against them (*Isaiah* 65; 66. *Enoch*); which tends to indicate that Jews, especially Cabalistic Jews, were the driving force behind the Reformation and Puritanism, which had as their main goal the destruction of Roman Catholicism.

The apocalyptic works derive from the flood story told in the legends of Gilgamesh, and the flood is a genocide meant to cleanse the Earth of the unclean mixture of the blood of angels which had commingled with humans and other animals through angelic miscegenation with women and animals. The character Enoch parallels Enmeduranna and Noah replicates Utnapishtim. In Jewish mythology, the angels brought evil to the world and taught humans to do wrong.

The ultimate source of the giant myths, as told in Gilgamesh, *Enoch*, the book of *Giants*, and in the legends of the Greeks, is probably to be found in dinosaur, elephant, mastodon and mammoth bones; which Adrienne Mayor has shown were kept in ancient temples and which were believed to be the bones of the legendary giant beasts and men.[354] Christian Messianic and apocalyptic mythology certainly derives not only from the prophecies found in the Old Testament, but also from the genocidal book of *Enoch*, with its "Elect" and "Elect One", and the book of *Giants*, which were valued by the Essenians who created, or at least contributed to, the Christian myth. These dangerous mythologies each teach their blind adherents to welcome genocide, and they provide a religious basis for the mass murder of our fellow human beings and the destruction of our natural environment.

In another instance of Jewish genocidal hatred, the book of *Enoch* calls for the extermination of the "seed of Cain",[355] which "race" descended from the evil mixture of the "angels" with women. This gives Jews religious license to

---

The Seedline Hypothesis under Scrutiny, Mission to Israel Ministries, Scottsbluff, Nebraska, (2000). M. Barkun, *Religion and the Racist Right: The Origin of the Christian Identity Movement*, Revised Edition, University of North Carolina Press, (1997).
352. *Genesis* 6:1-5; "The Book of the Watchers", I *Enoch*, Chapters 1-36.
353. A. Heidel, *The Gilgamesh Epic and Old Testament Parallels*, University of Chicago Press, (1949).
354. A. Mayor, *The First Fossil Hunters: Paleontology in Greek and Roman Times*, Princeton University Press, (2000).
355. I *Enoch* 22:7.

mercilessly mass murder any group which they oppose. It is interesting that the genocidal Nazis, as a philosophical movement, sprang forth from, and adopted, the ancient Jewish myths expressed in the book of *Enoch*. One also sees the book of *Enoch* in the legend of Faust, and the story of Mohammed and his flight with Gabriel.

William Winwood Reade published an influential book which applied Darwinist principles to history, *The Martyrdom of Man*, Trübner & Co., London, (1872). This work influenced Cecil John Rhodes, among many others. Reade discussed various revolutionary movements and concluded that they would next destroy religion—he very much wanted to destroy Christianity—a goal he had in common with Talmudic writers,

> "The anti-slavery movement, which we shall now briefly sketch, is merely an episode in that great rebellion against authority which began in the night of the Middle Ages; which sometimes assumed the form of religious heresy, sometimes of serf revolt; which gradually established the municipal cities, and raised the slave to the position of the tenant; which gained great victories in the Protestant Reformation, the two English Revolutions, the American Revolution, and the French Revolution; which has destroyed the tyranny of governments in Europe, and which will in time destroy the tyranny of religious creeds."

Reade saw that Communism was patterned after the Christian faith, with its homogenous and obedient followers; and the damnation of the wealthy, which led Christians to accept their own misery with joy,

> "A young man named Joshua or Jesus, a carpenter by trade, believed that the world belonged to the devil, and that God would shortly take it from him, and that he the Christ or Anointed would be appointed by God to judge the souls of men, and to reign over them on earth. In politics he was a leveller and communist, in morals he was a monk; he believed that only the poor and the despised would inherit the kingdom of God. All men who had riches or reputations would follow their dethroned master into everlasting pain. He attacked the churchgoing, sabbatarian ever-praying Pharisees; he declared that piety was worthless if it were praised on earth. It was his belief that earthly happiness was a gift from Satan, and should therefore be refused. If a man was poor in this world, that was good; he would be rich in the world to come. If he were miserable and despised, he had reason to rejoice; he was out of favour with the ruler of this world, namely Satan, and therefore he would be favoured by the new dynasty. On the other hand, if a man were happy, rich, esteemed, and applauded, he was for ever lost. He might have acquired his riches by industry; he might have acquired his reputation by benevolence, honesty, and devotion; but that did not matter; he had received his reward. So Christ taught that men should sell all that they had and give to the poor; that they should renounce all family ties; that they should let tomorrow take care of itself; that they should not trouble

about clothes: did, not God adorn the flowers of the fields? He would take care of them also if they would fold their hands together and have faith, and abstain from the impiety of providing for the future. The principles of Jesus were not conducive to the welfare of society; he was put to death by the authorities; his disciples established a commune; Greek Jews were converted by them, and carried the new doctrines over all the world. The Christians in Rome were at first a class of men resembling the Quakers. They called one another brother and sister; they adopted a peculiar garb, and peculiar forms of speech; the Church was at first composed of women, slaves, and illiterate artisans but it soon became the religion of the people in the towns. All were converted excepting the rustics (pagani) and the intellectual freethinkers, who formed the aristocracy. Christianity was at first a republican religion; it proclaimed the equality of souls; the bishops were the representatives of God, and the bishops were chosen by the people. But when the emperor adopted Christianity and made it a religion of the state, it became a part of imperial government, and the parable of Dives was forgotten. The religion of the Christians was transformed; its founder was worshipped as a god; there was a doctrine of the incarnation; they had their own holy books, which they declared to have been revealed; they established convents, and nunneries, and splendid temples, adorned with images, and served by priests with shaven heads, who repeated prayers upon rosaries, and who taught that happiness in a future state could best be obtained by long prayers and by liberal presents to the Church. In the Eastern or Greek world, Christianity in no way assisted civilisation, but in the Latin world it softened the fury of the conquerors, it aided the amalgamation of the races. The Christian priests were reverenced by the barbarians, and these priests belonged to the conquered people."

The Communists replaced the slavish dogma of Christianity with the slavish dogma of Marx, making their leaders the new gods and breaking the power of the emerging democracies of Europe, which had led to the assimilation of the Jews. Reade believed that war had many beneficial effects for humanity, though he predicted that weapons of mass destruction would eventually make war unthinkable,

"Thus war will, for long years yet to come, be required to prepare the way for freedom and progress in the East; and in Europe itself, it is not probable that war will ever absolutely cease until science discovers some destroying force, so simple in its administration, so horrible in its effects, that all art, all gallantry, will be at an end, and battles will be massacres which the feelings of mankind will be unable to endure."

The same principle of assured mutual destruction, with which we are all familiar in this age of nuclear weapons, had already appeared in the writings of Edward Bulwer-Lytton. In *The Coming Race*, Lord Lytton wrote, at least as early as 1848,

"But the effects of the alleged discovery of the means to direct the more terrible force of vril were chiefly remarkable in their influence upon social polity. As these effects became familiarly known and skillfully administered, war between the Vril-discoverers ceased, for they brought the art of destruction to such perfection as to annul all superiority in numbers, discipline, or military skill. The fire lodged in the hollow of a rod directed by the hand of a child could shatter the strongest fortress, or cleave its burning way from the van to the rear of an embattled host. If army met army, and both had command of this agency, it could be but to the annihilation of each. The age of war was therefore gone, but with the cessation of war other effects bearing upon the social state soon became apparent. Man was so completely at the mercy of man, each whom he encountered being able, if so willing, to slay him on the instant, that all notions of government by force gradually vanished from political systems and forms of law. It is only by force that vast communities, dispersed through great distances of space, can be kept together; but now there was no longer either the necessity of self-preservation or the pride of aggrandizement to make one State desire to preponderate in population over another."[356]

Henri de Saint-Simon[357] predicted the end of war due to the destructive force of modern technologies, in the early 1800's. He also argued for a United Nations, a world government, and a socialistic world in which science liberated mankind. His concepts were derivative of Francis Bacon's Seventeenth Century work, *New Atlantis*, which in turn is derivative of Campanella's *Civitas Solis* and Sir Thomas More's *Utopia*—all these Utopian works, including also Cicero's *De Republica* and Augustine's *City of God*, deriving from Plato's description of Atlantis in his *Timæus*, and Plato's *Republic*. It is interesting to note that Saint Simon of Trent was a young Christian boy who was murdered on 21 March 1475. It was alleged that a group of Rabbis had ritually murdered the boy in order to ridicule Christ and use the boy's blood in the matzoh for Passover. He was made a Saint and became one of the most popular Saints in history.

In 1913, H. G. Wells crafted a novel which envisioned many of the events which later took place in the First World War and in the Second World War. This novel was titled, *The World Set Free: A Story of Mankind*, Macmillan, London, (1914); also published in Leipzig, Germany, by B. Tauchnitz. Wells' story tells of a "world war" which ends when "atomic bombs" fall and a "world government" is formed. Wells later published *The Open Conspiracy; Blue Prints*

---

356. E. Bulwer-Lytton, *Rienzi: The Pilgrims of the Rhine; The Coming Race*, Brainard, New York, Continental Press, New York, (1848); and *Rienzi, Two Volumes in One; The Pilgrims of the Rhine; the Coming Race*, Boston, Dana Estes & Co., 1848; here quoted from: *The Pilgrims of the Rhine to which is Prefixed The Ideal World. The Coming Race*, Chapter 9, Estes and Lauriat, (1892), p. 277; which appears to be a reprint of: *The Pilgrims of the Rhine. The Coming Race*, Dana Estes & Co., Boston, (1849).
357. M. M. Dondo, *The French Faust: Henri de Saint-Simon*, Philosophical Library, New York, (1955).

for a World Revolution, V. Gollancz Ltd., London, (1928); and several other related works.[358] Wells' book inspired Michael Higger to publish a depiction of the rabbinical "Utopia" Zionists had planned for their domination of the Earth entitled: *The Jewish Utopia*, Lord Baltimore Press, Baltimore, (1932). Racist Zionist Albert Einstein used his fame to promote world government, an ideal which in the Old Testament, as well as Cabalistic and Talmudic writings, takes the form of universal Jewish rule and the subjugation and eventual extermination of the Gentiles.

Many believed and believe that the events of the Twentieth Century fulfilled many of the Bible's prophecies. Many of these persons do not recognize the willful intervention of groups who organized themselves for the expressed purpose of bringing these events about in order to "fulfill prophecy". When the "race war", the First World War, finally broke out, it was not easy to define just

---

358. H. G. Wells, *Imperialism and the Open Conspiracy*, Faber & Faber, London, (1929); **and** *The Way to World Peace*, E. Benn Ltd., London, (1930); **and** *What Are We to Do with Our Lives?*, Doubleday, Doran, Garden City, New York, (1931); **and** *The Work, Wealth and Happiness of Mankind*, Doubleday, Doran & Company, Inc., Garden City, New York, (1931); **and** *After Democracy: Addresses and Papers on the Present World Situation*, Watts & Co., London, (1932); **and** *What Should Be Done—Now: A Memorandum on the World Situation*, The John Day Company, New York, (1932); **and** *The Shape of Things to Come*, The Macmillan Co., New York, (1933); **and** *The New America, the New World*, Macmillan Co., New York, (1935); **and** *Things to Come: A Film Story Based on the Material Contained in His History of the Future "The Shape of Things to Come"*, Cresset Press, London, (1935); **and** *World Brain*, Methuen & Co., London, (1938); *and The Holy Terror*, Simon and Schuster, New York, (1939); **and** *The Fate of Homo Sapiens: An Unemotional Statement of the Things That Are Happening to Him Now, and of the Immediate Possibilities Confronting Him*, Secker and Warburg, London, (1939); **and** *The Fate of Man: An Unemotional Statement of the Things That Are Happening to Him Now, and of the Immediate Possibilities Confronting Him*, Longmans, New York, (1939); **and** *The New World Order: Whether it Is Attainable, How it Can Be Attained, and What Sort of World a World at Peace Will Have to Be*, Secker and Warburg, London, (1940); **and** *The Rights of Man; Or, What Are We Fighting For?*, Penguin Books Ltd., Harmondsworth, Middlesex, England, (1940); **and** *The Common Sense of War and Peace: World Revolution or War Unending*, Penguin Books, Harmondsworth, Middlesex, England, New York, (1940); **and** *Babes in the Darkling Wood, a Novel*, Alliance book Corp., New York, (1940); **and** *All Aboard for Ararat*, Secker & Warburg, London, (1940); **and** *You Can't be too Careful: A Sample of Life 1901-1951*, Secker & Warburg, London, (1941); **and** *The Outlook for Homo Sapiens: An Unemotional Statement of the Things That Are Happening to Him Now, and of the Immediate Possibilities Confronting Him*, Secker and Warburg, London, (1942); **and** *Science and the World Mind*, New Europe Pub. Co., London, (1942); **and** *Phoenix: How to Rebuild the World: A Summary of the Inescapable Conditions of World Reorganization*, Haldeman-Julius, Girard, Kansas, (1942); **and** *A Thesis on the Quality of Illusion in the Continuity of the Individual Life in the Higher Metazoa: With Particular Reference to the Species Homo Sapiens*, London, (1942). ***See also:*** L. V. Uspenskii, H. G. Wells, *The New Rights of Man: Text of Letter to Wells from Soviet Writer, Who Pictures the Ordeal and Rescue of Humanistic Civilization, H. G. Wells' Reply and Program for Liberated Humanity*, Haldeman-Julius Publications, Girard, Kansas (1942).

what "race" was fighting which other "race", and "races" came to defined by religious affiliation and language, as well as historical groupings and phenotypes.[359]

The Frankists, their cabalistic predecessors and their nihilistic descendants were successful in breaking apart Western Christendom under the Roman Catholics of the Holy Roman Empire. They worked to destroy the hegemony of Christianity and replace it with Judaic hegemony in fulfilment of ancient Judaic prophecies. This struggle played out in part Vienna during the *Kulturkampf*.

A common theme of many politicians was the notion that war must not result in changed borders—beyond the dissolution of empires. Both World Wars did little to change the map of Europe from its traditional complexion, other than to enhance segregation, and promote Bolshevism. Though many Zionists allegedly sponsored "Internationalism", they sought to segregate out the "Nationalities" which were disappearing under the empires and thereby causing Jews to assimilate in a spirit of true internationalism and integration. The Zionists, who were forbidden to practice their racism in the empires, sought to promote instead the rabid and racist Nationalism which led to the near destruction of Europe, without much changing its ethnic map. This was their short term goal, because it enabled the Zionists to justify their racism and to take the opportunity of peace talks which would follow a world war—at which talks small nations would appeal for independence—to ask for Palestine as an independent nation for the formation of a "Jewish State".

The *Kulturkampf* further complicated matters, because some Catholics desired to take Constantinople and Jerusalem from the Turks, who were Germany's allies, and make them Christian centers. The Zionists sought sympathy for their cause by promising Christians easy access to the Holy Land and by promoting Biblical prophecies, and more recent pseudo-Christian inventions like the "rapture", which they themselves did not believe. Many Catholics, as well as British and American Protestants, desired that Rome, Athens, Jerusalem and Constantinople forever be in Jewish and Christian hands. Greece had obtained its independence from the Turks with help of England, France and Russia and had long desired to reconquer all of the Byzantine Empire for Christendom. Greek Christians (doubtless many were crypto-Jews or the agents of Jews) managed the accounts of the Sultan and despite the prosperity the rise in cotton prices (which resulted from the American Civil War[360]) and other factors should have brought to the region, the Sultan was led towards bankruptcy.

Before political Zionism and Theodor Herzl, many "Christian" writers (doubtless many were crypto-Jews or the agents of Jews) and movements sought to reestablish a Jewish nation in Palestine allegedly in order to fulfill Biblical prophecy and hasten the second coming of Christ. Napoleon sought to destroy

---

359. See the forward of the first German edition of K. Kautsky, "Rasse und Judentum", *Ergänzungshefte zur Neuen Zeit*, Number 20, (1914/1915 Ausgeben am 30. Oktober 1914), pp. 1-5.
360. "Late Items of Foreign News", *The Chicago Tribune*, (8 November 1861), p. 3.

the Turkish Empire and take Palestine and give it to the Jews, believing himself to be the Messiah. Napoleon invaded Poland and Russia in order to emancipate the Jews—at the expense of his French soldiers and the Russian people, as well as many peoples in between the Russians and the French.

There were many Christian Zionists in the Nineteenth Century many of whom hoped to bring on the Apocalypse (whose loyalty had been bought with Rothschild money). These included Queen Victoria, Louis Way, the Christadelphians, William Blackstone, Charles Henry Churchill, Lord Anthony Ashley Cooper, the Earl of Shaftesbury, Lord Manchester, Lord Lindsay, Lord Palmerston, F. Laurence Oliphant, Holman Hunt, Sir Charles Warren, George Eliot, Hall Caine, George Gawler, Orson Hyde, John Nelson Darby, Jean Henri Dunant, and William Henry Hechler—who inspired and encouraged Theodor Herzl when he was feeling defeated, and who contacted Frederick the Grand Duke of Baden, Kaiser Wilhelm II, the Sultan of Turkey and Arthur Balfour on behalf of the Zionist cause.[361] David Lloyd George's Christianity made him favorable to Zionism.

Then, as now, England and America were the staunchest supporters of Zionism. English Protestants had been promoting the "restoration of the Jews" for centuries. Many English believed that the ancient Britons were of Jewish descent and that the Royal Family were direct descendants of King David—David who took Jerusalem and whose seed was prophesied to bear the Messiah. The Germans had hoped in both World Wars that the British and Americans would side with them against the Slavs, or remain neutral.

8.7 Cabalistic Jews Calling Themselves Christian Condition the British to Assist in Their Own Demise—Rothschild Makes an Open Bid to Become the Messiah

It is interesting to note that the Damascus Affair, which united Jews around the world, happened shortly after a broad based and well-publicized Zionist movement got underway in England in the 1830's. Both this movement to "restore Jews to Palestine" and the Damascus Affair received a great deal of press coverage in England. Was the Damascus Affair and the murder of Father Thomas the work of *agents provocateur* of the Lavon Affair[362] type today

---

361. Consult Herzl's Diaries and see: H. Ellern and B. Ellern, *Herzl, Hechler, the Grand Duke of Baden and the German Emperor, 1896-1904. Documents Found by Hermann and Bessi Ellern Reproduced in Facsimile*, Tel Aviv, (1961).
362.A. Golan, *Operation Susannah*, Harper & Row, New York, (1978). *See also:* D. Raviv, *Every Spy a Prince: The Complete History of Israel's Intelligence Community*, Houghton Mifflin, Boston, (1990). *See also:* V. Ostrovsky and C. Hoy, *By Way of Deception: A Devastating Insider's Portrait of the Mossad*, Stoddart, Toronto, (1990). V. Ostrovsky, *The Other Side of Deception: A Rogue Agent Exposes the Mossad's Secret Agenda*, Harper Paperbacks, New York, (1994). *See also:* I. Black and B. Morris, *Israel's*

celebrated in Israel?[363]

In an article entitled "The Jews", *The Knickerbocker; or New York Monthly Magazine*, Volume 53, Number 1, (January, 1859), pp. 41-51, at 50-51, wrote,

> "Of all Mussulmans the Egyptians doubtless regard the Jews with most aversion. In the year 1844 a young man belonging to a respectable family in Cairo, suddenly disappeared. Several of the resident Consuls, moved by the solicitations of the wretched mother, requested of the Viceroy a searching investigation into the circumstances of the case. It could only be discovered that the young man had gone to the Jews' quarter, from which no one had seen him return. He had been missed a few days before the feast of the Passover, and the terrible accusation was laid upon the Jews of having offered the blood of a human victim as a holocaust, instead of the blood of the paschal lamb.
>
> Had the Israelites not been protected by the Austrian Consul, it is probable that the infuriated and bigoted populace would have razed their quarter of the city level with the ground. Four years previous a similar event had occurred at Damascus. The Père Thomas, a Christian priest, greatly beloved by the people, was treacherously murdered in the house of an opulent Jew named Daout-Arari. The affair created much excitement even in Europe. Two celebrated French advocates were sent to Egypt to plead the cause of the accused before Mohammed Ali, then master of Syria. The intrigues of the Austrian Consul and other secret influences brought to bear, procured. an acquittal of the accused. But during the judicial investigation, several important revelations were obtained. Seven Israelites confessed the crime, and turned Mussulmans in order to claim the clemency of the Cadis. From them it was learned that a Jewish barber had murdered the Père Thomas in the house of Daout-Arari, and that the blood of the priest had been mixed with the unleavened bread. The same year the Jews of Rhodes were charged with a like offence. Similar accusations have been brought against the Israelites living in Germany and Hungary.
>
> The Greeks of Constantinople affirm that heretofore the Jews have been in the habit of purloining children, in order to sacrifice them as paschal lambs. This sacrilege was universally talked of and generally believed a few years ago in Pera and the Fanar, when the traditional enmity of the Jews and Greeks was at its height. During the Greek Revolution the Israelites assisted

---

Secret Wars: A History of Israel's Intelligence Services, Grove Weidenfeld, New York, (1991). ***See also:*** S. Teveth, *Ben-Gurion's Spy: The Story of the Political Scandal That Shaped Modern Israel*, Columbia University Press, New York, (1996). ***See also:*** J. Beinin, *The Dispersion of Egyptian Jewry: Culture, Politics, and the Formation of a Modern Diaspora*, University of California Press, Berkeley, (1998).

363. L. Fry, *Waters Flowing Eastward: The War Against the Kingship of Christ*, TBR Books, Washington, D. C., (2000), pp. 101-102. For a nearly contemporary portrayal of events, *see:* "The Modern Jews", *The North American Review*, Volume 60, Number 127, (April, 1845), pp. 329-368, at 340-342.

the Turks against the Hellenes; and when the venerable Greek Patriarch was hanged by the Moslems, the Jews volunteered to drag his corpse through the streets to the sea."

Sandwiched between the memorandum to the Protestant monarchs of Europe and the leader of the United States on the "Restoration of the Jews" which was published together with attendant correspondence,[364] and a story about the murder of Father Thomas which "occupies in a marked manner the whole journalism of Europe", were the following two Letters to the Editor of *The London Times* published on 26 August 1840 on page 6 (note the expression of tensions which led to WW I and WW II),

"TO THE EDITOR OF THE TIMES.

Sir,—Every right-minded person must feel gratified at the general expression of interest in the Jewish nation which has been elicited by the recent sufferings of their brethren at Damascus. It is to be hoped that the public feeling will not be allowed to evaporate in the mere expression of sympathy, but that some effectual measures may be adopted to prevent a recurrence of these atrocities, not merely in our own times, but in generations yet to come. We must not forget, when giving utterance to our indignation at the late transactions in the east, that but a few centuries have passed since our own country was the scene of similar enormities on a far larger scale. What reader of English history does not recall with shame and sorrow the wholesale tortures, executions, and massacres of the Jews who had sought shelter here, or who can estimate the amount of property seized and confiscated, or the number of hearts wrung by the endless repetition of cruelty and injustice? If in England they have till lately been thus treated, how can they look for more security elsewhere? Instead of wondering that they should become sordid and debased, the only cause for surprise is that any should rise to intelligence and respectability. Subject to the caprice and cruelty of any nation among whom they may dwell, fleeing from the persecutions of one only to meet with like treatment from another, having no city of refuge where they can be in safeguard, no single spot to call their own, they are in a more pitiable condition than the Indian of the forest, or the Arab of the desert.

'The wild bird hath her nest, the fox his cave,
'Mankind their country, Israel but the grave.'

Is this state of things always to continue? They think not. Though many hundreds of years of hope deferred might have been enough to quench the anticipations of the most sanguine, they still hope on, and turn with constant and earnest longing to the land of their forefathers. Their little children are

---

[364]. "Restoration of the Jews: Memorandum", *The London Times*, (26 August 1840), pp. 5-6. News of this "Memorandum" first appeared in *The London Times* on 9 March 1840, on page 3, under the title, "Restoration of the Jews".

taught to expect that they shall one day see Jerusalem. They purchase no landed property, and hold themselves in readiness at a few hours' notice to revisit what they and we tacitly agree to call 'their own land.' It is theirs by a right which no other nation can boast, for God gave it to them, and though dispossessed of it for so many ages, it is still but partially peopled, and held with a loose hand and a disputed title by a hostile power, as if in readiness for their return.

There are political reasons arising from the present aspect of affairs in Russia, Turkey, and Egypt, which would make it to the interest not only of England but of other European nations, either by purchase or by treaty, to procure the restoration of Judea to its rightful claimants. About a year since, I heard it it said by a German Jew, that a proposal had some time before been made by our (then) Government to the late Baron Rothschild, that he should enter into a negotiation for this purpose, and that he declined, assigning as a reason, 'Judea is our own; we will not buy it, we wait till God shall restore it to us.' The desirableness as well as the possibility of such a step seems daily to become more evident, but England has lately proved that she needs no selfish motives to induce her to discharge a debt of national honour and justice, or to perform an act of pure benevolence. The one now suggested would not, judging from appearances, cost 20,000,000*l.* of money, or be unaccomplished after 50 years of exertion, or be so vast and so laborious an undertaking as the extinction of slavery throughout the world. It would be a noble thing for a Christian nation to restore these wanderers to their homes again. It would be a crowning point in the glory of England to bring about such an event. The special blessings promised in the Scriptures to those who befriend the Jews would rest upon her, and her sons and daughters would sit down with purer enjoyment to their domestic comforts when they thought that the persecuted outcasts of so many ages had, through their agency, been replaced in homes as happy and secure as theirs.

Hoping that some master mind may be led to take up this subject in all its bearings, and to form some tangible plan for its accomplishment, and that some Wilberforce may be raised up to plead for it by all the powerful and heart-stirring arguments of which it is capable,

I am, Sir, your obedient servant,

AN ENGLISH CHRISTIAN.

TO THE EDITOR OF THE TIMES.

Sir,—The extraordinary crisis of Oriental politics has stimulated an almost universal interest and investigation, and the fate of the Jews seems to be deeply involved with the settlement of the Syrian dilemma now agitating every Court of Christendom.

You have well and wisely recommended that a system of peaceful umpirage and arbitration should be adopted as the proper *role* of Britain, France, Austria, Prussia, and Russia, and you have exposed the extreme absurdity which these Powers would commit if in their zeal for accommodating the quarrels of the Ottomans they should stir up bloody wars among themselves.

The peace of Europe and the just balance of its powers being therefore assumed as the grand desideratum, as the consummation most devoutly to be wished, I peruse with particular interest a brief article in your journal of this day relative to the restriction of the Jews in Jerusalem, because I imagine that this event has become practicable through an unprecedented concatenation of circumstances, and that moreover it has become especially desirable, as the exact expedient to which it is the interest of all the belligerent parties to consent.

The actual feasability of the return of the Jews is no longer a paradox; the time gives it proof. That theory of the restoration of the Jewish kingdom, which a few years ago was laughed at as the phantasy of insane enthusiasm, is now calculated on as a most practical achievement of diplomacy.

Let us view the question more nearly. It is granted that the Jews were the ancient proprietors of Syria; that Syria was the proper heart and centre of their kingdom. It is granted that they have a strong conviction that Providence will restore them to this Syrian supremacy. It is granted that they have entertained for ages a hearty desire to return thither, and are willing to make great sacrifices of a pecuniary kind to the different parties interested, provided they can be put in peaceful and secure possession.

It is likewise notorious, that since the Jews have been thrust out of Syria, that land has been a mere arena of strife to neighbouring Powers, all conscious that they had no legitimate right there, and all jealous of each other's intrusion.

Such having been the case, why, it may be asked, have not the Jews long ago endeavoured to regain possession of Syria by commercial arrangements? In reply it may be said, that though they have evidently wished to do so, and have made overtures of the kind, hitherto circumstances have mainly opposed their desires. For instance, they could not expect to purchase a secure possession of Syria from Turkey, while that empire, in the pride of insolent despotism, could have suddenly revoked its stipulations, and have seized on Jewish treasuries, none venturing to call it to account. Nor could the Jews have ventured to purchase Syria while the right to that country was vehemently disputed between Turkey and Egypt, without any powerful arbitrators to arrange the right at issue, and lend sanction and binding authority to diplomatic documents.

Now, however, these obstacles and hindrances are in a great measure removed; all the strongest Powers in Europe have come forward as arbitrators and umpires to arrange the settlement of Syria.

Under such potent arbitrators, pledged to the performance of any conditions finally agreed on, I have reason to believe that the Jews would readily enter into such financial arrangements as would secure them the absolute possession of Jerusalem and Syria.

If such an arrangement were formed, one great cause of dissension between France and England would be at once removed; for both the Porte and Egypt are decidedly in want of money, and will gladly sell their respective rights in the Syrian territory. They themselves begin to see the folly of enacting the part of the dog in the manger; they will drop the apple of discord if they can get fair compensation for their trouble.

I know no reason, under such powerful umpires, why the Hebrews should not restore an independent monarchy in Syria, as well as the Egyptians in Egypt, or the Grecians in Greece.

As a practical expedient of politics, I believe it will be easier to secure the peace of Europe and Asia by this effort to restore the Jews, than by any allotment of Syrian territories to the Turks or Egyptians, which will be sure to occasion fresh jealousies and discords.

In offering these remarks, I have viewed the question merely as a lawyer and a politician, and proposed the restoration of the Jews as a sort of *tertium quid*, calculated to win the votes of several of the parties at issue. But, Sir, there is a higher point of view from which many of the readers of *The Times* may wish to regard this topic of investigation. Whichever way the restoration of the Jews may finally be brought about, there is no doubt that it is a subject frequently illustrated by Biblical prophecies.

I will, therefore, if I may do so without the vain and presumptuous curiosity which some of the neologists have manifested, endeavour to detail the opinion of the church on this subject in the words of some of her most respectable writers.

It is generally supposed by Newton, Hales, Faber, and others, that the great prophetical period of 1,260 years is not very far from its termination. If they are right in this supposition, the period of the restoration of the Jews cannot be very remote.

These two contingencies are evidently connected by the prophet Daniel, who distinctly states that at the time of the end of this period there shall be great contests among the Eastern nations in Syria. And at that time (continues Daniel) shall Michael stand up, even the great Prince who standeth up for the children of the Jews, and there shall be a time of trouble such as never was since there was a nation, and at that time the Jews shall be delivered. (Daniel xii.)

Whatever this mysterious passage may imply, all the most learned expositors agree that it refers to the same crisis indicated by the author of the Apocalypse (Chapter xvi., verses 12, 16.) Most of these expositors seem to think that by the phrase 'drying up the great river Euphrates, that the way of the Kings of the East might be prepared,' we are to understand the diminution of the Turkish empire, that the Jews may regain their long lost kingdom of Syria.

I will not detain you by quoting a host of learned authorities in confirmation of this interpretation; but it may be important to hint, that the moral and intellectual position of the Jews in the present day, as well as their commercial connexions, has enabled them to assume a political sphere of activity at once lofty and extensive.

As to religion, they have of late years realized many of the predictions of Mendelssohn and D'Israeli. They have thrown off the absurd bigotry which once rendered them contemptible, and begin to give the New Testament and the writings of Christian divines that attention to which they are every way entitled among truth-searching and philosophic men. Though, perhaps, fewer positive conversions to Christianity have taken place than were expected by the clergy, still the Hebrew intellect has made within a few years past a wonderful approximation to that temper of impartial inquiry in which such books as *Grotius de Veritate* produce an indeliable impression.

I believe that the cause of the restoration of the Jews is one essentially generous and noble, and that all individuals and nations that assist this world-renounced people to recover the empire of their ancestors will be rewarded by Heaven's blessing. [It was and is commonplace for Zionists to appeal to the superstitions of Christians and others with the myth that Jews have supernatural connections which will bless those who help Jews and punish those who do not. The real forces at work are generally control over public opinion through media, planted rumor and gossip; sophisticated intelligence networks; and the might of higher education and investment capital, or lack thereof, which can raise a nation above others, or destroy it. Whoever controls news outlets and financial institutions is the first to learn of events and investments, and to profit from them, or prevent them.—CJB] Everything that is patriotic and philanthropic should urge Great Britain forward as the agent of prophetic revelations so full of auspicious consequence.

I dare not allow my mind to run into the enthusiasm on this subject which I find predominant among religious authors. I will, therefore, conclude with one quotation from *Hale's Analysis of Chronology*:—

'The situation of the new Jerusalem,' says this profound mathematician, 'as the centre of Christ's millennary kingdom in the Holy Land, considered in a geographical point of view, is well described by Mr. King in a note to his *Hymns to the Supreme Being*. How capable Syria is of a more universal intercourse than any other country with all parts of the world is most remarkable, and deserves to be well considered, when we read the numerous prophecies which speak of its future grandeur, when its people shall at length be gathered from all nations among whom they have wandered, and Sion shall be the joy of the whole earth.'
Your very obedient servant,
  Aug. 17. F. B."

The "Memorandum" was advertised in *The London Times* on 9 March 1840, on page 3,

"RESTORATION OF THE JEWS.—A memorandum has been addressed to the Protestant monarchs of Europe on the subject of the restoration of the Jewish people to the land of Palestine. The document in question, dictated by the peculiar conjuncture of affairs in the East, and the other striking 'signs of the times,' reverts to the original covenant which secures that land to the descendants of Abraham, and urges upon the consideration of the powers addressed what may be the possible line of duty on the part of Protestant Christendom to the Jewish people in the present controversy in the East. The memorandum and correspondence which has passed upon the subject have been published."

The "Memorandum to the Protestant Powers of the North of Europe and America" was published in *Memorials concerning God's Ancient People of Israel*. It was later republished together with attendant correspondence in *The London Times* on 26 August 1840 on pages 5-6. It is an attempt to persuade Protestant leaders to bring to fruition Biblical apocalyptic prophecy by forcing it to "come true" through less than divine willful human intervention. This was a tradition for the Christians which dates from the *Gospels*. For example, *Matthew* 21:1-11 states, referring to *Zachariah* 9:9,

"And when they drew nigh unto Jerusalem, and were come to Bethphage, unto the mount of Olives, then sent Jesus two disciples. Saying unto them, Go into the village over against you, and straightway ye shall find an ass tied, and a colt with her: loose *them*, and bring *them* unto me, And if any *man* say ought unto you, ye shall say, The Lord hath need of them; and straightway he will send them. All this was done, that it might be fulfilled which was spoken by the prophet, saying, Tell ye the daughter of Sion, Behold, thy King cometh unto thee, meek, and sitting upon an ass, and a colt the foal of an ass. And the disciples went, and did as Jesus commanded them, And brought the ass, and the colt, and put on them their clothes, and they set *him* thereon. And a very great multitude spread their garments in the way; others cut down the branches from the trees, and strawed *them* in the way. And the multitudes that went before, and that followed, cried, saying, Hosanna to the son of David: Blessed *is* he that cometh in the name of the Lord; Hosanna in the highest. And when he was come into Jerusalem, all the city was moved, saying, Who is this? And the multitude said, This is Jesus the prophet of Nazareth of Galilee."

Rothschild saw himself as the Messiah, but could not convince any large number of Jews of the fact. He could buy Palestine, but could not buy enough Jews to populate it. Rothschild could even buy the support of the governments of Europe, but there was only one means to persuade Jews to move to the desert—by mass murdering and otherwise terrorizing European assimilatory Jews. Both the Old Testament (*Leviticus* 26. *Deuteronomy* 4:24-27; 28:15-68; 30:1-3. II *Chronicles* 7:19-22. *Jeremiah* 29:1-7) and the Babylonian Talmud, *Tractate Kethuboth* (*also:* "Ketubot"), 111a, make it clear that the Jews must not hasten the coming of the Messiah and must wait for the Messiah to establish a Jewish state, before emigrating to Palestine in large numbers. Israel Shahak and Norton Mezvinsky wrote in their book *Jewish Fundamentalism in Israel*,

"The Haredi objection to Zionism is based upon the contradiction between classical Judaism, of which the Haredim are the continuators, and Zionism. Numerous Zionist historians have unfortunately obfuscated the issues here. Some detailed explanation is therefore necessary. In a famous talmudic passage in *Tractate Ketubot*, page 111, which is echoed in other parts of the Talmud, God is said to have imposed three oaths on the Jews. Two of these oaths that clearly contradict Zionist tenets are: 1) Jews should not rebel against non-Jews, and 2) as a group should not massively emigrate to Palestine before the coming of the Messiah. (The third oath, not discussed here, enjoins the Jews not to pray too strongly for the coming of the Messiah, so as not to bring him before his appointed time.) During the course of post-talmudic Jewish history, rabbis extensively discussed the three oaths. Of major concern in this discussion was the question of whether or not specific Jewish emigration to Palestine was part of the forbidden massive emigration. During the past 1,500 years, the great majority of traditional Judaism's most important rabbis interpreted the three oaths and the continued existence of the Jews in exile as religious obligations intended to expiate the Jewish sins that caused God to exile them."[365]

Christians believe that the Jews had broken the Covenant and that a new Covenant had been made between God and the Christians, thereby voiding the Covenant with the Jews (*Matthew* 12:30; 21:43-45. *Romans* 9; 11:7-8. *Galatians* 3:16. *Hebrews* 8:6-10).

The New-Yorker, Volume 9, Number 13, Whole Number 221, (13 June 1840), pp. 196-197; wrote of Rothschild's desires to be King of the Jews, and by the implications of Jewish prophecy, King of the World—and by the implications of Christian prophecy, the anti-Christ:

"RESTORATION OF THE JEWS.—On more than one occasion we have called attention to the signs, of one kind or another, by which the exiles of Israel are beginning to express their impatience for the accomplishment of the prophecies that point to their restoration; and the changes, physical and moral, which are gradually breaking down the barriers to the final fulfilment of the promise. These are curious and worth attention; and more significant in their aggregation, and with reference to the character of the people in question, than those of our readers who have looked at them hastily and separately, may have been prepared to suspect. The Malta letters brings accounts from Syria, in which some curious particulars are given of Sir Moses Montefiore's proceedings, during his late visit to the Holy Land. We remember rumors, which had currency some years ago, of the Jewish capitalist's (Rothschild's) design to employ his wealth in the purchase of Jerusalem, as the seat of a kingdom, and bring back the tribes under his own

---

365. I. Shahak and N. Mezvinsky, *Jewish Fundamentalism in Israel*, Pluto Press, London, (1999), p. 18.

guidance and sovereignty. If the scheme, amid its sublimity, savored sufficiently of the romantic to make the rumor suspicious, the positive acts of Sir Moses, at least, exhibit an anxiety to gather together the wanderers in the neighborhood of their ancient home and future hopes; that they may await events on the ground where they can best be made available to the fulfilment of the promise. During his pilgrimage he sought his way to the hearts of his countrymen, by giving a *talaris* (we believe about fifteen piastres) to every Israelite; and having instituted strict inquiries respecting the various biblical antiquities on his way, and ascertained the amount of duty which the sacred places and villages paid to the Egyptian Government to be about 64,000 purses (a purse being equal to fifteen talaris,) he proposed to the Viceroy of Egypt, that he (Sir Moses) should pay this revenue out of his own pocket, as the price of that prince's permission to him to colonize all those places with the Children of Israel. The offer has been, it is said, accepted, subject to the condition that the colony shall be considered national, and not under European protection. Athenæum."

Though the majority of Jews opposed political Zionism from its inception for the very reason that it was an artificial effort to do God's will in the absence of a Messiah, some modern Jewish and Christian Zionist groups are planning to artificially create the horrors of the Apocalypse, in order to artificially begin the Messianic Era—in their twisted dreams. Jessica Stern writes, referring to Judaism, Christianity and Islam; and citing the Bible at *Zechariah* 14:2-12, *Daniel* 12:1-2, *Revelation* 16:14-16, 20:1-6, and the Koran at Sura 14:48 and Sura 18:8:

"Millenarian Jews believe that at the End of Days, there will be a time of great troubles. Jerusalem will be taken in battle, but God will smite the enemies of the Jews. The wicked will act wickedly and not understand, while the knowledgeable will grow refined and radiant. The righteous among the dead will rise to eternal life, while others will be left to everlasting abhorrence. All three monotheistic traditions have a conception of an apocalypse, but each believes that its own group will prevail in the catastrophic events of the final days.[14] Some millenarians hope to bring on that very catastrophe, which they see as a necessary stage in the process of redemption. Evangelical Christians and Messianic Jews have developed a cooperative relationship, based on their common belief that rebuilding the Temple will facilitate the process of redemption, even though each believes its own group will ultimately triumph."[366]

The "Memorandum to the Protestant Powers of the North of Europe and America" was soon followed by the memorandum of Lord Ashley (Shaftesbury)

---

[366]. J. Stern, *Terror in the Name of God: Why Religious Militants Kill*, Ecco, New York, (2003), p. 95.

to British Foreign Secretary Lord Palmerston of 25 September 1840 and the memorandum to Palmerston of 2 March 1841.

Almost a century before the "Memorandum to the Protestant Powers of the North of Europe and America", another English "Christian", David Hartley, published his *Observations on Man* in 1749.[367] Hartley evinces the desire of a (recently reemerging) sect of philo-Semitic Christian Zionists for the destruction of Catholicism (in anticipation of the French Revolution and the *Kulturkampf*), the "restoration of the Jews to Palestine"; then Jewish world rule followed by the utter destruction of human kind, in anticipation of the First and Second (and Third?) World Wars. He tried to persuade his Christian readers to welcome despair, death and destruction in the hopes that it "may fit us for *the new Heavens, and new Earth.*" (*Isaiah* 65:16-17; 66:22-24). Hartley asked Christians to accept that this life must be miserable, while promising them a better afterlife—a promise he knew he would never be asked to honor.

In the Jewish dominated media of today we find many Jews preaching to the public that the end times are coming and that Christians ought to view their own destruction in a positive light as if it were the divine fulfillment of Jewish and Christian prophecy. Many Christians have been duped by these charlatans, be they psychics, pseudo-Christians preachers, UFO and ghost investigators, etc. These dupes must awaken to fact that the destruction of the world and its nations is occurring as a result of the deliberate intervention of immensely wealthy Jews, and not as the result of God's will. These Jewish leaders view the Hebrew Bible as a plan which they are deliberately fulfilling without their God's help and in violation of Christian principles and prophecy, unless it be Christian prophecy of the "anti-Christ" against whom Christians are duty bound to fight. Christianity, like Communism has always been used by Jews as a trap to destroy Europeans. It promises a Utopia if only the Europeans surrender their power to State authority and surrender their wealth to the Jews. In the meantime, Jews are taught that they need only obey God's laws and that they are duty bound to accumulate wealth, most especially gold and jewels. Under such a system, Christians cannot compete and the Jews have provided them with belief systems meant to destroy them. Whereas Christians are taught to surrender their struggle for individual survival to fatalism under the promise of a perfect afterlife, Jews are taught that immortality rests in the segregation and survival of their "race" and that the individual must struggle for the survival and segregation of the "Jewish race", and must also encourage all other "races" to destroy themselves, because they view the mere existence of other "races" as a threat to the survival of the "Jewish race", both because they sense the ever present danger that assimilation will dissolve them, and because they sense that Esau will someday take revenge on Jacob for its deliberate deceit, theft and genocide of non-Jews.

David Hartley was a Cabalistic Jew who wanted to bring ruin upon the

---

367. D. Hartley, *Observations on Man, His Frame, His Duty, and His Expectations in Two Parts*, Printed by S. Richardson for James Leake and Wm. Frederick, booksellers in Bath and sold by Charles Hitch and Stephen Austen, booksellers in London, London, (1749).

Gentiles by deceiving them with Christian mythology into mass murdering themselves for the benefit of the Jews. He was next in a long line of traitors who had come under the influence of wealthy Cabalistic Jewish mystics, a lineage which can be traced through Sir Isaac Newton to Henry More and beyond.

The genocidal Zionists attempted to justify their inhuman actions and plans as if divine manifestations of the Messianic myth of "hevlei Mashiah", or "the birth pangs of the Messiah".[368] This madness of self-destruction imposed on Christians by Jewish Zionists and their Cabalistic agents—including Henry More, Isaac Newton, Samuel Clarke and David Hartley—has culminated today, after two horrific world wars which they and their progeny planned and brought about—has culminated today in the apocalyptic desires of Dispensationalist Christians, who slavishly promote the evils of Israel and eagerly await a nuclear holocaust they intend to deliberately bring about, which will destroy human life on Earth.[369] These insane dupes of the racist Jewish Zionists have been taught that they will be raptured up into Heaven and that God will create a new heaven and Earth just for them. The racist Jewish Zionists use their media control and wealth to promote these pseudo-Christians in America in order to subvert the American political process and to lead America into World War Three with a dim-witted smile on its face.

David Hartley was influenced by Isaac Newton's student and defender, the quasi-Anglican Arian philosopher (cabalistic Jew) Samuel Clarke. Clarke's Arianism was in fact Judaic—he, Newton, and later Hartley, would not sign the Thirty-Nine Articles of the Church of England, which would have required them to affirm a belief in the Trinity. Clarke compiled a series of Bible quotations concerning the "restoration of the Jews".[370] Hartley apparently copied much from Clarke's *A Demonstration of the Being and Attributes of God And Other Writings*, without any attribution, including Clarke's space-time theory of

---

368. "Messianic Movements", *Encyclopaedia Judaica*, Volume 11 LEK-MIL, Encyclopaedia Judaica, Jerusalem, The Macmillan Company, New York, (1971), cols. 1417-1427, at 1418. G. Scholem, *Kabbalah*, New American Library, New York, (1974), p. 284. Compare to: *Matthew* 24:7-8. *Mark* 13:7-8.
369. G. Halsell, *Prophecy and Politics: Militant Evangelists on the Road to Nuclear War*, Lawrence Hill & Co., Westport, Connecticut, (1986); **and** *Prophecy and Politics: The Secret Alliance Between Israel and the U. S. Christian Right*, Lawrence Hill & Co., Westport, Connecticut, (1986); **and** *Forcing God's Hand: Why Millions Pray for a Quick Rapture—and Destruction of Planet Earth*, Crossroads International Pub., Washington, D.C., (1999), Amana Publications, Beltsville, Maryland, (2003); **Turkish:** M. Acar, H. Özmen, *et al.* translators, *Tanri'yi kiyamete zorlamak: Armagedon, Hristiyan kiyametçiligi ve Israil = Forcing God's Hand : Why Millions Pray for a Quick Rapture: And Destruction of Planet Earth*, Kim, Ankara, (2002).
370. S. Clarke, "The Conversion and Restoration of the Jews", *A Collection of the Promises of Scripture: or, The Christian's Inheritance*, Part 3, Section 10, American Tract Society, New York, and J. Buckland, London, (1750). **See also:** *A Discourse Concerning the Connexion of the Prophecies in the Old Testament, and the Application of Them to Christ. Being an Extract from the Sixth Edition of a Demonstration of the Being and Attributes of God, &c...* ., J. Knapton, London, (1725).

1705,[371] which anticipated the special theory of relativity by two-hundred years, and which had its origins in the Cabalistic space-time theories of Giordano Bruno,[372] Henry More,[373] John Locke,[374] and Isaac Newton—and the *Kabbala Denudata* which inspired all of these pseudo-Christians to destroy Christian society.[375] These men were Cabalists who denied the divinity of Jesus, and who were greatly influenced by prominent and wealthy Jewish mystics, and who also wrote about the "restoration of the Jews" and the conversion of Jews to Christianity which they argued would bring about the millennium, the destruction of the old world and the creation of a new world.[376] Again, it is important to stress, that we have as their legacy two world wars and a coming third.

Some Jews were spreading the message that in order for Christianity to

---

371. S. Clarke, *A Demonstration of the Being and Attributes of God And Other Writings*, Edited by E. Vialati, Cambridge University Press, (1998), pp. 19-20. *Cf.* Thomas Reid, *Essays on the Intellectual Powers of Man*, Essay III, Of Memory, CHAPTER III, OF DURATION, (1785); in *The Works of Thomas Reid, D.D. F.R.S. Edinburgh. Late Professor of Moral Philosophy in the University of Glasgow. With an Account of His Life and Writings*, Edited by D. Stewart, Volume 2, E. Duyckinck, Collins and Hannay, and R. and W. A. Bartow, New York, (1822), pp. 132-134.

372. G. Bruno, *De la causa, principio, et vno*, John Charleswood, London, (1584); English translation, *Cause, Principle, and Unity*, Multiple Editions; German translation, *Von der Ursache, dem Princip und dem Einen*, Multiple Editions; **and** *De l'Infinito Universo e Mondi*, John Charleswood, London, (1584); English translation, *Giordano Bruno, His Life and Thought. With Annotated Translation of his Work, On the Infinite Universe and Worlds*, Schuman, New York, (1950); German translation, *Zwiegespräche vom Unendlichen: All und den Welten*, E. Diedrich, Jena, (1892). Collected Works in German, *Gesammelte Werke*, E. Diedrich, Leipzig, (1904-1909).

373. H. More, *A COLLECTION Of Several Philosophical Writings OF Dr. HENRY MORE, Fellow of* Christ's-College *in* Cambridge, Joseph Downing, London, (1712); which contains: *AN ANTIDOTE AGAINST ATHEISM: OR, An Appeal to the Natural Faculties of the Mind of Man, Whether there be not a GOD*, The Fourth Edition corrected and enlarged: WITH AN APPENDIX Thereunto annexed, "An Appendix to the foregoing Antidote," Chapter 7, pp. 199-201.

374. J. Locke, *Essay Concerning Human Understanding*, Chapter 15, Section 12.

375. I. Newton, *Principia*, Book I, Definition VIII, Scholium; **and** Book III, General Scholium.

376. J. E. Force and R. H. Popkin, Editors, *The Millenarian Turn: Millenarian Contexts of Science, Politics, and Everyday Anglo-American Life in the Seventeenth and Eighteenth Centuries*, Kluwer Academic Publishers, Dordrecht, Boston, (2001). H. More, J. Flesher, *et al.*, *Conjectura Cabbalistica., Or, a Conjectural Essay of Interpreting the Minde of Moses, According to a Threefold Cabbala: viz., Literal, Philosophical, Mystical, Or, Divinely Moral*, Printed by James Flesher, and are to be sold by William Morden bookseller in Cambridge, London, (1653). F. M. v. Helmont, H. More, J. Gironnet, *et al.*, *Opuscula Philosophica: Quibus Continentur Principia Philosophiæ Antiquissimæ & Recentissimæ. Ac Philosophia Vulgaris Refutata. Quibus Subjuncta Sunt Cc. Problemata De Revolutione Animarum Humanarum*, Prostant Amstelodami, (1690). I. Newton, *Observations upon the Prophecies of Daniel, and the Apocalypse of St. John*, Printed by J. Darby and T. Browne and sold by J. Roberts etc., London, (1733).

succeed, Jews would have to convert Christianity. This gave them privilege and the power to amend Christianity so as to make it more palatable to Jews. It also prevented a backlash against Jews who would emigrate to Palestine and who would be seen by Christians as the minions of the anti-Christ were they not to feign Christian conversion.

Isaac Newton, like Clarke after him, disbelieved in the Trinity, wanted to see the Gentile nations laid to waste, and hoped that the Jews would rule the world from Jerusalem. Newton wrote, among other things,

> "For they understand not that ye final return of ye Jews captivity & their conquering the nations <of ye four Monarchies> & setting up a ~~peaceable~~ righteous & flourishing Kingdom at ye day of judgment is this mystery. Did they understand this they would end it in all ye old Prophets who write of ye last times as in ye last chapters of Isaiah where the Prophet conjoyns the new heaven & new earth wth ye ruin of ye wicked nations, the end of ~~all troubles~~ weeping & of all troubles, the return of ye Jews captivity & their setting up a flourishing & everlasting Kingdom."[377]

and,

> "'Tis in ye last days yt this is to be fulfilled & then ye captivity shall return & become a strong nation & reign over strong nations afar off, & ye Lord shal reign in mount Zion from thenceforth for ever, & many nations shal receive ye law of righteousness from Jerusalem, & they shall beat their swords into plow-shares & their spears into pruning hooks & nation shall not lift up a sword against nation, neither shal they learn war any more; all wch never yet came to pass."[378]

Stephen Snobelen wrote of Newton,

> "Newton had a profound interest in things Jewish. His library alone supplies ample evidence of this.[15] Newton owned five of the works of Maimonides,[16] and makes numerous references to them in his manuscripts. He also

---

[377]. S. Snobelen, "'The Mystery of the Restitution of All Things': Isaac Newton on the Return of the Jews", in J. E. Force and R. H. Popkin, Editors, *The Millenarian Turn: Millenarian Contexts of Science, Politics, and Everyday Anglo-American Life in the Seventeenth and Eighteenth Centuries*, Chapter 7, Kluwer Academic Publishers, Dordrecht, Boston, (2001), pp. 95-118, at 95. Snobelen cites: Jewish National and University Library (Jerusalem) Yahuda MS 6, f. 12r.

[378]. S. Snobelen, "'The Mystery of the Restitution of All Things': Isaac Newton on the Return of the Jews", in J. E. Force and R. H. Popkin, Editors, *The Millenarian Turn: Millenarian Contexts of Science, Politics, and Everyday Anglo-American Life in the Seventeenth and Eighteenth Centuries*, Chapter 7, Kluwer Academic Publishers, Dordrecht, Boston, (2001), pp. 95-118, at 101. Snobelen cites: Jewish National and University Library (Jerusalem) Yahuda MS 9.2, f. 143r.

possessed Christian Knorr von Rosenroth's *Kabbala denudata* (1677-84), which shows extensive signs of dog-earing,[17] along with an edition of the first-century Jewish philosopher Philo.[18] His writings reveal that he used the Talmud, the learning of which he accessed through Maimonides and other sources in his library.[19] Although he never acquired a competency in the language, Newton picked up a smattering of Hebrew and armed himself with an array of Hebrew lexicons and grammars.[20] He also owned and used a Hebrew Bible.[21] Much attention is given in Newton's writings to studies of the Jewish Temple and its rituals.[22] His fascination with these things was motivated in large part by the importance of understanding both the complexities of Jewish ritual and the design of the Temple for the interpretation of prophecy.[23] Newton owned a number of works on these subjects as well.[24] A further testimony to his research on the Temple exists in the physical evidence of his octavo Bible, the pages of which are heavily soiled in the section detailing the Temple of Ezekiel's prophecy.[25] This study also bore its fruit. Several scholars have pointed to Newton's appropriation of elements of Jewish theology. John Maynard Keynes famously characterized Newton as a 'Judaic monotheist of the school of Maimonides.'[26],[379]

The first known records of Christianity appeared after the destruction of the Temple and the dispersion of the Jews from Jerusalem. Religious Jews were fanatically concerned that the nation of the Jews be preserved. Christianity itself was probably nothing but a means to convert the Romans to Judaism so that the Romans would then restore the Jews to Palestine and force the Jews back to Judaism, which the Jews had largely abandoned. After, or as, the Jews were being restored to Judaism, Jews would then restore the Christians to Paganism. This appears to be the plan of treacherous Paul, who was born a Jew named Saul, and who set down this plan in *Romans* 9-11. The fulfilment of this plan occurred in the Twentieth Century, when Communism and Nazism largely destroyed the religion of European Christians and forced Jews to move to Palestine out of fear. The anti-religious doctrines of Communism are well known. The anti-religious doctrines of Nazism are discussed in Uriel Tal's introduction to J. M. Snoek's *The Grey Book*, Humanities Press, New York, (1970), pp. I-XXVI. Tal writes, *inter alia*,

> "[T]he Nazis appropriated the messianic structure of religion which they exploited to their own ideological and political ends[... ,] but which is designed to de-Christianize the German people[.] Anti-Semitism is not only called to combat religion and Christianity; its chief aim is to save the

---

[379]. S. Snobelen, "'The Mystery of the Restitution of All Things': Isaac Newton on the Return of the Jews", in J. E. Force and R. H. Popkin, Editors, *The Millenarian Turn: Millenarian Contexts of Science, Politics, and Everyday Anglo-American Life in the Seventeenth and Eighteenth Centuries*, Chapter 7, Kluwer Academic Publishers, Dordrecht, Boston, (2001), pp. 95-118, at 97.

German nation and the whole world from Jewish domination and from the moral depredation of the Jewish race. [*i. e.* to segregate and persecute Jews as the Zionists desired and to force them to Palestine, while destroying the Judaism of Gentiles—while destroying Christianity.] [***] The general tendency of this movement was directed against Christianity as an ecclesiastical institution, sometimes chiefly against the Catholic Church which was suspected of 'ultramontanist' sympathies for a foreign ecclesiastical power."

After making it appear that he was a neutral arbiter in Chapters 9 and 10, Paul, born Saul, warns Gentiles and apostate Jews of their ultimate fate when he writes in Chapter 11 of *Romans*,

"1 I say then, Hath God cast away his people? God forbid. For I also am an Israelite, of the seed of Abraham, *of* the tribe of Benjamin. 2 God hath not cast away his people which he foreknew. Wot ye not what the scripture saith of Elias? how he maketh intercession to God against Israel, saying, 3 Lord, they have killed thy prophets, and digged down thine altars; and I am left alone, and they seek my life. 4 But what saith the answer of God unto him? I have reserved to myself seven thousand men, who have not bowed the knee to *the image of* Baal. 5 Even so then at this present time also there is a remnant according to the election of grace. 6 And if by grace, then *is it* no more of works: otherwise grace is no more grace. But if *it be* of works, then is it no more grace: otherwise work is no more work. 7 What then? Israel hath not obtained that which he seeketh for; but the election hath obtained it, and the rest were blinded 8 (According as it is written, God hath given them the spirit of slumber, eyes that they should not see, and ears that they should not hear;) unto this day. 9 And David saith, Let their table be made a snare, and a trap, and a stumblingblock, and a recompence unto them: 10 Let their eyes be darkened, that they may not see, and bow down their back alway. 11 I say then, Have they stumbled that they should fall? God forbid: but rather through their fall salvation *is come* unto the Gentiles, for to provoke them to jealousy. 12 Now if the fall of them *be* the riches of the world, and the diminishing of them the riches of the Gentiles; how much more their fulness? 13 For I speak to you Gentiles, inasmuch as I am the apostle of the Gentiles, I magnify mine office: 14 If by any means I may provoke to emulation *them which are* my flesh, and might save some of them. 15 For if the casting away of them *be* the reconciling of the world, what *shall* the receiving *of them* be, but life from the dead? 16 For if the firstfruit *be* holy, the lump *is* also *holy*: and if the root be holy, so are the branches. 17 And if some of the branches be broken off, and thou, being a wild olive tree, wert graffed in among them, and with them partakest of the root and fatness of the olive tree; 18 Boast not against the branches. But if thou boast, thou bearest not the root, but the root thee. 19 Thou wilt say then, The branches were broken off, that I might be graffed in. 20 Well; because of unbelief they were broken off, and thou standest by faith. Be not highminded, but fear: 21 For if God spared not the natural branches, *take heed* lest he also spare not thee. 22 Behold therefore the goodness and severity of God: on them which fell, severity; but toward thee, goodness, if thou continue in *his* goodness: otherwise thou also shalt be cut off. 23 And they also, if they abide not still in unbelief, shall be graffed in: for God is able to graff them in again. 24 For if thou wert cut out of the olive tree which is wild by nature, and wert graffed contrary to nature into a good olive tree: how much more shall these, which be the natural *branches*, be graffed into their own olive tree? 25 For I would not, brethren, that ye should be ignorant of this mystery, lest ye should be wise in your own conceits; that blindness in part is happened to Israel, until the fulness of the Gentiles be come in. 26 And so all Israel shall be saved: as it is written, There shall come out of Sion the Deliverer, and shall turn away ungodliness from Jacob: 27 For this is my covenant unto them, when I shall take away their sins. 28 As concerning the gospel, *they are* enemies for your sakes: but as touching the election, *they are* beloved for the fathers' sakes. 29 For the gifts and calling of God *are* without repentance. 30 For as ye in times past have not believed God, yet have now obtained mercy through their unbelief: 31 Even so have these also now not believed, that through your mercy they also may obtain mercy. 32 For God hath concluded them all in unbelief, that he might have mercy upon all. 33 O the depth of the riches both of the wisdom and knowledge of God! how unsearchable *are* his judgments, and his ways past finding out! 34 For who hath known the mind of the Lord? or who hath been his counsellor? 35 Or who hath first given to him, and it shall be recompensed unto him again? 36 For of him, and through him, and to him, *are* all things: to whom be glory for ever. Amen."

Paul, born Saul, also warned his fellow Jews in I *Thessalonians* 2:15-16, where Paul stated,

"For ye, brethren, became followers of the churches of God which in Judaea are in Christ Jesus: for ye also have suffered like things of your own countrymen, even as they *have* of the Jews. Who both killed the Lord Jesus, and their own prophets, and have persecuted us; and they please not God, and are contrary to all men: Forbidding us to speak to the Gentiles that they might be saved, to fill up their sins alway: for the wrath is come upon them to the uttermost."

We see that "Jesus" is an allegory for Judaism, which the Romans had attacked, and which many Jews had abandoned. The name "Jesus" in the original means "Jew". The "life" of Jesus was concurrent with the life of Philo the Jew, who Hellenized Judaism—an act which made Judaism palatable to Romans; and who obliged the conversion of the Temple to the worship of the Roman Emperors after the Jews had exhibited religious intolerance against Rome. The parallels between the story of "Jesus" and the history of Judaism are many. The sale of Judaism by "Judas", which name is the same word as "Jesus" in the original and which means "Jew" as in *Philo Judæus*—the doubting of Thomas and the denial of Peter as Jews became more secular or pagan—the promise of everlasting life to a religion that was dying out[380]—the destruction of the Temple—twelve Apostles of "the Jew" judging the Twelve Tribes of Israel (*Matthew* 19:28. *Luke* 22:28-30)—forgiveness of the whore which had slept with Judah (*Genesis* 38), etc. What better act of vengeance could there have been for Caligula's desecration of the Temple and Titus' destruction of it, than to convert Romans to a Romanized and Hellenized branch of Judaism, which had the Romans worshiping "the Jew" and joyfully looking forward to their ultimate destruction?

In 1925, Bialik gave a speech at the inauguration of the "Hebrew University" and arrogantly spoke of the salvation of the pagan and the rôle Jesus played in conditioning Gentiles to accept the Jewish world view, that ultimately led to the Balfour Declaration.[381] The closing book of the Old Testament states (*Malachi* 1:11-14), in the context of the continual ruin of Edom—the continual ruin of the world of the Gentiles:

"11 For from the rising of the sun even unto the going down of the same my name *shall* be great among the Gentiles; and in every place incense *shall* be offered unto my name, and a pure offering: for my name *shall* be great among the heathen, saith the LORD of hosts. 12 But ye have profaned it, in that ye say, The table of the LORD *is* polluted; and the fruit thereof, *even* his meat, *is* contemptible. 13 Ye said also, Behold, what a weariness *is it!* and ye have snuffed at it, saith the LORD of hosts; and ye brought *that which was* torn, and the lame, and the sick; thus ye brought an offering: should I accept this of your hand? saith the LORD. 14 But cursed *be* the deceiver, which hath in his flock a male, and voweth, and sacrificeth unto the LORD a corrupt thing: for I *am* a great King, saith the LORD of hosts, and my name *is* dreadful among the heathen."

The stumblingblocks we face even today are many. Christianity, Islam and Judaism pose a great danger to our modern existence, with their suicidal hopes and apocalyptic dreams which are used to justify inhumanity and war and the selfishness and self-destructiveness of the "elect" (*Isaiah* 65; 66. *Enoch*). In the Twentieth Century, Marxism, Einsteinism and Freudism became dark dogmas rooted in ancient mythologies, which monopolized discourse, while far more enlightened views were suppressed. The Christian religion of obedience to the Jewish God of war and destruction has been one of the worst stumblingblocks Europe ("Rome") has faced—as those who fabricated the mythology probably intended (note that Jesus was effectively the Messiah of the Gentiles, not the

---

[380]. Refer to the "Third Letter" in: M. Hess, *Rom und Jerusalem: die letzte Nationalitätsfrage*, Eduard Wengler, Leipzig, (1862); English: *Rome and Jerusalem: A Study in Jewish Nationalism*, Bloch, New York, (1918).

[381]. H. N. Bialik, "Bialik on the Hebrew University", in A. Hertzberg, *The Zionist Idea*, Harper Torchbooks, New York, (1959), pp. 281-288, at 287.

Jews).[382] *Psalm* 69:22, may have inspired some Jews to trap the Romans with Christianity:

> "Let their table become a snare before them: *and that which should have been* for *their* welfare, *let it become* a trap."

The Jews, whose religion taught them to mercilessly destroy other peoples, had long seen religious conversion as a means to trap a people. *Deuteronomy* 7:2, 16-18 states:

> "7:2 And when the LORD thy God shall deliver them before thee; thou shalt smite them, *and* utterly destroy them; thou shalt make no covenant with them, nor shew mercy unto them: [***] 16 And thou shalt consume all the people which the LORD thy God shall deliver thee; thine eye shall have no pity upon them: neither shalt thou serve their gods; for that *will be* a snare unto thee. 17 If thou shalt say in thine heart, These nations *are* more than I; how can I dispossess them? 18 Thou shalt not be afraid of them: *but* shalt well remember what the LORD thy God did unto Pharaoh, and unto all Egypt;"

Where Christianity has been forcibly replaced by Communism, still worse mythologies have been imposed. Benjamin Disraeli, who was to become Britain's Prime Minister, wrote in 1852,

> "Nor is it indeed historically true that the small section of the Jewish race which dwelt in Palestine rejected Christ. The reverse is the truth. Had it not been for the Jews of Palestine the good tidings of our Lord would have been unknown for ever to the northern and western races. The first preachers of the gospel were Jews, and none else; the historians of the gospel were Jews, and none else. No one has ever been permitted to write under the inspiration of the Holy Spirit except a Jew. For nearly a century no one believed in the good tidings except Jews. They nursed the sacred flame of which they were the consecrated and hereditary depositories. And when the time was right to diffuse the truth among the ethnicks, it was not a senator of Rome or a philosopher of Athens who was personally appointed by our Lord for that office, but a Jew of Tarsus, who founded the seven churches of Asia. And that greater church, great even amid its terrible corruptions, that has avenged the victory of Titus by subjugating the capital of the Cæsars and has changed every one of the Olympian temples into altars of the God of Sinai and of Calvary, was founded by another Jew, a Jew of Galilee.
> [***]
> They may be traced in the last outbreak of the destructive principle in

---

382. *Deuteronomy* 18:15-19. *Psalm* 2:1-12, 69:22. *Isaiah* 8:14-15. *Luke* 2:34-35. *Romans* 9:33. I *Corinthians* 1:18, 23. 2 *Corinthians* 2:15-16. I *Peter* 2:8.

Europe. An insurrection takes place against tradition and aristocracy, against religion and property. Destruction of the Semitic principle, extirpation of the Jewish religion, whether in the Mosaic or in the Christian form, the natural equality of man and the abrogation of property, are proclaimed by the secret societies who form provisional governments, and men of Jewish race are found at the head of every one of them. The people of God co-operate with atheists; the most skilful accumulators of property ally themselves with communists; the peculiar and chosen race touch the hand of all the scum and low castes of Europe! And all this because they wish to destroy that ungrateful Christendom which owes to them even its name, and whose tyranny they can no longer endure.

When the secret societies, in February 1848, surprised Europe, they were themselves surprised by the unexpected opportunity, and so little capable were they of seizing the occasion, that had it not been for the Jews, who of late years unfortunately have been connecting themselves with these unhallowed associations, imbecile as were the governments the uncalled for outbreak would not have ravaged Europe. But the fiery energy and the teeming resources of the children of Israel maintained for a long time the unnecessary and useless struggle. If the reader throws his eye over the provisional governments of Germany, and Italy, and even of France, formed at that period, he will recognise everywhere the Jewish element. Even the insurrection, and defence, and administration of Venice, which, from the resource of statesmanlike moderation displayed, commanded almost the respect and sympathy of Europe, were accomplished by a Jew—Manini, who by the bye is a Jew who professes the whole of the Jewish religion, and believes in Calvary as well as Sinai, 'a converted Jew', as the Lombards styled him, quite forgetting, in the confusion of their ideas, that it is the Lombards who are the converts—not Manini.

[***]

Is it therefore wonderful, that a great portion of the Jewish race should not believe in the most important portion of the Jewish religion? As however the converted races become more humane in their behaviour to the Jews, and the latter have opportunity fully to comprehend and deeply to ponder over true Christianity, it is difficult to suppose that the result will not be very different. Whether presented by a Roman or Anglo-Catholic, or Geneveve, divine, by pope, bishop, or presbyter, there is nothing one would suppose very repugnant to the feelings of a Jew when he learns that the redemption of the human race has been effected by the mediatorial agency of a child of Israel; if the ineffable mystery of the Incarnation be developed to him, he will remember that the blood of Jacob is a chosen and peculiar blood, and if so transcendent a consummation is to occur he will scarcely deny that only one race could be deemed worthy of accomplishing it. There may be points of doctrine on which the northern and western races may perhaps never agree. The Jew, like them, may follow that path in those respects which reason and feeling alike dictate; but nevertheless it can hardly be maintained that there is anything revolting to a Jew to learn that a Jewess is the queen of heaven, or that the flower of the Jewish race are even now sitting on the right hand of the Lord God of Sabaoth.

Perhaps too in this enlightened age as his mind expands and he takes a comprehensive view of this period of progress, the pupil of Moses may ask himself, whether all the princes of the house of David have done so much for the Jews as that prince who was crucified on Calvary? Had it not been for Him, the Jews would have been comparatively unknown, or known only as a high oriental caste which had lost its country. Has not He made their history the most famous in the world? Has not He hung up their laws in every temple? Has not He vindicated all their wrongs? Has not He avenged the victory of Titus and conquered the Cæsars? What successes did they anticipate from their Messiah? The wildest dreams of their rabbis have been far exceeded. Has not Jesus conquered Europe and changed its name into Christendom? All countries that refuse the cross wither while the whole of the new world is devoted to the Semitic principle and its most glorious offspring the Jewish faith, and the time will come when the vast communities and countless myriads of America and Australia, looking upon Europe as Europe now looks upon Greece and wondering how so small a space could have achieved such great deeds, will still find music in the songs of Sion and solace in the parables of Galilee.

These may be dreams, but there is one fact which none can contest. Christians may continue to persecute Jews and Jews still persist in disbelieving Christians, but who can deny that Jesus of Nazareth, the Incarnate Son of the Most High God, is the eternal glory of the Jewish race?"[383]

The ancient Judeans prevailed in one sense against the Romans, whom they identified as their mortal enemy "Esau", they themselves being "Jacob". Jewish proselytes made Rome the new capital of the Jewish religion, where Roman gods were spat upon, where a Jewish son was worshiped as God, and where a Jewish woman, who the Jews claimed was a prostitute, was worshiped as the mother of God.

The *Encyclopaedia Judaica* writes in its article "Messianic Movements":

"One trend of Jewish messianism which left the national fold was destined 'to conquer the conquerors'—by the gradual Christianization of the masses throughout the Roman Empire. Through Christianity, Jewish messianism became an institution and an article of faith of many nations. Within the Jewish fold, the memory of glorious resistance, of the fight for freedom, of martyred messiahs, prophets, and miracle workers remained to nourish future messianic movements."[384]

---

[383]. B. Disraeli, *Lord George Bentinck: A Political Biography*, Chapter 24, Third Revised Edition, Colburn, (1852), pp. 485, 497-498, 505-507.
[384]. "Messianic Movements", *Encyclopaedia Judaica*, Volume 11 LEK-MIL, Encyclopaedia Judaica, Jerusalem, The Macmillan Company, New York, (1971), cols.

The story of Jesus appeared at a time when many Jews believed that God was punishing the Jews for a long list of transgressions including Solomon's marriage to the Pharaoh's daughter and subsequent idolatry (*Sabbath* 56*b*. I *Kings* 11. II *Chronicles* 7:19-23), as well as the transgressions of Aaron's worship of the Golden Calf, and the increase in "intermarriage" with the "daughter of a strange god" and apostasy (*Malachi* 2:10-12). The ten northern tribes were allegedly sent into captivity for impiety (II *Kings* 17), and the southern tribes, who remained unrepentant, soon followed into their own captivity (II *Kings* 18:13; 24:3-16; 25), Solomon's Temple was destroyed, thus beginning the age of Gentile domination and the yoke on Israel. II *Chronicles* 36:18-21, attributes the destruction of the First Temple, at least in part, to the failure of the Israelites and Judeans to maintain the Shemmitah (*Exodus* 23:10-11. *Leviticus* 25. *Deuteronomy* 15; 23:20; 31:10-13),

> "18 And all the vessels of the house of God, great and small, and the treasures of the house of the LORD, and the treasures of the king, and of his princes; all *these* he brought *to* Babylon. 19 And they burnt the house of God, and brake down the wall of Jerusalem, and burnt all the palaces thereof with fire, and destroyed all the goodly vessels thereof. 20 And them that had escaped from the sword carried he away to Babylon; where they were servants to him and his sons until the reign of the kingdom of Persia: 21 To fulfil the word of the LORD by the mouth of Jeremiah, until the land had enjoyed her sabbaths: *for* as long as *she* lay desolate she kept sabbath, to fulfil threescore and ten years."

Solomon was a magician and is said to have built the Temple with the assistance of *demons* and angels. Due to his evil, Solomon lost his Kingdom and ruled only his staff at the end of his life (*Sanhedrin* 20*b*). Some Jews believed that God would not permit the existence of the Temple, or send the Messiah, until the Jews had atoned for Solomon's sins and for the sins of Israel—some even viewed the Holocaust as atonement for the sins of Israel and justify their conclusion by pointing to the existence of Israel—others believe that Zionists instigated the Holocaust as an artificial atonement for the sin of worshiping the Golden Calf, which the Talmud asserted caused the Jews eternal suffering (*Sanhedrin* 102*a*). The very gift of the Covenant is tainted by Jacob's sins against Esau.[385] Moses iterated many curses which would befall the Jews if they were disobedient to God (*Leviticus* 26. *Deuteronomy* 4:24-27; 28:15-68; 30:1-3. II *Chronicles* 7:19-22. *Jeremiah* 29:1-7). Many Jews view the Diaspora, and their supposed eternal suffering, as God's retribution against them for the Jews' disobedience to God.

Mordecai Manuel Noah made it crystal clear in 1837 that powerful Jews had

---

1417-1427, at 1421.
[385] "Jacob and Esau", *The Jewish Chronicle*, (24 November 1911), p. 22.

decided to take the atonement of the Jews, and the "restoration of the Jews to Palestine", into their hands. Noah stated in 1837,

> "Firmly as I believe the American Indian to have been descended from the tribes of Israel, and that our continent is full of the most extraordinary vestiges of antiquity, there is one point, a religious as well as a historical point, in which you may possibly continue to doubt, amidst almost convincing evidences.
> If these are the remnants of the nine and a half tribes which were carried into Assyria, and if we are to believe in all the promises of the restoration, and the fulfillment of the prophecies, respecting the final advent of the Jewish nation, what is to become of these our red brethren, whom we are driving before us so rapidly, that a century more will find them lingering on the borders of the Pacific ocean?
> Possibly, the restoration may be near enough to include even a portion of these interesting people. Our learned Rabbis have always deemed it sinful to compute the period of the restoration; they believe that when the sins of the nation were atoned for, the miracle of their redemption would be manifested. My faith does not rest wholly in miracles—Providence disposes of events, human agency must carry them out. That benign and supreme power which the children of Israel had never forsaken, has protected the chosen people amidst the most appalling dangers, has saved them from the uplifted sword of the Egyptians, the Assyrians, the Medes, the Persians, the Greeks and the Romans, and while the most powerful nations of antiquity have crumbled to pieces, we have been preserved, united and unbroken, the same now as we were in the days of the patriarchs—brought from darkness to light, from the early and rude periods of learning to the bright reality of civilization, of arts, of education and of science.
> The Jewish people must now do something for themselves, they must move onward to the accomplishment of that great event long foretold—long promised—long expected; and when they DO move, that mighty power which has for thousands of years rebuked the proscription and intolerance shown to the Jews, by a benign protection of the *whole* nation, will still cover them with his invincible standard.
> My belief is, that Syria will revert to the Jewish nation by *purchase*, and that the facility exhibited in the accumulation of wealth, has been a providential and peculiar gift to enable them, at a proper time, to re-occupy their ancient possessions by the purse-string instead of the sword.
> We live in a remarkable age, and political events are producing extraordinary changes among the nations of the earth.
> Russia with its gigantic power continues to press hard on Turkey. The Pacha of Egypt, taking advantage of the improvements and inventions of men of genius, is extending his territory and influence to the straits of Babelmandel on the Red sea, and to the borders of the Russian empire; and the combined force of Russia, Turkey, Persia and Egypt, seriously threaten the safety of British possessions in the East Indies. An intermediate and

balancing power is required to check this thirst of conquest and territorial possession, and to keep in check the advances of Russia in Turkey and Persia, and the ambition and love of conquest of Egypt. This can be done by restoring Syria to its rightful owners, not by revolution or blood, but as I have said, by the purchase of that territory from the Pacha of Egypt, for a sum of money too tempting in its amount for him to refuse, in the present reduced state of his coffers. Twelve or thirteen millions of dollars have been spoken of in reference to the cession of that interesting territory, a sum of no consideration to the Jews, for the good will and peaceable possession of a land, which to them is above all price. Under the cooperation and protection of England and France, this re-occupation of Syria within its old territorial limits, is at once reasonable and practicable.

By opening the ports of Damascus, Tripoli, Joppa, Acre, &c., the whole of the commerce of Turkey, Egypt, and the Mediterranean will be in the hands of those, who even now in part, control the commerce of Europe. From the Danube, the Dneister, the Ukraine, Wallachia and Moldavia, the best of agriculturalists would revive the former fertility of Palestine. Manufacturers from Germany and Holland; an army of experience and bravery from France and Italy; ingenuity, intelligence, activity, energy and enterprise from all parts of the world, would, under a just, a tolerant and a liberal government, present a formidable barrier to the encroachments of surrounding powers, and be a bulwark to the interests of England and France, as well as the rising liberties of Greece.

Once again unfurl the standard of Judah on Mount Zion, the four corners of the earth will give up the chosen people as the sea will give up its dead, at the sound of the last trumpet. Let the cry be Jerusalem, as it was in the days of the Saracen and the lion-hearted Richard of England, and the rags and wretchedness which have for eighteen centuries enveloped the persons of the Jews, crushed as they were by persecution and injustice, will fall to the earth; and they will stand forth, the richest, the most powerful, the most intelligent nation on the face of the globe, with incalculable wealth, and holding in pledge the crowns and sceptres of kings. Placed in possession of their ancient heritage by and with the consent and co-operation of their Christian brethren, establishing a government of peace and good will on earth, it may then be said, behold the fulfillment of prediction and prophecy: behold the chosen and favoured people of Almighty God, who, in defence of his unity and omnipotence, have been the outcast and proscribed of all nations, and who for thousands of years have patiently endured the severest of human sufferings, in the hope of that great advent of which they never have despaired:—and then when taking their rank once more among the nations of the earth, with the good wishes and affectionate regards of the great family of mankind, they may by their tolerance, their good faith, their charity and enlarged liberal views, merit what has been said in their behalf by inspired writers, 'Blessed are they who bless Israel.'"[386]

The Zionists put Hitler into power in order to bring about an unprecedented human sacrifice, which would finally atone for the Jews' sins against God, through their own treachery to the Jewish People. Dualist, or Satanist, Jews see Jacob's treachery against his brother as his greatest strength. They argue that evil deeds are rewarded many times in the Old Testament. The Satanic Cabalistic cults believe that evil triumphs over good. Jewish Dualist cults seek the combined power of both good and evil, but tend to fear the Devil more than God, and so are eager to do the Devil's bidding. These genocidal Jews found divine authority for their actions throughout the Hebrew Bible, which calls for the mass murder of assimilatory Jewry.

Early Christians created their myth of Judas Iscariot's betrayal of Jesus from the Old Testament story of Judah's sale of Joseph, which betrayal saved Josehp's life. Many suspect that the New Testament stories of Jesus and Judas evolved from the myth of Joseph and his brother Yehuda (Judah, or "Judas"), who betrayed him by selling him for 20 pieces of silver in order to save his life (*Genesis* 37:26-28).[387] Yehuda (Judas) of Galilee is thought by some to have been the source for the legend of the crucifixion, the legend of "Jesus". Hence, the legend of "Yehuda", "Judah", "Judas Iscariot", who fulfills prophecy and saves through betrayal; and "Yehuda", "Judah", "Judas of Galilee", "Jesus of Nazareth" (in the every instance in modern English texts), who is crucified as a human sacrifice that offers eternal life and ultimate atonement putting an end to animal sacrifices. Judas (the Jew) as Jesus, and Judas (the Jew) as Judas Iscariot, each individually and together as a dialectic dualism, symbolized the salvation of the Jewish nation, or "Judah", from the Romans, through deliberate betrayal which artificially fulfilled the prophecies, and which betrayal offers the promise of eternal life for the Jewish nation (*John* 3:16). Why was it important that Jesus be a Galilean? Isaiah prophesied that the Messiah would come from Galilee (*Isaiah* 9). The authors of the Gospels acknowledged the importance of this prophecy (*Matthew* 4:12-15).

The New Testament book of Acts speaks of both "Judas the Galilean" (*Acts* 5:37) and "Judas Iscariot" (*Acts* 1, with reference to *Psalm* 69:25 and 109:8). It appears that "Jesus" is a corruption of "Judas"="Judah"="Yehuda"="Jew" through the confusion of terms "Bar-Jesus" ("son of salvation") and "Elymas" (another name for Bar-Jesus, which means "magician", or "sorcerer", *Acts* 13:6-8; *cf.* E. F. Harrison, Editor, The Wycliffe Bible Commentary, Moody Press, Chicago, (1962), p. 1147). Jesus was viewed by many as a magician, and it is your author's contention that "Judas the Galilean" morphed into "Jesus the Nazarene" and that the English "Jesus of Nazareth", therefore ultimately refers

---

386. M. M. Noah, *Discourse on the Evidences of the American Indians Being the Descendants of the Lost Tribes of Israel: Delivered Before the Mercantile Library Association*, Clinton Hall: J. Van Norden, New York, (1837), pp. 37-40.

387. J. S. Spong, *Liberating the Gospels: Reading the Bible with Jewish Eyes: Freeing Jesus from 2,000 Years of Misunderstanding*, Harper, San Francisco, (1996).

to a historical character named "The Jew of Galilee"—to the Jewish nation. Judas Iscariot and Judas of Galilee are the good and evil of Jewish Dualism, and the ancient book *The Gospel of Judas* celebrates Judas Iscariot as a necessary figure in the sacrifice and salvation as was Jesus.

Christians called for Jews to atone for the death of Jesus Christ, and some will not be satisfied unless Israel evaporates beneath a storm of mushroom clouds and rains human ash upon the desert. Jews, especially assimilated Jews, have not only Christian mythology to fear, but Judaic mythology, as well. The Jews killed off many of their fellow Jews in the Holocaust in the belief that they were fulfilling Old Testament prophecies. Their campaign is not over and will not end until all assimilated Jews and all Gentiles are dead.

Some Jews, the same type of racist tribal Jews who caused the Holocaust, want to kill off all Christians and all assimilated Jews. They believe that all anti-Semitism stems from Esau's pledge to destroy the seed of Jacob, and that God insisted that the Jews exterminate the seed of Amalek, grandson of Esau—and all assimilated Jews. Rather than fault Jacob for his vile treachery, racist Jews excuse their immoral hatred of Esau, by faulting Esau for being angry at Jacob for stealing the Covenant. Note that Esau was Jacob's brother and that the genocidal Jews believe in pruning off whole branches from their own family tree and exterminating whole lines of Jewish ancestral blood. Note further that Jews believe in treachery against their own blood as a means of maintaining the Covenant, for after all they are told again and again that only a remnant of Jews will survive in the end times, and racist Jews are convinced that that means them and that they have right to kill off assimilated Jews and Gentiles. This was one of the ways in which the racist Zionist Jews justified their mass murder of fellow Jews to themselves during the Holocaust.

The success of the story of Jesus led the Jews into another dilemma, in that Christians asserted that Jews must convert to Christianity as stated in *Romans* 9-11, though Saul, a. k. a. Paul, was probably only asking Jews to remain Jews at a time when many Jews were becoming secular. After more than a thousand years of antagonism, something had to give, and some Jews sought to undermine Christianity by converting it to Judaism, while pretending to convert Jews to Christianity. In many waves, over many centuries, swarms of Talmudists, Cabalists and false Messiahs have swept across Europe literally peddling social, spiritual and medical panaceas. Zionist anti-Catholic ministers preached the conversion and restoration of the Jews to Palestine and readied their gullible Christian brethren for their own demise.

The Jews had another reason to feign Christian conversion before colonizing Palestine. They knew that the Christians would see the Biblical implications of Jewish financiers using their corruptly gotten gains to take Jerusalem from its rightful ancient inhabitants as the manifestation of the "anti-Christ". The Jews feared that the Christians would join forces with Islam to crush the "anti-Christ" Jewish King and with him the Jews. Moses Hess quoted Ernest Laharanne, *La nouvelle question d'Orient: Empires d'Egypte et d'Arabie. Reconstitution de la nationalité juive*, E. Dentu, Paris, (1860):

"I may, therefore, recommend this work, written, not by a Jew, but by a French patriot, to the attention of our modern Jews, who plume themselves on borrowed French humanitarianism. I will quote here, in translation, a few pages of this work, *The New Eastern Question*, by Ernest Laharanne.[*Footnote:* See note IX at end of book.]

'In the discussion of these new Eastern complications, we reserved a special place for Palestine, in order to bring to the attention of the world the important question, whether ancient Judæa can once more acquire its former place under the sun.

'This question is not raised here for the first time. The redemption of Palestine, either by the efforts of international Jewish bankers, or the nobler method, of a general subscription in which all the Jews should participate, has been discussed many times. Why is it that this patriotic project has not as yet been realized? It is certainly not the fault of pious Jews that the plan was frustrated, for their hearts beat fast and their eyes fill with tears at the thought of a return to Jerusalem.[*Footnote:* My friend, Armond L., who traveled for several years through the Danube Principalities, told me that the Jews were moved to tears when he announced to them the end of their suffering, with the words 'The time of the return approaches.' The more fortunate Occidental Jews do not know with what longing the Jewish masses of the East await the final redemption from the two thousand year exile. They know not that the patriotic Jew cannot suppress his cry of anguish at the length of the exile, even in the midst of his festive songs, as, for instance, the patriotic poem which is read on Chanukah, closes with the mournful call:

'For salvation is delayed for us and there is no end to the days of evil.'

'They asked me,' continued my friend, 'what are the indications that the end of the exile is approaching?' 'These,' I answered, 'that the Turkish and the papal powers are on the point of collapse.']

'If the project is still unrealized, the cause is easily cognizable. The Jews dare not think of the possibility of possessing again the land of their fathers. Have we not opposed to their wish our Christian veto? Would we not continually molest the legal proprietor when he will have taken possession of his ancestral land, and in the name of piety make him feel that his ancestors forfeited the title to their land on the day of the Crucifixion?

'Our stupid Ultramontanism has destroyed the possibility of a regeneration of Judæa, by making the present of the Jewish people barren and unproductive. Had the city of Jerusalem been rebuilt by means of Jewish capital, we would have heard preachers prophesying, even in our progressive nineteenth century, that the end of the world is at hand and predictions of the coming of the Anti-Christ. Yes, we have lived to see such a state of affairs, now that Ultramontanism has made its last stand in oratorical eloquence. In the sacred beehive of religion, we still hear a continuous buzzing of those insects who would rather see a mighty sword in the hands of the barbarians, than greet the resurrection of nations and hail the revival of a free and great thought inscribed on their banner. This is undoubtedly the reason why Israel did not make any attempt to become master of his own flocks, why the Jews, after wandering for two thousand years, are not in a position to shake the dust from their weary feet. The continuous, inexorable demands that would be made upon a Jewish settlement, the vexatious insults that would be heaped upon them and which would finally degenerate into persecutions, in which fanatic Christians and pious Mohammedans would unite in brotherly accord—these are the reasons, more potent than the rule of the Turks, that have deterred the Jews from attempting to rebuild the Temple of Solomon, their ancient home, and their State."[388]

The Christians believed that the Jews had only one way to save themselves from ultimate annihilation—to convert to Christianity. Christians believe that only a small remnant of the Jews will convert and survive. They plan to slaughter the others. Even those Gentiles who were willing to help the Jews to take Palestine from the Turks believed that the Jews would be attacked by Christians unless they pretended to convert to Christianity. The Jews also believed that the Moslems would attack them and many toyed with the idea of massive feigned Christian conversion so that the Jews in Palestine would have Christendom as an artificial ally against Islam. Hence the countless books that were published by "Christians" calling for the "restoration of the Jews to Palestine" concurrently called for the conversion of the Jews, so as to protect the Jews from the Christians and grant them Christian protection from Islam.

The Holocaust had the effect of making the Jews appear impotent and vulnerable—non-threatening. Centuries of Jewish intrigues and propaganda eventually had the effect of weakening Christianity and subverting its beliefs such that the threat of a negative Christian response to massive Jewish emigration to Palestine has greatly diminished, though the possibility that the Jews will find themselves in a trap of their own making persists.

The numbing pain inspired by the shocking images of the victimization of the Jews in the Holocaust has been abused by racist Jews to shield themselves from criticism, such that their arrogance makes them an open menace which tarnishes the image of all Jews. As has always happened in the past when leading Jews grow insufferably arrogant and hypocritical, it might some day come about that true Christians will feel that they have been betrayed by "evil Jewish leadership" and will retaliate against the "anti-Christ" and the Zionists—pseudo-Christian and Jew, who have misled them. Real Christians may join forces with Islam and crush a foe which has been attacking them from the beginning, and which views the Hebrew Bible as a plan they intend to fulfill with their own deliberate actions. It is possible that the Christians and Moslems will learn from Jewish racists and adopt Jewish inhumanity and religious intolerance.

Very early on, Cyprian stated in his Twelfth Treatise, "Three Books of

---

[388]. M. Hess, "Eleventh Letter", *Rom und Jerusalem: die letzte Nationalitätsfrage*, Eduard Wengler, Leipzig, (1862); English: *Rome and Jerusalem: A Study in Jewish Nationalism*, Bloch, New York, (1918/1943), pp. 141-159, at 150-152.

Testimonies Against the Jews", First Book, Testimony 24, that the Jews had but one option to atone for the death of Christ,

> "24. That by this alone the Jews can receive pardon of their sins, if they wash away the blood of Christ slain, in His baptism, and, passing over into His Church, obey His precepts.
> In Isaiah the Lord says: 'Now I will not release your sins. When ye stretch forth your hands, I will turn away my face from you; and if ye multiply prayers, I will not hear you: for your hands are full of blood. Wash you, make you clean; take away the wickedness from your souls from the sight of mine eyes; cease from your wickedness; learn to do good; seek judgement; keep him who suffers wrong; judge for the orphan, and justify the widow. And come, let us reason together, saith the Lord: and although your sins be as scarlet, I will whiten [*Footnote:* 'Exalbabo.'] them as snow; and although they were as crimson, I will whiten [*Footnote:* 'Inalbabo.'] them as wool. And if ye be willing and listen to me, ye shall eat of the good of the land; but if ye be unwilling, and will not hear me, the sword [Esau] shall consume you; for the mouth of the Lord hath spoken these things. [*Footnote:* Isa. i. 15-20.]"[389]

The Zionists who wanted to remain openly practicing Jews had to carefully nurture an antagonism over the course of many centuries in Europe against the Pope, and depict him as the anti-Christ, and against Catholicism as the evil ecumenical Church of the Apocalypse, and against Islam and the Turks as heathens; so that "reformed" Christians would not see the Jews and Judaism as the evil ecumenical Church of the Apocalypse headed by the anti-Christ; and so that the English Esau, or some other European force, would take Palestine from the Turks and give it to the Jews, who could then regulate the trade of the world. The best means to accomplish this feat was to create anti-Catholic "reformations" and "second reformations" creating the Protestant and Puritan Churches which mirrored the Jewish faith, and for the Jews to pretend to convert to these Judaised Churches and form an alliance with Gentile Christians against Islam, while destroying Catholic Christianity.

Cabalist Giordano Bruno influenced Queen Elizabeth, and a short time later an interest in the *Kabbala Denudata*, edited by Christian Knorr von Rosenroth and Francis Mercury van Helmont, appeared in England.[390] Franciscus

---

[389]. Cyprian, Twelfth Treatise, "Three Books of Testimonies Against the Jews", First Book, Testimony 24, *The Anti-Nicene Fathers: Translations of the Writings of the Fathers down to A.D. 325*, Volume 5, Christian Literature Publishing Company, New York, (1886), p. 514-515.

[390]. T. Vaughan, T. W. and H. Blunden, *Magia Adamica, Or, the Antiquitie of Magic, and the Descent Therof from Adam Downwards, Proved: Whereunto Is Added, a Perfect, and Full Discoverie of the True Coelum Terrae, or the Magician's Heavenly Chaos, and First Matter of All Things*, Printed by T.W. for H. Blunden, London, (1650). *See also:* T. Vaughan, H. Blunden, R. Vaughan, et al., *Lumen De Lumine: Or, a New Magicall Light,*

Mercurius van Helmont[391] promoted cabalistic reformist dogma in England. Van

---

Printed for H. Blunden at the Castle in Corne-Hil, London, (1651). *See also:* H. C. Agrippa von Nettesheim, *Three Books of Occult Philosophy*, Printed by R.W. for Gregory Moule, London, (1651). *See also:* R. Fludd, *Mosaicall Philosophy: Grounded upon the Essential Truth or Eternal Sapience Written First in Latin, and Afterwards Thus Rendered into English*, Printed for Humphrey Moseley, at the Prince's Armes in St. Paul's Church-Yard, London, (1659). *See also:* J. Brinsley, The *Christians Cabala, Or, Sure Tradition Necessary to Be Known and Believed by All That Will Be Saved: A Doctrine Holding Forth Good Tidings of Great Joy, to the Greatest of Penitent Sinners : with a Character of One That Is Truly Such: As it Was Lately Held Forth to the Church of God at Great Yarmouth*, Printed for George Sawbridge, London, (1662).
391. F. M. v. Helmont, H. More, J. Gironnet, *et al.*, *Opuscula philosophica: quibus continentur principia philosophiæ antiquissimæ & recentissimæ. Ac philosophia vulgaris refutata. Quibus subjuncta sunt cc. Problemata de revolutione animarum humanarum*, Prostant Amstelodami, (1690). *See also:* P. Buchius, F. M. v. Helmont, and Philanglus, *The Divine Being and its Attributes Philosophically Demonstrated from the Holy Scriptures, and Original Nature of Things According to the Principles of F.M.B. of Helmont*, Printed and are to be sold by Randal Taylor, London, (1693). *See also:* J. B. v. Helmont and F. M. v. Helmont, *Joannis Baptistæ van Helmont... Opuscula Medica Inaudita. I. De Lithiasi. II. De Febribus. III. De Humoribus Galeni. IV. De Peste*, Apud Ludovicum Elzevirium, Amsterodami, (1648). *See also:* Y-Worth, F. M. v. Helmont, Paracelsus, *et al.*, *Trifertes Sagani, or Immortal Dissolvent: Being a... Discourse of the Matter and Manner of Preparing the Liquor Alkahest of Helmont, the Great Hilech of Paracelsus, the Sal Circulatum minus of Ludovicus De Comit: or Our Fiery Spirit of the Four Elements. Together with its Use in Preparing Magisteries, Arcanas, Quintessences and Other Secret Medicines of the Adepts*, W. Pearson for T. Ballard, London, (1705). *See also:* J. B. v. Helmont and F. M. v. Helmont, *Ortus medicinæ: Id est, initia physicæ inaudita : progressus medicinae novus, in morborum ultionem, ad vitam longam*, Apud Ludovicum Elsevirium, Amsterodami, (1648). *See also:* J. B. v. Helmont, W. Charleton and F. M. v. Helmont, *A Ternary of Paradoxes: The Magnetick Cure of Wounds, Nativity of Tartar in Wine, Image of God in Man*, Printed by James Flesher for William Lee, London, (1650). *See also:* G. Starkey and J. B. v. Helmont, *Natures Explication and Helmont's Vindication, Or, a Short and Sure Way to a Long and Sound Life Being a Necessary and Full Apology for Chymical Medicaments, and a Vindication of Their Excellency Against Those Unworthy Reproaches Cast on the Art and its Professors... by Galenists, Usually Called Methodists*, Printed by E. Cotes for Thomas Alsop, London, (1657). *See also:* J. B. v. Helmont and F. M. v. Helmont, *Oriatrike, Or, Physick Refined: the Common Errors Therein Refuted, and the Whole Art Reformed & Rectified : Being a New Rise and Progress of Philosophy and Medicine for the Destruction of Diseases and Prolongation of Life*, Printed for L. Loyd, London, (1662). *See also:* J. B. v. Helmont, J. C., and F. M. v. Helmont, *Van Helmont's Works Containing His Most Excellent Philosophy, Physick, Chirurgery, Anatomy : Wherein the Philosophy of the Schools Is Examined, Their Errors Refuted, and the Whole Body of Physick Reformed and Rectified : Being a New Rise and Progresse of Philosophy and Medicine, for the Cure of Diseases, and Lengthening of Life*, Printed for Lodowick Lloyd, London, (1664). *See also:* J. B. v. Helmont and F. M. v. Helmont, *Opera Omnia*, Sumptibus Johannis Justi Erythropili, Typis Johannis Philippi Andreae, Francofurti, (1682). *See also:* J. B. v. Helmont, C. Knorr von Rosenroth and F. M. v. Helmont, *Aufgang der Artzney-Kunst: Das ist, noch nie erhörte Brund-Lehren von der Natur, zu einer neuen Beförderung der Artzney-*

Sachen, sowol die Kranckheiten zu vertreiben als ein langes Leben zu erlangen, In Verlegung Johann Andreæ Endters Sel. Söhne: Gedruckt bey Johann Holst, Sultzbach, (1683). ***See also:*** F. M. v. Helmont and J. B. v. Helmont, *A. T. Limojon de St. Didier, et al., One Hundred Fifty Three Chymical Aphorisms... : Done by the Labour... of Eremita Suburbanus*, Printed for the Author: And Are to Be Sold by W. Cooper at the Pelican in Little Britain: And D. Newman at the Kings-Arms in the Poultry, London, (1688). ***See also:*** F. M. v. Helmont, G. Janssonius Van Waesberghe, and J. Janssonius Van Waesberge, *Cliii aphorismi chemici: Ad quos quicquid est scientiae chemicae commodè referri potest*, Apud Janssonio-Waesbergios, Amstelaedami, (1688). ***See also:*** F. M. v. Helmont, P. Buchius, J. C. Amman, *Francisci Mercuri ab Helmont Observationes circa Hominem Ejusque Morbos: Certissimis Sanae Rationis & Experientiae Superstructae*, Apud Joannem Wolters, Amstelaedami, (1692). ***See also:*** P. B. Buchius, Philanglus, and F. M. v. Helmont, *The Divine Being and its Attributes Philosophically Demonstrated from the Holy Scriptures, and Original Nature of Things According to the Principles of F.m.b. of Helmont*, : Printed and Are to Be Sold by Randal Taylor, London,(1693). ***See also:*** F. M. v. Helmont and D. Kellner, *Kurtze Vorstellung der zur edlen chymie gehörigen Wissenschafft bestehend in CLIII. Aphorismis oder kurtzen Sätzen, dahin alles, was zur Alchymie gehöret, gar füglich gezogen werden kan*, A.m. Hynitzsch, Nordhausen, (1699). ***See also:*** F. M. v. Helmont and J. Clark, *Seder Olam, Or, the Order, Series or Succession of All the Ages, Periods, and Times of the Whole World Is Theologically, Philosophically and Chronologically Explicated and Stated Also the Hypothesis of the Pre-existency and Revolution of Humane Souls Together with the Thousand Years Reign of Christ on Earth... : to Which Is Also Annexed Some Explanatory Questions of the Book of the Revelations... : and an Appendix Containing Some Emendations and Explanations of Divers Passages in the Two Foregoing Treatises, out of the Author's Original Manuscripts and Papers*, Printed for Sarah Howkins, London, (1694). ***See also:*** F. M. v. Helmont, *A Cabbalistical Dialogue in Answer to the Opinion of a Learned Doctor in Philosophy and Theology, That the World Was Made of Nothing As it Is Contained in the Second Part of the Cabbala Denudata & Apparatus in Lib. Sohar, P. 308 &C., To Which Is Subjoyned a Rabbinical and Paraphrastical Exposition of Genesis I. Written in High-dutch by the Author of the Foregoing Dialogue, First Done in Latin, by Now Made English*, Printed for Benjamin Clark, London, (1682); **and** *The Spirit of Diseases, Or, Diseases from the Spirit Laid Open in Some Observations Concerning Man and His Diseases : Wherein Is Shewed How Much the Mind Influenceth the Body in Causing and Curing of Diseases : the Whole Deduced from Certain and Infallible Principles of Natural Reason and Experience*, Printed for Sarah Howkins, London, (1694); **and** *The Paradoxal Discourses of F.m. Van Helmont Concerning the Macrocosm and Microcosm, Or, the Greater and Lesser World and Their Union*, Printed by J.C. and Freeman Collins for Robert Kettlewel, London, (1685); **and** *Two Hundred Queries Moderately Propounded Concerning the Doctrine of the Revolution of Humane Souls and its Conformity to the Truths of Christianity*, Printed for Rob. Kettlemell, London, (1684); **and** *An Hundred and Fifty-three Chymical Aphorisms To Which, What-Ever Relates to the Science of Chymistry May Fitly Be Referred*, Printed for Awnsham Churchill, London, (1690); **and** *Franciscii Mercurii Freyherrn von Helmont Paradoxal Discourse: oder, ungemeine Meynüngen von dem Macrocosmo und Microcosmo, das ist: von der grossen und kleinern Welt und verselben Vereinigung mit Einander... auf der Englischen in die hochteutsche Sprache übersetset*, Gottfried Liebernickel, Hamburg, (1691); and *Alphabeti verè naturalis hebraici brevissima delineatio: quae simul methodum suppeditat, juxta quam qui surdi nati sunt sic informari possunt, ut non alios saltem loquentes intelligant, sed &*

Helmont taught an ecumenical religion which converted Christianity into Judaism. The Inquisition accused him of Judaising Europe. He was a good friend of Leibnitz.

Van Helmont disseminated his message in England though Anne Conway[392]

---

ipsi ad sermonis usum perveniant, Typis Abrahami Lichtenthaleri, Sulzbaci, (1657); and *Seder olam, sive ordo seculorum: historica enarratio doctrinae*, Leyden, (1693); **and** *Seder Olam, Or, the Order, Series, or Succession of All the Ages, Periods, and Times of the Whole World Is Theologically, Philosophically, and Chronologically Explicated and Stated ; Also the Hypothesis of the Pre-Existency and Revolution of Humane Souls ; Together with the Thousand Years Reign of Christ on the Earth*, Printed for Sarah Hopkins, in George-Yard, Lumbard-Street, London, (1694); **and** *Kurtzer Entwurff des eigentlichen Natur-Alphabets der heiligen Sprache: Nach dessen Anleitung man auch taubgeborne verstehend und redend machen kan*, Abraham Lichtenthaler, Sultzbach, (1667); **and** *Een Zeer Korte Afbeelding Van Het Ware Natuurlyke Hebreuwse A.B.C. Welke Te Gelyk De Wyse Vertoont, Volgens Welke Die Doof Geboren Syn, Sodanig Konnen Onderwesen Werden, Dat Sy Niet Alleenig Andere Die Spreken Konnen, Verstaan, Maar Selfs Tot Het Gebruik Van Spreken Komen*, Pieter Rotterdam, Amsterdam, (1697); **and** *Eenige Voor-Bedagte En Over-Wogene Bedenkingen: Over De Vier Eerste Kapittelen Des Eersten Boeks Moysis, Genesis Genaamt*, Pieter Rotterdam, T'Amsterdam, (1698); **and** *Some Premeditate and Considerate Thoughts upon the First Four Chapters of the First Book of Moses Called Genesis*, London, (1701); and *Einige Gedancken über die vier ersten Capitel des ersten Buchs Mosis, Genesis genannt*, (1698); and *Quædam Præmeditatae & consideratae cogitationes super quatuor priora capita libri primi Moysis, Genesis nominati*, Prostant Apud Henr. Wetstenium, Amstelodami, (1697); **and** *C.LIII Aphorismes Chymiques: Ausquels on Peut Facilement Rapporter Tout Ce Qui Regarde La Chymie : Mis En Ordre Par Les Soins & Le Travail De L'hermite du Fauxbourg*, Laurent D'houry, Paris, (1692); **and** *Admonitio de rationis humanae deceptione in spiritualibus fugiendâ*, Regio-Monti, (1646); **and** *Brevissima Descrizione Dell'alfabeto Veramente Naturale Ebraico. Essa Può Fornire Insieme Il Metodo Col Quale Coloro Che Sono Nati Sordi Possono Essere Istruiti Così Da Comprendere non Solo Gli Altri Parlanti, Ma Essi Stessi Giungere All'uso Del Linguaggio*, Scuola Tipografica Sordomuti, Siena, (1667/1960); **and** *Autobiographical Memoirs of F. M. van Helmont: Or, Extraordinary Passages from His Life*, (1600); **and** *Eenige Gedakten*, Hendrik Jansse, T'Amsterdam, (1690); and *Taina tvoreniia, po vidimym I nevidimym chudesam ego, iz bozhestvennago magicheskago tsentral'nago svieta: pokazannaia chadam premudrosti ot niekiikh prosvieshchennykh magov: nynie dlia razmnozheniia poznaniia v pervyi raz izdannaia na sviet iz drevnikhrukopisei*, V Tipografii i. Lopukhina, Moskva, (1785); **and** *Aanmerkingen*, Pieter Rotterdam, T'Amsterdam, (1692); **and** *Thesaurus Novus Experientiae Medicae Aureus: Oder, Guldener Artzney-Schatz neuer niemels entdeckter Medicamenten wider allerhand Leibs-Kranckheiten auss den fürtrefflichen Schriften*, Bey Eman. Und Joh. Rud. Thurneysen, Basel, (1723); **and** *Ortvs medicinae: Id est, initia physicae inavdita. Progreffus medicinae novus, in morborvm vltionem ad vitam longam*, Apud Ludovicum Elzevirium, Amsterodami, (1652).

392. A. Conway, *The Principles of the Most Ancient and Modern Philosophy Concerning God, Christ and the Creatures... Being a Little Treatise Published since the Author's Death, Translated out of the English into Latin, with Annotations Taken from the Ancient Philosophy of the Hebrews, and Now Again Made English*, Printed in Latin at Amsterdam by M. Brown, and reprinted at London, (1692). A. Conway, et al., *Conway letters: The Correspondence of Anne, viscountess Conway, Henry More, and Their Friends, 1642-*

to Henry More, Robert Boyle, John Locke, Isaac Newton, etc. Van Helmont also published on medicine and chemistry, subjects which would later interest David Hartley. The ecumenical Protestants, Puritans, and Arians like Isaac Newton, Samuel Clarke and David Hartley, converted Christians to Judaism under the guise of converting Jews to Christianity.

Frankist Jews converted to Christianity in order to destroy it. Some Jews used the institution of Freemasonry as a means to bring about the conversion of Christians to Judaism. As predicted in Biblical prophecy, they sought to make Jerusalem the capital of the ecumenical church of Judaism, which would replace the supposedly "Universal" or "Catholic" Church seated in Rome. Zionist Moses Hess wrote in his treatise published in 1862, *Rome and Jerusalem*,

> "You have certainly heard of Joseph Salvador, the author of the work entitled *History of the Mosaic Institutions and of the Hebrew People*. This same author recently published a work entitled *Paris, Rome and Jerusalem*, in which he clearly shows that even among our enlightened brethren, there are dreamers who wish for a rebuilding of the Temple of Jerusalem. But he attaches to this rebuilding conditions that are acceptable neither to pious nor to progressive Christians and Jews. If I understand the author correctly, he expects his New Jerusalem to become the world capital of the fusionists. Salvador, furthermore, seems to cherish the curious idea that the Jews ought first to turn Christians, so that they may be the better able to convert the Christians afterward to Judaism. This work is, in reality, not as new as Salvador thinks; it began eighteen hundred years ago. It seems, however, that the Judaism of which Salvador is thinking is as new as his Christianity.
>
> More reasonable are the attempts of those fusionists who, like my friend Hirsch, of Luxemburg, are utilizing freemasonry as a means to amalgamate all the historical cults into one. The Luxemburg Rabbi, the antipode of his namesake, the Frankfort Rabbi Hirsch, developed the idea of fusion so thoroughly in the excellent lectures which he delivered at the Luxemburg Lodge, and later published under the title *Humanity as a Religion*, that, according to him, the matter may be considered closed. All that remains for the rabbis to do is to close up their reform temples and send the school children to the masonic temples. In truth, the logical consequences of reform have long since led those who took the sermons of the reform rabbis seriously, toward making such a step; as you, being a resident of Frankfort, well know. In vain did they afterward ornament their fusionist sermons with Talmudic quotations. It was too late and they had to be satisfied to preach to empty pews.
>
> Jewish rationalists, who have as little reason to remain within the fold of Judaism as have Christian rationalists for clinging to Christianity are, like their Christian friends, very energetic in discovering new grounds for the existence of a religion which, according to them, has no longer any reason

---

1684, Oxford University Press, (1930).

to exist. According to them, the dispersion of the Jews was merely a preliminary step to their entering upon their great mission. What great things are the Jews in exile to accomplish in their opinion? First of all, they are to represent 'pure' theism, in contradistinction to Christianity. In the next place, tolerant Judaism is to teach intolerant Christianity the principles of humanitarianism. Furthermore, it is the function of exilic Judaism to take care that morality and life, which in the Christian world are severed from each other, should become one. And lastly, the Jews must also act as industrial and commercial promoters—be the leaven of such activities among the civilized nations in whose midst they live. I have even heard it remarked quite seriously, that the Indo-Germanic race must improve its quality by mingling with the Jewish race!

But, mark you, from all these real or imaginary benefits which the Jews in dispersion confer upon the world, none will be diminished even after the restoration of the Jewish State. For just as at the time of the return from the Babylonian exile not all the Jews settled in Palestine, but the majority remained in the lands of exile, where there had been Jewish settlements since the dispersion of Israel and Judah, so need we not look forward to a larger concentration of Jews at the future restoration. Besides, it seems to me that those benefits which the Jews in exile confer upon the world are exaggerated, 'for the sake of the cause.' I consider it an anachronism to assign to the Jews those missions which they certainly performed in antiquity, and to some extent also in mediaeval times, but which, at present, no longer belong peculiarly to them. As to affecting the unity of life and theory, it is only possible with a nation which is politically organized; such a nation alone is able to realize it practically by embodying it in its institutions.

Again, what section of world-Jewry is to teach the Christians tolerance and humanity? You will surely say the enlightened Jews. But is not the enlightened Christian entitled to repeat to the enlightened Jew the words which Lessing, in his *Nathan the Wise*, puts into the mouth of the liberal Christian in his answer to the liberal Jew: 'What makes me a Christian in your eyes, makes you a Jew in mine.'

Or, on the other hand, should the enlightened Jew say to the orthodox Christian, 'Your beliefs are mere superstitions, your religion only fanaticism,' may the enlightened Christian not turn to the orthodox Jew and make similar remarks in defense of his own religion? Our cultured Jews who accuse Christians of possessing a persecution mania, reason as fallaciously as does Bethmann Hollweg when he charges the Jews with the same trait. History can neither be explained nor changed in its course by such explanations.

From the viewpoint of enlightenment, I see as little reason for the continuation of the existence of Judaism as for Christianity. It is better for the Jew who does not believe in the national regeneration of his people, to labor, like the enlightened Christian, for the dissolution of his religion. I understand how one can hold such an opinion. But what I do not understand is, how it is possible to believe simultaneously in 'enlightenment' and in a Jewish Mission in exile; in other words, in the ultimate dissolution and in the continued existence of Judaism at the same time."[393]

Christianity itself was a movement to convert Gentiles to Judaism in the guise of Liberalism, and to take the hatred and menacing nature of the creator God of the Old Testament out of Judaism so as to make it more palatable for Gentile consumption. A new call for "fusion" reappeared in the Zionism of Protestants, who often wrote of converting Jews to "Christianity"—while calling for the restoration of the Jews to Palestine, so as to make Jerusalem (as opposed to Rome) the seat of a new international despotism that was based on Judaism, which treachery against Christians signified the terror and devastation of the prophecies, the mysticism of the gnostics who were influenced by the East, and the despotism and deceit of the worst of the Talmud. Protestantism itself takes a large step towards converting Christianity back into original Judaism, with all its horrors and inhumanity.

8.7.1 The "British-Israel" Deceit

Biblical prophecies require that in order for the millennium to begin all of the Tribes of Israel must return to Palestine (*Isaiah* 11:11-12. *Jeremiah* 23:8; 30:3. *Ezekiel* 37:21. *Hosea* 3:4-5). The ten northern tribes were missing (never existed), though some were believed to have been found in the mid-1800's. Some in England had long believed that the English descended from one of the ten lost tribes of Israel which had allegedly traveled to England on Phoenician ships in ancient times.

The belief that the British were a lost tribe of Israel was promoted in Russia as evidence that England might be a place of respite for the anti-Christ—especially since British Royals claimed to be descended from King David and the Zionists published countless books in England and America calling for the "restoration of the Jews to Palestine" and concurrently seeking to foment a war with Rome, Russia and Turkey by calling the Pope, the Czar and the Sultan the "anti-Christ". There is little doubt but that it was Jews in England who inspired the belief that the Pope in Rome, the Russian Czar and the Turkish Sultan were the anti-Christ, because the propaganda which popularized these beliefs served the perceived self-interests of the Jews. It was also likely the Jews in Russia who inspired a belief there that the English King was the anti-Christ. It was not mere

---

[393]. M. Hess, *Rom und Jerusalem: die letzte Nationalitätsfrage*, Eduard Wengler, Leipzig, (1862); English translation by M. Waxman: *Rome and Jerusalem: A Study in Jewish Nationalism*, Bloch, New York, (1918/1943), pp. 114-117. Hess refers to: J. Salvador, *Histoire des institutions de Moïse, et du peuple hébreu*, Ponthieu et Cie., Paris, Ponthieu, Michelsen et Cie., Leipzig, (1828); German: *Geschichte der mosaischen Institutionen und des jüdischen Volks*, Hoffmann und Campe, Hamburg, (1836); **and** *Paris, Rome, Jérusalem; ou, La question religieuse au XIXe siècle*, M. Lévy, Paris, (1860).

coincidence that this antagonistic propaganda calling for wars on all sides amongst the empires uniformly called for the "restoration of the Jews to Palestine" and uniformly stigmatized an artificial enemy as the "anti-Christ". The Jews had been trying to provoke a world war through their hateful and intolerant propaganda for centuries.

A vast movement existed in England and the Commonwealth Nations during and after Queen Victoria's reign, which called itself "Anglo-Israel" or "British-Israel". They claimed that the English descended from Israel, that Queen Victoria descended from the House of King David, and that the Jews should be restored to Palestine.[394] It is likely that all movements which call for the "restoration of the Jews to Palestine" were begun by Jews, overtly or covertly.

The prophecies require that the Messiah be descended from David (II *Samuel* 7; 22:44-51; 23:1-5. *Isaiah* 9:6-7. *Jeremiah* 23:5; 33:15, 17). In an attempt to avoid Christian suspicion and persecution, many Jewish groups spread the myth that their ancestors had left Israel before the crucifixion of Christ, or had opposed it.[395] The question naturally arises, was the entire British-Israel movement, which began more than one thousand years ago, initiated by

---

394. The literature of the "Anglo-Israel" or "British-Israel" movement is extensive. To name but a few of the hundreds of titles published from the early 1800's to the present: R. Weaver, *Monumenta antiqua: Or, The Stone Monuments of Antiquity Yet Remaining in the British Isles, Particularly as Illustrated by Scripture. Also a Dissertation on Stonehenge: Together with a Compendious Account of the Druids. To Which Are Added Conjectures on the Origin and Design of the Pyramids of Egypt, and of the Round Towers of Ireland*, J.B. Nichols & Son, London, (1840). *See also:* E. Hine, *The English Nation Identified with the Lost House of Israel by Twenty-Seven Identifications*, J. Heywood, Birmingham, R. Davies, Manchester, (1871). *See also:* W. Carpenter, *The Israelites Found in the Anglo-Saxons: The Ten Tribes Supposed to Have Been Lost, Traced from the Land of Their Captivity to Their Occupation of the Isles of the Sea: With an Exhibition of Those Traits of Character and National Characteristics Assigned to Israel in the Books of the Hebrew Prophets*, G. Kenning, London, (1874). *See also:* W. H. Poole, *Anglo-Israel, or, The British Nation the Lost Tribes of Israel*, Toronto, (1879); and *Anglo-Israel: Or, The Saxon Race, Proved to Be the Lost Tribes of Israel. In nine Lectures*, W. Briggs, Toronto, (1889). *See also:* F. R. A. Glover, *England the Remnant of Judah and the Israel of Ephraim*, London, (1881). *See also:* T. R. Howlett, *Anglo-Israel and the Jewish problem. The Ten Lost Tribes of Israel Found and Identified in the Anglo-saxon Race. The Jewish Problem Solved in the Reunion of Israel and Judah, and Restoration of the Israelitish Nation*, Spangler & Davis, Philadelphia, (1892). *See also:* P. S. McKillop, *Britain and America, the Lost Israelites: Or, the Ten Tribes Identified in the Anglo-Celtic Race*, St. Albans, Vermont, (1902). *See also:* R. Harris, *The Lost Tribes of Israel*, S. W. Partridge, London, (1907). *See also:* J. L. Thomas, *The Restoration of Israel*, Marshall, London, New York, (1922). *See also:* S. A. Brown, *The House of Israel: Or, The Anglo-Saxon*, Pub. for S.A. Brown by Boyer Print. & advertising Co., Portland, Oregon, (1925). *See also:* M. Barkun, *Religion and the Racist Right: The Origin of the Christian Identity Movement*, Revised Edition, University of North Carolina Press, (1997).

395. "The Modern Jews", *The North American Review*, Volume 60, Number 127, (April, 1845), pp. 329-368, at 333-334, 355.

Jews who sought to distance themselves from the crucifixion of Christ? Some Jews asserted that America was the new Israel and that Jews were important members and sponsors of Christopher Columbus' voyage to America—even that Christopher Columbus was himself a Jew searching for a new homeland for the Jews.[396] In America, Judge Noah, a Jewish Zionist, argued that the American Indians were descended from the Israelites, and Noah sought to privilege Jews in America on this basis.

John Spargo was quoted in *The New York Times* on 22 February 1921 on page 10, referring to the publication of the *Protocols of the Learned Elders of Zion* in English translation, in an article entitled "Spargo Condemns Racial Antagonism"

> "In 1895 a book was published in France which attempted to prove the existence of a world-wide conspiracy against Christian civilization. In that book the theory was advanced that the English people are all of the Jewish race, and that the British Government is the central force of this worldwide Jewish conspiracy. In his book Nilus reproduced this fantastic theory but, recognizing that it would cause the protocols to be laughed out of court, The Dearborn Independent, The London Morning Post and all the other publishers of the protocols in England and America have carefully deleted this part of the book by Nilus. The reason for the deletion is as obvious as the dishonor of it."

Spargo was mistaken if he would assert that there was no belief among the British themselves that they had descended from the "Israelites" and that this belief was instead concocted in Russia in 1895 in order to discredit the Jews and the British. The belief that the British descended from the "Israelites" was very old and enduring, as was the belief that they descended from Noah.[397] William

---

396. A. Leroy-Beaulieu, *Israel chez les nations: Les Juifs et l'antisémitisme*, C. Lévy, Paris, (1893); English translation by F. Hellman, *Israel among the Nations: A Study of the Jews and Antisemitism*, G. P. Putnam's Sons, New York, W. Heinemann, London, (1895), p. 356. H. N. Casson, "The Jew in America", *Munsey's Magazine*, Volume 34, Number 4, (January, 1906), pp. 381-395. B. J. Hendrick, "The Jews in America: I How They Came to This Country", *The World's Work*, Volume 44, Number 2, (December, 1922), pp. 144-161.

397. B. W. Tuchman, *Bible and Sword: England and Palestine from the Bronze Age to Balfour*, New York University Press, New York, (1956). *See also:* C. Duvernoy, *Le prince et le prophète*, Le département des publications de l'Agence juive, Jérusalem, (1966); English translation: *The Prince and the Prophet*, Land of Promise Productions, Paradise, California, (1973); Christian Action for Israel, (1979). *See also:* E. Newman, "Non-Jewish Pioneers of Zionism", in A. W. Kac, Editor, *The Messiahship of Jesus: What Jews and Jewish Christians Say*, Moody Press, Chicago, (1980), pp. 291-297. *See also:* M. Ould-Mey, "The Non-Jewish Origin of Zionism", *The Arab World Geographer / Le Géographe du monde arabe*, Volume 5, Number 1, (2002), pp. 34-52: <http://mama.indstate.edu/users/mouldmey/The%20Non-Jewish%20Origin%20of%20Zionism.PDF>

Camden in his *Britannia* of 1586,[398] and Theophilus Evans in his *Drych y prif oesoedd* of 1716,[399] told of the Welsh legend that the Ancient Britons, the Welsh, had descended from Noah's grandson Gomer. Camden's view even found its way onto the 1606 English edition of the map *ANGLIÆ, SCOTIÆ ET HIBERNIÆ, SIVE/BRITANNICAR: INSVLARVM DESCRIPTIO*:

"The first Inhabitants which settled here not long after the universal Flood and the Confusion of Babel came here from France, considering its Proximity, Similarity of language, Manners, Government, Customs and Name, as is stated by the learned Clarencieux Camden, the only light shining on our histories, as demonstrated in his treatise called Britannia. For to this day the ancient Britans, the Welshmen, call themselves CUMRI, (not Cambri), derived from Gomer, the son of Iaphet (called by the Romans Cimber) from whom the Celtæ or Gauls are descended."[400]

Circa 800AD, Nennius wrote that the British descended from Noah in his *Historia Britonum*.[401] Aylett Sammes published *Britannia Antiqua Illustrata*[402] in 1676, in which he argued that the British descended from the Phoenicians. Henry Rowlands[403] argued in 1723 that the ancient Druids were the descendants of Noah. In the 1740's, William Stukeley held that the British were the children

---

398. W. Camden, *Britannia sive Florentissimorvm regnorvm, Angliæ, Scotiæ, Hiberniae, et invlarvm adiacentium ex intima antiquitate chorographica descriptio*, Per Radulphum Newbery. Cum gratia & priuilegio Regiæ Maiestatis, Londini, (1586).
399. T. Evans, G. H. Hughes, *Drych y prif oesoedd: yn ôl yr argraffiad cyntaf, 1716*, Gwasg Prifysgol Cymru, Caerdydd [Wales], (1961).
400. English translation from the Latin by Dr. Marcel van den Broecke, Cartographica Neerlandica Background for Ortelius Map No. 16 ANGLIÆ, SCOTIÆ ET HIBERNIÆ, SIVE/BRITANNICAR: INSVLARVM DESCRIPTIO, English Edition of 1606:
http://www.orteliusmaps.com/book/ort16.html
http://www.orteliusmaps.com/book/ort_text16.html
401. B. W. Tuchman, *Bible and Sword: England and Palestine from the Bronze Age to Balfour*, New York University Press, New York, (1956), p. 3.
402. A. Sammes, *Britannia Antiqua Illustrata: Or, the Antiquities of Ancient Britain, Derived from the Phoenicians: Wherein the Original Trade of this Island Is Discovered, the Names of Places, Offices Dignities, as Likewise the Idolatry, Language, and Customs of the Primitive Inhabitants Are Clearly Demonstrated from That Nation... Together with a Chronoloical History of this Kingdom, from the First Traditional Beginning, until the Year of Our Lord 800... Faithfully Collected out of the Best Authors ... with the Antiquities of the Saxons, as Well as Phoenicians, Greeks, and Romans. The First Volume*, Printed by T. Roycroft, for the Author, London, 1676
403. H. Rowlands, *Mona Antiqua Restaurata. An Archæological Discourse on the Antiquities, Natural and Historical, of the Isle of Anglesey, the Antient Seat of the British Druids. In Two Essays. With an Appendix, Containing a Comparative Table of Primitive Words, and the Derivatives of Them in Several of the Tongues of Europe; with Remarks upon Them. Together with Some Letters, and Three Catalogues, Added Thereunto. I. of the Members of Parliament from the County of Anglesey, Ii. Of the High-sheriffs; and Iii. Of the Beneficed Clergy Thereof*, Printed by A. Rhames, Dublin, (1723).

of Abraham.[404] Queen Victoria believed that she was descended from King David, which also meant that Victoria's grandson Kaiser Wilhelm II was also believed by the family to be descended from David. In 1924, Laurence Austine Waddell published *The Phoenician Origin of Britons, Scots & Anglo-Saxons Dicovered by Phoenician & Sumerian Inscription in Britain, by Pre-Roman Briton Coins & a Mass of New History*.[405]

8.7.2 For Centuries, England is Flooded with Warmongering Zionist Propaganda

Zionism appeared early and often in England and America.[406] For example, in addition to the works cited above, Thomas Brightman published his *Apocalypsis Apocalypseos* in 1585.[407] In 1585, Francis Kett, like Martin Luther, declared that the Pope was the "Beast" prophesied in *Revelation* and the man foretold to

---

[404]. W. Stukeley, *Stonehenge: A Temple Restor'd to the British Druids*, : Printed for W. Innys and R. Manby, London, (1740); **and** *Abury, a Temple of the British Druids with Some Others, Described, Wherein Is a More Particular Account of the First and Patriarchal Religion, and of the Peopling the British Islands*, Printed for the Author, London, (1743).

[405]. L. A. Waddell, *The Phoenician Origin of Britons, Scots & Anglo-Saxons Dicovered by Phoenician & Sumerian Inscription in Britain, by Pre-roman Briton Coins & a Mass of New History*, Williams and Norgate, London, (1924).

[406]. B. W. Tuchman, *Bible and Sword: England and Palestine from the Bronze Age to Balfour*, New York University Press, New York, (1956).

[407]. T. Brightman, *Brightmans predictions and prophecies vvritten 46 yeares since: concerning the three churches of Germanie, England, and Scotland : fore-telling the miserie of Germanie, the fall of the pride of bishops in England by the assistance of the Scottish Kirk: all which should happen, as he foretold, between the yeares of 36 and 41, &c.*, (1641); **and** *A reuelation of the Reuelation: that is, the Reuelation of St. John opened clearly with a logicall resolution and exposition : wherein the sense is cleared, out of the Scripture, the euent also of thinges foretold is discussed out of the church-historyes*, Amsterdam, (1615); **and** *The revelation of S. Iohn illustrated with an analysis & scholions Where in the sence is opened by the scripture, & the euent of things fore-told, shewed by histories*, Class [on van Dorpe], Leyden, (1616); **and** *A revelation of the Apocalyps, that is, the Apocalyps of S. Iohn illustrated vvith an analysis & scolions where the sense is opened by the scripture, & the events of things foretold, shewed by histories. Hereunto is prefixed a generall view: and at the end of the 17. chapter, is inserted a refutation of R. Bellarmine touching Antichrist, in his 3. book of the B. of Rome*, Iudocus Hondius & Hendrick Laurenss, Amsterdam, (1611); **and** *Apocalypsis apocalypseos: id est Apocalypsis D. Joannis analysi et scholiis illustrata; ubi ex scriptura sensus rerumque praedictarum ex historiis eventus discutiuntur. Huic Synopsis praefigitur universalis, et refutatio Rob. Bellarmini de antichristo libro tertio de Romano Pontifice ad finem capitis decimi septimi inseritur*, Heidelberg, (1612); **and** *Een Grondighe ontdeckinghe ofte duydelijcke uytlegginghe, met een logicale ontknoopinghe, over de gantsche openbaringe Iohannis des Apostels: waer in de sin uyt de Schriftuere verklaert, ende insghelijchs de uytkomsten der dinghen die voorseyt waren, met de kerchelijcke historien aenghewesen worden*, Jan Evertsz Cloppenburch, boeckvercooper, 'tAmstelredam, (1621).

pretend to be God in the Temple.[408] In 1585, Kett envisioned Jerusalem as the heavenly seat of the new Kingdom of Christ.[409] Kett was burned at the stake in 1589 for declaring that the Bible prophesied that the Jews would be restored to Palestine. The "Eastern Question" arose again and again in apocalyptic literature and the authors frequently discussed scenarios that eventually played out—Russia's wars against Turkey, Napoleon and the East, Greek independence, the Crimean War, the Congress of Berlin, World War I and World War II, etc. It is no coincidence that the works which called for the "restoration of the Jews to Palestine" correctly foretold the wars the Jews deliberately caused to further their goal of creating a "Jewish State".

"Lord Protector" Oliver Cromwell, Queen Victoria, Prime Minister Benjamin Disraeli, Prime Minister Arthur Balfour, Prime Minister David Lloyd George and Prime Minister Winston Churchill were each outspoken and long-term Zionists.[410] This remarkably high percentage of Zionist leaders in England is especially odd given that only a very small percentage of Jews were Zionists and there was never a large Jewish population in England. This oddity is explained by the grossly disproportionate influence of Cabalistic Jews and Jewish bankers in England over the course of many centuries.

Puritans, like Oliver Cromwell, were ardent Zionists and carried out a "second reformation" in order to attack the Catholics—whom the Jews hated. Many Puritans migrated to America. Though American schools teach that they came for religious freedom, the truth of the matter is that they migrated to America so that they would have the freedom to practice extreme intolerance.

Puritans sought to forcefully convert Christians to Judaism while pretending to seek to convert Jews to Christianity. Like many of the Protestants of Germany, they generally named their children with names taken from the Old Testament, not the New. In Amsterdam, English Puritan Zionists Joanna and Ebenezer Cartwright issued a Zionist petition in 1649 calling on the English and the Dutch to lead the Jews back to Palestine. Zionist Cabalist Franciscus Mercurius van Helmont traveled from Amsterdam to England to spread Cabalistic Judaism and Zionism among the intellectual elite of England—and he was quite successful. Rabbi Manassah ben Israel, of Amsterdam, persuaded Oliver Cromwell to readmit Jews into England on the premise that the Biblical pronouncement that the Jews shall be scattered to the ends of the Earth meant that they shall enter

---

408. F. Kett, *An Epistle [S]ent to Divers [Pa]pistes in England Prouing [Th]e Pope to Bee the Beast in the [1]3 of the Reuelations, and to Be the Man Exalted in the Temple of God, as God, Thess. 2.2*, Henry Marsh, London, (1585).
409. F. Kett, *The Glorious and Beautifull Garland of Mans Glorification. Containing the Godlye Misterie of Heauenly Ierusalem, the Helmet of Our Saluation. The Comming of Christ in the Fleshe for Our Glorie, and His Glorious Com[m]ing in the End of the World to Crowne Men with Crownes of Eternall Glorie. Beeing an Heauenly Adamant to Drawe Thee to Christ and a Spirituall Rod to Mortifie Thy Life. Made and Set Foorth by Frauncis Kett, Doctor of Phisick*, Roger Ward, London, (1585).
410. R. Sharif, "Christians for Zion, 1600-1919", *Journal for Palestine Studies*, Volume 5, Number 3/4, (Spring-Summer, 1976), pp. 123-141.

England, which would trigger the "ingathering" of the Jews to Palestine.[411] The Old Testament instructed the Jews to enter every nation and the affluent Jews of Amsterdam no doubt recognized the benefits of gaining inroads into the affairs of England and of profiting from its wealth. Jews were famous for gathering political and economic intelligence from around the world.[412]

Regina Sharif wrote in her article "Christians for Zion, 1600-1919",

> "Nowhere in Europe has support for Zionism been as widespread and popular over the ages as in England. It was there that the idea of Jewish restoration in Palestine became prominent and developed into a doctrine that lasted well over three centuries.[1] Nahum Sokolow, the well-known Jewish historian of the Zionist movement, commented on this permanent connection between England and Zionism: 'English Christians taught the underlying principles of Jewish nationality.'[2] He expressed his gratitude to the many 'English thinkers, men of letters arid poets throughout the ages,' who championed the Zionist cause through many generations. 'For nearly three centuries Zionism was a religious as well as a political idea which great Christians and Jews, chiefly in England, handed down to posterity.'[3]
> [***] Weizmann's skills in international diplomacy and persuasion, however great they might have been, would have remained fruitless had not English culture been conditioned to Zionism long before the time of Herzl or Weizmann and had not the seeds of Zionism been sown and cultivated in England by non-Jewish Zionists long before the appearance of Herzl's *Judenstaat*."[413]

See also: Eliyahu Tal, *You Don't Have to be Jewish to be a Zionist: A Review of 400 Years of Christian Zionism*, International Forum for a United Jerusalem, Tel Aviv, (2000).

In the 1500's and continuing through the 1800's and beyond, a great many books were published in Great Britain and in America advocating: (1) the overthrow of the Pope, who was called "the Beast", and the destruction of the Catholic Church; (2) the destruction of the Turkish Empire, and of Mohammedanism; (3) the destruction of the French and German Empires; (4)

---

411. R. Sharif, "Christians for Zion, 1600-1919", *Journal for Palestine Studies*, Volume 5, Number 3/4, (Spring-Summer, 1976), pp. 123-141.

412. "The State and Prospect of the Jews", *The London Times*, (24 January 1839), p. 3. B. W. Tuchman, *Bible and Sword: England and Palestine from the Bronze Age to Balfour*, New York University Press, New York, (1956), p. 89.

413. R. Sharif, "Christians for Zion, 1600-1919", *Journal for Palestine Studies*, Volume 5, Number 3/4, (Spring-Summer, 1976), pp. 123-141, at 123-124. Sharif cites: (1) F. Kobler, *The Vision was There: Of the British Movement for the Restoration of the Jews to Palestine*, Published for the World Jewish Congress, British Section, by Lincolns-Prager, London, (1956), p. 7. (2) N. Sokolow, "Introduction", *History of Zionism, 1600-1918*, Volume 1, Longmans, Green and Co., London, (1919), p. xxvi. (3) N. Sokolow, "Introduction", *History of Zionism, 1600-1918*, Volume 1, Longmans, Green and Co., London, (1919), p. xxvii.

world war; (5) the "restoration of the Jews to Palestine", the rebuilding of the Temple and turning Jerusalem into the capital of a new world government—many of which objectives Jewish leaders accomplished through the Russian Revolution and the First World War.

There were many advocates of these beliefs, including Thomas Drake, who published *The Calling of the Jews* in 1608. Henry Finch published *The Worlds Great Restauration. Or the Calling of the Ievves and (With Them) of All the Nations and Kingdomes of the Earth, to the Faith of Christ* in 1621.[414] Manasseh ben Israel's work was translated into English as: *The Hope of Israel*, Printed by R.I. for Hannah Allen, London, (1650); and *The Great Deliverance of the Whole House of Israel: What it Truly Is, by Whom it Shall Be Performed, and in What Year... in Answer to a Book Called the Hope of Israel, Written by a Learned Jew of Amsterdam Named Menasseh ben Israel*, Printed by M.S., London, (1652). John Milton published *Paradise Regained* in 1671.[415] In 1747, John Collet published *A Treatise of the Future Restoration of the Jews and Israelites to Their Land: with Some Account of the Goodness of the Country, and Their Happy Condition There, till They Shall Be Invaded by the Turks : with Their Deliverance from All Their Enemies, When the Messiah Will Establish His Kingdom at Jerusalem, and Bring in the Last Glorious Ages*.[416] Joseph Eyre published *Observations upon the Prophecies Relating to the Restoration of the Jews: with an Appendix in Answer to the Objections of Some Late Writers* in 1771.[417] After winning an award for his work on Zionism in 1795 while a divinity student at Cambridge, Charles Jerram published *An Essay Tending to Shew the Grounds Contained in Scripture for Expecting a Future Restoration of the Jews* in 1796.[418]

Scores of such works appeared in Britain, America, and elsewhere advocating world war, the "restoration of the Jews to Palestine" and the destruction of heaven and Earth; including: G. Fletcher, *The Policy of the Turkish Empire. The First Booke*, Printed by Iohn Windet for W[illiam]

---

[414]. H. Finch, *The Worlds Great Restauration. Or the Calling of the Ievves and (With Them) of All the Nations and Kingdomes of the Earth, to the Faith of Christ*, William Gouge, London, (1621).

[415]. J. Milton, *Paradise Regained*, Printed by J.M. for John Starkey, London, (1671).

[416]. J. Collet, *A Treatise of the Future Restoration of the Jews and Israelites to Their Land: with Some Account of the Goodness of the Country, and Their Happy Condition There, till They Shall Be Invaded by the Turks : with Their Deliverance from All Their Enemies, When the Messiah Will Establish His Kingdom at Jerusalem, and Bring in the Last Glorious Ages*, Printed for J. Highmore, M. Cooper and G. Freer, London, (1747).

[417]. J. Eyre, *Observations upon the Prophecies Relating to the Restoration of the Jews: with an Appendix in Answer to the Objections of Some Late Writers*, Printed for T. Cadell, London, (1771).

[418]. C. Jerram, *An Essay Tending to Shew the Grounds Contained in Scripture for Expecting a Future Restoration of the Jews*, Printed by J. Burges and sold by W.H. Lunn, J. Deighton, and J. Nicholson, Cambridge, (1796). *See also:* C. Jerram, *A Sermon Preached at the Parish Church of St. Paul, Covent Garden, on Thursday Evening, May 7, 1829, Before the London Society for Promoting Christianity Amongst the Jews*, (1829).

S[tansby] and are to be soulde at Powles Wharfe at the signe of the Crosse Keyes, London, (1597); and *Of the Rvsse Common Wealth, Or, Maner of Gouernement by the Russe Emperour, (Commonly Called the Emperour of Moskouia): With the Manners, and Fashions of the People of That Countrey*, Thomas Charde, London, (1591); and *De literis antiquae Britanniae, regibus praesertim qui doctrinâ claruerunt, quíque Collegia Cantabrigiae fundârunt*, Ex Academiae celeberrimae typographeo, Cantabrigiae, (1633); and *Israel Redux: Or the Restauration of Israel, Exhibited in Two Short Treatises. The First Contains an Essay upon Some Probable Grounds, That the Present Tartars Near the Caspian Sea, Are the Posterity of the Ten Tribes of Israel. The Second, a Dissertation Concerning Their Ancient and Successive State, with some Scripture Evidences of Their Future Conversion, and Establishment in Their Own Land*, Printed by S. Streater for John Hancock, London, (1677); and *The English Works of Giles Fletcher, the Elder*, University of Wisconsin Press, Amsterdam, (1964). See also: T. Draxe, *The VVorldes Resurrection, or the Generall Calling of the Iewes a Familiar Commentary Vpon the Eleuenth Chapter of Saint Paul to the Romaines, According to the Sence of Scripture, and the Consent of the Most Iudicious Interpreters, Wherein Aboue Fiftie Notable Questions Are Soundly Answered, and the Particular Doctrines, Reasons and Vses of Euery Verse, Are Profitable and Plainly Deliuered*, Iohn Wright, London, (1608); and *The Earnest of Our Inheritance Together with a Description of the New Heauen and the New Earth, and a Demonstration of the Glorious Resurrection of the Bodie in the Same Substance*, George Norton, London, (1613); and *An Alarum to the Last Iudgement. Or an Exact Discourse of the Second Comming of Christ and of the Generall and Remarkeable Signes and Fore-Runners of It Past, Present, and to Come; Soundly and Soberly Handled, and Wholesomely Applyed. Wherein Diuers Deep Mysteries Are Plainly Expounded, and Sundry Curiosities Are Duely Examined, Answered and Confuted*, Matthew Law, London, (1615). See also: J. Mede, *Clauis apocalyptica ex innatis et insitis visionum characteribus eruta et demonstrata. Ad eorum usum quibus deus amorem studiúmq[ue] indiderit prophetiam illam admirandam cognoscendi scrutandíque*, T. and J. Buck, Cantabrigiæ, (1627); English translation by R. B. Cooper, *A Translation of Mede's Clavis Apocalyptica*, Rivington, London, (1833). See also: J. Archer, *The Personall Reigne of Christ upon Earth: In a Treatise Wherein Is Fully and Largely Laid Open and Proved, That Jesus Christ, Together with the Saints Shall Visibly Possesse a Monarchicall State and Kingdome in this World*, Benjamin Allen, London, (1643). See also: T. Brightman, *The Revelation of Saint John: Illustrated with Analysis and Scholions, Wherein the Fence Is Opened by the Scripture, and the Events of Things Foretold Showed by Histories, Together with a Most Comfortable Exposition of the Last and Most Difficult Part of the Prophecy of Daniel, Wherein the Restoring of the Jews, and Their Calling to the Faith of Christ, after the Utter Overthrow of Their Three Last Enemies, Is Set Forth in Lively Colours*, Printed by Thomas Stafford, Amsterdam, (1644); and *The Workes of That Famous, Reverend, and Learned Divine, Mr. Tho. Brightman viz., a Revelation of the Apocalyps, Containing an Exposition of the Whole Book of the Revelation of Saint John, Illustrated with Analysis and*

Scholions : Wherein the Sense Is Opened by the Scripture, and the Event of Things Foretold, Shewed by History : Whereunto Is Added, a Most Comfortable Exposition of the Last and Most Difficult Part of the Prophesie of Daniel : Wherein the Restoring of the Jews, and Their Calling to the Faith of Christ, after the Utter Overthrow of Their Three Last Enemies, Is Set Forth in Lively Colours : Together with a Commentary on the Whole Book of Canticles, or Song of Salomon, Printed by John Field for Samuel Cartwright, London, (1644). See also: R. J., *Compunction or Pricking of Heart with the Time, Meanes, Nature, Necessity, and Order of It, and of Conversion; with Motives, Directions, Signes, and Means of Cure of the Wounded in Heart, with Other Consequent or Concomitant Duties, Especially Self-Deniall, All of Them Gathered from the Text, Acts 2.37. And Fitted, Preached, and Applied to His Hearers at Dantzick in Pruse-land, in Ann. 1641. And Partly 1642. Being the Sum of 80. Sermons. With a Post-Script Concerning These Times, and the Sutableness of this Text and Argument to the Same, and to the Calling of the Jews. By R. J. Doctor of Divinity,* Printed by Ruth Raworth for Thomas Whitaker, and are to be sold at his shop, at the Kings Armes in Pauls Church-Yard, London, (1648). See also: S. Gott, *Novæ solymæ libri sex: sive Institutio Christiani 1. De pueritia. 2. De creatione mundi. 3. De juventute. 4. De peccato. 5. De virili ætate. 6. De redemptione hominis,* Johannis Legati, Londini, (1649); English translation: *Nova Solyma, the Ideal City; Or, Jerusalem Regained,* London, J. Murray, (1902). See also: T. Thorowgood, J. Dury, Manasseh ben Israel, *Digitus dei: Nevv Discoveryes with Sure Arguments to Prove That the Jews (A Nation) or People Lost in the World for the Space of near 200 Years, Inhabite Now in America; How They Came Thither; Their Manners, Customs, Rites and Ceremonies; the Unparallel'd Cruelty of the Spaniard to Them; and That the Americans Are of That Race. Manifested by Reason and Scripture, Which Foretell the Calling of the Jewes; and the Restitution of Them into Their Own Land, and the Bringing Back of the Ten Tribes from All the Ends and Corners of the Earth, and That Great Battell to Be Fought. With the Removall of Some Contrary Reasonings, and an Earnest Desire for Effectuall Endeavours to Make Them Christians. Whereunto Is Added an Epistolicall Discourse of Mr John Dury, with the History of Ant: Monterinos, Attested by Manasseh Ben Israell, a Chief Rabby. By Tho: Thorowgood, B:D, :* Printed for Thomas Slater, and are to be sold at his shop at the signe of the Angell in Duck-Lane, London, (1652). See also: E. Hall, *He apostasia, ho antichristos, Or, a Scriptural Discourse of the Apostasie and the Antichrist, by Way of Comment, upon the Twelve First Verses of 2 Thess. 2 under Which Are Opened Many of the Dark Prophecies of the Old Testament, Which Relate to the Calling of the Jews, and the Glorious Things to Be Affected at the Seventh Trumpet Through the World : Together with a Discourse of Slaying the Witnesses, and the Immediate Effects Thereof : Written for the Consolation of the Catholike Church, Especially the Churches of England, Scotland, and Ireland,* London, (1653). See also: E. Lane, *Look unto Jesus, Or, An Ascent to the Holy Mount to See Jesus Christ in His Glory Whereby the Active and Contemplative Believer May Have the Eyes of His Understanding More Inlightned to Behold in Some Measure the Eternity and Immutability of the Lord Jesus Christ: At the End of the Book Is an Appendix, Shewing the Certainty of the Calling of the Jews*, Printed by Thomas Roycroft for the Authour, and are to be sold by Humphrey Tuckey, and by William Taylor, London, (1663). See also: R. R., *The Restauration of the Jevves: Or, a True Relation of Their Progress and Proceedings in Order to the Regaining of Their Ancient Kingdom. Being the Substance of Several Letters viz. from Antwerp, Legorn, Florence, &c.*, A. Maxwell, London, (1665). See also: J. A. Comenius, *The Way of Light*, Hodder & Stoughton, Ltd., London, (1668/1938). See also: G. Ben Syrach, *Nevvs from the Jews, or a True Relation of a Great Prophet in the Southern Parts of Tartaria; Pretending Himself to Be Sent to Gather Together the Jews from All Parts: as Well the Ten Tribes That Have So Long Abscronded Themselves from the World; as the Known Tribes of Judah and Benjamin: Promising to Them the Restoration of the Land of Canaan, and All That They Formerly Enjoyed in the Time of King Solomon. As it Was Communicated to Rabbi Josuah Ben Eleazar, Merchant in Amsterdam, by a Letter from Adrianople. Faithfully Translated into English, by Josephus Philo-Judæus, Gent. With Allowance*, Printed for A.G., London, (1671). See also: W. Alleine, *The Mystery of the Temple and City Described in the Nine Last Chapeters of Ezekiel, Unfolded Also These Following Particulars Are Briefly Handled, 1. The Calling of the Jews, 2. The Restitution of All Things, 3. The Description of the Two Beasts, Rev. 13, 4. The Day of Judgment, and the World Perishing by Fire, 5. Some Signs of the Times When the Fall of Babylon Is Near, 6. Some Advantages Which the Knowledge of These Truths Will Afford, 7. The Conclusion of All in Some Counsels and Directions*, Printed for E. Harris: And are to be sold by T. Wall, London, (1677). See also: "Lover of His Country's Peace", *The Mystery of Ambras Merlins, Standardbearer Wolf, and Last Boar of Cornwal With Sundry Other Misterious Prophecys, Both Ancient and Modern, Plainly Unfolded in the Following Treatise, on the Signification and Portent of That Prodigious Comet, Seen by Most Part of the World, Anno 1680, with the Blazing Star Anno 1682, and the Conjunctions of Saturn and Jupiter in October Following and since : All Which Do Purport Many Sad Calamitys to Befall Most Parts of the Europian Continent in General Before the Year 1699, ... the Ruin of the House of Austria, Vienna, and the Empire of Germany : with Rome, Italy, and the Pope and Papicy, the King and Kingdom of France, with Several Other Countrys in Europe, and the Danger of an Invation in England by the Turks, and Then the Convertion of the Said Nation to the Christian Faith, Before this Present Expedition of the Turks into Hungary and Germany Be Over, Which Will Be Followed, (1) with the Calling of the Jews, (2) the Reducing of All Wayes of Religious Worship into One by Which an Universal Peace Will Ensue to All the Earth*, Printed for Benj. Billingsley, London, (1683). See also: R. Baxter, *The Glorious Kingdom of Christ, Described and Clearly Vindicated, Against the Bold Asserters of a Future Calling and Reign of the Jews, and 1000 Years Before the Conflagration. And the Asserters of the 1000 Years Kingdom after the Conflagration. Opening the Promise of the New Heaven and Earth, and the Everlastingness of Christ's Kingdom, Against Their Debasing It, Who Confined it to 1000 Years, Which with the Lord Is but as One Day*, Printed by T. Snowden, for Thomas Parkhurst at the

Bible and Three Crowns, the lower end of Cheapside, London, (1691). *See also:* "Lay Hand", *The Great Signs of the Times Giving a True Account of the Universal Change That Is Now Expected: With a Preface Concerning Prophecies, and an Introduction Wherein the Right Notion of the Calling of the Jews and the Kingdome of Christ, So Much Obscur'd, Is True and Faithfully Declar'd*, Printed for the author, and are to be sold by J. Nutt, London, (1699). *See also:* S. Willard, *The Fountain Opened, Or, the Great Gospel Priviledge of Having Christ Exhibited to Sinfull Men: Wherein Also Is Proved That There Shall Be a National Calling of the Jews from Zech. XIII, 1*, Printed by B. Green and J. Allen for Samuel Sewall, Junior, Boston in New-England, (1700). *See also:* R. Fleming, *Apocalyptical Key: an Extraordinary Discourse on the Rise and Fall of Papacy, Or, the Pouring out of the Vials, in the Revelation of St. John, Chap. XVI: Containing Predictions Respecting the Revolutions of France, the Fate of It's Monarch, the Decline of Papal Power, Together with the Fate of the Surrounding Nations, the Destruction of Mahometanism, the Calling in of the Jews, the Restoration and Consummation of All Things, &c. &c.*, Printed for G. Terry, London, (1701/1793). *See also:* S. Clarke, "The Conversion and Restoration of the Jews", *A Collection of the Promises of Scripture: or, The Christian's Inheritance*, Part 3, Section 10, American Tract Society, New York, and J. Buckland, London, (1750); and *A Discourse Concerning the Connexion of the Prophecies in the Old Testament, and the Application of Them to Christ. Being an Extract from the Sixth Edition of a Demonstration of the Being and Attributes of God, &c...*, J. Knapton, London, (1725). *See also:* W. Whiston, *An Essay on the Revelation of Saint John, So Far as Concerns the past and Present Times: To Which Are Added Two Dissertations, the One upon Mark II. 25, 26. The Other upon Matthew XXIV. And the Parallel Chapters: With a Collection of Scripture-Prophecies Relating to the Times after the Coming of the Messiah*, Cambridge: Printed at the University-Press; for B. Tooke, London, (1706); and *The Accomplishment of Scripture Prophecies: Being Eight Sermons Preach'd at the Cathedral Church of St. Paul, in the Year MDCCVII, at the Lecture Founded by the Honourable Robert Boyle Esq.: With an Appendix, to Which Is Subjoin'd a Dissertation, to Prove That Our Savior Ascended into Heaven on the Evening after His Resurrection*, Cambridge : Printed at the University-Press for B. Tooke, London, (1708); and *Historical Memoirs of the Life of Dr. Samuel Clarke Being a Supplement to Dr. Sykes's and Bishop Hoadley's Accounts. Including Certain Memoirs of Several of Dr. Clarke's Friends*, London, Fletcher Gyles, (1730); and *Memoirs of the Life and Writings of Mr. William Whiston: Containing, Memoirs of Several of His Friends Also. Written by Himself*, J. Whiston and B. White, London, (1753). *See also:* T. Burnet, *De statu mortuorum et resurgentium tractatus: adjicitur: Appendix de futurâ Judaeorum restauratione*, J. Hooke, Londini, (1727). *See also:* I. Newton, *Observations upon the Prophecies of Daniel, and the Apocalypse of St. John*, Printed by J. Darby and T. Browne and sold by J. Roberts etc., London, (1733). *See also:* T. Newton, *Dissertations on the Prophecies; Which Have Remarkably Been Fulfilled, and at this Time Are Fulfilling in the World*, William Butler, Northhampton, Massachusetts, (1746). *See also:* T. Newans, *A Key to the Prophecies of the Old and New Testaments: Shewing the Approaching Invasion of England, the Desolation of Germany ..., the Destruction of Rome, the Expulsion of the Mahometans, the Extirpation of Popery ..., the Restoration of the Jews to Their Own Land, the Rebuilding of the Temple at Jerusalem, the Fulness of the Gentiles, and the Glorious and Triumphant Estate of Christ's Church upon Earth*, London, (1747). See also: J. Collet, *A Treatise of the Future Restoration of the Jews and Israelites to Their Land: With Some Account of the Goodness of the Country, and Their Happy Condition There, till They Shall Be Invaded by the Turks : with Their Deliverance from All Their Enemies, When the Messiah Will Establish His Kingdom at Jerusalem, and Bring in the Last Glorious Ages*, J. Highmore, M. Cooper and G. Freer, London, (1747). See also: R. Clayton, *An Enquiry into the Time of the Coming of the Messiah, and the Restoration of the Jews*, Printed for J. Brindley, London, 1751); and *An Impartial Enquiry into the Time of the Coming of the Messiah, Together with an Abstract of the Evidence on Which the Belief of the Christian Religion Is Founded: In Two Letters from Robert, Lord Bishop of Clogher, to an Eminent Jew*, J. Brindley, London, (1751). See also: Archaicus, *The Rejection and Restoration of the Jews, According to Scripture Declar'd: With Indications of the Means by Which, And, Nearly, of the Time When, the Latter of These Great Events Is to Be Brought to Pass. To Which Are Added, Some Intimations That Neither Is this Time Yet Nigh at Hand, Nor Will Any Extraordinary Civil Privileges Indulg'd to That People Conduce to Accelerate, but Rather to Retard It, and for What Reasons*, R. Baldwin, London, (1753). See also: Presbyter of the Church of England, *An Explanation of Some Prophecies Contained in the Book of Daniel, Wherein the Particular Times of the Destruction of the Mahometans, and of the Restoration of the Jews, Are Pointed Out*, Printed by E. Say and sold by R. Baldwin, London, (1753). See also: W. Torrey, *A Brief Discourse Concerning Futurities or Things to Come Viz. The Next, of Second Coming of Christ. Of the Thousand Years of Chrrst's Kingdom. Of the First Resurrection. Of the New Heavens and New Earth; and of the Burning of the Old. Of the New Jerusalem. Of Gog and Magog. Of the Calling of the Jews. Of the Pouring out of the Spirit on All Flesh. Of the Greatest Battle That Ever Was, or Shall Be Fought in the World. And Many Other Things Coincident with These Things. Together with Some Useful Consideration upon the Whole Discourse*, Prince, Thomas, Publication, Printed and sold by Edes and Gill, at their printing-office, next to the prison, in Queen-Street, Boston, (1757). See also: J. Inglis, *By the Way of a Scripture Interpretation. Theism: a Prophecy: Or, Prophetical Dissertation. Predicting and Declaring the Coming of the Expected Messiah, in the Character of Lord and King; the Setting up of a National Theocracy, in the Calling of the Jews, and Redemption of the Gentile Church. Part I. Consisting of an Astro-theological Unfolding of Certain Formerly Obscure, but Highly-interesting and Capital Points of Doctrine. Adapted to the Present Crisis of Affairs*, Printed for the author by William Dunlap, Philadelphia, (1763). See also: J. Inglis and W. Dunlap, et al., *The Little Book Open [Double Dagger]: A Prophecy, Or, Prophitical Dissertation. Predicting and Declaring the Coming of the Expected Redeemer, in the Character of Lord and King; the Setting up of a National Theocracy, in the*

Calling of the Jews, and Redemption of the Gentile Church. Part I. Consisting of an Astro-Theological Unfolding of Certain Formerly Obscure, but Highly-Interesting and Capital Points of Doctrine. Adapted to the Present Crisis of Affairs, William Dunlap, Philadelphia, (1763). See also: J. Eyre, *Observations upon the Prophecies Relating to the Restoration of the Jews: with an Appendix in Answer to the Objections of Some Late Writers*, T. Cadell, London, (1771). See also: R. Hurd, *An Introduction to the Study of the Prophecies Concerning the Christian Church: And in Particular Concerning the Church of Papal Rome: in Twelve Sermons, Preached in Lincoln's-Inn-Chapel, at the Lecture of the Right Reverend William Warburton*, Thomas Ewing, Dublin, (1772). See also: C. Love, *The History of the Holy Bible Containing the Old and New Testaments, by Question and Answer, Giving, I., an Account of the Remarkable Events and Transactions of the Antideluvian and Patriarchal Ages Before and after the Flood: as Also, Several Very Curious Critical Remarks and Practical Observations upon the Lives of the Patriarchs ; II., a Minute Description of the Jews, from the Calling of Abraham to Their Settlement in the Promised Land: with Suitable Remarks upon the Messages of the Prophets Sent to That People; III., and Lastly, the History of Our Lord and Saviour Jesus Christ, and His Apostles, from the Birth of John the Baptist, to the Conclusion of the Canon of Scripture; for the Benefit of Every Real Christian*, Printed and sold by Patrick Mair, Falkirk, (1783). See also: E. W. Whitaker, *A Dissertation on the Prophecies Relating to the Final Restoration of the Jews*, J. Rivington and Sons, London, (1784). See also: J. Priestley, *Letters to the Jews: Inviting Them to an Amicable Discussion of the Evidences of Christianity*, Pearson and Rollason, Birmingham, (1787); and *Letters to the Jews; Part II: Occasioned by Mr. David Levi's Reply to the Former Letters*, Pearson and Rollason, Birmingham, (1787); and *The Evidence of the Resurrection of Jesus Considered: In a Discourse First Delivered in the Assembly-room, at Buxton, on Sunday, September 19, 1790. To Which Is Added, an Address to the Jews*, J. Thompson, Birmingham, (1791); and *An address to the Jews*, Birmingham, (1791); and *A Comparison of the Institutions of Moses with Those of the Hindoos and Other Ancient Nations With Remarks on Mr. Dupuis's Origin of All Religions, the Laws and Institutions of Moses Methodized, and an Address to the Jews on the Present State of the World and the Prophecies Relating to It*, A. Kennedy, Northumberland, Pennsylvania, (1799). See also: J. Bicheno, *A Friendly Address to the Jews: Stating the Motives to Serious Inquiry into the Cause of Their Dispersion... : To Which Is Added, a Letter to Mr. D. Levi, Containing Remarks on His Answer to Dr. Priestley's Letters to the Jews*, Buckland, London, (1787); and *The Signs of the Times, Or, the Overthrow of the Papal Tyranny in France, the Prelude of Destruction to Popery and Despotism, but of Peace to Mankind*, Carter and Wilkinson, Providence, Rhode Island, (1794); and *The Restoration of the Jews, the Crisis of All Nations, Or, an Arrangement of the Scripture Prophesies Which Relate to the Restoration of the Jews... : Drawn from the Present Situation and Apparent Tendencies of Things, Both in Christian and Mahomedan Countries*, Printed by Bye and Law, London, (1800); and *The Restoration of the Jews. The Crisis of All Nations; to Which Is Now Prefixed, a Brief History of the Jews, from Their*

First Dispersion, to the Calling of Their Grand Sanhedrim at Paris, October 6th, 1806, and an Address on the Present State of Affairs, in Europe in General, and in this Country in Particular, J. Barfield, London, (1807). See also: D. Levi and J. Priestley, *Letters to Dr. Priestley, in Answer to His Letters to the Jews, Part. II. Occasioned by Mr. David Levi's Reply to the Former Part. Also Letters 1. To Dr. Cooper, in Answer to His "One Great Argument in Favour of Christianity from "A Single Prophecy." 2. To Mr. Bicheno, 3. To Dr. Krauter... Occasioned by Their Remarks on Mr. David Levi's Answer to Dr. Priestley's First Letters to the Jews*, London, (1789). See also: R. Beere, *An Epistle to the Chief Priests and Elders of the Jews: Containing an Answer to Mr. David Levis Challenge to Christians of Every Denomination ... Predictive of the Time of the First Coming and Crucifiction of the Messiah. To Which Is Added an Investigation and Computation of the Exact Time of Their Final Restoration... Together with an Accurate Chronology of the World... Confirmed by Astronomical Observations*, D. Brewman, London, (1789). See also: "Watchman", *A Divine Call to That Highly Favoured People the Jews: Justice and Mercy Opening Now the Way for Their Restoration*, Frederick Green, Anapolis, Maryland, (1790). See also: J. A. Comenius, *The Lives, Prophecies, Visions and Revelations, of Christopher Kotterus, and Christian Poniatonia: Two Eminent Prophets in Germany ; Containing Predictions Concerning the Pope, the King of France, and the Roman Emmpire, with the Sudden Destruction of the Papal Power, the Miraculous Conversion of the Turks, the Calling in of the Jews, and the Uniting All Religions into One Universal Visible Church ; Many of Which Prophecies Being Desired by the Then King of Bohemia, Were by the Learned Comenius Presented to Him*, Printed for G. Terry, London, (1794). See also: W. Ashburnham, *Restoration of the Jews: A Poem*, London, (1794). See also: F. Wrangham, *The Restoration of the Jews: A Poem*, R. Edwards, London, (1795). See also: R. Brothers, *A Revealed Knowledge of the Prophecies and Times Book the First. Wrote under the Direction of the Lord God, and Published by His Sacred Command; it Being the First Sign of Warning for the Benefit of All Nations. Containing, with Other Great and Remarkable Things, Not Revealed to Any Other Person on Earth, the Restoration of the Hebrews to Jerusalem, by the Year 1798; under Their Revealed Prince and Prophet Richard Brothers*, Robert Campbell, Philadelphia, (1795); and *A Revealed Knowledge of the Prophecies and Times Particularly of the Present Time, the Present War, and the Prophecy Now Fulfilling. The Year of the World 5913. Book the Second. Containing, with Other Great and Remarkable Things, Not Revealed to Any Other Person on Earth, the Sudden and Perpetual Fall of the Turkish, German, and Russian Empires*, Robert Campbell, Philadelphia, (1795). See also: N. B. Halhed, *A Revealed Knowledge of the Prophecies and Times. Wrote under the Direction of the Lord God, and Published by His Sacred Command; it Being the First Sign of Warning for the Benefit of All Nations. Containing, with Other Great and Remarkable Things, Not Revealed to Any Other Person on Earth, the Restoration of the Hebrews to Jerusalem, by the Year of 1798, under Their Revealed Prince and Prophet. To Which Is Added, the Testimony of the Authenticity of the Prophecies of Richard Brothers, and of His Mission to Recall the Jews / Book the First*, Dublin, (1795); and *Testimony of the Authenticity of the Prophecies of Richard Brothers, and of His Mission to Recall the Jews*, London : Printed for H.D. Symonds, (1795). *See also:* C. Jerram, *An Essay Tending to Shew the Grounds Contained in Scripture for Expecting a Future Restoration of the Jews*, J. Burges, Cambridge, (1796). *See also:* D. Levi, *Dissertations on the Prophecies of the Old Testament: Part I Contains All Such Prophecies as Are Applicable to the Coming of the Messiah: the Restoration of the Jews, and the Resurrection of the Dead: Whether So Applied by Jews or Christians. Part Ii Contains All Such Prophecies as Are Applied to the Messiah by Christians Only, but Which Are Shewn Not to Be Applicable to the Messiah*, D. Levi, London, (1796-1800). *See also:* C. J. Ligne, *Mémoire sur les Juifs*, (1797); reprinted *Oeuvres du Prince de Ligne*, Volume 1, F. van Meenen, Bruxelles, L. Van Bakkenes, Amsterdam, (1860). *See also:* E. King, *Remarks on the Signs of the Times*, George Nicol, London, (1798); and *A Supplement to the Remarks on the Signs of the Times: With Many Additional Remarks*, George Nicol, London, (1799). *See also:* H. Kett, *History the Interpreter of Prophecy, Or, a View of Scriptural Prophecies and Their Accomplishment in the past and Present Occurrences of the World; with Conjectures Respecting Their Future Completion*, Hanwell and Parker, and J. Cooke, Oxford, (1799). *See also:* T. Witherby, *Observations on Mr. Bicheno's Book, Entitled the Restoration of the Jews, the Crisis of All Nations: Wherein the Revolutionary Tendency of That Publication Is Shown to Be Most Inimical to the Real Interest of the Jews... Together with an Inquiry Concerning Things to Come*, S. Couchman, London, (1800); and *An Attempt to Remove Prejudices Concerning the Jewish Nation: By Way of Dialogue*, Stephen Couchman, London, (1804); and *A Vindication of the Jews: By Way of Reply to the Letters Addressed to Perseverans to the English Israelite ; Humbly Submitted to the Consideration of the Missionary Society, and the London Society for Promoting Christianity among the Jews*, Stephen Couchman, London, (1809). *See also:* D. Lewis, *An Address to the Jews; Shewing the Time of Their Obtaining the Knowledge of the Messiah, and Their Restoration to the Land of God's Promise to Abraham... to Which Is Added, an Address to the Nations, Shewing the Origin of Apostacy; Their Continuance Therein; and the Time of Their Delivery Therefrom. Also, a Few Observations on the Plan of a Modern Utopia*, H. D. Symonds, London, (1800). *See also:* L. Mayer, *Restoration of the Jews: Being an Extract from an Entire New Work, Intended to Be Published by Subscription Entitled "Truth Dispelling the Clouds of Error, by the Fulfilment of the Prophecies": Addressed to the Jews*, London, (1803); and *Bonaparte the Emperor of the Gauls, Considered as the Lucifer and Gog of Isaiah and Ezekiel: And the Issue of the Present Contest Between Great Britain and France Represented According to Divine Revelation, with an Appeal to Reason on the Errors of Commentators*, C. Stower, London, (1804); and *Restoration of the Jews: Containing an Explanation of the Prophecies in the Books of Daniel and the Revelations, That Relate to the Period When Their Restoration Will Be Accomplished. With an Illustration, Applicable to the Jews, of the Two Olive Trees, and the Two Candlesticks, That Are Said to Stand Before the God of the Earth, and the Two Witnesses, Who Were to Prophesy, Clothed in Sackcloth,*

1260 days. Addressed to the Jews, London, (1806); and *Peace with France, and Catholic Emancipation: Repugnant to the Command of God*, London, (1806); and *The Important Period, and Long Wished for Revolution, Shewn to Be at Hand, When God Will Cleanse the Earth by His Judgments*, Williams & Smith, London, (1806); and *The Prophetic Mirror; Or, a Hint to England: Containing an Explanation of Prophecy That Relates to the French Nation, and the Threatened Invasion; Proving Bonaparte to Be the Beast That Arose out of the Earth, with Two Horns like a Lamb, and Spake as a Dragon, Whose Number is 666. Rev. XIII*, London, (1806); and *Bonaparte the Emporor of the French, Considered as the Lucifer and Gog of Isaiah and Ezekiel: And the Issues of the Present Contest Between Great Britain and France, Represented According to Divine Revelation with an Appeal to Reason, on Prophecy, and the Errors of Commentators... Also an Hieroglyphic Published in 1804, of the Destiny of Europe, the Fate of the German Empire, and the Fall of Russia. And a New Explanation of Daniel's Seventy Weeks*, London, (1806); and *Truth Dispelling the Clouds of Error: Containing a New Explanation of Nebuchadnezzar's Great Image and the Prophecies of Balaam, Which Relate to the Total Destruction of the Antichristian Powers, and the Annihilation of the Turkish and Persian Empires. Part I*, W. Nicholson for Williams & Smith, London, (1807); and *Death of Bonaparte, and Universal Peace: A New Explanation of Nebuchadnezzar's Great Image, and Daniel's Four Beasts*, W. Nicholson, London, (1809). See also: J. Rathbun, *A Sign, with a Looking-glass, Or, a Late Vision Opened and Explained, in the Light of the Prophecies and Revelations: In Which Is Shown, the Sudden Destruction of the Draggon, and Beast, and False-church, and the Sudden Gathering in of the Jews, into Their Own Land, and Their Final Restoration to Christ ; and the Curse Taken off from the Earth, and the Glory of the Millennium ; Also, the Sudden Second Coming of Christ, Which Will Be like the Opening of the Eyelids of the Morning to All Nations, When Every Man May Sit down under His Own Vine and Fig Tree, and None Shall Hurt Them*, Phinehas Allen, Pittsfield, (1804). See also: G. White and H. Witsius, *The Restoration of the Jews: An Extract from Herman Witsius*, Printed for Williams & Smith, by W. Heney, London, (1806). See also: Hunter, *The Rise, Fall, and Future Restoration of the Jews: To Which Are Annexed, Six Sermons, Addressed to the Seed of Abraham by Several Evangelical Ministers : Concluding with an Elaborate Discourse, by the Late Dr. Hunter, Entitled, 'The Fullness of the Gentiles Coeval with the Salvation of the Jews'*, W. Button, London, (1806). See also: G. S. Faber, *A Dissertation on the Prophecies, That Have Been Fulfilled, Are Now Fulfilling, or Will Hereafter Be Fulfilled, Relative to the Great Period of 1260 Years; the Papal and Mohammedan Apostasies: the Tyrannical Reign of Antichrist, or the Infidel Power; and the Restoration of the Jews*, Printed for F.C. and J. Rivington, London, (1806). See also: *Sanhedrin Hadashah, and, Causes and Consequences of the French Emperor's Conduct Towards the Jews: Including Official Documents and the Final Decisions of the Grand Sanhedrin : a Sketch of the Jewish History since Their Dispersion, Their Recent Improvements in the Sciences and the Polite Literature upon the Continent : and the Sentiments of Their Principal Rabbins, Fairly Stated and*

Compared with Some Eminent Christian Writers, upon the Restoration, the Rebuilding of the Temple, the Millennium, &C. ; *with Considerations on the Question: "Whether There Is Any Thing in the Prophetic Records That Seems to Point Particularly to England?"*, Printed by Day & co., for M. Jones, London, (1807). See also: W. Ettrick, *The Second Exodus; Or, Reflections on the Prophecies, Relating to the Rise, —Fall, —and Perdition of the Great Roman Beast of the 1260 Years and His Last Head, and Their Connection with the Long Captivity and Approaching Restoration of the Jews*, J. Graham, Sunderland, England, (1814). See also: J. M'Donald, *Isaiah's Message to the American Nation: A New Translation of Isaiah, Chapter XVIII, with Notes Critical and Explanatory: A Remarkable Prophecy, Respecting the Restoration of the Jews, Aided by the American Nation, with a Universal Summons to the Battle of Armageddon, and a Description of That Solemn Scene*, Printed by E. & E. Hosford, Albany, New York, (1814). See also: C. Maitland, *A Brief and Connected View of Prophecy: Being an Exposition of the Second, Seventh, and Eighth Chapters of the Prophecy of Daniel Together with the Sixteenth Chapter of Revelation : to Which Are Added, Some Observations Respecting the Period and Manner of the Restoration of the Jews*, J. Hatchard, London, (1814). See also: M. M. Noah, *Call to America to Build Zion*, Arno Press, New York, (1814/1977); and *Discourse Delivered at the Consecration of the Synagogue of [K. K. She`erit Yisra`el] in the City of New-York on Friday, the 10th of Nisan, 5578, Corresponding with the 17th of April, 1818*, Printed by C.S. Van Winkle, New-York, (1818); and *Discourse on the Evidences of the American Indians Being the Descendants of the Lost Tribes of Israel: Delivered Before the Mercantile Library Association*, Clinton Hall, J. Van Norden, New York, (1837); and *Discourse on the Restoration of the Jews: Delivered at the Tabernacle, Oct. 28 and Dec. 2., 1844*, Harper, New York, (1845); and *The Jews, Judea, and Christianity: A Discourse on the Restoration of the Jews*, Hugh Hughes, London, (1849). See also: W. Ettrick, *The Season and Time, Or, an Exposition of the Prophecies Which Relate to the Two Periods of Daniel Subsequent to the 1260 Years Now Recenter Expired: Being the Time of the Seventh Trumpet... Together with Remarks upon the Revolutionary Anti-Christ Proposed by Bishop Horsley and the Rev. G. S. Faber*, Longman, Hurst, Rees, Orne, and Brown, London, (1816). See also: N. L. Moore, *The Restoration of Sodom, Samaria and Judah, Or, the Return of the Jews to Their Former Estate: A Sermon*, Printed by John B. Johnson, Hamilton, New York, (1817). See also: "Citizen of Baltimore", *The Return of the Jews, and the Second Advent of Our Lord, Proved to Be a Scripture Doctrine*, Printed by Richard J. Matchett, Baltimore, (1817). See also: W. Witherby and J. Eyre, *A Review of Scripture in Testimony of the Truth of the Second Advent, the First Resurrection and the Millennium*, W. Marchant for Longman, Hurst, Rees, Orme, and Brown, London, (1818). See also: H. McNeile, *The Church of Rome the Apostasy, and the Pope the Man of Sin and Son of Perdition. With an Appendix*, Presbyterian Board of Publication, Philadelphia, (1818/1841); and *Popular Lectures on the Prophecies Relative to the Jewish Nation*, J. Hatchard, London, (1830); and *The Relative Position Occupied by the Jewish Nation in the Revealed Purposes of*

Jehovah, Towards Our World: A Sermon Preached on Behalf of the Philo-Judaean Society at the Church of St. Clement Danes, on Tuesday Evening, April 27th, 1830, Hatchard & Son, London, (1830); and *Nationalism in Religion: A Speech Delivered at the Annual Meeting of the Protestant Association, Held in Exeter Hall, on Wednesday, May 8, 1839*, (1839); and *Jezebel: A Type of Popery: A Speech*, New Irish Pulpit Office, Dublin, (1840); and *The Papal Antichrist. Church of Rome Proved to Have the Marks of Antichrist: A Speech, March 7, 1843*, Hatchards, London, (1843); and *A Sermon Preached at the Parish Church of the United Parishes of Christ Church, Newgate-Street, at St. Leonard, Foster-Lane, on Thursday, May 7, 1846 Before the London Society for Promoting Christianity Amongst the Jews*, London Society, London, (1846); and *The Covenants Distinguished: A Sermon, on the Restoration of the Jews, Preached in the Parish Church of St. George's, Bloomsbury, on Thursday, the 22d of November, 1849, and Published by Request*, J. Hatchard and Son, London, Arthur Newling, Liverpool, (1849); and *The Rev. Dr. M'Neile's Speech on the Papal Aggression: Delivered at Exeter Hall, on Tuesday, December 17th, 1850*, C. Westerton, London, (1850); and *The Jews and Judaism. A Lecture by the Rev. Hugh M'Neile, D.d., St.paul's, Liverpool, Delivered Before the Young Men's Christian Association, in Exeter Hall, February 14, 1854*, James Nisbet, London, (1854); and *The English Reformation, a Re-Assertion of Primitive Christianity. A Sermon, Preached in Christ Church, Newgate Street, on the 17th of November, 1858, the Tercentenary Commemoration of the Accession of Queen Elizabeth*, A. Holden, Liverpool, (1858). See also: P. Fisk, L. Parsons, et al., *Holy Land Missions and Missionaries*, Arno Press, New York, (1819-1977). See also: P. Fisk and L. Parsons, *Sermons of Rev. Messrs. Fisk & Parsons, Just Before Their Departure on the Palestine Mission*, Samuel T. Armstrong, Boston, (1819). See also: L. Parsons, *The Dereliction and Restoration of the Jews: A Sermon Preached in Park Street Church, Boston, Sabbath, Oct. 31, 1819, Just Before the Departure of the Palestine Mission*, S. T. Armstrong, Boston, (1819). See also: A. Power, *An Appeal to the Jewish Nation in Particular, and the Infidel in General: With an Endeavour to Prove the Pyramid to Be the Ensign or Beacon of Isaiah, for the Call and Restoration of all Jews, &c.*, G. & W.B. Whittaker, London, (1822). See also: "Jerusalem", *An Account of the Siege and Destruction of Jerusalem; with Some Observations on the Present State of the Jews, and on Their Future Restoration to Former Privileges*, Edmond Barber, Cork, Brown-Street, (1822). See also: J. P. Haven, *Israel's Advocate, Or, the Restoration of the Jews Contemplated and Urged*, Serial Publication Published for the American Society for Meliorating the Condition of the Jews by John P. Haven, New York, (1823-1827). See also: J. Wilson, *A Dissertation on the Future Restoration of the Jews, the Overthrow of the Papal Civil Authority, and on Other Interesting Events of Prophecy, in Two Sections*, H. H. Brown, Providence, Rhode Island, (1828). See also: J. Burridge, *The Budget of Truth: Relative to the Present Aspect of Affairs in the Religious and the Political World, Especially to the Existing State of Christendom: To Which Are Added, Observations on the Restoration of the Jews, and "The Holy Alliance," Being a Development of the Prophecies of Daniel & John, with an Appendix Containing*

Curious Official Correspondence, &c., London, (1830). *See also:* J. Tyso, *An Inquiry after Prophetic Truth Relative to the Restoration of the Jews and the Millenium: Containing a Map of the Countries to Be Possessed by the Restored Tribes, and Ground Plans of the New City and Temple to Be Built, According to the Patterns Showed to Ezekiel in the Mount: Addressed to the Jews and Gentiles*, Holdsworth and Ball, London, (1831). *See also:* G. H. Wood, *The Believer's Guide to the Study of Unfulfilled Prophecy. Containing the Scripture Testimony Respecting the Gentile Apostacy, the Second Advent of Christ in Judgment, His Personal Reign on Earth with All His Saints, the Restoration of the Jews, the Restitution of All Things, Hades, or the Intermediate State of Departed Spririts, and Other Important Subjects, with an Appendix, Containing the Testimony of the Fathers, Reformers, &C. To the Truth of the above Doctrines*, J. Nisbet, London, (1831). *See also:* B. Disraeli, *The Wondrous Tale of Alroy. The Rise of Iskander*, Saunders and Otley, London, (1833); and *Tancred, or, The New Crusade*, Henry Colburn, London, (1847); and *Die jüdische Frage in der orientalischen Frage*, Wien, (1877); reproduced in: N. M. Gelber, *Tokhnit ha-medinah ha-Yehudit le-Lord Bikonsfild (Binyamin Deyizra'eli)*, Ts. Lainman, Tel-Aviv, (1946), pp. 61-91; also attributed to Disreali in: N. H. Frankel and T. H. Gaster, *Unknown Documents on the Jewish Question: Disraeli's Plan for a Jewish State (1877)*, The Schlesinger Pub. Co., Baltimore, (1947); on attribution to Disraeli *see:* C. Roth, *Benjamin Disraeli, Earl of Beaconsfield*, Philosophical Library, New York, (1952). *See also:* J. Gregg, *Elisama; or, The Captivity and Restoration of the Jews: Including the Period of Their History from the Year 606 to 408, B.C.*, American Sunday-School Union, Philadelphia, (1835). *See also: Remarks on the Expatriation of the Jews from Judea: and the Probability of Their Restoration to That Country*, B. Fellowes, London, (1836). *See also:* P. Colby, *The Conversion and Restoration of the Jews: A Sermon Delivered at Randolph, Mass., Before the Palestine Missionary Society, June 17, 1835*, (1836). *See also:* J. S. C. F. Frey, *Judah and Israel, or, The Restoration and Conversion of the Jews and Ten Tribes: To Which Is Added Essays on the Passover*, T. Ward & Co., London, (1837). *See also:* E. Bickersteth, *The Way of Christ Prepared: An Address Both to Christians and Jews, on the Duty and Blessedness of Removing Their Mutual Stumbling-Blocks: Being the Substance of a Sermon Preached to the Jews in the Episcopal Jews' Chapel, in London, March 12, and at St. Augustines, in Liverpool, Sept. 27, 1837*, Seeley & Co., London, (1837); and *The Time to Favour Zion, Or, an Appeal to the Gentile Churches in Behalf of the Jews: Being the Substance of Four Sermons Preached in the Episcopal Churches of St. James, Trinity, and St. John, in Edinburgh, on Whit-Sunday, May 19, 1839, and the Following Wednesday ; with the Proceedings on the Formation of the Edinburgh Auxiliary to the London Society for Promoting Christianity Among the Jews*, John Lindsay, Edinburgh, (1839); and *The Future Destiny of Israel*, O. Rogers, Philadelphia, (1840); and *The Restoration of the Jews to Their Own Land: In Connection with Their Future Conversion and the Final Blessedness of Our Earth*, R. B. Seeley and W. Burnside, London, (1841); and *Scriptural Studies Relating to the Conversion and Restoration of the Jews*, London

Society's Office, London, (1843); and *The Way of the Jewish People to Be Prepared: A Sermon, Preached at the Parish Church of St. Clement Danes, Strand, on Tuesday Evening, May 8, 1834, Before the London Society for Promoting Christianity Amongst the Jews*, Sold at the London Society's House, London (1844); and *The Mind of Christ Respecting the Jews*, H. B. Pratt, Boston, (1845); and *Israel's Sins, and Israel's Hopes: Being Lectures Delivered During Lent, 1846, at St. George's, Bloombury*, James Nisbet and Co., London, (1846); and *The Forty-Eight Report of the London Society for Promoting Christianity among the Jews: With an Appendix Containing a List of Subscribers and Benefactors, and a Statement of Accounts to March 31, 1856; to Which Is Prefixed the Annual Sermon Preached Before the Society on May 8, 1856, at the Church of St. Dunstan-in-the-West, Fleet Street*, London Society for Promoting Christianity amongst the Jews, London, (1856). See also: A. McCaul, *The Conversion and Restoration of the Jews: Two Sermons, Preached Before the University of Dublin*, B. Wertheim, London, (1837); and *Equality of Jew and Gentile in the New Testament Dispensation: A Sermon Preached at the Parish Church of St. Clement Danes, Strand, on Thursday Evening May 2, 1833, Before the London Society for Promoting Christianity Amongst the Jews*, B. Wertheim, London, (1838); and *The Conversion and Restoration of the Jews: A Lecture Delivered on Tuesday Evening October 28 1845*, J. Nisbet London, (1845); and *New Testament Evidence to Prove That the Jews Are to Be Restored to the Land of Israel*, Sold at the London Society's House, London, (1850). See also: A. C. L. Crawford, a. k. a. Lord Lindsay, "Letters on Egypt, Edom, and the Holy Land", *The Quarterly Review*, Volume 125, (December, 1838), pp. 166-192. See also: W. Aldis, *The Holy Prophecies, Visions and Life of the Prophet Enoch: Quoted by Saint Jude's Epistle on Christ's Millennium Reign. Introduced by an Epistle on Church Union, for the Jews' Conversion, and Restoration of the Twelve Tribes of Israel. Preached to Vast Multitudes in England and Scotland*, R. Menzies, Edinburgh, (1839). See also:"Restoration of the Jews", *The New-Yorker: A Weekly Journal of Literature, Politics and General Intelligence* (H.Greeley & Co., New York), Volume 9, Number 13 (13 June 1840), pp. 196-197. See also: J. Litch, *An Address to the Clergy on the near Approach of the Glorious, Everlasting Kingdom of God on Earth: As Indicated by the Word of God, the History of the World, Signs of the Present Times, the Restoration of the Jews, &c.*, Dow & Jackson, Boston, (1840). See also: J. W. Brooks, *The Testimony of Prophecy Concerning the Conversion of the Gentiles and the Restoration of the Jews: An Address Delivered to the Clergy of Bath and its Vicinity, and the Members of the Bath and East Somerset Auxiliary Society for Promoting Christianity Amongst the Jews, Assembled at Breakfast at Bath, Preparatory to the Anniversary Meeting, April 12, 1842*, Printed for the Society, by George Wood & Sons, Bath, (1842). See also: C. Elizabeth, *Judah's Lion*, M. W. Dodd, New York, (1843). See also: R. H. Herschell, *The National Restoration of the Jews to Their Fatherland, and Consequent Fulfilment of the Promise to the Patriarchs. A Sermon*, London, (1843). See also: O. Bacheler, *Restoration and Conversion of the Jews*, Potter, Pawtucket, (1843). See also: A. Keith, *The Land of Israel, According to the Covenant with Abraham, with Isaac, and with Jacob*, William Whyte, Edinburgh, (1843); and *Isaiah as It Is: Or, Judah and Jerusalem the Subjects of Isaiah's Prophesying*, William Whyte and Co., Edinburgh, (1850). *See also:* G. Bush, *The Valley of Vision, Or, the Dry Bones of Israel Revived: An Attempted Proof from Ezekiel, Chap. XXXVII. 1-14 of the Restoration and Conversion of the Jews*, Saxton & Miles, New York, (1844). *See also:* Abram-François Pétavel, *La fille de Sion, ou, le rétablissement d'Israel: Poème en sept chants, avec notes et éclaircissemens Bibliques*, Chez Gerster, Neuchatel, (1844); and Israël peuple de l'Avenir: Discours prononc'e a l'assembl'ee g'en'erale des Chr'etiens 'evang'eliques de tout pays, à Paris, Librairie de Grassart, Paris, (1861). *See also:* L. Gaussen, *Geneva and Jerusalem. The Gospel at Length Preached to the Jews, and Their Restoration at Hand. A Discourse Delivered at a Missionary Meeting at Geneva, March 12, 1843*, W. H. Dalton, London, (1844). *See also:* J. L. Rhees, *A Scriptural View of the Restoration of the Jews, the Second Advent of the Lord Jesus and Some of the Leading Circumstances of That Glorious Event*, King & Baird, Philadelphia, (1844). *See also:* L. Gaussen, *Geneva and Jerusalem. The Gospel at Length Preached to the Jews, and Their Restoration at Hand. A Discourse Delivered at a Missionary Meeting at Geneva, March 12, 1843*, W.H. Dalton, London, (1844). *See also:* E. Winchester, H. Ballou, *et al., Select Theological Library: Containing Valuable Publications Principally Treating of the Doctrine of Universal Salvation*, Gihon, Fairchild, Philadelphia, (1844). *See also:* S. A. Bradshaw, *A Tract for the Times, Being a Plea for the Jews*, (1844); and *Modus Operandi in Political, Social, and Moral Forecast Concerning the East*, (1884). *See also:* G. Gawler, *Tranquillization of Syria and the East: Observations and Practical Suggestions, in Furtherance of the Establishment of Jewish Colonies in Palestine, the Most Sober and Sensible Remedy for the Miseries of Asiatic Turkey*, T. & W. Boone, London, (1845); and *The Emancipation of the Jews Indispensable for the Maintenance of the Protestant Profession of the Empire; and, in Other Respects, Most Entitled to the Support of the British Nation*, Boone, London, (1847); and *Syria and Its near Prospects: The Substance of an Address Delivered in the Young Men's Christian Association Lecture Room, Derby, on Tuesday, 25th January, 1853. With an Appendix*, Hamilton, Adams, London, (1853). *See also:* R. W. Johnson, *The World Enlightened by the Restoration of Judah's Palace: A Sermon Preached on the 9th of March, 1845, at St. Anne's Chapel, Wandsworth, Surrey, in Behalf of the Society for Promoting Christianity among the Jews*, Simpkin and Marshall, London, (1845). *See also:* P. Fairbairn, *The Typology of Scripture; Or, the Doctrine of Types Investigated in its Principles, and Applied to the Explanation of the Earlier Revelations of God, Considered as Preparatory Exhibitions of the Leading Truths of the Gospel. With an Appendix on the Restoration of the Jews*, T. Clark, Edinburgh, (1845). *See also:* S. Hawley, *The Fulness of the Jews: The Restoration of the Jews and Subsequent Probation to the Gentiles Demonstrated from Romans Eleventh*, H. B. Pratt, Boston, (1845). *See also:* L. M. Auerbach, *Claims of the Jews in Two Parts: I. Claims of the Jews on Christians and Their Obligations to the Jews, a Discourse Delivered on 25th, Dec. 1845 in the City Hall, Glasgow at the Request of Christians Who Seek the Good of God's Ancient*

People; Ii. The True Nature and Character of the Returning Exiles the House of Israel from the Land of Strangers to Their Fatherland and Second Advent, Reign, and Personal Ministry of the Lord Jesus Christ on Earth over the House of Israel in Their Fatherland with a Few Hebrew Anthems Translated into English, as Relating to the Restoration of Israel, Glasgow, (1846). See also: J. Thomas, *Elpis Israel: A Book for the Times: Being an Exposition of the Kingdom of God ; with Reference to the Time of the End, and the Age to Come*, London, (1849); and *The Coming Struggle among the Nations of the Earth, Or, the Political Events of the Next Fifteen Years, Described in Accourdance with Prophecies in Ezekiel, Daniel, and the Apocalypse: Showing Also the Important Position Britain Will Occupy During, and at the End of, the Awful Conflict*, T. Maclear, Toronto, (1853); and *Anatolia: Or Russia Triumphant and Europe Chained: Being an Exposition of Prophecy: Showing the Inevitable Fall of the French and Ottoman Empires: The Occupation of Egypt and the Holy Land by the British... : And Consequent Establishment of the Kingdom of Israel*, Mott Haven, New York, (1854); and *Phanerosis: An Exposition of the Doctrine of the Old and New Testaments, Concerning the Manifestation of the Invisible Eternal God in Human Nature : Being Alike Subversive of Jewish Rabbinical Tradition and the Theology of Romish and Protestant Sectarianism*, R. Roberts, Birmingham, (1869); and *Destiny of the British Empire, as Revealed in the Scriptures*, G. J. Stevenson, London, (1871). See also: A. G. H. Hollingsworth, *The Holy Land Restored; Or, an Examination of the Prophetic Evidence for the Restitution of Palestine to the Jews, in Twelve Dissertations*, Seeleys, London, (1849); and *Remarks upon the Present Condition and Future Prospects on the Jews in Palestine and the Duty of England to That Nation*, Seeleys, London, (1853). See also: W. Ashburnham, *The Restoration of the Jews, and Other Poems*, R. Bentley, London, (1849). See also: W. W. Ewbank, *The National Restoration of the Jews to Palestine Repugnant to the Word of God : A Speech, Delivered... in Liverpool at the Anniversary Meeting of the Auxiliary Society for Promoting Christianity amongst the Jews, Oct. 21, 1849*, Deighton and Laughton, Liverpool, (1849). See also: W. W. Ewbank and H. M. Villiers, *A Distinction Without a Difference: a Letter to the Hon. & Rev. H. Montagu Villiers, M.a., Rector of St. George's Bloomsbury, on a Sermon Lately Preached in His Church, in Favour of the Restoration of the Jews, and Entitled, "The Covenants Distinguished."*, Deighton and Laughton, Liverpool, F. and J. Rivington, London, (1850). See also: W. H. Johnstone, *Israel After the Flesh: The Judaism of the Bible, Separated from its Spiritual Religion*, John W. Parker, London, (1850); and *Israel in the World: Or, the Mission of the Hebrews to the Great Military Monarchies*, J. F. Shaw, London, J. Menzies, Edinburgh, J. Robertson, Dublin, (1854). See also: B. Musolino, *Gerusalemme ed il popolo ebreo*, La Rassegna mensile d'Israel, Roma, (1851/1951). See also: E. Avery, *A Few Thoughts Taken from the Word of God, In Favor of Christ's Body Being of a Divine Nature, He Being the Son of God and Not the Eternal Father. The End of Idolatry and the Restoration of the Jews* , (1851). See also: S. Lewis, *The Restoration of the Jews, with the Political Destiny of the Nations of the Earth, as Foretold in the Prophecies of Scripture*, J.S. Redfield, New York, (1851). See *also:* J. Wright, *Christianity and Commerce the Natural Results of the Geographical Progression of Railways; Or, a Treatise on the Advantage of the Universal Extension of Railways in Our Colonies and Other Countries, and the Probability of Increased National Intercommunication Leading to the Early Restoration of the Land of Promise to the Jews*, Dolman, London, (*ca.* 1850). *See also:* S. M. M., *Remarks on the Prophecies Relating to the Restoration of the Jews*, W.E. Painter, London, (1852). *See also:* D. D. Buck, *An Original Harmony and Exposition of the Twenty-fourth Chapter of Matthew: and the Parallel Passages in Mark and Luke, Comprising a Review of the Common Figurative Theories of Interpretation, with a Particular Examination of the Principal Passages Relating to the Second Coming of Christ, the End of the World, the New Creation, the Millennium, the Resurrection, the Judgment, the Conversion and Restoration of the Jews, the Final Gathering of the Elect, etc., etc.*, Henry W. Derby, Cincinnati, (1853); and *Our Lord's Great Prophecy, and its Parallels Throughout the Bible, Harmonized and Expounded: Comprising a Review of the Common Figurative Theories of Interpretation. With a Particular Examination of the Principal Passages Relating to the Second Coming of Christ, the End of the World, the New Creation, the Millennium, the Resurrection, the Judgment, the Conversion and Restoration of the Jews, and a Synopsis of Josephus' History of the Jewish War*, Miller, Orton & Mulligan, New York and Auburn, (1856). *See also:* R. Browning, *Holy-Cross Day: on Which the Jews Were Forced to Attend an Annual Christian Sermon in Rome*, Poem of 1855 reproduced in many of Browning's works. *See also: Expected Restoration of the Jews; and the Millennium: Being the Seventh Lecture of a View of the Scripture Revelations Concerning a Future State*, J.W.Parker, London, (1859). *See also:* E. Hanes, *The Observer of the Signs of the Times, Including the Final Restoration of the Jews and the Messiah's Reign*, Pierce, Armstrong Co., Pennsylvania, (1860). *See also:* E. Laharanne, *La nouvelle question d'Orient: Empires d'Egypte et d'Arabie. Reconstitution de la nationalité juive*, E. Dentu, Paris, (1860). *See also:* J. C. M'Causland, *The Hope of Israel; Or, the Testimony of Scripture to the National Restoration and Conversion of the Jews*, Hodges, Smith & Co., Dublin, (1860). *See also:* R. Raine, *The Restoration of the Jews: And the Duties of English Churchmen in That Respect*, London, (1860). *See also:* D. Brown, *The Restoration of the Jews: The History, Principles, and Bearings of the Question*, A. Strahan & Co., Edinburgh, (1861). *See also:* E. B. Eaton, *The Signs of the Times, or What Things Are Coming on the Earth: The Downfall of Monarchy in Europe, the Restoration of the Jews, Second Advent of Christ-Jesus the Messiah, the Millenium, the Whole World a Republican Comm-Union of Continental and Adjacent Insular Unions of States*, R.J. Trumbull, San Francisco, (1868). *See also:* S. Henn, *The Return of the Jews: Or, The restoration of Israel*, Worcestershire, (*ca.* 1870). *See also:* E. R. Talbot, *The Mystery of the Jew, as Revealed by St. Paul in Romans XI.; Being an Expository Paraphrase of the Scope and Argument of the Chapter, with Four Lectures on the Leading Features of the Revelation as to the Future National Restoration and Conversion of the Jews. To Which Is Added, a Refutation of the Theory as to the Identity of the English Nation with the Lost House of Israel*, W. Macintosh,

London, (1872). *See also:* C. Warren, *The Land of Promise: Or, Turkeys Guarantee*, George Bell & Sons, London, (1875). *See also:* G. Eliot (Mary Ann Evans), *Daniel Deronda*, William Blackwood and Sons, Edinburgh, London, (1876). *See also:* L. Glueckstein, *The Eastern Question and the Jews*, P. Vallentine, London, (1876). *See also:* C. H. Spurgeon, *The Restoration and Conversion of the Jews*, Sovereign Grace Advent Testimony, Chelmsford, Essex. *See also:* Philadelphos, *The Coming Trouble: Certain Fate of Turkey ; the World's Tribulation; and Time of the End, Or, the Eastern Question and the Turkish Revolution Viewed in the Light of Prophecy, Showing the Certain Fate of the Turkish Empire, the Return of the Jews, the Destruction of the Papacy*, J.G. Berger, London. *See also:* H. Folbigg, *Millennial Glory, Or, the Doom of Turkey and the Battle of the Nations: The Restoration of the Jews, &c.*, London, (1877). *See also:* J. Neil, *Palestine Re-Peopled: Or, Scattered Israel's Gathering, a Sign of the Times*, J. Nisbet, London, (1877). *See also:* R. Roberts, *Prophecy and the Eastern Question: Being an Exhibition of the Light Shed by the Scriptures of Truth on the Matters Involved in the Crisis That Has Arrived in Eastern Affairs, Showing the Approaching Fall of the Ottoman Empire, War Between England and Russia; the Settlement of the Jews in Syria under British Protectorate*, F. Pitman, London, (1877). *See also:* E. Cazalet, *The Eastern Question: An Address to Working Men*, Edward Stanford, London, (1878); and *The Berlin Congress and the Anglo-Turkish Convention*, Edward Stanford, London, (1878); and *England's Policy in the East: Our Relations with Russia and the Future of Syria*, Edward Stanford, London, (1879). *See also:* J. P. Henderson, *The Destiny of Russia as Foretold by God's Prophets: Together with an Outline of the Future Movements and Destiny of England, Germany, Persia, Africa, and the Jews*, Thomas Wilson, Chicago, (1878). *See also:* L. Oliphant, *The Land of Gilead, with Excursions in the Lebanon*, W. Blackwood and Sons, Edinburgh, London, (1880). *See also:* A. Cairns, *The Jews: Their Fall and Restoration: Two Discourses, Preached in Chalmer's Church, on September the 3rd, 1854, in Behalf of the Suffering Jews of Palestine*, Hutchinson, Melbourne, (1881). *See also:* T. H. Dawson, *The Restoration of the Jews at the Second Coming of Christ: A Lecture*, Bosqui Engraving & Print. Co., San Francisco, (1885). *See also:* C. W. Meiter, *The Restoration of the Jews, and the Re-Building of King Solomon's Temple*, London, (1887). *See also:* A. W. Miller, *The Restoration of the Jews*, Constitution Pub. Co., Atlanta, (1887). *See also:* W. E. Blackstone, *Palestine for the Jews*, W. Blackstone, Oak Park, Illinois, (1891); and *Christian Protagonists for Jewish Restoration*, Arno Press, New York, (1891/ 1977). *See also:* A. C. Tris, *The Restoration of Israel, the Jews in Canaan, Jehovah Jesus, Their King: A Word to All*, Iowa Print. Co., Des Moines, (1895). *See also:* B. H. Charles, *Lectures on Prophecy: An Exposition of Certain Scriptures with Reference to the History and End of the Papacy; the Restoration of the Jews to Palestine, Their Repentance and Enlargement under the Reign of the Son of David; and the New State in the Millennium*, Fleming H. Revell Company, New York, (1897). *See also:* Cheiro, a. k. a. Count Louis Hamon, *Cheiro's World Predictions: the Fate of Europe, the Future of the U.S.A., the Coming War of Nations, the Restoration of the Jews*, The London Pub. Co.,

London, (1928).

Jewish forces in England who wanted to destroy Catholicism and attack the Pope and the Turks in order to "restore the Jews to Palestine" fabricated prophecies meant to win converts to their cause. In 1641, a pamphlet appeared in England purporting to be the prophecies of one Ursula Shipton, a. k. a. Mother Shipton, a. k. a. Agatha Shipton, a. k. a. Ursula Sontheil (ca. 1488-1561). This six page pamphlet entitled *The Prophesie of Mother Shipton in the Raigne of King Henry the Eigth Foretelling the Death of Cardinall Wolsey, the Lord Percy and Others, as Also What Should Happen in Insuing Times*[419] printed several statements of fact in 1641, which purported to be predictions of events yet to occur in Shipton's lifetime, but which had already occurred by 1641. There were no extant records proving that any such woman as "Mother Shipton" ever existed.

The pamphlet was political propaganda issued by those who wished to rid England of Catholicism and justify revolution and murder. It was so successful, that new prophecies allegedly written by "Mother Shipton" began to appear referring to the Pope, the Turks and the "calling of the Jews".[420] "Mother Shipton" predicted terrible wars, which had not yet occurred when these new prophecies appeared, but which had been in the plans of the Protestants who would overthrow the Pope and take Palestine from the Turks in order to give it to the Jews. Numerous later and expanded editions appeared. In 1862, Charles Hindley[421] lent greater credibility to the hoax by adding passages about

---

[419]. "Mother Shipton", *The Prophesie of Mother Shipton in the Raigne of King Henry the Eigth Foretelling the Death of Cardinall Wolsey, the Lord Percy and Others, as Also What Should Happen in Insuing Times*, Printed for Richard Lounds, London, (1641).

[420]. U. Shipton, *The Second Part of Mother Shiptons Prophecies: With Sixteen Others... Not Onely Concerning the Kingdome of England, but Also of the Turks Invading Germany, the Downfall of the Pope, and the Calling of the Jews*, Printed for Joshua Coniers in the Long-Walke near Christ-church hospital, London, (1651). *See also:* R. Head, *The Life and Death of the Famous Mother Shipton; Containing, an Account of Her Strange Birth, and the Most Important Passages of Her Life; Also Her Prophecies Now Newly Collected and Explained, and Illustrative of Some of the Most Wonderful Events That Have Happened, or Are to Come to Pass. Taken from a Very Scarce Copy*, Published Upwards of Two Hundred Years Since, Dean and Munday, London. *See also:* R. Head, *The Life and Death of Mother Shipton. Being not only a true Account of her Strange BIRTH, and most Important Passages of her LIFE, but also of her Prophesies: Now newly Collected. and Historically Experienced, from the time of her Birth, in the Reign of KING HENRY the VII, until this present year 1684, Containing the most Important Passages of State during the reign of these Kings and Queens of England following, viz. Henry the VIII. King James. Edward the VI. King Charles the I. Queen Mary. King Charles the II. Queen Elizabeth. Whom God Preserve. Strangly Preserved amongst other writings belonging to an Old Monastry in York-shire, and now published for the Information of Posterity. To which are added some other Prophesies yet unfulfil'd. As also Mr. Folwells's Predictions concerning the Turk, Pope, and French King, With Reflections thereupon*, Printed for Benj. Harris, at the Stationers-Armes and Anchor under the Piazza of the Royal Exchange, London, (1684).

[421]. C. Hindley, *The Life, Prophecies, and Death of the Famous Mother Shipton :*

machines, which did not exist in the period of 1488-1561, but which had since been invented. He also infamously added the prediction that the world would end in 1881, but later admitted that these additions were the products of his own imagination.[422]

8.7.3 As a Good Cabalist Jew, David Hartley Conditions Christians to Welcome Martyrdom for the Sake of the Jews

Though he came from humble beginnings, David Hartley was well-connected and had married into the immensely wealthy family of his second wife Elizabeth Packer in 1735. His plea for the destruction of the Christian Temple of Europe—principally *Roman* Catholicism, and the diaspora of Christendom—smack of revenge for the Jewish Diaspora brought on by the Romans. It is amazing that some Christians, to this day, are gullible enough to destroy themselves and humanity for the sake of ancient Jewish prophecies, for the sake of modern Jews. Their leaders are well paid.

In an odd twist on the Crusader culture of the English, Hartley tried to make one feel un-Christian if one did not support world revolution, Zionism and Jewish world rule after the intentional destruction of Christendom. Anti-Semitic Christian Zionists worked the other end of the political spectrum, but issued the same ultimate message, *i. e.* they promoted world revolution, Zionism and the destruction of Christendom.[423] It is interesting to note that the founder of Protestantism—the founder of the Gentile movement to destroy Catholicism and label the Pope the "Beast of the Apocalypse"—was an expressed philo-Semite, Martin Luther, a "Reformer" who appeared to seek the cooperation of the Jews to end the religious hegemony of Catholicism—Luther who had published *That Jesus Christ was Born a Jew* in 1523.[424] Long after creating a divide in European Christians, Martin Luther forwarded the Zionist agenda by taking an anti-Semitic stand. He published *On the Jews and Their Lies* in 1543.[425] Luther, with his close contacts with the Jewish community, may well have been an agent for Zionists and Protestantism was a device to divide and destroy Christendom. It might also

---

Being Not Only a True Account of Her Strange Birth and Most Important Passages of Her Life, but Also All Her Prophecies, J. Buck, Brighton, (1862).
422. W. H. Harrison, "Spiritualist", *Mother Shipton Investigated: the Result of Critical Examination in the British Museum of the Literature Relating to the Yorkshire Sibyl*, Norwood Editions, Norwood, Pennsylvania, (1881/1976).
423. J. M. Snoek, *The Grey Book*, Humanities Press, New York, (1970), pp. I-XXVI.
424. M. Luther, *Das Ihesus Christus ain geborner Jude sey*, Melchior Ramminger, Wittemberg, (1523); ***also:*** *Das Jhesus Christus eyn geborner Jude sey*, Cranach u. Döring, 1523; English translation in: "That Jesus Christ was Born a Jew", *Luther's Works*, Volume 45, Muhlenberg Press, Philadelphia, (1962), pp. 199-229.
425. M. Luther, *Von den Juden und ihren Lügen*, Hans Lufft, Wittenberg, (1543); Reprinted, Ludendorffs, München, (1932); English translation by Martin H. Bertram, "On the Jews and Their Lies", *Luther's Works*, Volume 47, Fortress Press, Philadelphia, (1971), pp. 123-306.

be that near the end of his life Luther eventually sickened of killing Christians and was sincerely revolted by the Jews' plans to exterminate all Gentile races. The ultimate motives behind the Crusades and the persecution of Jews during the Crusades are also open to question.

Some have taken the view that Protestantism created Zionism in its quest for an ally against the Catholic Church—and in England with the purpose of securing trade routes to India and China (and later oil).[426] Though these forces were no doubt in play during the movement—at the instigation of Jews, it would appear far more likely that Zionists created Protestantism as a means to destroy the Roman Catholic Church they so hated, than that the Protestants created Zionism—given the fact that Zionism pervades the Old Testament. The Rothschilds had no small amount of influence in England and in France—they helped to put Disraeli and the Napoleons in power—and the alleged trade advantages of securing Palestine for the Jews would profit Jewish financiers, as well as the British or French. It was always the Jews who were whispering of these alleged advantages into the ears of the Christians. It was the Jews who went from one country to another preaching this same message. It was the Jews who alleged that only Jews could secure European interests in the region, which was not only a patently false message, it was absurd and the exact opposite of the truth.

Abbé Barruel alleged that the Jacobins, who instigated the French Revolution, were a current manifestation of a very old revolutionary conspiracy of the Freemasons to establish a world government through world revolution. In 1806, Barruel produced a letter he received from A. J. B. Simonini, which he alleged proved a Jewish conspiracy to destroy Christendom and rule the world.[427] At about the same time, George Stanley Faber[428] alleged that the Pope and Islam

---

426. M. Ould-Mey, "The Non-Jewish Origin of Zionism", *The Arab World Geographer / Le Géographe du monde arabe*, Volume 5, Number 1, (2002), pp. 34-52: <http://mama.indstate.edu/users/mouldmey/The%20Non-Jewish%20Origin%20of%20Zionism.PDF>

427. A. J. B. Simonini to A. Barruel Simonini of 1 August 1806, *Le Contemporain* (Paris), (July, 1878), pp. 58-61; **also:** N. Deschamps, *Les sociétés secrètes et la société, ou, Philosophie de l'histoire contemporaine*, Volume 3, Seguin Frère, Avignon (1881), pp. 658-661; *also* A. Nechvolodov, *L'empereur Nicolas II et les Juifs: Essais sur la révolution russe dans ses rapports avec l'activité universelle du judaisme contemporain*, Étienne Chiron, Paris, (1924), pp. 231-234.

428. G. S. Faber, *Thoughts on the Calvinistic and Arminian Controversy*, Printed for F.C. and J. Rivington, London, (1803); **and** *A Dissertation on the Prophecies That Have Been Fulfilled, Are Now Fulfilling, or Will Hereafter Be Fulfilled Relative to the Great Period of 1260 Years, the Papal and Mohammedan Apostacies, the Tyrannical Reign of Antichrist, or the Infidel Power, and the Restoration of the Jews: to Which Is Added, an Appendix*, Andrews and Cummings, Boston, (1808); **and** *A General and Connected View of the Prophecies, Relative to the Conversion, Restoration, Union, and Future Glory of the Houses of Judah and Israel; the Progress, and Final Overthrow, of the Antichristian Confederacy in the Land of Palestine; and the Ultimate General Diffusion of Christianity*, Published by William Andrews. T.B. Wait & Co. Printers, Boston, (1809); **and** *A*

were an evil conspiracy, which stood in the way of the "restoration of the Jews" and the fulfilment of prophecy. Faber proposed the destruction of the Turkish Empire, and the destruction of Catholicism, in preparation for the "restoration of the Jews". In this period we find such fanatical titles as: W. Ettrick, *The Season and Time, Or, an Exposition of the Prophecies Which Relate to the Two Periods of Daniel Subsequent to the 1260 Years Now Recenter Expired: Being the Time of the Seventh Trumpet... Together with Remarks upon the Revolutionary Anti-Christ Proposed by Bishop Horsley and the Rev. G. S. Faber*, Longman, Hurst, Rees, Orne, and Brown, London, (1816). There was a large and long-lived religious movement in Protestant England and America to bring about World War in order to "restore the Jews to Palestine". This had no benefits for Christians.

The Bolsheviks under Trotsky sought the destruction of religion and a world government. These were expressed Jewish objectives. The Bolsheviks mass murdered tens of millions of Christians and plunged many millions more into

---

Dissertation on the Prophecy Contained in Daniel IX 24-27: Generally Denominated the Prophecy of the Seventy Weeks, F.C. and J. Rivington, London, (1811); **and** *The conversion of the Jews to the Faith of Christ*, A. Macintosh, London, (1822); **and** *The Fourteenth Report of the London Society for Promoting Christianity Amongst the Jews: with an Appendix Containing Extracts of Correspondence, and a List of Subscribers and Benefactors to March 31, 1822; to Which Is Prefixed a Sermon Preached Before the Society on April 18, 1822 at the Parish Church of St. Paul, Covent Garden, by the Rev. George Stanley Faber*, London Society for Promoting Christianity amongst the Jews, London, (1822); **and** *The Conversion of the Jews to the Faith of Christ the True Medium of the Conversion of the Gentile World: a Sermon Preached Before the London Society for Promoting Christianity Amongst the Jews, on Thursday, April 18, 1822, at the Parish Church of St. Paul, Convent Garden*, Published for the Baltimore Auxiliary Society for Meliorating the Condition of the Jews, Baltimore, (1823); **and** *Protestantism & Catholicism: the Catholic Question: to the editor of the St. James's Chronicle*, St. James Chronicle, Edinburgh, (1829); **and** *The Difficulties of Romanism*, Towar & Hogan, Philadelphia, (1829); **and** *Recapitulated Apostasy, the True Rationalè of the Concealed Apocalyptic Name of the Roman Empire*, Printed for J.G. & F. Rivington, London, (1833); **and** *Difficulties of Infidelity*, Rivington, London, (1833); **and** *The Primitive Doctrine of Justification Investigated: Relatively to the Several Definitions of the Church of Rome and the Church of England; and with a Special Reference to the Opinions of the Late Mr. Knox, as Published in His Remains*, R.B. Seeley and W. Burnside, London, (1839); **and** *Views of Daniel VIII; Extracted from His Dissertations on the Prophecies*, Sold by I. Wilcox, Providence, (1844); **and** *Rome and the Bible*, Society for Promoting Christian Knowledge, London, (1845); **and** *A Sermon Preached Before the London Society for the Promotion of Christianity Amongst the Jews*, Justus Cobb, Printer, Middlebury, (1847); **and** *Facts and Assertions, Or, a Brief and Plain Exhibition of the Incongruity of the Peculiar Doctrines of the Church of Rome: With Those Both of the Sacred Scriptures and of the Early Writers of the Christian Church Catholic*, Printed for the Society for Promoting Christian Knowledge, London, (1851); **and** *The Revival of the French Emperorship Anticipated from the Necessity of Prophecy*, T. Bosworth, London, (1853); **and** *The Predicted Downfall of the Turkish Power: The Preparation for the Return of the Ten Tribes*, T. Bosworth, London, (1853).

misery. The British and Americans eventually succeeded in destroying the Turkish Empire, "restoring the Jews to Palestine" and securing their access to the Orient and to oil. The Jews had their way, at a horrible cost to humanity, which we continue to pay.

One hundred years before Marx published his *Manifesto*, Hartley called for world revolution and the destruction of the Christian Churches and of European civil institutions so as to cause suffering Christians to disperse throughout the world and evangelize—just as the Roman dispersion of the Jews into Diaspora caused Jews to roam and proselytize. In Hartley's day, many governments had both "evangelical and civil" power—the Church and the State were often one institution with two faces.

At that time, the Roman Catholic Church was one of the most powerful institutions in the world and stood in the way of the Old Testament prophecy that the "Jewish Nation" should attain political and religious hegemony, and rule the world after the other nations had been obliterated. The Catholics pretended to the Jewish throne as the elect, as the chosen of God. The Catholics asserted the doctrine that the Catholic Church is the "Mystical Body of Christ", which has divine dominion over the nations. The Jews believed that their Messianic prophecies gave them this divine right.

Herzl believed that he would not receive the support of the Pope and the Catholic Church and he was correct. Jews also had many other reasons to hate Catholics. Romans destroyed the Jewish Nation and Rome was the seat of Catholicism. Catholics had committed numerous atrocities against Jews, including the Ghetto system and the Inquisition—the Ghetto of Rome was an especially degrading system.

In Europe, absolute hegemony had always been the goal of empires and churches—and the cause of numerous wars. Jews were by no means alone in their quest for hegemony. In addition, the Catholic Popes had sought to take Palestine in the Crusades, supposedly not in hopes of the "restoration of the Jews to Palestine", but in the hopes of taking the Holy Land for the Christians. This made Catholicism an obstacle to Zionism. Catholicism had long been the chief obstacle to Jewish religious hegemony. It also sought hegemony over the Christian faith, for example, in the case of the Council of Trent.

The Old Testament, the Talmud[429] and the Cabalistic writings led Jews like the Frankists to believe that they had a right and an obligation to enslave the rest of mankind to serve them, that evil was good, and that the only means to bring about the reign of God was to destroy all competitive religions and governments and bring about absolute suffering throughout the world. Ironically and paradoxically, a major part of the Jewish religious plan is the objective of making Jews irreligious. Jewish leaders believe that the prophets commanded that Jews fall away from God and that they are duty bound to see to it that two thirds of Jews perish as a result (*Isaiah* 48:10. *Ezekiel* 5:12. *Zechariah* 13:8-9). They believe that the Messiah will only come when Jews have embraced heresy and

---

429. *Sanhedrin* 97*a*, 97*b*, 98*a*, 98*b*, 99*a*, 99*b*. Compare to *Job* 12.

have made the world evil (*Sanhedrin* 97a).

These Shabbataian/Frankist Cabalistic Dualistic sects among Jews even promoted anti-Semitism—even Blood Libel accusations—in order to promote their political agendas in an unbroken chain of revolutionaries from the Frankists to the Marxists to the Zionists. They preached reincarnation and taught that their leaders were incarnations of the Messiah. It is no coincidence that Newton, Clarke, Hartley and the other British "Christians" who rejected the divinity of Christ preached the message that Christians must destroy themselves with a world revolution and "restore the Jews to Palestine". These treacherous men were obviously serving the interests of the Cabalistic Jews who led them.

8.7.3.1 Jewish Revolutionaries and Napoleon the Messiah Emancipate the Jews

Pragmatically, in order for the Jews to obtain emancipation throughout the world, the governments which held them as chattel would have to be overthrown. In order for the governments to be overthrown, the basic structures of society had to be destroyed so as to promote misery, gross dissatisfaction and revolution. Satisfied people tend to preserve the *status quo*. The last vestiges of the Holy Roman Empire and the Turkish Empire had to be eliminated in order for the Jews to obtain Palestine.

Jewish revolutionaries seek to tear down society so that the common people will have no option but to revolt. Though they pretend to work for the interests of the common people, the Jewish revolutionaries covertly do everything in their power to make the people suffer. When the revolution occurs, Jewish revolutionaries deliberately throw the nation into chaos and economic disaster. Jewish revolutionaries then use their power over the press to spread the myth that only a dictator can restore order, the order the Jews covertly and deliberately subverted. After the Jewish revolutionaries have their puppet dictator in place, they attack religion and mass murder Christians and especially attack the intellectual elite so as to ruin the genetic heritage of the Gentile peoples and prevent counter-revolution—prevent Gentile self-determination. This "revolutionary" process is the fulfillment of Judaism.

The Jacobins used pro-democracy propaganda to install the dictator Robespierre in the French Revolution. After Robespierre failed, the Jews put Napoleon, a dictator who considered himself to have been the Messiah, into power. Napoleon almost achieved the Jews goals. However, when Napoleon's success in emancipating the Jews led to assimilation, Jewish leaders turned against him for having helped the Jews. Jewish leaders preferred oppressive segregation to assimilation.

Liberal apostate Jews began to treat Napoleon as if something of a god. On 4 April 1806, Napoleon mandated a single catechism for the entire Empire, which included the statements that Napoleon was "the image of God on Earth" and the "the Lord's anointed", *i. e.* the "Messiah".[430] Napoleon instituted the Feast of St. Napoleon on 15 August 1806 in honor of Neopolas and mixed the cipher of Napoleon and Josephine with the unutterable name of Jehovah and placed the imperial eagle higher than the Ark of the Covenant on his official crest. Before Napoleon, the French Revolution had largely lost favor with Catholics and religious Jews when Robespierre attacked Judaic and Christian beliefs and instituted the Cult of the Supreme Being and pretended that he was himself a god. Napoleon, the Messiah, emancipated Jews wherever he could, tried to take Palestine for the Jews, re-instituted the Sanhedrin, laid much of the foundation for reform Judaism, etc.[431] *The North American Review* wrote in 1845,

> "The performance of Racine's tragedy of 'Esther' is said to have excited Napoleon's sympathy for the Jews; and he intended at once to improve their condition, and win them to his own interests. In 1806, their usurious practices led to complaint, and serious question, whether their rights, under the decree of 1791, should not be withdrawn. Whereupon, the emperor convened at Paris an assembly of the principal French Jews, to whom he proposed questions respecting their opinions and practices, with measures for establishing their brethren throughout the kingdom in honest and useful professions. The questions were answered, for the most part, to the satisfaction of the emperor; and he called a grand sanhedrim of seventy-one members, to convert the doctrinal explanations of the first assembly into authoritative decrees; hoping that the Jews out of the kingdom, also, would send representatives, and thus Paris would be made the centre of a powerful influence to unite and control the Jews throughout the world. The sanhedrim assembled at Paris in 1807,—a truly venerable body. A few foreign deputies attended; but its authority has never been recognized out of France, nor by all in that country; where, however, it seems to have been productive of benefit, in turning many Jews from dishonest and sordid to respectable and useful employments. Indeed, the decrees of this assembly contained a submissive renunciation of many firm Judaic principles. They declared, that France was the only 'fatherland' of the French Jews, that intermarriage with

---

430. "Napoleon I (Bonaparte)", *The Catholic Encyclopedia*, Volume 10, Robert Appleton Company, (1911).
431. A. Lemoine, *Napoléon Ier* [i.e. Premier] *et les Juifs*, : F. Frères, Paris, (1900). *See also:* E. Kahn, *Napoleon and the Jews*, Rabbinical Dissertation, Hebrew Union College, Ohio, (1902). *See also:* R. Anchel, ... *Napoléon et les Juifs*, Les Presses Universitaires de France, Paris, (1928). *See also:* K. E. Zeis, *Napoleon and the Jews*, Masters Thesis, Univ of Wisconsin, Madison, (1939). *See also:* B. Weider, *Napoleon et les Juifs*, Souvenir Napoléonien du Canada Montreal, (1971). *See also:* F. Kobler, *Napoleon and the Jews*, Schoken Books, New York, (1975). S. Schwarzfuchs, *Napoleon, the Jews, and the Sanhedrin*, Routledge & Kegan Paul, London, Boston, (1979). *See also:* J. Lémann, *Napoléon et les juifs*, Avalon, Paris, (1989).

Christians was lawful, and that no trades were prohibited."[432]

When Napoleon sought the "restoration of the Jews to Palestine", Czar Alexander, under the influence of religious Jewish leadership, called Napoleon the anti-Christ and declared that he was out to destroy Christendom. Jewish leaders used their influence around the world to prevent the complete emancipation of the Jews, which they believed would lead to assimilation and the loss of their power over the Jewish People. The Holy Synod of Moscow proclaimed,

> "In order to bring about a debasement of the Church he [Napoleon] has convened to Paris the Jewish synagogues, restored the dignity of the rabbis and founded a new Hebrew Sanhedrin, the same infamous tribunal which once dared to condemn our Lord and Saviour Jesus Christ to the cross. And now he has the impudence to contemplate the unification of the Jews whom God in His wrath has dispersed over the surface of the earth and to organize all of them for the destruction of the Church of Christ to the purpose — oh, unspeakable audacity surpassing all the misdeeds! — that they may proclaim the Messiah in the person of Napoleon."[433]

The Jews exerted their influence in England as well as in Russia. Lewis Mayer, who desired the "restoration of the Jews" and who sought the annihilation of Catholicism, and the German, Turkish and Russian Empires, also declared that Napoleon was the anti-Christ in 1806.[434] During Napoleon's reign,

---

432. "The Modern Jews", *The North American Review*, Volume 60, Number 127, (April, 1845), pp. 329-368, at 344.
433. F. Kobler, *Napoleon and the Jews*, Schoken Books, New York, (1975), p. 166. *See also:* A. C. L. Crawford, a. k. a. Lord Lindsay, "Letters on Egypt, Edom, and the Holy Land", *The Quarterly Review*, Volume 125, (December, 1838), pp. 166-192. *See also:* J. D. Klier, *Russia Gathers Her Jews: The Origins of the "Jewish Question" in Russia, 1772-1825*, Northern Illinois University Press, Dekalb, Illinois, (1986).
434. L. Mayer, *Restoration of the Jews: Being an Extract from an Entire New Work, Intended to Be Published by Subscription Entitled "Truth Dispelling the Clouds of Error, by the Fulfilment of the Prophecies": Addressed to the Jews*, London, (1803); **and** *Bonaparte the Emperor of the Gauls, Considered as the Lucifer and Gog of Isaiah and Ezekiel: And the Issue of the Present Contest Between Great Britain and France Represented According to Divine Revelation, with an Appeal to Reason on the Errors of Commentators*, C. Stower, London, (1804); **and** *Restoration of the Jews: Containing an Explanation of the Prophecies in the Books of Daniel and the Revelations, That Relate to the Period When Their Restoration Will Be Accomplished. With an Illustration, Applicable to the Jews, of the Two Olive Trees, and the Two Candlesticks, That Are Said to Stand Before the God of the Earth, and the Two Witnesses, Who Were to Prophesy, Clothed in Sackcloth, 1260 Days. Addressed to the Jews*, London, (1806); **and** *Peace with France, and Catholic Emancipation: Repugnant to the Command of God*, London, (1806); **and** *The Important Period, and Long Wished for Revolution, Shewn to Be at Hand, When God Will Cleanse the Earth by His Judgments*, Williams & Smith, London, (1806); **and** *The Prophetic Mirror; Or, a Hint to England: Containing an Explanation of*

some Jews betrayed him and encouraged all Jews to side against Napoleon and with an "anti-Semitic" Czar, because they feared that Napoleon's emancipation of the Jews was leading to assimilation,[435] and one must wonder if Russian anti-Semitism was the work of such Jews and if the anti-Semitism of the Czars came at the request of Jewish leaders. A powerful Jewish leader of the time, Shneur Zalman, who hated Gentiles, reasoned that,

> "If Bonaparte wins, the wealth of the Jews will increase and their positions will be raised. But their hearts will be estranged from their Father in Heaven. However, if Czar Alexander wins, then although the poverty of the Jews will increase and their position will be lower, their hearts will cleave to and be bonded with their Father in Heaven."[436]

Napoleon III was also seen by some as the anti-Christ, who would reign over America and England and persecute and destroy Christendom.[437] When

---

Prophecy That Relates to the French Nation, and the Threatened Invasion; Proving Bonaparte to Be the Beast That Arose out of the Earth, with Two Horns like a Lamb, and Spake as a Dragon, Whose Number is 666. Rev. XIII, London, (1806); **and** *Bonaparte the Emporor of the French, Considered as the Lucifer and Gog of Isaiah and Ezekiel: And the Issues of the Present Contest Between Great Britain and France, Represented According to Divine Revelation with an Appeal to Reason, on Prophecy, and the Errors of Commentators... Also an Hieroglyphic Published in 1804, of the Destiny of Europe, the Fate of the German Empire, and the Fall of Russia. And a New Explanation of Daniel's Seventy Weeks,* London, (1806); **and** *Truth Dispelling the Clouds of Error: Containing a New Explanation of Nebuchadnezzar's Great Image and the Prophecies of Balaam, Which Relate to the Total Destruction of the Antichristian Powers, and the Annihilation of the Turkish and Persian Empires. Part I,* W. Nicholson for Williams & Smith, London, (1807); **and** *Death of Bonaparte, and Universal Peace: A New Explanation of Nebuchadnezzar's Great Image, and Daniel's Four Beasts,* W. Nicholson, London, (1809).

435. A. M. Dershowitz, *The Vanishing American Jew: In Search of Jewish Identity for the Next Century,* Little, Brown and Company, Boston, New York, Toronto, London, (1997), pp. 2-3.

436. *From:* A. Nadler, "Last Exit to Brooklyn: The Lubavitcher's Powerful and Preposterous Messianism", *The New Republic,* (4 May 1992), pp. 27-35, at 34. Nadler appears to quote from: N. Loewenthal, *Communicating the Infinite: The Emergence of the Habad School,* University of Chicago Press, (1990).

437. M. P. Baxter, "The Coming Battle and the Appalling National Convulsions Foreshown in Prophecy Immediately to Occur During the Period 1861-67 / by an Episcopal Minister", *Second Advent Library* (Jenks), Volume 105, Number 2, W. Harbert, Philadelphia, (1860); **and** "End of the World about 1864-69 : as Held and Clearly Demonstrated by More than Fifty Expositors ... Whose Predictions of Coming Calamities Are Verified by the Present American Commotion Which Is Only a Prelude to the Dreadful Wars, Famines, Pestilences, and Earthquakes, That Will Prevail until the End in 1869 ... : and Then the Subjugation of England about 1864-5 by Napoleon the Antichrist ... / by the Rev. M. Baxter ...", *Second Advent Library* (Jenks), Volume 105, Number 1, E. Dutton, Boston, (1861); **and** *Louis Napoleon, the Destined Monarch of the World, and Personal Antichrist, Foreshown in Prophecy to Confirm a Seven Years' Convenant with*

Napoleon Bonaparte's attempt to capture Palestine for the Jews failed, he sought to bring Jews from around the world to France—only five hundred Jews lived in Paris in 1789,[438] and there were only 40,000 Jews in all of France.[439] If Napoleon had defeated the British, it would have meant the hegemony of the Jews over Christendom as Hartley had desired.

Napoleon Bonaparte told Barry Edward O'Meara,

> "I wanted to make them leave off usury, and become like other men. There were a great many Jews in the countries I reigned over; by removing their disabilities, and by putting them upon an equality with Catholics, Protestants, and others, I hoped to make them to become good citizens, and conduct themselves like the rest of the community. I believe that I should have succeeded in the end. My reasoning with them was, that as their rabbis explained to them that they ought not practise usury against their own tribes, but were allowed to practise it with Christians and others, that, therefore, as I had restored them to all their privileges, and made them equal to my other subjects, they must consider me like Solomon or Herod, to be the head of their nation, and my subjects as brethren of a tribe similar to theirs. Consequently, they were not permitted to deal usuriously with them or me, but to treat us as if we were of the tribe of Judah. Enjoying similar privileges to my subjects, they were, in like manner, to pay taxes, and submit to the laws of conscription, and to other laws. By this I gained many soldiers. Besides, I should have drawn great wealth to France, as the Jews were very numerous and would have flocked to a country where they enjoyed such

---

the Jews About, or Soon after 1863, and Then, (After the Resurrection and the Translation of the Wise Virgins Has Taken Place Two Years and from Four to Six Weeks after the Convenant,) Subsequently to Become Completely Supreme over England and Most of America, and All Christendom, and Fiercely to Persecute Christians During the Latter Half of the Seven Years, until He Finally Perishes at the Descent of Christ at the Battle of Armageddon, about or Soon after 1870: Including an Examination of the Views of the Revs. G. S. Faber, Edward Irving, E. Bickersteth, T. Birks, C. Maitland, Sir E. Denny, Lord Congleton, Major Phillips, Judge Strange, Dr. Tregelles, Etc.: with Seven Diagrams and Two Maps... / by the Rev. M. Baxter, of the Episcopal Church..., Third Enlarged Edition, Wm. S. & A. Martien, Philadelphia; : D. Appleton & Co., New York; Sheldon & Co. ; J. E. Tilton & Co., Boston; S. C. Griggs & Co., Chicago; W. C. Chewett & Co., Toronto, (1863); **and** *Forty Future Wonders Predicted in Daniel and Revelation: Between 1906 and the End of this Age in Passover Week, 1929 or 1931, as Foreshown in the Prophecies of Daniel and Revelation... ; with Quotations from the Expositions of Archbishop Cyprian, the Late Duke of Manchester, Lord Cavan... ; with Five Appendices and 50 Illustrations (Some Being from Ancient Bibles in the British Museum / by Rev. M. Baxter (Founder of the "Christian Herald" and "Prophetic News")*, Eleventh Edition, M. Baxter, London, (1903).

438. A. Muhlstein, *Baron James: The Rise of the French Rothschilds*, Vendome Press, New York, (1982), p. 41.

439. S. Schwarzfuchs, *Napoleon, the Jews, and the Sanhedrin*, Routledge & Kegan Paul, London, Boston, (1979), p. 1.

privileges. Moreover, I wanted to establish a universal liberty of conscience and thought to make all men equal, whether Protestants, Catholics, Mohammedans, Deists, or others; so that their religion should have no influence in getting them employment under government. In fact, that it should neither be the means of serving, nor of injuring them: and that no objection should be made to a man's getting a situation on the score of religion, provided he were fit for it in other respects. I made everything independent of religion."[440]

In August of 1806, the Venetian representative of the Viennese Court stated that the assembly of the Notables of France and Italy "aimed at the realization of far-reaching plans and 'even to the gathering of the Jews in a particular Kingdom'."[441] On 24 September 1806, Metternich wrote to Standion of Napoleon, the Messiah,

"The impulse has been given: the Israelites of all the lands have their eyes turned to the Messiah who seems to free them from the yoke under which they find themselves; the aim of so many sentences (as it is only that much) is not at all to give full licence to the citizens professing this religion in the lands submitted to French rule, but the desire to prove to the whole nation that its real fatherland is France."[442]

If France were to become the Jewish homeland, as Napoleon desired after his failure to take Palestine for the Jews, that would have made Napoleon the King of the Jews, the Jewish Messiah—the "anti-Christ". Napoleon's uncle, Cardinal Joseph Fesch, purportedly said to him,

"Do you want indeed to bring about the end of the world? Do you not know that the Holy Scriptures predict the end of the world for the moment when the Jews will be recognized as a corporate nation?"[443]

Israel Jacobson published *Les premiers pas de la nation juive vers le bonheur sous des auspices du Grand Monarche Napoléon*, Paris, (1806); which treated of Napoleon as if he were the Messiah.

8.7.3.2 Hitler Accomplishes for the Zionists What Napoleon Could Not

---

[440]. F. Kobler, *Napoleon and the Jews*, Schoken Books, New York, (1975), pp. 174-175.
[441]. S. Schwarzfuchs, *Napoleon, the Jews, and the Sanhedrin*, Routledge & Kegan Paul, London, Boston, (1979), p. 165.
[442]. S. Schwarzfuchs, *Napoleon, the Jews, and the Sanhedrin*, Routledge & Kegan Paul, London, Boston, (1979), p. 166.
[443]. F. Kobler, *Napoleon and the Jews*, Schoken Books, New York, (1975), p. 162. Kobler cites: J. A. C. Chaptal, *Mes Souvenirs sur Napoléon*, E. Plon, Nourrit et Cie, Paris, (1893), pp. 242ff. Kobler believes the quotation is apocryphal.

Later, the Nazis, with their dictator, and the Bolsheviks, with their many dictators, sought to destroy all religions in Europe—sought to destroy Europe, itself. Hitler called for a millennium of Nazism. Much of this revolutionary and nihilistic fervor in Europe stemmed from the Reformation as a revolution against Catholic corruption and in this period revolutions were commonly justified based upon scripture.[444] The Illuminati sought revolution, the elimination of private property and religion. Even more revolutionary and nihilistic was the Jewish reformatory movement of Frankism.[445] This Jewish sect encouraged its members to join other religions in order to destroy them—to become leaders in government in order to subvert society—to practice the mafia creed of *Omerta* silence and to lie and deceive.

Hitler, whose political career in many ways reflects Napoleon's and in many ways was the polar opposite of Napoleon's though meant to fulfill the same ends Napoleon failed to achieve—Hitler tells us of his apocalyptic visions that Nature

---

[444]. J. Somers, D. Defoe, J. Dunton, G. Burnet, T. Harrison, *Vox populi, vox Dei: being true maxims of government : proving I. That all kings, governours, and forms of government proceed from the people, II. The nature of our constitution is fairly stated, with the original contract between king & people, and a journal of the late revolution, III. That resisting of tyrannical power is allow'd by scripture and reason, IV. That the children of Israel did often resist and turn out their evil princes, and that God Almighty did approve of resistance, V. That the primitive Christians did often resist their tyrannical emperors, and that Bishop Athanasius, St. Chrysostom, Luther, and Melancthon, &c. did approve of resistance, VI. That the Protestants in all ages did resist their evil and destructive princes, VII. Together with a historical account of the depriving of kings for their evil government in Israel, France, Spain, Scotland, &c. and in England before and since the Conquest, VIII. That absolute passive-obedience is a damnable and treasonable doctrine. By contradicting the glorious attributes of God, and incouraging of rebellion, usurpation and tyranny : To which no answer will be made, or dare be made, or can be made, without treason, not to be behind Mr. Lesley, or any Jacobite assurance*, Printed for the Author, and are to be sold by T. Harrison ..., London, (1709). T. Paine, *Common sense: addressed to the inhabitants of America, on the following interesting subjects : I. Of the origin and design of government in general, with concise remarks on the English Constitution. II. Of monarchy and hereditary succession. III. Thoughts on the present state of American affairs. IV. Of the present ability of America, with some miscellaneous reflections*, W. and T. Bradford, Philadelphia, (1776).

[445]. See: "Frank, Jacob, and the Frankists", *Encyclopaedia Judaica*, Volume 7 Fr-Ha, Encyclopaedia Judaica, Jerusalem, The Macmillan Company, New York, (1971), cols. 55-71. ***See also:*** G. Scholem, "The Holiness of Sin", *Commentary* (American Jewish Committee), Volume 51, Number 1, (January, 1971), pp. 41-70; reprinted: G. Scholem, "Redemption Through Sin", *The Messianic Idea in Judaism and Other Essays on Jewish Spirituality*, Schocken Books, New York, (1971), pp. 78-141; **and** *Sabbatai Sevi: The Mystical Messiah, 1626-1676*, Princeton University Press, (1973); **and** *Kabbalah*, New American Library, New York. ***See also:*** Rabbi M. S. Antelman, *To Eliminate the Opiate*, Volume 1, Chapter 10, Zahavia, New York, (1974). ***See also:*** H. Graetz, *Popular History of the Jews*, Volume 5, Fifth Edition, Hebrew publishing Company, New York, (1937), p. 245-259.

might have chosen the Jews. The pledge of a thousand year empire, *ein tausendjähriges Reich*, is reminiscent of the prophesy of the millennium of Christ (*Revelation* 20:1-7). Hitler, the Bolshevik who did what he could do to destroy Europe—Hitler, who ultimately called on the German People to admit their defeat and kill themselves at the close of the war in Europe, who wrote in *Mein Kampf*, after complaining of the francophilia of the Viennese press and stating that Zionism had convinced him to finally accept anti-Semitism, Hitler stated,

> "Just once more — and this was the last time — fearful, oppressive thoughts came to me in profound anguish.
> When over long periods of human history I scrutinized the activity of the Jewish people, suddenly there rose up in me the fearful question whether inscrutable Destiny, perhaps for reasons unknown to us poor mortals, did not with eternal and immutable resolve, desire the final victory of this little nation.
> Was it possible that the earth had been promised as a reward to this people which lives only for this earth?
> Have we an objective right to struggle for our self-preservation, or is this justified only subjectively within ourselves?
> As I delved more deeply into the teachings of Marxism and thus in tranquil clarity submitted the deeds of the Jewish people to contemplation, Fate itself gave me its answer.
> The Jewish doctrine of Marxism rejects the aristocratic principle of Nature and replaces the eternal privilege of power and strength by the mass of numbers and their dead weight. Thus it denies the value of personality in man, contests the significance of nationality and race, and thereby withdraws from humanity the premise of its existence and its culture. As a foundation of the universe, this doctrine would bring about the end of any order intellectually conceivable to man. And as, in this greatest of all recognizable organisms, the result of an application of such a law could only be chaos, on earth it could only be destruction for the inhabitants of this planet.
> If, with the help of his Marxist creed, the Jew is victorious over the other peoples of the world, his crown will be the funeral wreath of humanity and this planet will, as it did millions' of years ago, move through the ether devoid of men.
> Eternal Nature inexorably avenges the infringement of her commands.
> Hence today I believe that I am acting in accordance with the will of the Almighty Creator: *by defending myself against the Jew, I am fighting for the work of the Lord.*"[446]

---

[446]. A. Hitler, English translation by Ralph Manheim, *Mein Kampf*, Houghton Mifflin, Boston, New York, (1971), pp. 64-65.

Jewish Dualists believed that the millenium could be brought about by committing monumental acts of evil. They believed that by betraying the Jewish People, as Judas betrayed Jesus—Jew betrayed Jew, anti-Semitic Jews could fulfill the Jewish prophecies. They believed in Hitler.

English Protestant Zionists, vile traitors under the direction and influence of Jewish Zionist financiers, planned the destruction of European society, which they planned would result in the "restoration of the Jews to Palestine" and the downfall of Christianity—ultimately the destruction of Heaven and Earth by fire. The Socialist ideology that almost brought this about was promoted by the anti-Semite Karl Marx and his good friend, the eager assimilationist—turned anti-Semite—turned racist Zionist, Moses Hess—who, together with Ghillany, Bauer, and others, provided the anti-Semitic Socialistic dogma that gave rise to Dühring and eventually to Adolf Hitler. Such Socialists had always used anti-Semitism to bring themselves into power and their goal was always to destroy the social institutions of Europe to make it ripe for revolution, which revolution would emancipate the Jews, then expel them to Palestine.

In 1749, with the English Revolution of 1688 against Catholicism in fairly recent memory, Hartley had iterated these goals in three corollaries to his 83rd Proposition in the second volume of his *Observations on Man*:

> "COR. 1. May not the two Captivities of the *Jews*, and their two Restorations, be Types of the first and second Death, and of the first and second Resurrections?
>
> COR. 2. Does it not appear agreeable to the whole Analogy both of the Word and Works of God, that the *Jews* are Types both of each Individual in particular, on one hand, and of the whole World in general, on the other? May we not therefore hope, that, at least after the second Death, there will be a Resurrection to Life eternal to every Man, and to the whole Creation, which groans, and travails in Pain together, waiting for the Adoption, and glorious Liberty, of the Children of God?
>
> COR. 3. As the Downfal of the *Jewish* State under *Titus* was the Occasion of the Publication of the Gospel to us Gentiles, so our Downfal may contribute to the Restoration of the *Jews*, and both together bring on the final Publication and Prevalence of the true Religion; of which I shall treat in the next Proposition. Thus the Type, and the Thing typified, will coincide; the First-fruits, and the Lump, be made holy together."

Hartley called for the destruction of the Christian Temple—principally *Roman* Catholicism. Jews hated Romans and that hatred carried over to the Pope and Catholicism. Gustaf Dalman wrote of the Talmud, which is riddled with hateful comments,

> "In the Talmud no people have a name so hated as the Romans, who destroyed the Jews' holy city and took from them the last remnant of

independence."[447]

In Proposition 84, Hartley calls for a Christian diaspora to serve the interests of the Jews by spreading Jewish monotheism to all the peoples of the Earth and by making it easy for the Jews to monopolize trade and take all the wealth of the Gentiles, which objectives fulfill Jewish Messianic prophecy,

> "Fifthly, The Downfal of the Civil and Ecclesiastical Powers, mentioned in the 81st and 82d Propositions, must both be attended with such public Calamities, as will make Men serious, and also drive them from the Countries of *Christendom* into the remote Parts of the World, particularly into the *East* and *West-Indies*; whither consequently they will carry their Religion now purified from Errors and Superstitions.
> Sixthly, The Restoration of the *Jews*, mentioned in the last Proposition, may be expected to have the greatest Effect in alarming Mankind, and opening their Eyes. This will be such an Accomplishment of the Prophecies, as will vindicate them from all Cavils. Besides which, the careful Survey of *Palæstine*, and the neighboring Countries, the Study of the *Eastern* Languages, of the Histories of the present and antient Inhabitants, &c. (which must follow this Event) when compared together, will cast the greatest Light upon the Scriptures, and at once prove their Genuineness, their Truth, and their Divine Authority."

Hartley concludes his many fallacies by asserting that Christendom should rejoice in its own deliberate self-destruction and the annihilation of the Earth, because destroying itself proves its faith in, and the truth of, the Jews' prophecies, by artificially and willfully bringing them about,

> "One ought also to add, with St. *Peter*, as the practical Consequence of this Proposition, that the Dissolution of this World by Fire is the strongest Motive to an Indifference to it, and to that holy Conversation and Godliness, which may fit us for *the new Heavens, and new Earth*."

The Dispensationalist "Christians" are the modern version of the Hartleys and the Newtons. They have nuclear bombs at their disposal and intend to bring about the destruction of life on Earth in the vain and suicidal hope that Jesus will fabricate them a new heaven and Earth. These religious fanatics are a menace to mankind and are under the direct control of modern Jewish leadership, who have fabricated their mythologies and promoted them. They are slaves to Israel who intend to deliberately destroy humankind. They are psychopathic and have no sympathy for others, nor respect for the self-determination of others, nor any regard for human life. They are the ideal slaves of Israel.

It is interesting that the New Testament contains in its creed the seeds of the

---

[447]. G. Dalman, *Jesus Christ in the Talmud, Midrash, Zohar, and the Liturgy of the Synagogue*, Deighton Bell, Cambridge, (1893), p. 22.

self-destruction of the enemies of the Jews prophesied in the Old Testament, and converts the enemies of the Jews to a mythology that results in their own demise (*see, for example: Romans* 11). The apocalyptic book of *Enoch* contains many of the mythologies found in the apocalyptic aims of "Christian" Zionists, who seem to wish to stamp out the "seed of Cain"[448]—the seed of the fair—the seed of the European Gentiles. Hartley and countless others readied Christians to joyfully accept war and their own extermination.

History's most highly regarded theological expert on Judaism, Johannes Buxtorf alleged that Jews were readying to destroy Christianity and to take the Christians remaining after the devastation as slaves—as is prophesied in the Old Testament, in the apocalyptic books of Qumran, and in the Talmud and Cabalistic writings. Buxtorf reiterated the intentions of some Jews as told in the 14th Century Jewish author Machir of Toledo's (this is perhaps a false name and the work may have been fabricated by Turkish Jews) *Avkat Rokhel*, Constantinople/Istanbul, (1516). Machir's *Avkat Rokhel* was and is a very influential work, which was translated from Hebrew to Yiddish, and which has been republished many times in both Hebrew and Yiddish. The Jews wrote of Hitler and the persecutions of the Third Reich centuries before they came about. The Zionists put Adolf Hitler into power to fulfil these plans. The Jews also wrote of world government and of the league of nations following world war, centuries before they came about. The Zionists have agitated for both World Wars in order to fulfil these plans, and are today agitating for a Third World War.

The book of I *Enoch* taught Jews many apocalyptic lessons. It is interesting to note that being victims of oppressive laws, Jews had experience with "excessive laws, tyrannical rulers," etc. and one is struck by how these methods were applied by Bolsheviks and Nazis under Zionist control, and are today used against the Palestinians in the illegally occupied territories. Some Jewish writers knew that such oppression could make peoples lackadaisical, defeatist and lose their will to fight back, or be involved in politics, which they would degrade into vicious combat—especially vulnerable were peoples who had been conditioned by Jewish mythology to welcome their own demise, like Christians were conditioned to exterminate themselves by vile traitors like David Hartley. Jewish writers told that chemical and biological weapons, as well as environmental degradation and psychological warfare, would decimate Gentiles and apostate Jews, while antidotes spared pious Jews, who prospered from the destruction of their neighbors. Jewish writers predicted dictators like Napoleon and Hitler who would ask their people to worship them as gods—there being no better means to defeat Roman Catholicism in Europe. Jewish writers often spoke of the extermination of assimilatory Jews, like those of Europe in the mid-Twentieth Century.

Buxtorf, a renowned expert on Judaism and the life of the Jews who were his contemporaries, and with whom he had an extensive correspondence (his son

---

[448]. I *Enoch* 22:7.

corresponded with Manasseh ben Israel), wrote in his *Synagoga Judaica: Das ist Jüden Schul ; Darinnen der gantz Jüdische Glaub und Glaubensubung... grundlich erkläret*, Basel, (1603), as translated in the 1657 English edition, *The Jewish Synagogue: Or An Historical Narration of the State of the Jewes, At this Day Dispersed over the Face of the Whole Earth*, Printed by T. Roycroft for H. R. and Thomas Young at the Three Pidgeons in Pauls Church-Yard, London, (1657), (margin notes appear here in {braces}):

"CHAP. XXXVI.

Touching the Jews Messias who is yet for to come.

THat a Messias was promised unto the Jews, they all with one mouth acknowledge; hereupon petitioning in their daily prayers that he would come quickly; before the houreglass of their life be run out. The only scruple is of the time when, and the state in which he shall appear.

They generally beleeve, that this their future Messias shall be a simple man, yet nevertheless far exceeding the whole generation of mortals in all kinde of vertues: who shall marry a wife and beget children, to sit upon the throne of his kingdom after him. When therefore the Scripture mentioneth a twofold Messias, the one plain, poor, and meek, subject to the stroke of death: the other illustrious, powerful, highly advanced and exalted: the Jews forge unto themselves two of the same sort, one which they call by the name of Messias the son of *Joseph* that poor and simple one, yet an experienced and valiant leader for the warrs; Another whom they entitle Messias the son of David that true Messias who is to be king of Israel, and to rule over them in their own land. About whose coming they are among themselves altogether disagreeing.

Those ancient Jews who lived before Christs incarnation, did not much miss the marke, when *Elias* said that the world should continue six thousand years, whereof two thousand were to be void and without force, that is, without the law of God, the other two thousand under the law: and the last under the Messias.

Their hope was therefore this, that foure thousand years after the worlds creation fully expired, their Messias should come in the flesh: in which their errour was small or none at all; for according to the vulgar account of us Christians, Christ the true Messias was borne in the 3963. year of the world, but according to the Jews computation in the year 3761, we and they differing 202 years. And now because Christ came not unto them in great power, a king of glorious state (such as were *David* and *Solomon*) to deliver them from the tyranny of that usurping *Herod*, and Roman cruelty, neither with a rod of iron to break in pieces and destroy their enemies: but only began his kingdom over them with the spiritual scepter of his doctrine, even for this very cause they would not receive him for the true Messias, though some few did acknowledge and embrace him, and at that time the most ancient and approved men amongst them did expect his coming: {Luk.

2.25.} thus we finde a *Simeon* waiting for the consolation of Israel, {Ib. v. 38.} and *Anna* that old Prophetess speaking of him to all that hoped for deliverance in Jerusalem. The very same that the Apostle *Paul* witnesseth in his Epistle to the *Romans*, {Rom. 11.5.} that though the Jews were most ingrateful, *yet is there a remnant of them according to the election of grace.* Yea, when all kingly power, sacerdotal honour and dignity was taken from them, the city Jerusalem made a ruinous heap, and their beauty the temple turned into ashes, every one now begins to suspect the time of the coming of the Messias to be past. Hence it was that in the 52. years after the destruction of the Temple, a certain proud and haughty Jew boasting that he was the true Messias, feared not to affirme himself the same of whom *Balaam* prophesied in these words: {Num. 24.17, 18.} *I shall see him, but not now, I shall behold him, but not nigh: there shall come a star out of Jacob, and a Scepter shall rise out of Israel, and he shall smite the corners of Moab, and destroy all the children of Sheth. And Edom shall be a possession, Seir also shall be a possession for his enemies, and Israel shall do valiantly.* Others understood this prophesie of the then newly begun kingdom of the Christians. But the Jews even at this day determine their Messias as yet to come, and to fulfil those things which *Balaam* foretold, according to their substance. That the said Jew should proclaim himself the Messias, was most grateful unto them: who presently in their own conceits can nourish hopes, that they should become the conquerours of the Romans, who a little before had destroyed their City and Temple. This Seducer following the letter of the prophesie, names himself, *Ben Chocab*, which is by interpretation, the son of a Star. His chief follower, who at the very first clave unto him, was *Rabbi Akibha*, a man of great learning, who had under his tuition twenty four thousand Scholars, proclaiming him to be *Malka Meschiccha*, Christ the King. By this means much people went after him; insomuch, that he chused unto himself the City *Bittera* for the seat of his kingdom. But when that *Adrian* the Roman Emperour, had after a siege of three years and an half taken and killed this their Messias, and together with this beautiful Star had miserably slaughtered more then four hundred thousand Jews, then the remnant of so great a massacre perceiving themselves led astray by this their Star, turn Anabaptists, and call him from that day to this *Barcozabh*, that is, the son of a lye, a lying and bastardly Messias. Yet neverthelesse, many since have lived who would be reputed for the Messias, as you may read in a book called *Schebhet Jehudah**. {*Schebet Jehudah*, the tribe of Judah. A historical book of the many afflictions, martyrdoms of the Iews, as also of their disputes with the Christians in *Spain*, and *Italy*. It was printed at *Crncovia* in Germany. An. d.1591.}

The issue of all is this; that the Jews convicted in their own consciences, will they, nill they, [willy nilly] are forced to confesse that the time in which the Messias was to come, is already past.

When therefore they had despised and rejected Christ the true Messias, and no other appeared, they falsified the above mentioned tradition of *Elias* (which was that the Messias should come about the four thousandth year of the world) by annexing unto it this Comment; that the time was prolonged for their offences. But when at length no reason could be pretended of this long delay, neither could they define the time of his coming: their onely evasion is, to smite with this curse the head of him that should determine a certain season for his coming, *Tippach ruchan atzman schel mechasschebhe Kitzin**, {Sanhedrin c.II.p.97.} *Which is, Let their soul and body burst with a swolling Rupture, who peremptorily set down the time*; that time (I say) in which the Messias is expressly for to come. Yet this not at all pondered, and nothing set by, many of them moved by the prophesies of the men of God concerning the coming of the Messias, have in their souls and consciences confessed, that the time of his coming was already past; and therefore in their writings they acknowledge that he is born indeed; but for their sins and impenitent life, not as yet revealed. And at this instant all the Jews dwelling amongst us are of the same opinion. Hereupon *Rabbi Solomon Jarchi* saith, that according to their ancestors, the Messias was born in that day in which Ierusalem was last of all destroyed, but where he hath so long been hid, to be uncertain. Some of them think that he lies in Paradise, bound to the womans hair, grounding upon these words in the Song of *Solomon:* {Cant. 7.5.} *Thy head upon thee is like Carmel, and the hair of thy head like purple, the King is bound in the Galleries*. By King understanding the *Messias*, and by Galleries, paradise. *Rabbi Solomon* follows this exposition of these ancient Rabbines. The *Talmudists* write, {Sanhedrin c.II. p. 98.} that he lies in *Rome* under a gate among sick folks and Lepers, perswaded by the words of *Esay* [Isaiah], who saith, {Esay 53.3.} that he is one *despised and rejected of men, a man of sorrows, and acquainted with grief*. Others forge other lies and tales.

 Well, let all these things fall out according to their own desire; yet they still believe he is to come. {The miracles before Christs coming.} First then before his coming shall happen ten notable miracles, by which every one shall be admonished and incited to an accurate preparation for his coming, and also be warned to conceive that he shall not come so poor and privately as Christ came. These ten miracles I mean here to present in the same words that the Rabbines have commended them to posterity, in a little book called *Abkas Rochel*. {*Abkas rochel pulvis aromatarius*, the author *Rimchar* a little book in *octavo* it hath 3 parts, the first of the miracles, before the coming of the *Messias*, two of the soule, and the state of it after this life. The third of *Moses* his tradition about Mount Sinai, mans creation, &c. It was printed at *Venice anno Dom*. 1597.}

 The first miracle, God shall stir up and produce three kings, who proving traitors to their own faith, shall also turn Apostates: so living before men as though they served the true God: yet in very deed practising nothing less; seducing silly souls, and after such a manner tormenting their consciences, that they may abjure God and their own faith, even so that many of the sinners of Israel shall utterly despair of redemption, being ready to deny God, and forsake his fear. Concerning these things *Isaiah* speaketh,

c. 59. 14,15. *Judgment is turned away backward, and justice standeth afar off; for truth is fallen in the street, and equity cannot enter, yea truth faileth.* What? All they why shall love the truth shall flee in troops, and flying hide themselves in the caves and holes of the earth, and shall be massacred by the great, and mighty, and tyrannical persecutors. At that time shall be no king in Israel, as it is written. {Hos. 3.4.} *The children of Israel shall abide many dayes without a King, and without a Prince, and without a sacrifice, and without an Image, and without an Ephod, and without a Teraphin:* There shall not be any more *Rosch Ieschibhah* (b) {*Jascbhah.* [***] a Synagogue from [***] to sit or rest.} that is head of the Synagogue, no faithful teachers who may feed the people with the word of God, no merciful and holy, no famous and eminent persons shall remain. The heaven shall be shut up and food shall fail; these three kings shall enact laws so many, so burdensome, and so tyrannical, pronounce such heavie judgments upon men, that but a very few shall be left, because they had rather die, then living deny their maker. Yet these three kings by Gods ordinance and disposition shall only reign three moneths.

In the time of their reign, they shall double the ordinary tribute, so that who formerly paied only eight pieces only eight pieces, shall then pay eighty, he who formerly paied ten, shall then be forced to give an hundred. He that hath nothing at all to give, shall be punished with the loss of his head: yea also, the longer they shall reign, the greater and heavier will the burdens be which they shall impose upon the children of Israel. There shall also come certain men from the ends of the earth, so black and abominable, that if any man look upon them he will die through fear. Every one of them shall have two heads, and eight eyes, shining like a flame of fire. They shall run as nimbly and swiftly as an hart. Then shall Israel cry out, woe unto us, woe unto us, the frighted little ones cry alass alass, dear father what shall we doe? then shall the father answer, the deliverance of Israel is now at hand, and even at the door.

{The second miracle.} The second miracle, God shall make the sun to exceed in heat, that many burning feavers, plagues, and other diseases shall be scattered abroad upon the earth, by reason of which, a thousand thousand of the Gentiles and people of the world shall die daily. Hereupon, the Gentiles at length weeping, shall bitterly cry out, woe and alass whither shall we turn our selves? where shall we hide us? Thus with expedition they shall goe and dig their own graves, wish for death, and oppressed with thirst and grief, hide themselves in the Caves and Dens of the Earth. But this great heat shall be as physick and a refreshing to them that are just and good in Israel, as it is written, {Mal. 4.2.} *unto you that fear my name shall the sun of righteousness arise with healing in his wings, and ye shall go forth and grow up as calves of the stall;* by this sun of righteousness understanding that in the heavens. {Num. 24.23.} *Balaam* (say they) also prophesied of this; saying, *alass who shall live when the Lord hath brought it to pass.*

{The third miracle.} The third miracle, God shall make a dew of blood to fall upon the earth: which all Christians and people of the earth thinking to be watery and most delightful, shall take and drink, and drinking die. The Reprobate also in Israel who despaired of redemption, shall also die by drinking of it, but it shall not be hurtful to them who are just among the Iews, who in true faith firmly cleaving unto God, do persevere in the same, as it is written. {Dan. 12.3.} *They that be just shall shine as the brightness of the firmament, and they that turne many to righteousness, as the stars for ever and ever:* again, the whole world for three dayes space shall be full of blood; according to that which is written: {Joel 2.30.} *I will give signes in heaven and in earth, blood and fire and pillars of smoke.*

{The fourth miracle.} The fourth miracle, God shall send a wholsome dew upon the earth. They shall drink of this who are indifferent honest: It shall serve as a salve to them who were made sick by drinking of the former, as it is written. {Hos. 14.5.} *I will be as a dew to Israel, he shall grow as the lillie, and cast forth his root as Lebanon.*

{The fifth miracle.} The fifth miracle. God shall turn the sune into so thick a darkness, that it shall not shine for the space of thirty dayes, as it is written, *The sun shall be turned into darkness, and the moon into blood, before the great and terrible day of the Lord come.* At the end of thirty dayes God shall restore its light; as it is written, {Es. 24.22.} *They shall be gathered as prisoners are gathered in the pit, and shall be shut up in prison, and after many dayes they shall be visited.* The Christians being sore afraid to see these things, they shall be confounded with shame, and acknowledg that all these things come to pass for Israels sake: yea, many of them shall embrace the Jewish religion: as it is written, {Jon. 2.8.} *They that observe lying vanities forsake their own mercy.*

{The sixth miracle.} The sixth miracle, God shall permit the kingdom of *Edom* (to whit that of the Romans) to bear rule over the whole world. One of whose Emperours shall reign over the whole earth nine moneths, who shall bring many great kingdoms to desolation, whose anger shall flame towards the people of Israel, exacting a great tribute from them, and so bringing them into much misery and calamity. Then shall Israel after a strange manner be brought low and perish, neither shall they have any helper: of this time *Esay* prophesied, {Esa. 59.16.} *And he saw that there was no man, and wondered that there was no intercessor: therefore his arm brought salvation unto him.* After the expiration of these nine moneths, God shall send the *Messias son of Joseph*, who shall come of the stock of *Joseph*, whose name shall be *Nehemiah*, the son of *Husiel*. He shall come with the stem of Ephraim, *Benjamin* and *Manasses*; and with one part of the sons of *Gad*. As soon as the Israelites shall hear of it, they shall gather unto him out of every City and nation, as it is written: {Jer. 3.14.} *Turn ye backsliding children saith the Lord, for I will reign over you, I will take you one of a City, and two of a tribe, and bring you to Sion.*

Then shall *Messias* the son of *Joseph*, make great war against the king of *Edom*, or the *Pope of Rome*, and being conqueror shall kill a great part of his army, and also cut the throat of the king of *Edom*, make desolate the Roman Monarchie, bring back some of the holy vessels to Jerusalem, which are treasured up in the house of *Ælianus*. Moreover the king of Egypt shall enter into league with Israel, and shall kill all the men inhabiting about Jerusalem, Damascus, and Ascalon: which thing once noised over the whole earth, a horrid dread and astonishment shall overwhelm the inhabitants thereof.

{The Seventh miracle.} The seventh miracle. They say that at *Rome* there is a certain piece of marble, in shape resembling a Virgin, so framed and fashioned, not by mans workmanship, but by the Lords hand. To this Image shall all the wicked livers in the world gather themselves, and burning in lust towards it, shall commit incest with it. Hereupon, in the same marble will the Lord forme an infant, which by a certain rupture shall issue out of it. This infant shall be called *Armillus Harascha*, *Armillus* the wicked, and shall be the same which the Christians call Antichrist. His length and bredth shall be tenn els, the space betwixt his eyes and the palm cross wise. His hollow eyes red, his hair yellow like gold, the soles of his feet green; and to make his deformity compleat, he shall have two heads. He coming to the wicked king of Rome, shall affirm himself to be the *Messias* and god of the Romans, to whom they easily give credit: and make him king over them. All the sons of *Esau* shall love and stick fast unto him. He shall bring under his yoak the whole Roman Monarchie, and to all *Esaus* ofspring glorying in the name of Christian, he shall say, bring me the law which I gave unto you. Which they shall presently deliver, together with their book of Common-prayer, which he shall receive as true and legitimate, acknowledging that he gave that law and book unto them, desiring that they will beleeve in him.

These things once finished, he shall send his Embassadors to Jerusalem to *Nehemiah* the son of *Husiel*, and to all the Congregation of Israel; with this mandate to bring their law unto him: and confess him to be God: At the report of this, fear and wonder assault their souls: and *Nehemias* accompanied with three hundred thousand voluntiers of the tribe of *Ephraim*, carrying also the book of the law with him, shall come unto *Armillus*, and out of it read him this sentence, {Exod. 20.} *I am the Lord thy God, thou shalt have none other Gods before me*. To whom *Armillus* making answer, shall deny any such sentence to be extant in their law, and that therefore they ought to acknowledg him for a God, following the example of the Christians, and other people of the earth. Then shall *Nehemiah* the son of *Husiel* in that instant command his followers to binde *Armillus*, and entering the field with thirty thousand armed Nobles, shall put to the sword two hundred thousand of his assistants. For this cause *Armillus* greatly enraged, shall gather together all his forces in a deep valley to fight against Israel, and to destroy no small number of *Jacobs* posterity. There shall *Messias* the son of *Joseph* breath his last, whom the holy Angels shall take, hide, and casket up with other Patriarks of the world. The Israelites shall be struck with such astonishment, their hearts shall fleet like water; but *Armillus* himself shall not know of the death of their Messias, who otherwise would not leave one of them alive.

Then shall all the Nations of earth banish the Jews out of their dominions, no way permitting them any longer to be their co-inhabitants. Moreover, such trouble and distresse shall at that time perplex the Jews, as hath not been from the beginning of the world.

{The coming of Michael.} Then shall *Michael* come and fan away the wicked in *Israel*, as it is written; {Dan. 12.1.} *At that time shall Michael stand up, the great Prince, which standeth for the children of thy people, and there shall be a time of trouble, such as never was since there was a nation even to that same time.* Then the remnant shall flee into the wildernesse, where God shall try and purge them after the same manner that silver and gold is tried in the Furnace. For the Lord saith, {Exek. 20.38.} *I will purge out from among you the Rebels, and them that transgresse against me.* And again, {Dan. 12.10.} *Many shall be purified, made white, and tryed; but the wicked shall do wickedly, and none of the wicked shall understand: but the wise shall understand.* Then shall the whole remainder of Israel be in the wildernesse for forty five days, the chief of their fare being grasse, leaves, and herbs; and that Scripture shall be fulfilled in their ears, {Hos. 2.14.} *I will allure her, and bring her into the wildernesse, and speak comfortably unto her.* The truth of this appears out of that of the Prophet, *From the time that the daily sacrifice shall be taken away, and the abomination that maketh desolate, set up, there shall be a thousand two hundred and ninety dayes. Blessed is he that cometh to the thousand three hundred and five and thirty dayes. But goe thee thy way till the end be: for thou shalt rest, and stand in the lot at the end of the dayes.*

Conceive that forty five being added to the precedent number of ninty, the last number of 1335 daies doth arise. In that time all the wicked in Israel shall perish; who are unworthy to be copartners in such a deliverance. Finally, *Armillus* invading Egypt with great power shall subdue it, as it is written: {Dan. 11.42.} *The land of Egypt shall not escape.* From Egypt he shall muster his forces for Jerusalem, striving with might and main once more to make it a desolate heap. {Dan. 11.45.} *And he shall plant the tabernacle of his palace, between the Seas, in the glorious holy mountain, yet he shall come to his end, and shall help him.*

{The eighth miracle.} The eighth miracle. The Archangel *Michael* shall arise, and shall thrice winde a mighty trumpet, as it is written; {Jsa. 27.13.} *It shall come to pass in that day, that the great trumpets shall be blowen, and they shall come that were ready to perish in the land of Assyria, and the outcasts in the land of Egypt, and shall worship the Lord in the holy mount at Jerusalem.* Again, {Zech. 9.14.} *The Lord God shall blow the trumpet, and shall goe with the whirlewinds of the South.* At the sound of this trumpet the true *Messias the son of David*, and the Prophet, *Elias* shall appear and manifest themselves to the devout Israelites inhabiting the wilderness of Judea. Then shall they receive incouragement, the weary hands shall be lifted up, and strength shall visit the feeble knees. All the Jews also wheresoever dispersed over the whole earth shall hear the sound of the trumpet, and at last confess, that God in mercy hath visited his people, and by a plenary deliverance hath been gracious to his inheritance, and all the captives of *Ashur* shall be gathered together. But the sound of this trumpet shall blast the Christians and people of the world with fear and astonishment, casting them into horrid maladies, Then shall the Jews gird up their loins, and with many a weary journey seek to revisite their Jerusalem. Messias also the son of *David*, together with his harbinger *Elias*, and all the faithfull his followers in Israell with great joy shall come into Jerusalem. So soon as this pierceth the ears of wicked *Armillus:* he will babble out, how long will this abject and base people thus behave themselves? and shall once more with a great army of Christians hasten to Jerusalem to give battel to to their newly inaugurated soveraign. But God shall not permit that the Israelites should fall out of the fire into the pit, but speaking unto the *Messias* shall say unto him, Come thou and sit at my right hand, and to the children of Israel, sit you still, hold your peace, and quietly expect that great deliverance which the Lord this day will impart unto you. Then shall the Lord rain from heaven fire and brimstone, as it is recorded, {Ezech. 38.22.} *I will plead against him with pestilence, and with blood, and I will rain upon him, and upon his bands, and upon the many people that are with him, an overflowing rain, and great hailstones, fire and brimstone.* Then shall *Armillus* with his whole army die, and the Atheistical Edomites (the Christians they mean) who laid waste the house of our God, and led us captive into a strange land, shall miserably perish; then shall the Jews be revenged upon them, as it is written, {Obad. 18} *The house of* Jacob *shall be a fire, and the house of* Joseph *a flame, and the house of Esau* (that is, we Christians, as the Jews interpret, whom they Christen Edomites) *shall be for stubble.* This stubble the Jews shall set in fire, that nothing be left to us Edomites which shall not be burnt and turned into ashes.

{The ninth miracle.} The ninth miracle. At the second blast of *Michael* his trumpet being long and loud, all the graves in Jerusalem shall open, and the dead arise, *Messias* also the son of *David* together with *Elias* the Prophet shall restore to life *Messias* that good son of *Joseph* reserved under a certain gate.

At the same time shall all the Congregation of Israel send *Messias* the son of *David* as an Embassador to the remnant of the Jews superviving the last slaughter, dispersed here and there among the Christians and other people of the earth, to summon them to Jerusalem. Then shall the kings of the nations without delay, carry the Jews inhabiting their quarters, upon their shoulders, and in Chariots unto Sion. I think this will come to pass much about the Greek Calends. {I. never.}

{The tenth miracle.} The tenth miracle. At what time the Angel *Michael* shall blow the trumpet the third time, then shall God bring them forth who border upon the rivers *Gosane Lachlacke, Chabore,* and also inhabited the cities of Juda, and they in number infinite and immesurable, together with their infants shall enter into *Moses* Paradise; the earth before and behinde them shall be nothing but a flame of fire, which shall consume all which is needful for the preservation of life among the Christians and other people.

When the ten tribes of Israel shall return out of the land of their captivity, then the pillar of the cloud of the divine glory and majesty shall encompass them, as it is written: {Micah 2.13.} the breaker up is to come before them: *they have broken up, and have passed through the gate, and are gone out by it, and their king shall pass before them, and the Lord on the head of them.* Moreover God shall open unto them fountains flowing out of the tree of life, wherewith he shall refresh them in their journey, lest at any time thirst should annoy them. For the Lord saith, {Jsa. 41.18.} *I will open rivers in high places, and fountains in the midst of the vallies: I will make the wilderness a pool of water, and dry land springs of water.* Again, {Jsa. 49.10.} *They shall not hunger nor thirst, neither shall the heat nor sun smite them, for he that hath mercy on them, shall lead them, even by the springs of water shall he guide them.* {The Jews ten fould comfort against the foresaid signes.} To comfort them against these ten signes foregoing the coming of the *Messias*, the most of which pretend great calamity and affliction to the Jews, they have a tenfold consolation. {Consol. 1.} The first is, that the *Messias* is certainly yet for to come: according to that of the Prophet, {Zach. 9.9.} *Behold thy king cometh &c.*

{The 2. Cons.} The second that he shall again gather them together being dispersed over the face of the whole earth, as it is written: *I will bring them from the north country, and gather them from the coasts of the earth, and with them the blinde and the lame, the women with childe, and her that travelleth with childe together, a great company shall return thither:* From which place we may learn thus much, that if any went unto his grave blind or lame, the same shall God raise up cloathed with the same imperfections: that one may more easily know another, yet the Lord shall so perfectly cure the lame, that they shall skip like Roes, as the Scripture witnesseth, {Esa. 35.6} *Then shall the lame man leap as an hart, and the tongue of the dumbe sing; for in the wilderness shall the waters break out, and streams in the desert.*

{The 3. Cons.} The third is; that God shall raise up the dead: as it is written; *Many that sleep in the dust of the earth shall arise: these to life eternal, they to shame and everlasting contempt.*

{The 4. Cons.} The fourth is, that God shall build them up a third temple, according to that plat-form and fashion which *Ezekiel* hath described *cap. 41. ver.* 1, 2, 3.

{The 5. Cons.} The fift is, that the people of Israel shall be the sole Monarchs of the whole world, their dominion stretching from one end of the earth unto the other, according to that of *Esay* 60.12. *The nation and kingdom that will not serve thee shall perish: yea, these nations shall be utterly wasted.* Yea, the whole world being turned unto the Lord shall be subject to his law, as it is recorded, {Zeph. 3.9.} *For then will I turn to the people a pure language, that they may all call upon the name of the Lord, to serve him with one consent.*

{The 6. Cons.} The sixth is, that God at that time shall defeat and destroy all the enemies of his people (that is, the Christians) and mightily to revenge himselfe upon them: as it is written, {Ezek. 25.14.} *I will lay vengeance upon Edom by the hand of my people Israel, and they shall do in Edom according to mine anger.*

{The 7. Cons.} The seventh is, that God shall take away all diseases and maladies from among the people of Israel, according to that; {Jsa. 33.24.} *The inhabitants shall not say I am sick: the people that dwell therein shall be forgiven their iniquitie.*

{The 8. Cons.} The eight is, God shall prolong the dayes and yeares of the life of the Israelites. So that they shall live as long as the oake or other of that kinde: {Jsa. 65.22.} *for* saith the lord, *as the dayes of a tree are the dayes of my people, and my elect shalt long enjoy the works of their hands*, and againe, *there shall be no more thence an infant of dayes, nor an old man that hath not filled his dayes: for the child shall die an hundred yeares old, but the Sinner being an hundred years old shall be accursed*, which is as much as to say, {See *Reschaim* in the Talmud c. 6. p. 68.} if any die at an hundred years of age, it shall be said of him, that he died as a little infant, or in his infancy: for at that time the years of life of the Israelites shall be equal to them of the fathers from *Adam* to *Noah*, as *Abenezra* comments upon the place.

{Ninth Cons.} The ninth is, that God shall so clearly manifest himself to the Israelites, that they shall see him face to face. As it is recorded: {Isa. 40.5.} *The glory of the Lord shall be revealed, and all flesh shall see together: because the mouth of the Lord hath spoken it.* Yea, all the Lords people shall be Prophets, as it is written: *It shall come to pass afterward that I will powr out my spirit upon all flesh, and your sons and your daughters shall prophesie: your old men shall dream dreams, your yong men shall see visions.*

{Tenth Cons.} The last degree of comfort is, that God shall quite root out of them all imbred lusts, and inclinations unto evil, as it is written: {Ezek. 36.26.} *A new heart also will I give you, and a new spirit will I put within you, and I will take away the stony heart out of your flesh, and I will give you an heart of flesh.* Hitherto we have delivered what we promised out of the book called *Abhkas rochel*, in which though it be summarily set down what the Jews beleeve concerning their *Messias*, as also the manner how he is to bring them back to Jerusalem: yet I think not impertinent in this place a litle more largely to declare with what solemnities their *Messias* shal give them intertainement in their own land, and with what happiness and felicitie they shall lead their lives under him.

{The feast which the *Messias* shall make unto the Jews at his coming.} When then the Messias hath gathered all the Jews together out of all the nations under heaven and from the foure winds of the earth, and hath brought them unto the land of Canaan flowing with milk and hony; then shall he cause to be prepared a sumptuous and delicate banquet, inviting and friendly welcoming unto it all the Jews with great pomp and joy inexpressible.

At this banquet shall be dished up and served in, the greatest beasts, fishes and fouls that ever God created. The worst wine that they shall drink shall be whose grape had its growth in paradise, and hath been barrel'd up and reserved in *Adams* Cellar unto that time.

{The first dish. *Behemoth*. Job. 4.10.} The first dish in this feast shall be that huge oxe described in the book of *Job*, to be of such great strength and magnitude, named *Behemoth*. This is the Rabbines affirme to be the same oxe whereof *David* makes mention in his 50 *Psalm* and 10 verse. *All the beasts of the forrest are mine, and the cattel (Behemoth) feeding on a thousand hills,* that is to say, which every day eateth up the grass of a thousand hils. But a man will aske what at length would have become of this oxe, if he had lived so long, seeing he had long since eaten up all his fodder. The Rabbines (*a*) {(a) *Rabbi Sal: Jarchi, & Rabhuenski.*} learnedly answer that this oxe is stall-fed, and remains always in the same place, and that whatsoever he eateth on the day grows again upon the night in the same length and forme.

{The 2. dish. *Leviathan.*} The second dish adorning the table shall be that vast whale, *Leviathan*, (according to the Jewish tone Pronounced Lipiasan) who is also described in the book of Job, and mentioned in other places of holy writ.

Concerning these two beasts there hath bin handsomly compiled this tradition by the wit and ingenuity of the solid pated Rabbins in their Talmud, {*Babha Basra*. c. 5. p. 74.} it runs thus, Rabbi Jehudah saith that what thing soever God created in the world he created it male and female, and that without all doubt; for he created the Leviathan yet least the he and she Leviathan: by engendring should augment the number, and at length by there monstrous magnitude and multitude destroy the whole world, God gelded the male, and killed the female, reserving her in pickle to be meat for them that are just in Judah and feared him, in the dayes of the Messias, as it is written: {Jsa. 27.1} *In that day will the lord with a sore and great and strong sword punish Leviathan the piercing serpent, even Leviathan that crooked serpent, and he shall slay the dragon that is in the sea.* In the same manner he created that great ox called *Behemoth* feeding on a thousand hils male and female: yet lest by multiplying they might fill and destroy the earth, he gelded the male and killed the female, reserving it for the Jewes diet in time to come, as it is written: {Job 40.16.19} *Loe now his strength is in his loynes, and his force in the navell of his belly, he that made him can make his sword to approach unto him.*

{The third dish. *Barinchue*.} The third dish in this banquet as Elias Levita in his dictionarie named Tesbi out of the Rabbins reports, shall be that horrible huge bird called *Barinchue* which killed and unboweld shall then be rosted. Concerning this bird it is written in the Talmud {*Bechoros* c. *ult.* p. 57} she cast an Egge out of her nest by whose fall three hundred tall Cedars were broken down, and the Egge breaking in the full drowned three score villages. By this relation it is easie to conceive this bird to have been little inferiour in greatnes to the forementioned oxe and fish; whence we may also collect how glorious a dish the Messias is to make of it for his guests, and when there are many such birds (Guls I think) found in the land of Judah, none ought to think that which is reported of this to be fabulous.

{The Crow. *Babha basra.* c. 5. p. 72.} In the forementioned book of the Talmud, we read of a certain great crow which was seen of a Rabbine, worthy to be credited. The relation runs thus. *Rabbi barchannah* saith, At a certain time I saw a frog, which is as great as the village *Akra* in *Hagronia*, well how big was the village? It consisted of no fewer then threescore houses. Then came a mighty serpent and swallowed up this frog. Instantly upon this, a great crow flying that way picked up as a small morsel both the frog and the serpent; and taking him to flight sat upon a Tree, now think with your selves how great and strong this tree must be. To which *Rabbi papa* the son of *Samuel* making answer, unless I had been in the place, and with these mine eyes seen the very tree, I would not have beleeved it. Thus much the Talmudist. Who dare give the lie to this *Rabbine?* When that good man *Kimchi* commenting on the fifty Psalm, {The great bird. ziz.} and explaining the word Ziz hath there witnessed that *Rabbi Judah* the son of *Simeon* did avouch Ziz to be a bird of that bigness, that when he spreads abroad his wings he hides the body of the sun, and wraps the world in darkness. Furthermore, on a certain time, a certain *Rabbine* was upon the sea in a little ship, in the middle of which he saw a bird standing of such an height, that water came only to her knees: {Talmud in the same place.} which the *Rabbine* observing, bespeaks his companions that there they might wash themselves seeing the water was not deep. But a voice from heaven hindred the attempt, saying unto the *Rabbine*, see that thou do it not: for now seven whole years are gone and past, since a certain man let a hatchet fall in this very place, which hath been ever since a falling, and is not as yet come to the bottom. By which a man may easily gather how long legs this bird had, and how great her body ought to be in proportion to her feet. Without doubt these birds keep their residence in the wood *Ela*, in which, a Lion is reported to live of such an unheard of portraicture, that only to relate would strike a man with astonishment. {The great Lion *Chohn*. cap. 3. p. 59.} Of this Lion the Talmud thus fables. When upon a certain time the Emperor of *Rome* asked Rabbi *Joshua* the son of *Hananiah*, what the reason was why their God compared himself unto a Lion; and whether he was of so great strength that he could kill a Lion? the *Rabbine* made answer, that God did not compare himself unto an ordinary Lion, but unto such an one as lived in the wood *Ela:* to whom the Prince replied, shew me that Lion. Then the *Rabbine* by prayer obtained of God that the lion should leave the wood, and come, when hs was yet four hundred miles distant from the Emperour, he roared so terribly, that all the women with child in Rome became abortive, and the walls of the City fell flat unto the ground. When he had come an hundred miles nearer, he the second time roared so fearefully that all the teeth of the Romanes fell out of their heads, & the Emperour falling from his throne, lay prostrate upon the earth half dead; who with vehement entreaties begs of the Rabbin to send back the Lion; which was likewise put in execution. But these fables draw us too far from the smell of that feast which the Messias hath provided for the Jews in the land of promise.

The flesh of the foresaid Behemoth and Leviathan will not digest well without a Cup of older wine; therefore the Messias shall broach that wine and give it unto his guests, {The wine for the feast.} which was made in Paradise, and was kept from the begining of the world to that time in *Adams* Cellar, as it is written: {Esa. 27.2.3.} *In that day sing you unto her a vineyard of red wine. I the Lord do keep it, I do water it every moment: lest any hurt it, I will keep it night and day:* again, {Psal. 75.9.} *There is a cup in the hand of the Lord, and the wine thereof is red: it is full mixt; he shall poure it out, and the dregs thereof all the ungodly of the earth shall drink and suck them up.*

{The sports where with the Messias will delight the Jews.} Before the supper be served in, the *Messias* after the manner of Kings, and Princes, and others celebrating Festivals and Marriages, shall present the Jews with pleasant sports and plaies to make them merry. He will cause *Behemoth* and *Leviathan* to meet in some spacious place, and there they shall play before the *Messias* to pass away the time, and for his minds refreshing, as it is written: {Job 40.20.} *Surely the Mountains bring him forth food, where all the beasts of the field play.* And again, {Psal. 104.26.} *There go the ships, there is that Leviathan whom thou hast made to play therein.* Then the oxe running hither and thither shall bend his hornes against the *Leviathan*; which will greatly affect the *Messias*, according to that, {Psal. 69.32.} *It will be more grateful to the Lord then a bullock that hath horns and hoofs.* The *Leviathan* also shall come to encounter the oxe, armed with his fins as an helmet, not easie to be seen, as it is written: {Job 40.14.15.} *Who can open the doors of his face, his teeth are terrible round about. His scales are his pride, shut up together as with a close seal.* Here shall be the summons to the battle, and the first encounter begin most hot and furious, but to small purpose, for they being of equall strength neither can overcome the other, but at last wearied out both shall fall upon the ground. Then the Messias drawing out his sword shall slay them both, as it is written: {Jsa. 27.1.} *At that day will the Lord with a sore, great and strong sword punish Leviathan.*

Now comes the Cooks part, nothing but boyling and roasting: and great provision for this sumptuous supper, as it is recorded: {Esa. 25.6} *The Lord of hosts shall make unto all people in this mountain a feast of fat things, of fat things full of marrow.* The fish shall be served up in parcels to the guests, which done, every one shall greatly rejoyce, as it is written: {Job 41.6} *shall thy companions make a banquet of him? shall they part him among the merchants.*

{The marriage of the Messias} This donative supper being ended, the *Messias* shall marry a wife: the Scripture being witness: {Ps. 45.10.} *Kings daughters were among thy honourable women: upon thy right hand stood the Queen in a vesture of gold.* So the Jews themselves interpret: {*Schegal* [***] properly signifieth the wife of a King from [***] *Shagal* which is to exercise the very act of venery.} and the meaning is this, as *Kimchi* professeth in his great gloss: Among the honourable women which the

Messias shall have, shall be the daughters of Kings. For every King of the earth shall esteem himself highly graced, so that he may give his daughter in marriage unto the *Messias*. But the genuine and rightly so named wife of the *Messias* (properly signified by the word *Schegal*) shall be one of the most eminent beauties among the daughters of Israel; she shall sit at his right hand, without intermission abide in the Kings closet: whereas the other shall stay in the supping room, or house of the women: not approaching the King, unless it be his pleasure to send for them. In this bond of Wedlock the *Messias* shall beget children; after he shall die as other mortals, and his children shall sit upon his throne after him, as it is written: {Isa. 53.10.} *He shall see his seed, he shall prolong his dayes, and the will of the Lord shall prosper in his hands*, that is, as a *Rabbine* expounds it, The *Messias* shall live to a good old age, and at last shall be brought to his grave with great solemnity: and his son shall reign after him, and after his death his posterity shall possess his seat.

{The manner of life the Jews shall have under their Messias.} For the manner of life which the Jews shall have under their *Messias*. First of all the remnant of the Christians and other people which fell not by the hand of the Jews shall make hast and build the Jews houses and Cities, not for hire, but of free accord, till their ground, plant them vineyards, yea, bestow their very goods upon them; moreover Kings and Princes shall be their servants whom they have subdued. They themselves shall be cloathed in costly aray: all their Priests anointed shall be holiness to the Lord; as it is written: {Jsa. 60.10,11,12.} *The sons of strangers shall build up thy walls, and their Kings shall minister unto thee: for in my wrath I smote thee, but in mercy have I had favour on thee, therefore thy gates shall be open continually, they shall not be shut day nor night, that men may bring unto thee the forces of the Gentiles, and that their Kings may be brought, for the nation and kingdom that will not serve thee shall perish, yea those nations shall be utterly wasted*, and again {Jsa. 61.5.6.} *strangers shall stand and feed your flocks, and all the sons of the alien shall be your plow-men, and your vine-dressers. But you shall be named the Priests of the Lord, men shall call you the Ministers of our God: you shall eat the riches of the Gentiles, and in their glory shall you boast your selves.* (Oh here with hunger and thirst how are the Jews opprest? Although some of them satisfie and appease both, without the sweat of their own brows gaining many a million: for which many a poor Christian suffers toile and vexation.)

{The 2 benefit.} 2. They shall have a new and wholsome aire, as it is written: {Jsa. 65.17.} *Behold I create a new heaven and a new earth, the former shall not be thought upon*, by the benefit of this aire they shall enjoy their health and prolong their life, even as the men before the flood. In their hoary old age their strength and agility shall not forsake them, but remain in the same temper as in their youth, as it is written, {Psal. 92.14,15.} *They who are planted in the house of our God, shall flourish in the courts of the Lord, they shall bring forth more fruit in their age, they shall be fat and well liking.*

{The 3 Benefit} 3. The seed once sown shall for ever grow up, increase, and ripen of its own accord: after the manner of Vines which require but one plantation, as it is written, {Hos. 14.8.} *They shall revive as wheat, flourish like a vine, his smell is like Lebanon.*

Whensoever any one shall desire rain for the watering of any particular Field, Garden, or the smallest herb therein, the Lord will pour out upon that place, and on that onely, without delay: for saith the Prophet, *Ask you rain of the Lord, and he shall create lightnings, and give you showres of rain.* Then shall they gather their fruits and wine with great quietnesse and security, and shall not be molested by any enemy: as it is written, The Lord hath sworn by his right hand, and by the arm of his strength, {Isay 62.8,9.} *I will no more give thy corn to be meat for thine enemies, and the sons of strangers shall not drink thy wine for the which thou hast laboured, but they that have gathered it shall eat it.*

{The 4 Benefit} 4. No war nor rumour of war shall any more be heard in the land: and there shall be a firm and secure peace established, not only between man and man, but also between man and beast; as it is written, *I will make a covenant for them in that day with the beasts of the field, with the fowls of heaven, and creeping things of the earth: I will put away the bow and the sword and war from the earth, and make them to sleep secure. And I will espouse thee unto me for ever and ever: I will marry thee in justice and judgement, in mercy and commiseration.* Again, {Esay 11.17.} *The Cow and the Bear shall feed: their young ones shall lie down together, and the Lion shall eat straw with the Ox. The Wolf shall lie down with the Lamb, and the Leopard with the Kid: and the Calf and the young Lion and the fatling together, and a little childe shall lead them.*

{The 5 Benefit} 5. When any war or discord ariseth among the Gentiles, then the Messias shall reconcile them, and renew the league amongst them: so that there shall be no more mutiny; as it is written, {Isay 2.4} *He shall judge among the nations, and rebuke many people; he shall beat their swords into plowshares, and their spears into pruning-hooks: nation shall not lift up sword against nation, nor learn war any more.* Then shall the Iews live in everlasting joyes, make new marriages, sing praise and glory to God without ceasing: shall be full of the wisdom and knowledge of the Lord: as it is written, *In this place of which you say that it is forsaken, shall again be heard the voice of joy, the voice of exultation, the voice of the Bride and the Bridegroom, the voice of them that say, Give thanks to the Lord of hosts.* And again, *the earth shall be full of the knowledge of the Lord as the sea is full of water.*

Briefly, the happiness of this holy people shall at that time be so immeasurable, that neither can the heart of man conceive it, or the tongue yeeld the least expression thereof. Which things thus ordered and declared, leaving the Iewes in this their prosperous estate, I will put a period to my labours, and hide the secret of their faith from the Christians; seeing I have attempted more then they themselves, if they could have ruled the matter, would have permitted. What I have done already will not be pleasing unto them, in which I have exposed to every mans eye the full anatomy of their life and belief.

The Christian Reader may easily perceive by that which hath been said, that the faith of the Jews and their whole religion, is not grounded upon *Moses*, but upon meer lies, false and forged constitutions, fables of the Rabbines, and inventions of seduced Pharisees. And that therefore it ought no more to issue out of the mouth of a Christian, that the Jewes stand for the Law of *Moses*, but rather with *Jeremy*, {Jer. 8.} that they are strong defendants of the false worship of the true God, not suffering themselves any way to be drawn from it. And with our Saviour to affirm, that {Matth. 15.5} *they make the Commandments of God of none effect by their traditions*; in vain they worship him, when they teach nothing but the mandates of men: honouring him with their lips, but in their hearts are far from him. In their words they professe to know God, but in their works they deny him: {Titus 1.} these are the men whom the Lord abhors, who being disobedient unto his word are unto every good work reprobate, as the Apostle *Paul* hath recorded. By which it is more manifest then the light of the Sun at noontide, that the punishment is now fallen heavie upon them wherewith *Moses* threatened them: that {Deut. 28.} *the Lord should smite them with madnesse, blindnesse, and astonishment of heart, that they should grope at noon day as the blinde gropeth in darknesse*. And this appears most clearly, and is more then evident from this, that they miserably pervert, and contrary to all reason with an impudent front invested with a dull ignorance expound and interpret the word of God.

O merciful God, who hast vouchsafed to impart this gracious favour unto us Christians, that we being warned by such an horrible example of the divine wrath, should with awe and reverence embrace his holy word, lest the same things should befal us, and so our Candlestick should be removed for our ingratitude, God of his mercy grant, that the Sun of his justice may alwayes shine in our hearts until perfect day, and by the illumination of his good Spirit conduct us unto all truth. Amen."

Interestingly, Charles Taze Russell determined in 1876 that the reign of the Gentile governments would end in 1914—which is the year World War I began—and that the Jews would then take over the world. Russell supposedly made his prediction based on Scripture, and his followers spread his message widely. In an article, "Gentile Times: When Do They End?", *The Bible Examiner*, Volume 21, Number1, Whole Number 313, (October, 1876), pp. 27-28; Charles Taze Russell wrote,

> "'Jerusalem shall be trodden down of the Gentiles until the times of the Gentiles are fulfilled.'—Luke 21:24.
>
> Doubtless our Lord intended to communicate to His disciples some knowledge, and possibly it was addressed more to the disciples in our day, than to the early church.
>
> Let us then search what times the prophecy, which was in Christ, did

signify. Of course, if it be one of the secret things of God, we cannot find out; but if a secret, why should Jesus mention it? If, on the contrary, it is revealed it *belongs* to us. Shall we guess and suppose? No: let us go to God's treasure-house; let us search the Scriptures for the key.

Jesus does not *foretell* its treading under foot of the Gentiles, as Rome had her foot upon them at that time. He does tell us, however, how long it will continue so, even the disciples thought 'that it was he which should have DELIVERED Israel.'

We believe that God has given the key. We believe He doeth nothing but he revealeth it unto His servants. Do we not find part of the key in Lev. 26:28, 33 'I, even I will chastise you seven times for your sins: ... and I will bring your land into desolation ... and will scatter you among the heathen.' Israel did not hearken unto the Lord, but disobeyed him, and this prophecy is now being fulfilled, and has been since the days of Zedekiah, when God said, 'Remove the diadem, take off the crown, ... I will overturn, overturn, overturn it, ... until He comes whose right it is, and I will give it unto Him.' Comparing these Scriptures, we learn, that God has scattered Israel for a period of seven times, or until 'he comes whose right' the Government is, and puts an end to Gentile rule or government. This gives us a clue at least, as to how long until the Jews are delivered. Further, Nebuchadnezzar, king of Babylon, the head of gold, is recognized by God as the representative of the beast, or Gentile Governments. 'A king of kings and wheresoever the children of men dwell, the beasts of the field, and the fowls of the air, hath God given into his hand.' Dan. 2:38. God had taken the crown off Zedekiah and declared the Image, of which Nebuchadnezzar is the head, ruler of the world until the kingdom of God takes its place (smiting it on its feet); and, as this is the same time at which Israel is to be delivered, (for 'Jerusalem shall be trodden down of the Gentiles until the times of the Gentiles are fulfilled'), we here get our second clue, viz.: these two events, noted of the Scriptures of truth—'Times of Gentiles,' and 'Treading of Jerusalem,' are parallel periods, commencing at the same time and ending at the same time; and, as in the case of Israel, their degradation was to be for seven times, so with the dominion of the Image; it lasts seven times; for, when in his pride the 'Head of Gold' ignored 'The God of heaven,' the glory of that kingdom (which God gave him, as a representative of the Image,) departed, and it took on its beastly character, which lasts seven times. Dan 4:23—and, (prefigured by the personal degradation for seven years, of Nebuchadnazzar, the representative) until the time comes when they shall acknowledge, and 'give honor to the Most High, whose Kingdom is an everlasting Kingdom.' Dan 4:34: for all the ends of the earth shall remember and turn unto the Lord when He is the Governor among the nations.

Our next question naturally, is, How long are seven times? Does God in his word, furnish us any clue from which to determine the length of that period? Yes, in Revelations we learn that three and one-half times, 42 months, and 1260 prophetic days, literal years, are the same (it has for years been so accepted by the church,) and it was so fulfilled: if three and one-half times are 1260 years, seven times would be twice as much, i.e., 2520 years. At the commencement of our Christian era, 606 years of this time had passed, (70 years captivity, and 536 from Cyrus to Christ) which deducted from 2520, would show that the seven times will end in A.D. 1914; when Jerusalem shall be delivered forever, and the Jew say of the Deliverer, 'Lo, this is our God, we have waited for Him and He will save us.' When Gentile Governments shall have been dashed to pieces; when God shall have poured out of his fury upon the nation, and they acknowledge, him King of Kings and Lord of Lords.

But, some one will say, 'If the Lord intended that we should know, He would have told us plainly and distinctly how long.' But, no, brethren, He never does so. The Bible is to be a *light* to God's children;—to the world, foolishness. Many of its writings are solely for *our* edification upon whom the ends of the world are come. As well say that God should have put the gold on top instead of in the bowels of the earth it would be too common; it would lose much of its value. So with truth; but, 'to you it is given to know the mysteries of the kingdom.

We will ask, but not now answer, another question: If the Gentile Times end in 1914, (and there are many other and clearer evidences pointing to the same time) and we are told that it shall be with fury poured out; at time of trouble such as never was before, nor ever shall be; a day of wrath, etc., how long before does the church escape? as Jesus says, 'watch, that *ye* may be accounted worthy to *escape* those things coming upon the world.'

Brethren, the taking by Christ of His Bride, is evidently, one of the first acts in the Judgment; for judgment must begin at the house of God.

W. *Philadelphia.*"

The World, of New York, wrote on 30 August 1914,

"The terrific war outbreak in Europe has fulfilled an extraordinary prophecy. For a quarter of a century past, through preachers and through press, the 'International Bible Students,' best known as 'Millenial Dawners,' have been proclaiming to the world that the Day of Wrath prophesied in the Bible would dawn in 1914. 'Look out for 1914! has been the cry of the hundreds of traveling evangelists."[449]

Were Cabalistic Jews working with Russell and conditioning Gentiles to surrender their rights to Jews? Did Cabalistic Jews simply time the war based on the same premises as Russell used to arrive at his predictions, or was it the other way around? Cabalistic Jews have long practiced numerology.

Ben Justin Martyr alleged that Jews murdered and defamed Christians from the very beginning of the Christian movement—as did other sources, including

---

449. Quoted in: *Revelation: It's Climax at Hand!*, Watchtower Bible and Tract Society of New York, Inc., International Bible Students Association, Brooklyn, New York, (1988), p. 105.

Biblical sources. Gustaf Dalman wrote,

> "SINCE everyone has not the writings of Justin at hand, we venture to offer some important extracts from them bearing on this subject. We quote in accordance with the edition of J. C. Th. Otto, Jena, 1843:—'The Jews regard us as foes and opponents, and kill, and torture us, if they have the power. In the lately-ended Jewish war Bar Kokh'ba, the instigator of the Jewish revolt, caused Christians alone to he dragged to terrible tortures, whenever they would not deny and revile Jesus Christ [*Footnote: Apology*, I. chap. 31.].' 'The Jews hate us, because we say that Christ is already come, and because we point out that He, as had been prophesied, was crucified by them [*Footnote: Ibid.* chap. 36].'—'Therefore we pray both for you Jews and for all other men who hate us, that you place yourselves in company with us, and against those, whom His works, and the miracles now still wrought through the invoking of His Name, and His teaching, as well as the prophecies concerning Him as wholly undefiled and blameless, all unite to admonish that they should vomit forth no revilings against Jesus Christ, but believe on Him [*Footnote: Dialogue with Trypho*, chap. 35.].' 'The high-priests of your nation and your teachers have caused that the name of Jesus should be profaned and reviled through the whole world [*Footnote: Ibid.* chap. 117].'—' Ye have killed the Just and His prophets before Him. And now ye despise those, who hope in Him and in God, the King over all and Creator of all things, who has sent Jesus; ye despise and dishonour them, as much as in you lies, in that in your synagogues ye curse those who believe in Christ. Ye only lack the power, on account of those who hold the reins of government, to treat us with violence. But as often as ye have had this power, ye have also done this [*Footnote: Ibid.* chap. 16].' 'In your synagogues ye curse all who have become Christians, and the same is done by the other nations, who give a practical turn to the curse, in that when any one merely acknowledges himself a Christian, they put him to death [*Footnote: Dialogue with Trypho*, chap. 96.].' 'Nay, ye have added thereto, that Christ taught those impious, unlawful, horrible actions, which ye disseminate as charges above all, against those who acknowledge Christ as Teacher and as the Son of God [*Footnote: Ibid.* chap. 108].' 'Yet revile not the Son of God, and hearken not to the Pharisees as teachers, that after prayer ye should ill-treat the King of Israel with scoffs, as they have been taught you by the rulers of the synagogue [*Footnote: Ibid.* chap. 137.].' — 'As far as depends on you and the rest of mankind, each Christian is driven not only from his possession, but completely out of the world: ye permit no Christian to live [*Footnote: Ibid.* chap. 110.].'—'Your hand is stretched out for ill-doing. For instead of experiencing repentance for having put Christ to death, ye hate us who through Him believe on God and the Father of all things, and ye put us to death as often as ye have the power, and ye continually curse Christ and His adherents, whereas we all pray for you as in general for all men' (after the wording of Matt. v. 44; Luke vi. 27 f.) [*Footnote: Ibid.* chap. 133.],—'Your teachers exhort you to permit

yourselves no conversation whatever with us [*Footnote: Ibid.* chap. 112.].'—'There does not press upon other nations so heavy an offence against us and Christ as upon you, who are the originators of the preconceived evil opinion, which the nations cherish concerning Christ and us, His disciples. For since ye have attached Him the only blameless and righteous One to the Cross, ye have not only made no amends for your atrocious action, but at that time ye sent forth chosen men from Jerusalem, to proclaim throughout the world, that there is a new sect, namely, the Christians, arisen, which reverence no God, and to spread abroad what all who know us not maintain concerning us. It was your most earnest endeavour that bitter, dark, unjust charges should be put into circulation throughout the whole world against that sole spotless and righteous Light, which was sent from God to men [*Footnote: Ibid.* chap. 17.].'—'The Jews make war against the Christians as against a foreign nation, and the Greeks (*i.e.* the Gentiles) persecute them; but their enemies can allege no ground of hostility [*Footnote: Letter to Diognetus*, chap. 5.].'"[450]

Jews massacred Christians in Palestine. James Parkes wrote,

"The day of a brief relief and revenge was, however, approaching. Justinian's grandiose dreams of imperial magnificence, and his passion for building—including several churches in Palestine—had heavily overstrained the empire's weak economic resources. His successors could not possibly maintain what he had so rashly conquered; and the empire fell a prey to disorder. Then occurred a repetition of the superstitious fears which had led Valerian and Diocletian to persecute the Christians, only this time the infidels who were said to be angering the Almighty were the Jews. Phocas (602-610) and his successor Heraclius (610-641) were said to have been warned that the empire was menaced by 'the circumcised', and both in consequence ordered the Jews of the empire to accept baptism. What numbers submitted we have no means of knowing. In any case their submission was probably of short duration, for in 611 the Persians swept through the eastern provinces, and in 614 they took Jerusalem after a siege lasting only twenty days. There is no doubt that the Persians received substantial help from the Jews of Galilee. One chronicler mentions a figure of 20,000 Jewish soldiers, another 26,000. While the actual figures are as unreliable as all ancient figures, there is no reason to question the fact that the Jews aided the Persians with all the men they could muster, and that the help they gave was considerable. Once Jerusalem was in Persian hands a terrible massacre of Christians took place, and the Jews are accused of having taken the lead in this massacre. It would not be surprising if the accusation were true, even though the fantastic stories told of Jewish

---

450.G. Dalman, *Jesus Christ in the Talmud, Midrash, Zohar, and the Liturgy of the Synagogue*, Deighton Bell, Cambridge, (1893), pp. 99-100.

revenge by Christian chroniclers are certainly exaggerated. The Jews seem to have hoped that the Persians would allow them the full possession of Jerusalem, and even the re-establishment of an autonomous state. But the Persian occupation was too short for such plans to develop. It lasted only fifteen years.

In those fifteen years, however, changes occurred which centuries were not to repair. The country had been desolated by the Persian armies; agriculture had come to a standstill; cities were empty, while their inhabitants had fled to the mountains; churches and monasteries were in ruins, and much of Jerusalem itself was burnt. All the treasures collected in its shrines, including the 'true cross' itself, had been taken away, and the patriarch Zacharias sent with thousands of others as prisoners to Persia. It was a half empty country filled with ruins which, by a supreme effort, Heraclius managed to reoccupy in 629. Though he himself seems to have been inclined to spare the Jews for the part they had played as allies of the Persians, the clergy of Jerusalem thought only of revenge; and as bloody a massacre took place of Jews as had, previously taken place of Christians. But that—and the recovery of the 'true cross'—was all the satisfaction that the Christians got. A far more powerful enemy was approaching. In the year in which Heraclius regained Palestine, Muhammad was completing his conquest of Mecca. In 636 his followers entered the country; in 640 Caesarea surrendered and Byzantine rule was at an end."[451]

One can expect that when the Jews anoint their Messiah, he will be especially vicious to Christians, because he will resent their belief that Jesus was the Messiah and not him.

In 1802, in the context of Hartley's and Napoleon's Zionism, Johann Gottfried Herder believed that Hartley and his ilk were trying to "restore the Jews to Palestine" in order to make the world safe for a Jewish monopolization of trade among the Continents, because Palestine itself could not provide the Jews with the great wealth they needed to fund the dominance Hartley had planned for them. If the Christians were ruined and dispersed, as Hartley planned, Judaized "Christian" settlements could provide the Jews with infrastructure around the world, and Christian armies could "civilize" and dominate lands the Jews could not, and Christian navies could secure Jewish trade. It was obvious that Hartley had called for a Christian diaspora, based on the model of the Jewish Diaspora, in order to forward the interests of the Jews, not the interests of the Christians. He wanted Christians to become Jews and then spread Judaism around the globe. Hartley, who was an agent of the Cabalistic Jews, would accomplish these ends by teaching the Christians to welcome their demise at the hands of Jewish revolutionaries.

The Jewish revolutionaries accomplished their goals in France and Poland.

---

451. J. Parkes, *A History of Palestine from 135 A. D. to Modern Times*, Oxford University Press, New York, (1949), pp. 81-82. See also: E. Horowitz, *Reckless Rites: Purim and the Legacy of Jewish Violence (Jews, Christians, and Muslims from the Ancient to the Modern World)*, Princeton University Press, (2006).

In 1899, Edouard Drumont wrote, *inter alia*,

> "During the Revolution, [Jewish money power] was with us; then it supported Bonaparte; in 1815, it was clearly against him, and, at Waterloo, with Rothschild it fought as energetically as Wellington. [***] After having been, at its birth, the apotheosis of Power, France culminates in the apotheosis of Money. It had two masters; Napoleon, in the beginning; Rothschild, personification of the Jewish Conquest, at its decline. [***] Already in 1875 a Jew who is mostly forgotten today but who was then almost famous and who was, in any case, a most interesting and very curious spirit, Alexandre Weill,[11] explained to me that France was obliged to undergo the same fate as Poland and that it would be good, in the best interests of Humanity, that the French, dispersed and countryless like the Poles, would go and spread throughout the world the general truths of civilization and progress"[452]

Drumont recounted in 1899, that Alexandre Weill, an elderly and supposedly prophetic man of Jewish descent, had told him that France would end up in a diaspora like Poland, which had been devastated, divided and dispersed by Frankist Jews. In fact, both Poland and France, two predominantly Catholic nations which at one time had led European culture, were battlegrounds in both World Wars—in the case of France, just as Weill and Drumont had predicted. In the early 1790's, Poland suffered under Russian tyranny after the Frankist Jews had undermined the Polish Government. Many Polish intellectuals, philosophers, poets, artists, political theorists, etc. fled to places like France,[453] which was embroiled in a revolution, and carried with them their sophisticated knowledge and ways. Weill looked forward to another destruction of France which he hoped would result in a similar migration of talent and wisdom—all of which recalls David Hartley's desires that Christianity be destroyed and dispersed so as to spread Judaism around the world—which reminds one of *Exodus* 1:7-12,

> "¶7 And the children of Israel were fruitful, and increased abundantly, and multiplied, and waxed exceeding mighty; and the land was filled with them. 8 Now there arose a new king over Egypt, which knew not Joseph. 9 And he said unto his people, Behold, the people of the children of Israel *are* more and mightier than we: 10 Come on, let us deal wisely with them; lest they

---

452. E. A. Drumont, *Les juifs contre la France une nouvelle Pologne*, Librairie Antisémite, Paris, (1899), pp. 36-48; English translation in: R. S. Levy, *Antisemitism in the Modern World: An Anthology of Texts*, D. C. Heath and Company, Toronto, (1991), pp. 107-112, at 107, 111. *See also:* E. A. Drumont, A. de Rothschild and A. L. Burdeau, *Burdeau-Rothschild contre Drumont; Le proces de la libre parole, debats complets*, Paris, (1892).

453. G. Hosking, *Russia and the Russians: A History*, Harvard University Press, (2001), p. 258.

multiply, and it come to pass, that, when there falleth out any war, they join also unto our enemies, and fight against us, and *so* get them up out of the land. 11 Therefore they did set over them taskmasters to afflict them with their burdens. And they built for Pharaoh treasure cities, Pithom and Raamses. 12 But the more they afflicted them, the more they multiplied and grew. And they were grieved because of the children of Israel."

Racist Jews and Reformed Jews believed that the Diaspora of the Jews had benefitted the world by dispersing the Jews, who then spread knowledge of Judaism around the world. The Jews shattered French and Polish society, in part, so that the intellectuals of these highly advanced and sophisticated nations would travel the world spreading modernity and Jewish monotheism, which would make the way easier for Jewish infiltration of the rest of the world, which would fulfill the forecasts of the Jewish prophets who predicted the demise of the Gentiles and the rise of the Jews. Many German scientists left Germany after the Second World War, taking their knowledge of weapons, war and politics to the United States and the Soviet Union.

Jewish and Christian investors and merchants had long profited from trade with the colonial new world, in slaves,[454] furs, sugar, etc. Many great fortunes that were made, were made with inside information and manipulation in the money, commodity and stock markets—especially during wars. Rothschild made a fortune from Napoleon's adventures.[455] As Smedley D. Butler said, "war is a racket."[456]

Herder wrote, in 1802, shortly after Napoleon commenced his Zionist campaigns,

"Good luck to [World Jewry], if a Messiah-Bonaparte may victoriously lead

---

454. Historical Research Department of the Nation of Islam (Chicago), *The Secret Relationship between Blacks and Jews*, Chicago, Latimer Associates, (1991). **For counter-argument, *see:*** H. D. Brackman, *Ministry of Lies: The Truth behind the Nation of Islam's The Secret Relationship between Blacks and Jews*, Four Walls Eight Windows, New York, (1994); **and** "Jews Had Negligible Role in Slave Trade", *The New York Times*, (14 February 1994), p. A16. **Contrast these with Brackman's own statements in his PhD dissertation:** H. D. Brackman, PhD Dissertation, University of Californian, Los Angeles, *The Ebb and Flow of Conflict—History of Black-Jewish Relations Through 1900*, University Microfilms International (Dissertation Services), Ann Arbor, Michigan, (1977); **and *see:*** T. Martin, *The Jewish Onslaught: Despatches from the Wellesley Battlefront*, Majority Press, Dover, Massachusetts, (1993). ***See also:*** L. Brenner, Letter to the Editor, *The New York Times*, (28 February 1994), p. A16; **and** "Harold Brackman Believes in Recycling Garbage", *New York Amsterdam News*, (11 March 1995). ***See also:*** M. A. Hoffman II, *Judaism's Strange Gods*, Independent History and Research, Coeur d'Alene, Idaho, (2000), pp. 66-67.
455. B. M. Baruch, *Baruch: My Own Story*, Henry Holt and Company, New York, (1957), pp. 107-108. A. Muhlstein, *Baron James: The Rise of the French Rothschilds*, Vendome Press, New York, (1982).
456. S. D. Butler, *War Is a Racket*, Round Table Press, New York, (1935).

them there, good luck to them in Palestine! But it will be difficult for this richly competitive nation to live in a narrow Palestine if they cannot there take over the general middle trade of both the old and the new world. For the old world would be convenient to their land. Fine sharp-witted race, wonder of the ages! One of the brilliant glosses of their rabbis yokes together a complaining Esau and Israel [Jacob]. Both suffer from the kiss, but they cannot separate themselves."[457]

In the 1830's, Godfrey Higgins suspected that Napoleon viewed himself as the Messiah of the Jews,

"To what I have said in Vol. I. p. 688, respecting Napoleon, I think it expedient to add a well-known anecdote of him. When his uncle, Cardinal Fesch, once expostulated with him, and expressed his belief that he must one day sink beneath that universal hatred with which his actions were surrounding his throne, he led his uncle to the window, and, pointing upwards, said, *"Do you see yonder star?"* *"No sire,"* was the reply. *"But I see it,"* answered Napoleon, and abruptly dismissed him.[*Footnote:* J. T. Baker, of Deptford, to Ed. of Morn. Chron., Oct. 12, 1832.] What are we to make of this? Here we have the star of Jacob, of Abraham, of Cæsar. Here we have a star, probably from the East. The whole of Napoleon's actions in the latter part of his life bespeak mental alienation. I believe that he continued to retain expectations and hopes of restoration to the empire of the world, till the day of his death. Many circumstances unite to persuade me that he was latterly the victim of monomania. I cannot help suspecting that Napoleon was tainted with a belief that he was *the promised one*. [***] Victor Cousin says, "You will remark, that all great men have, in a greater or less degree, been fatalists: *the error is in the form, not at the foundation of the thought*. They feel that, in fact, they do not exist on their own account: they possess the consciousness of an immense power, and being unable to ascribe the honour of it to themselves, they refer it to a higher power which uses them as its instruments, in accordance with its own ends."[*Footnote:* For. Quar. Review, No. XXIII. July 1833, p.202.] With the exception of the words in Italics, which I do not understand, I quite agree with M. Cousin. But how completely it bears me out in the assertion I have made, that the belief in each person that *he was the great one that was for to come* has led either to his success or to his destruction! It led Julian into the dessert—Napoleon to Moscow."[458]

---

457. P. L. Rose, *Revolutionary Antisemitism in Germany from Kant to Wagner*, Princeton University Press, (1990), pp. 104-105. Rose cites: J. H. Herder, "Bekehrung der Juden", *Adrastea* (Leipzig), Volume 4, (1802); Reprinted: J. H. Herder, *Sämtliche Werke*, Collected Works in 33 Volumes Edited by Bernhard Ludwig Suphan, Volume 24, Georg Olms, Hildesheim, (1877 Reprinted 1967), p. 67.
458. G. Higgins, *Anacalypsis: An Attempt to Draw Aside the Veil of the Saitic Isis : Or, an Inquiry into the Origin of Languages, Nations, and Religions*, Volume 2, Book 5,

Hartley and later Shaftesbury iterated themes repeated again and again by English Christian Zionists through to the time of Winston Churchill, and beyond. The same themes reappear today in the beliefs of evangelical Dispensationalist Christian Zionists and neo-conservative Zionists in the United States. On the other end of the Protestant political spectrum, the anti-Semites followed Martin Luther's call for the expulsion of the Jews and the destruction of Catholicism, all of which forwards the Jewish Zionist agenda.

Hartley wrote in 1749, and his work is but one of thousands of such examples of Jewish Zionist propaganda published by pseudo-Christian traitors,

"PROP. 41.

The Divine Authority of the Scriptures may be inferred from the superior Wisdom of the Jewish *Laws, considered in a political Light; and from the exquisite Workmanship shewn in the Tabernacle and Temple.*

ALL these were Originals amongst the *Jews,* and some of them were copied partially and imperfectly by ancient Heathen Nations. They seem also to imply a Knowledge superior to the respective Times. And I believe, that profane History gives sufficient Attestation to these Positions. However, it is certain from Scripture, that *Moses* received the whole Body of his Laws, also the Pattern of the Tabernacle, and *David* the Pattern of the Temple, from God; and that *Bezaleel* was inspired by God for the Workmanship of the Tabernacle. Which Things, being laid down as a sure Foundation, may encourage learned Men to inquire into the Evidences from profane History, that the Knowledge and Skill to be found amongst the *Jews* were superior to those of other Nations at the same Period of Time, *i. e.* were supernatural.

[***]

SECT. II.

Of the Expectation of Bodies Politic, the Jews *in particular, and the World in general, during the present State of the Earth.*

PROP. 81.

It is probable, that all the present Civil Governments will be overturned.

THIS may appear from the Scripture Prophecies, both in a direct way, *i. e.* from express Passages, such as those concerning the Destruction of the Image, and Four Beasts, in *Daniel*; of Christ's *breaking all Nations with a Rod of Iron, and dashing them in Pieces like a Potter's Vessel, &c.* and from the Supremacy and universal Extent of the Fifth Monarchy, or Kingdom of the Saints, which is to be set up.

We may conclude the same Thing also from the final Restoration of the *Jews,* and the great Glory and Dominion promised to them, of which I shall speak below.

And it adds some Light and Evidence to this, that all the known Governments of the World have the evident Principles of Corruption in

---

Chapter 2, Longman, Rees, Orme, Brown, Green, and Longman, London, (1836), p. 358.

themselves. They are composed of jarring Elements, and subsist only by the alternate Prevalence of these over each other. The Splendor, Luxury, Self interest, Martial Glory, &c. which pass for Essentials in Christian Governments, are totally opposite to the meek, humble, self-denying Spirit of Christianity; and whichsoever of these finally prevails over the other, the present Form of the Government must be dissolved. Did true Christianity prevail throughout any Kingdom intirely, the Riches, Strength, Glory, &c. of that Kingdom would no longer be an Object of Attention to the Governors or Governed; they would become a Nation of Priests and Apostles, and totally disregard the Things of this World. But this is not to be expected: I only mention it to set before the Reader the natural Consequence of it. If, on the contrary, worldly Wisdom and Infidelity prevail over Christianity, which seems to be the Prediction of the Scriptures, this worldly Wisdom will be found utter Foolishness at last, even in respect of this World; the Governments, which have thus lost their Cement, the Sense of Duty, and the Hopes and Fears of a future Life, will fall into Anarchy and Confusion, and be intirely dissolved. And all this may be applied, with a little Change, to the *Mahometan* and *Heathen* Governments. When Christianity comes to be propagated in the Countries where these subsist, it will make so great a Change in the Face of Affairs, as must shake the Civil Powers, which are here both externally and internally opposite to it; and the Increase of Wickedness, which is the natural and necessary Consequence of their Opposition, will farther accelerate their Ruin.

The Dissolution of antient Empires and Republics may also prepare us for the Expectation of a Dissolution of the present Governments. But we must not carry the Parallel too far here, and suppose that as new Governments have arisen out of the old ones, resembling them in great measure, subsisting for a certain time, and then giving place to other new ones, so it will be with the present Governments. The Prophecies do not admit of this; and it may be easily seen, that the Situation of Things in the Great World is very different from what it has ever been before. Christianity must now either be proved true, to the intire Conviction of Unbelievers; or, if it be an Imposture, it will soon be detected. And whichsoever of these turns up, must make the greatest Change in the Face of Affairs. I ought rather to have said, that the final Prevalence and Establishment of Christianity, which, being true, cannot but finally prevail, and be established, will do this. But it may perhaps be of some Use just to put false Suppositions.

How near the Dissolution of the present Governments, generally or particularly, may be, would be great Rashness to affirm. Christ will come in this Sense also *as a Thief in the Night*. Our Duty is therefore to watch, and to pray; to be faithful Stewards; to give Meat, and all other Requisites, in due Season, to those under our Care; and to endeavour by these, and all other lawful Means, to preserve the Government, under whose Protection we live, from Dissolution, seeking the Peace of it, and submitting to every Ordinance of Man for the Lord's sake. No Prayers, no Endeavours of this Kind, can fail of having some good Effect, public or private, for the Preservation of ourselves or others. The great Dispensations of Providence are conducted by Means that are either secret, or, if they appear, that are judged feeble and inefficacious. No man can tell, however private his Station may be, but his fervent Prayer may avail to the Salvation of much People. But it is more peculiarly the Duty of Magistrates thus to watch over their Subjects, to pray for them, and set about the Reformation of all Matters Civil and Ecclesiastical, to the utmost of their Power. Good Governors may promote the Welfare and Continuance of a State, and wicked ones must accelerate its Ruin. The sacred History affords us Instances of both Kinds, and they are recorded there for the Admonition of Kings and Princes in all future Times.

It may not be amiss here to note a few Instances of the Analogy between the Body Natural, with the Happiness of the Individual to which it belongs, and the Body Politic, composed of many Individuals, with its Happiness, or its flourishing State in respect of Arts, Power, Riches, &c. Thus all Bodies Politic seem, like the Body Natural, to tend to Destruction and Dissolution, as is here affirmed, through Vices public and private, and to be respited for certain Intervals, by partial, imperfect Reformations. There is no complete or continued Series of public Happiness on one hand, no utter Misery on the other; for the Dissolution of the Body Politic is to be considered as its Death. It seems as romantic therefore for any one to project the Scheme of a perfect Government in this imperfect State, as to be in Pursuit of an universal Remedy, a Remedy which should cure all Distempers, and prolong human Life beyond Limit. And yet as Temperance, Labour, and Medicines, in some Cases, are of great Use in preserving and restoring Health, and prolonging Life; so Industry, Justice, and all other Virtues, public and private, have an analogous Effect in respect of the Body Politic. As all the Evils, which Individuals suffer through the Infirmity of the mortal Body, and the Disorders of the external World, may, in general, contribute to increase their Happiness even in this Life, and also are of great Use to others; and as, upon the Supposition of a future State, Death itself appears to have the same beneficial Tendency in a more eminent Degree than any other Event in Life, now considered as indefinitely prolonged; so the Distresses of each Body Politic are of great Use to this Body itself, and also of great Use to all neighbouring States; and the Dissolution of Governments have much promoted the Knowledge of true Religion, and of useful Arts and Sciences, all which seem, in due time and manner, intended to be intirely subservient to true Religion at last. And this affords great Comfort to benevolent and religious Persons, when they consider the Histories of Former Times, or contemplate the probable Consequences of Things in future Generations.

PROP. 82
It is probable, that the present Forms of Church-Government will be dissolved.

THIS Proposition follows from the forgoing. The Civil and Ecclesiastical Powers are so interwoven and cemented together, in all the Countries of *Christendom*, that if the first fall, the last must fall also.

But there are many Prophecies, which declare the Fall of the Ecclesiastical Powers of the Christian World. And through each Church seems to flatter itself with the Hopes of being exempted; yet it is very plain, that the prophetical Characters belong to all. They have all left the true, pure, simple Religion; and teach for Doctrines the Commandments of Men. They are all Merchants of the Earth, and have set up a Kingdom of this World, abounding in Riches, temporal Power, and external Pomp. They have all a dogmatizing Spirit, and persecute such as do not receive their own Mark, and worship the Image which they have set up. They all neglect Christ's Command of preaching the Gospel to all Nations, and even of going to *the lost Sheep of the House of Israel*, there being innumerable Multitudes in all Christian Countries, who have never been taught to read, and who are, in other respects also, destitute of the Means of saving Knowledge. It is very true, that the Church of *Rome* is *Babylon the Great, and the Mother of Harlots, and of the Abominations of the Earth*. But all the rest have copied her Example, more or less. They have all received Money, like *Gehazi*; and therefore the Leprosy of *Naaman* will cleave to them, and to their Seed for ever. And this Impurity may be considered not only as justifying the Application of the Prophecies to all the Christian Churches, but as a natural Cause for their Downfal. The corrupt Governors of the several Churches will ever oppose the true Gospel, and in so doing will bring Ruin upon themselves.

The Destruction of the Temple at *Jerusalem*, and of the Hierarchy of the *Jews*, may likewise be considered as a Type and Presage of the Destruction of that *Judaical* Form of Rites, Ceremonies, and human Ordinances, which takes place, more or less, in all Christian Countries.

We ought, however, to remark here,

First, That though the Church of Christ has been corrupted thus in all Ages and Nations, yet there have been, and will be, in all, many who receive the Seal of God, and worship him *in Spirit, and in Truth*. And of these as many have filled high Stations, as low ones. Such Persons, though they have concurred in the Support of what is contrary to the pure Religion, have, however, done it innocently, with respect to themselves, being led thereto by invincible Prejudices.

Secondly, Nevertheless, when it so happens, that Persons in high Stations in the Church have their Eyes enlightened, and see the Corruptions and Deficiences of it, they must incur the prophetical Censures in the highest Degree, if they still concur, nay, if they do not endeavour to reform and purge out these Defilements. And though they cannot, according to this Proposition, expect intire Success; yet they may be blessed with such a Degree, as will abundantly compensate their utmost Endeavours, and rank them with the Prophets and Apostles.

Thirdly, As this Corruption and Degeneracy of the Christian Church has proceeded from the fallen State of Mankind, and particularly of those Nations to whom the Gospel was first preached, and amongst whom it has been since received; so it has, all things being supposed to remain the same, suited our Circumstances, in the best Manner possible, and will continue to do so, as long as it subsists. God brings Good out of Evil, and draws Men to himself in such manner as their Natures will admit of, by external Pomp and Power, by things not good in themselves, and by some that are profane and unholy. He makes use of some of their Corruptions as Means of purging away the rest. The Impurity of Mankind is too gross to unite at once with the strict Purity of the Gospel. The *Roman* Empire first, and the *Goths* and *Vandals* afterwards, required, as one may say, some Superstitions and Idolatries to be mixed with the Christian Religion; else they could not have been converted at all.

Fourthly, It follows from these Considerations, that good Men ought to submit to the Ecclesiastical *Powers that be*, for Conscience-sake, as well as to the Civil ones. They are both from God, as far as respects Inferiors. Christ and his Apostles observed the Law, and walked orderly, though they declared the Destruction of the Temple, and the Change of the Customs established by *Moses*. Both the *Babylonians*, who destroyed *Jerusalem* the first time, and the *Romans*, who did it the second, were afterwards destroyed themselves in the most exemplary Manner. And it is probable, that those who shall hereafter procure the Downfal of the Forms of Church-Government, will not do this from pure Love, and Christian Charity, but from the most corrupt Motives, and by Consequence bring upon themselves, in the End, the severest Chastisements. It is therefore the Duty of all good Christians to obey both the Civil and Ecclesiastical Powers under which they were born, *i. e.* provided Disobedience to God be not injoined, which is seldom the Case; to promote Subjection and Obedience in others; gently to reform and rectify, and to pray for the Peace and Prosperity of, their own *Jerusalem*.

PROP. 83.

It is probable, that the Jews *will be restored to* Palæstine.

THIS appears from the Prophecies, which relate to the Restoration of the *Jews* and *Israelites* to their own Land. For,

First, These have never yet been fulfilled in any Sense agreeable to the Greatness and Gloriousness of them. The Peace, Power, and Abundance of Blessings, temporal and spiritual, promised to the *Jews* upon their Return from Captivity, were not bestowed upon them in the Interval between the Reign of *Cyrus*, and the Destruction of *Jerusalem* by *Titus*; and ever since this Destruction they have remained in a desolate State.

Secondly, The Promises of Restoration relate to the Ten Tribes, as well as the Two of *Judah* and *Benjamin*. But the Ten Tribes, or *Israelites*, which were captivated by *Salmaneser*, have never been restored at all. There remains therefore a Restoration yet future for them.

Our Ignorance of the Place where they now lie hid, or Fears that they are so mixed with other Nations, as not to be distinguished and separated, ought not to be admitted as Objections here. Like Objections might be made to the Resurrection of the Body; and the Objections both to the one, and the other, are probably intended to be obviated by *Ezekiel*'s Prophecy concerning the dry Bones. It was one of the great Sins of the *Jews* to call God's Promises in Question, on account of apparent Difficulties and Impossibilities; and the *Sadduces*, in particular, erred concerning the Resurrection, because *they knew not the Scriptures, nor the Power of God*. However, it is our Duty to inquire, whether the Ten Tribes may not remain in the Countries where they were first settled by *Salmaneser*, or in some others.

Thirdly, A double Return seems to be predicted in several Prophecies.

Fourthly, The Prophets who lived since the Return from *Babylon*, have predicted a Return in similar Terms with those who went before. It follows therefore, that the Predictions of both must relate to some Restoration yet future.

Fifthly, The Restoration fo the *Jews* to their own Land seems to be predicted in the New Testament.

To the Arguments, drawn from Prophecy, we may add some concurring Evidences, which the present Circumstances of the *Jews* suggest.

First, then, The *Jews* are yet a distinct People from all the Nations amongst which they reside. They seem therefore reserved by Providence for some such signal Favour, after they have suffered the due Chastisement.

Secondly, They are to be found in all the Countries of the known World. And this agrees with many remarkable Passages of the Scriptures, which treat both of their Dispersion, and of their Return.

Thirdly, They have no Inheritance of Land in any Country. Their Possessions are chiefly Money and Jewels. They may therefore transfer themselves with the greater Facility to *Palæstine*.

Fourthly, They are treated with Contempt and Harshness, and sometimes with great Cruelty, by the Nations amongst whom they sojourn. They must therefore be the more ready to return to their own Land.

Fifthly, They carry on a Correspondence with each other throughout the whole World; and consequently must both know when Circumstances begin to favour their Return, and be able to concert Measures with one another concerning it.

Sixthly, A great Part of them speak and write the *Rabbinical Hebrew*, as well as the Language of the Country where they reside. They are therefore, as far as relates to themselves, actually possessed of an universal Language and Character; which is a Circumstance that may facilitate their Return, beyond what can well be imagined.

Seventhly, The *Jews* themselves still retain a Hope and Expectation, that God will once more restore them to their own Land.

COR. 1. May not the two Captivities of the *Jews*, and their two Restorations, be Types of the first and second Death, and of the first and second Resurrections?

COR. 2. Does it not appear agreeable to the whole Analogy both of the Word and Works of God, that the *Jews* are Types both of each Individual in particular, on one hand, and of the whole World in general, on the other? May we not therefore hope, that, at least after the second Death, there will be a Resurrection to Life eternal to every Man, and to the whole Creation, which groans, and travails in Pain together, waiting for the Adoption, and glorious Liberty, of the Children of God?

COR. 3. As the Downfal of the *Jewish* State under *Titus* was the Occasion of the Publication of the Gospel to us Gentiles, so our Downfal may contribute to the Restoration of the *Jews*, and both together bring on the final Publication and Prevalence of the true Religion; of which I shall treat in the next Proposition. Thus the Type, and the Thing typified, will coincide; the First-fruits, and the Lump, be made holy together.

PROP. 84.

The Christian Religion will be preached to, and received by, all Nations.

THIS appears from the express Declarations of Christ, and from many of his Parables, also from the Declarations and Predictions of the Apostles, and particularly from the *Revelation*. There are likewise numberless Prophecies in the Old Testament, which admit of no other Sense, when interpreted by the Events which have since happened, the Coming of Christ, and the Propagation of his Religion.

The Truth of the Christian Religion is an Earnest and Presage of the same Thing, to all who receive it. For every Truth of great Importance must be discussed and prevail at last. The Persons who believe can see no Reasons for their own Belief, but what must extend to all Mankind by degrees, as the Diffusion of Knowledge to all Ranks and Orders of Men, to all Nations, Kindred, Tongues, and People, cannot now be stopped, but proceeds ever with an accelerated Velocity. And, agreeably to this, it appears that the Number of those who are able to give a Reason for their Faith increases every Day.

But it may not be amiss to set before the Reader in one View some probable Presumptions for the universal Publication and Prevalence of the Christian Religion, even in the way of natural Causes.

First, then, The great Increase of Knowledge, literary and philosophical, which has been made in this and the Two last Centuries, and continues to be made, must contribute to promote every great Truth, and particularly those of Revealed Religion, as just now mentioned. The Coincidence of the Three remarkable Events, of the Reformation, the Invention of Printing, and the Restoration of Letters, with each other, in Time, deserves particular Notice here.

Secondly, The Commerce between the several Nations of the World is inlarged perpetually more and more. And thus the Children of this World are opening new Ways of Communication for future Apostles to spread the glad Tidings of Salvation to the uttermost Parts of the Earth.

Thirdly, The Apostasy of nominal Christians, and Objections of Infidels, which are so remarkable in these Days, not only give Occasion to search out and publish new Evidences for the Truth of Revealed Religion, but also oblige those who receive it, to purify it from Errors and Superstitions; by which means its Progress amongst the yet Heathen Nations will be much forwarded. Were we to propagate Religion, as it is now held by the several Churches, each Person would propagate his own Orthodoxy, lay needless Impediments and Stumbling-blocks before his Hearers, and occasion endless Feuds and Dissensions amongst the new Converts. And it seems as if God did not intend, that the general Preaching of the Gospel should be begun, till Religion be discharged of its Incumbrances and Superstitions.

Fourthly, The various Sects, which have arisen amongst Christians in late Times, contribute both to purify Religion, and also to set all the great Truths of it in a full Light, and to shew their practical Importance.

Fifthly, The Downfal of the Civil and Ecclesiastical Powers, mentioned in the 81st and 82d Propositions, must both be attended with such public Calamities, as will make Men serious, and also drive them from the Countries of *Christendom* into the remote Parts of the World, particularly into the *East* and *West-Indies*; whither consequently they will carry their Religion now purified from Errors and Superstitions.

Sixthly, The Restoration of the *Jews*, mentioned in the last Proposition, may be expected to have the greatest Effect in alarming Mankind, and opening their Eyes. This will be such an Accomplishment of the Prophecies, as will vindicate them from all Cavils. Besides which, the careful Survey of *Palæstine*, and the neighboring Countries, the Study of the *Eastern* Languages, of the Histories of the present and antient Inhabitants, &c. (which must follow this Event) when compared together, will cast the greatest Light upon the Scriptures, and at once prove their Genuineness, their Truth, and their Divine Authority.

Seventhly, Mankind seem to have it in their Power to obtain such Qualifications in a natural way, as, by being conferred upon the Apostles in a supernatural one, were a principal Means of their Success in the first Propagation of the Gospel.

Thus, as the Apostles had the Power of Healing miraculously, future Missionaries may in a short time accomplish themselves with the Knowledge of all the chief practical Rules of the Art of Medicine. This Art is wonderfully simplified of late Years, has received great Additions, and is improving every Day, both in Simplicity and Efficacy. And it may be hoped, that a few theoretical Positions, well ascertained, with a moderate Experience, may enable the young Practitioner to proceed to a considerable Variety of Cases with Safety and Success.

Thus also, as the Apostles had the Power of speaking various Languages miraculously, it seems possible from the late Improvements in Grammar, Logic, and the History of the human Mind, for young Persons, by learning the Names of visible Objects and Actions in any unknown barbarous Language, to improve and extend it immediately, and to preach to the Natives in it.

The great Extensiveness of the *Rabbinical Hebrew*, and of *Arabic*, of *Greek* and *Latin*, of *Sclavonic* and *French*, and of many other Languages, in their respective ways, also of the *Chinese* Character, ought to be taken into Consideration here.

And though we have not the Gift of Prophecy, yet that of the Interpretation of Prophecy seems to increase every Day, by comparing the Scriptures with themselves, the Prophecies with the Events, and, in general, the Word of God with his Works.

To this we may add, that when Preachers of the Gospel carry with them the useful manual Arts, by which human Life is rendered secure and comfortable, such as the Arts of Building, tilling the Ground, defending the Body by suitable Cloathing, &c. it cannot but make them extremely acceptable to the barbarous Nations; as the more refined Arts and Sciences, Mathematics, natural and experimental Philosophy, &c. will to the more civilized ones.

And it is in an additional Weight in favour of all this Reasoning, that the Qualifications here considered may all be acquired in a natural way. For thus they admit of unlimited Communication, Improvement, and Increase; whereas, when miraculous Powers cease, there is not only one of the Evidences withdrawn, but a Recommendation and Means of Admittance also.

However, far be it from us to determine by Anticipation, what God may or may not do! The natural Powers, which favour the Execution of this great Command of our Saviour's, to preach the Gospel to all Nations, ought to be perpetual Monitors to us to do so; and as we now live in a more adult Age of the World, more will now be expected from our natural Powers. The *Jews* had some previous Notices of Christ's First Coming, and good Persons were thereby prepared to receive him; however, his Appearance, and intire Conduct, were very different from what they expected; so that they stood in need of the greatest Docility and Humility, in order to become Disciples and Apostles. And it is probable, that something analogous to this will happen at Christ's Second Coming. We may perhaps say, that some Glimmerings of the Day begin already to shine in the Hearts of all those, who study and delight in the Word and Works of God.

PROP. 85
It is not probable, that there will be any pure or complete Happiness, before the Destruction of this World by Fire.

THAT the Restoration of the *Jews*, and the universal Establishment of the true Religion, will be the Causes of great Happiness, and change the Face of the World much for the better, may be inferred both from the Prophecies, and from the Nature of the Thing. But still, that the great Crown of Glory promised to Christians must be in a State ulterior to this Establishment, appears for the following Reasons.

First, From the express Declarations of the Scriptures. Thus St. *Peter* says, that the Earth must be burnt up, before we are to expect *a new Heaven, and new Earth, wherein dwelleth Righteousness*; and St. *Paul*, that *Flesh and Blood cannot inherit the Kingdom of God*; the celestial, glorious Body, made like unto that of Christ, at the Resurrection of the Dead, being requisite for this Purpose.

Secondly, The present disorderly State of the natural World does not permit of unmixed Happiness; and it does not seem, that this can be rectified in any great Degree, till the Earth have received the Baptism by Fire.

But I presume to affirm nothing particular in relation to future Events. One may just ask, whether Christ's Reign of a Thousand Years upon Earth does not commence with the universal Establishment of Christianity; and whether the Second Resurrection, the new Heavens, and new Earth, &c. do not coincide with the Conflagration.

One ought also to add, with St. *Peter*, as the practical Consequence of this Proposition, that the Dissolution of this World by Fire is the strongest Motive to an Indifference to it, and to that holy Conversation and Godliness, which may fit us for *the new Heavens, and new Earth*."[459]

Note Hartley's statement,

"First, then, The *Jews* are yet a distinct People from all the Nations amongst which they reside. They seem therefore reserved by Providence for some such signal Favour, after they have suffered the due Chastisement."

Many Christian Zionists and many Jewish Zionists tried to justify the Holocaust as "due Chastisement". Politically powerful Dispensationalist Christians and their Jewish handlers are today actively promoting nuclear war and an apocalyptic holocaust which will kill us all, because they believe that God will create a new Earth after they have destroyed the old Earth. It is a new heaven and a new Earth which will only sustain the "elect", the "chosen", the Jews. *Isaiah* 65 states (*see also: Enoch*),

"1 I am sought of *them that* asked not *for me;* I am found of *them that* sought me not: I said, Behold me, behold me, unto a nation *that* was not called by my name. 2 I have spread out my hands all the day unto a rebellious people, which walketh *in* a way *that was* not good, after their own thoughts; 3 A people that provoketh me to anger continually to my face; that sacrificeth in gardens, and burneth incense upon altars of brick; 4 Which remain among the graves, and lodge in the monuments, which eat swine's flesh, and broth of abominable *things is in* their vessels; 5 Which say, Stand by thyself, come not near to me; for I am holier than thou. These *are* a smoke in my nose, a fire that burneth all the day. 6 Behold, *it is* written before me: I will not keep silence, but will recompense, even recompense into their bosom, 7 Your iniquities, and the iniquities of your fathers together, saith the LORD, which have burned incense upon the mountains, and blasphemed me upon the hills: therefore will I measure their former work into their bosom. 8 Thus saith the LORD, As the new wine is found in the cluster, and *one* saith, Destroy it

---

459. D. Hartley, *Observations on Man, His Frame, His Duty, and His Expectations in Two Parts*, Volume 2, Printed by S. Richardson for James Leake and Wm. Frederick, booksellers in Bath and sold by Charles Hitch and Stephen Austen, booksellers in London, London, (1749), pp. 184, 366-381.

not; for a blessing *is* in it: so will I do for my servants' sakes, that *I* may not destroy them all. 9 And I will bring forth a seed out of Jacob, and out of Judah an inheritor of my mountains: and mine elect shall inherit it, and my servants shall dwell there. 10 And Sharon shall be a fold of flocks, and the valley of Achor a place for the herds to lie down in, for my people that have sought me. 11¶ But ye *are* they that forsake the LORD, that forget my holy mountain, that prepare a table for *that* troop, and that furnish the drink offering unto *that* number. 12 Therefore will I number you to the sword, and ye shall all bow down to the slaughter: because when I called, ye did not answer; when I spake, ye did not hear; but did evil before mine eyes, and did choose *that* wherein I delighted not. 13 Therefore thus saith the Lord GOD, Behold, my servants shall eat, but ye shall be hungry: behold, my servants shall drink, but ye shall be thirsty: behold, my servants shall rejoice, but ye shall be ashamed: 14 Behold, my servants shall sing for joy of heart, but ye shall cry for sorrow of heart, and shall howl for vexation of spirit. 15 And ye shall leave your name for a curse unto my chosen: for the Lord GOD shall slay thee, and call his servants by another name: 16 That he who blesseth himself in the earth shall bless himself in the God of truth; and he that sweareth in the earth shall swear by the God of truth; because the former troubles are forgotten, and because they are hid from mine eyes. 17¶ For, behold, I create new heavens and a new earth: and the former shall not be remembered, nor come into mind. 18 But be ye glad and rejoice for ever *in that* which I create: for, behold, I create Jerusalem a rejoicing, and her people a joy. 19 And I will rejoice in Jerusalem, and joy in my people: and the voice of weeping shall be no more heard in her, nor the voice of crying. 20 There shall be no more thence an infant of days, nor an old man that hath not filled his days: for the child shall die an hundred years old; but the sinner *being* an hundred years old shall be accursed. 21 And they shall build houses, and inhabit *them;* and they shall plant vineyards, and eat the fruit of them. 22 They shall not build, and another inhabit; they shall not plant, and another eat: for as the days of a tree *are* the days of my people, and mine elect shall long enjoy the work of their hands. 23 They shall not labour in vain, nor bring forth for trouble; for they *are* the seed of the blessed of the LORD, and their offspring with them. 24 And it shall come to pass, that before they call, I will answer; and while they are yet speaking, I will hear. 25 The wolf and the lamb shall feed together, and the lion shall eat straw like the bullock: and dust *shall be* the serpent's meat. They shall not hurt nor destroy in all my holy mountain, saith the LORD."

Isaiah 66:22-24 states,

"22 For as the new heavens and the new earth, which I *will* make, *shall* remain before me, saith the LORD, so shall your seed and your name remain. 23 And it shall come to pass, *that* from one new moon to another, and from one sabbath to another, shall all flesh come to worship before me, saith the LORD. 24 And they shall go forth, and look upon the carcases of the men that have transgressed against me: for their worm shall not die, neither shall their fire be quenched; and they shall be an abhorring unto all flesh."

Note Hartley's pronouncement, which became a policy of inhumanity for the Zionists, both Christian and Jewish, who allied themselves with the anti-Semites and funded the anti-Semites' rise to political power in hopes that the persecution of assimilated Jews would force them to Zionism,

"Fourthly, They are treated with Contempt and Harshness, and sometimes with great Cruelty, by the Nations amongst whom they sojourn. They must therefore be the more ready to return to their own Land."

Further note Hartley's statement,

"Fifthly, The Downfal of the Civil and Ecclesiastical Powers, mentioned in the 81st and 82d Propositions, must both be attended with such public Calamities, as will make Men serious, and also drive them from the Countries of *Christendom* into the remote Parts of the World, particularly into the *East* and *West-Indies*; whither consequently they will carry their Religion now purified from Errors and Superstitions."

In 1899, anti-Semite Edouard Drumont alleged that Protestants and Jews had united to corrupt and destroy the predominantly Catholic nation of France. Drumont also predicted that Jewish financiers would unite with the German Government to destroy Russia, years before Jacob Schiff boasted of his success in destroying the Russian People. Drumont also alleged that "Jews" would build up the economy of a nation only to then use corrupt influence in its thriving markets, artificially enhanced by an influx of Jewish investment capital, to deplete the nation of its wealth. He argued that Jews made money the controlling factor in society, and then corruptly obtained control over the fortunes of nations. Drumont believed that Napoleon had been put in power to serve the interests of Jewish wealth accumulation; and that when there was little left to take, the Jews turned against Napoleon, in particular, Rothschild turned against Napoleon.

Many have alleged that Jewish Liberalism was a farce that led to tyranny and absolute Jewish dominance. They further asserted that wealthy Jewish materialistic Capitalists deliberately destroyed all virtue in Gentile society, so as to turn God against the Gentile world and back towards the Jews. Anti-Semitic political movements often concluded that they, not liberal or capitalistic Jews, represented the genuine interests of the working class; which the Jews only desired to deceive and exploit.[460] It was a pattern of general vilification of all Jews that suited the Zionists well, in that it segregated the wealthier Jews from

---

[460]. *See, for example:* H. Bielohlawek, "Yes, We Want to Annihilate the Jews!" in R. S. Levy, *Antisemitism in the Modern World: An Anthology of Texts*, D. C. Heath and Company, Toronto, (1991), pp. 115-120.

the societies into which they were otherwise comfortably and wilfully assimilating.

8.7.3.3 Zionists Develop a Strategy Which Culminates in the Nazis and the Holocaust as Means to Attain the "Jewish State"

In the 1640's, Orthodox Ukrainian Bohdan Chmielnicki alleged that Jews and Polish Catholics had enslaved the peoples who were under Polish control. In retaliation, Chmielnicki allegedly slaughtered large numbers of Jews. Some Jews saw this holocaust as the punishment which signaled the coming of the Messiah. Some Jews believed that God would not allow the existence of the Temple, or send the Messiah, until the Jews had atoned for Solomon's marriage to the Pharaoh's daughter and subsequent idolatry (*Sabbath* 56b. I *Kings* 11); which became associated with the "sin" of assimilation. There was also a perceived need to finally atone for Aaron's worship of the Golden Calf (*Sanhedrin* 102a) and the impiety of the ten Northern Tribes, and the impiety of southern tribes of Judah and Benjamin. Note that the Jews correlated a Jewish Holocaust with the redemption of Israel through the arrival of the Messiah, whose primary task was to "restore the Jews to Palestine".

Many had predicted that the year 1666 would mark the arrival of the Messiah. For the Christians, this meant the second coming of Christ, for the Jews, the arrival of the Jewish King. After the Chmielnicki holocaust, which some saw as the sacrifice of masses of Jewish lives as an act of atonement, Shabbatai Zevi declared himself to be the Jewish Messiah and a large Messianic sect followed him. He traveled to Palestine, as a good Messiah would, and attracted a large Jewish following. While traveling through Turkey, Shabbatai Zevi was taken prisoner and was forced to feign conversion to Islam in order to save his life.

A branch of the Shabbataian sect of crypto-Jews, called the *Dönmeh*, formed in Turkey. They pretended to convert to Islam, but practiced Judaism in secret. For centuries this sect of crypto-Jewish Turks have bred subversive crypto-Jewish agents who have been sent around the world to prepare the way for Jewish world domination. They created a secret society in Paris and eventually led a revolt from Salonika. They were the hidden masters of the "Young Turks" and flooded Turkey with revolutionary propaganda defaming the Sultan. Their reach extended across the globe.

The Shabbataians believed that Shabbatai Zevi's Messianic spirit passed from one Jewish King to the next in a process of Metempsychosis. They argued that the line of David was a dynasty, which would not end when any given King of the Jews died, but rather the spirit of the Messiah would leave one body of the Jewish King and enter the next, sort of like a kosher Dalai Lama. In the form of the "Young Turks", the *Dönmeh* eventually succeed in overthrowing the Sultan whose ancestors had shamed Shabbatai Zevi. They also destroyed Turkish culture and committed genocide against the Armenian Christians. Shabbatai Zevi was a bizarre individual, a bit of a "flake". He wore a bride's dress and wedded himself to the Torah.

Jacob Frank—a Polish Jew who was born Jacob Leibowitz, or Jacob Ben Judah Leib, whose father belonged to the Messianic sect of Shabbatai Zevi—joined the *Dönmeh* in Turkey. Frank declared himself to be the successor of Shabbatai Zevi and the then present Messiah. Frank opposed the Talmud and convinced prominent Catholic leaders that his sect would convert Jews to Christianity. The Frankist reformation, as well as Moses Mendelssohn's and Napoleon's reforms, set the stage for reformed Judaism, which, it was alleged, would lead to better relations between Christians and Jews, and which would afford revolutionary Jews with a means by which they could subvert Gentile society.[461] The Talmud, with its anti-Christian passages, had long been a source of anti-Christian and anti-Semitic tensions. Though Ashkenazi Jews had lain greater emphasis on the Talmud than even the Pentateuch, Sephardic Jews had a greater respect for the original books of Judaism and viewed the Talmud as the mere commentary it is. The Sephardic Jews developed Cabalism as an outgrowth of original Judaism with less emphasis on the Rabbinical authority of Talmudism—unless it happened to be convenient at any given time to quote a Talmudic authority.

Just as the Rabbis used the Talmud to justify their power and authority over Jewry, the Cabalists used the anti-Gentile and anti-Christian passages of the Talmud as a weapon against the Rabbis, to usurp their authority, and to bring them into conflict with Christians. Cabalistic writings are also severally anti-Christian and anti-Gentile, and the attacks were hypocritical, but the Cabalists survived their hypocrisy by becoming crypto-Jews who pretended to Christian and Moslem conversion. The Talmud, in *Tractate Kethuboth* 111a, prevents the Jews from forcing the Messianic Era and from emigrating to Palestine in large numbers before the coming of the Messiah. The Cabalists opposed this stance and had a powerful Messianic message and model, by which they used politics and wealth accumulation to carry out the Messianic prophecies, and anointed their own false messiahs at will.

The Jewish descendants of the Frankists became leading figures in Poland. Granted special privileges by the elite of Europe, they pretended to convert to Catholicism, but the Frankist conversions to Catholicism and Islam were instead efforts to subvert both religions and the Jews secretly carried on as Jews. The Frankists had many reasons for attacking Rabbinical culture. The Rabbis opposed any "artificial" establishment of a Jewish State, and the Catholic Church would likely have ended its opposition to "the restoration of the Jews to Palestine" if the Jews professed to be Christians and accepted the "new Covenant of Christ". The New Testament calls for a "remnant of Jews" to convert and live in Palestine.

The Frankists advocated many of the same beliefs as the Illuminati—and Communism and Bolshevism. The leadership elements of each of these groups

---

461. A. Leroy-Beaulieu, *Israel chez les nations: Les Juifs et l'antisémitisme*, C. Lévy, Paris, (1893); English translation by F. Hellman, "The Jew is the product of His Tradition and His Law", *Israel among the Nations: A Study of the Jews and Antisemitism*, Chapter 6, G. P. Putnam's Sons, New York, W. Heinemann, London, (1895), pp. 123-147.

are notable both for their disproportionate Jewish influence and for their highly perverse sexual deviancy. The Frankists believed that if they could destroy all Gentile religions, the Gentiles would be left without gods to protect them and their Jewish God would reign supreme. The Frankists also believed that evil is good and found many passages in the Old Testament to support their view that the Messiah would only be successful when evil ruled the Earth. They did everything they could to infiltrate and overthrow governments and sought world revolution. They wormed their way into the leadership of governments through pretended conversions and through intermarriage and did what they could to cause calamities, starvation and war.

Shabbatai Zevi, Jacob Frank and the Frankists had a long relationship with Turkey, as did Adolf Hitler's Hungarian Jewish patron Moses Pinkeles, a. k. a. Ignatius Trebitsch-Lincoln, and Adam Alfred Rudolf Glauer, a. k. a. Rudolf Glandeck Freiherr von Sebottendorf, both of whom helped to put Adolf Hitler into power—there were also the genocidal Young Turks of Jewish descent,[462] of *Dönmeh* descent, and there have been many Israeli leaders with intimate involvements in Turkey, including David Ben-Gurion.[463]

The belief that the God of the Old Testament sponsored evil was not new. The Talmud contains passages indicating that evil must reign before the Messiah will appear.[464] Some of the earliest Christians saw the creator God of the Old Testament as an evil force, who was supplanted by the supreme God who was the Father of Jesus. Marcion[465] believed that Jesus was not the Messiah of the Old Testament God, who was in Marcion's view the evil creator God who would restore the Jews, but was instead the Messiah of a good God, a supreme God who reigned over the *many* gods referred to in the Old Testament—for example in *Genesis* 3:5, 22,

> "For God doth know that in the day ye eat thereof, then your eyes shall be opened, and ye shall be as gods, knowing good and evil [***] And the LORD God said, Behold, the man is become as one of us, to know good and evil: and now, lest he put forth his hand, and take also of the tree of life, and eat, and live for ever:"

and *Psalm* 82:1,

---

462. I. Zangwill, *The Problem of the Jewish Race*, Judaen Publishing Company, New York, (1914), pp. 9, 11; which was first published as an article, "The Jewish Race", *The Independent*, Volume 71, Number 3271, (10 August 1911), pp. 288-295, at 290-291. J. Prinz, *The Secret Jews*, Random House, New York, (1973), pp. 111-112.
463. Y. Küçük, *Şebeke = Network*, YGS Yayinlari, Kadiköy, Istanbul, (2002).
464. *Sanhedrin* 97a-99b.
465. Tertullian, *Adversus Marcionem*, Clarendon Press, Oxford, (1972); English translation: *The Five Books of Quintus Sept. Flor. Tertullianus Against Marcion*, T. & T. Clark, Edinburgh, (1868). "Marcionites", *The Catholic Encyclopedia*, Volume 9, Robert Appleton Company, New York, (1910), pp. 645-649.

"God standeth in the congregation of the mighty; he judgeth among the gods."

and *Jeremiah* 10:10-11,

"10 But the LORD *is* the true God, he *is* the living God, and an everlasting king: at his wrath the earth shall tremble, and the nations shall not be able to abide his indignation. 11 Thus shall ye say unto them, The gods that have not made the heavens and the earth, *even* they shall perish from the earth, and from under these heavens."

Marcion believed that the Jew's Messiah was yet to appear and Marcion shunned the Old Testament creator God as an evil force and sought to keep the Christian faith from falling into the belief that Jesus was the Messiah of the Jews. *The Catholic Encyclopedia* wrote of Marcion, among other things,

"II. DOCTRINE AND DISCIPLINE.—We must distinguish between the doctrine of Marcion himself and that of his followers. Marcion was no Gnostic dreamer. He wanted a Christianity untrammeled and undefiled by association with Judaism. Christianity was the New Covenant pure and simple. Abstract questions on the origin of evil or on the essence of the Godhead interested him little, but the Old Testament was a scandal to the faithful and a stumbling-block to the refined and intellectual gentiles by its crudity and cruelty, and the Old Testament had to be set aside. The two great obstacles in his way he removed by drastic measures. He had to account for the existence of the Old Testament and he accounted for it by postulating a secondary deity, a demiurgus, who was god in a sense, but not the supreme God; he was just, rigidly just, he had his good qualities, but he was not the good god, who was Father of Our Lord Jesus Christ. The metaphysical relation between these two gods troubled Marcion little; of divine emanation, æons, syzygies, eternally opposed principles of good and evil, he knows nothing. He may be almost a Manichee in practice, but in theory he has not reached absolute consistency as Mani did a hundred years later. Marcion had secondly to account for those passages in the New Testament which countenanced the Old. He resolutely cut out all texts that were contrary to his dogma; in fact, he created his own New Testament, admitting but one gospel, a mutilation of St. Luke, and an Apostolicon containing ten epistles of St. Paul. The mantle of St. Paul had fallen on the shoulders of Marcion in his struggle with the Judaisers. The Catholics of his day were nothing but the Judaisers of the previous century. The pure Pauline Gospel had become corrupted and Marcion not obscurely hinted that even the pillar Apostles, Peter, James and John, had betrayed their trust. He loves to speak of 'false apostles', and lets his hearers infer who they were. Once the Old Testament has been completely got rid of, Marcion has no further desire for change. He makes his purely New Testament Church as like the Catholic Church as possible, consistent with his deep-seated Puritanism. The first description of Marcion's doctrine dates from St. Justin: 'With the help of the devil Marcion has in every country contributed to blasphemy and the refusal to acknowledge the Creator of all the world as God. He recognizes another god, who, because he is essentially greater (than the World-maker or Demiurge) has done greater deeds than he (ὡς ὄντα μείζονα τὰ μείζονα παρὰ τοῦτον πεποιηκέναι). The supreme God is ἀγαθός, good, kind; the inferior god is merely δίκαιος, just and righteous. The good God is all love, the inferior god gives way to fierce anger. Though less than the good God, yet the just god, as world-creator, has his independent sphere of activity. They are not opposed as Ormuzd and Ahriman, though the good God interferes in favour of men, for He alone is all-wise and all-powerful and loves mercy more than punishment. All men are indeed created by the Demiurge, but by special choice he elected the Jewish people as his own and thus became the god of the Jews.

His theological outlook is limited to the Bible, his struggle with the Catholic Church seems a battle with texts and nothing more. The Old Testament is true enough, Moses and the Prophets are messengers of the Demiurge, the Jewish Messias is sure to come and found a millennial kingdom for the Jews on earth, but the Jewish Messias has nothing whatever to do with the Christ of God. The Invisible, Indescribable, Good God (ἀόρατος, ἀκατάνομαστος, ἀγαθὸςθεός), formerly unknown to the creator as well as to his creatures, has revealed Himself in Christ. How far Marcion admitted a Trinity of persons in the Supreme Godhead is not known; Christ is indeed the Son of God, but he is also simply 'God' without further qualification; in fact, Marcion's Gospel began with the words; 'In the fifteenth year of the Emperor Tiberius God descended in Capharnaum and taught on the Sabbaths'. However daring and capricious this manipulation of the Gospel text, it is at least a splendid testimony that in Christian circles of the first half of the second century the Divinity of Christ was a central dogma. To Marcion however Christ was God Manifest, not God Incarnate. His Christology is that of the Docetæ (q.v.) rejecting the inspired history of the Infancy, in fact any childhood of Christ at all; Marcion's Savior is a 'deus ex machina' of which Tertullian mockingly says: 'Suddenly a Son, suddenly Sent, suddenly Christ!' Marcion admitted no prophecy of the Coming of Christ whatever; the Jewish prophets foretold a Jewish Messias only, and this Messias had not yet appeared. Marcion used the story of the three angels, who ate, walked and conversed with Abraham and yet had no real human body, as an illustration of the life of Christ (Adv. Marc., III, ix). Tertullian says (*ibid.*) that when Apelles and seceders from Marcion began to believe that Christ had a real body indeed, not by birth but rather collected from the elements, Marcion would prefer to accept even a putative birth rather than a real body. Whether this is Tertullian's mockery or a real change in Marcion's sentiments we do not know. To Marcion matter and flesh are not indeed essentially evil, but are contemptible things, a mere production of the Demiurge, and it was inconceivable that God should really have made them His own. Christ's life on earth was a continual contrast to the conduct of the Demiurge. Some of the contrasts are cleverly staged: the Demiurge sent bears to devour children for puerile merriment (Kings)—Christ bade children come to Him and He fondled and blessed them; the Demiurge in his law declared lepers unclean and banished them—but Christ touched and healed them. Christ's putative passion and death was the work of the Demiurge, who in revenge for Christ's abolition of the Jewish law delivered Him up to hell. But even in hell Christ overcame the Demiurge by preaching to the spirits in Limbo, and by His Resurrection He founded the true Kingdom of the good God. Epiphanius (Haer., xlii, 4) says that Marcionites believed that in Limbo Christ brought salvation to Cain, Core, Dathan and Abiron, Esau and the Gentiles, but left in damnation all Old Testament saints. This may have been held by some Marcionites in the fourth century, but it was not the teaching of Marcion himself, who had no Antinomian tendencies. Marcion denied the resurrection of the body, 'for flesh and blood shall not inherit the Kingdom of God', and denied the second coming of Christ to judge the living and the dead, for the good God, being all goodness, does not punish those who reject Him; He simply leaves them to the Demiurge, who will cast them into everlasting fire.

With regard to discipline, the main point of difference consists in his rejection of marriage, i.e. he baptized only those who were not living in matrimony: virgins, widows, celibates, and eunuchs (Tert., 'Adv. Marc.', I, xxix); all others remained catechumens. On the other hand the absence of division between catechumens and baptized persons in Marcionite worship, shocked orthodox Christians, but it was emphatically defended by Marcion's appeal to Gal., vi, 6. According to Tertullian (Adv. Marc., I, xiv) he used water in baptism, anointed his faithful with oil and gave milk and honey to the catechumens and in so far retained the orthodox practices, although, says Tertullian, all these things are 'beggarly elements of the Creator.' Marcionites must have been excessive fasters to provoke the ridicule of Tertullian in his Montanist days. Epiphanius says they fasted on Saturday out of a spirit of opposition to the Jewish God, who made the Sabbath a day of rejoicing. This however may have been merely a western custom adopted by them."[466]

The Frankists wanted to be the Messiahs not of the creator God of the Old Testament whom they also called evil, but of Marcion's good God, whom they recast into the ultimate and supreme God of Israel. The Frankist Jews believed that they could accomplish this end by being apostates, nihilists and deceivers, who achieved God's will by doing evil, and who did evil by hiding their true intentions. The Messiah himself would have to be crypto-Jew who would torment other Jews—like Adolf Hitler. The Frankist Jews tried to force God to restore them to Israel as he promised to do after punishing them for their evil

---

466. "Marcionites", *The Catholic Encyclopedia*, Volume 9, Robert Appleton Company, New York, (1910), pp. 645-649, at 646-647.

acts. They believed that they had to first perform said divine evil on an unprecedented scale and thereby hasten the punishment of the Jews in a horrific holocaust, which would also hasten the arrival of the Messianic Era. The Frankist sophists thereby converted the action of doing evil into a divine act of obedience to God. They set about to destroy the world as an invitation to God to punish them and begin the Messianic Age. Adolf Hitler was their apostate Messiah, who restored the Jews to Palestine by punishing the Jews and committing gross acts of deliberate evil.

These Frankist Jews quickly became the guiding force behind world leadership. Gershom Scholem encapsulated their beliefs as follows:

"1) The belief in the necessary apostasy of the Messiah and in the sacramental nature of the descent into the realm of the *kelipot*. 2) The belief that the 'believer' must not appear to be as he really is. 3) The belief that the Torah of *atzilut* must be observed through the violation of the Torah of *beriah*. 4) The belief that the First Cause and the God of Israel are not the same, the former being the God of rational philosophy, the latter the God of Religion. 5) The belief in three hypostates of the God-head, all of which have been or will be incarnated in human form."[467]

Scholem wrote,

"According to Frank, the 'cosmos' (*tevel*) or 'earthly world' (*tevel ha-gashmi*) as it was called by the sectarians in Salonika, is not the creation of the Good or Living God, for if it were it would be external and man would be immortal, whereas as we see from the presence of death in the world this is not at all the case."[468]

Scholem quotes Frankist doctrine:

"This much I tell you: Christ, as you know, said that he had come to redeem the world from the hands of the devil, but I have come to redeem it from all the laws and customs that have ever existed. It is my task to annihilate all this so that the Good God can reveal Himself. [***] Wherever Adam trod a city was built but wherever I set foot all will be destroyed, for I came into this world only to destroy and to annihilate. But what I build will last forever. [***] I did not come into this world to lift you up but rather to cast you down to the bottom of the abyss. further than this it is impossible to descend, nor can one ascend again by virtue of one's own strength, for only the Lord can raise one up from the depths by the power of His hand."[469]

---

[467]. G. Scholem, "The Holiness of Sin", *Commentary* (American Jewish Committee), Volume 51, Number 1, (January, 1971), pp. 41-70, at 63.
[468]. G. Scholem, "The Holiness of Sin", *Commentary* (American Jewish Committee), Volume 51, Number 1, (January, 1971), pp. 41-70, at 64.
[469]. G. Scholem, "The Holiness of Sin", *Commentary* (American Jewish Committee),

Jacob Frank gave out his wife and daughter for sexual favors in order to gain converts and influence the influential. He accused his fellow Jews of ritual sacrifice for personal political gain, and otherwise tried to appeal to the mythologies and aspirations of Moslem and Christian leaders. Frank's agents and their descendants have corrupted the Gentile world with Communist, Masonic and Illuminati-style leaders, who bought into the mythologies he promulgated, and who have done his bidding. The Hasidic Jews seem very earnest in carrying out his objectives and some practice his perversions. Frank's ultimate goal was to destroy life on Earth, and the means to accomplish that end today exist. The Nazis and Communists, under crypto-Jewish leadership, inflicted terrible harm on humanity. For the Frankist Jews, there is still worse evil yet to be done.

It is interesting to note that Baal worshipers practiced the prostitution of women and homosexual men in their temples to gain converts and as an expression of their fertility religion, and some Jewish temples were used for Baal worship by Jews. The Gnostics also used communal women and homosexual sex to lure in converts. The dissemination of insincere Liberalism was another tactic some Zionists have used to undermine the structure of Gentile societies.

What the Italian mafia called *Omerta*, the code of silence, Frank called *massa dumah*. The *Encyclopaedia Judaica* writes in its article, "FRANK, JACOB, AND THE FRANKISTS,"

> "The motto which Frank adopted here was *massa dumah* (from Isa. 21:11), taken to mean 'the burden of silence'; that is, it was necessary to bear the heavy burden of the hidden faith in the abolition of all law in utter silence, and it was forbidden to reveal anything to those outside the fold. Jesus of Nazareth was no more than the husk preceding and concealing the fruit, who was Frank himself. Although it was necessary to ensure an outward demonstration of Christian allegiance, it was forbidden to mix with Christians or to intermarry with them, for in the final analysis Frank's vision was of a Jewish future, albeit in a rebellious and revolutionary form, presented here as a messianic dream. [***] Frank went with his daughter to Vienna in March of 1775 and was received in audience by the empress and her son, later Joseph II. Some maintain that Frank promised the empress the assistance of his followers in a campaign to conquer parts of Turkey, and in fact over a period of time several Frankist emissaries were sent to Turkey, working hand in glove with the Doenmeh, and perhaps as political agents or spies in the service of the Austrian government. During this period Frank spoke a great deal about general revolution which would overthrow kingdoms, and the Catholic Church in particular, and he also dreamed of the conquest of some territory in the wars at the end of time which would be the

---

Volume 51, Number 1, (January, 1971), pp. 41-70, at 65.

Frankist dominion."[470]

It is difficult to believe that it is merely a coincidence that this religion of revolution and Nihilism was heavily promoted in England at the same time in the writings of David Hartley—and can be traced back to the Cabalist Van Helmont. It was their intention to destroy and corrupt; and the Frankists relied upon passages in the Old Testament and the Lurian Cabalah to justify deceit, lying, world revolution, destruction, evil and atheism among Gentiles—passages such as *Isaiah* 45:7; and 59:15-16:

> "I form the light, and create darkness: I make peace, and create evil: I the LORD do all these *things*. [***] Yea, truth faileth; and he *that* departeth from evil maketh himself a prey. And the LORD saw *it*, and it displeased him that *there was* no judgment. And he saw that *there was* no man, and wondered that *there was* no intercessor: therefore his arm brought salvation unto him; and his righteousness, it sustained him."

and *Job* Chapter 12,

> "And Job answered and said, 2 No doubt but ye *are* the people, and wisdom shall die with you. 3 But I have understanding as well as you; I *am* not inferior to you: yea, who knoweth not such things as these? 4 I am *as* one mocked of his neighbor, who calleth upon God, and he answereth him: the just upright *man is* laughed to scorn. 5 He that is ready to slip with *his* feet *is as* a lamp despised in the thought of him that is at ease. 6 The tabernacles of robbers prosper, and they that provoke God are secure; into whose hand God bringeth *abundantly*. 7 But ask now the beasts, and they shall teach thee; and the fowls of the air, and they shall tell thee: 8 or speak to the earth, and it shall teach thee; and the fishes of the sea shall declare unto thee. 9 Who knoweth not in all these that the hand of the LORD hath wrought this? 10 In whose hand *is* the soul of every living thing, and the breath of all mankind. 11 Doth not the ear try words? and the mouth taste his meat? 12 With the ancient *is* wisdom; and in length of days understanding. 13 With him *is* wisdom and strength, he hath counsel and understanding. 14 Behold, he breaketh down, and it cannot be built again: he shutteth up a man, and there can be no opening. 15 Behold, he withholdeth the waters, and they dry up: also he sendeth them out, and they overturn the earth. 16 With him *is* strength and wisdom: the deceived and the deceiver *are* his. 17 He leadeth counselors away spoiled, and maketh the judges fools. 18 He looseth the bond of kings, and girdeth their loins with a girdle. 19 He leadeth princes away spoiled, and overthroweth the mighty. 20 He removeth away the speech of the trusty, and taketh away the understanding of the aged. 21 He

---

470. "Frank, Jacob, and the Frankists", *Encyclopaedia Judaica*, Volume 7 Fr-Ha, Encyclopaedia Judaica, Jerusalem, The Macmillan Company, New York, (1971), cols. 55-71, at 60, 68.

poureth contempt upon princes, and weakeneth the strength of the mighty. 22 He discovereth deep things out of darkness, and bringeth out to light the shadow of death. 23 He increaseth the nations, and destroyeth them: he enlargeth the nations, and straiteneth them *again*. 24 He taketh away the heart of the chief of the people of the earth, and causeth them to wander in a wilderness *where there is* no way. 25 They grope in the dark without light, and he maketh them to stagger like a drunken man."

It was a long road for the Frankist Nihilists to the end of time, which came during, and shortly after, the Second World War. The Frankists chose Jacob Frank's nephew, Moses Dobrushka a. k. a. Junius Frey, a. k. a. Franz Thomas von Schoenfeld, as Jacob Frank's successor—it was a Frankist-Shabbataian tradition to change names, and give the appearance of changing religions, in order to gain the confidence of Gentiles so as to enable the Frankists to more easily destroy them and subvert their societies. Moses Dobrushka became a Jacobin, a leader of Freemasonry and a powerful influence in the French Revolution. It is interesting that Robespierre and Napoleon saw themselves as Messiahs, as had Shabbatai Zevi and Jacob Frank, and as Adolf Hitler later would. Frankist mythologies asserted that Messiahdom was a generational passage—a matter of reincarnation. Frankism primarily took root in Poland, which has been at the epicenter of the destruction of Catholic Europe.

According to Edouard Drumont, Alexandre Weill found good in the destruction and dismemberment of Poland and the planned destruction of France and diaspora of the French—recall that David Hartley had wished for the fall of Christendom and the diaspora of Christians. Drumont wrote that Weill had told him in 1875 that,

> "[...]France was obliged to undergo the same fate as Poland and that it would be good, in the best interests of Humanity, that the French, dispersed and countryless like the Poles, would go and spread throughout the world the general truths of civilization and progress."[471]

Roman Dmowski iterated a Polish Gentile's view of the First World War in his article *The Jews and the War* of 1924.[472] He noted that many of the Jews who had supported the Central Powers in the beginning of the war changed sides to the Allies in early 1917. Dmowski believed in 1924 that Jews intended to make Poland a new Palestine. Great masses of Jews were deported to Poland in both world wars by both sides of the conflict. Poland was the epicenter of the Jewish

---

[471]. E. A. Drumont, *Les juifs contre la France une nouvelle Pologne*, Librairie Antisémite, Paris, (1899), pp. 36-48; English translation in: R. S. Levy, *Antisemitism in the Modern World: An Anthology of Texts*, D. C. Heath and Company, Toronto, (1991), pp. 107-112, at 111.
[472]. R. Dmowski, "The Jews and the War", in R. S. Levy, Editor, J. Kulczycki, translator, *Antisemitism in the Modern World: An Anthology of Texts*, D. C. Heath and Company, Lexington, Massachusetts, Toronto, (1991), pp. 182-189.

Holocaust.

The initial plan was evidently to concentrate them for deportation to Palestine, which neither a majority of the Jews, nor many of the world's nations, desired. It is interesting to note that Hitler was allegedly surprised by the reaction of the British when Germany invaded Poland in a quest for *Lebensraum* for Germans and for a place in which to segregate the Jews to the East and prevent their assimilation while preparing them for forced deportation to Palestine. The English had obstructed the Nazis' attempts to deport Jews to Palestine and then declared war on Germany when the Nazis invaded Poland—the ultimate destination for millions of Jews, many of whom perished under the crypto-Jewish Zionist Nazi leaders Adolf Eichmann and Hans Frank. After World War II, the Allies allowed the Soviet Union to take over Poland. The Soviets tried for another forty years to destroy religion in Poland—primarily Catholicism. The Jews were forced to suffer through the war in Poland so that enough Jewish blood would be spilled to justify the theft of Palestine and frighten the Jews into moving there and staying.

David Hartley's work followed the works of Thomas Brightman, who published his *Apocalypsis Apocalypseos* in 1585;[473] and Henry Finch, who published *The Worlds Great Restauration. Or the Calling of the Ievves and (With Them) of All the Nations and Kingdomes of the Earth, to the Faith of Christ* in 1621.[474]

---

[473]. T. Brightman, *Brightmans predictions and prophecies vvritten 46 yeares since: concerning the three churches of Germanie, England, and Scotland : fore-telling the miserie of Germanie, the fall of the pride of bishops in England by the assistance of the Scottish Kirk: all which should happen, as he foretold, between the yeares of 36 and 41, &c.*, (1641); **and** *A reuelation of the Reuelation: that is, the Reuelation of St. John opened clearely with a logicall resolution and exposition : wherein the sense is cleared, out of the Scripture, the euent also of thinges foretold is discussed out of the church-historyes*, Amsterdam, (1615); **and** *The revelation of S. Iohn illustrated with an analysis & scholions Where in the sence is opened by the scripture, & the euent of things fore-told, shewed by histories*, Class [on van Dorpe], Leyden, (1616); and *A revelation of the Apocalyps, that is, the Apocalyps of S. Iohn illustrated vvith an analysis & scolions where the sense is opened by the scripture, & the events of things foretold, shewed by histories. Hereunto is prefixed a generall view: and at the end of the 17. chapter, is inserted a refutation of R. Bellarmine touching Antichrist, in his 3. book of the B. of Rome*, Iudocus Hondius & Hendrick Laurenss, Amsterdam, (1611); **and** *Apocalypsis apocalypseos: id est Apocalypsis D. Joannis analysi et scholiis illustrata; ubi ex scriptura sensus rerumque praedictarum ex historiis eventus discutiuntur. Huic Synopsis praefigitur universalis, et refutatio Rob. Bellarmini de antichristo libro tertio de Romano Pontifice ad finem capitis decimi septimi inseritur*, Heidelberg, (1612); **and** *Een Grondighe ontdeckinghe ofte duydelijcke uytlegginghe, met een logicale ontknoopinghe, over de gantsche openbaringe Iohannis des Apostels: waer in de sin uyt de Schriftuere verklaert, ende insghelijchs de uytkomsten der dinghen die voorseyt waren, met de kerchelijcke historien aenghewesen worden*, Jan Evertsz Cloppenburch, boeckvercooper... , 'tAmstelredam, (1621).

[474]. H. Finch, *The Worlds Great Restauration. Or the Calling of the Ievves and (With Them) of All the Nations and Kingdomes of the Earth, to the Faith of Christ*, William Gouge, B. of D. and Preacher of Gods word in Black-Fryers, London, Printed by Edvvard

The Zionists had the vocal support of prominent Protestant Christians who hoped to bring about the Apocalypse through active political intervention—as opposed to waiting for God to do what He promised to do. More modern Jewish Zionists repeated much of the rhetoric and tactics the Christian Zionists used, which was originally covertly crafted by Cabalistic Jews. It was a strange cycle, whereby Jews learned their Zionism from the Christians who had secretly learned it from Jews.

All that the modern Jewish Zionists lacked was the widespread support of Jews, which they only received after the end of the Second World War—after the Frankist Jews had done their dirty deeds. It took the Zionists two world wars and Adolf Hitler to change the Jews' collective mind to embrace secular Zionism, which led many to realize that Zionists and their Protestant supporters had agitated for both world wars and had created and continually sponsored Adolf Hitler's rise to power.

In 1933, Zionist Horace Mayer Kallen blamed the First World War on the Germans, and stated,

> "The formation of the League of Nations on the initiative and insistence of a great American President, Woodrow Wilson, was fruit of this War, and an explicit, if weak, acknowledgment of this interdependence. Mr. Wilson's successor of today just as frankly acknowledges it stresses it."[475]

Kallen goes on to quote Zionist Franklin Delano Roosevelt's message to the World Economic Conference and the Disarmament Conference of 16 May 1933.[476]

Like Wilson, Roosevelt later lied to the American public in order to be elected and told them that he was against American involvement in the war in Europe. Tyler Gatewood Kent[477] documented President Roosevelt's secret

---

Griffin for William Bladen, and are to be sold at his shop neare the great north dore of Pauls, at the signe of the Bible, (1621).

475. H. M. Kallen quoted in A. Hertzberg, *The Zionist Idea*, Harper Torchbooks, New York, (1959), p. 532.

476. *Peace and War: United States Foreign Policy, 1931-1941*, United States Department of State, Publication 1983, U.S. Government Printing Office, Washington, D. C., (1943), pp. 178-185.

477. A. H. P. Kent, *Petition to Members of the Seventy-eighth Congress of the United States for the Redress of Grievances Suffered by My Son, Tyler Kent, a Loyal Citizen of the United States*, Washington, D. C., (1944). *See also:* J. S. Snow, *The case of Tyler Kent*, Published jointly by Domestic and Foreign Affairs and Citizens Press, Chicago, New York, (1946). *See also:* R. Whalen, "The Strange Case of Tyler Kent", *Diplomat*, (November, 1965), pp. 16-19, 62-64. *See also:* W. F. Kimball, "Churchill and Roosevelt: The Personal Equation", *Prologue*, Volume 6, (Fall, 1974), pp. 169-82. *See also:* F. L. Loewenheim, H. D. Langley, and M. Jonas, *Roosevelt and Churchill: Their Secret Wartime Correspondence*, Saturday Review Press, New York, (1975). *See also:* J. Leutze, "The Secret of the Churchill-Roosevelt Correspondence: September 1939-May 1940", *Journal of Contemporary History*, Volume 10, (1975), pp. 465-91. *See also:* J. P.

communications to Zionist Winston Churchill beginning in October of 1939, in which Roosevelt assured Churchill that America would not be truly neutral and would rescue the British. This emboldened the British in their declaration of war on Germany, and revealed Roosevelt's duplicity.

When this correspondence began, Neville Chamberlain was Prime Minister of England and Churchill was head of the British Navy. Roosevelt went behind Chamberlain's back and apparently knew ahead of time that Churchill would succeed Chamberlain. Just when Kent had accumulated all the evidence needed to prove Roosevelt a lair, and a criminal, and as Kent was preparing to send this evidence to the American Congress; British authorities arrested him, seized his records and in violation of his American diplomatic immunity, which they conspired to have waived, imprisoned him for the duration of the war. Just as the American Wilson Administration passed several laws which enabled them to imprison dissenters, and the Roosevelt Administration used the Sedition Act to persecute its critics; the British had in place Regulation 18B, which enabled authorities to arrest and detain anyone they wanted to keep quiet, including Captain Archibald Henry Maule Ramsay.

Like Zionist President Wilson, Zionist President Roosevelt betrayed the American blacks who initially helped to put him in office; and, unlike his wife Eleanor, Franklin Delano Roosevelt opposed the anti-lynching bill. While it is obviously a good thing that the Russians, Americans and British defeated the Nazis, it obviously would have been a better thing if the Zionists had not caused both of the world wars.

Zionists and their supporters often spoke of Wilson's "New World Order" following the "war to end all wars." The concept of the "war to end all wars" is a prophetic and apocalyptic one foretold in *Isaiah* 2:1-4:

"1 The word that Isaiah the son of Amoz saw concerning Judah and Jerusalem. 2 And it shall come to pass in the last days, *that* the mountain of the LORD's house shall be established in the top of the mountains, and shall be exalted above the hills; and all nations shall flow unto it. 3 And many people shall go and say, Come ye, and let us go up to the mountain of the LORD, to the house of the God of Jacob; and he will teach us of his ways, and we will walk in his paths: for out of Zion shall go forth the law, and the word of the LORD from Jerusalem. 4 And he shall judge among the nations, and shall rebuke many people: and they shall beat their swords into plowshares, and their spears into pruninghooks: nation shall not lift up

---

Lash, *Roosevelt and Churchill 1939-1941: The Partnership that Saved the West*, Norton, New York, (1976). *See also:* D. Irving, "Tyler Gatewood Kent: The Many Motives of a Misguided Cypher Clerk", *Focal Point*, (23 November 1981), pp. 3-10. *See also:* Kimball, Warren F., and Bartlett, Bruce. "Roosevelt and Prewar Commitments to Churchill: The Tyler Kent Affair", *Diplomatic History*, Volume 5, Number 4, (Fall, 1981), pp. 291-312. *See also:* R. Harris, "The American Tearoom Spy", *The London Times*, (4 December 1982), p. 6. *See also:* R. Bearse and A. Read, *Conspirator: the Untold Story of Churchill, Roosevelt and Tyler Kent, Spy*, Macmillan, London, (1991).

sword against nation, neither shall they learn war any more."

In 1943, Zionist Rabbi Abba Hillel Silver saw this new world order of "justice" as the allegedly just action of taking Palestine from its majority population and giving it to the Zionists. Referring to Americans of Jewish descent, the Rabbi asked them in 1943 to give their approval to Zionism and to pressure American politicians,

> "with the same sympathy and the same understanding as the Presidents of the United States from Wilson down, and the Congress of the United States, helped [the Yishuv] in the earlier years."[478]

In 1944, while the Nazis were massacring innocent and helpless Slavs, Jews, Gypsies, etc., Zionist David Ben-Gurion stated,

> "One Degania [resident of the first communal settlement of Zionists in Palestine] is worth more than all the 'Yevsektzias' [Jewish Bolsheviks who sought to secularize Jews] and assimilationists in the world."[479]

Ben-Gurion boasted,

> "This people was the first to prophesy about 'the end of days,' the first to see the vision of a new human society. [***] Our small and land-poor Jewish people, therefore, lived in constant tension between the power and influence of the neighboring great empires and its own seemingly insignificant culture—a culture poor in material wealth and tangible monuments, but rich and great in its human and moral concepts and in its vision of a universal 'end of days.'"[480]

Christopher Sykes wrote, "[...]Zionist leaders were determined at the very outset of the Nazi disaster to reap political advantage from the tragedy."[481] David Ben-Gurion stated in 1932,

---

[478]. A. H. Silver, *Vision and Victory*, Zionist Organization of America, New York, (1949); in A. Hertzberg, *The Zionist Idea*, Harper Torchbooks, New York, (1959), pp. 592-600, at 599.
[479]. D. Ben-Gurion, *Ba-Maarachah*, Volume 3, Tel-Aviv, (1948), pp. 200-211, English translation in A. Hertzberg, *The Zionist Idea*, Harper Torchbooks, New York, (1959), pp. 606-619, at 616.
[480]. D. Ben-Gurion, *Ba-Maarachah*, Volume 3, Tel-Aviv, (1948), pp. 200-211, English translation in A. Hertzberg, *The Zionist Idea*, Harper Torchbooks, New York, (1959), pp. 606-619, at 607-608.
[481]. K. Polkehn, "The Secret Contacts: Zionism and Nazi Germany, 1933-1941", *Journal of Palestine Studies*, Volume 5, Number 3/4, (Spring-Summer, 1976), pp. 54-82, at 58; citing C. Sykes, *Crossroads to Israel*, London, (1965); *Kreuzwege nach Israel; die Vorgeschichte des jüdischen Staates*, C. H. Beck, München, (1967), p. 151.

"What Zionist propaganda for years and years could not do, disaster has done overnight. Palestine is today the fiery question for the Jews of East and West, and the New World as well."[482]

Ben-Gurion also stated,

"The disaster facing European Jewry is not directly my business."[483]

and,

"The First World War brought us the Balfour Declaration. The Second ought to bring us the Jewish State."[484]

and,

"It is the job of Zionism not to save the remnant of Israel in Europe but rather to save the land of Israel for the Jewish people and the yishuv."[485]

The majority of Jews did not want the desert the Zionists wanted for them, until the Nazis had mass murdered European Jews. Racist Zionist leader Chaim Weizmann stated in 1914, before the First World War began,

"We cannot take Palestine yet, even if it were given to us. Even if the great miracle had happened and we had obtained the Charter, we should have to wait for the greater miracle—for the Jews to know how to make use of this Charter."[486]

Weizmann admitted in 1927 that,

"We Jews got the Balfour Declaration quite unexpectedly; or, in other words, we are the greatest war profiteers. [***] The Jews, they knew, were against us; we stood alone on a little island, a tiny group of Jews with a foreign past."[487]

---

[482]. C. Weizmann, "The Key to Immigration", *Rebirth and Destiny of Israel*, Philosophical Library, New York, (1954), p. 41.
[483]. T. Segev, *The Seventh Million: The Israelis and the Holocaust*, Hill and Wang, New York, (1993), p. 98.
[484]. M. Bar-Zohar, *Ben-Gurion: The Armed Prophet*, Prentice-Hall, Englewood Cliffs, New Jersey, (1967), p. 69.
[485]. T. Segev, *The Seventh Million: The Israelis and the Holocaust*, Hill and Wang, New York, (1993), p. 129.
[486]. C. Weizmann, *Chaim Weizmann*, V. Gollanez, London, (1945); quoted in A. Hertzberg, *The Zionist Idea*, Harper Torchbooks, New York, (1959), pp. 575-578, at 576.
[487]. C. Weizmann, *Chaim Weizmann*, V. Gollanez, London, (1945); quoted in A. Hertzberg, *The Zionist Idea*, Harper Torchbooks, New York, (1959), pp. 578-583, at 581.

David Ben-Gurion stated,

"The First World War brought us the Balfour Declaration. The Second ought to bring us the Jewish State."[488]

Countless millions died as the Zionists depended on both world wars to bring them Palestine.

The London Times had published on the Protestant Zionist movement on 24 January 1839 on page 3, quoting extensively from *The Quarterly Review*[489] of January, 1839,

"THE STATE AND PROSPECT OF THE JEWS.

(From a Correspondent.)
'What is to become of the Jews?' is a question that must as often occur to the reflecting statesman as to the reader of the ancient prophecies. Wherever he turns his eye he beholds a people exiled and scattered, persecuted and despised, as a body ground almost to powder by the iron hand of poverty; and yet, everywhere intelligent, learned, and possessed of unbounded influence, and, however paradoxical it may sound, of immense wealth; inhabitants of all countries, but at home in none; apparently a mass of disjointed fragments, but in reality knit together in the most intimate religious and national union, and in continual and rapid communication with their brethren in all parts of the world. What, then, is to become of them? Some of the continental statesmen solve the enigma by an attempt at amalgamation, and think that the ties of religion and nationality, which have stood the wear and tear of 18 centuries, are to be rent asunder by the simple process of naturalization. Very similar is the expectation of the church of Rome and of most sectarians. Looking upon their own little communion as the church and people of God, they appropriate to themselves the promises of future glory which Hebrew prophets have announced to the Hebrew people, and think that by the process of conversion the Jews will gradually melt down and be lost in the Christian church. The great writers of the Anglican church, adopting an interpretation more worthy of their faith and their scholarship, trace out for the children of Abraham a destiny more congenial to their hitherto marvellous history, the main features of which are ably delineated in an article on Lord Lindsay's travels in the last number of the *Quarterly Review.*

The writer, treading in the steps of Bishops Lowth, Butler, Horsley, and

---

[488]. M. Bar-Zohar, *Ben-Gurion: The Armed Prophet*, Prentice-Hall, Englewood Cliffs, New Jersey, (1967), p. 69.

[489]."Letters on Egypt, Edom, and the Holy Land" by A. C. L. Crawford, a. k. a. Lord Lindsay, *The Quarterly Review*, Volume 63, Number 125, (January, 1839), pp. 166-192.

Van Mildert, has turned the public attention to the claims which the Jewish people still have upon the land of Israel as their rightful inheritance, and their consequent political importance in the progress of that great struggle which has already commenced in the East, and which threatens soon to absorb the regards and energies of the old world, possibly of the new also. The subject may be new to many of our readers, but it is one deserving the solemn consideration of a people possessing an oriental empire of such vast extent. The article breathes also a spirit of kindness towards a deeply injured people, and a freedom from prejudice which does honour to the author. No people on the face of the earth has been so little understood and so grossly misrepresented as the Jewish, but no wonder, for no people ever did so much to misrepresent and caricature themselves as the Jews have done in the maxims and legends of the Talmud. A new era is, however, commencing. The Jews themselves, in London as elsewhere, are taking steps to abdicate the follies and intolerance of Rabbinism, and Christians at the same moment begin to renounce their most unchristian prejudice.

The following extracts from a journal so highly respectable as the *Quarterly Review* must tend to prove to the Jews that the feelings of those whose opinions are worth having are those of kindliness and good will.

After a notice of Lord Lindsay's work, the author thus proceeds:—

'We have alluded, in the commencement of this article, to the growing interest manifested in behalf of the Holy Land. This interest is not confined to the Christians—it is shared and avowed by the whole body of the Jews, who no longer conceal their hope and their belief that the time is not far distant when 'the Lord shall set his hand again the second time to recover the remnant of his people which shall be left, from Assyria, and from Egypt, and from Pathros, and from Cush, and from Elam, and from Shinar, and from Hamath, and from the islands of the sea; and shall set up an ensign for the nations, and shall assemble the outcasts of Israel, and shall gather together the dispersed of Judah from the four corners of the earth.' Isaiah xi., 11.

'Doubtless, this is no new settlement among the children of the dispersion. The novelty of the present day does not lie in the indulgence of such a hope by that most venerable people; but in their fearless confession of the hope, and in the approximation of spirit between Christians and Hebrews, to entertain the same belief of the future glories of Israel, to offer up the same prayer, and look forward to the same consummation. To most former periods a development of religious feeling has been followed by a persecution of the ancient people of God; from the days of Constantine to Leo XII, the disciples of Christ have been stimulated to the oppression of the children of Israel; and Heaven alone can know what myriads of that suffering race fell beneath the *piety* of the Crusaders, as they marched to recover the sepulchre of their Saviour from the hands of the infidels. But a mighty change has come over the hearts of the Gentiles; they seek now the temporal and eternal peace of the Hebrew people; societies are established in England and Germany to diffuse among them the light of the gospel; and the increasing accessions to the parent institution in London attest the public estimation of its principles and services. * * * *

'But a more important undertaking has already been begun by the zeal and piety of those who entertain an interest for the Jewish nation. They have designed the establishment of a church at Jerusalem, if possible on Mount Zion itself, where the order of our service and the prayers of our liturgy shall daily be set before the faithful in the Hebrew language. A considerable sum has been collected for this purpose; the missionaries are already resident on the spot; and nothing is wanting but to complete the purchase of the ground on which to erect the sacred edifice. Mr. Nicolayson, having received ordination at the hands of the Bishop of London, has been appointed to the charge; and Mr. Pieritz, a Hebrew convert, is associated in the duty. The service meanwhile proceeds, though 'the ark of God is under curtains;' and a small but faithful congregation of proselytes hear daily the Evangelical verities of our church on the mount of the Holy City itself, in the language of the prophets, and in the spirit of the apostles. To anyone who reflects on this event it must appear one of the most striking that have occurred in modern days, perhaps in any days since the corruptions began in the church of Christ. It is well known that for centuries the Greek, the Romanist, the Armenian, and the Turk, have had their places of worship in the city of Jerusalem, and the latitudinarianism of Ibrahim Pasha had lately accorded that privilege to the Jews. The pure doctrines of the Reformation, as embodied and professed in the church of England, have alone been unrepresented amidst all these corruptions; and Christianity has been contemplated both by Mussulman and Jew as a system most hateful to the creed of each, a compound of mummery and image-worship.

'It is surely of vital importance to the cause of our religion that we should exhibit it in its pure and apostolical form to the children of Israel. We have already mentioned that they are returning in crowds to their ancient land; we must provide for the converts an orthodox and spiritual service, and set before the rest, whether residents or pilgrims, a worship as enjoined by our Saviour himself, 'a worship in spirit and in truth,'—its faith will then be spoken of through the whole world. A great benefit of this nature has resulted from the Hebrew services of the London Episcopal Chapel; it has not only afforded instruction and opportunity of worship to the converted Israelite, but has formed a point of attraction to foreign Jews on a visit to this country, and has been largely and eagerly commented on in many of the Hebrew journals published in Germany. In the purity of our worship they confess our freedom from idolatry; and in the sound of the language of Moses and the prophets, they forget that we are Gentiles. But if this be so in London, what will it be in the Holy City? They will hear the Psalms of David, in the very words that fell from his inspired lips, once more chanted on the holy hill of Zion; they will see the whole book of the law and the prophets laid before them, and hear it read at the morning and evening oblation; they will admire the church of England, with all its comprehensive fulness of doctrine, truth, and love, like a pious and humble daughter, doing final homage to that church first planted at Jerusalem, which is the mother of us all. Our soul-stirring and soul-satisfying liturgy—in Hebrew—its deep and tender devotion—the evangelical simplicity of its ritual—will form, in the mind of the Jew, an inviting contrast to the idolatry and superstition of the Latin and eastern churches; its enlarged charity will affect his heart, and its scriptural character demand his homage. It is surely a high privilege reserved to our church and nation to plant the true cross on the holy hill of Zion; to carry back the faith we thence received by the apostles; and uniting, as it were, the history, the labours, and the blood of the primitive and Protestant martyrs, 'light such a candle in Jerusalem as by God's blessing shall never be put out.'

'But this privilege will not be unaccompanied by practical benefits to the character and position of our own establishment. Whatever promotes the study and reverence of the Hebrew Scriptures promotes, in a similar degree, the honour and stability of the church of England. Her appointed orders, her liturgical services, her decent splendour, her national endowments, are 'according to the pattern that God showed us in the Mount.' The principle of an establishment then received the august sanction of the divine wisdom; and whether we look back to the earliest periods of the Jewish history, or forwards to the day of their future glory, as displayed in the concluding chapters of Ezekiel, we shall find that a national and established church is ever a main portion of the polity of the people of God. The arch-assailants of our Zion are well aware of this truth, and seek, therefore, to disparage the Old Testament by a contemptuously exclusive preference of the New!—irreverently excluding from their 'Christian' catalogue the 'Law, the Prophets, and the Psalms,' they ascribe to the Gospels and Epistles *alone* the title of the *Christian Scriptures!* And they are wise in their generation,—perceiving, as they do, that the co-ordinate authority and mutual dependence of all parts of the written word would manifest that the Saviour of Mankind, no less in the temporal than in the spiritual necessities of his church, 'came not destroy, but to fulfil.'

'The growing interest manifested for those regions, the larger investment of British capital, and the confluence of British travellers and strangers from all parts of the world, have recently induced the Secretary of State for Foreign Affairs to station there a representative of our Sovereign, in the person of a vice-consul. This gentleman set sail for Alexandria at the end of last September—his residence will be fixed at Jerusalem, but his jurisdiction will extend to the whole country within the ancient limits of the Holy Land; he is thus accredited, as it were, to the former kingdom of David and the twelve tribes. The soil and climate of Palestine are singularly adapted to the growth of produce required for the exigencies of Great Britain; the finest cotton may be obtained in almost unlimited abundance; silk and madder are the staple of the country, and olive-oil is now, as it ever was, the very fatness of the land. Capital and skill are alone required: the presence of a British officer, and the increased security of property which his presence will confer, may invite them from these islands to the cultivation of Palestine; and the Jews, who will betake themselves to agriculture in no other land,* having found in the English Consul a mediator between their people and the Pasha, will probably return in greater numbers, and become once more the husbandmen of Judea and Galilee.

'This appointment has been conceived and executed in the spirit of true wisdom. Though we cannot often commend the noble Lord's official proceedings, we must not withhold our meed of gratitude for the act, nor of praise for the zeal with which he applied himself to great preliminary difficulties, and the ability with which he overcame them. It is truly a national service: at all times it would have been expedient, but now it is necessary. To pass over commercial advantages—which the country will best perceive in the experience of them—we may discern a manifest benefit to our political position. We have done a deed which the Jews will regard as an honour to their nation, and have thereby conciliated a body of well-wishers in every people under heaven. Throughout the East they nearly monopolize the concerns of traffic and finance, and maintain a secret but uninterrupted intercourse with their brethren in the West. Thousands visit Jerusalem in every year from all parts of the globe, and carry back to their respective bodies that intelligence which guides their conduct and influences their sympathies. So rapid and accurate is their mutual communication, that Frederick the Great confessed the earlier and superior intelligence obtained through the Jews of all affairs of moment. Napoleon knew well the value of an Hebrew alliance, and endeavoured to reproduce in the capital of France the spectacle of the ancient Sanhedrim, which, basking in the sunshine of imperial favour, might give laws to the whole body of the Jews throughout the habitable world, and aid him, no doubt, in his audacious plans against Poland and the East. His scheme, it is true, proved abortive, for the mass of the Israelites were by no means inclined to merge their hopes in the destinies of the empire—exchange Zion for Montmartre, and Jerusalem for Paris. The few liberal unbelievers whom he attracted to his view ruined his projects with the people by their impious flattery, and averted the whole body of the nation by blending, on the 15th of August, the cipher of Napoleon and Josephine with the unutterable name of Jehovah, and elevating the imperial eagle above the representation of the ark of the covenant. A misconception, in fact, of the character of the people has vitiated all the attempts of various sovereigns to better their condition; they have sought to amalgamate them with the body of their subjects, not knowing or not regarding the temper of the Hebrews, and the plain language of Scripture, that 'the people shall dwell alone, and shall not be reckoned among the nations.'

'That which Napoleon designed in his violence and ambition, thinking 'to destroy nations not a few,' we may wisely and legitimately undertake for the maintenance of our empire. The affairs of the East are lowering on Great Britain, but it is singular and providential that we should at this moment have executed a measure which will almost assure us the co-operation of the eastern Jews, and kindle in our behalf the sympathies of nearly 2,000,000 in the heart of the Russian dominions. These hopes rest on no airy foundation; but, pleasing as they are, we cannot disguise our far greater satisfaction that, in the step just taken, in the appointment just made, England has attained the praise of being the first of the Gentile nations that has ceased to '*to tread down Jerusalem!*' This is, indeed, no more than justice, since she was the first to set the evil and cruel example of banishing the whole people in a body from her inhospitable bosom. France next, and then Spain, aped our unchristian and foolish precedent. Spain may have exceeded us in barbarity; but we invented the oppression, and preceded her in the infliction of it.'

*Dr. Henderson says of the Polish Jews—'Comparatively few of the Jews learn any trade, and most of those attempts which have been made to accustom them to agricultural habits have proved abortive. [Later political Zionists were anxious to persuade Jews to take up farming so as to cease to be, in their minds, "parasites". They did not want foreign workers to live in Israel and, in their minds, pollute their gene pool and corrupt their culture. Jacob worked the field. Esau wielded the sword. Cain was a farmer who slew Abel. "Abel was a keeper of sheep." (*Genesis* 4:2) The Talmud taught Jews that agricultural was the lowest form of work (*Yebamoth* 63a).—CJB] Some of those who are in circumstances of affluence possess houses and other immoveable property; but the great mass of the people seem destined to sit loose from every local tie, and are waiting, with anxious expectation, for the arrival of the period when, in pursuance of the Divine promise, they shall be restored to what they still consider *their own land*. Their attachment indeed to Palestine is unconquerable.'—*Biblical Researches and Travels in Russia*, 1826."

The Zionists often attempted to draw the might of the British Empire into the Middle East, so that the British citizens could sacrifice their lives for the sake of Israel, just as the French had done under Napoleon. The Zionists flattered and tempted the British, just as they had done to the French, with promises of Messiahdom, the Messianic Age, wealth and millions of Jewish allies against the Russians in the heart of the Russian Empire. Disraeli would later draw the British into the swamp of the Suez and Queen Victoria, the Queen of the House of David, became "Empress of India", in an effort to defend British interests from an imagined Russian and Turkish threat through the trade routes of the Middle East. While pretending to solve these "problems", the Jews created and agitated them. Zionists persuaded the British to die to take Palestine in order to curry favor with Russian Jews, and Zionists brought America into the war in exchange for the Balfour Declaration—to this day Americans are killing Moslems in pursuit of the Zionists' perceived self-interests. As they had done to the British and French, Jews covertly and artificially create disasters for America, and then offer up greater destruction as a solution, a solution which benefits them and destroys all others.

In 1839, *The Quarterly Review* pitched Zionism to the British by appealing to their sympathies, and to their greed,

"That which Napoleon designed in his violence and ambition, thinking 'to destroy nations not a few,' we may wisely and legitimately undertake for the maintenance of our Empire. The affairs of the East are lowering on Great Britain—but it is singular and providential that we should, at this moment, have executed a measure, which will almost assure us the co-operation of the Eastern Jews, and kindle, in our behalf, the sympathies of nearly two millions in the heart of the Russian dominions. [*Footnote:* 'Look to their present state of suffering in Poland and Russia, where they are driven from place to place, and not permitted to live in the same street where the so-called Christians reside! It not unfrequently happens, that when one or more wealthy Jews have built commodious houses in any part of a town, not hitherto prohibited, this affords a reason for proscribing them; it is immediately enacted that no Jew must live in that part of the city, and they are forthwith driven from their houses, without any compensation for their loss being given them'... ... ... 'they are oppressed on every side, yet dare not complain; robbed and defrauded, yet obtain no redress'... ... 'in the walk of social life, insult, and contempt, meet them at every turning.'— *Herschel's Sketch*, p. 7.] These hopes rest on no airy foundation; but pleasing as they are, we cannot disguise our far greater satisfaction that, in the step just taken, in the appointment just made, England has attained the praise of being the first of the Gentile nations that has ceased *'to tread down Jerusalem!'* This is, indeed, no more than justice, since she was the first to set the evil and cruel example of banishing the whole people in a body from her inhospitable bosom. France next, and then Spain, aped our unchristian and foolish precedent. Spain may have proceeded us in barbarity; but we invented the oppression, and preceded her in the infliction of it."[490]

The majority of Jews wanted nothing of the Protestant movement to banish them to the deserts of Palestine in the hopes that Jesus might return in the form of a Rothschild. *The London Times* published the following set of queries on 17 August 1840 on page 3,

"SYRIA.—RESTORATION OF THE JEWS.
(From a Correspondent.)

The proposition to plant the Jewish people in the land of their fathers, under the protection of the five Powers, is no longer a mere matter of speculation, but of serious political consideration. In a Ministorial paper of the 31st of July an article appears bearing all the characteristics of a feeler on this deeply interesting subject. However, it has been reserved for a noble lord opposed to Her Majesty's Ministers to take up the subject in a practical

---

[490]."Letters on Egypt, Edom, and the Holy Land" by A. C. L. Crawford, a. k. a. Lord Lindsay, *The Quarterly Review*, Volume 63, Number 125, (January, 1839), pp. 166-192, at 190.

and statesmanlike manner, and he is instituting inquiries, of which the following is a copy:—

QUERIES.

'1. What are the feelings of the Jews you meet with respect to their return to the Holy Land?

'2. Would the Jews of station and property be inclined to return to Palestine, carry with them their capital, and invest it in the cultivation of the land, if by the operation of law and justice life and property were rendered secure?

'3. How soon would they be inclined and ready to go back?

'4. Would they go back entirely at their own expense, requiring nothing further than the assurance of safety to person and estate?

5. Would they be content to live under the Government of the country as they should find it, their rights and privileges being secured to them under the protection of the European powers?

'Let the answers you procure be as distinct and decided and detailed as possible: in respect as to the inquiries as to property, it will of course be sufficient that you should obtain fair proof of the fact from general report.'

The noble Lord who is instituting these inquiries has given deep attention to the matter, and is well known as the writer of an able article in the *Quarterly* on the subject, in December, 1838.

In connexion with this, a deeply interesting discovery has been made on the south-west shores of the Caspian, enclosed in a chain of mountains, of the remnant of the Ten Tribes, living in the exercise of their religious customs in a primitive manner, distinct from the customs of modern Judaism. The facts which distinguish them as the remnant of that branch of the Jewish family are striking and incontrovertible, and are about to be given to the world. An intrepid missionary, the Rev. Mr. Samuel, of Bombay, has made the discovery, and resided amongst this people several months, under permission from the Russian Government, who directed him to institute inquiry concerning them."

9 THE PRIORITY MYTH

It is well known in the Physics community that Albert Einstein was a career plagiarist. Immediately after the Annalen der Physik *published the Einsteins' 1905 paper on the theory of relativity, which wanted for a single reference to the published work of the Einsteins' predecessors, Walter Kaufmann dubbed the special theory of relativity the "Lorentz-Einstein" theory. Kaufmann was overly generous to Einstein at the expense of the Frenchman Henri Poincaré.*

> "The secret to creativity is knowing how to hide your sources."—ALBERT EINSTEIN

> "All this was maintained by Poincaré and others long before the time of Einstein, and one does injustice to truth in ascribing the discovery to him."—CHARLES NORDMANN

9.1 Introduction

It is easily proven that Albert Einstein did not originate the special theory of relativity in its entirety, or even in its majority.[491] The historic record is readily

---

[491]. *See:* P. Langevin, "Le Physicien", *Revue de Métaphysique et de Morale*, Volume 20, Number 5, (September, 1913), pp. 675-718. *See also:* H. A. Lorentz, "Deux mémoires de Henri Poincaré sur la physique mathématique", *Acta Mathematica*, Volume 38, (1921), pp. 293-308; reprinted in *Œuvres de Henri Poincaré*, Volume 9, Gautier-Villars, Paris, (1954), pp. 683-695; **and** Volume 11, (1956), pp. 247-261. *See also:* W. Pauli, "Relativitätstheorie", *Encyklopädie der mathematischen Wissenschaften*, Volume 5, Part 2, Chapter 19, B. G. Teubner, Leipzig, (1921), pp. 539-775; English translation by G. Field, *Theory of Relativity*, Pergamon Press, London, Edinburgh, New York, Toronto, Sydney, Paris, Braunschweig, (1958). *See also:* H. Thirring, "Elektrodynamik bewegter Körper und spezielle Relativitätstheorie", *Handbuch der Physik*, Volume 12 ("Theorien der Elektrizität Elektrostatik"), Springer, Berlin, (1927), pp. 245-348, *especially* 264, 270, 275, 283. *See also:* S. Guggenheimer, *The Einstein Theory Explained and Analyzed*, Macmillan, New York, (1929). *See also:* J. Mackaye, *The Dynamic Universe*, Charles Scribner's Sons, New York, (1931). *See also:* J. Le Roux, "Le Problème de la Relativité d'Après les Idées de Poincaré", *Bulletin de la Société Scientifique de Bretagne*, Volume 14, (1937), pp. 3-10. *See also:* Sir Edmund Whittaker, *A History of the Theories of Aether and Electricity*, Volume II, Philosophical Library Inc., New York, (1954), *especially* pp. 27-77; **and** "Albert Einstein", *Biographical Memoirs of Fellows of the Royal Society*, Volume 1, (1955), pp. 37-67. *See also:* G. H. Keswani, "Origin and Concept of Relativity, Parts I, II & III", *The British Journal for the Philosophy of Science*, Volume 15, Number 60, (February, 1965), pp. 286-306; Volume 16, Number 61, (May, 1965), pp.19-32; Volume 16, Number 64, (February, 1966), pp. 273-294; **and** Volume 17, Number 2, (August, 1966), pp. 149- 152; Volume 17, Number 3, (November, 1966), pp. 234-236. *See also:* G. H. Keswani and C. W. Kilmister, "Intimations of Relativity before Einstein", *The British Journal for the Philosophy of Science*, Volume 34, Number 4, (December, 1983), pp. 343-354. *See also:* G. B. Brown, "What is Wrong with Relativity?", *Bulletin of the Institute of Physics and the Physical Society*, Volume 18, Number 3, (March, 1967), pp. 71-77. *See also:* C. Cuvaj, "Henri Poincaré's Mathematical Contributions to

Relativity and the Poincaré Stresses", *American Journal of Physics*, Volume 36, (1968), pp. 1109-1111. ***See also:*** C. Giannoni, "Einstein and the Lorentz-Poincaré Theory of Relativity", *PSA: Proceedings of the Biennial Meeting of the Philosophy of Science Association*, Volume 1970, (1970), pp. 575-589. JSTOR link:

<http://links.jstor.org/sici?sici=0270-8647%281970%291970%3C575%3AEATLTO%3E2.0.CO%3B2-Z>

See also: J. Mehra, *Einstein, Hilbert, and the Theory of Gravitation*, Reidel, Dordrecht, Netherlands, (1974). ***See also:*** W. Kantor, *Relativistic Propagation of Light*, Coronado Press, Lawrence, Kansas, (1976). ***See also:*** R. McCormmach, "Editor's Forward", ", *Historical Studies in the Physical Sciences*, Volume 7, (1976), pp. xi-xxxv. ***See also:*** H. Ives, D. Turner, J. J. Callahan, R. Hazelett, *The Einstein Myth and the Ives Papers*, Devin-Adair Co., Old Greenwich, Connecticut, (1979). ***See also:*** J. Leveugle, "Henri Poincaré et la Relativité", *La Jaune et la Rouge*, Volume 494, (April, 1994), pp. 29-51; **and** *La Relativité, Poincaré et Einstein, Planck, Hilbert: Histoire véridique de la Théorie de la Relativité*, L'Harmattan, Paris, (2004). ***See also:*** A. A. Logunov, *On the Articles by Henri Poincaré ON THE DYNAMICS OF THE ELECTRON*, Publishing Department of the Joint Institute for Nuclear Research, Dubna, (1995); **and** *The Theory of Gravity*, Nauka, Moscow, (2001); **and** Анри Пуанкаре и ТЕОРИЯ ОТНОСИТЕЛЬНОСТИ, Наука, Москва, (2004). An English translation of this book will soon appear as: *Henri Poincaré and the Theory of Relativity*. ***See also:*** E. Gianetto, "The Rise of Special Relativity: Henri Poincaré's Works before Einstein", *ATTI DEL XVIII CONGRESSO DI STORIA DELLA FISICA E DELL'ASTRONOMICA*, pp. 172-207; URL:

<http://www.brera.unimi.it/Atti-Como-98/Giannetto.pdf>

See also: S. G. Bernatosian, *Vorovstvo i obman v nauke*, Erudit, St. Petersburg, (1998), ISBN: 5749800059 . *See also:* U. Bartocci, *Albert Einstein e Olinto De Pretto: La vera storia della formula piu famosa del mondo*, Societa Editrice Andromeda, Bologna, (1999). ***See also:*** Jean-Paul Auffray, *Einstein et Poincaré: sur les Traces de la Relativité*, Le Pommier, Paris, (1999). ***See also:*** Y. Brovko, "Einshteinianstvo—agenturnaya set mirovovo kapitala", Molodaia Gvardiia, Number 8, (1995), pp. 66-74, at 70. Юрий Бровко, "Эйнштейнианство — агентурная сеть Мирового капитала", Молодая гвардия, № 8, (1995), сс. 66-74; **and** Y. Brovko, "Razgrom einshteinianstvo", Priroda i Chelovek. Svet, Number 7, (2002), pp. 8-10. Юрий Бровко, "Разгром эйнштейнианства", Природа и Человек. Свет, № 7, (2002), сс. 8-10. URL:

<http://medograd.narod.ru/einstein.html>

For counter-argument, see: R. Dugas, *A History of Mechanics*, Dover, New York, (1988), pp. 463-501. ***See also:*** T. Hirosige, "Electrodynamics before the Theory of Relativity, 1890-1905", *Japanese Studies in the History of Science*, Volume 5, (1966), pp. 1-49; **and** "Theory of Relativity and the Ether", *Japanese Studies in the History of Science*, Volume 7, (1967), pp. 37-53; **and** "Origins of Lorentz' Theory of Electrons and the Concept of the Electromagnetic Field", *Historical Studies in the Physical Sciences*, Volume 1, (1969), pp. 151-209; **and** "The Ether Problem, the Mechanistic Worldview, and the Origins of the Theory of Relativity", *Historical Studies in the Physical Sciences*, Volume 7, (1976), pp. 3-82. ***See also:*** S. Goldberg, "Henri Poincare and Einstein's

Theory of Relativity", *American Journal of Physics*, Volume 35, (1967), pp. 934-944; **and** "The Lorentz Theory of Electrons and Einstein's Theory of Relativity", *American Journal of Physics*, Volume 35, (1969), pp. 982-994; **and** "Poincare's Silence and Einstein's Relativity: The Role of Theory and Experiment in Poincaré's Physics", *The British Journal for the History of Science*, Volume 5, Number 17, (1970), pp. 73-84; **and** *Understanding Relativity*, Birkhäuser, Boston, Basel, Stuttgart, (1984). *See also:* A. Pais, *Subtle is the Lord*, Oxford University Press, Oxford, New York, Toronto, Melbourne, (1982). *See also:* A. I. Miller, "A Study of Poincaré's 'Sur la Dynamique de l'Électron'", *Archive for History of Exact Sciences*, Volume 10, (1972), pp. 207-328; *American Journal of Physics*, Volume 45, Number 11, (November, 1977), pp. 1040-1048; **and** *Albert Einstein's Special Theory of Relativity, Emergence (1905) and Early Interpretation (1905-1911)*, Addison-Wesley Publishing Company, Inc., (1981). *See also:* W. Rindler, *American Journal of Physics*, Volume 38, (1970), pp. 1111-1115. *See also:* G. Holton, "On the Origins of the Special Theory of Relativity", *Thematic Origins of Scientific Thought*, Harvard University Press, Cambridge, Massachusetts, London, (1973 revised 1988), pp. 191-236; **and** "On the Origins of the Special Theory of Relativity", *American Journal of Physics*, Volume 28, Number 7, (October, 1960), pp. 627-636, *especially* 633-636; Volume 30, Number 6, (June, 1962), 462-469; **and** *Special Relativity Theory: Selected Papers*, American Institute of Physics, New York, (1963), pp. 1-8.

available. Ludwig Gustav Lange,[492] Woldemar Voigt,[493] Oliver Heaviside,[494]

---

[492]. Lange introduced the term "Inertialsystem" and defined the concept: L. Lange, "Über die wissenschaftliche Fassung der Galilei'schen Beharrungsgesetzes", *Philosophische Studien*, Volume 2, (1885), pp. 266-297, 539-545; L. Lange, "Ueber das Beharrungsgesetz", *Berichte über die Verhandlungen der Königlich Sächsischen Gesellschaft der Wissenschaften zu Leipzig, mathematisch-physische Classe*, Volume 37, (1885), pp. 333-351; **and** *Die geschichtliche Entwickelung des Bewegungsbegriffes und ihr voraussichtliches Endergebniss. Ein Beitrag zur historischen Kritik der mechanischen Principien von Ludwig Lange*, Wilhelm Engelmann, Leipzig, (1886); **and** "Die geschichtliche Entwicklung des Bewegungsbegriffes und ihr voraussichtliches Endergebniss", *Philosophische Studien*, Volume 3, (1886), pp. 337-419, 643-691; **and** "Das Inertialsystem vor dem Forum der Naturforschung: Kritisches und Antikritisches", *Philosophische Studien*, Volume 20, (1902), pp. 1-71; **and** *Das Inertialsystem vor dem Forum der Naturforschung*, Wilhelm Engelmann, Leipzig, (1902); **and** "Mein Verhältnis zu Einsteins Weltbild" ("My Relationship to Einstein's Conception of the World"), *Psychiatrisch-neurologische Wochenschrift*, Volume 24, (1922), pp. 116, 154-156, 168-172, 179-182, 188-190, 201-207.

For references to Lange, and analysis of his work, *see:* B. Thüring, "Fundamental-System und Inertial-System", *Methodos; rivista trimestrale di metodologia e di logica simbolica*, Volume 2, (1950), pp. 263-283; **and** *Die Gravitation und die philosophischen Grundlagen der Physik*, Duncker & Humblot, Berlin, (1967), pp. 75-77, 234-240; *See also:* M. v. Laue, "Dr. Ludwig Lange", *Die Naturwissenschaften*, Volume 35, Number 7, (1948), pp. 193-196; *See also:* E. Mach, *The Science of Mechanics*, Open Court, (1960), pp. 291-297; *See also:* E. Gehrcke, *Kritik der Relativitätstheorie*, Hermann Meusser, Berlin, (1924), pp. 17, 30-34; **and** "Über den Sinn der Absoluten Bewegung von Körpern", *Sitzungsberichten der Königlichen Bayerischen Akademie der Wissenschaften*, Volume 12, (1912), pp. 209-222; **and** "Über die Koordinatensystem der Mechanik", *Verhandlungen der Deutschen Physikalischen Gesellschaft*, Volume 15, (1913), pp. 260-266; **and** A. Müller, "Das Problem des absoluten Raumes und seine Beziehung zum allgemeinen Raumproblem", *Die Wissenschaft*, Volume 39, Friedr. Vieweg & Sohn, Braunschweig, (1911); *See also:* H. Seeliger, "Kritisches Referat über Lange's Arbeiten", *Vierteljahrsschrift der Astronomischen Gesellschaft*, Volume 22, (1886), pp. 252-259; **and** "Über die sogenannte absolute Bewegung", *Sitzungsberichte der mathematische-physikalische Classe der Königlich Bayerische Akademie der Wissenschaften zu München*, (1906), Volume 36, pp. 85-137.

For Lange's immediate predecessors, *see:* C. Neumann, *Ueber die Principien der Galilei-Newton'schen Theorie*, B. G. Teubner, Leipzig, (1870); "The Principles of the Galilean-Newtonian Theory", *Science in Context*, Volume 6, (1993), pp. 355-368; *See also:* William Thomson and P. G. Tait, *Treatise on Natural Philosophy*, Volume 1, Part 1, §§ 245, 249, 267, Cambridge University Press, (1879); *See also:* H. Streintz, *Die physikalischen Grundlagen der Mechanik*, B.G. Teubner, Leipzig, (1883); *See also:* James Thomson, "On the Law of Inertia; the Principle of Chronometry; and the Principle of Absolute Clinural Rest, and of Absolute Rotation", *Proceedings of the Royal Society of Edinburgh*, Volume 12, (November 1883-July 1884), pp. 568-578; **and** "A Problem on Point-motions for Which a Reference-frame Can So Exist as to Have the Motions of the Points, Relative to It, Rectilinear and Mutually Proportional", *Proceedings of the Royal Society of Edinburgh*, Volume 12, (November 1883-July 1884), pp. 730-742; *See also:* P. G. Tait, "Note on Reference Frames", *Proceedings of the Royal Society of*

Edinburgh, Volume 12, (November 1883-July 1884), pp. 743-745.
493. W. Voigt, "Ueber das Doppler'sche Princip", *Nachrichten von der Königlichen Gesellschaft der Wissenschaften und der Georg-Augusts-Universität zu Göttingen*, (1887), pp. 41-51; reprinted *Physikalische Zeitschrift*, Volume 16, Number 20, (15 October 1915), pp. 381-386; English translation, as well as very useful commentary, are found in A. Ernst and Jong-Ping Hsu (W. Kern is credited with assisting in the translation), "First Proposal of the Universal Speed of Light by Voigt in 1887", *Chinese Journal of Physics* (The Physical Society of the Republic of China), Volume 39, Number 3, (June, 2001), pp. 211-230; URL's:

<http://psroc.phys.ntu.edu.tw/cjp/v39/211/211.htm>

<http://psroc.phys.ntu.edu.tw/cjp/v39/211.pdf>

See also: W. Voigt, "Theorie des Lichtes für bewegte Medien", *Nachrichten von der Königlichen Gesellschaft der Wissenschaften und der Georg-Augusts-Universität zu Göttingen*, (1887), pp. 177-238.

Lorentz acknowledged Voigt's priority, and suggested that the "Lorentz Transformation" be called the "Transformations of Relativity": H. A. Lorentz, *Theory of Electrons*, B. G. Teubner, Leipzig, (1909), p. 198 footnote; **and** "Deux mémoires de Henri Poincaré sur la physique mathématique", *Acta Mathematica*, Volume 38, (1921), p. 295; reprinted in *Œuvres de Henri Poincaré*, Volume 9, Gautier-Villars, Paris, (1954), pp. 683-695; **and** Volume 11, (1956), pp. 247-261.

Minkowski also acknowledged Voigt's priority: *The Principle of Relativity*, Dover, New York, (1952), p. 81; **and** *Physikalische Zeitschrift*, Volume 9, Number 22, (November 1, 1908), p. 762. J. Le Roux proposed the nomenclature "Voigt-Lorentzian Transformation" and "Voigt-Lorentzian Group" in "Der Bankrott der Relativitätstheorie", H. Israel, *et al*, Eds., *Hundert Autoren Gegen Einstein*, R. Voigtländer, Leipzig, (1931), pp. 20-27, at 22.

For further discussion of Voigt's relativistic transformation, *see:* F. Hund, "Wer hat die Relativitätstheorie geschaffen?", *Physikalische Blätter*, Volume 36, Number 8, (1980), pp. 237-240. *See also:* W. Schröder, "Hendrik Antoon Lorentz und Emil Wiechert (Briefwechsel und Verhältnis der beiden Physiker)", *Archive for History of Exact Sciences*, Volume 30, Number 2, (1984), pp. 167-187. *See also:* R. Dugas, *A History of Mechanics*, Dover, New York, (1988), pp. 468, 484, 494. *See also:* A. Pais, *Subtle is the Lord*, Oxford University Press, Oxford, New York, Toronto, Melbourne, (1982), pp. 121-122.

494. O. Heaviside, "The Electromagnetic Effects of a Moving Charge", *The Electrician*, Volume 22, (1888), pp. 147-148; **and** "On the Electromagnetic Effects due to the Motion of Electricity Through a Dielectric", *Philosophical Magazine*, Volume 27, (1889), pp. 324-339; **and** "On the Forces, Stresses and Fluxes of Energy in the Electromagnetic Field", *Philosophical Transactions of the Royal Society*, Volume 183A, (1892), p. 423; **and** "A Gravitational and Electromagnetic Analogy", *The Electrician*, Volume 31, (1893), pp. 281-282, 359; **and** *The Electrician*, Volume 45, (1900), pp. 636, 881.

Heinrich Rudolf Hertz,[495] George Francis FitzGerald,[496] Joseph Larmor,[497]

---

495. H. R. Hertz, "Über sehr schnelle electrische Schwingungen", *Annalen der Physik und Chemie*, Volume 31, (1887), pp. 421-449; English translation in: *Electric Waves, Being Researches on the Propagation of Electric Action with Finite Velocity Through Space*, London, New York, Macmillan, (1893), p. 29ff.; **and** "Über einen Einfluß des ultravioletten Lichtes auf die electrische Entladung", *Annalen der Physik und Chemie*, Volume 31, (1887), pp. 983-1000; English translation in: *Electric Waves, Being Researches on the Propagation of Electric Action with Finite Velocity Through Space*, London, New York, Macmillan, (1893), p. 63ff.; **and** *Sitzungsberichte der Königlich Preussischen Akademie der Wissenschaften zu Berlin*, (1887), pp. 487ff.; **and** "Über die Einwirkung einer geradlinigen electrischen Schwingung auf eine benachbarte Strombahn", *Annalen der Physik und Chemie*, Volume 34, (1888), pp. 155-171; **and** "Über die Ausbreitungsgeschwindigkeit der electrodynamischen Wirkungen", *Annalen der Physik und Chemie*, Volume 34, (1888), pp. 551-569; **and** "Über elektrodynamische Wellen im Lufträume und deren Reflexion", *Annalen der Physik und Chemie*, Volume 34, (1888), pp. 609-623; **and** "Ueber die Grundgleichungen der Elektrodynamik für ruhende Körper", *Nachrichten von der Königlichen Gesellschaft der Wissenschaften und der Georg-Augusts-Universität zu Göttingen*, (1890), pp. 106-149; reprinted *Annalen der Physik und Chemie*, Volume 40, (1890), pp. 577-624; reprinted *Untersuchung über die Ausbreitung der Elektrischen Kraft*, Johann Ambrosius Barth, Leipzig, (1892), pp. 208-255; translated into English by D. E. Jones, as: "On the Fundamental Equations of Electromagnetics for Bodies at Rest", *Electric Waves, Being Researches on the Propagation of Electric Action with Finite Velocity Through Space*, London, New York, Macmillan, (1893), pp. 195-239; and "Ueber die Grundgleichungen der Elektrodynamik für bewegte Körper", *Annalen der Physik und Chemie*, Volume 41, (1890), pp. 369-399; reprinted *Untersuchung über die Ausbreitung der Elektrischen Kraft*, Johann Ambrosius Barth, Leipzig, (1892), pp. 256-285; translated into English by D. E. Jones, as: "On the Fundamental Equations of Electromagnetics for Bodies in Motion", *Electric Waves, Being Researches on the Propagation of Electric Action with Finite Velocity Through Space*, London, New York, Macmillan, (1893), pp. 241-268.; **and** "Über den Durchgang der Kathodenstrahlen durch dünne Metallschichten", *Annalen der Physik und Chemie*, Volume 45, (1892), pp. 28-32.
496. G. F. FitzGerald, "On the Electromagnetic Theory of the Reflection and Refraction of Light", *Philosophical Transactions of the Royal Society*, Volume 171, (1880), pp. 691-711; **and** "On Electromagnetic Effects Due to the Motion of the Earth", *The Scientific Transactions of the Royal Dublin Society*, Series 2, Volume 1, (Read May 5th, 1882, published 1883), pp. 319-324; **and** "The Ether and Earth's Atmosphere (Letter to the Editor)", *Science*, Volume 13, Number 328, (1889), p. 390; **and** "Boltzmann on Maxwell", *Nature*, Volume 49, (1894), pp. 381-382. *See also:* O. Lodge, "Aberration Problems", *Philosophical Transactions of the Royal Society of London A*, Volume 184, (1893), p. 184.
497. J. Larmor, "A Dynamical Theory of the Electric and Luminiferous Medium", *Philosophical Transactions of the Royal Society of London A*, Volume 185, (1894), pp. 719-822; Volume 186, (1895), pp. 695-743; Volume 188, (1897), pp. 205-300; **and** "Dynamical Theory of the Ether I & II", *Nature*, Volume 49, (11 January 1894), pp. 260-262; **and** (18 January 1894), pp. 280-283; **and** *Philosophical Magazine*, Series 5, Volume 44, (1897), p. 503; **and** *Aether and Matter*, Cambridge University Press, (1900); **and** "Can Convection Through the Æther be Detected Electrically?", *The Scientific Writings of the Late George Francis FitzGerald*, Longmanns, Green & Co., London, (1902), pp.

566-569; **and** "On the Influence of Convection on Optical Rotary Polarization", *Philosophical Magazine*, Series 6, Volume 4, (September, 1902), pp. 367-370; **and** "On the Intensity of the Natural Radiation from Moving Bodies and its Mechanical Reaction", *Philosophical Magazine*, Series 6, Volume 7, (May, 1904), pp. 578-586; **and** "On the Ascertained Absence of Effects of Motion through the Æther, in Relation to the Constitution of Matter, and the FitzGerald-Lorentz Hypothesis", *Philosophical Magazine*, Series 6, Volume 7, (June, 1904), pp. 621-625; **and** "Æther and Absolute Motion", *Nature*, Volume 76, (18 July 1907), pp. 269-270.

Hendrik Antoon Lorentz,[498] Jules Henri Poincaré,[499] Paul Drude,[500] Paul

---

498. H. A. Lorentz, "Over den Invloed, Dien de Beweging der Aarde op de Lichtverschijnselen Uitoefent", *Koninklijke Akademie van Wetenschappen, Afdeeling Natuurkunde, Verslagen en Mededeelingen* (Amsterdam), Volume 2, (1886), pp. 297-372; French translation, "De l'Influence du Mouvement de la Terre sur les Phénomènes Lumineux", *Archives Néerlandaises des Sciences Exactes et Naturelles*, First Series, Volume 21, (1887), pp. 103-176; reprinted *Collected Papers*, Volume 4, M. Nijhoff, The Hague, (1935-1939), pp. 153-214; **and** "Over de terugkaatsing van licht door lichamen die zich bewegen", *Verslagen der Zittingen de Wis- en Natuurkundige Afdeeling der Koninklijke Akademie van Wetenschappen* (Amsterdam), Volume 1, (1892), pp. 28-31; reprinted in English "On the Reflection of Light by Moving Bodies", *Collected Papers*, Volume 4, pp. 215-218; **and** "De Relatieve Beweging van de Aarde en den Aether", *Verslagen der Zittingen de Wis- en Natuurkundige Afdeeling der Koninklijke Akademie van Wetenschappen* (Amsterdam), Volume 1, (1892/1893), pp. 74-79; translated in English, "The Relative Motion of the Earth and the Ether", *Collected Papers*, Volume 4, pp. 219-223; **and** "La Théorie Électromagnétique de Maxwell et son Application aux Corps Mouvants", *Archives Néerlandaises des Sciences Exactes et Naturelles*, First Series, Volume 25, (1892), pp. 363-552; reprinted *Collected Papers*, Volume 2, pp. 164-343; **and** *Versuch einer Theorie der electrischen und optischen Erscheinungen in bewegten Körpern*, E. J. Brill, Leiden, (1895); unaltered reprint by B. G. Teubner, Leipzig, (1906); reprinted *Collected Papers*, Volume 5, pp. 1-137; **and** "Concerning the Problem of the Dragging Along of the Ether by the Earth", *Verslagen der Zittingen de Wis- en Natuurkundige Afdeeling der Koninklijke Akademie van Wetenschappen* (Amsterdam), Volume 6, (1897), p. 266; reprinted *Collected Papers*, Volume 4, pp. 237-244; **and** "Optical Phenomena Connected with the Charge and Mass of the Ions", *Verslagen der Zittingen de Wis- en Natuurkundige Afdeeling der Koninklijke Akademie van Wetenschappen* (Amsterdam), Volume 6, (1898), p. 506, 555; reprinted *Collected Papers*, Volume 3, pp. 17-39; **and** "Vereenvoudigde Theorie der Electrische en Optische Verschijnselen in Lichamen die Zich Bewegen", *Verslagen van de gewone vergaderingen der wis- en natuurkundige afdeeling, Koninklijke Akademie van Wetenschappen te Amsterdam*, Volume 7, (1899), pp. 507-522; English translation, "Simplified Theory of Electrical and Optical Phenomena in Moving Bodies", *Proceedings of the Section of Sciences, Koninklijke Akademie van Wetenschappen te Amsterdam*, Volume 1, (1899), pp. 427–442; reprinted in K. F. Schaffner, *Nineteenth Century Aether Theories*, Pergamon Press, New York, Oxford, (1972), pp. 255-273; French translation, "Théorie Simplifiée des Phénomènes Électriques et Optiques dans des Corps en Mouvement", *Verslagen der Zittingen de Wis- en Natuurkundige Afdeeling der Koninklijke Akademie van Wetenschappen* (Amsterdam), Volume 7, (1899), p. 507; **and** *Archives Néerlandaises des Sciences Exactes et Naturelles*, Volume 7, (1902), pp. 64-80; reprinted *Collected Papers*, Volume 5, pp. 139-155; **and** "La Théorie de l'Aberration de Stokes dans l'Hypothèse d'un Éther n'Ayant pas Partout la Même Densité", *Verslagen der Zittingen de Wis- en Natuurkundige Afdeeling der Koninklijke Akademie van Wetenschappen* (Amsterdam), Volume 7, (1899), p. 523; *Proceedings of the Royal Academy of Sciences at Amsterdam* (*Noninklijke Nederlandse Akademie van Wetenschappen te Amsterdam*), Volume 1, (1899), p. 425 *et seq.*; reprinted *Collected Papers*, Volume 4, pp. 245-251; **and** "Considérations sur la Pesantuer", *Verslagen der Zittingen de Wis- en Natuurkundige Afdeeling der Koninklijke Akademie van Wetenschappen* (Amsterdam), Volume 8, (1900), p. 603; *Archives Néerlandaises des Sciences Exactes et Naturelles*, Series 2, Volume 7, (1902), pp. 325-342; reprinted *Collected Papers*, Volume 5, pp. 198-215; **and**

"Über die scheinbare Masse der Ionen", *Physikalische Zeitschrift*, Volume 2, Number 5, (November, 1900), pp. 78-80; reprinted *Collected Papers*, Volume 3, pp. 113-116; **and** "Théorie des Phénomènes Magnéto-Optiques Récemment Découverts", *Rapports présentés au Congrès international de physique réuni à Paris en 1900 sous les auspices de la Société française de physique, rassemblés et publiés par Ch.-Éd. Guillaume et L. Poincaré*, Volume 3, Gauthier-Villars, Paris, (1900-1901), pp. 1-33; **and** "De Electronen-Theorie", *Nederlandsch Natuur en Geneeskundig Congres, Verhandelingen*, Volume 8, (April 12, 1901), p. 35; reprinted in *Collected Papers*, Volume 9, pp. 102-111; **and** "Sur la Méthode du Miroir Tourant pour la Détermination de la Vitesse de la Lumière", *Archives Néerlandaises des Sciences Exactes et Naturelles*, Series 2, Volume 6, (1901), pp. 303-318; reprinted in *Collected Papers*, Volume 4, 104-118; **and** *Sichtbare und unsichtbare Bewegungen*, F. Vieweg, Braunschweig, (1902); **and** *Encyklopädie der mathematischen Wissenschaften*, Volume 5, Part 2, B. G. Teubner, Leipzig, (1904), "Maxwells elektromagnetische Theorie" (submitted June, 1903), Chapter 13, pp. 63-144, and "Weiterbildung der Maxwellschen Theorie, Elektronentheorie" (submitted December, 1903), Chapter 14, pp. 145-280; **and** "Electromagnetische Verschijnselen in een Stelsel dat Zich met Willekeurige Snelheid, Kleiner dan die van Het Licht, Beweegt", *Koninklijke Akademie van Wetenschappen te Amsterdam, Wis- en Natuurkundige Afdeeling, Verslagen van de Gewone Vergaderingen*, Volume 12, (23 April 1904), pp. 986-1009; translated into English, "Electromagnetic Phenomena in a System Moving with any Velocity Smaller than that of Light", *Proceedings of the Royal Academy of Sciences at Amsterdam (Noninklijke Nederlandse Akademie van Wetenschappen te Amsterdam)*, 6, (May 27, 1904), pp. 809-831; reprinted *Collected Papers*, Volume 5, pp. 172-197; a redacted and shortened version appears in *The Principle of Relativity*, Dover, New York, (1952), pp. 11-34; a German translation from the English, "Elektromagnetische Erscheinung in einem System, das sich mit beliebiger, die des Lichtes nicht erreichender Geschwindigkeit bewegt," appears in *Das Relativitätsprinzip: eine Sammlung von Abhandlungen*, B. G. Teubner, Leipzig, (1913), pp. 6-26; **and** *Ergebnisse und Probleme der Elektronentheorie*, Julius Springer, Berlin, (published 1905, received on 20 December 1904; second revised edition published 1906); reprinted *Collected Papers*, Volume 8, pp. 79-124; **and** *Abhandlungen über theoretische Physik*, B. G. Teubner, Leipzig, (1907); **and** *The Theory of Electrons and Its Applications to the Phenomena of Light and Radiant Heat*, B.G. Teubner; Leipzig, G.E. Stechert & Co., New York, (1909); Second Edition, B. G. Teubner, Leipzig, (1916); reprinted Dover, New York, (1952); **and** "Alte und neue Fragen der Physik", *Physikalische Zeitschrift*, Volume 11, (1910), pp. 1234-1257; **and** *Het Relativiteitsbeginsel voor eenparige translaties (1910–1912)*, Brill, Leiden; **and** *Het Relativiteitsbeginsel; drie Voordrachten Gehouden in Teyler's Stiftung*, Erven Loosjes, Haarlem, (1913); German translation, *Das Relativitätsprinzip; drei Vorlesungen gehalten in Teylers Stiftung zu Haarlem*, B. G. Teubner, Leipzig-Berlin, (1914/1920); **and** "Considérations Élémentaires sur le Principe de Relativité", *Revue Générale des Sciences Pures et Appliquées*, Volume 25, (1914), pp. 179-201; reprinted *Collected Papers*, Volume 7, pp. 147-165; **and** "Die Maxwellsche Theorie und die Elektronentheorie", *Die Kultur der Gegenwart*, E. G. Warburg, Ed., *Physik*, B. G. Teubner, Berlin, Leipzig, (1915), p. 311; **and** "The Michelson-Morley-Experiment and the Dimensions of Moving Bodies", *Nature*, Volume 106, (1921), pp. 793-795; **and** *Lessen Over Theoretische Natuurkunde aan de Rijks-Universiteit te Leiden Gegeven*, (1922); German translation, *Vorlesungen über theoretische physik an der Universität Leiden*, Akademische Verlagsgesellschaft M. B. H., Leipzig, (1927-1931); English translation, *Lectures on Theoretical Physics*, Macmillan, London, (1931); **and**

Problems of Modern Physics. A Course of Lectures Delivered in the California Institute of Technology, Ginn and Company, Boston, (1927).
499. J. H. Poincaré, URL:

<http://gallica.bnf.fr/metacata.idq?Bgc=&Mod=&CiRestriction=%28@_Auteur%20henri%26poincare%29&RPT=>

"Sur les Hypothèses Fondamentales de la Géométrie", *Bulletin de la Société Mathématique de France*, Volume 15, (1887), pp. 203-216; **and** *Théorie Mathématique de la Lumière*, Naud, Paris, (1889); **and** "Les Géométries Non-Euclidiennes", *Revue Générale des Sciences Pures et Appliquées*, Volume 2, (1891), pp. 769-774; **and** "A Propos de la Théorie de M. Larmor", *L'Éclairage électrique*, Volume 3, (April 6, May 18, 1895), pp. 5-13, 289-295; Volume 5, (October 5, November 8, 1895) pp. 5-14, 385-392; reprinted *Œuvres de Henri Poincaré*, Volume 9, Gautier-Villars, Paris, (1954), pp. 369-426; **and** "La Mesure du Temps", *Revue de Métaphysique et de Morale*, Volume 6, (January, 1898) pp. 1-13; English translation by G. B. Halsted, *The Value of Science*, The Science Press, New York, (1907), pp. 26-36; **and** "La Théorie de Lorentz at le Principe de Réaction", *Archives Néerlandaises des Sciences Exactes et Naturelles*, Series 2, Volume 5, (1900), pp. 252-278; reprinted *Œuvres*, Volume IX, pp. 464-488; **and** "RELATIONS ENTRE LA PHYSIQUE EXPÉRIMENTALE ET LA PHYSIQUE MATHÉMATIQUE", *RAPPORTS PRÉSENTÉS AU CONGRÈS INTERNATIONAL DE PHYSIQUE DE 1900*, Volume I, Gauthier-Villars, Paris, (1900), pp. 1-29; translated into German "Über die Beziehungen zwischen der experimentellen und mathematischen Physik", *Physikalische Zeitschrift*, Volume 2, (1900-1901), pp. 166-171, 182-186, 196-201; English translation in *Science and Hypothesis*, Chapters 9 and 10; **and** *Electrité et Optique*, Gauthier-Villars, Paris, (1901), *especially* Chapter 6, pp. 516-536; **and** *La Science et l'Hypothèse*, E. Flammarion, Paris, (1902); translated into English *Science and Hypothesis*, Dover, New York, (1952), which appears in *The Foundations of Science*; translated into German *with substantial notations* by Ferdinand and Lisbeth Lindemann *Wissenschaft und Hypothese*, B. G. Teubner, Leipzig, (1904); **and** Poincaré's St. Louis lecture from September of 1904, *La Revue des Idées*, 80, (November 15, 1905); "L'État Actuel et l'Avenir de la Physique Mathématique", *Bulletin des Sciences Mathématique*, Series 2, Volume 28, (1904), p. 302-324; English translation, "The Principles of Mathematical Physics", *The Monist*, Volume 15, Number 1, (January, 1905), pp. 1-24; **and** *La Valeur de la Science*, E. Flammarion, Paris, (1905); English translation by G. B. Halsted, *The Value of Science*, The Science Press, New York, (1907), pp. 26-36; the *The Value of Science*, itself, appears in *The Foundations of Science; Science and Hypothesis, The Value of Science, Science and Method*, The Science Press, Garrison, New York, (1913/1946), University Press of America, Washington D. C., (ca. 1982); **and** "Sur la Dynamique de l'Électron", *Comptes rendus hebdomadaires des séances de L'Académie des sciences*, Volume 140, (1905), pp. 1504-1508; reprinted in H. Poincaré, *La Mécanique Nouvelle: Conférence, Mémoire et Note sur la Théorie de la Relativité / Introduction de Édouard Guillaume*, Gauthier-Villars, Paris, (1924), pp. 77-81 URL:

<http://gallica.bnf.fr/scripts/ConsultationTout.exe?E=0&O=N029067>

reprinted *Œuvres de Henri Poincaré*, Volume 9, Gautier-Villars, Paris, (1954), pp. 489-493; English translations appear in: G. H. Keswani and C. W. Kilmister, "Intimations of Relativity before Einstein", *The British Journal for the Philosophy of Science*, Volume

34, Number 4, (December, 1983), pp. 343-354, at pp. 350-353; **and**, translated by G. Pontecorvo with extensive commentary by A. A. Logunov, *On the Articles by Henri Poincaré ON THE DYNAMICS OF THE ELECTRON*, Publishing Department of the Joint Institute for Nuclear Research, Dubna, (1995), pp. 7-14; **and** "Sur la Dynamique de l'Électron", *Rendiconti del Circolo matimatico di Palermo*, Volume 21, (1906, submitted July 23rd, 1905), pp. 129-176; reprinted in H. Poincaré, *La Mécanique Nouvelle: Conférence, Mémoire et Note sur la Théorie de la Relativité / Introduction de Édouard Guillaume*, Gauthier-Villars, Paris, (1924), pp. 18-76 URL:

<http://gallica.bnf.fr/scripts/ConsultationTout.exe?E=0&O=N029067>

reprinted *Œuvres*, Volume IX, pp. 494-550; redacted English translation by H. M. Schwartz with modern notation, "Poincaré's Rendiconti Paper on Relativity", *American Journal of Physics*, Volume 39, (November, 1971), pp. 1287-1294; Volume 40, (June, 1972), pp. 862-872; Volume 40, (September, 1972), pp. 1282-1287; English translation by G. Pontecorvo with extensive commentary by A. A. Logunov with modern notation, *On the Articles by Henri Poincaré ON THE DYNAMICS OF THE ELECTRON*, Publishing Department of the Joint Institute for Nuclear Research, Dubna, (1995), pp. 15-78; **and** "La Dynamique de l'Électron", *Revue Générale des Sciences Pures et Appliquées*, Volume 19, (1908), pp. 386-402; reprinted *Œuvres*, Volume IX, pp. 551-586; English translation: "The New Mechanics", *Science and Method*, Book III, which is reprinted in *Foundations of Science*; **and** *Science et Méthode*, E. Flammarion, Paris, (1908); translated in English as *Science and Method*, numerous editions; *Science and Method* is also reprinted in *Foundations of Science*; **and** "La Mécanique Nouvelle", *Comptes Rendus des Sessions de l'Association Française pour l'Avancement des Sciences*, Conférence de Lille, Paris, (1909), pp. 38-48 ; *La Revue Scientifique*, Volume 47, (1909), pp. 170-177; reprinted in H. Poincaré, *La Mécanique Nouvelle: Conférence, Mémoire et Note sur la Théorie de la Relativité / Introduction de Édouard Guillaume*, Gauthier-Villars, Paris, (1924), pp. 18-76 URL:

<http://gallica.bnf.fr/scripts/ConsultationTout.exe?E=0&O=N029067>

and **28 April 1909 Lecture in Göttingen:** "La Mécanique Nouvelle", *Sechs Vorträge über der reinen Mathematik und mathematischen Physik auf Einladung der Wolfskehl-Kommission der Königlichen Gesellschaft der Wissenschaften gehalten zu Göttingen vom 22.-28. April 1909*, B. G. Teubner, Berlin, Leipzig, (1910), pp. 51-58; "The New Mechanics", *The Monist*, Volume 23, (1913), pp. 385-395; **13 October 1910 Lecture in Berlin:** "Die neue Mechanik", *Himmel und Erde*, Volume 23, (1911), pp. 97-116; *Die neue Mechanik*, B. G. Teubner, Berlin, Leipzig, (1911); **and** *Dernières Pensées*, E. Flammarion, Paris, (1913); translated in English as *Mathematics and Science: Last Essays*, Dover, New York, (1963); **and** *The Foundations of Science; Science and Hypothesis, The Value of Science, Science and Method*, The Science Press, Garrison, New York, (1913/1946), University Press of America, Washington D. C., (ca. 1982).
500. P. Drude, "Ueber Fernewirkungen", *Annalen der Physik und Chemie*, Volume 62, (1897), pp. 693, I-XLIX; **and** *Lehrbuch der Optik*, S. Hirzel, Leipzig, (1900); translated into English *The Theory of Optics*, Longmans, Green and Co., London, New York, Toronto, (1902), *see especially* pp. 457-482; **and** "Zur Elektronentheorie der Metalle. I & II", *Annalen der Physik*, Series 4, Volume 1, (1900), pp. 566-613; Volume 3, (1900), pp. 369-402; **and** "Optische Eigenschaften und Elektronentheorie, I & II", *Annalen der*

Langevin,[501] and many others, slowly developed the theory, step by step, and based it on thousands of years of recorded thought and research. Einstein may have made a few contributions to the theory, such as the relativistic equations for aberration and the Doppler-Fizeau Effect;[502] though he also rendered an

---

Physik, Series 4, Volume 14, (1904), pp. 677-725, 936-961; **and** "Die Natur des Lichtes" in A. Winkelmann, *Handbuch der Optik*, Volume 6, Second Edition, J. A. Barth, Leipzig, (1906), pp. 1120-1387; **and** *Physik des Aethers auf elektromagnetischer Grundlage*, F. Enke, Stuttgart, (1894), Posthumous Second Revised Edition, W. König, (1912).
501. P. Langevin, *Comptes Rendus Hebdomadaires des Séances de l'Académie des Sciences*, Volume 139, (1904), p. 1204; **and** "La Physique des Électrons", *Revue Générale des Sciences Pures et Appliquées*, Volume 16, (1905), pp. 257-276; **and** *Bulletin des Séances de la Société Française de Physique: Résumé des Communications*, (1905), p. 13; **and** "Sur une Formule Fondamentale de la Théorie Cinétique", *Comptes Rendus Hebdomadaires des Séances de l'Académie des Sciences*, Volume 140, (1905), pp. 35-38; **and** "Sur l'Impossibilité Physique de Mettre en Évidence le Mouvement de Translation de la Terre", *Comptes Rendus Hebdomadaires des Séances de l'Académie des Sciences*, Volume 140, (1905), pp. 1171-1173; **and** "Magnétisme et Théorie des Électrons", *Annales de Chimie et de Physique*, Volume 5, (1905), pp. 70-127; **and** "Une Formule Fondamentale de la Théorie Cinétique", *Annales de Chimie et de Physique*, Volume 5, (1905), pp. 245-288; **and** "La Physique des Électrons", Revue Générale des Sciences et Appliquées, Volume 16, (1905), pp. 257-276; **and** "Sur la Théorie du Mouvement Brownien", *Comptes Rendus Hebdomadaires des Séances de l'Académie des Sciences*, Volume 146, (1908) pp. 530-533; **and** "L'Évolution de l'Espace et du Temps", *Scientia*, Volume 10, (1911), pp. 31-54; **and** *La Théorie Rayonnement et les Quanta. Rapports et Discussions de la Réunion Tenue à Bruxelles, du 30 Octobre au 3 Novembre 1911, sous les Auspices de M. E. Solvay. Publiés par MM. P. Langevin et M. de Broglie*, Gauthier-Villars, Paris, (1912), 393-406; **and** "L'Inertie de l'Énergie et ses Conséquences", *Journal de Physique*, Volume 3, (1913), pp. 553-591; **and** "Le Physicien", *Revue de Métaphysique et de Morale*, Volume 20, Number 5, (September, 1913), pp. 675-718; **and** *Die Theorie der Strahlung und der Quanten : Verhandlungen auf einer von E. Solvay einberufenen Zusammenkunst, 30. Oktober bis 3. November 1911 : mit einem Anhange über die Entwicklung der Quantentheorie vom Herbst 1911 bis zum Sommer 1913 / in deutscher Sprache herausgegeben von A. Eucken*, Verlag Chemie, Berlin, (1914), pp. 318-329; **and** *Oeuvres Scientifiques de Paul Langevin*, Paris, Centre National de la Recherche Scientifique, (1950).
502. Aberration: J. Bradley, "An Account of a New Discovered Motion of the Fixed Stars", *Philosophical Transactions of the Royal Society of London*, Volume 35, Number 406, (1729), p. 637; F. Fresnel, *Annales de Chimie et de Physique*, Series 2, Volume 9, (1818), pp. 57-66. See also: W. Veltmann, "Fresnel's Hypothese zur Erklärung der Aberrationserscheinungen", *Astronomische Nachrichten*, Volume 75, (1870), pp. 145-160; **and** "Ueber die Fortpflanzung des Lichtes in bewegten Medien", Volume 76, (1870), pp. 129-144; **and** "Ueber die Fortpflanzung des Lichtes in bewegten Medien", *Annalen der Physik und Chemie*, Volume 150, (1873), pp. 497-534.
Doppler-Fizeau Effect: C. Doppler, "Über das farbige Licht der Doppelsterne u. einiger anderer Gestirne des Mimmels. Versuch einer das Bradleysche Abberrationstheorem als integrierenden Teil in sich schliessenden allgemeinen Theorie", *Neuere Abhandlungen der Königlichen Böhmischen Gesellschaft der Wissenschaften*, Series 2, Volume 5, (1842, Jahrgang 1841-42), pp. 462-482, (1845), p. 419ff.; H. Fizeau, *Annales de Chimie et de Physique*, Volume 19, (1870), p. 211. M. Einstein-Marity and

incorrect equation for the transverse mass of an electron, which, when corrected, becomes Lorentz' equation.[503]

Albert Einstein's first work on the theory of relativity did not appear until 1905. There is substantial evidence that Albert Einstein did not write this 1905 paper[504] on the "principle of relativity" alone. His wife, Mileva Einstein-Marity, may have been co-author, or the sole author, of the work.[505]

A. Einstein, "Zur Elektrodynamik bewegter Körper", *Annalen der Physik*, Volume 17, (1905), pp. 910-912.
503. *See:* M. Planck, "Das Prinzip der Relativität und die Grundgleichungen der Mechanik", *Verhandlungen der Deutschen Physikalischen Gesellschaft*, Volume 8, (1906), pp. 136-141. *See also:* W. Kaufmann, "Über die Konstitution des Elektrons", *Annalen der Physik*, Volume 19, (1906), pp. 530-531. *See also:* E. G. Cullwick, "Einstein and Special Relativity: Some Inconsistencies in his Electrodynamics", *The British Journal for the Philosophy of Science*, Volume 32, Number 2, (June, 1981), pp. 167-176.
504. M. Einstein-Marity and A. Einstein, "Zur Elektrodynamik bewegter Körper", *Annalen der Physik*, Volume 17, (1905), pp. 891-921.
505. D. Trbuhović-Gjurić, *Im Schatten Albert Einsteins, Das tragische Leben der Mileva Einstein-Marić*, Paul Haupt, Bern, (1983). *See also:* D. Krstic, Matica Srpska (Novi Sad), Collected Papers. *Natural Sciences*, Volume 40, (1971), p. 190, note 2; **and** "The Wishes of Dr. Einstein", *Dnevnik* (Novi Sad), Volume 28, Number 9963, (1974), p. 9; **and** "The Education of Mileva Marić-Einstein, the First Woman Theoretical Physicist, at the Royal Classical High School in Zagreb at the End of the 19th Century", *Collected Papers on History of Education* (Zagreb), Volume 9, (1975), p. 111; **and** "The First Woman Theoretical Physicist", *Dnevnik* (Novi Sad), Volume 30, VIII/21, (1976); **and** *Mileva and Albert Einstein: Love and Joint Scientific Work*, Diodakta, (1976); and D. Krstic, "Mileva Einstein-Marić", in E. R. Einstein, *Hans Albert Einstein: Reminiscences of His Life and Our Life Together*, Appendix A, Iowa Institute of Hydraulic Research, University of Iowa, Iowa City, Iowa, (1991), pp. 85-99, 111-112. *See also:* T. Pappas, *Mathematical Scandals*, Wide World Publishing/Tetra, San Carlos, California, (1997), pp. 121-129. *See also:* M. Maurer, "Weil nicht sein kann, was nicht sein darf... 'DIE ELTERN' ODER 'DER VATER' DER RELATIVITÄTSTHEORIE? Zum Streit über den Anteil von Mileva Marić an der Entstehung der Relativitätstheorie", *PCnews*, Number 48 (Nummer 48), Volume 11 (Jahrgang 11), Part 3 (Heft 3), Vienna, (June, 1996), pp. 20-27; reprinted from *Dokumentation des 18. Bundesweiten Kongresses von Frauen in Naturwissenschaft und Technik vom 28.-31*, Birgit Kanngießer, Bremen, (May, 1992), not dated, pp. 276-295; an earlier version appeared, co-authored by P. Seibert, *Wechselwirkung*, Volume 14, Number 54, Aachen, (April, 1992), pp. 50-52 (Part 1); Volume 14, Number 55, (June, 1992), pp. 51-53 (Part 2). URL:

http://rli.at/Seiten/kooperat/maric1.htm

See also: E. H. Walker, "Did Einstein Espouse his Spouses Ideas?", *Physics Today*, Volume 42, Number 2, (February, 1989), pp. 9, 11; **and** "Mileva Marić's Relativistic Role", *Physics Today*, Volume 44, Number 2, (February, 1991), pp. 122-124; **and** "Ms. Einstein", *AAAS* [American Association for the Advancement of Science] *Annual Meeting Abstracts for 1990*, (February 15-20, 1990), p. 141; **and** "Ms. Einstein", *The Baltimore Sun*, (30 March 1990), p. 11A. *See also:* S. Troemel-Ploetz, "Mileva Einstein-Marić: The Woman Who did Einstein's Mathematics", *Women's Studies International Forum*, Volume 13, Number 5, (1990), pp. 415-432; *Index on Censorship*, Volume 19,

9.2 Opinions of Einstein and "His" Work

If Albert Einstein did not originate the major concepts of the special theory of relativity, how could such a historically significant fact have escaped the attention of the world for nearly a century? The simple answer is that it did not.
   Some called Einstein's priority into question almost immediately. As early

---

Number 9, (October, 1990), pp. 33-36. *See also:* A. Pais, *Subtle is the Lord*, Oxford University Press, New York, (1982), p. 47. *See also:* W. Sullivan, "Einstein Letters Tell of Anguished Love Affair", *The New York Times*, (3 May 1987), pp. 1, 38. *See also:* "Did Einstein's Wife Contribute to His Theories?", *The New York Times*, (27 March 1990), Section C, p. 5. *See also:* S. L. Garfinkel, "First Wife's Role in Einstein's Work Debated", *The Christian Science Monitor*, (27 February 1990), p. 13. *See also:* D. Overbye, "Einstein in Love", *Time*, Volume 135, Number 18, (30 April 1990), p. 108; **and** *Einstein in Love : A Scientific Romance*, Viking, New York, (2000). *See also:* "Was the First Mrs Einstein a Genius, too?", *New Scientist*, Number 1706, (3 March 1990), p. 25. *See also:* A. Gabor, *Einstein's Wife: Work and Marriage in the Lives of Five Great Twentieth-Century Women*, Viking, New York, (1995). *See also:* J. Haag, "Einstein-Marić, Mileva", *Women in World History: A Biographical Encyclopedia*, Volume 5, Yorkin Publications, (2000), pp. 77-81. *See also:* M. Zackheim, *Einstein's Daughter: The Search for Lieserl*, Riverhead Books, Penguin Putnam, New York, (1999). *See also:* Television Documentary, *Einstein's Wife: The Life of Mileva Maric-Einstein*, URL:

http://www.pbs.org/opb/einsteinswife/

   For counter-argument, *see:* J. Stachel, "Albert Einstein and Mileva Marić: A Collaboration that Failed to Develop", found in *Creative Couples in the Sciences*, Rutgers University Press, New Brunswick, New Jersey, (1996), pp. 207-219; **and** Stachel's reply to Walker, *Physics Today*, Volume 42, Number 2, (February, 1989), pp. 11, 13. *See also:* A. Fölsing, "Keine 'Mutter der Relativitätstheorie'", *Die Zeit*, Number 47, (16 November 1990), p. 94. *See also:* A. Pais, *Einstein Lived Here*, Oxford University Press, New York, (1994), pp. 14-16.
   For the Einstein's correspondence, *see:* J. Stachel, Editor, *The Collected Papers of Albert Einstein*, Volume 1, Princeton University Press, (1987); English translations by A. Beck, *The Collected Papers of Albert Einstein*, Volume 1, Princeton University Press, (1987). *See also:* J. Renn and R. Schulmann, Editors, *Albert Einstein/Mileva Maric: The Love Letters*, Princeton University Press, (1992). *See also:* M. Popović, *In Albert's Shadow: The Life and Letters of Mileva Maric, Einstein's First Wife*, The Johns Hopkins University Press, (2003).

as the years 1905-1907, Max Planck,[506] Walter Kaufmann,[507] Paul Ehrenfest,[508] Jakob Laub,[509] Max von Laue,[510] Hermann Minkowski, and Albert Einstein,[511] himself, referred to the Einsteins' theory as being a mere interpretation and generalization of Hendrik Antoon Lorentz' principle of relativity, which interpretation and generalization was first accomplished by Henri Poincaré,[512] and later became known as the "Special Theory of Relativity".

In 1905, immediately after the appearance of the Einsteins' first paper on the principle of relativity, which did not contain any references to previous works, Walter Kaufmann coined the term "Lorentz-Einstein" for the theory, in recognition of Lorentz' priority,

> "Finally, there is a recently published theory of electrodynamics by Mr. A. Einstein, which leads to consequences which are formally identical to those of Lorentz' theory, and for which, therefore, the second equation applies, as well. [***] (Lorentz-Einstein) [***] The above results speak decidedly against the correctness of the Lorentzian, and, therefore, also the Einsteinian, fundamental assumption. If one considers this basic assumption as thereby disproved, then the attempt to base the whole of Physics including electrodynamics and optics on the principle of relative motion

---

506. M. Planck, "Das Prinzip der Relativität und die Grundgleichungen der Mechanik", *Verhandlungen der Deutschen Physikalischen Gesellschaft*, Volume 8, (1906), pp. 136-141; "Die Kaufmannschen Messungen der Ablenkbarkeit der ß-Strahlen in ihrer Bedeutung für die Dynamik der Elektronen", *Physikalische Zeitschrift* Volume 7, (1906), pp. 753-759, with a discussion on pp. 759-761.
507. W. Kaufmann, "Über die Konstitution des Elektrons", *Sitzungsberichte der Königlich Preussischen Akademie der Wissenschaften zu Berlin*, (1905), pp. 949-956, *especially* p. 954; **and** "Über die Konstitution des Elektrons", *Annalen der Physik*, Volume 19, (1906), pp. 487-553; " Nachtrag zu der Abhandlung: 'Über die Konstitution des Elektrons'", *Annalen der Physik*, Volume 20, (1906), pp. 639-640.
508. P. Ehrenfest, "Die Translation deformierbarer Elektronen und der Flächensatz", *Annalen der Physik*, 23, (1907), pp. 204-205.
509. J. Laub, "Zur Optik der bewegten Körper", *Annalen der Physik*, 23, (1907), pp. 738-744.
510. M. v. Laue, "Die Mitführung des Lichtes durch bewegte Körper nach dem Relativitätsprinzip", *Annalen der Physik*, 23, (1907), pp. 989-990.
511. A. Einstein, "Über das Relativitätsprinzip und die aus demselben gezogenen Folgerung", *Jahrbuch der Radioaktivität und Elektronik*, Volume 4, (1907), pp. 411-462.
512. *Cf.* S. Mohorovičić, *Die Einsteinsche Relativitätstheorie und ihr mathematischer, physikalischer und philosophischer Charakter*, Walter de Gruyter & Co., Berlin, Leipzig, (1923), pp. 29-30; **and** "Über die räumliche und zeitliche Translation", "»Bulletin« d. süslaw. Akad. D. Wiss." (*Jugoslovenska Akademija Znanosti i Umjetnosti*), Volume 6/7, Zagreb, (1916-1917), p. 48; **and** "Die Folgerung der allgemeinen Relativitätstheorie und die Newtonsche Physik", *Naturwissenschaftliche Wochenschrift*, New Series, Volume 20, Jena, (1921), pp. 737-739; **and**, as cited by Mohorovičić: E. Guillaume and C. Willigens, "Über die Grundlagen der Relativitätstheorie", *Physikalische Zeitschrift*, Volume 22, (1921), pp. 109-114; **and** E. Guillaume, "Graphische Darstellung der Optik bewegter Körper", *Physikalische Zeitschrift*, Volume 22, (1921), pp. 386-388.

must be considered a failure."

"Endlich ist noch eine von Hrn. A. Einstein² kürzlich publizierte Theorie der Elektrodynamik zu erwähnen, die zu Folgerungen führt, die mit denen LORENTZschen Theorie formell identisch sind, und für die deshalb auch die zweite Gleichung in Anwendung kommt. [***] (LORENTZ-EINSTEIN) [***] Die vorstehenden Ergebnisse sprechen entschieden gegen die Richtigkeit der Lorentzschen und somit auch der Einsteinschen Grundannahme. Erachtet man diese Grundannahme als hierdurch widerlegt, so würde der Versuch, die ganze Physik, einschließlich der Elektrodynamik und der Optik auf das Prinzip der Relativbewegung zu gründen, einstweilen als mißglückt zu bezeichnen sein."[513]

Kaufmann again used the phrase "Lorentz-Einstein" in 1906, and reiterated the formal identity of the two authors' works,

"Einstein's theory leads to the same formula as Lorentz'[.]"

"Die Einsteinsche Theorie führt zu derselben Formel wie die Lorentzsche[.]"[514]

Max Planck stated in the early spring of 1906,

"The'principle of relativity' recently introduced by H. A. Lorentz¹) and more generally worded by A. Einstein²)[.]"

"Das vor kurzem von H. A. Lorentz¹) und in noch allgemeinerer Fassung von A. Einstein²) eingeführte „Prinzip der Relativität''[.]"[515]

In 1906, Planck referred to the theory of relativity as the Lorentz-Einstein theory and referenced Poincaré,

"I have only done the calculations for those two theories, which are the most developed at this point: Abraham's [*Footnote:* M. Abraham, Ann. d. Phys.

---

[513]. W. Kaufmann, "Über die Konstitution des Elektrons", *Sitzungsberichte der Königlich Preussischen Akademie der Wissenschaften zu Berlin*, (1905), pp. 949-956, at 954 and 956.
[514]. W. Kaufmann, "Über die Konstitution des Elektrons", *Annalen der Physik*, Volume 19, (1906), pp. 487-553, at 530.
[515]. M. Planck, "Das Prinzip der Relativität und die Grundgleichungen der Mechanik", *Verhandlungen der Deutschen Physikalischen Gesellschaft*, Volume 8, (1906), pp. 136-141, at 136. Planck repeated himself in a very similar statement in 1907, "Zur Dynamik bewegter Systeme", *Sitzungsberichte der Königlich Preussischen Akademie der Wissenschaften, Sitzung der physikalisch-mathematischen Classe*, Volume 13, (13 June 1907), pp. 541-570, at 546.

(4) 10, 105, 1903.], according to which the electron has the form of a rigid sphere, and Lorentz-Einstein's [*Footnote:* H. A. Lorentz, Versl. Kon. Akad. v. Wet. Amsterdam 1904, S. 809. A. Einstein, Ann. d. Phys. (4) 17, 891, 1905. Also confer with H. Poincaré, C. R. 140, 1504, 1905.], according to which the 'principle of relativity' is rigorously valid. In order to be concise, I will dub the first theory 'theory of the sphere', and the second 'theory of relativity'. [***] The Lorentz-Einstein theory is based upon the postulate that no absolute translation is provable."

"Ich habe die Rechnungen nur für diejenigen beiden Theorien durchgeführt, welche bis jetzt die meiste Ausbildung erfahren haben: die Abrahamsche [*Footnote:* M. Abraham, Ann. d. Phys. (4) 10, 105, 1903.], wonach das Elektron die Form einer starren Kugel hat, und die Lorentz-Einsteinsche [*Footnote:* H. A. Lorentz, Versl. Kon. Akad. V. Wet. Amsterdam 1904, S. 809. A. Einstein, Ann. d. Phys. (4) 17, 891, 1905. Vgl. auch H. Poincaré, C. R. 140, 1504, 1905.], wonach das „Prinzip der Relativität" genaue Gültigkeit besitzt. Zur Abkürzung werde ich im folgenden die erste Theorie als Kugeltheorie, die zweite als „Relativtheorie" bezeichnen. [***] Der Lorentz-Einsteinschen Theorie liegt auch ein Postulat zugrunde, nämlich, daß keine absolute Translation nachzuweisen ist."[516]

Relativistic theories were commonplace at the time. Friedrich Kottler wrote an article entitled "Gravitation and the Theory of Relativity" in 1903.[517]

Albert Einstein believed he had a right to plagiarize these ideas of Lorentz, and others, if he could put a new spin on them. He asserted this "privilege" in 1907, and note that in order for Einstein to assert that his viewpoint is "new" he must have known what the "old" viewpoint was,

"It appears to me that it is the nature of the business that what follows has already been partly solved by other authors. Despite that fact, since the issues of concern are here addressed from a new point of view, I believe I am entitled to leave out what would be for me a thoroughly pedantic survey of the literature, all the more so because it is hoped that these gaps will yet be filled by other authors, as has already happened with my first work on the principle of relativity through the kind efforts of Mr. *Planck* and Mr. *Kaufmann*."

"Es scheint mir in der Natur der Sache zu liegen, daß das Nachfolgende zum Teil bereits von anderen Autoren klargestellt sein dürfte. Mit Rücksicht

---

516. M. Planck,"Die Kaufmannschen Messungen der Ablenkbarkeit der β-Strahlen in ihrer Bedeutung für die Dynamik der Elektronen", *Physikalische Zeitschrift* Volume 7, Number 21, (1906), pp. 753-759, with a discussion on pp. 759-761, at 756 and 761.
517. S. Oppenheim and F. Kottler, *Kritik des Newton'schen Gravitationsgesetzes; mit einem Beitrag: Gravitation und Relativitätstheorie von F. Kottler*, Deutsche Staatsrealschule in Karolinenthal, Prag, (1903).

darauf jedoch, daß hier die betreffenden Fragen von einem neuen Gesichtspunkt aus behandelt sind, glaubte ich, von einer für mich sehr umständlichen Durchmusterung der Literatur absehen zu dürfen, zumal zu hoffen ist, daß diese Lücke von anderen Autoren noch ausgefüllt werden wird, wie dies in dankenswerter Weise bei meiner ersten Arbeit über das Relativitätsprinzip durch Hrn. Planck und Hrn. Kaufmann bereits geschehen ist."[518]

Rather than claim independence from Lorentz' work, in 1907, Einstein endorsed Kaufmann's and Planck's declarations that his work was merely an extension of Lorentz' prior work. In 1907, Einstein wrote a review article on the principle of relativity for the *Jahrbuch der Radioaktivität und Elektronik*, and again declared that his work was an interpretation of Lorentz' 1904 paper on electromagnetic phenomena in moving systems—though Einstein would later lie about this point.

In 1907, Einstein wrote to Johannes Stark, who edited the *Jahrbuch der Radioaktivität und Elektronik*, that the only work by Lorentz related to the special theory of relativity which he knew of was Lorentz' 1904 paper (which contains the "Lorentz transformation").[519] This alone would indicate that when the Einsteins spoke of "Lorentzian electrodynamics" in their 1905 paper, they were speaking of Lorentz' work of 1904—a position held by Prof. G. H. Keswani. However, Einstein's statement is contradicted by a letter from Albert Einstein to Mileva Marić, written in 1901 in which Albert pledges to delve into the work of Lorentz.[520]

Einstein stated on 19 December 1952,

"I learned of [the Michelson-Morley experiment] through H. A. Lorentz' decisive investigation of the electrodynamics of moving bodies, with which I was acquainted before developing the special theory of relativity."[521]

However, Albert Einstein lied to R. S. Shankland on 4 February 1950 and stated,

---

518. A. Einstein, "Über die vom Relativitätsprinzip gefordterte Trägheit der Energie", *Annalen der Physik*, Series 4, Volume 23, (1907), pp. 371-384, at 373.
519. A. Einstein to J. Stark, *The Collected Papers of Albert Einstein*, Volume 5, Document 58, Princeton University Press, (1995), p. 42.
520. A. Einstein to M. Marić, *The Collected Papers of Albert Einstein*, Volume 1, Document 131, (1987), p. 189.
521. R. S. Shankland, "The Michelson-Morley Experiment", *Scientific American*, Volume 211, Number 5, (1964), pp. 107-114, at 114. See also: R. S. Shankland, "Conversations with Albert Einstein. II", *American Journal of Physics*, Volume 41, Number 7, (July, 1973), pp. 895-901. R. S. Shankland, "Conversations with Albert Einstein", *American Journal of Physics*, Volume 31, Number 1, (January, 1963), pp. 47-57. R. S. Shankland, "Michelson-Morley Experiment", *American Journal of Physics*, Volume 32, Number 1, (January, 1964), pp. 16-35.

"[I] had become aware of [the Michelson-Morley experiment] through the writings of H. A. Lorentz, but only after 1905 had it come to [my] attention."[522]

In Einstein's famous lecture of 1922 in Kyoto, Japan, he recounts that he derived inspiration from "Michelson's experiment":

"While I was thinking of this problem in my student years, I came to know the strange result of Michelson's experiment. Soon I came to the conclusion that our idea about the motion of the earth with respect to the ether is incorrect, if we admit Michelson's null result as a fact. This was the first path which led me to the special theory of relativity."[523]

On 21 September 1909, Einstein stated the "principle of relativity" is the generalization of the empirical result of the Michelson experiment,

"Michelson's experiment suggested the assumption that, relative to a coordinate system moving along with the earth, and, more generally, relative to any system in nonaccelerated motion, all phenomena proceed according to exactly identical laws. Henceforth, we will call this assumption in brief 'the principle of relativity.'"[524]

R. S. Shankland recorded a letter Einstein had sent him in 1952, in which Einstein stated,

"I learned of [the Michelson-Morley experiment] through H. A. Lorentz' decisive investigation of the electrodynamics of moving bodies, with which I was acquainted before developing the special theory of relativity."[525]

Assuming Einstein did not intend to lie to Stark, one must further assume that when Einstein stated in the 1905 paper that,

"[T]he electrodynamic foundation of Lorentz's theory of the

---

[522]. R. S. Shankland, "Conversations with Albert Einstein. II", *American Journal of Physics*, Volume 41, Number 7, (July, 1973), pp. 895-901, at 896. See also: R. S. Shankland, "Conversations with Albert Einstein", *American Journal of Physics*, Volume 31, Number 1, (January, 1963), pp. 47-57, at 48; **and** R. S. Shankland, "Michelson-Morley Experiment", *American Journal of Physics*, Volume 32, Number 1, (January, 1964), pp. 16-35, at 35.
[523]. A. Einstein, translated by Y. A. Ono, "How I Created the Theory of Relativity", *Physics Today*, Volume 35, Number 8, (August, 1982), pp. 45-47, at 46.
[524]. A. Einstein, translated by A. Beck, "On the Development of our Views Concerning the Nature and Constitution of Radiation", *The Collected Papers of Albert Einstein*, Volume 2, Document 60, Princeton University Press, (1989), pp. 379-394, at 383.
[525]. R. S. Shankland, "The Michelson-Morley Experiment", *Scientific American*, Volume 211, Number 5, (1964), pp. 107-114, at 114.

electrodynamics of moving bodies is in agreement with the principle of relativity."[526]

Einstein must have been alluding to Lorentz' 1904 paper, which paper he did not cite in 1905, but which paper he correctly found the most relevant of Lorentz' writings at the time. Prof. G. H. Keswani has arrived at this same conclusion on other grounds.[527] Keswani avers that the Einsteins' 1905 paper's assertion of conformity between the relativity principle and Lorentzian electrodynamics could only have referred to Lorentz' paper of 1904, and that Lorentz' earlier efforts were not in conformity with the principle of relativity, according to Keswani, and Max Born would seemingly have agreed,

"In the new theory of Lorentz the principle of relativity holds, in conformity with the results of experiment, for all electrodynamic events."[528]

Albert Einstein clearly lied when he told Carl Seelig,

"There is no doubt, that the special theory of relativity, if we regard its development in retrospect, was ripe for discovery in 1905. LORENTZ had already observed that for the analysis of MAXWELL's equations the transformations which later were known by his name are essential, and POINCARÉ had even penetrated deeper into these connections. Concerning myself, I knew only LORENTZ' important work of 1895—'La théorie électromagnétique de Maxwell' [sic (1892)] and 'Versuch einer Theorie der electrischen und optischen Erscheinungen bewegten Körpern'—but not LORENTZ' later work, nor the consecutive investigations by POINCARÉ. In this sense my work of 1905 was independent."[529]

It is obvious that Einstein not only contradicted himself, but lied to both Johannes Stark and Carl Seelig regarding Lorentz' work. Einstein probably lied to Stark in 1907 in order emphasize the freshness of Lorentz' 1904 work in 1905, thereby emphasizing the novelty of the work, and likely lied to Seelig many years later in order emphasize the distinction of Lorentz' earlier works from Lorentz' 1904 paper, and hence the Einsteins' 1905 paper, which contained the perfected form of the Lorentz Transformation the Einsteins had plagiarized from Lorentz and Poincaré. When Albert Einstein published the article Stark had requested in 1907 for the *Jahrbuch der Radioaktivität und Elektronik*, Einstein

---

526. *The Principle of Relativity*, Dover, New York, (1952), p. 60.
527. G. H. Keswani, "Origin and Concept of Relativity, Part I", *The British Journal for the Philosophy of Science*, Volume 15, Number 60, (February, 1965), pp. 286-306, at 299-300.
528. M. Born, *Einstein's Theory of Relativity*, Methuen & Co. Ltd., London, (1924), pp. 188.
529. M. Born, *Physics in my Generation*, second revised edition, Springer-Verlag, New York, (1969), p. 104.

emphasized the fact that his work of 1905 was an extension of Lorentz' 1904 paper, and that his 1907 article would heal any wounds which existed between Lorentz' 1904 paper and the Einsteins' 1905 paper. When the Einsteins' 1905 paper was reproduced in the book *Das Relativitätsprinzip* in 1913 together with Lorentz' prior work of 1895 and 1904, Arnold Sommerfeld annotated the Einsteins' paper, which so obviously parroted Lorentz' prior work, with the following footnote—which we know, based on the above facts, to be untrue,

> "Die im Vorhergehenden abgedruckte Arbeit von H. A. Lorentz war dem Verfasser noch nicht bekannt."[530]

We know from Maurice Solovine that Einstein had studiously read Poincaré's books *Science and Hypothesis* of 1902 and *The Value of Science* of 1904, which reprinted Poincaré's famous St. Louis lecture of 1904 and his 1898 work on relative simultaneity. We know from Einstein's citations that he was familiar with Poincaré's 1900 paper on the theory of Lorentz, which contained the clock synchronization procedure Einstein parroted, and which implicitly contained the formula $E = mc^2$ which Einstein also plagiarized from Poincaré. Therefore, Albert Einstein's statement to Carl Seelig that in 1905 he was unfamiliar with Poincaré's works, which followed from Lorentz' work of 1892 and 1895, was a deliberate lie.

Einstein stated in a lecture in Kyoto, Japan, on 14 December 1922, that,

> "At that time I firmly believed that the electrodynamic equations of Maxwell and Lorentz were correct. Furthermore, the assumption that these equations should hold in the reference frame of the moving body leads to the concept of the invariance of the velocity of light, which, however, contradicts the addition rule of velocities used in mechanics. Why do these concepts contradict each other? I realized that this difficulty was really hard to resolve. I spent almost a year in vain trying to modify the idea of Lorentz in the hope of resolving this problem."[531]

Said "year in vain" was the year from Lorentz' work of 1904 to the Einsteins' 1905 paper, and the missing link required to "modify the idea of Lorentz" was supplied by Poincaré months before Mileva and Albert's 1905 paper appeared in print. Poincaré corrected the defects in Lorentz' theory, before the Einsteins, and thus rendered simultaneity fully relative from the additions of velocity perspective, perfecting the Lorentz group, and attaining full reciprocity for all inertial systems and the covariance of the laws of physics, without a preferred reference frame.[532] Poincaré also went far beyond this, and asserted

---

[530]. A. Sommerfeld as annotator, *Das Relativitätsprinzip*, B. G. Teubner, Berlin, Leipzig, (1913), p. 27.
[531]. A. Einstein, *Physics Today*, Volume 35, Number 8, (August, 1982), p. 46.
[532]. *See*, H. A. Lorentz, "Deux Mémoires de Henri Poincaré sur la Physique Mathématique", *Acta Mathematica*, Volume 38, (1921), pp. 293-308; **and** compare it to

that gravity propagates at light speed, and introduced the four-dimensional interpretation of the Lorentz group, before Minkowski or Einstein.

This new spin on the principle of relativity for which Einstein claimed sole credit, had already been spun in the papers of Henri Poincaré, and Einstein failed to acknowledge this fact in his 1907 review article, which was the perfect opportunity for Einstein to have made amends for the sins of his wife's and his 1905 paper, which lacked any references to, or even mention of, the work of Henri Poincaré. It appears that Einstein never gave Poincaré due credit for the extension of the principle of relativity to electrodynamics; or for the light postulate; or for the concept of, and the exposition on, relative simultaneity; or for the first covariant relativistic theory of gravity based on the presupposition that gravitational effects propagate at light speed; or for the introduction of four-dimensional space-time into the theory of relativity. Einstein was deeply indebted to Poincaré for these ideas, and failed to specifically credit him for them, though Einstein knew that they were Poincaré's ideas, not his.

In 1908, Alfred Heinrich Bucherer published a paper titled, "The Experimental Verification of the Lorentz-Einstein Theory".[533] In 1909, Philipp Frank wrote of the "principle of relativity according to Lorentz" and "The Lorentzian theorem of relativity" and also employed the designation "Lorentz-Einstein".[534] Walther Ritz, who once coauthored a paper with Albert Einstein,[535] spoke of the "Lorentz-Einstein Theory of Relativity".[536] Erich Hupka wrote of the "Lorentz-Einstein theory" and W. Heil wrote of the "Lorentz-Einstein

---

Lorentz' commentary from 1906, and notes from 1909, H. A. Lorentz, *Theory of Electrons*, B. G. Teubner, Leipzig, (1909), pp. 223-230, 328-329; in which he calls Albert Einstein's proxy theory of Poincaré's corrections of Lorentz' theory "artificial", a fallacy of *Petitio Principii*. Lorentz, who attributed to Einstein, what he knew to belong to Poincaré, may have done so out of resentment for Poincaré, as some have suggested—after all, Poincaré completed gracefully that which Lorentz had long struggled after—*and/or Lorentz may have given Einstein credit for Poincaré's work so that he, Lorentz, could attack its fallacious nature without offending Poincaré.*

533. A. H. Bucherer, "Messungen an Becquerelstrahlen. Die experimentelle Bestätigung der Lorentz-Einsteinschen Theorie", *Physikalische Zeitschrift*, Volume 9, Number 22, (November 1, 1908), pp. 755-762.

534. P. Frank, "Die Stellung des Relativitätsprinzips im System der Mechanik und der Elektrodynamik", *Sitzungsberichte der mathematisch-naturwissenschaftlichen Klasse der Kaiserlichen Akademie der Wissencahften in Wien*, Volume 118, (1909), pp. 373-446; at 373, 376, 420, and 442. Frank introduced the term "Group of the Galilean Transformations" in this paper, at page 382.

535. W. Ritz and A. Einstein, "Zum gegenwärtigen Stand des Strahlungsproblems", *Physikalische Zeitschrift*, Volume 10, Number 9, (1909), pp. 323-324; reprinted in *The Collected Papers of Albert Einstein*, Volume 2, Document 57; republished in W. Ritz, *Gesammelte Werke: Œuvres Publiées par la Société Suisse de Physique*, Gauthier-Villars, Paris, (1911), pp. 507-508.

536. W. Ritz, "Das Prinzip der Relativität in der Optik. (Antrittsrede zur Habilitation.)", *Gesammelte Werke: Œuvres Publiées par la Société Suisse de Physique*, Gauthier-Villars, Paris, (1911), pp. 509-518, at 516.

relativity theory" in 1910.[537] Max Born wrote in 1910 and 1911 of the "Lorentz-Einstein principle of relativity".[538] Richard Hiecke wrote of the "Lorentz-Einstein Theory of Relativity" in 1914.[539] George Braxton Pegram spoke of the "Lorentz-Einstein relativity theory in 1917.[540] The designation "Lorentz-Einstein" was quite common at least through the 1920's, and was found in the writings of Emil Cohn, Ferdinand Lindemann, Arvid Reuterdahl, Erwin Freundlich and Hans Reichenbach, among many others.[541] Hermann Weyl wrote of "Lorentz's Theorem of Relativity" and of the "Lorentz-Einstein Theorem of Relativity", in 1921.[542]

While the theory was known most commonly as the "Lorentz-Einstein theory of relativity", it was really Hermann Minkowski who gave the theory its sex appeal based on Poincaré's innovations; and probably Minkowski, more than Larmor, Lorentz, Einstein and even Poincaré, created a stir for the special theory of relativity outside the small circle of theoretical physicists of the day—that is, before the media circus surrounding the eclipse observations of 1919 made Einstein internationally famous. Minkowski, in dramatic style, elevated the theory from an absurd proposition to an intriguing possibility in the eyes of many of his contemporary mathematicians, physicists and philosophers.

Minkowski acknowledged Woldemar Voigt's priority for the "Lorentz Transformation", the mathematical backbone of the special theory of relativity,

"In the interest of history, I want yet to add, that the transformations which play the main rôle in the principle of relativity were first mathematically formulated by Voigt, in the year 1887."

---

537. *Cf.* A. I. Miller, *Albert Einstein's Special Theory of Relativity, Emergence (1905) and Early Interpretation (1905-1911)*, Addison-Wesley Publishing Company, Inc., (1981), pp. 376-377.

538. *Cf.* A. I. Miller, *Albert Einstein's Special Theory of Relativity, Emergence (1905) and Early Interpretation (1905-1911)*, Addison-Wesley Publishing Company, Inc., (1981), p. 270. *See, for example:* M. Born, "Zur Kinematik des starren Körpers im System des Relativitätsprinzips", *Nachrichten von der Königlichen Gesellschaft der Wissenschaften und der Georg-Augusts-Universität zu Göttingen*, (1910), pp. 161-179, at 161.

539. R. Hiecke, "Über das Relativitätsprinzip", *Verhandlungen der Deutschen Physikalischen Gesellschaft*, Volume 16, Number 13, (1914), p. 569.

540. *Cf.* A. I. Miller, *Albert Einstein's Special Theory of Relativity, Emergence (1905) and Early Interpretation (1905-1911)*, Addison-Wesley Publishing Company, Inc., (1981), p. 295.

541. *See:* E. Cohn, "Physikalisches über Raum und Zeit", *Himmel und Erde*, Volume 23, (1911), pp. 117ff.; *Physikalisches über Raum und Zeit*, B. G. Teubner, Leipzig, Berlin, (1911). E. Freundlich, *The Foundations of Einstein's Theory of Gravitation*, Second Edition, Methuen & Co., London, (1924). A. Reuterdahl, *Scientific Theism Versus Materialism*, The Devin-Adair Company, New York, (1920), pp. 174, 267, and 268. F. A. Lindemann, "Introduction" dated "March, 1920" in M. Schlick, *Space and Time in Contemporary Physics*, Oxford University Press, New York, (1920), p. iv. H. Reichenbach, *The Philosophy of Space & Time*, Dover, New York, (1958), p. 161.

542. H. Weyl, *Space-Time-Matter*, Dover, New York, (1952), pp. 165, 169, 172, 327.

"Historisch will ich noch hinzufügen, daß die Transformationen, die bei dem Relativitätsprinzip die Hauptrolle spielen, zuerst mathematisch von Voigt im Jahre 1887 behandelt sind."[543]

Minkowski named Lorentz, Planck and Poincaré, together with Einstein,[544] as the developers of the principle of relativity,

"H. A. Lorentz has found out the 'Relativity theorem' and has created the Relativity-postulate as a hypothesis that electrons and matter suffer contractions in consequence of their motion according to a certain law."[545]

and,

"The credit for the development of the general principle [of relativity] belongs to Einstein, Poincaré and Planck, upon whose works I shall presently expound."

"Verdienste um die Ausarbeitung des allgemeinen Prinzips haben Einstein, Poincaré und Planck, über deren Arbeiten ich alsbald Näheres sagen werde."[546]

Planck[547] and Poincaré attributed the principle of relativity to H. A. Lorentz,

"Will not the principle of relativity, as conceived by Lorentz, impose upon us an entirely new conception of space and time and thus force us to abandon some conclusions which might have seemed established? [***] What, then, is the revolution which is due to the recent progress of physics? The principle of relativity, in its former aspect, has had to be abandoned; it is replaced by the principle of relativity according to Lorentz. It is the transformations of 'the group of Lorentz' which do not falsify the differential equations of dynamics. [***] No, it was the mechanics of Lorentz, the one dealing with the principle of relativity; the one which,

---

[543]. *Physikalische Zeitschrift*, Volume 9, Number 22, (November 1, 1908), p. 762.
[544]. S. Walter, "Minkowski, Mathematicians, and the Mathematical Theory of Relativity", in H. Goenner, et al., Editors, *The Expanding Worlds of General Relativity*, Birkauser, Boston, (1999), pp. 45-86.
[545]. *The Principle of Relativity: Original Papers by A. Einstein and H. Minkowski Translated into English by M. N. Saha and S. N. Bose*, University of Calcutta, (1920), H. Minkowski, "Principle of Relativity", translated by Dr. Meghnad N. Saha, p. 2.
[546]. H. Minkowski, *Annalen der Physik*, 47, (1915), p. 928.
[547]. M. Planck, *Verhandlungen der Deutschen Physikalischen Gesellschaft*, Volume 8, (1906), p. 136; "Die Kaufmannschen Messungen der Ablenkbarkeit der ß-Strahlen in ihrer Bedeutung für die Dynamik der Elektronen", *Physikalische Zeitschrift* Volume 7, (1906), pp. 753-759, with a discussion on pp. 759-761.

hardly five years ago, seemed to be the height of boldness. [***] In all instances in which it differs from that of Newton, the mechanics of Lorentz endures. We continue to believe that no body in motion will ever be able to exceed the speed of light; that the mass of a body is not a constant, but depends on its speed and the angle formed by this speed with the force which acts upon the body; that no experiment will ever be able to determine whether a body is at rest or in absolute motion either in relation to absolute space or even in relation to the ether. [***] This is easy; we have only to apply Lorentz' principle of relativity."[548]

---

548. H. Poincaré, *Dernières Pensées*, Flammarion, Paris, (1913); English translation *Mathematics and Science: Last Essays*, Dover, New York, (1963), pp. 15, 23, 75, 99.

Poincaré makes clear his attribution of the formulation, which he completed and corrected, stems from Lorentz in "Sur la Dynamique de l'Électron", *Comptes rendus hebdomadaires des séances de L'Académie des sciences*, Volume 140, (1905), pp. 1504-1508; reprinted in H. Poincaré, *La Mécanique Nouvelle: Conférence, Mémoire et Note sur la Théorie de la Relativité / Introduction de Édouard Guillaume*, Gauthier-Villars, Paris, (1924), pp. 77-81 URL:

<http://gallica.bnf.fr/scripts/ConsultationTout.exe?E=0&O=N029067>

reprinted *Œuvres de Henri Poincaré*, Volume 9, Gautier-Villars, Paris, (1954), pp. 489-493; English translations appear in: G. H. Keswani and C. W. Kilmister, "Intimations of Relativity before Einstein", *The British Journal for the Philosophy of Science*, Volume 34, Number 4, (December, 1983), pp. 343-354, at pp. 350-353; **and**, translated by G. Pontecorvo with extensive commentary by A. A. Logunov, *On the Articles by Henri Poincaré ON THE DYNAMICS OF THE ELECTRON*, Publishing Department of the Joint Institute for Nuclear Research, Dubna, (1995), pp. 7-14; **and** "Sur la Dynamique de l'Électron", *Rendiconti del Circolo matimatico di Palermo*, Volume 21, (1906, submitted July 23rd, 1905), pp. 129-176; reprinted in H. Poincaré, *La Mécanique Nouvelle: Conférence, Mémoire et Note sur la Théorie de la Relativité / Introduction de Édouard Guillaume*, Gauthier-Villars, Paris, (1924), pp. 18-76 URL:

<http://gallica.bnf.fr/scripts/ConsultationTout.exe?E=0&O=N029067>

reprinted *Œuvres*, Volume IX, pp. 494-550; redacted English translation by H. M. Schwartz with modern notation, "Poincaré's Rendiconti Paper on Relativity", *American Journal of Physics*, Volume 39, (November, 1971), pp. 1287-1294; Volume 40, (June, 1972), pp. 862-872; Volume 40, (September, 1972), pp. 1282-1287; English translation by G. Pontecorvo with extensive commentary by A. A. Logunov with modern notation, *On the Articles by Henri Poincaré ON THE DYNAMICS OF THE ELECTRON*, Publishing Department of the Joint Institute for Nuclear Research, Dubna, (1995), pp. 15-78; **and** "La Dynamique de l'Électron", *Revue Générale des Sciences Pures et Appliquées*, Volume 19, (1908), pp. 386-402; reprinted *Œuvres*, Volume IX, pp. 551-586; English translation: "The New Mechanics", *Science and Method*, Book III, which is reprinted in *Foundations of Science*.

Note that in the latter papers, Poincaré address relativistic gravitation, the propagation of gravitation at *celeritas*, and the perihelion of Mercury, long before Einstein published his version of the general theory of relativity in 1915/16. In this context

In 1911, Max von Laue wrote of, "the principle of relativity of classical mechanics," and of, "the principle of relativity of the Lorentz Transformation."[549] Lorentz, himself, attributed the principle of relativity to Poincaré,

"For certain of the physical magnitudes which enter in the formulas I have not indicated the transformation which suits best. This has been done by Poincaré, and later by Einstein and Minkowski. [***] I have not established the principle of relativity as rigorously and universally true. Poincaré on the contrary, has obtained a perfect invariance of the electromagnetic equations, and he has formulated the 'postulate of relativity,' terms which he was the first to employ."[550]

Albert Einstein stated,

"The term relativity refers to time and space. [***] This led the Dutch professor, Lorentz, and myself to develop the special theory of relativity."[551]

---

see also: H. A. Lorentz, "Alte und neue Fragen der Physik", *Physikalische Zeitschrift*, Volume 11, Number 26, (December 15, 1910), pp. 1234-1257; reprinted in part in H. A. Lorentz, A. Einstein, and H. Minkowski, *Das Relativitätsprinzip: eine Sammlung von Abhandlungen*, B. G. Teubner, Leipzig & Berlin, (1913), pp. 74-89, though not in later editions; see also: *Das Relativitätsprinzip, Drei Vorlesungen gehalten in Teylers Stiftung zu Haarlem*, B. G. Teubner, Leipzig, Berlin, (1914/1920), pp. 19-20.

549. M. v. Laue, *Die Relativitätsprinzip der Lorentztransformation*, Friedr. Vieweg & Sohn, Braunschweig, (1921), pp. 12, 48.

550. H. A. Lorentz, "Deux mémoires de Henri Poincaré sur la physique mathématique", *Acta Mathematica*, Volume 38, (1921), pp. 293-308; reprinted in *Œuvres de Henri Poincaré*, Volume 9, Gautier-Villars, Paris, (1954), pp. 683-695; **and** Volume 11, (1956), pp. 247-261; taken from H. E. Ives' translation, "Revisions of the Lorentz Transformation", *Proceedings of the American Philosophical Society*, Volume 95, Number 2, (April, 1951), p. 125; reprinted R. Hazelett and D. Turner Editors, *The Einstein Myth and the Ives Papers, a Counter-Revolution in Physics*, Devin-Adair Company, Old Greenwich, Connecticut, (1979), p. 125.

Contrast Lorentz' statement of Poincaré's priority, with the fact that Lorentz participated in the production of: H. A. Lorentz, A. Einstein, and H. Minkowski, *Das Relativitätsprinzip: eine Sammlung von Abhandlungen*, B. G. Teubner, Leipzig & Berlin, (1913), which did not include any work by Poincaré, and Lorentz' statement published in 1913, "The principle of relativity, for which we have Einstein to thank, ... " "Das Relativitätsprinzip, welches wir Einstein verdanken, ... "—H. A. Lorentz, *Das Relativitätsprinzip; drei Vorlesungen gehalten in Teylers Stiftung zu Haarlem*, B. G. Teubner, Leipzig-Berlin, (1920). p. 1; "Het relativiteitsbeginsel, dat wij aan EINSTEIN te danken hebben, ... "—H. A. Lorentz, *Het Relativiteitsbeginsel; drie Voordrachten Gehouden in Teyler's Stiftung*, Erven Loosjes, Haarlem, (1913), p. 1; and it appears that Lorentz succumbed to some pressure, real or perceived, to award Einstein undeserved credit, while on other occasions, he told the truth.

551. H. A. Lorentz, *The Einstein Theory of Relativity*, Brentano's, New York, (1920), pp.

Einstein, who knew that Lorentz had the power to end Einstein's masquerade at any time, wrote to Lorentz,

> "My feeling of intellectual inferiority with regard to you cannot spoil the great delight of [our] conversation, especially because the fatherly kindness you show to all people does not allow any feeling of despondency to arise."[552]

Einstein was grateful to Lorentz, for his theory and for his tact,

> "Lorentz is a marvel of intelligence and exquisite tact. A living work of art! In my opinion he was the most intelligent of the theorists present".[553]

At the 1953 centennial celebration of Lorentz' birthday, Einstein stated,

> "At the turn of the century, H. A. Lorentz was regarded by theoretical physicists of all nations as the leading spirit; and this with the fullest justification. No longer, however, do physicists of the younger generation fully realise, as a rule, the determinant part which H. A. Lorentz played in the formation of the basic principles of theoretical physics."[554]

Robert Shankland records that,
> "[Einstein] repeatedly praised H. A. Lorentz and at our last meeting he told me: 'People do not realize how great was the influence of Lorentz on the development of physics. We cannot imagine how it would have gone had not Lorentz made so many great contributions.'"[555]

Abraham Pais recounts that,

> "As [Einstein] told me more than once, without Lorentz he would never have been able to make the discovery of special relativity."[556]

11-12.
[552]. A. Einstein, quoted in *The Expanded Quotable Einstein*, Princeton University Press, Princeton, Oxford, (2000), p. 91.
[553]. A. Einstein, quoted in *The Expanded Quotable Einstein*, Princeton University Press, Princeton, Oxford, (2000), p. 91.
[554]. A. Einstein, "H. A. Lorentz, His Creative Genius and His Personality", in G. L. De Haas-Lorentz, Ed., *H. A. Lorentz: Impressions of His Life and Work*, North-Holland Publishing Company, Amsterdam, (1957), p. 5.
[555]. R. S. Shankland, *Einstein, a Centenary Volume*, Harvard University Press, (1979), p. 39. *See also:* R. S. Shankland, "Conversations with Albert Einstein", *American Journal of Physics*, Volume 31, Number 1, (January, 1963), pp. 47-57, at 57.
[556]. A. Pais, *Subtle is the Lord*, Oxford University Press, (1982), p. 13.

Adriaan D. Fokker wrote,

> "This transposition received the name of the Lorentz transformation of co-ordinates and time. After Einstein the same theory came to be known as the theory of relativity. [***] The invariance of the laws of nature had already been postulated by [Lorentz] in 1892."[557]

Einstein stated in 1912,

> "To fill this gap, I introduced the principle of the constancy of the velocity of light, which I borrowed from H. A. Lorentz's theory of the stationary luminiferous ether, and which, like the principle of relativity, contains a physical assumption that seemed to be justified only by the relevant experiments (experiments by Fizeau, Rowland, etc.)."[558]

Einstein professed in 1935, that it is the Lorentz Transformations which are fundamental in deducing the "two postulates" of special relativity, not the other way around, which means that the "postulates" are in fact corollaries, and that those who first induced the Lorentz transformation ought to be considered the founders of the special theory of relativity,

> "The special theory of relativity grew out of Maxwell electromagnetic equations. So it came about that even in the derivation of the mechanical concepts and their relations the consideration of those of the electromagnetic field has played an essential role. The question as to the independence of those relations is a natural one because the Lorentz transformation, the real basis of the special relativity theory, in itself has nothing to do with the Maxwell theory".[559]

Einstein also stated,

> "This rigid four-dimensional space of the special theory of relativity is to some extent a four-dimensional analogue of H. A. Lorentz's rigid three-dimensional æther."[560]

---

[557]. A. D. Fokker, "The Scientific Work", in G. L. De Haas-Lorentz, Ed., *H. A. Lorentz: Impressions of His Life and Work*, North-Holland Publishing Company, Amsterdam, (1957), p. 78-79.

[558]. A. Einstein, translated by A. Beck, "Relativity and Gravitation: Reply to a Comment by M. Abraham", *The Collected Papers of Albert Einstein*, Volume 4, Document 8, (1996), p. 131.

[559]. A. Einstein, "Elementary Derivation of the Equivalence of Mass and Energy", *Bulletin of the American Mathematical Society*, Series 2, Volume 41, (1935), pp. 223-230, at 230.

[560]. A. Einstein, *Relativity, The Special and the General Theory*, Crown Publishers, Inc., New York, (1961), pp. 150-151.

and,

> "I think, that the ether of the general theory of relativity is the outcome of the Lorentzian ether, through relativation."[561]

and,

> "The four men who laid the foundations of physics on which I have been able to construct my theory are Galileo, Newton, Maxwell, and Lorenz."[562]

Einstein's sycophantic behavior towards Lorentz may well explain why Lorentz did not take a stronger stance against Einstein's plagiarism. Another factor in Lorentz' reluctance to discuss Einstein's plagiarism may have been that Lorentz, together with Einstein, stood much to lose in a priorities dispute, and Lorentz owed much of his fame to Einstein's promotion. Lorentz owed a great debt of acknowledgment (which he most often paid prior to Einstein's sycophantic adoration) to Weber, Mossotti, Zöllner, Gerber, Mewes, Tisserand, Voigt, Heaviside, Hertz, FitzGerald, Poincaré and Larmor, among others—many others. That no articles from these men appeared in the 1913 book *Das Relativitätsprinzip* is a moral crime, one in which Hendrik Antoon Lorentz fully participated.

Lorentz, like Einstein, was a pacifist, even before World War I,[563] and found an ally in Einstein against war and against Germany. In a letter to Einstein dated 28 October 1920, Max Born charged Lorentz with plagiarism, and with committing a gross injustice against Max Planck in order to curry favor with Lorentz' "well-fed friends amongst the Allies"—this at a time when Germans were starving.[564] Max Born called Lorentz dishonest and ignoble.

Beyond all of this, Lorentz shared another character flaw with Einstein—supreme arrogance. At a conference in California, Lorentz stated, near the end of his life,

> "As to the second-order effect, the situation was much more difficult. The experimental results could be accounted for by transforming the co-ordinates in a certain manner from one system of co-ordinates to another. A transformation of the time was also necessary. So I introduced the conception of a local time which is different for different systems of

---

561. A. Einstein, *Sidelights on Relativity*, translated by: G. B. Jeffery and W. Perret, Methuen & Co., London, (1922); *republished, unabridged and unaltered:* Dover, New York, (1983), p. 20.
562. A. Einstein quoted in "Einstein, too, is puzzled; It's at Public Interest", *The Chicago Tribune*, (4 April 1921), p. 6.
563. G. L. De Haas-Lorentz, "Reminiscences", *H. A. Lorentz: Impressions of His Life and Work*, North-Holland Publishing Company, Amsterdam, (1957), pp. 113-115.
564. M. Born, *The Born-Einstein Letters*, Walker and Company, New York, (1971), p. 44.

reference which are in motion relative to each other. But I never thought that this had anything to do with the real time. This real time for me was still represented by the old classical notion of an absolute time, which is independent of any reference to special frames of co-ordinates. There existed for me only this one true time. I considered my time transformation only as a heuristic working hypothesis. So the theory of relativity is really solely Einstein's work. And there can be no doubt that he would have conceived it even if the work of all his predecessors in the theory of this field had not been done at all. His work is in this respect independent of the previous theories."[565]

If he in fact uttered these words, Lorentz' statement is not only supremely arrogant—he took it upon himself to deny the legacies of many scientists, philosophers and mathematicians (most notably Voigt who introduced "local time" before Lorentz and Lorentz knew it), knowing that his legacy was secure—Lorentz' statement is also irrational. One usually gives the credit and honor of priority to she or he who originated the subject idea, and one does not give credit for the evolution of a theory to someone who later summarizes it.

Furthermore, Lorentz was under the gun when he made this statement, in that the special theory of relativity had been discredited by Miller, who also spoke at the gathering at which Lorentz made his statement. Lorentz was careful to distance himself from "Einstein's theory", while cautiously promoting himself, knowing he was widely considered the forefather of this theory, such that whether the special theory of relativity won or lost the day, Lorentz' legacy would remain intact. It is shameful that Lorentz took credit for Voigt's "Ortszeit" and gave Einstein credit for Poincaré's renouncement of the concept of absolute time and the assertion of relative simultaneity, and gave Einstein undue credit for Michelson's experimental results, if Lorentz in fact made the last of the above comments, which were published almost two years after the conference, and after Lorentz' death. Perhaps Lorentz' lecture notes have survived and will show that he did make the statements. Lorentz also must have known that Poincaré's work was vastly superior to the Einsteins'.

Lorentz also had political interests in promoting Einstein. Both were pacifists and Lorentz was interested in the success of the eclipse expeditions in 1919 because he hoped it would promote the interests of *rapprochement*. Lorentz delighted in Einstein's celebrity for many reasons. Lorentz wanted Einstein to come to Leyden, but Einstein knew that Lorentz would discover that Einstein had no talent. Lorentz must have known that Einstein was very well connected and had numerous important contacts in the press and in the publishing business.

Though the press claimed that Einstein was the greatest and most original thinker the world had ever seen. Einstein wrote to Lorentz on 19 January 1920,

"Nevertheless, unlike you, nature has not bestowed me with the ability to

---

565. "Conference on the Michelson-Morley Experiment", *The Astrophysical Journal*, Volume 68, Number 5, (December, 1928), pp. 341-402, at 350.

deliver lectures and dispense original ideas virtually effortlessly as meets your refined and versatile mind. [***] This awareness of my limitations pervades me all the more keenly in recent times since I see that my faculties are being quite particularly overrated after a few consequences of the general theory stood the test."[566]

Paul Ehrenfest, who was close to Lorentz and Einstien, already knew this about Einstein and wrote to Einstein on 2 September 1919,

"No one here expects any accomplishments, all simply want you nearby."[567]

In 1905 and 1906, Paul Ehrenfest considered Lorentz' 1904 paper on special relativity and Poincaré's Rendiconti paper on space-time as the most significant work (both historically and scientifically) on the subject of the principle of relativity. Paul Ehrenfest and his wife Tatiana attended David Hilbert's 1905 Göttingen seminars on electron theory, which described Lorentz' and Poincaré's work on special relativity. They knew that Einstein did not create the theory of relativity. Paul Ehrenfest wrote to Albert Einstein on 9 December 1919,

"I hear, for ex., that your accomplishments are being used to make propaganda, with the 'Jewish Newton, who is simultaneously an ardent Zionist' (I personally haven't *read* this yet, but only *heard* it mentioned). [***] But I cannot go along with the propagandistic fuss with its *inevitable* untruths, precisely *because* Judaism is at stake and *because* I feel myself so thoroughly a Jew."[568]

As for the alleged inevitability of Einstein's hypothetical genesis of the theory of relativity sans all predecessors, Einstein wrote in late 1907,

"That the supposition made here, which we want to call the 'principle of the constancy of the velocity of light', is actually met in Nature, is by no means self-evident, nevertheless, it is—at least for a system of coordinates in a definite state of motion—rendered probable through its verification, which Lorentz' theory based upon an absolutely resting æther has ascertained through experiment."

---

566. Letter from A. Einstein to H. A. Lorentz of 19 January 1920, English translation by A. Hentschel, *The Collected Papers of Albert Einstein*, Volume 9, Document 265, Princeton University Press, (2004), p. 220.
567. Letter from P. Ehrenfest to A. Einstein of 2 September 1919, English translation by A. Hentschel, *The Collected Papers of Albert Einstein*, Volume 9, Document 98, Princeton University Press, (2004), pp. 81-82, at 82.
568. Letter from P. Ehrenfest to A. Einstein of 9 December 1919, English translation by A. Hentschel, *The Collected Papers of Albert Einstein*, Volume 9, Document 203, Princeton University Press, (2004), pp. 173-175, at 174.

"Daß die hier gemachte Annahme, welche wir „Prinzip von der Konstanz der Lichtgeschwindigkeit" nennen wollen, in der Natur wirklich erfüllt sei, ist keineswegs selbstverständlich, doch wird dies — wenigstens für ein Koordinatensystem von bestimmtem Bewegungszustande — wahrscheinlich gemacht durch die Bestätigungen, welche die, auf die Voraussetzung eines absolut ruhenden Äthers gegründete Lorentzsche Theorie durch das Experiment erfahren hat."[569]

The "supposition" was, in Einstein's eyes, not a self-evident truth, but an empirical observation—not *a priori*, but *a posteriori*. In fact, Einstein depended upon the Michelson-Morley result, which he later cited in this 1907 paper as compelling a change in Lorentz' theory of 1895 *and 1904*, which change Einstein argues was the result of the merger of Lorentz' theory with the principle of relativity, a merger made by Poincaré before the Einsteins. Einstein makes clear in this 1907 article that his 1905 work on the principle of relativity was an evolution of Lorentz' 1904 paper, and Einstein told Shankland that he learned of Michelson's experiments in Lorentz' work, before 1905.

The so-called "Lorentz Transformation" which is contained in Lorentz' 1904 paper, first appeared Joseph Larmor's work before Lorentz adopted it. The theory of relativity was not a popular theory among scientists in the early part of the twentieth century, and Lorentz was likely glad to have Einstein on the team to help popularize the unpopular theory. Making much of Einstein's plagiarism would have entailed the risk that Lorentz' theoretical work would itself have been blackened by the scandal. Planck and Kaufmann forced Einstein to acknowledge Lorentz early on, and Lorentz' legacy was thus secured.

Poincaré died in 1912. He is not known to have mentioned Einstein in the context of the theory of relativity in any positive sense. Of course, it would have been ludicrous for Poincaré to have referenced Einstein when describing his own work, which Einstein plagiarized. It is disappointing that Lorentz did not do more to restore Poincaré's legacy, though he did credit Poincaré with perfecting his theory, before Einstein and Minkowski.

While Einstein was demonstrably a sycophant, he had another side to his personality, as sycophants often do. Einstein would not hesitate to arrogantly express ruthless disdain for those who had nothing to offer him and those whom he wished to smear in order to avoid scandal and criticism. This is abundantly clear in Einstein's letters and statements. Einstein's smear tactics and his infamous cowardly avoidance of criticism, as well as his reticence in response to accusations of his plagiarism have already been addressed.

In 1912 Johannes Stark accused Einstein of plagiarism. Einstein did not deny the charge, but arrogantly held,

"J. Stark has written a comment on a recently published paper of mine[1] for

---

[569]. A. Einstein, "Über das Relativitätsprinzip und die aus demselben gezogenen Folgerung", *Jahrbuch der Radioaktivität und Elektronik*, 4, (1907), p. 416.

the purpose of defending his intellectual property.² I will not go into the question of priority that he has raised, because this would hardly interest anyone, all the more so because the law of photochemical equivalence is a self-evident consequence of the quantum hypothesis.³"[570]

The "self-evident" ploy was one of Einstein and his coterie's favorite tactics to manipulate credit for the ideas of others through fallacy of *Petitio Principii*. Knowing the published results others had derived, Einstein and his friends would assert the results, *later*, as "natural consequences" of "their" subsequent theory, which conclusions they had also irrationally presumed in their premises, as if this gave them priority for the thoughts others had published before them, because they would falsely claim that they had derived what others were forced to hypothesize.[571] Einstein would turn the deductive synthetic scientific theories of his predecessors on their heads and argue the same theories inductively, as if that gave him the right to take credit for them. He would do this without making reference to the works of his predecessors and then would later lie and claim that he had had no knowledge of the prior works.

Einstein had a very different attitude when it came to his alleged priority. Contrary to the impression some would have us believe, that Einstein was oblivious to the issue of priority, Einstein had written to Stark on 17 February 1908,

"I find it somewhat strange that you do not recognize my priority regarding the connection between inertial mass and energy."[572]

Einstein and his followers often promoted the theory of relativity as if revolutionary, a supposedly unprecedented departure from all that came before it. The issue of priority was very important to Einstein and to his supporters. Had it not been, Einstein would have been more honest and forthcoming when he wrote his papers and when he described the history of the theory of relativity.

But others had not forgotten Poincaré. In 1912, shortly after Poincaré's untimely death, Vito Volterra wrote in a tribute to Poincaré,

"But a celebrated experiment was performed by Michelson and Morley which kept account of the terms depending on the square of the aberration, and even this experiment, as is well known, gave a negative result.

In a famous paper of 1904 Lorentz showed that this result could be explained by introducing the hypothesis that all bodies are subjected to a contraction in the direction of the motion of the earth.

---

570. A. Einstein, *Annalen der Physik*, Volume 38, (1912), p. 888; English translation by A. Beck, *The Collected Papers of Albert Einstein*, Volume 4, Document 6, Princeton University Press, (1996), p. 125.
571. *See*, for example, H. Weyl, *Space-Time-Matter*, Dover, New York, (1952), p. 324, note 35.
572. A. Einstein to J. Stark, translated by A. Beck, *The Collected Papers of Albert Einstein*, Volume 4, Document 85, Princeton University Press, (1995), p. 58.

This paper was the point of departure for the later investigations. The results of Poincaré, Einstein and Minkowski followed closely that of Lorentz. In 1905 Poincaré published a summary of his ideas in the 'Comptes Rendus' of the French Academy of Sciences. An extended memoir on the same subject appeared shortly afterwards in the 'Rendiconti' of Palermo.

The basic idea in this set of investigations is founded upon the principle that no experiment could show any absolute motion of the earth. That is what is called the *Postulate of Relativity*. Lorentz showed that certain transformations, called now by his name, do not change the equations that hold for an electromagnetic medium; two systems, one at rest, the other in motion, are thus the exact images each of the other, in such a way that we can give every system a motion of translation without affecting any of the apparent phenomena."[573]

In 1913, Arthur Gordon Webster wrote in his memorial to Poincaré,

"The development of Maxwell's electromagnetic theory that has taken place in the last twenty-five years has led to a theory that has attracted the greatest interest among mathematical physicists and has, in fact, become in certain parts of the world no less than a mania. I refer to the so-called principle of relativity, a name which was given to it first, if I am not mistaken, by Poincaré. This principle is no less than a fundamental relation between time and space, intended to explain the impossibility of determining experimentally whether a system, say the earth, is in motion or not. In an elaborate paper published in 1905 in the *Palermo Rendiconti* entitled, 'Sur la dynamique de l'électron,' he defines the principle of relativity by means of what he calls the Lorentz transformation. If the coordinates and the time receive the following linear transformation

$$x' = kl(x + \varepsilon t), \quad t' = kl(t + \varepsilon x), \quad y' = ly,$$
$$z' = lz, \quad k = \frac{1}{\sqrt{1 - \varepsilon^2}}$$

the function $x^2 + y^2 + z^2 - t_1^2$ and the equations of electric propagation will remain invariant. From this follows the impossibility of determining absolute motion. Poincaré then submits the Lorentz transformation, which he shows belongs to a group, to an examination with regard to the principle of least action, which he shows holds for the principle of relativity. He

---

[573]. V. Volterra, "Henri Poincaré", a lecture delivered at the inauguration of The Rice Institute, *The Book of the Opening of the Rice Institute*, Volume 3, Houston, Texas, (1912), pp. 919-920; reprinted in: *The Rice Institute Pamphlets* Volume 1, Number 2, (May, 1915), pp. 153-154; reprinted in: *Saggi scientifici*, N. Zanichelli, Bologna, (1920), pp. 119-157.

further shows that by aid of certain hypotheses gravitation can be accounted for and shown to be propagated with the velocity of light."⁵⁷⁴

In 1913, Ernst Gehrcke wrote,

"The theory of relativity is nothing but a completely novel interpretation of the theory of the electrodynamics and optics of bodies in motion, which Lorentz had already developed. The theory of relativity is *not* distinguished by the creation of substantially new equations, but by a substantially new *interpretation* of the known transformation equations of Lorentz. The arguments made against this *interpretation* condemn it, not the equations themselves, which, as was stated, are not Einstein's, but rather Lorentz' equations, and still stand intact today."

"Die Relativitätstheorie ist nichts anderes, als eine völlig neuartige Interpretation einer schon von LORENTZ entwickelten Theorie der Elektrodynamik und Optik bewegter Körper. Das Charakteristikum der Relativitätstheorie besteht *nicht* in der Aufstellung wesentlich neuer Gleichungen, sondern in der Aufstellung einer wesentlich neuen *Interpretation* der bekannten Transformationsgleichungnen von LORENTZ. Gegen diese *Interpretation* richten sich die gemachten Einwände, nicht gegen die Gleichungen selbst, die, wie gesagt, keine EINSTEINschen, sondern LORENTZsche Gleichungen sind und die bis heute unangegriffen dastehen."⁵⁷⁵

Alfred Arthur Robb spoke to the issue in 1914,

"Although generally associated with the names of Einstein and Minkowski, the really essential physical considerations underlying the theories are due to Larmor and Lorentz."⁵⁷⁶

Einstein had already conceded this fact in early 1911,
"In fact, there are no fundamental differences between Minkowski's and Lorentz's theory."⁵⁷⁷

---

574. A. G. Webster, "Henri Poincaré as a Mathematical Physicist", *Science*, Volume 38, Issue 991, (December 26, 1913), p. 907.
575. E. Gehrcke, "Die gegen die Relativitätstheorie erhobenen Einwände", *Die Naturwissenschaften*, Volume 1, Number 3, (January 17, 1913), pp. 62-66; reprinted in *Kritik der Relativitätstheorie*, Hermann Meusser, Berlin, (1924), p. 20.
576. A. A. Robb, *A Theory of Time and Space*, Cambridge University Press, (1914), p. 1.
577. A. Einstein, translated by Anna Beck, "Die Relativitäts-Theorie", *Vierteljahrsschrift der Naturforschenden Gesellschaft in Zürich*, Volume 56, (1911), pp. 1-14; in *The Collected Papers of Albert Einstein*, Volume 3, Princeton University Press, Princeton, New Jersey, (1993), p. 355.

Einstein saw the only difference between the two as being a "top down" versus "bottom up" approach to the *same* problem with the *same* results, as in inductive versus deductive reasoning of the same problem with the same solution.

Harry Bateman asserted his priority over Albert Einstein, in 1918,

> "The appearance of Dr. Silberstein's recent article on 'General Relativity without the Equivalence Hypothesis'[578] encourages me to restate my own views on the subject. I am perhaps entitled to do this as my work on the subject of General Relativity was published before that of Einstein and Kottler,[579] and appears to have been overlooked by recent writers."[580]

---

[578]. L. Silberstein, "General Relativity without the Equivalence Hypothesis", *Philosophical Magazine*, Series 6, Volume 36, (July, 1918), pp. 94-128.

[579]. F. Kottler, "Über die Raumzeitlinien der Minkowski'schen Welt", *Sitzungsberichte der mathematisch-naturwissenschaftlichen Klasse der Kaiserlichen Akademie der Wissenschaften in Wien* (Wiener Sitzungsberichte), Volume 121, (1912), pp. 1659-1759; **and** "Relativitätsprinzip und beschleunigte Bewegung", *Annalen der Physik*, Volume 44, (1914), pp. 701-748; **and** "Fallende Bezugssysteme vom Standpunkt des Relativitätsprinzip", *Annalen der Physik*, Volume 45, (1914), pp. 481-516; **and** "Beschleunigungsrelative Bewegungen und die konforme Gruppe der Minkowski'schen Welt", *Sitzungsberichte der mathematisch-naturwissenschaftlichen Klasse der Kaiserlichen Akademie der Wissenschaften in Wien* (Wiener Sitzungsberichte), Volume 125, (1916), pp. 899-919; **and** "Über Einsteins Äquivalenzhypothese und die Gravitation", *Annalen der Physik*, Volume 50, (1916), pp. 955-972; **and** "Über die physikalischen Grundlagen der Einsteinschen Gravitationstheorie", *Annalen der Physik*, Series 4, Volume 56, (1918), pp. 401-462; **and** F. Kottler, *Encyklopädie der mathematischen Wissenschaften*, 6, 2, 22a, pp. 159-237; **and** "Newton'sches Gesetz und Metrik", *Sitzungsberichte der mathematisch-naturwissenschaftlichen Klasse der Kaiserlichen Akademie der Wissenschaften in Wien* (Wiener Sitzungsberichte), Volume 131, (1922), p. 1-14; **and** "Maxwell'schen Gleichungen und Metrik", *Sitzungsberichte der mathematisch-naturwissenschaftlichen Klasse der Kaiserlichen Akademie der Wissenschaften in Wien* (Wiener Sitzungsberichte), Volume 131, (1922), pp. 119-146.

[580]. H. Bateman, "On General Relativity", *Philosophical Magazine*, Series 6, Volume 37, (1919), pp. 219-223, at 219. *See also:* H. Bateman, "The Conformal Transformations of a Space of Four Dimensions and their Applications to Geometrical Optics", *Proceedings of the London Mathematical Society*, Series 2, Volume 7, (1909), pp. 70-89; **and** *Philosophical Magazine*, Volume 18, (1909), p. 890; **and** "The Transformation of the Electrodynamical Equations", *Proceedings of the London Mathematical Society*, Series 2, Volume 8, (1910), pp. 223-264, 375, 469; **and** "The Physical Aspects of Time", *Memoirs and Proceedings of the Manchester Literary and Philosophical Society*, Volume 54, (1910), pp. 1-13; **and** *American Journal of Mathematics*, Volume 34, (1912), p. 325; **and** *The Mathematical Analysis of Electrical and Optical Wave-Motion on the Basis of Maxwell's Equations*, Cambridge University Press, (1915); **and** "The Electromagnetic Vectors", *The Physical Review*, Volume 12, Number 6, (December, 1918), pp. 459-481; **and** *Proceedings of the London Mathematical Society*, Series 2, Volume 21, (1920), p. 256. **Confer:** E. Whittaker, *Biographical Memoirs of Fellows of the Royal Society*, Volume 1, (1955), pp. 44-45; **and** *A History of the Theories of Aether and Electricity*, Volume 2, Philosophical Library, New York, (1954), pp. 8, 64, 76, 94, 154-156, 195. *See also:* W. Pauli, *Theory of Relativity*, Pergamon Press, New York, (1958), pp. 81, 96, 199.

In 1920, Johannes Riem stated,

"Auf Wunsch der Schriftleitung soll hier der Versuch gemacht werden, zu zeigen, worum es sich eigentlich bei dem jetzt so viel genannten und mit so großer Reklame verbreiten Prinzip handelt, das an sich so merkwürdig und allen Erfahrungssätzen so sehr widersprechend ist, daß Einstein selber erzählt, er habe erst Monate lang darüber nachgedacht, ehe er dahinter gekommen sei, daß es kein Unsinn sei. Dabei ist zu betonen, daß es nicht etwa fertig aus Einsteins Kopfe entsprungen ist. Zunächst ist der berühmte Mathematiker Riemann zu nennen, dessen Habilitationsschrift von 1854 die Gedanken gibt, die weiter geführt, zu Einstein führen, indem Riemann zeigte, daß Physik und Geometrie zusammengehören. Erheblich später hat dann Lorentz 1895 und Minkowski 1907 die Lehre weiter ausgebaut, letzterer führte schon die Verbindung von Raum und Zeit als Weltpostulat ein und benutzte es dazu, die elektrodynamischen Grundgleichungen für bewegte Materie abzuleiten. Endlich hat dann Einstein alle diese Gedankengänge in mathematischer Weise vertieft und einen die ganze Mechanik, Physik und Astronomie umfassenden Bau daraus gemacht, freilich in einer Weise, die der elementaren Darstellung durchaus spottet. Gleichzeitig mit dieser Entwicklung ist dann eine zweite gegangen, die, von gleichen Gedanken ausgehend, zu anderen Folgerungen kommt und sich daher Einstein gegenüber kritisch verhält, seine Schlüsse zum Teil ablehnt. Das sind die Entwicklungen von Rudolf Mewes in Berlin, der schon 1889 in einem Aufsatz über das Wesen der Materie und des Naturerkennens die Relativität der Materie und der von einander untrennbaren Begriffe Raum und Zeit nachweist. Fußend auf dem Weberschen Grundgesetz und dem Dopplerschen Prinzip, hat er schon drei Jahre vor Lorentz eine Relativitätstheorie aufgestellt, welche außer der relativen Bewegung der Körper zueinander auch noch deren Drehbewegung berücksichtigt, ein Umstand, der bei Einstein nicht vorhanden ist.

Wir kommen so zu einem nach Einsteins Meinung ganz allgemeinen Grundgesetz der Natur, dessen Aufstellung ihn nach der Behauptung der Tagespresse mit Newton auf eine Stufe stelle oder noch darüber. Dem gegenüber ist nicht scharf genug zu betonen, daß erstens sein Prinzip nicht von ihm aufgefunden ist, sondern nur erweitert, und daß ferner der Streit für und wider noch weit davon entfernt ist, ein Ende zu haben. See hat in den „Astronomischen Nachrichten" soeben mehrere Aufsätze erscheinen lassen, die sich scharf gegen Einstein wenden, seine Leugnung des Äthers

---

See also: E. Bessel-Hagen, "Über der Erhaltungssätze der Elektrodynamik", *Mathematische Annalen*, Volume 84, (1921), pp. 258-276. ***See also:*** F. D. Murnaghan, "The Absolute Significance of Maxwell's Equations", *The Physical Review*, Volume 17, Number 2, (February, 1921), pp. 73-88. ***See also:*** G. Kowalewski, "Über die *Bateman*sche Transformationsgruppe", *Journal für die reine und angewandte Mathematik*, Volume 157, Number 3, (1927), pp. 193-197.

als unsinnig bezeichnen, dagegen betonen, wie die amerikanischen Physiker und Astronomen Einstein ablehnen, und Michelson selber sich dagegen verwahrt, seinen Versuch so zu deuten, wie es Einstein tut."[581]

Charles Nordmann averred, in 1921,

"The only time of which we have any idea apart from all objects is the psychological time so luminously studied by M. Bergson: a time which has nothing except the name in common with the time of physicists, of science.
  It is really to Henri Poincaré, the great Frenchman whose death has left a void that will never be filled, that we must accord the merit of having first proved, with the greatest lucidity and the most prudent audacity, that time and space, as we know them, can only be relative. A few quotations from his works will not be out of place. They will show that the credit for most of the things which are currently attributed to Einstein is, in reality, due to Poincaré. [***] I venture to sum up all this in a sentence which will at first sight seem a paradox: in the opinion of the Relativists it is the measuring rods which create space, the clocks which create time. All this was maintained by Poincaré and others long before the time of Einstein, and one does injustice to truth in ascribing the discovery to him."[582]

On 28 March 1921, *The New York Times* reported that Edmund Noble claimed to have anticipated the deductions Einstein made from the theory of relativity. Noble published a relativistic article in the journal *The Monist* in 1905,[583] which set forth a research program for a unified field theory, a relational

---

581. J. Riem, "Das Relativitätsgesetz", *Deutsche Zeitung* (Berlin), Number 286, (26 June 1920). See also: "Gegen den Einstein Rummel!", *Umschau*, Volume 24, (1920), pp. 583-584; "Amerika über Einstein", *Deutsche Zeitung*, (1 July 1921 evening edition); **and** "Zu Einsteins Amerikafahrt. Stimmen amerikanischer Blätter und die Antwort Reuterdahls." *Deutsche Zeitung*, (13 September 1921); **and** "Ein amerikanisches Weltanschauungsbuch", *Der Reichsbote* (Berlin), Number 463, (4 October 1921); **and** "Um Einsteins Relativitätstheorie", *Deutsche Zeitung*, (18 November 1921); **and** "Die astronomischen Beweismittel der Relativitätstheorie", *Hellweg Westdeutsche Wochenschrift für Deutsche Kunst*, Volume 1, (1921), pp. 314-316; **and** "Keine Bestätigung der Relativitätstheorie", *Naturwissenschaftliche Wochenschrift*, Volume 36, (1921), p. 420; **and** "Lenards gewichtige Stimme gegen die Relativitätstheorie", *Naturwissenschaftliche Wochenschrift*, Volume 36, (1921), p. 551; "Neues zur Relativitätstheorie", *Naturwissenschaftliche Wochenschrift*, Volume 37, (1922), pp. 13-14; **and** "Rotverschiebung und Michelsonscher Versuch", *Naturwissenschaftliche Wochenschrift*, Volume 37, (1922), p. 717; **and** "Beobachtungstatsachen zur Relativitätstheorie", *Umschau*, Volume 27, (1923), pp. 328-329; **and** "Relativitätstheorie und Konstitution der Materie", *Umschau*, Volume 29, (1925), pp. 908-910.
582. C. Nordmann, *Einstein et l'universe*, (1921), translated by Joseph McCabe as *Einstein and the Universe*, Henry Holt & Co., New York, (1922), pp. 10-11, 16.
583. E. Noble, "The Relational Element in Monism", *The Monist*, Volume 15, Number 3, (1905), pp. 321-337.

theory of a finite (of necessity) universe in which space and time exist only as the universe itself, etc. Though Noble does not note the fact, it is interesting that the article which follows his in *The Monist* was written by David Hilbert,[584] from whom Einstein plagiarized the generally covariant field equations of gravitation of the general theory of relativity, and from whom Einstein plagiarized the unified field theory concept.

This volume of *The Monist* of 1905 also contains an English translation of Poincaré's famous St. Louis lecture of 1904,[585] which iterated so many of the essential elements of the special theory of relativity, before Einstein, and which lecture Einstein must have read when reading Poincaré's book *The Value of Science*. Poincaré and Hilbert were frequent contributors to *The Monist*, an Open Court publication—a publishing house under the direction of Paul Carus, which helped bring Ernst Mach's works to the English speaking audience. Monistic[586] and Anti-Kantian philosophy defined the research program of the general theory of relativity in the Nineteenth Century. Einstein considered himself an "Anti-Kantian", and certainly pursued the reasoning of Bolliger, who iterated "Mach's principle" in terms of a Boscovichian dynamistic unified field theory.[587]

On 3 April 1921, *The New York Times* quoted Chaim Weizmann,

> "When [Einstein] was called 'a poet in science' the definition was a good one. He seems more an intuitive physicist, however. He is not an experimental physicist, and although he is able to detect fallacies in the conceptions of physical science, he must turn his general outlines of theory over to some one else to work out."[588]

Einstein told Leopold Infeld, "I am really more of a philosopher than a physicist."[589]

On 27 April 1921, Gertrude Besse King wrote in *The Freeman* of New York,

> "ALADDIN EINSTEIN. THE popular interest in America in Professor Einstein's theories has astonished the professor. The public who does not know whether the theory of relativity has accounted for the alteration of

---

[584]. D. Hilbert, "On the Foundations of Logic and Arithmetic", *The Monist*, Volume 15, Number 3, (1905), pp. 338-352.
[585]. H. Poincaré, "The Principles of Mathematical Physics", *The Monist*, Volume 15, Number 1, (January, 1905), pp. 1-24.
[586]. An interesting clip appeared in *The New York Times* on 27 January 1932, page 20, quoting Henry White Callahan's letter to the editor in the context of the theory of relativity on the monist view that "the whole thing is one!" See also: D. M. Y. Sommerville, *The Elements of Non-Euclidean Geometry*, G. Bell, London, (1914), p. 201.
[587]. A. Bolliger, *Anti-Kant oder Elemente der Logik, der Physik und der Ethik*, Felix Schneider, Basel, (1882), *esp.* pp. 336-354.
[588]. *The New York Times*, (3 April 1921), pp. 1, 13, at 13.
[589]. L. Infeld, *Quest—An Autobiography*, Chelsea, New York, (1980), p. 258.

mercury or of Mercury, waylays his steps, and delights, with the exception of a mere alderman or two, to do him honour. Gifted newspaper-reporters herald him as the originator of the theory of relativity, which, by the way he is not, and question him as to the ultimate nature of space, though only a mathematical physicist who is also a philosopher could understand the professor's answers.

This general interest in an extremely difficult science is not quite what it seems. Probably Professor Einstein does not realize how sensationally and cunningly he has been advertised. From the point of view of awakening popular curiosity, his press-notices could hardly have been improved. The newspapers first announced his discovery as revolutionizing science. This sounds well, but its meaning, after all, is rather vague. Then they printed a series of entertaining oddities, supposedly deducible from his hypothesis, although most of them could have been equally well deduced from the conclusions of Lorentz or Poincaré: for example, moving objects are shortened in the direction of their motion. This is a gay novelty until one learns the proportion of the reduction, which is calculated to divest the statement of interest to any but scientists. Further, our newspapers told us that if we were to travel from the earth with the speed of light, and could see the clock we left behind, it would always remain at the same moment, permanently pausing, unable to reach the next tick. But we should be unable to travel at the rate of light for a number of reasons, the most interesting and perhaps the most decisive being that such a speed would cause our mass to be infinite! Finally, our informants assert that no point in space, no moment of time can serve as a permanent base for measurement; we can measure only the relations of space, the relations of time, never absolute space or time; and even to measure space-relations, we have to take into account time! What a fascinating dervish-dance of what we used to regard as immutable fixities! Is it possible that these delicious contradictions are serious and accredited doctrines among those who know? Yet so they appear, for though Professor Einstein is always careful in stating that his hypothesis enjoys as yet only a tentative security, his methods are vouched for by the experts, his procedure is according to Hoyle, and the crowd is at liberty to gorge its appetite for marvels untroubled by the ogres of scientific orthodoxy.

Aside from the fact that Professor Einstein comes as a distinguished and somewhat mysterious foreigner to partake of our insatiable hospitality, his popular welcome is to be accounted for by the spell of wizardry that the press has cast upon his interpretations. For it is the necromancy of these strange theories, not their science, that catches the gaping crowd. Reporters are often good, practical psychologists. Instinctively they have divined the public eagerness for miracles, without grasping the factors that feed this taste. They know that most of us are essentially children still clamouring for fairy tales. Man is congenitally restless with the prison-house of this too, too solid world. He is always looking for short-cuts to power. Since he can not find them to his mental satisfaction as once he could through the miracles and divine dispensations of the Church, or through the magic and occultism that were his legitimate resources in the Middle Ages, he now turns to the wonders of science and philosophy. Here, even in theories that he does not understand, he can find release for his cramped position, here he can taste the intoxicating freedom of a boundless universe, and renew his sense of personal potency. [...]"[590]

Arvid Reuterdahl wrote in *The Bi-Monthly Journal of the College of St. Thomas*, Volume 9, Number 3, (July, 1921):

"Einstein and the New Science.
BY
ARVID REUTERDAHL
HISTORICAL NOTE ON THE NEW SCIENCE.

A New Science has been born, a science in which metaphysics and philosophy find a prominent place. This statement conjures before your vision the internationally celebrated figure of Professor Dr. Albert Einstein, who was born in 1879 in the town of Ulm, Wurtemberg, Germany. Although Dr. Einstein, through his colossal and unprecedented advertising campaign, has done more than any other man to bring this New Science before the world, nevertheless, the year of birth of this new departure in scientific thought cannot be considered as coincident with the appearance, in 1905, of Dr. Einstein's first contribution to the subject of Relativity.

On the contrary, we must look back to the year 1887 as the proper birth year of the New Science, which bids fair to inaugurate a new era in intellectual thought. In that year the famous Michelson-Morley experiment was performed at Cleveland, Ohio. At the time Dr. Albert A. Michelson was Professor of Physics at the Case School of Applied Science. Dr. Edward W. Morley was Professor of Chemistry at the same institution. The writer, because of the far reaching significance of this experiment, considers the year 1887 as marking the birth of the New Science.

THE PIONEERS OF THE NEW SCIENCE.

The New Science was, in part, foreshadowed by the work of Baron Karl von Reichenbach in the years 1844 and 1856. Reichenbach, in his various works, laid the foundation to the theory of radiation. He also held that physiological organisms exhibited characteristics of a decidedly electrical nature.

In the years 1870 and 1871, Aurel Anderssohn of Breslau, Germany, announced the theory that there is no force of attraction extant in the universe. He maintained that gravitation, that is, universal attraction of material systems, is not due to a force but is a mutual effect produced by radiation from bodies.

---

[590]. "Aladdin Einstein", *The Freeman* (New York), Volume 3, Number 59, (27 April 1921), pp. 153-154.

Dr. Johannes Zacharias expanded the limited principle of Anderssohn into practically universal proportions. The results of the earliest work of Zacharias were presented in a lecture before the Physical Society of Breslau in the year 1882. In the hands of a capable and a prodigious worker as Zacharias the elementary suggestion of Anderssohn grew into The Mass-Pressure Theory of Electricity and Magnetism. The essential principle of this theory was publicly demonstrated in Berlin (November, 1908) by means of a colossal rotating electromagnet. The careful and exhaustive experimental work of Zacharias confirmed the vision of Anderssohn that the force of gravitation is merely fictitious.

'Kinertia,' during the period of time from 1877 to 1881, convinced himself that the so-called attractive force of gravitation was an illogical inference not warranted by facts. (For more complete details refer to the author's article, 'Kinertia Versus Einstein' which appeared in The Dearborn Independent, April 30, 1921.) On the 27th day of June 1903, 'Kinertia' filed with the 'Kgl. Preussische Akademie Der Wissenschaften' a description of a mechanical device and an account of an experiment by which 'gravity' could be produced experimentally. (The writer is in possession of the original acknowledgment of the receipt of this deposition.) The 'gravity machine' of 'Kinertia', when water only is used, generates a spiral vortex in space similar to the vortex of a spiral nebulae. When lead balls are projected from the machine by means of either water or compressed air, then the balls describe elliptical orbits, like the planets, while advancing along the neutral axis of rotation. The resultant path, in the latter case, is therefore an elliptical spiral. Many years later (1911-1915 inclusive) Dr. Einstein presented this same theory to a then receptive scientific world with the result that he was subsequently proclaimed a 'greater than Newton.'

'Kinertia' concludes that the effects formerly attributed to the action of a 'force' called gravitation are due to acceleration. He includes a dynamic principle in his concept.

It is an incontrovertible fact, therefore, that 'Kinertia' announced the now famous, 'Principle Of Equivalence,' many years before the alleged discoverer Einstein won the excessive plaudits of the over-enthusiastic scientific world. The work of 'Kinertia,' however, is free from the erroneous sophistical solipsism of Einstein.

Dr. J. Henry Ziegler, of Zurich, Switzerland, in the year 1902, laid the foundation of a cosmic theory in a lecture entitled, 'Die Universelle Weltformel und ihre Bedeutung fur die wahre Erkenntnis aller Dinge.' This theory is of basic and far reaching significance to the New Science. Ziegler's cosmology is based upon the fundamental conception that the world is a unitary structure generated from the universal trinity of space, time and force. Ziegler does not commit the solipsistic error of Einstein by omitting the inclusion of a genuine Absolute Principle in his system. Any cosmological theory which endeavors to construct the universe upon purely relative fundamentals leads to the ultimate verdict 'ignorabimus'. Absolute truth becomes impossible and knowledge is merely a matter of individual opinion. Einstein's system is of this latter type and the name "solipsism" is therefore a proper and fitting designation for the Einsteinian Theory of Relativity. The significant and universal relationship of light to the physical and chemical manifestations of matter led Ziegler to regard light as an absolute essential in physical phenomena. In Ziegler's theory we find, therefore, the root of the only absolute in Einstein's entire system. The fact that Dr. Einstein lived in Bern, Switzerland, at the time when Ziegler's theory of light was a topic of general discussion, leads one to justly question the extraordinary claims to originality of the founder of the Theory of Relativity.

In the same year (1902) that Dr. Ziegler first announced his theory to the world, the writer presented a brief outline of his Space-Time Potential and Theory of Interdependence. At the Inaugural Meeting of the American Electrochemical Society, held at Philadelphia, April 5, 1902, the writer presented his conclusions in a lecture entitled 'The Atom of Electrochemistry.'

In this lecture the writer showed that the physical universe is ultimately reducible to centers of activity (action point-instants) which undergo compensating changes and displacements in conformity with the requirements of the whole cosmos regarded as a unitary, interacting, and interdependent system of multiplicity. This unitary multiplicity system is its own continuum. Action-at-a-distance between its ultimates is not only postulated as inevitably necessary between the primordial centers regarded as discrete (which is an incontrovertible fact of experience), but is also inherent in the fundamental concept of a unitary continuum whose principal constituents are space, time and interdependent interaction.

The writer, consequently, found it possible at that time (1902), to dispense with the old inconsistent ether hypothesis. Moreover, he took occasion, in this lecture, to protest against the attempts of the pangeometers to mathematically manufacture reality by conceptual extensions of actuality. The mythical edifice erected by Minkowski and Einstein, based upon the merely speculative mathematical contributions of the non-Euclideans, has not caused him to feel any necessity whatsoever to modify his views of 1902. This lecture has been fully developed in a work published by the Devin-Adair Company of New York City, bearing the title, 'Scientific Theism Versus Materialism, The Space-Time Potential.'

The great contribution of Ziegler has afforded the writer profound pleasure. Ziegler, working independently in Switzerland, evolved the theory of the unitary triune, Space, Time, and Force. The present author developed his Space-Time Potential in the United States without being aware of the conclusions of Ziegler. The word 'Potential' was used merely to emphasize the fact that the Space-Time Chart is potentially receptive to the play of energy or substance. Ziegler and the present writer are at one in their emphasis upon the dynamic element in the universe which has been so blatantly omitted in the system of Einstein.

It is with utmost pleasure that I here call particular attention to the fact that Ziegler was the first to advance a complete theory of light from the standpoint of the New Science. Einstein has nowhere in his works referred to the work of Ziegler, despite the fact that the much heralded Doctor, undoubtedly, owes a great debt to the illustrious Swiss savant.

Dr. H. Fricke completed his investigations concerning the nature of gravitation and space in the year 1914. The war delayed the publication of this work which finally appeared in 1919 under the title 'Eine neue und einfache Deutung der Schwerkraft'. For Fricke the old ether disappears, but he replaces it with a field of force. The static or stationary ether of Lorentz gives way before the energetic and mobile medium of Fricke which, however, like space, with which it is identified, retains abiding properties. Gravitation (Schwerkraft) is regarded, in the theory of Fricke, as a continuous stream of energy which acts as a *concurrent* system in the equilibration of the *excitant* systems in the universe. Cosmic bodies exhibit outgoing radiational and ingoing gravitational fields of force and all fields of activity are, in their last analysis, moving fields of force. Dr. Einstein, who, it seems, is not in the habit of extending recognition to the deserving, has nevertheless, reluctantly admitted that this theory of Fricke is both highly significant and original. Fricke has announced a New Cosmic Law of far reaching consequences. This epoch-making law may be briefly stated as follows: In vacuous space, if we disregard all other disturbing influences, a definite temperature pertains to every gravitational field. It follows that the temperature of cosmic space does not correspond to the absolute zero, but it is proportional to the gravitational field present in each particular location. Fricke, moreover, concludes that the work done by gravitation is not only changed into heat, but, in part, appears as a directed motion of cosmic bodies.

The homogeneity of inertial resistance and gravitation is a basic principle with Fricke. In cases of inertia the medium of Fricke has a decelerating action toward ponderable masses in conformity with the same laws which govern the accelerative force in the case of gravitation. This conception plays an important role in the theory of Einstein which, however, lacks even the semblance of an explanation. The Pressure Theory of Fricke not only affords an explanation of this cosmic phenomenon but also obviates the difficulties, ably pointed out by Maxwell, in a mechanical theory of gravitation.

In the United States we find Dr. Robert T. Browne in the front rank of the new scientific movement. In his great work 'The Mystery of Space', Dr. Browne emphasizes the actuality of a genuine dynamic element in space. He fully appreciates the weakness and danger of the Relativistic position. For him the universe is inexplicable without an Absolute Principle.

Dr. Charles F. Brush, the world famous electrical engineer and scientist of Cleveland, Ohio; with a series of carefully conducted experiments has challenged the investigation of the Hungarian Baron, Eötvos, performed with a torsion-balance in the year 1890. The issue involved in both investigations is the equivalence or non-equivalence of the inertial and the gravitational mass of a body. Eötvos concludes that the two are equivalent. The General Theory of Relativity of Einstein relies upon the correctness of the conclusions of Eötvos. (See, Relativity, by A. Einstein, pages 80 to 83 inclusive.) The conclusions of Eötvos may be stated in another manner: The magnitude of the effect of gravitation does not depend upon the kind of the material. According to Eötvos, one unit of mass of bismuth should be affected in precisely the same manner as one mass unit of zinc by the gravitational influence. Dr. Brush, on the contrary, asserts that his experiments indicate that the gravitational field exerts a greater influence upon the same mass of bismuth than it does upon precisely the same mass of zinc. The inference from Dr. Brush's experiments is that gravitation takes cognizance, as it were of those subtle differences in matter which we ordinarily group under the term 'qualities'.

The significance of the issue here involved is almost staggering when one reflects upon its far reaching import to the New Science. The old school of science built its stupendous edifice upon the assumption of 'sameness' in its ultimates. Diversity is the result of differences in the number of identical ultimates. For many years the writer has been of the opinion that the physical universe cannot be constructed from mere number. On the contrary, it is my firm conviction, grounded in reason and experience, that observable diversity owes its being to genuine and individually different characteristics in the ultimate particles out of which material aggregates are formed (See author's 'Scientific Theism Versus Materialism, The Space-Time Potential,' paragraph 82, page 44).

Einstein's elaborate speculative edifice falls to the ground, if the momentous experimental results of Dr. Brush are completely substantiated. Practically all the foundation stones of Einstein's structure are composed of unproven, volatile material.

J. G. A. Goedhart, Officer of the Royal Netherlands Navy, Retired, in his work 'L'Orbite En Spirale Dans La Mecanique Celeste' (The Spiral Orbit In Celestial Mechanics, Amsterdam, Netherlands, 1921) presents Six Principal Laws pertaining to the movements of celestial bodies. At this time Goedhart's Second Law is of particular interest because of its relation to the work of 'Kinertia' and the alleged originality of Einstein's conclusions. Goedhart's Second Law is: 'Secondary celestial bodies revolve around the centers of gravitation of planetary systems in eccentric logarithmic spiral orbits, the asymptotes of which are ellipses'.

The work of Goedhart is of unique significance to the scientific world at the present time because it proves conclusively that the spiral orbit, in the case of a planet, can be derived without recourse to the Minkowski-Einstein, four-dimensional, Space-Time speculative product.

Dr. Sten Lothigius of Stockholm, Sweden, in a brochure entitled 'Esquisse D'Une Theorie Nouvelle De La Lumiere' (Sketch Of a New Theory Of Light; Stockholm, 1920) presents a 'Thread-Theory' of light. In his theory of light Lothigius gives a more tangible significance to the usual term 'ray of light'. For him a light-ray is a continuous and coherent structure. Along the axis of transmission undulatory crests may therefore appear without the auxiliary assistance of an ether. Referring to the hypothetical ether, Lothigius states: 'Here lies someone who lived long although he never existed'. It would be difficult to condense a criticism of so vast a subject into fewer words.

Professor P. Lenard, the illustrious physicist, whose brilliant investigations concerning the behavior and properties of certain types of radiations or rays, formed, in part, the basis of the award of a Nobel Prize, has rendered the New Science a service of immeasurable value in stabilizing its formative tendencies during the disruptive attack of Einsteinism.

Lenard's fearless attack on the theory of the 'Zauberkünstler' (Z. K.) (Einstein) has had an exceptionally wholesome influence in preserving the dignity and sanity of the scientific world.

In this connection the forceful exposures of 'Z. K.' by Paul Weyland, E. Gehrcke, H. Fricke, E. Guillaume, and A. Patscke, deserve particular mention.

Dr. Lenard's work, 'Uber Relativitätsprinzip, Äther, Gravitation,' is of such profound import that it cannot be lightly set aside by the mere flippant gesture of Einstein.

Professor Lenard is now preparing a work, exposing the errors of Einsteinism.

Dr. Rudolf Mewes, the distinguished physicist and engineer, with his contribution, 'Raumzeitlehre oder Relativitätstheorie in Geistes—und Naturwissenschaft und Werkkunst', has rendered a lasting service to the New Science. His first work on Space, Time, and Relativity appeared in 1884, thus antedating Einstein by twenty one years.

Camille Flammarion, the eminent French astronomer, writing in the 'Revue Mondiale', calls attention to the fact that Denis Diderot was undoubtedly the first to present an outline of a theory of relativity. Flammarion repudiates the Space-Time Combination of Minkowski and Einstein.

Professor Henri Poincaré, the famous French physicist and mathematician, advisedly ignores the name of Einstein in his lectures on 'Relativity'.

In this short resume it has been impossible to do justice to the momentous issues brought before the intellectual world by the Pioneers of the New Science. Many names have, undoubtedly, been omitted, not intentionally, however, but because of lack of first hand information.

CONTRIBUTORS TO THE THEORY OF RELATIVITY

(In Einstein's works passing references are found to the influential contributions of Cristoffel, Riemann, Ricci, Levi-Civita, Gauss, and Hamilton in mathematics; and to Galilei, Newton, Minkowski, and Lorentz in physics.)

In the year 1869, E. B. Christoffel laid the basis for a new type of calculus which was later used by Einstein in his speculative development of the Theory of Relativity. (See Crelle's Journal fur die Math., Vol. LXX,

1869). Riemann developed the work of Christoffel. In the hands of Ricci and Levi-Civita these contributions took the form of the Absolute Differential Calculus, used by Einstein in his mathematical treatment of Relativity.

Certain functions developed by Sir William R. Hamilton, and known as the Hamiltonian Functions, were also used by Einstein. It would, indeed, have been difficult for Einstein to avoid the references to these men which are found in 'Die Grundlage der allgemeinen Relativitätstheorie', Annalen der Physik, Band 49, No. 7, 1916, (see pages 782, 799, and 804) It so happens that their names have been permanently associated with particular mathematical devices.

Professor Einstein mentions the work of Newton and Galilei, merely in passing, and by way of contrast with his own system which, by means of this delicate stratagem, is thereby made to assume far greater significance than the work of Galilei and Newton, because of its alleged inclusive universality. Einstein seems to be reasonably certain that no serious competition can arise from the graves of the great.

The great German mathematician Karl Friedrich Gauss, receives post-mortem glorification by having his name associated with the four-dimensional system of coordinates which has proved a useful instrument for Einstein.

Without the 'Space Time' contribution of Hermann Minkowski, the electrodynamics of Maxwell-Lorentz, and the H. A. Lorentz Transformation, the Einsteinian tower would reduce to a mere excavation. Consequently, conservative references are made, always in passing, however, to the work of these men.

ADDITIONAL BASIC CONTRIBUTIONS TO THE THEORY OF RELATIVITY.

(Einstein either advisedly ignores or is unaware of the contributions of Anderssohn, Zacharias, 'Kinertia', Larmor, Gerber, Palagyi, Ziegler, Reuterdahl, Mewes, Fricke, and Varicak).

Anderssohn paved the way to a new conception of gravitation (1870-1871).

Zacharias extended the principle of Anderssohn to include electrical and magnetic phenomena (1882).

'Kinertia' developed the Principle Of Equivalence (1877-1881) many years before its announcement by Einstein (1911-1915).

Larmor's work, 'Aether And Matter', was published in the year 1900. Einstein's dissociation of the name of Larmor from Lorentz is incomprehensible.

Paul Gerber, in the year 1898, developed a formula descriptive of the perturbed motion of the Planet Mercury. (See, Zeitschrift fur Mathematik und Physik, 1898). Professor Dr. E. Gehrcke fully realizing the great importance of this work of Gerber, arranged for its reprinting in Annalen der Physik (1917, Vol. 52, page 415). Einstein made his calculations for the motion of the perihelion of Mercury in the year 1915.

Melchior Palagyi published, in the year 1901, a contribution entitled 'Neue Theorie des Raumes und der Zeit' (Engelmanns Verlag in Leipzig) which contained the essentials of the Minkowski-Einstein Space-Time conception. Minkowski's first paper appeared in 'Der Göttinger Mathematischen Gesellschaft,' Nov. 5, 1907. In the following year his Cölner lecture, entitled 'Raum und Zeit', was delivered. This was reprinted in Annalen Der Physik, Vol. 47, No. 15, page 927; June 15, 1915.

Zeigler, in the year 1902 announced his new cosmic theory involving the unitary triune, Space, Time, and Force, together with Light as the universal, physical absolute.

Einstein's first paper bears the date September 1905. It was written in Bern, Switzerland where Ziegler's theory was much discussed.

The present writer's first paper was published in the year 1902. This paper briefly outlined the basic elements of his complete work 'Scientific Theism Versus Materialism, The Space-Time Potential', which appeared in 1920.

The present author's direct and simple method of calculating the deflection of light, due to the Sun, is a closer approximation to the observed 'bending' than the result obtained by the more indirect and involved method of Einstein. (See work cited, pages 271 and 272).

Rudolf Mewes' contributions, when they appear in a collected form, will exert exceptional influence upon the position of Lorentz. In fact, the older works of both Mewes and Gerber will then attain unique significance.

H. Fricke, in his work (1914-1919), presented a physical basis for the Principle Of Equivalence which was arbitrarily announced by Einstein from purely speculative reasons. Fricke's researches on the relation of heat to gravitation are certain to open fruitful fields of investigation for the New Science.

Varicak, the mathematician, was the first (1915) to point out that the Principle of Relativity leads directly to the formulae of non-Euclidean geometry. (See, Sitzungsberichte der Berliner Akademie 1915, page 847). Einstein's fabric is woven from the fibers supplied by the metageometers.

A retrospective view of the above facts can result in but one question: What original contribution has Einstein made which warrants the, now common, verdict that he is 'a greater than Newton'?

The Scientific American (May 14, 1921), in an unwarranted, sarcastic editorial attack on the present writer, answers this question as follows: 'He (Einstein) has formulated mathematically and as a concrete whole ideas which have had a rather nebulous existence before him, cementing the structure with ideas to which he has himself given birth. His crowning achievement is the precise mathematical formulation; this has never been approached or approximated in any way.' This is surely an extraordinary claim, especially in view of the fact, that the editorial itself was called forth because the writer demanded, in the name of justice, that credit be given to the originators of those 'nebulous' ideas, without which the Theory of Relativity would have been an impossibility.

The case is analogous to that of the builder who appropriated sufficient bricks to build a house, and when payment was demanded, replied: 'I have furnished the cement, which binds the bricks into a structure, therefore I owe you nothing for the bricks.'

THE 'MAIN MEMBERS' OF EINSTEIN'S STRUCTURE.

The relations of the 'main members' in Einstein's structure may, most readily, be illustrated by a reinforced concrete arch bridge composed of two ribs or segments, hinged at the crown (center of span) and at the abutments. A reinforced concrete floor, laid in the bed of the stream, connects the two thrust-resisting abutments. The left abutment in Einstein's arch is the Lorentz Transformation. Non-Euclidean Space-Time (Minkowski) constitutes the right abutment, and the Michelson-Morley Experiment is the connecting floor between the abutments.

The left arch rib is The Absolute Velocity of Light, while the Principle Of Equivalence constitutes the right arch rib. The three alleged experimental verifications of Einstein's theory form the three hinges. From left to right we may think of these hinges as being, 1st, The Perturbations of the Planet Mercury; 2nd, Displacement of the Spectral Lines towards the Red; and 3rd, The Deflection of Light in a Gravitational Field.

TESTS OF THE MEMBERS.

The limited scope of this article prevents a full discussion of the structural value of the members. 'The Fallacies Of Einstein,' now in preparation by the writer, will consider these and other matters in detail.

RIGHT ABUTMENT.-NON-EUCLIDEAN SPACE-TIME.

It can be shown that the Minkowski-Einstein Space-Time is a mathematically camouflaged type of four-dimensional space. In the invariant form of the General Quadratic Differential, which is basic to Relativity, the last term is formed by multiplying the velocity of light by time. Velocity is reducible to length divided by time. Therefore time is eliminated from this term, leaving it as a pure spatial expression. Consequently we have here nothing but a new version of four-dimensional space which is not a physical reality. The writer challenges the relativists and the metageometers to construct a model of the four co-ordinate axes required by this conceptual space. This demand can be satisfied in the case of three dimensional space, which is our only real space.

Conclusion.

The right abutment of Einstein's arch bridge is merely mythical and not a physical reality. From the standpoint of engineering this verdict is sufficient to condemn the entire structure. Certainly pure science ought not to be less exacting in its demands than the applied science of engineering.

THE FLOOR. - MICHELSON - MORLEY - MILLER EXPERIMENT.

This experiment involves the ether, and the possibility of relative motion between the earth and the ether. The constancy of the velocity of light was assumed in the experiment.

Known Facts.

The motion and velocity of the earth.
The constant velocity of light.
Unknown Facts.
(The experiment assumed the existence of the ether. This assumption takes too much for granted.)
Does the medium called 'The Ether' exist?
Assuming the existence of the ether, then in regard to its possible motion, only two assumptions can be made: viz.,
1st. The ether is stationary,
2nd. The ether is in motion.

Michelson and Morley, using an interferometer, failed to detect any relative motion between the earth and the ether. Miller and Morley, with a much larger interferometer, were unable to detect any relative motion.

Sir Oliver Lodge, assuming that the ether was carried along with moving bodies, experimented with rapidly rotating discs only to obtain negative results. Both of the above possibilities proved futile in the attempt to determine the earth's motion in respect to the ether.

At the time when these experiments were performed science was not prepared to abandon the ether because of its *conceptual* usefulness in explaining the phenomena of light and electro-magnetics.

In the New Science the old inconsistent ether is being replaced by interactional vehicles and interdependent activities.

Classical mechanics and the theory of relativity as held by Newton took cognizance of relative velocities computed by reference to arbitrary systems of co-ordinates. If the Michelson-Morley experiment had yielded a positive result, indicating that the earth's velocity could be calculated in reference to a stationary ether, then the measurement of so-called 'absolute motion' would have been possible. The ether would then have constituted a universal and fixed reference system.

Because of the negative results of these experiments, Einstein expanded the older notions of relativity to include these later results. He concludes therefore, that an *absolute* determination of *uniform* motion is impossible, and he holds that the ether cannot be used as a reference system by which relative uniform motion may be detected.

It should be recalled that Archimedes failed to find a fixed point in the universe. Michelson, Morley, Miller, and Lodge also failed in the more modern case of the ether. Then the voice of the Prophet of Relativity was heard crying from the house-tops, 'There is no Absolute! Everything is conditioned and relative! Truth itself is variable!' We listen but we are not convinced. We conclude that the relativists have sought the philosopher's stone in vain, for they have searched for a static point in the dynamic world. They have tried to achieve the impossible. We become content then and decide to continue making observational references from 'fixed' points that move.

The negative result of the Michelson-Morley experiment may, with confidence, be regarded as a conclusive proof that there is no ether. If this position is entertained, then interactional vehicles, acting in conjunction with Space-Time (properly interpreted), must be introduced in order to function in the cases of light and electro-magnetic phenomena. Einstein, however, has not allowed this phase of the problem to disturb his equanimity. On the contrary, he has seized upon the Larmor-Lorentz Transformation as the only way out of the difficulty.

Conclusion.

At the present time the results of the Michelson-Morley-Miller experiments must be accepted as experimental facts. The abuse of these results by Larmor, Lorentz, and Einstein, in no way influences the previous statement. Notwithstanding its Eisteinian misapplication and abuse, the floor must certainly be pronounced as structurally safe.

The experiment actually proves that the time required for light to travel from an initial to a final observation point, in a closed vectorial configuration is independent of the path.

The result of the Michelson-Morley trial, therefore, substantiates the writer's theory of light. (See discussion in this article under caption, 'A New Theory Of Light.')

LEFT ABUTMENT. THE FITZGERALD-LARMOR-LORENTZ TRANSFORMATION.

Assuming that the ether exists, Fitzgerald conceived the idea that if the material composing a body contracted along lines and planes parallel to the direction of motion through the ether, then the negative results of the Michelson-Miller experiment could be explained. According to the modern view matter is composed of electrons which are identical in size and deterministic characteristics. This is merely an arbitrary assumption which is not warranted by the great diversity manifest in the physical universe. However, the assumption, it appeared, would obviate such difficulties as would arise from the differences in the structural material of the interferometer. Moreover, it would serve to generalize this entire class of phenomena—a generalization purchased with a sacrifice of truth.

Larmor and Lorentz, adopting the suggestion of Fitzgerald, conceived its mathematical form. The amount of the contraction in the direction of motion must be something definite. Moreover, it must agree with that space and time coefficient for moving bodies which now constitutes an important element in the left and right abutments of Einstein's structure. The 'plot' was a master stroke of ingenious imagination. Since we have bound ourselves to refrain from mathematical developments in this article we are forced merely to label this Space-Time Motion expression for purposes of discussion. We shall designate it as the 'Space-Time Coefficient.' The writer, in his work, has referred to this expression as the Fundamental Scalar of Einstein's Relativity. It is used in both Scalar and Vector Analysis.

With this expression known, (derived by *Euclidean* geometry from Space and Time considerations and not from experimental evidence) Larmor and Lorentz could readily *speculatively determine* the amount of the contraction in the direction of motion. If then the diameter of an electron at 'rest' is known, its contracted diameter could be calculated by multiplying the 'rest-diameter' by the Space-Time Coefficient. This is a pathetic illustration of the fact that the Relativists, whilst disclaiming any knowledge of 'fixed' points, persistently employ moving points (electron in this case) as 'fixed'. They are continually cutting the eternally moving infinite chain of relativities in order to 'fix' a point. As a speculation their theory is interesting. Practically it cannot be consistently applied.

Knowing the mass of the electron at 'rest' its so-called 'transverse mass' can be *speculatively determined* by introducing this known mass into the Space-Time Coefficient. The 'transverse mass' is therefore based upon that diameter of the electron which is *parallel* or *coincident* with its direction of motion.

If we align one arm of the Michelson interferometer in the direction of the earth's motion, then the time required by light (according to Lorentz, a type of motion in the stationary ether) to travel from a point to a mirror and back again *ought to be* greater than if light travels an *equivalent* distance (twice the sum of the distance from the point to the mirror) in a *continuous and unreversed* path. This statement assumes the constancy of the velocity of light. The Michelson-Morley-Miller experiment showed *no difference* in time. In fact the other arm of the interferometer, constructed at right angles to the first, gave the same time interval. The two arms were identical in all essential details. Moreover, *no difference* in the time period could be detected by swinging the interferometer on its axis into any position whatsoever.

If that interferometer arm which was parallel to the direction of the earth's motion would only be sufficiently accommodating to always contract in length that precise amount which would compensate the theoretical excess in the time period, then all would be well, because light would then travel over a shorter path and the time-excess would disappear. The science of mathematics is a boon to the modern school of scientific speculators. By its manipulations we can produce the most gratifying compensations and accommodations. It does not seem to be particularly important if the alleged compensations are actual physical facts. The principal issue is, the derivation of a satisfactory *mathematical* result.

In any event, if physical experiment should fail to cope with the situation, it must be determined mathematically and then imposed upon our long-suffering physical universe. Fortunately for Relativity, Larmor and Lorentz, in their Space-Time Coefficient, had the *mathematically, built-to-order*, instrument of precision which unerringly could *speculatively* annihilate the alleged difficulty.

Hence, if the length of that arm of the interferometer which moved in the earth's direction of motion, was multiplied by the Space-Time Coefficient (a reduction expression) then this length would be sufficiently decreased to compensate for the supposed time-excess.

This contraction theory of Larmor and Lorentz, in the hands of Einstein, became a means of producing a confusing pyrotechnic display designed to intellectually impress the masses. Serious calculations were made concerning the diminution of a human being due to motion. The poor victim, we are told, is totally unaware of the change in his dimensions because his associates are all *suffering* diminution in the same relative proportion. Everything in motion contracts in the same relative ratio. One cannot even physically *determine* the actual amount of the alleged contraction. It always eludes you. This fact is an extraordinarily ingenious protective element inserted, inadvertently perhaps, into the Theory of Relativity. Nothing can be verified experimentally. Reality has been dethroned and mathematics has become the final creator, director, judge, jury, and arbiter of the type and destiny of a physical universe which, no longer, is permitted a voice in these matters.

By way of summarizing the results of this discussion of the contraction theory, the writer desires merely to restate that which is now self-evident.

The Larmor-Lorentz contraction theory is purely a mathematical device designed to meet an emergency. It has not been shown by physical experiment that an electron contracts in the manner claimed by the theory. The relativists themselves take great delight, it seems, in pointing out that, from the standpoint of their own theory the affair is beyond proof or disproof. One must conform, without murmur, to the precepts laid down in the Relativistic Koran. If this work, however, is regarded as the product of a fallible mind, then we may venture into that real world which lies beyond the confines of Relativity and there discover facts which serve, like dynamite, to cause the collapse of this speculative structure.

The experiments of Kaufmann and others have shown that the mass of an electron increases as its velocity increases. (Below a certain limiting velocity the mass remains practically constant.) As this velocity approaches the velocity of light, the mass increases towards an infinite amount. Lorentz and Einstein employed the same expression to mathematically describe this experimentally observed increase that was used in the calculation of the contraction of the electronic diameter in the direction of its motion. The writer desires to call particular attention to the dangerous dilemma which arises from this maneuver.

Left Horn.

If the mass of the electron at rest is divided by the Space-Time Coefficient, in which the velocity of the electron equals that of light, then the expression indicates a resulting infinite mass for the moving electron.

It should be noted that the Kaufmann Effect is an observed fact and that the mathematical expression is merely an attempt to describe an actuality. Therefore, a true scientist, in contradistinction to a mathematical speculator, will abide by the result of an experiment whenever mathematical speculation and actual observation disagree.

Right Horn.

If the diameter of the electron at rest is multiplied by the Space-Time Coefficient, in which the velocity of the electron equals that of light, the expression indicates a zero value for the diameter. In other words, the electron will have no diameter at all. In the absence of any statement to the contrary on the part of the Relativists we are at liberty to asume that a similar fate befalls all lines of the electron which are parallel to the direction of motion. It follows that, if the Larmor-Lorentz contraction hypothesis is true, the mass of the electron reduces, in this case, to zero.

The two horns of this dilemma have been presented with complete recognition of a somewhat similar expression for the so-called 'longitudinal mass.'

Between these two horns, the proper choice is apparent at once, if facts and not speculation shall be our guide. Therefore, we discard the right horn as untenable because it is incompatible with the left horn which is based upon facts. Moreover, we demand that the advocates of the contraction theory, if they desire serious consideration for their claims, prove their contentions by an experiment. We cannot accept the subterfuge that this is not possible. Whatever we accept as truthfully descriptive of the physical world must be verifiable by experimental observation. Any theory which cannot meet this requirement is not worthy of serious consideration.

The Space-Time Coefficient owes its origin in Relativity to mathematical speculations concerning Space, Time, and Motion, depicted in the terms of Euclidean geometry. Nowhere do we find even a trace, in Relativity, of its source in an actual dynamic world. It is not surprising that it is continually misapplied by the Relativists. If the Relativists had first probed for its supporting source in the physical universe, then this very origin would have served as an unerring guide in its future application. In his investigations concerning Interdependent and Independent Motion the writer has shown that its origin is grounded in the facts of dynamic action which exhibit interdependent motion. (See Scientific Theism etc., pages 273-280).

The contraction hypothesis is a flagrant case of the misapplication of a mathematical product to physical reality. The Larmor-Lorentz contribution to the Theory of Relativity must be discarded because it is not only contrary to known facts, but it is also incapable of experimental verification.

Conclusion. The Left Abutment of Einstein's arch is not only inadequate to withstand the thrust, but also non-existent as a genuine physical fact.

LEFT ARCH RIB. THE CONSTANCY OF THE VELOCITY OF LIGHT.

Without entering into refined particulars we may state that, 'The Second Postulate of Einstein's Theory' maintains that the velocity of light, in a vacuum, is the same for all observers and is independent of the relative motion existing between the observer and the source of light.

Einstein regards this constancy of light as a necessary assumption in his theory. In this he has shown unusual caution. The reason for his prudence is that he cannot suggest even a semblance of an explanation of this glaring exception to his world-scheme of Relativity. This situation is not devoid of humor. That member in his arch which is indisputably sound he regards as a postulate. The Michelson-Morley Experiment, of course, is exempt from the previous implied criticism for the reason that the result of this experiment must be regarded, at the present time, as an *experimental fact*. The interpretation, however of this result, is an entirely different matter. The Relativistic version is palpably fallacious. Another test has been proposed. It is evident from the nature of the Einsteinian arch that the outcome of this proposed test can have no beneficial bearing upon the stability of Einstein's structure.

We have already shown that the Arch fails for a negative result. It is self-evident that a positive result, being fatally inimical to the Larmor-Lorentz Contraction, is of less value, and therefore cannot prevent the collapse of the structure.

Although Einstein has failed utterly to find even a clew to that greatest of all world mysteries, The Constancy of Light, nevertheless, the New Science stands ready with a solution.

A NEW THEORY OF LIGHT.

The New Science has found it necessary to abandon the ether hypothesis in its inconsistent and antequated form. The only physically known is differential matter in motion in the sympathetic presence of the compensating integrator Space-Time. This conception is the root of the author's Space-Time Potential in which Space and Time are regarded as the Intermutational Matrices of Reality. The writer has failed to find the word which will adequately express the thought which he desires to convey by the word 'Intermutational.' The idea cannot be expressed by the word 'Interactional' because action, in the physical universe, is always associated with matter. Space and Time are not material essences, therefore the word 'action, in any form whatsoever, cannot properly be associated with these basic matrices of reality. The 'inter-play' of Space and Time, although not genuine action, nevertheless *suggests* action, foreshaddows it from its very essence as a hope which can be realized in the presence of dynamic substance. This is the thought which the writer has, so inadequately, attempted to express by the word 'Intermutational.' If there exists one characteristic in either Space or Time (real not conceptual) which is totally different and not found in the other, then Space and Time are not mere phases of a single entity Space-Time. That such distinct features exist becomes apparent upon reflection. Real Time is irreversible. Space is reversible. The limited scope of this article prohibits a detailed discussion of this phase of the subject. The mere hint, here given, is sufficient to prove that the Einsteinian single entity Space-Time is not grounded in experience. That actually existing, though shadowy phase of the 'inter-play' of Space and Time, which we have here termed 'intermutation' is impossible in a single entity Space-Time.

The writer has sought for an explanation of that greatest of all cosmic mysteries, the constancy of the velocity of light, at the very fountain-head of reality, the 'Intermutation of Space and Time.' An explanation cannot be found anterior to the fountain-head. The solution is therefore startling in its directness and simplicity. Only a brief exposition of the author's theory is possible in this article.

In order to make the content of this theory clear, let us erect a straight line, in any desirable direction, in Space. We will call this line the 'Space-Directrix.' It is evident that we can erect an infinite number of such directrices. Erect a plane perpendicular to the Space-Directrix. Regard this plane of sufficiently great dimensions to include all elements under investigation. Consider matter as the 'Now of Substance.' No other conception of matter conforms with observed reality. There is a 'Now' and a 'Future' for every kern (mass-acceleration unit) of reality. The 'Now' can be depicted in our plane, provided that we identify our consciousness with it. When this is done we will designate our plane as the 'Now Plane'. The 'Future' (substance) of every kern can be depicted as a *kern-extension filament* reaching beyond the Now Plane into Space. We give the name 'Cosmic Filaments' to all such extension filaments. This picture of the Cosmos is merely pictorially symbolic of a reality which defies the most profound attempts of finite represeiitation. In our picture, Time is represented by the Now Plane. In the Intermutative background of the Cosmos, Time corresponds to the 'dynamic urge' of substance and may therefore be regarded as an *Underlying* Principle of Motion, which in conjunction with Space, Substance, and all the Categorical and Empirical Determinations manifest as cosmic phenomena.

With Time regarded as an Underlying Principle of Motion, the question immediately arises: 'Have we any precise experiential knowledge of the 'Motion' of Time?' The answer is close at hand. So close, indeed that it has completely escaped the notice of both science and philosophy. The writer has found the answer at the very fountain-head of existence in the matrices of Time and Space whose intermutational motion is the underlying basis of the known constancy of the velocity of light. There is, of course, a material side to this primordial relation, but this material phase of the problem must ultimately be grounded in the intermutational matrices which form the responsive equilibrating background of all physical reality. If we refer again to the Cosmic Model, presented ahove, the thought here outlined becomes clear. Let the Now Plane moves in such a manner that it is continuously perpendicular to a Space-Directrix. In cosmic phenomena, such as light and gravitation, the Now Plane moves, in reference to any initial point of reference in the Space-Directrix, with a velocity of 300,000 kilometers per second. In my Space-Time Potential I have given the name 'Kosmometer' to this cosmic unit of length. (Scientific Theism, page 173.) The cosmic unit of time is therefore that time period (one second) which is required for the Now Plane to travel a distance of one Kosmometer along a Space-Directrix. It is understood that the Now Plane in all its positions is perpendicular to the arbitrarily selected Space-Directrix. The velocity which arises in this manner is an Absolute Cosmic Velocity because it is the *constant cosmic ratio* of intermutation of the matrices of Time and Space. The converse is also true. The generation of this *constant cosmic ratio* is possible because

Time and Space are distinct, though intermutational matrices. Einstein erroneously considers Space and Time as merely subjective precipitates from the single entity Space-Time.

As the Now Plane moves with this constant velocity, it continuously intersects the Cosmic Filaments whose 'Now Sections' responsively adjust themselves in such relative positions and configurations which conform with their inmost interactional nature and also with the co-responsive Cosmic Mold, Space-Time. Thus it is seen that intermutational Space-Time constitutes a cosmic chart capable of (the '*becoming-kinetic*' of substance) exhibiting deterministic future action. Herein lies the essence of the author's use of the word Potential in his 'Space-Time Potential.' The entire system is thus both interacting and unitary, and individually distinct forces, regarded as entities separate from matter, have no meaning. (Thus the 'force of gravitation', regarded as a separate and distinct entity is meaningless.) In such a unitary system the objections usually entertained against 'action-at-a distance,' completely disappear.

The material side of the phenomenon of light is in perfect harmony with its intermutational aspect in Space-Time. (See Scientific Theism, pages 172, 274, 275, and 276.) Here we deal with transverse and longitudinal displacements arising during the interaction of the *excitant* and *concurrent material systems*. The *ratio* of the velocities of the excitant and concurrent material system is that constant velocity which is known as the velocity of light. The concurrent system is composed of gyratory groups of monons which are interactionahly responsive to the presence of the constituent units of the excitant system. The latter travels in straight lines, unless subjected to the deflective interactional influence of other material systems. Normally to the direction of motion of the excitant units, the gyrational groups, constituting the concurrent system, undergo a cyclical augmentation in their orbital radii. The result is a genuine physical light-wave which cannot be even conceived in a continuum like the ether which contains no real discrete parts.

The writer desires to point out a few of the results which follow from his theory.

1st. Cosmic Space and Time become genuine primordial realities. The Cosmic Now Plane moves with an Absolute, and known, velocity in reference to Space. The velocity of Cosmic Phenomena (light, gravitation etc.,) becomes known as a universal Cosmic Constant (that of Light).

2nd.

The discrepancy between the sum of the component vectors and the resultant vector in the ordinary velocity and force triangle is completely accounted for by this theory. The truth of this assertion follows from the fact that a displacement in this Now Plane is inseperably associated with, and actually impossible without, a coordinated displacement of the Now Plane itself along a Space-Directrix. Thus for every vector which is not perpendicular to the Now Plane two components are inevitable. Here then we have the ultimate source of the vector triangle and also the root of the above mentioned discrepancy.

3rd.

It follows, that in a closed vectorial configuration the time period between an initial and final point is independent of the path. Since the paths are unequal in length, it follows that the velocities also will be unequal.

In the case of the interferometer experiment, if we regard the Space-Directrix as parallel or coincident with the direction of the earth's motion, it follows that the observed time period, referred to the Now Plane, is independent of the path of the light-ray. The time period required for a net displacement of the ray *along the Space-Directrix*, is the same whether the path be a curve, a continuous broken line, two adjoining hypotenuses of right triangles, or the net resultant of a simple forward and backward motion. The governing element is the initial and final position of the Now Plane in reference to the Space-Directrix.

The interpreters of the Michelson-Morley experiment have not given due consideration to the fact that light is a *continuously generated* phenomenon. It is a generated (dynamic) vector *subject to interdependent interaction*.

The writer's theory permits variability in the velocity of both the *excitant* and the *concurrent* systems. It is only the *ratio* of these variable velocities which *remains constant*. The excitant system is actively responsive to interactional intensities. The concurrent system is continuously equilibrated and therefore exists in a neutral action phase. The excitant system is subject to acceleration in the presence of interacting fields. This obviates the dilemmas (like that arising out of the Doppler effect), difficulties, and omissions which are constitutionally inherent in Einstein's system.

The excitant system, when passing near the Sun, will be subject to its direct interactional influence, and also to the refractive effect due to its corona. The combined result will be a retardation of the velocity of the excitant system.

The difficulties and possibilities of observational errors involved in this class of physico-astronomical investigations are both numerous and large. Such allowances for the combined influences which can be made are consequently, rather in the nature of assumptions than precise determinations. These effects, however, cannot be ignored. It would seem that Einstein, in his calculations, has taken no cognizance whatever of these combined influences. In the writer's calculations an attempt was made to make allowance for these disturbing effects by a reduction in the velocity of the excitant system. In the near future, comparatively accurate information may make it possible to substitute precise data for enforced assumption.

At the present time, therefore, the main significance of the writer's calculation lies in the directness and simplicity of the method employed, together with the additional important feature that whenever reliable information is available concerning the retardational influences mentioned above, a precise determination can be made without recourse to the unnecessarily complicated, and basically erroneous, hypothesis of a curved space.

It is important to note that Cosmic phenomena involve the Cosmic Now Plane whose movement in reference to any Space-Directrix is describable ia terms of the velocity of light.

The motion, however, of a discrete material system is describable in terms of a particular Now Plane which may be regarded as associated with the system.

4th.

The perfect harmonious agreement between the dynamical behavior of substance and the cooperative responsiveness of Space-Time is evident from the fact that the same vectorial relations also arise from a study of the basic dynamic laws of the universe. These dynamic relations were first developed by the writer in the year 1904. They are treated in his work under the caption, 'Fundamental Physico-mathematical Relations of the Space-Time Potential' (pages 261-268 inclusive). The relations which pertain to Interdependent and Independent Motion follow directly from these basic dynamic relations. (See Scientific Theism, pages 278-279 inclusive). The Interaction Coefficient for Light was developed by the writer from these dynamic relations, which were based directly upon experimental facts. The author's Interaction Coefficient is identical in form with the Space-Time Coefficient of Larmor, Lorentz and Einstein. Not comprehending the nature of its interdependent source in both Space-Time and real dynamic action (not mathematically speculative) the Relativists misinterpreted its significance and grossly abused its use. The unsound Larmor-Lorentz Contraction hypothesis is only one of their many misapplications.

RIGHT ARCH RIB. THE PRINCIPLE OF EQUIVALENCE.

Stated briefly, the Principle Of Equivalence asserts the equivalence of acceleration and gravitation. Einstein was not the first one to announce this Principle, even if he was the first to misinterpret it in order to link the mutilated product into his system.

Everyone grants that acceleration can be produced by *mechanical* means. Energy must then be expended in its production. The effects produced by acceleration mechanically-generated are precisely the same as the effects which result from its *cosmic generation*, generally described by the term 'gravitation.' The 'dynamic urge' in the case of gravitational acceleration is hidden in its cosmic generation. It is, however, just as much a reality as the energy which must be supplied to generate acceleration mechanically. The 'pure acceleration' of Einstein can therefore never be the equivalent of an actual and physically real 'dynamic urge.' 'Mere motion' is purely theoretical. The attempts of Einstein to account for physically real gravitation by means of the convenient substitute of 'pure acceleration,' can therefore result in nothing but complete failure.

The substitution of a term, empty of dynamic being, does not warrant Einstein's claim for Equivalence. This artifice is on par with many other similar sophistical half-truths emanating from the Father of Relativity. The affair is nothing more than a clever shift of terms in two causal series. Acceleration produced mechanically is an *effect* arising from the application of power. The word 'gravitation' invariably refers to that *cosmic cause* which is capable of producing *gravitational acceleration* as an *effect*. Therefore it follows, that acceleration is not equivalent to gravitation. No one has ever disputed, however, that both mechanically produced acceleration and gravitational acceleration can be discussed analytically under the general term 'acceleration.'

Sophistical half-truths are always productive of dilemmas. If Einstein claims that cause and effect (in the case cited) are equivalent, then it follows, with equal show of sanity, that black is white, evil is good, error is truth, etc., etc. On the other hand, if Einstein claims extraordinary originality in having made the 'astounding' discovery that gravitational and mechanical acceleration are types of acceleration, then we must freely concede the truth of the latter statement whilst marvelling at the unparalleled audacity of tbe claim.

It is here pertinent to call attention again, now, however, by way of contrast, to the substantial work of Anderssohn, Zacharias, 'Kinertia,' Mewes, and Fricke, whose serious endeavors to probe the phenomenon of gravitation to its ultimate source, constitute lasting records in the history of science.

In the light of all this we are utterly unable to account for the wave of enthusiasm which swept the scientific world when Einstein announced 'his great discoveries.'

Conclusion.

Einstein's type of the 'Principle Of Equivalence' is a mere quibble and inversion of words, which is another illustration of 'Much ado about Nothing.' The right arch rib must consequently be declared worthless, because of the failure of this 'Principle' to establish a real 'dynamical' equivalence.

LEFT ABUTMENT HINGE. PERTURBED MOTION OF MERCURY.

It must be admitted that this 'hinge' is the strongest auxiliary member in Einstein's Arch. Dr. William H. Pickering has shown that a discrepancy of about 10% exists between the observed advance of the perihelion and the amount calculated by Einstein. (Dr. Pickering made allowance for the fact that the sun is an oblate spheroid. Einstein assumed it to be a perfect sphere.)

In this connection Jeffreys points out a serious weakness in Einstein's theory. There is, accoring to Jeffreys, 'no abritrary constituent (in Einstein's theory) capable of adjustment to suit empirical facts.'

Moreover, it is a significant fact that Einstein's theory is not successfully applicable to such other well known cases of perturbation as the secular acceleration of the Moon. In fact, his theory fails utterly in universal application. This fatal weakness in Einstein's theory has been revealed by the able work of Professor C. L. Poor.

We have previously mentioned the formula of Paul Gerber (1898) which covered this ground in a much simpler manner. Therefore, even here where the theory is the strongest, it is not indispensable. Its speculative complexity is a serious fault. Since three of the four 'main members' of the structure have collapsed, this auxiliary hinge cannot save the Arch of Einstein from complete destruction. This verdict is in complete harmony with Einstein's expressed opinion concerning his own theory: 'If any deduction from it should prove untenable, it (the theory in its entirety) must be given up. A modification of it seems impossible without destruction of the whole.'

It still remains for Einstein to admit the priority of Gerber.

Conclusion.

This hinge cannot save the Einsteinian structure. It is based upon a fallacious theory. It is not universal in its application. A simple and consistent substitute is available. Therefore this hinge must be discarded.

RIGHT ABUTMENT HINGE. DEFLECTION OF LIGHT.

As far as the results of the calculations are concerned no legitimate criticism can be presented. The percentage of error is comparatively small when the observational difficulties are considered. Here again, however, the theory is not indispensable. The deflection can be calculated with greater precision and by a more direct and simple method. Attention has already been drawn to this fact in the preceding.

This work of Einstein is, moreover, open to severe criticism on the ground of perversion of facts. The 'bending' of light-rays by the sun is used to strengthen the 'Curved Space' phase of this theory. The rays are supposed to follow the *geodesic lines* of Einstein's Curved World-Frame, and again we loose sight, in his theory, of the genuine cause of the phenomenon. In every phase of his system we encounter a departure from the direct and simple. Repeatedly the actual causes of a phenomenon are obscured by a complex fabric spun in the looms of speculative mathematics. So in this case Einstein prefers the devious to the direct, perversion instead of simplicity, and *unreal* curved space becomes the all-important feature of the 'bending' instead of the simple interactional influence of the sun upon particles of matter. The direct attact of the problem should be as follows: Light-rays are composed of matter (since matter is the only known physical reality). The Sun is a great aggregate of matter. Groups of matter interact causing mutual deflections, whose relative amounts depend upon the magnitudes of the interacting groups. Therefore, light-rays being composed of very small particles of matter will be deflected when they pass near a great mass like the Sun.

Einstein's omission of the effects due to the Sun's atmosphere is a serious error.

The New Science has enough problems of real import with which to grapple without accepting the *unnecessary* burdens inconsiderately created by speculative mathematics. Any theory of unnecessary complexity must be regarded as a useless burden to the New Science. We have already seen the Einsteinian Arch crumble into dust. Whatever consideration we give to the hinges must be considered merely as formal and indulgent courtesy.

Conclusion.

The design of this hinge is based upon erroneous assumptions. The details of construction involve unnecessary and inconsistent complexity together with serious omissions. Since a simple and consistent substitute is available, this hinge must be rejected.

CROWN HINGE. DISPLACEMENT OF SPECTRAL LINES TOWARD THE RED.

The average result of all the experiments made, fails to support Einstein's theory. Einstein, while in the United States, publicly stated that he is willing to hazard the truth of his entire theory on the results of this experiment. Up to date the average result has been decidedly against The Theory of Relativity. It is, indeed, strange if science accepts the implied mandate of Einstein in regard to this 'hazard.' The risk involved in such an acceptance is enormous because the displacement is exceedingly small. Moreover, only a limited number of lines can be used in the experiment. Excessively large displacements are likely to occur because of the rapid motion in the line of sight. This excess will vitiate the entire experiment unless absolute allowance for it becomes possible.

Now Einstein is willing to risk the truth of his theory upon this slight probability of apparent experimental confirmation. It would seem that either he has the faith of one obsessed, or even now, he realizes that his theory has no basis in fact. In the latter event his proposal would tend to delay the arrival of that, for him, most potent moment when he would be forced to confess to the world that his intricately spun fabric is worthless.

Dr. Pickering points out that St. John, in an experiment conducted at Mt. Wilson, found a displacement for the cyanogen lines of only +0.0018A, whereas the displacement predicated by Einstein, from his Theory of Relativity, should be +0.0080A. The actual discrepancy is +0.0062A which represents an excess of 344 per cent. If Einstein had genuine confidence in the alleged affirmative results obtained by Grebe, Bachem, Schwarzschild, and Evershed he would not have made, while in the United States, the public statement cited above. In his work 'Relativity' he refers to this experiment as follows: 'It is an open question whether or not this effect exists ... . At all events, a definite decision will be reached during the next few years.' He is like a man who uses the technical machinery of the courts to delay the final and inevitable verdict.

In view of the above one can but marvel at the extraordinary reception, a mounting to a triumphal ovation, which was accorded a theory, built upon a foundation of quicksand, and 'hinging', according to its originator, upon an experiment yet to be proved.

Conclusion.

This hinge must be rejected because it is not only unsafe, but also non-existent.

SUMMARY OF RESULTS OF TESTS.
RIGHT ABUTMENT. NON-EUCLIDEAN SPACE-TIME.

Non-Euclidean Space-Time is based upon unsound and erroneous departures from, and extensions of, Euclidean geometry. The Minkowski and Einstein version of four-dimensional Space-Time reduces to a type of four-dimensional space which is not a reality. Therefore, the right abutment is structurally non-existent. The Space-Time idea is not even original with Minkowski and Einstein. Palagyi, in 1901, expounded the essentials of this theory.

THE FLOOR.-MICHELSON-MORLEY-MILLER EXPERIMENT.

At the present time the results must be regarded as experimental facts. The significance of these results has been misconstrued by Larmor, Lorentz, and Einstein. Relativity is based upon a wrong interpretation of these results. We, however, must pronounce the floor as structurally safe.

LEFT ABUTMENT. THE FITZGERALD-LARMOR-LORENTZ TRANSFORMATION.

The Larmor-Lorentz contraction theory is a purely speculative mathematical device designed to meet an emergency. Its contentions have not been substantiated experimentally. Moreover, the Relativists (including Einstein) maintain that an experimental proof is impossible. Therefore we are forced to conclude that the left abutment is conceptually unsound, experimentally unverifiable, and structurally nonexistent. This transformation is not due to Einstein but is the work of Larmor and Lorentz based upon a suggestion by Fitzgerald.

LEFT ARCH RIB. THE CONSTANCY OF THE VELOCITY OF LIGHT.

The constancy of the velocity of light is an experimentally established fact. The left arch rib is, therefore, a sound and safe structural member. This experimental fact was not discovered by Einstein.

RIGHT ARCH RIB. THE PRINCIPLE OF EQUIVALENCE.

As defined by Einstein it is a mere quibble and inversion of words. It is an erroneous substitution of effect for cause, followed by a claim of 'Equivalence' of the reversed product and its causal source. This is a pure case of 'Much ado about Nothing.' The experimentally sound feature of the principle has been misinterpreted by Einstein, and as such, becomes a self-destructive member of his Arch. The discovery of the real facts, which were perverted by Einstein, are not even due to him but must be accredited to Anderssohn, Zacharias, 'Kinertia,' and Fricke.

LEFT ABUTMENT HINGE. PERTURBED MOTION OF MERCURY.

Here we have the best agreement of Einstein's theory with observed facts. The unnecessary complexity of Einstein's method of calculation, however, eliminates the result from serious consideration. This element in his unstable structure cannot save a theory which so blatantly lacks internal consistency and external verification. Einstein's theory is here impossible because it lacks universal applicability. The priority of Gerber here removes all ground for claims to originality on the part of Einstein.

RIGHT ABUTMENT HINGE. DEFLECTION OF LIGHT.

Einstein's calculated deflection is in comparatively close agreement with the observed amount. The calculated is less than the observed by about 11 per cent. Einstein's deflection is twice the amount obtained by the use of

Newton's gravitational expression. Newton's is less than the observed by about 56%. This, then, is the status of the calculations which brought Einstein into prominent opposition to Newton. Einstein has committed a serious error in neglecting to allow for the retardational effect of the Sun's atmosphere.

We have previously mentioned that a closer approximation can be derived by simple methods founded upon the readily verifiable laws of dynamics. Therefore this attempt of Einstein is merely historically interesting. Moreover, the basic assumptions of the Einsteinian calculations are erroneous, being founded upon an untenable theory. This hinge must therefore, be removed from the world structure because it is both lacking in possible precision, and also involves unnecessary complexity in design.

THE CROWN HINGE. DISPLACEMENT OF SPECTRAL LINES TOWARD THE RED.

Einstein hazards the stability of his whole structure upon this hinge. Experimental evidence, now at hand, is decidedly damaging to Einstein's position. He admits that his contentions have not been verified. This is borne out by his own statements, recently made, in the United States.

The proposed experiment involves extremely small displacements. Varying pressure in the solar atmosphere together with the rapid motion in the line of sight, constitute decidedly detrimental extraneous influences which increase the inevitable inaccuracy of the experiment. Therefore, whatever may be the result of this proposed experiment, its significance will be open to challenge. It is never safe to base, even a less important theory, upon such dangerous experimental ground-work.

We must therefore, even now, discard the future result of this experiment as having significant bearing upon the validity of Einstein's theory.

In view of these facts we draw the inevitable conclusion that the crown hinge is not only unsafe but also non-existent.

CRITICAL WORKS ON EINSTEIN'S RELATIVITY.

E. Gehrcke—Die Relativitätstheorie, Berlin, 1920.

H. Fricke—Der Fehler in Einsteins Relativitätstheorie, Berlin, 1920.

Edouard Guillaume—La Theorie de la Relativite, Lausanne, Switzerland, 1921.

Edouard Guillaume—La Theorie De La Relativite Et Sa Signification. (Revue de Metaphysique et de Morale.)

A. N. Whitehead—An Enquiry Concerning The Principles of Natural Knowledge.

Cambridge University Press, 1919.

A. Patscke—Umsturz der Einsteinschen Relativitätstheorie.

P. Lenard is preparing a work on the errors of Einsteinism.

Rudolf Mewes—Raumzeitlehre oder Relativitätstheorie. (Berlin, 1921. The Collected Works of Mewes dating from 1884 to 1899 inclusive.)

William H. Pickering—The Einstein Theories. (Scientific American Monthly, April, 1921.)

John T. Blankart—Relativity or Interdependence. (The Catholic World, Feb., 1921.)

J. E. Turner—Some Philosophic Aspects of Scientific Relativity. (The Journal of Philosophy, April 14, 1921.)

Arvid Reuterdahl—Scientific Theism Versus Materialism, The Space-Time Potential (1920).

The Fallacies Of Einstein. (Now in preparation.)

The writer has, in this brief article, presented facts not previously available in collected form. He will feel amply repaid for his labors if their presentation will further the cause of justice and truth."

From the *St. Paul Pioneer Press*, 21 August 1921,

"REVIEWER SAYS REUTERDAHL'S NEW BOOK CLEARLY DRIVES EINSTEIN TO THE ROPES

William Wyckoff Clark of St. Paul, graduate of the University of Minnesota in its earlier days, has made a clear study of the theory of relativity, and an article by him entitled 'Divine Relativity,' discussing a metaphysical aspect of the theory, will appear soon in the Homoletic Magazine, a leading scientific journal. Prof. Arvid Reuterdahl is dean of the department of engineering and architecture at the College of St. Thomas, and is widely known as a scientist. He challenged Einstein to a debate, some weeks ago, but never has had a reply from him.

Prof. Reuterdahl is receiving daily letters and telegrams of commendation of his book attacking the Einstein theory. They have come from Berlin, where Einstein is at present, from Prague, from Jugo-Slavia and Switzerland as well as from scientists in America.

Dr. T. J. J. See, director of the United States Naval observatory at Mare Island, Calif., writes: 'I am glad that you have punctured Einstein's bubble, which justifies the remark that 'Einstein is the Doctor Cook of physical science'.'

By William Wyckoff Clark

EINSTEIN AND THE NEW SCIENCE, Dean Arvid Reuterdahl, College of St. Thomas, St. Paul.

In this article, appearing in the Journal of the College of St. Thomas and reprinted for general circulation, Dean Reuterdahl does three things creditably: First, he makes an accurate notation of the sources from which, it is claimed, Einstein acquired the various ideas composing the theory of relativity; offers a concise, vigorous and scholarly criticism of the theory; and, third, introduces an outline of his own striking and strikingly original Time-Space Potential, in so far as it is akin to relativistic principles.

The St. Paul mathematician is the most fearless and unrelenting foe of Einstein's relativity that has, up to the present time, voiced his criticism of the theory in the English language. German and French scientists have flayed Einstein and his teachings and his methods unmercifully, but for some reason or other, those English and American scientists, who have not joined the relativistic ranks, have maintained a very polite and kindly silence anent the theory. Many of them reject it, many of them adopt the Scotch verdict of 'not proven,' but few indeed are they who have taken pen in hand to write for publication even the mildest sort of adverse comment. Reuterdahl, therefore, enters an almost empty field. That he does so willingly and even joyfully no one who has read the very brief comments on relativity contained in his book, 'Scientific Theism,' will for a moment doubt. He is a fighter, but withal fair and dignified.

Leaves Case With Jury.

Without any waving of arms or shouting of 'plagiarist,' 'thief,' 'robber,' Reuterdahl introduces his evidence and leaves the case with the jury. He gives the names of scientists and mathematicians, with the titles and dates of publication of their various works, periodicals, etc., from which, it is claimed, Einstein obtained the data and the very ideas composing the theory of relativity, specifying accurately the subject matter appropriated. To such an extent and so thoroughly does Reuterdahl perform this work of exhibiting 'parallelism' that the possibility of honest, independent origination by Einstein is made to appear very remote and the burden is clearly placed upon his friends to show any original work of value by him in connection with the theory. Practically all that the author concedes to Einstein is a limited amount of grouping of ideas and an unlimited amount of self-glorifying advertisement.

In his criticism of relativity Reuterdahl is fair and discriminating, conceding merit to certain parts of the theory and acknowledging as authentic a number of its important postulates. He very rightly regards Minkowski's Space-Time composite as one of the abutments on which the arch of relativity must necessarily rest. Minkowski is the man who coupled space and time together in an inseparable 'bund,' and then, in the ecstacy of delight over his achievement, made use of that expression which finds a place in every treatise on relativity, viz.: 'Henceforth Space in itself and Time in itself sink into mere shadows, and only a kind of bund of the two can be maintained as self-existent.' And it was Minkowski who worked out the process, on which all relativistic mathematics rests, in which time is treated as functionally equivalent to a fourth dimension of space.

'High Brow' Camp Annoyed.

The school of relativity is divided into two camps, on embracing those who frankly believe in a four-dimensional space with time actually one of the dimensions, and one embracing those who would merely assert that under certain conditions time enters the mathematics of relativity as quantitatively equivalent to a dimension of space. The latter group consider themselves the 'high-brows' of relativity and are much annoyed by the success which has attended the members of the other camp in conveying the impression to the public that relativity sponsors four-dimensional space.

In his consideration of the subject, Reuterdahl starts out with the new, original and highly important demonstration that Minkowski's mathematics really gives a four-dimensional space. From page 11 I quote:

> It can be shown that the Minkowski-Einstein space-time is a mathematically camouflaged type of four-dimensional space. In the invariant form of the General Quadratic Differential, which is basic to Relativity, the last term is formed by multiplying the velocity of light by time. Velocity is reducible to length divided by time. Therefore time is eliminated from this term, leaving it as a pure spatial expression. Consequently we have here nothing but a new version of four-dimensional space which is not a physical reality.

Relativists Put on Defensive.

Unless relativists are able to show that Reuterdahl is mistaken in this analysis of Minkowski's Time-Space mathematics, the theory is left tied up with and bearing the burden of a four-dimensional space: and relativity is seriously handicapped by Reuterdahl's initial attack.

Reuterdahl next takes up the celebrated Michelson-Morley experiment and concedes that at the present time its results must be accepted as experimental facts. He agrees with relativity in regarding the experiment as conclusive proof that there is no ether. Relativity fails to provide any substitute for the ether and thereby lays itself open to the charge of incompleteness in providing no medium for the transmission of light or other electro-magnetic waves.

Reuterdahl, however, avoids that mistake; his 'concurrent system' offers a satisfactory substitute for ether and one which is free from the inconsistent and even contradictory properties ascribed by the scientists of the last generation to that medium.

Consideration is next given to the Fitzgerald-Lorentz contraction hypothesis and to the Larmor-Lorentz transformation equations. The importance of these matters to the theory of relativity amply warrants the extended space devoted to them by the writer, but his treatment is too technical to authorize an extended review of it at this time.

Reuterdahl considers the contraction theory to be a 'purely mathematical device designed to meet an emergency.' It has not been and cannot be confirmed by experiment; it is a 'flagrant case of the misapplication of a mathematical product to physical reality.'

Taking up the subject of the constancy of the velocity of light, Reuterdahl accepts the second postulate of Einstein's theory that this velocity, in a vacuum, is the same to all observers and is independent of the relative motion existing between the observer and the source of light. This is the startling postulate holding that, whether an observer were rapidly approaching a light source, or relatively at rest with it, or rapidly receding from it, in each of the three cases the waves of light would reach him with the same velocity. Paradoxical as this may seem, scientists in general accept it, although with great reluctance. In a recent letter to the writer of this review a former president of the American Academy for the Advancement of Science says:

> I quite agree with you that the postulate of relativity as to constancy

of the velocity of light without reference to the motion of the stars is unsatisfactory, and I hope that at some time the experimental grounds for this assumption will be found to be less compelling than seems to me to be the case at present.

Under the circumstances, therefore, Reuterdahl's acceptance of this postulate is undoubtedly justified, especially so in view of the fact that he immediately points out the misinterpretation and misuse of the postulate by relativity.

Then follows that portion of the article which, to the philosophically inclined reader, will be found most intensely interesting, i. e., Reuterdahl's own theory of the velocity of light together with an altogether too brief outline of his Space-Time Potential. The reviewer has tried, but without success, to contract within the space at his disposal an understandable resume of this work. Any more concise presentation of it than the author himself gives would necessarily be incomplete. An understanding of it involves an acquaintance with the author's former work, 'Scientific Theism.' It is therefore with great regret that we dismiss the topic with the totally inadequate comment that Reuterdahl's Space-Time Potential and theory of light transmission are strikingly original, scientifically and philosophically consistent and worthy of the profoundest study.

Among the most outstanding features of relativity is Einstein's much-heralded 'Principle of Equivalence' between gravitation and acceleration. Reuterdahl performs a very important bit of work in showing that the identifying of the two by relativity is a confusion of cause and effect which robs the 'principle' of all heuristic value, indeed of all verity. Acceleration is an 'effect,' one that can be produced mechanically or by the action of that 'cosmic cause' which we call 'gravitation.' But acceleration, an effect, and gravitation, a cause, can not be identical or equivalent.

We must pass over without adequate consideration Einstein's proposed three tests for his theory, the perturbed motion of Mercury, the deflection of light and the displacement of spectral lines. Regarding the first, Reuterdahl admits the accuracy of the relativistic calculations to within about 10 per cent, but shows that the same system of computation, applied to other cases of perturbation, produces inconsistent results.

Regarding Einstein's calculated deflection of light the author concedes its approximate correctness, a variance of about 11 per cent being shown, but points out that 'the deflection can be calculated with greater precision and by a more direct and simple method.'

The third test, the displacement of spectral lines toward the red, not being claimed by relativity to have been confirmed, is dismissed by Reuterdahl with but little more than the passing comment that Einstein is taking long chances on resting the validity of his entire theory on this doubtful base.

'Einstein and the New Science' is a valuable addition to relativistic literature. Students of the subject, whether favorably inclined to the theory or otherwise, can not afford to miss reading it."

On 24 August 1921, *The New York Times* reported on page 2,

"*CALL FITZGERALD FATHER OF RELATIVITY. English Writer Gives Him Credit for the Genesis of the Einstein Theory.* Special Cable to THE NEW YORK TIMES. LONDON, Aug. 23.—Referring to the conferring by the Royal Society of its fellowship on Dr. Robb for his work on relativity, a scientific correspondent of The Daily Chronicle says that the credit for the evolution of the theories of time and space is due to the initiative of three Irishmen: Professor G. F. Fitzgerald of Trinity College, Dublin; Sir Joseph Larmor, who is Lucasian Professor of Mathematics at Cambridge, and Dr. Robb.

Robb has admitted his indebtedness to Larmor, but, says, the correspondent, the theory of relativity owes its origin to Fitzgerald's explanation, as far back as 1888, of Nicholson's [*sic*] and Morley's failure to detect any relative motion between earth and ether.

He showed that if all bodies contracted in the same proportion in the direction of their motion we should have no fixed standard of length, as measuring rules and all scientific instruments would likewise change their dimesions [*sic*]. Hence we could not ascertain the exact size of things, nor detect their motion relatively to fixed absolute space.

This was known as Fitzgerald's contraction theory, which, in the hands of Larmor and Sovenx [*sic:* Lorentz?] of Leyden has led up to the remarkable theories of space and time since developed by Robb and Einstein."

In 1921, Wolfgang Pauli set the record straight in the *Encyklopädie der mathematischen Wissenschaften*,

"The metamorphoses in physical concepts brought about by the theory of relativity was a long time in the making. As far back as 1887, Voigt observed in one of his works [***] that it is mathematically possible to introduce a time of position t' into a moving reference system, whose origin is a linear function of the spatial coordinates, while the unit of time, however, is taken to be constant. Whereby, one can assert, of course, that the wave equation

$$\Delta \varphi - \frac{1}{c^2} \frac{\partial^2 \varphi}{\partial t^2} = 0$$

also remains valid in the moving system. [***] We now come to a review of the three works of *Lorentz, Poincaré* and *Einstein*, which contain the thoughts and developments which are the foundation of the theory of relativity. Lorentz' work led the way. Above all, it furnished the proof that Maxwell's equations are invariant under the transformation of coordinates

[*Lorentz Transformation equations deleted*] provided that one at the same time suitably selects the field intensity in the primed system."[591]

Pauli argues that Lorentz holds priority for the proof of invariance. Pauli next addresses Poincaré's contribution,

"The formal gaps left by Lorentz's work were filled by Poincaré. He stated the relativity principle to be generally and rigourously valid. Since he, in common with the previously discussed authors, assumed Maxwell's equations to hold for the vacuum, this amounted to the requirement that all laws of nature must be covariant with respect to the 'Lorentz transformation' [The terms 'Lorentz transformation' and 'Lorentz group' occurred for the first time in this paper by Poincaré—*notation found in the original*]. The invariance of the transverse dimensions during the motion is derived in a natural way from the postulate that the transformations which affect the transition from a stationary to a uniformly moving system must form a group which contains as a subgroup the ordinary displacements of the coordinate system. Poincaré further corrected Lorentz's formulae for the transformations of charge density and current and so derived the complete covariance of the field equations of electron theory. We shall discuss his treatment of the gravitational problem, and his use of the imaginary coordinate *ict*, at a later stage (see §§ 50 and 7)."[592]

After giving Poincaré his due credit, and acknowledging that Einstein holds no priority for the special theory of relativity, Pauli, half-heartedly, pays the seemingly obligatory homage to Einstein, the then recently emerged celebrity,

"It was Einstein, finally, who in a way completed the basic formulation of this new discipline."[593]

And it appears that Pauli was forced, or felt compelled, to praise Einstein with additional inappropriate and, evidently, insincere comments.

Einstein's work was not so well-received, nor so perfect, as his present day sycophantic advocates would have us believe.[594] Louis Essen wrote,

---

[591]. W. Pauli, *Encyklopädie der mathematischen Wissenschaften*, Volume 5, Part 2, Section 19, B. G. Teubner, Leipzig, (1921), pp. 539-775, at 543-545.
[592]. W. Pauli, *Theory of Relativity*, Pergamon Press, London, Edinburgh, New York, Toronto, Sydney, Paris, Braunschweig, (1958), p. 3.
[593]. W. Pauli, *Theory of Relativity*, Pergamon Press, London, Edinburgh, New York, Toronto, Sydney, Paris, Braunschweig, (1958), p. 3.
[594]. *See, for example:* A. Lynch, *The Case Against Einstein*, P. Allan, London, (1932). H. Dingler, *Die Grundlagen der Physik; synthetische Prinzipien der mathematischen Naturphilosophie*, Second Edition, Walter de Gruyter & Co., Berlin, (1923); **and** *Physik und Hypothese Versuch einer induktiven Wissenschaftslehre nebst einer kritischen Analyse der Fundamente der Relativitätstheorie*, Walter de Gruyter & Co., Berlin,

"But there have always been its critics: Rutherford treated it as a joke: Soddy called it a swindle: Bertrand Russell suggested that it was all contained in the Lorentz transformation equations and many scientists commented on its contradictions. These adverse opinions, together with the fact that the small effects predicted by the theory were becoming of significance to the definition of the unit of atomic time, prompted me to study Einstein's paper. I found that it was written in imprecise language, that one assumption was in two contradictory forms and that it contained two serious errors."[595]

John T. Blankart stated in 1921,

"The 'Kinertia' articles offer food for thought when considered in connection with the colossal claims made by Einstein's supporters concerning his almost super-human originality. In fact, one begins to doubt the justice of these claims and to wonder if the charges made by a fast growing group of German scientists who, like E. Gehrcke, P. Lenard, and Paul Weyland, hold that Einstein is both a plagiarist and a sophist, are not, after all, true. We have done little justice in the above to the rare dialectic skill with which Dr. Einstein has applied his intellectual anæsthesia to the minds of his readers. All intellectual obstructions have been removed, and the reader is prepared to venture forth boldly into the mysterious realm of 'curved' space *whose geometrical properties depend upon the matter present*. This most curious inference of Einstein is the master stroke in his skillful massing of inconsistent sophistries."[596]

Sydney T. Skidmore wrote, in 1921,

"THE MISTAKES OF DR. EINSTEIN
By SYDNEY T. SKIDMORE

WE begin this essay by saying that Einsteinism is an erudite elaboration of sophistry and is closely akin to, if indeed it does not spring from, the same root as classic sophistry. The tap root of that system of philosophy developed in the fifth century before the Christian era, and consisted in a denial of the existence

---

Leipzig, (1921); **and** "Kritische Bemerkungen zu den Grundlagen der Relativitätstheorie", *Physikalische Zeitschrift*, Volume 21, (1920), pp. 668-669. H. Nordenson, *Relativity, Time and Reality: A Critical Investigation of the Einstein Theory of Relativity from a Logical Point of View*, Allen and Unwin, London, (1969).

595. L. Essen, "Relativity — Joke or Swindle?", *Electronics and Wireless World*, (February, 1988), pp. 126-127. URL:

<http://www.cfpf.org.uk/articles/scientists/essen.html>

596. J. T. Blankart, "Relativity or Interdependence", *Catholic World*, Volume 112, (February, 1921), pp. 588- 610, at 606.

of objective truth. Its thought and attitude can only become intelligible from a presentation of what 'objective truth' is, and for this, a little tax must be imposed on the reader's patience.

Its definition is simple enough. It consists of, and includes, the being of all created things and their relativities. It is objective because its essence is independent of subjective thinking which can apprehend it in part—can pick up pebbles of it from an ocean strand—and assemble what is gathered as knowledge. Since it inheres in the essence of created things it is coinstant with their creation.

Creation is originate; and all created things must have a beginning. The first creative act necessitated a 'where' for its occurrence, and that where has existed ever since as a changeless objective truth. Each creative act likewise necessitated a where, and the aggregate of all wheres, or whereness, constitutes a changeless, undistortable, frame of objective truth to which the term Space has been applied. Objective truth or 'isness' pertains to the wheres or loci in space, and since the loci are fixed, it also pertains to the changeless relations of loci.

The first creative act not only required a where, it also required a when—an instant—for its occurrence. Each creative act likewise required an instant, and the aggregate of all whens or whenness, constitutes another frame of objective truth, to which the term Time has been applied. Unlike loci, instants are not simultaneous, they are sequential, and their objective truth pertains to a procession rather than to a distribution.

In each creative event, therefore, three orders of objective truth are present, viz., cause, locus and instant. Since history is composed of events, and experience is concerned with them, the foregoing analysis may serve to show what the nature of objective truth is, and also that the objective truths, cause, space and time, supply and equip the generative arena of events, i.e., of physical phenomena.

Objective truths are presented in every fact and may be apprehended in all phenomena. They are not thoughts but they are thinkables, and are cognized by each mind according to its scope. Now, because the Eleatics failed to formulate it or define it as an abstract oneness, the Sophists denied that it had any existence whatever.

Since abstraction plays an important part in this discussion it must receive some attention. Abstraction consists in withdrawing attributes, or qualities, from their home correlatives in nature, and installing them in a psychical abode for mental contemplation. As the word stands it means the separation of something from something; but never a separation of something from nothing. Inception is usually the word for that. There must always be a residue from which the final abstraction is made. The relativity of attributes in and with a thing, although they are mentally withdrawn, is still codestructible only with the thing itself. An abstraction of qualities does not annihilate the residue; nor can a sound philosophy be constructed from the relativities of attributes alone, with the residue ignored. We give the following statement prodigious emphasis because it is so much involved in the reasoning farther on.

No amount of abstraction can resolve a thing to a philosophical nullity nor psychalize it into nonexistence. The residue with its relativities still persists as objective truth.

The relativities of abstractions by themselves are subjective, mental, and may be correct, but are usually incorrect owing to the imperfection of mental action. True science is a developed knowledge of what *is* as revealed by discovery in wide open objectivity, and false or pseudo science is a knowledge of what *seems to be* as revealed by apprehendings in the inclusions of subjective recesses.

Since the Sophists denied the existence of objective truth they could not make it an objective goal of human endeavor. They must by necessity adopt a subjective goal, such as excellence, success, or victory. Truth, with them, was inherent in triumph. Whatever prevailed was true and true because it prevailed and truth had no other significance. It is easy to understand how such a philosophy as that should become reduced by human ambition, selfishness, and deceit, to the direct degradation. The success most esteemed by the Greeks was victory in debate, and after two centuries, Sophistry became such a system of thin verbal trickery that it fell into disrepute, and a stigmatum attached to its name.

Wherever the supreme goal of endeavor has been placed in things other than debate; and smartness of any kind has been substituted for objective truth, as an end anywhere, sophistry works the same degradation. While it appears to be always present as an inseparable corrupter, there have been some well marked epochs in which it acquired such dominance as to shape legislation and thinking and openly display its fruits. This occurred in the ancient sophistry of Greece as such; in medieval sophistry as Scholasticism; and in modern sophistry as Commercialism, Pragmatism, and Education. In war and politics it appears respectable as Strategy. In commercialism, somewhat less so as shrewdness; while in pragmatism and education it often wears the mask of efficiency.

Objective truths are distent and gloriously free. Subjective truths are stifled in mentality and subordinate to the ends of victorious achievement. Apprehendings of objective truths are obtained from objective things and, if incorrect, they may be checked up and corrected by reference to the things. Apprehendings of subjective truths are mental constructions, apart from things, and uncorrectable since subjectivity is not apt to correct itself. If they are crazed by mental inaccuracy the relativities of such truths are incurably queered likewise.

This presentation of sophistry as a system of thought, seems necessary to establish, by comparison, the validity of the statement made in the beginning of this essay; for we shall try to show that Einsteinism is sophistry, both in its nature, and in its dialectic construction.

It is purely subjective and Protagorean in that it ignores the objective truth of all steadfastness, and all relativity of steadfastness in general being.

There are two orders of relativity; that of the steadfast with changeables; and that of changeables with each other. Einstein relativity is exclusively of the second order. We are not aware that Einstein anywhere formally denies the existence of steadfastness as objective truth, but since it cannot be psychalized he everywhere ignores it, and all arguments for Einstein relativity are based on its non-existence; and it is Einstein relativity, with its astounding pretensions, that we are criticising.

The primary positional steadfasts in nature are the loci (points) in space. The earth and all things in it move, but space units do not. All things in the earth have a first order relativity with the points of space, and a first or second order with each other according as their motions are alike or unlike each other. Now because the points of space are ultra to experience, imperceptible and unsubjective, together with their relativities, their being is summarily denied by sophists and ignored by Einsteinism; and all semblance of steadfastness, like that of car seats in a moving car, or houses on a moving earth, have no steadfast relativity with anything; it is only subjective thinking.

Einsteinism claims to open a vast extension of physics but, if adopted and followed, it would tend to a collapse of physics because it works from a psychological rather than from a physical basis. The two are in reversion. Physics stimulates discovery by trailing the scent of objective truths occluded in the unknown. Einsteinism represses discovery by holding truth corralled within subjectivity. Even Space and Time, the fundamental containers of those objective truths which physicists are continually transferring from the unknown to the known, are said to be 'devoid of the last vestige of physical objectivity.' (Schlick, pages 53, 76. Eddington, page 34). [*Footnote:* We shall quote in this paper from Schlick's 'Space and Time', and from Eddington's 'Space, Time and Gravitation', because both these books are recognized as authoritative in Einstein literature and they are somewhat more definite and explicit than Einstein's own writing.]

Physical relativities are of the first order; Einstein relativities are of the second order and pertain to the relations of fluxing events as they are observed. Words such as cause, potential, and force, which are leaders in physics are of rare occurrence in Einstein literature and when used are slipped in edgewise. The relativity of physical effects with their causes is slightly discussed, but the relativity of mental states induced in observers when differently conditioned abounds, and forms the body of argument, and the plenitude of discussion.

Another citation, which shows how completely truth is restricted to the realm of subjective apprehendency, appears in the interpretations given to the Michelson and Morley experiment.

Those investigators truly assumed that if a non-viscous static aether existed, an aether wind opposite to the earth's motion must blow through the moving earth; and that the velocity of light would be different when moving against this wind, than when moving at right angles with it. A very delicate and crucial experiment showed that the earth's motion had no effect whatever on the velocity of light. Now what? Something must be wrong, either with the aether belief, or with the motion of light; and the mathematicians proceeded to explain it, as they usually do, by tinkerings at space and time. Fitzgerald and Lorenz [sic] devised that everything in the line of motion transforms and contracts, and so increased time was exactly compensated by shortened distance, and the velocity of light, as shown by simultaneous arrival, was apparently unchanged.

This saved a clumsily apprehended aether belief from Michelson and Morley extinction; but Einstein proposed a different explanation. Quite indifferent to the fate of current aether belief, he found the difficulty lurking in the relativities of motion. All things, relatively at rest in a system, maintain that relativity whether the system, as a whole, is moving or not. The motion of a system, moving relatively with objects external to it, has zero effect on the relativity of things within it. The relative direction of city streets abides when their direction from the sun changes continually. Street cars run a mile east in the same time as when running north, although the earth rushes westward one thousand one hundred miles per minute, and northward not at all. The interferometer, mirrors, and source of light, in the Michelson experiment, were all in the same Earth system and therefore the light moved between them through equal distances in equal times, whatever the direction might be. This neither proves nor disproves the existence of an aether, but it does show that if an aether exist it is of such a character that currents and whirls in it do not perceptibly affect the velocity of light. It is not an externality by which the relativity of light movement with it can be sensibly apprehended. Now, because a static aether of a particular character does not exist, the reasoning dialectically pussyfoots into an assumption that there is neither aether nor staticity. The aether is of small consequence in the case, but it is essential to Einstein relativity to put out of existence the principle of staticity as an objective truth and the ultimate physical reference basis of all motion.

Whatever may be true in metaphysics it is certainly true, that in physics such a principle does and must exist, as a physical necessity. A bird does not take the air along with it in flight; a ship does not take the ocean with it in sailing; a moving car does not take the ties of the road bed with it, and no moving thing takes space with it. Air, ocean, and ties have a static relativity with the moving objects mentioned. Whatever moves has changing relations with everything that does not move precisely as it does; and static relations with everything that does: but a truce to such platitudes. Space contains all moving things which therefore have a shifting relativity with it, because it does not move like them. It is the physical ultimate of staticity since nothing physical exists external to it to which its motion can be referred.

The changing relativity of things with the points of space or instants of time is of the first order (primary) and all changing relativity of things with each other is of the second order (casual).

Einstein relativity is exclusively of the second order. The expounders of it deny that there is any other, and back up the denial by ignoring the staticity of space; but this they cannot do without postulating something in metaphysics external to space which does not move as space does; and this they cannot do; so, to abolish its staticity, they must abolish space itself and replace it by a subjective creation.

Staticity must be removed from the space world to permit the entrance of Einstein curios and non-Euclidian queers. While it abides lineality abides. Forms in space are outlined in it by moveless points, and are differentiated from it just as an island boundary is different from the surrounding ocean. Points of space are located by rectilinear coordinates, and all other coordinates whether Gaussian, polar, or zigzag, only serve to locate places on the surface of a form in space, like the longitude and latitude circles on the surface of a terrestrial globe. They do not locate points of space; they merely locate points with reference to other points on the surface of a form in space. Hence arises the non-Euclidian sophistry of spherics, or eliptical space, and the Einstein sophistry of space curved and twisted around material bodies, like a swaddling striate aura, and the further sophistry that bodies moving through such space are impelled by inertia along curved rather than straight lines in accordance with a 'Principle of Least Action' that the longest way round is the shortest way home, because straight lines would lead across curving hurdles (Eddington, page 105).

Space as such has no form whatever. It is neither curved, flat, nor otherwise. The pure forms of things (the abstract residues) are defined in space by the fixed relativity of its moveless points. This statement squarely contradicts Einsteinism. It is based on logical inherences in objective creation, while its antithesis is grown from subjective apprehendings of shifting things. Whichever is truth, the other is devoid of truth and the choice is yours.

Staticity has been discussed at some length because it illustrates the attitude of Einstein relativity towards all objective truth. Because such truths, when postulated are imperceptible and make no psychic impression, words sophistically used present them as unreals, and cause them to appear as 'ambiguities and unnecessary thought elements', (Schlick, page 5) which should be thrown aside as meaningless and obstructive to a path that leads not to truth but to victory; not to amendment and improvement by new tributes of knowledge; but to a revolution of fundamental concepts which throws down an older and erects a new intellectual throne.

This revolution (when achieved) is a promise of something which will cause Newton and Copernicus to seem like infantile prattlers; 'inasmuch as the deepest foundations of our knowledge concerning physical nature have to be remodeled much more radically than after the discovery of Copernicus.' (Schlick, page 5.)

The signs of such an approaching revolution at present are not very auspicious. While one out of twenty, or possibly fifty, of savants are filling the world with a sounding applause of it, all the rest are waiting, silent, dubious, and withholding allegiance. Still it may come; for the human world delights in sophistry and dotes on truths of its own creation. Impressionism which is so powerful in Art may also yet prevail in Philosophy.

That Einsteinism presents a revel in such truths is made evident by Eddington in Chapter XII: 'The conclusion is that the whole of those laws of nature which have been woven into a unified scheme—mechanics, optics, gravitation, electro-dynamics—have their origin not in any special mechanism of nature but in the workings of the mind.'

'Give me matter and motion,' said Descartes, 'and I will construct the universe.' 'The mind reverses this,' says Einstein. 'Give me a world in which there are relations, and I will construct matter and motion.' The world thus is what it is conceived to be; is what we think it is. That is precisely what Descartes and Einstein each professed to do. Both are subjectivists—sophists. One would replace the objective truths of real relations, by such queered relations as he could mentally construct from observed things, and the other would replace the objective truths of real things, by such queered things as he could mentally construct from observed relations. Both alike substitute their psychical apprehending of nature's content, for the content itself, and then call it truth.

Recent writings in current literature suggest that many inquiries are baffled in attempts to comprehend Einsteinism. They read about it and think there must be something in it, and so there is, but it is a something not included in their somethingness. It is shapen from non-Euclidian, or what is sometimes termed meta, geometry. This consists entirely of mental constructions that are purely subjective and correspond to nothing in nature. In fact it prides itself on a disbelief or at least a disregard of the existence of objective truth, and boasts that 'mathematicians are never so happy as when talking about something of which they know nothing.' (Eddington, page 14.) Really it is no geometry at all, for it measures nothing and disallows all mental standards. It is a fantastic jazz of mathematical symbols, devoid of quanta, in a dance hall, floored by a parquetry of ifs, supposings, and assumptions.

The attitude of Einsteinism toward physics, and the fate of physics by occlusion in this thing, misnamed geometry, is well stated by Eddington (page 183). 'As the geometry becomes more complex, the physics becomes simpler, until it finally almost appears that the physics has been absorbed into the geometry.' While parading the attractive banner of a 'New Physics' or a 'New Philosophy,' Einsteinism is really nothing but a special chapter in psychology, which is offered as a new style of incubator for hatching nature's eggs.

In popular discussion two things are mixed up in Einsteinism as if they belonged to it, but they do not. One of these is the prediction that space and time will have an end. This is nothing new. It is a philosophical deduct of long standing that whatever has a beginning is finite, and must have a boundary and an ending; and that space and time which began with creation will cease to be when created things become non-existent. The other is a scientific derivative from the electronic theory, and preceded Einstein by a number of years. That theory changed the definition of mass from 'quantity of matter' in a body to 'quantity of force' in a body. The matter in a body is its mass or force in statu; the motion of a body is its mass or force in motu. Matter and motion together constitute the mass of a body and each is force with a modal difference. Mass and inertia are one and the same thing to which different names are given when differently apprehended.

This was all worked out physically before the time of Einstein and is no part of Einsteinism. If wonderful, it is a wonder of physical discovery and not a marvel of psychical geometry.

A peculiar feature of Einsteinism is that the crux of its doctrines is deeply submerged in mathematical obscurity. If one asks for proof he is told that it lies in mathematical profundities, quite beyond the reach of anyone other than an adept; and the unintelligible character of Einstein literature fully sustains the statement. Now the English language, with its rich vocabulary, direct idiom, and classic verbal quarries, is quite capable of expressing anything that has a meaning, and of rounding out the proof of any statement that admits proof. To understanding it is a wide open Bible; and cloistered secrets doled out by initiates for aweing the credulous are unnecessary. Proofs that vest in mathematical cryptograms are dubious. Mathematicians choose their own assumptions and, according to the assumption taken, they can prove that truth is truth; or falsehood is falsehood, or truth is falsehood, or falsehood is truth, with equal facility. Mathematics supplied cranks, cycles, and epicycles to Ptolemaic astronomy just as readily as it supplied ellipses, parabolas, and hyperbolas to Copernican. Cryptogramists follow rules of interpretation and have but slight regard for rules of philosophic sense.

A mathematician can only be trusted as far as he can be seen, or objectively checked up. Unlike space but quite like that of a political conscience, the mathematic psychology warps and twists in quaintest fashion to attain an end when left to its own devices. According to Einstein device, Space and Time are inseparable from matter. 'Space and Time determinations will henceforth be inseparably connected with matter and will have meaning only when connected with it.' (Schlick, page 4.) 'Time and Space can be dissociated from matter only by abstraction, i.e., mentally; the combination or oneness of space, time, and things is alone reality; each by itself is an abstraction' (a mental figment). (Schlick, page 6.) 'In this way Space and Time are deprived of the last vestige of physical objectivity, to use Einstein's words.' (Schlick, page 53.) 'Exactly so; Space is an abstraction of the extensional relations of matter.' (Eddington, page 8.) What matter has extensional relations with, is not stated; if it be with other matter, the thing that sustains the relationship is not stated; and you may find out if you can, but not from Einsteinism.

Since Space and Time as thus stated are mental investitures of matter, a bunch of it when moving must either take its space and time along with it as personal property, like clothes, color, or shape; or else find it as a place endowment wherever it goes. We would much like to know whether space is regarded as the mental baggage of travelling matter, or is an omnipresent mental continuum which forms a 'oneness' with matter wherever the matter happens to be. We are not told which it is because that would resolve a psychologic mystery that can be handily employed in discussion. It is sometimes convenient to take it one way and sometimes the other.

The matter in other stars is assumed to be rather similar to that of the earth; but it is bunched together quite differently; and that would create different kinds of space and time. That presents no difficulty, however, because 'there are different kinds of possible space to choose from, no one of which can be regarded more likely than any other.' (Eddington, page 15.) The difficulty becomes serious, however, if it be true that space and time are purely mental determinations. Indeed it becomes an open question whether or not the stars have any space or time worth mentioning. Our mental determination of Arcturian space is restricted to a point; and unless there be a developed mentality in Arcturus, or somewhere else, the poor star has no space other than a point, and no time other than what is marked by star drift. Moreover, if there be any system of physics in Arcturus, it must be quite different from ours, unless the Arcturians have minds like ours, for, according to Eddington, as previously quoted, 'the laws of nature ... have their origin, not in any special mechanism of nature, but in the workings of the mind.'

The vice of Einsteinism is that it transfers sense deception from ordinary things which check it up, to space, time, motion, and energy, which do not check it up, because their nature is ultra to experience.

From a puny bunch of relativity as psychologically impressed on differently conditioned observers, a mathematical explosive has been prepared for deranging established foundations of thought. A petty scheme of psychalized relativity is given as interpretative of a grand world universe filled with objective relativities that have not as yet been psychalized. Its nature is purely subjective and sophistical.— Q. E. D."[597]

There were many others who publicly opposed Einstein, the theory of relativity, and the deception of the general public by the pro-Einstein press on

---

[597]. S. T. Skidmore, "The Mistakes of Dr. Einstein", *The Forum*, Volume 66, (August, 1921), pp. 119-131.

similar grounds, including: Adler,[598] Weinmann,[599] Mohorovičić,[600] Bergson,[601]

---

598. F. Adler, *Ortzeit, Systemzeit, Zonenzeit und das ausgezeichnete Bezugssystem der Elektrodynamik. Eine Untersuchung über die Lorentzsche und Einsteinsche Kinematik*, Wiener Volksbuchhandlung, (1920).
599. R. Weinmann, *Gegen Einsteins Relativierung von Zeit und Raum (gemeinverständlich)*, München, Berlin, Oldenbourg, (1922); "Kommt der Relativitätstheorie philosohische Bedeutung zu?", *Philosophie und Leben*, Volume 2, (1923), pp. 154-159; *Anti-Einstein*, Hillmannn, Leipzig, (1923); "Anti-Einstein Quintessenz", *Archiv für Systematische Philosphie und Soziologie*, Volume 30, (1927), pp. 263-270; *Widersprüche und Selbstwidersprüche der Relativitätstheorie*, Hillmann, Leipzig, (1925); *Versuch einer endgültigen Widerlegung der speziellen Relativitätstheorie*, Hillmann, Leipzig, (1926); "Der Widersinn und die Überflüssigkeit der speziellen Relativitätstheorie", *Annalen der Philosophie und philosophischen Kritik*, Volume 8, (1929), pp. 46-57; "Die Unhaltbarkeit der speziellen Relativitätstheorie", *Natur und Kultur*, Volume 27, (1930), pp. 121-125.
600. S. Mohorovičić, *Die Einsteinsche Relativitätstheorie und ihr mathematischer, physikalischer und philosophischer Charakter*, Walter de Gruyter & Co., Berlin, Leipzig, (1923); "Raum, Zeit und Welt", in K. Sapper, Editor, *Kritik und Fortbildung der Relativitätstheorie*, Akademische Druck- u. Verlagsanstalt, Graz, (1958/1962), Part 1 in Volume 1, (1958), pp. 168-281; Part 2 in Volume 2, (1962), pp. 219-352.
601. H. Bergson, *Durée et Simultanéité, à Propos de la Théorie d'Einstein*, English translation by L. Jacobson, *Duration and simultaneity, with Reference to Einstein's Theory*, The Library of Liberal Arts, Bobbs-Merrill, Indianapolis, (1965); which contains a bibliography at pages xliii-xlv.

Guillaume,[602] Patschke,[603] Dingle,[604] Dingler,[605] Strasser,[606] Guggenheimer,[607]

---

602. E. Guillaume's letter, translated by A. Reuterdahl, "Guillaume, Barred in Move To Debate Einstein, Calls Meeting Political Reunion", *Minneapolis Journal*, (14 May 1922), p. 14; reprinted with slight modifications, "The Origin of Einsteinism", *The New York Times*, (12 August 1923), Section 7, p. 8. *See also:* "Einstein Faces in Paris Grave Blow at Theory", *The Chicago Tribune*, (31 March 1922). *See also:* "Dr. Guillaume's Proofs of Einstein Theory's Fallacy Revealed to the Journal", *Minneapolis Journal*, (9 April 1922). *See also:* E. Guillaume, "Un Résultat des Discussions de la Théorie d'Einstein au Collège de France", *Revue Générale des Sciences Pures et Appliquées*, Volume 33, Number 11, (15 June 1922), pp. 322-324. *See also:* "Les Bases de la Physique moderne", *Archives des Sciences Physiques et Naturelles*, Series 4, Volume 43, (1917), pp. 5-21, 89-112, 185-198; **and** "Sur le Possibilité d'Exprimer la Théorie de la Relativité en Fonction du Temps Universel", *Archives des Sciences Physiques et Naturelles*, Series 4, Volume 44, (1917), pp. 48-52; **and** "La Théorie de la Relativité en Fonction du Temps Universel", *Archives des Sciences Physiques et Naturelles*, Series 4, Volume 46, (1918), pp. 281-325; **and** "Sur la Théorie de la Relativité", *Archives des Sciences Physiques et Naturelles*, Series 5, Volume 1, (1919), pp. 246-251; **and** "Représentation et Mesure du Temps", *Archives des Sciences Physiques et Naturelles*, Series 5, Volume 2, (1920), pp. 125-146; **and** "La Théorie de la Relativité et sa Signification", *Revue de Métaphysique et de Morale*, Volume 27, (1920), pp. 423-469; **and** "Relativité et Gravitation", *Bulletin de la Société Vaudoise des Sciences Naturelles*, Volume 53, (1920), pp. 311-340; **and** "Les Bases de la Théorie de la Relativité", *Revue Générale des Sciences Pures et Appliquées*, (15 April 1920) pp. 200-210; **and** C. Willigens, "Représentation Géométrique du Temps Universel dans la Théorie de la Relativité Restreinte", *Archives des Sciences Physiques et Naturelles*, Series 5, Volume 2, (1920), p. 289; **and** E. Guillaume, *La Théorie de la Relativité. Résumé des Conférences Faites à l'Université de Lausanne au Semestre d'été 1920*, Rouge & Co., Lausanne, (1921); **and** E. Guillaume and C. Willigens, "Über die Grundlagen der Relativitätstheorie", *Physikalische Zeitschrift*, Volume 22, (1921), pp. 109-114; **and** E. Guillaume, "Graphische Darstellung der Optik bewegter Körper", *Physikalische Zeitschrift*, Volume 22, (1921), pp. 386-388; **and** Guillaume's Appendix II, "Temps Relatif et Temps Universel", in L. Fabre, *Une Nouvelle Figure du Monde: les Théories d'Einstein*, Second Edition, Payot, Paris, (1922); **and** E. Guillaume, "Y a-t-il une Erreur dans le PremierMémoire d'Einstein?", *Revue Générale des Sciences Pures et Appliquées*, Volume 33, (1922), pp. 5-10; **and** "La Question du Temps d'après M. Bergson, à Propos de la Théorie d'Einstein", *Revue Générale des Sciences Pures et Appliquées*, Volume 33, (1922), pp. 573-582; **and** Guillaume's introduction in H. Poincaré, *La Mécanique Nouvelle: Conférence, Mémoire et Note sur la Théorie de la Relativité / Introduction de Édouard Guillaume*, Gauthier-Villars, Paris, (1924), pp. V-XVI; **and** H. Bergson, *Durée et Simultanéité, à Propos de la Théorie d'Einstein*, English translation by L. Jacobson, *Duration and simultaneity, with Reference to Einstein's Theory*, The Library of Liberal Arts, Bobbs-Merrill, Indianapolis, (1965); which contains a bibliography at pages xliii-xlv. *See also:* P. Painlevé, "La Mécanique Classique et la Théorie de la Relativité", *Comptes rendus hebdomadaires des séances de L'Académie des sciences*, Volume 173, (1921), pp. 677-680. *See also:* S. Mohorovičić, "Raum, Zeit und Welt. II Teil", in K. Sapper, Editor, *Kritik und Fortbildung der Relativitätstheorie*, Akademische Druck- u. Verlagsanstalt, Graz, Volume 2, (1962), pp. 219-352, at 273-275. *See also:* K. Hentschel, *Interpretationen und Fehlinterpretationen der speziellen und der allgemeinen Relativitätstheorie durch Zeitgenossen Albert Einsteins*, Birkhäuser, Basel, Boston, Berlin, (1990). *See also:* A. Genovesi, *Il Carteggio tra Albert Einstein ed*

Edouard Guillaume. *"Tempo Universale" e Teoria della Relativtà Ristretta nella Filosofia Francese Contemporanea*, Franco Angeli, Milano, (2000). *See also:* Letter from A. Einstein to E. Guillaume of 24 September 1917, *The Collected Papers of Albert Einstein*, Volume 8, Part A, Document 383, Princeton University Press, (1998). *See also:* Letter from E. Guillaume to A. Einstein of 3 October 1917, *The Collected Papers of Albert Einstein*, Volume 8, Part A, Document 385, Princeton University Press, (1998). *See also:* Letter from A. Einstein to E. Guillaume of 9 October 1917, *The Collected Papers of Albert Einstein*, Volume 8, Part A, Document 387, Princetone University Press, (1998). *See also:* Letter from E. Guillaume to A. Einstein of 17 October 1917, *The Collected Papers of Albert Einstein*, Volume 8, Part A, Document 392, Princeton University Press, (1998). *See also:* Letter from A. Einstein to E. Guillaume of 24 October 1917, *The Collected Papers of Albert Einstein*, Volume 8, Part A, Document 394, Princeton University Press, (1998). *See also:* Letter from E. Guillaume to A. Einstein of 25 January 1920, *The Collected Papers of Albert Einstein*, Volume 9, Document 280, Princeton University Press, (2004). *See also:* Letter from M. Grossmann to A. Einstein of 5 February 1920, *The Collected Papers of Albert Einstein*, Volume 9, Document 300, Princeton University Press, (2004). *See also:* Letter from A. Einstein to E. Guillaume of 9 February 1920, *The Collected Papers of Albert Einstein*, Volume 9, Document 305, Princeton University Press, (2004). *See also:* Letter from E. Guillaume to A. Einstein of 15 February 1920, *The Collected Papers of Albert Einstein*, Volume 9, Document 316, Princeton University Press, (2004). *See also:* Letter from A. Einstein to M. Grossmann of 27 February 1920, *The Collected Papers of Albert Einstein*, Volume 9, Document 330, Princeton University Press, (2004). *See also:* Letter from A. Einstein to P. Oppenheim of 29 April 1920, *The Collected Papers of Albert Einstein*, Volume 9, Document 399, Princeton University Press, (2004).

603. A. Patschke, *Umsturz der Einsteinschen Relativitätstheorie. 4 mathematische Geburtsfehler. Einführung in die einheitliche Erklärung und Mechanik der Naturkräfte. Kreuzigung und Auferstehung des Lichtäthers*, Berlin-Wilersdorf, (1922).

604. H. Dingle, *Science at the Crossroads*, Martin Brian & O'Keeffe, London, (1972).

605. H. Dingler, *Die Grundlagen der Physik; synthetische Prinzipien der mathematischen Naturphilosophie*, Second Edition, Walter de Gruyter & Co., Berlin, (1923); **and** *Physik und Hypothese Versuch einer induktiven Wissenschaftslehre nebst einer kritischen Analyse der Fundamente der Relativitätstheorie*, Walter de Gruyter & Co., Berlin, Leipzig, (1921); **and** "Kritische Bemerkungen zu den Grundlagen der Relativitätstheorie", *Physikalische Zeitschrift*, Volume 21, (1920), pp. 668-669.

606. H. Strasser, *Die Transformationsformeln von Lorentz und die „Transformationsformeln'' der Einsteinschen speziellen Relativitätstheorie*, Ernst Bircher, Bern, Leipzig, (1924).

607. S. Guggenheimer, *The Einstein Theory Explained and Analyzed*, Macmillan, New York, (1929).

Lynch,[608] Mackaye,[609] Nordenson,[610] Essen,[611] Theimer,[612] Gut,[613] etc. Early bibliographies appear in Gehrcke's *Kritik der Relativitätstheorie*, Hermann Meusser, Berlin, (1924), pp. 95-98; and in H. Israel, *et al.*, editors., *Hundert Autoren Gegen Einstein*, R. Voigtländer, Leipzig, (1931), pp. 75-78.

In 1922, Stjepan Mohorovičić acknowledged what Albert Einstein did not,

> "I must point out what is little known, that the French physicist H. Poincaré had already called attention to the fact that the Lorentz Transformations form a group, he had already shown in 1900 (therefore 5 years before Einstein) [*Footnote:* See the book, which is cited in note 22 {M. Abraham, *Theorie der Elektrizität*, Volume 2, Fourth Edition, Leipzig, Berlin, 1920}, S. 359. It appears that Poincaré did not mention Einstien even once in his lecture '*The New Mechanics*' (Leipzig, Berlin, 1911) for this reason.], how one can set clocks by means of light signals to Lorentz' local time. [***] Therefore we must understand the method of signaling (which, as we have stressed, H. Poincaré had already applied in 1900) only as an interpretation of Lorentz' formulas."

> "Ich muß darauf hinweisen, was weniger bekannt ist, daß schon der französische Physiker H. Poincaré darauf aufmerksam gemacht hat, *daß die Lorentzschen Transformationen eine Gruppe bilden*; er hat schon 1900 (also 5 Jahre vor Einstein) gezeigt [*Footnote:* Siehe das Buch, welches in Anmerkung 22 zitiert ist {M. Abraham, *Theorie der Elektrizität*. II. Bd. 4. Aufl. Leizig-Berlin 1920}, S. 359. Es scheint, daß deswegen Poincaré in seinem Vortrage »*Die neue Mechanik*« (Leipzig-Berlin 1911) Einstein nicht einmal erwähnt.], wie man die Uhren mittels der Lichtsignale auf die Lorentzsche Ortszeit richten kann. [***] [D]eswegen müssen wir die Methode der Signalisierung (welche — wie wir betont haben — schon H. Poincaré 1900 aufgebracht hat), nur als eine Interpretation der Lorentzschen Formeln auffassen[29])."[614]

---

608. A. Lynch, *The Case Against Einstein*, P. Allan, London, (1932).
609. J. Mackaye, *The Dynamic Universe*, Charles Scribner's Sons, New York, (1931).
610. H. Nordenson, *Relativity, Time and Reality: A Critical Investigation of the Einstein Theory of Relativity from a Logical Point of View*, Allen and Unwin, London, (1969).
611. L. Essen, *The Special Theory of Relativity: A Critical Analysis*, Clarendon Press, Oxford, (1971); "Relativity — Joke or Swindle?", *Electronics and Wireless World*, (February, 1988), pp. 126-127. <http://www.cfpf.org.uk/articles/scientists/essen.html>
612. W. Theimer, *Die Relativitätstheorie, Lehre-Kritik-Wirkung*, Francke, Munich, (1977).
613. B. J. Gut, *Immanent-logische Kritik der Relativitätstheorie*, Kugler, Oberwil, Switzerland, (1981).
614. S. Mohorovičić, *Die Einsteinsche Relativitätstheorie und ihr mathematischer, physikalischer und philosophischer Charakter*, Walter de Gruyter & Co., Berlin, Leipzig, (1923), pp. 23-24, 30.

Stjepan Mohorovičić acknowledged Poincaré's priority for realizing that the Lorentz Transformations form a group. Mohorovičić cites Max Abraham's acknowledgment of Poincaré's priority for the clock synchronization method with light signals,[615] and asserts that Poincaré did not mention Einstein even once in his lecture *Die neue Mechanik* (*La mécanique nouvelle* = *The New Mechanics*),[616] because Einstein had plagiarized Poincaré's method of synchronizing clocks with light signals, which method is but an interpretation of Lorentz' "Ortszeit", and Poincaré's assertion of the group properties of the Lorentz Transformation.[617]

Felix Klein had made similar assertions in a private letter to Wolfgang Pauli on 8 March 1921, that Poincaré was the first to recognize that the Lorentz Transformations form a group and that Poincaré felt an animosity towards Einstein, and this was the only explanation for the fact that Poincaré snubbed Einstein in Poincaré's Göttingen lecture on the new mechanics. Klein wrote,

"Es ist nun doch einmal so, daß Poincarés erste Note in den Comptes Rendus 140 vor Einstein liegt und er im Anschluß daran (in den Rendiconti di Palermo) zuerst zeigte, daß es sich bei Lorentz um eine *Gruppe* von Transformationen handele. Von da aus ein Gegensatz, der allein es verständlich macht, daß P[oincaré] 1911 in seinem Göttinger Vortrag „sur

---

615. M. Abraham, *Theorie der Elektrizität*, Fourth Edition, Volume 2 ("Elektromagnetische Theorie der Strahlung"), Leipzig, Berlin, B. G. Teubner, (1920), pp. 350-390, at 359. The Third Edition of 1914 credits Poincaré at pp. 365-368.

616. H. Poincaré, "La Mécanique Nouvelle", *Comptes Rendus des Sessions de l'Association Française pour l'Avancement des Sciences*, Conférence de Lille, Paris, (1909), pp. 38-48; *La Revue Scientifique*, Volume 47, (1909), pp. 170-177; reprinted in H. Poincaré, *La Mécanique Nouvelle: Conférence, Mémoire et Note sur la Théorie de la Relativité / Introduction de Édouard Guillaume*, Gauthier-Villars, Paris, (1924), pp. 18-76 URL:

<http://gallica.bnf.fr/scripts/ConsultationTout.exe?E=0&O=N029067>

and **28 April 1909 Lecture in Göttingen:** "La Mécanique Nouvelle", *Sechs Vorträge über der reinen Mathematik und mathematischen Physik auf Einladung der Wolfskehl-Kommission der Königlichen Gesellschaft der Wissenschaften gehalten zu Göttingen vom 22.-28. April 1909*, B. G. Teubner, Berlin, Leipzig, (1910), pp. 51-58; "The New Mechanics", *The Monist*, Volume 23, (1913), pp. 385-395; **13 October 1910 Lecture in Berlin:** "Die neue Mechanik", *Himmel und Erde*, Volume 23, (1911), pp. 97-116; *Die neue Mechanik*, B. G. Teubner, Berlin, Leipzig, (1911);

617. S. Mohorovičić, "Über die räumliche und zeitliche Translation", »Bulletin« d. süslaw. Akad. D. Wiss." (*Jugoslovenska Akademija Znanosti i Umjetnosti*), Volume 6/7, Zagreb, (1916-1917), p. 48; **and** "Die Folgerung der allgemeinen Relativitätstheorie und die Newtonsche Physik", *Naturwissenschaftliche Wochenschrift*, New Series, Volume 20, Jena, (1921), pp. 737-739. *See also:* as cited by Mohorovičić: E. Guillaume and C. Willigens, "Über die Grundlagen der Relativitätstheorie", *Physikalische Zeitschrift*, Volume 22, (1921), pp. 109-114; **and** E. Guillaume, "Graphische Darstellung der Optik bewegter Körper", *Physikalische Zeitschrift*, Volume 22, (1921), pp. 386-388.

la nouvelle mécanique" den Namen Einstein überhaupt nicht nennt."[618]

Poincaré's silence also caught the attention of Max Born, who stated,

"One of these series of lectures was given by Henri Poincare, April 22nd-28th 1909[.] The sixth lecture had the title 'La mécanique nouvelle.' It is a popular account of the theory of relativity without any formulae and with very few quotations. EINSTEIN and MINKOWSKI are not mentioned at all, only MICHELSON, ABRAHAM and LORENTZ. But the reasoning used by POINCARÉ was just that, which EINSTEIN introduced in his first paper of 1905, of which I shall speak presently. Does this mean that POINCARÉ knew all this before EINSTEIN? It is possible, but the strange thing is that this lecture definitely gives you the impression that he is recording LORENTZ' work."[619]

Arvid Reuterdahl also was aware that Poincaré resented Einstein,

"Professor Henri Poincaré, the famous French physicist and mathematician, advisedly ignores the name of Einstein in his lectures on 'Relativity'."[620]

And Johannes Riem reiterated the fact,

"Neben dieser Aufklärung durch die Presse ging dann eine wissenschaftliche Bekämpfung Einsteins, vor allem durch den Mathematiker und Ingenieur *Reuterdahl* am St. Thomas College, der selbst schon *vor* Einstein über Relativität gearbeitet und Einstein zu einer öffentlichen Aussprache aufgefordert hat, bei der dieser das Richterscheinen vorzog. Reuterdahl hat eine kleine leicht lesbare Broschüre im Journal seines College erscheinen lassen „Einstein und die neue Wissenschaft". Hierin untersucht er physikalisch die Grundlagen der neuen Lehre. Er zeigt seinen Landsleuten, wie schon lange vor Einstein zahlreiche Gelehrte das Richtige der Relativitätstheorie gefunden und diesem als Quelle gedient haben, ohne daß dieser auf diese seine Vorgänger hinwiese, so daß es ganz falsch ist, die Relativitätstheorie immer auf Einstein zurückzuführen, wie dies meist geschieht. Es ist dies so wenig berechtigt, daß z. B. Poincaré in seinen Vorlesungen über Relativität Einstein überhaupt nicht erwähnt. Quellenmäßig wird dann von Reuterdahl gezeigt, wie bedeutende Gelehrte die Einsteinsche Fassung der Relativitätstheorie als falsch bekämpfen und

---

[618]. F. Klein to W. Pauli, *Wissenschaftlicher Briefwechsel mit Bohr, Einstein, Heisenberg, u.a.* = *Scientific correspondence with Bohr, Einstein, Heisenberg, a.o.*, Document 10, Springer, New York, (1979), p. 27.

[619]. M. Born, *Physics in my Generation*, second revised edition, Springer-Verlag, New York, (1969), pp. 102-103.

[620]. A. Reuterdahl, *Einstein And The New Science*, Reprint from *The Bi-Monthly Journal of the College of St. Thomas*, Volume 9, Number 3, (July, 1921), p. 8.

ganz andere Ueberlegungen and die Stelle setzen, wie *Lenard, Gehrcke, Fricke, Mewes* es tun. Endlich untersucht er das Einsteinsche Gebäude selbst auf seine Zusammensetzung, seine Grundlagen und Haltbarkeit, und findet, daß es ein Spiel mit Worten und Begriffen ist, denen in der Physik nichts tatsächliches entspricht. Es wäre sehr lohnend, die kleine Schrift von 26 Seiten zu übersetzen."[621]

Alexander Moszkowski was very confused by the letter of recommendation Poincaré allegedly wrote for Einstein in 1911—which letter makes no mention of the theory of relativity.[622] Moszkowski saw this as a reversal of the animosity Poincaré demonstrated towards Einstein in Berlin in 1910. Moszkowski wrote in 1921, describing his belief that Poincaré had come to recognize the "lasting importance of Einstein's researches[,]" and had overcome any doubts about the accumulating number of hypotheses in the new mechanics,

On the 13th October 1910 a memorable event took place in the Berlin Scientific Association: Henri Poincaré, the eminent physicist and mathematician, had been announced to give a lecture in the rooms of the institute 'Urania'; an audience of rather meagre dimensions assembled. [***] It was at this lecture that we heard the name Albert Einstein pronounced for the first time. Poincaré's address was on the New Mechanics [***] At that time, early in 1916, only a few members of the Literary Society divined who it was that was enjoying their hospitality. In the eyes of Berlin, Einstein's star was beginning its upward course, but was still too near the horizon to be visible generally. My own vision, sharpened by the French lecture and by a friend who was a physicist, anticipated events, and already saw Einstein's star zenith, although I was not even aware at that time that Poincaré had in the meantime overcome his doubts and had fully recognized the lasting importance of Einstein's researches."[623]

Poincaré did not mention Einstein in his lecture and Moszkowski must have heard Einstein's name from his friend. Poincaré's resentment of Einstein had nothing to do with the *ad hoc* hypotheses of the new mechanics, which he attributed to Lorentz, but was instead purely a product of Einstein's plagiarism,

---

[621]. J. Riem, "Zu Einsteins Amerikafahrt", *Deutsche Zeitung*, (13 September 1921).
[622]. This letter is reproduced in French in C. Seelig, *Albert Einstein: Eine dokumentarische Biographie*, Europa Verlag, Zürich, Stuttgart, Wien, (1954), p. 163; English translation by M. Savill, *Albert Einstein: A Documentary Biography*, Staples Press, London, (1956), pp. 134-135. R. S. Shankland stated that the letter was in the Einstein Archives in Princeton in 1973, *cf.* "Conversations with Albert Einstein. II", *American Journal of Physics*, Volume 41, Number 7, (July, 1973), pp. 895-901, at 895. A partial English translation is found in A. Moszkowski, *Einstein: The Searcher*, Chapter 6, E. P. Dutton, New York, (1921), p. 231.
[623]. A. Moszkowski, *Einstein: The Searcher*, Chapter 6, E. P. Dutton, New York, (1921), pp. 1, 3.

which fact was acknowledged by the experts Felix Klein and Stjepan Mohorovičić.

Moszkowski was simply lying to his reading audience. He knew quite well that Poincaré, himself, was the father of the new mechanics and that Einstein had plagiarized it from Poincaré, though in 1904, Poincaré had generously attributed the "new mechanics" to Lorentz, before the Einsteins had published on the subject. Poincaré famously stated in 1904,

> "From all these results, if they are confirmed, would arise an entirely new mechanics, which would be, above all, characterised by this fact, that no velocity could surpass that of light, any more than any temperature could fall below the zero absolute, because bodies would oppose an increasing inertia to the causes, which would tend to accelerate their motion; and this inertia would become infinite when one approached the velocity of light."[624]

Moszkowski failed to emphasize the fact, which was known to him, that Poincaré was himself the father of this new mechanics and had coined the term in 1904. Poincaré did object to the growing number of *ad hoc* hypotheses, but Poincaré nevertheless created the special theory of relativity, and the Einsteins plagiarized the theory from him. The fact that Poincaré was aware of the fatal flaws in the theory, while the Einsteins irrationally pretended them away by deliberately confusing induction with deduction, does not change the fact that Poincaré created the theory and the Einsteins copied it directly from him. This proves that the Einsteins were not only opportunistic plagiarists, but that they were also incompetent and dishonest scientists.

Moszkowski wrote,

> "For the theory asks us to brush aside habits of thought that have claimed an hereditary position in pre-eminent minds. One of the foremost physicists, Henri Poincaré, had confessed as late as 1910 that it caused him the greatest effort to find his way into Einstein's new mechanics. Another whole year passed before he gave up his last doubts. Then he passed with flying colours into Einstein's camp, and recommended Einstein's appointment to the Professorship at Zürich, in conjunction with the discoverer of radium, Madame Curie, in an exuberant letter which may add its note of appreciation here:
> 'Herr Einstein,' so wrote the great Poincaré, 'is one of the most original minds that I have ever met. In spite of his youth he already occupies a very honourable position among the foremost savants of his time. What we marvel at in him, above all, is the ease with which he adjusts himself to new

---

[624]. H. Poincaré's St. Louis lecture from September of 1904, *La Revue des Idées*, 80, (November 15, 1905); "L'État Actuel et l'Avenir de la Physique Mathématique", *Bulletin des Sciences Mathématique*, Series 2, Volume 28, (1904), p. 302-324; English translation, "The Principles of Mathematical Physics", *The Monist*, Volume 15, Number 1, (January, 1905), pp. 1-24, at 16.

conceptions and draws all possible deductions from them. He does not cling tightly to classical principles, but sees all conceivable possibilities when he is confronted with a physical problem. In his mind this becomes transformed into an anticipation of new phenomena that may some day be verified in actual experience... . The future will give more and more proofs of the merits of Herr Einstein, and the University that succeeds in attaching him to itself may be certain that it will derive honour from its connexion with the young master.'"

Moszkowski simply lied when he claimed that Poincaré had a difficult time understanding the theory Poincaré himself had created. Moszkowski simply lied when he attributed the theory Henri Poincaré had created to his plagiarist friend, who promised to make him rich, Albert Einstein.

A letter of recommendation would have been a matter of course and found no counterpart in Poincaré's published works. This alleged recommendation of Einstein was never met with public or private praise in the context of the theory of relativity, and it was Poincaré's nature to give such praise, which he so lavished on an undeserving Lorentz at every opportunity. Moszkowski made no such attack on Poincaré until after Poincaré had died and Moszkowski, who was a career sycophant, had made it his life's work to promote Einstein as a cult figure and in so doing promote himself and make his fortune. Alexander Moszkowski was biased and sought desperately to promote Einstein to the public. He wrote to Albert Einstein on 1 February 1917,

"Regardless of what happens, I would like to continue the 'cult'; for you it is secondary, for me it is of paramount importance in life. Additionally, I have the encouraging feeling that, with my modest writing abilities, I may also serve the cause once in a while."[625]

We know that Moszkowski's book of 1921 was deliberately deceitful, because he expressed very different feelings towards Poincaré in 1916 and 1917.[626] Moszkowski's more immediate impression of Poincaré's lecture, in 1911, is on record,

"Am humansten verfährt eigentlich noch Henri Poincaré, und unter den Büchern mit sieben Siegeln, die er sonst zu schreiben pflegt, ist seine Schrift über „Die neue Mechanik" noch das offenste. Anstatt von vornherein mit dem Geschütz unheimlicher Differentialgleichungen vorzurücken, vermenschlicht er die Aufgabe durch Einführung jenes Beobachters „Lumen", der uns zuerst von Camille Flammarion vorgestellt worden ist.

---

625. A. Moszkowski to A. Einstein, translated by A. M. Hentschel, *The Collected Papers of Albert Einstein*, Volume 8, Document 292, Princeton University Press, (1998), p. 281.
626. A. Fürst and A. Moszkowski, "Der Herr Lumen", *Das Buch der 1000 Wunder*, Section 187, Albert Langen, München, (1916), pp. 254-257. A. Moszkowski, *Der Sprung über den Schatten*, Albert Langen, München, (1917), pp. 213-219.

Mit diesem Lumen, „wie ich ihn sehe" wollen wir uns zunächst ein wenig beschäftigen."[627]

Though much has been made of Einstein's allegedly kinematic versus Poincaré's allegedly dynamic expositions of length contraction, which some assert indicates that Poincaré failed to understand the special theory of relativity, the facts are that Poincaré originated Einstein's plagiarized "kinematic" descriptions of length contraction and Poincaré went further by attempting a dynamic exposition of length contraction. This proves that Poincaré was the greater mind of the two, with the greater insight into the problem. Physics, as opposed to purely illustrative abstraction, compels a dynamic explanation for the physical dynamic interactions of matter in relative motion. To speak in terms of space and time without referring to physical bodies is scientifically meaningless.

It was Poincaré who first provided the quadri-dimensional exposition of length contraction, which Minkowski adopted, and which Einstein opposed for some time, and further which is truly the modern method of the theory of relativity as a mathematical formalism—a method of exposition which Einstein failed to understand for years, then when Minkowski published it in a form Einstein could almost understand, Einstein still opposed it for many years. Poincaré provided the conventionalist pseudo-kinematic exposition, the operational procedure and the space-time definition of length contraction, before Einstein and Minkowski manipulated credit for his ideas; and in 1909 Mittag-Leffler wrote to Poincaré that Ivar Fredholm recognized Poincaré's priority.[628] The fact that Poincaré actually attempted to interject Physics back into this mathematical formalism, Metaphysics, conventionalism and operationalism, does not eradicate his proven priority for the rest of the theory, nor would a change of mind erase what he had once stated from the historic record or the minds of the plagiarists.

Those who deny Poincaré's priority based on perceived flaws in his theories which allegedly do not render the "perfect" theory of special relativity, *i. e.* the Einsteins' "two postulate" fallacy of *Petitio Principii*, do not deny Einstein's priority even when it is pointed out to them that the Einsteins' 1905 paper is not the modern form of the theory and contains numerous mistakes. These apologists for Einstein operate on a double standard. They also fail to realize that the special theory of relativity is an evolving theory and has yet to be perfected, and no arbitrary point can be selected along this evolution and legitimately be called the first publication of the special theory of relativity.

Long before Einstein, Poincaré recognized the group properties of the Lorentz Transformation, perhaps as early as 1904, and wrote to Lorentz about his findings in a letter which is reproduced in Arthur I. Miller's *Albert Einstein's*

---

627. A. Moszkowski, "Das Relativitätsproblem", *Archiv für systematische Philosophie*, New Series, Volume 17, Number 3, (1911), pp. 255-281, at 258-259.
628. S. Walter, "Minkowski, Mathematicians, and the Mathematical Theory of Relativity", in H. Goenner, et al., Editors, *The Expanding Worlds of General Relativity*, Birkauser, Boston, (1999), pp. 45-86.

Special Theory of Relativity: Emergence (1905) and Early Interpretation, 1905-1911, Addison-Wesley, Reading, Massachusetts, (1981), p. 81. Poincaré almost certainly wrote to the Einsteins, because it is highly doubtful that the Einsteins knew what a group was. Poincaré published this mathematical discovery in the *Comptes Rendus* on 5 June 1905 before the Einsteins had submitted their paper to the *Annalen der Physik*, and long before the final paper of the Einsteins was published—perhaps published with modifications. It was ludicrous for Moszkowski to claim that Poincaré failed to grasp what he had created and what Albert Einstein had openly opposed.

Olivier Darrigol stated in 1996,

"The physicist-historian and the philosopher-historian usually argue that Einstein's new kinematics was an extremely important innovation that overthrew previous physical and philosophical concepts of time; and they tend to interpret Poincaré's, Lorentz's, and others' fidelity to the ether as a failure to understand Einstein's superior point of view. On the contrary, the social historian would argue that in 1905 Einstein's relativity had no stabilized meaning, that it could be read and used in various manners depending on the receiving local culture, and that it acquired a precise meaning only at the end of a complex, social structuring process."[629]

In 1922, Ludwig Lange, who had fought so hard for so long against so many, sought, without success, for acknowledgment of his parentage of the inertial system concept, which he published some twenty years before the Einsteins' absolutism. Lange wrote, *inter alia*,

"Als ich 1886 meine fünf Jahre lang fortgesetzen Forschungen über den Bewegungsbegriff abgeschlossen, in denen ich die relativistische Weiterentwicklung richtig vorausgesagt, im wesentlichen so, wie sie seitdem sich vollzogen hat, da harrte ich mit große Spannung, aber jahrelang vergeblich auf die werktätige Teilnahme der Physikerwelt. [***] Als ich nunmehr 1902 in der *Wundt-Festschrift* meine Revision des Systems der Inertialbegriffe herausgebracht hatte, überkam mich ein wohltuendes Gefühl der Befreiung, wie ich mir denke, daß es einer umfassenden und dabei nicht im mindesten zerknirschten Beichte auch sonst folgen mag. Von diesem Zeitpunkt an mußten aber immer noch drei weitere Jahre verstreichen, ehe mit Albert Einstein eine Denkrichtung unter den Physikern sich Bahn zu brechen begann, welche, wenn auch nur indirekt, auf

---

629. O. Darrigol, "The Electrodynamic Origins of Relativity Theory", *Historical Studies in the Physical and Biological Sciences: HSPS*, Volume 26, Number 2, (1996), pp. 241-312; which is reprinted as Chapter 9 of Darrigol's *Electrodynamics from Ampère to Einstein*, Oxford University Press, (2000); as quoted in S. Abiko, "On Einstein's Distrust of the Electromagnetic Theory: the Origin of the Light Velocity Postulate", *Historical Studies in the Physical and Biological Sciences: HSPS*, Volume 33, Number 2, (2003), pp. 193-215, at 200.

verwandten Gedankengängen aufzubauen unternahm, und ein viertes Jahr mußte hinzukommen, bis H. v. Seeliger (1906) in der Astronomie meine Nomenklatur „Inertialsystem" mit dem erfolg einführte, daß sie sich seitdem bei seinen Fachgenossen nahezu völligdurchgesetz zu haben scheint, während in der Physikfreilich erst die Ansätze dazu wahrzunehmen sind; denn Einstein selber und sein Anhang sträuben sich aus unverständlichen Gründen immer noch dagegen, eine so bequeme und charakteristische Bezeichnungsweise anzuwenden. Nun, die Zeit wird kommen, wo man mich als den Vater jener Nomenklatur und als den sorgfältigen Analysator des Sprachgebrauches der Mechanik, der die Wichtigkeit der relativistischen Richtung für die Physik besonders früh erkannte, nach Verdienst schätzen wird."[630]

Friedrich Kottler, author of *Gravitation und Relativitätstheorie*[631] in 1903, revealed on March 31st, 1922, through the prestigious, widely read and well-respected *Encyklopädie der mathematischen Wissenschaften*,

"*H. Poincaré,* Palermo Rend. Circ. Math. 21 (1906), p. 129-175, especially p. 175, Formula (14). — This work of *Poincaré's* is dated July 23, 1905 and is the elaboration of a memorandum by the same title in the Parisian C. R. 140 (June 5, 1905), pp. 1504-8. The 'postulate' of relativity was enunciated here for the first time, *before Einstein.*"

"*H. Poincaré,* Palermo Rend. Circ. Math. 21 (1906), p. 129-175, insbes. p. 175, Formel (14). — Diese Arbeit *Poincarés* stammt vom 23. Juli 1905 und ist die Ausarbeit einer Note gleichen Titels aus den Paris C. R. 140 (5. Juni 1905), p. 1504-8. Hier wurde zum erstenmal, *vor Einstein*, das „Postulat" der Relativität ausgesprochen."[632]

In 1923, Einstein's plagiarism became an international scandal, and some called for the revocation of his Nobel Prize. Thomas Jefferson Jackson See made a statement on 12 April 1923 picked up by the Associated Press and published in *The New York Times*,

"Professor Westin charges Einstein with downright plagiarism, saying: 'From these facts the conclusion seems inevitable that Einstein cannot be regarded as a scientist of real note. He is not an honest investigator.' Thus Westin protested to the Directorate of the Nobel Foundation against the

---

630. L. Lange, "Mein Verhältnis zu Einstein's Weltbild", *Psychiatrisch-neurologische Wochenschrift*, Volume 24, Number 29/30, (1922), pp. 188-189.
631. S. Oppenheim and F. Kottler, *Kritik des Newton'schen Gravitationsgesetzes; mit einem Beitrag: Gravitation und Relativitätstheorie von F. Kottler*, Deutsche Staatsrealschule in Karolinenthal, Prag, (1903).
632. *Encyklopädie der mathematischen Wissenschaften*, 6, 2, 22a, p. 171, note (13).

reward of Einstein."[633]

T. J. J. See published numerous articles accusing Albert Einstein of plagiarism.[634]

---

633. T. J. J. See, quoted in, "Prof. See Attacks German Scientist, Asserting That His Doctrine Is 122 Years Old", *The New York Times*, 13 April 1923, Section 1, p. 5.
634. T. J. J. See, "Einstein's Theory of Gravitation", *The Observatory*, Volume 39, (1916), pp. 511-512; *See also:* J. Riem, "Das Relativitätsgesetz", *Deutsche Zeitung*, Number 286, (26 June 1920). *See also:* "Prof. See Attacks German Scientist, Asserting That His Doctrine Is 122 Years Old", *The New York Times*, (13 April 1923), p. 5; **and** T. J. J. See, "Einstein a Second Dr. Cook?", *The San Francisco Journal*, (13 May 1923), pp. 1, 6; **and** (20 May 1923), p. 1; "Einstein a Trickster?", *The San Francisco Journal*, (27 May 1923); response by R. Trumpler, "Historical Note on the Problem of Light Deflection in the Sun's Gravitational Field", *Science*, New Series, Volume 58, Number 1496, (1923), pp. 161-163; reply by See, "Soldner, Foucault and Einstein", *Science*, New Series, Volume 58, (1923), p. 372; rejoinder by L. P. Eisenhart, "Soldner and Einstein", *Science*, New Series, Volume 58, Number 1512, (1923), pp. 516-517; rebuttal by A. Reuterdahl, "The Einstein Film and the Debacle of Einsteinism", *The Dearborn Independent*, (22 March 1924), p. 15; **and** T. J. J. See, "New Theory of the Ether", *Astronomische Nachrichten*, Volume 217, (1923), pp. 193-283. *See also:* "Is the Einstein Theory a Crazy Vagary?", *The Literary Digest*, (2 June 1923), pp. 29-30. *See also:* R. Morgan, "Einstein Theory Declared Colossal Humbug by U.S. Naval Astronomer", *The Dearborn Independent*, (21 July 1923), p. 14. *See also:* "Prof. See Attacks German Scientist Asserting that his Doctrine is 122 Years Old", *The New York Times*, Section 1, (13 April 1923), p. 5. *See also:* "Einstein Geometry Called Careless", *The San Francisco Journal*, (14 October 1924). *See also:* T. J. J. See, "Is Einstein's Arithmetic Off?", *The Literary Digest*, Volume 83, Number 6, (8 November 1924), pp. 20-21. *See also:* "Navy Scientist Claims Einstein Theory Error", *The Minneapolis Morning Tribune*, (13 October 1924). Ironically, Reuterdahl accused See of Plagiarizing his exposure of Einstein's plagiarism in America, first recognized by Gehrcke and Lenard in Germany! "Reuterdahl Says See Takes Credit for Work of Others", *The Minneapolis Morning Tribune*, (14 October 1924); **and** "A Scientist Yields to Temptation", *The Minneapolis Journal*, (2 February 1925). *See also:* "Prof. See declares Einstein in Error. Naval Astronomer Says Eclipse Observations Fully Confirm Newton's Gravitation Theory. Says German began Wrong. A Mistake in Mathematics is Charged, with 'Curved Space' Idea to Hide it." *The New York Times*, (14 October 1924), p. 14; responses by Eisenhart, Eddington and Dyson, *The New York Times*, (16 October 1924), p. 12. *See also:* "Captain See vs. Doctor Einstein", *Scientific American*, Volume 138, (February 1925), p. 128; **and** T. J. J. See, *Researches in Non-Euclidian Geometry and the Theory of Relativity: A Systematic Study of Twenty Fallacies in the Geometry of Riemann, Including the So-Called Curvature of Space and Radius of World Curvature, and of Eighty Errors in the Physical Theories of Einstein and Eddington, Showing the Complete Collapse of the Theory of Relativity*, United States Naval Observatory Publication: Mare Island, Calif. : Naval Observatory,(1925). *See also:* "See Says Einstein has Changed Front. Navy Mathematician Quotes German Opposing Field Theory in 1911. Holds it is not New. Declares he himself Anticipated by Seven Years Relation of Electrodynamics to Gravitation", *The New York Times*, Section 2, (24 February 1929), p. 4. See refers to his works: *Electrodynamic Wave-Theory of Physical Forces*, Thos. P. Nichols, Boston, London, Paris, (1917); **and** *New Theory of the Aether*, Inhaber Georg Oheim, Kiel,

See's quote originates from Arvid Reuterdahl's article in *The Dearborn Independent* of 6 January 1923, in which Reuterdahl gives the fuller translation,

> "From these facts the conclusions seem inevitable that Einstein cannot be regarded as a scientist of real note; that he is not an honest investigator; and that no valid reason can be assigned for awarding him the Nobel premium. It behooves the Nobel directorate carefully to examine all the charges of plagiarism made against him before taking an irrevocable step which later may be regretted."

In 1923, Arvid Reuterdahl published two long letters in *The New York Times* spelling out the case against Einstein and declared,

> "No unprejudiced person can deny that, in the absence of direct and incontrovertible proofs establishing his innocence, Einstein must, in view of the circumstantial evidence previously presented, stand convicted before the world as a plagiarist."[635]

Reuterdahl also published numerous articles accusing Einstein of plagiarism, the plagiarism of Reuterdahl's works, as well as those of others.[636] Reuterdahl challenged Einstein to a debate over his priority and the soundness of the theory of relativity.[637] Reuterdahl's challenge was heavily covered by the international

---

(1922); **and** "New Theory of the Ether", *Astronomische Nachrichten*, Volume 217, (1923), pp. 193-283.

635. A. Reuterdahl, "The Origin of Einsteinism", *The New York Times*, Section 7, (12 August 1923), p. 8. Reply to F. D. Bond's response, "Reuterdahl and the Einstein Theory", *The New York Times*, Section 7, (15 July 1923), p. 8. Response to A. Reuterdahl, "Einstein's Predecessors", *The New York Times*, Section 8, (3 June 1923), p. 8. Which was a reply to F. D. Bond, "Relating to Relativity", *The New York Times*, Section 9, (13 May 1923), p. 8. Which was a response to H. A. Houghton, "A Newtonian Duplication?", *The New York Times*, Section 1, Part 1, (21 April 1923), p. 10. *See also:* A. Reuterdahl, "Einstein and the New Science", *Bi-Monthly Journal of the College of St. Thomas*, Volume 11, Number 3, (July, 1921).

636. A. Reuterdahl, "The Origin of Einsteinism", *The New York Times*, Section 7, (12 August 1923), p. 8. Reply to F. D. Bond's response, "Reuterdahl and the Einstein Theory", *The New York Times*, Section 7, (15 July 1923), p. 8. Response to A. Reuterdahl, "Einstein's Predecessors", *The New York Times*, Section 8, (3 June 1923), p. 8. Which was a reply to F. D. Bond, "Relating to Relativity", *The New York Times*, Section 9, (13 May 1923), p. 8. Which was a response to H. A. Houghton, "A Newtonian Duplication?", *The New York Times*, Section 1, Part 1, (21 April 1923), p. 10. ***See also:*** A. Reuterdahl, "Einstein and the New Science", *Bi-Monthly Journal of the College of St. Thomas*, Volume 11, Number 3, (July, 1921). ***See also:*** J. T. Blankart, "Relativity of Interdependence; Reuterdahl's Theory Contrasted with Einstein's", *Catholic World*, Volume 112, (February, 1921), pp. 588-610.

637. "Challenges Prof. Einstein: St. Paul Professor Asserts Relativity Theory Was Advanced in 1866", *The New York Times*, (10 April 1921), p. 21.

press at the time. Einstein refused to accept the challenge.[638]

Reuterdahl made public the priority of Johann Heinrich (aka J. Henri) Ziegler over Einstein. Ziegler lectured in Switzerland while Einstein lived there and while Einstein was developing his copy of Lorentz' theory. Ziegler asserted his priority over Einstein and accused Einstein of plagiarizing his work,

> "Now if it was already suspicious that the antedated 'hypothesis' of the constancy of the speed of light appears in Einstein's theory, then the new Einsteinian discovery of the replacement of the nonsensical æther by the integral primal atom of light and empty space must now appear to us beyond any doubt as an instance of plagiarism, though admittedly based on poor understanding. One can compare the premature, purely mathematical plagiarism to the copying of a Raphael painting by a modern cubist, where only the sharpest eye is still able to discover the resemblance with the original, but in the present case it was an attempt at an exact copy by a dull-witted incompetent."

> "War nun schon jene „Annahme'' von der Konstanz der Lichtgeschwindigkeit in Einstein's Theorie verdächtig, so muß uns jetzt die neue Einstein'sche Entdeckung von der Ersetzbarkeit des sinnlosen Äthers durch die vollen Urlichtatome und den leeren Raum als ein ganz zweifelloses Plagiat erscheinen, aber allerdings als ein immer noch schlecht verstandenes. Das frühere, rein mathematische Plagiat kann man mit der Kopie eines Raphael'schen Gemäldes durch einen modernen Kubisten vergleichen, bei der nur schärfste Auge noch eine Ähnlichkeit mit dem Original zu entdecken vermag, das jetzige dagegen gleicht bereits einer gut gemeinten Kopie durch einen Stümper."[639]

In 1927, Hans Thirring wrote,

> "H. Poincaré had already completely solved the problem of time several years before the appearance of Einstein's first work (1905). Beginning with an article in Revue de Métaphysique et de Morale which appeared in 1898 (later reprinted in his book 'The Value of Science' as a chapter on the concept of time), Poincaré settled the general problem of time from the physical standpoint and had already there referred to the fact that the principle of the constancy of the velocity of light serves as a basis for a definition of time. Poincaré, in his work 'La Théorie de Lorentz et le Principe de Réaction' [*Relevant citations and quotations found in*

---

638. "Challenges Prof. Einstein: St. Paul Professor Asserts Relativity Theory Was Advanced in 1866", *The New York Times*, (10 April 1921), p. 21. *See also:* "Einstein Charged with Plagiarism", *New York American*, (11 April 1921). *See also:* "Einstein Refuses to Debate Theory", *New York American*, (12 April 1921).
639. J. H. Ziegler, „*Das Ding an sich*'' *und das Ende der sog. Relativitätstheorie*, Weltformel-Verlag, Zürich, (1923), pp. 31-32.

*endnote*⁶⁴⁰], then defined Lorentz' local time (Fig. 23) as time, which time is to be measured with clocks synchronized by light signals."

"Die Klärung des Zeitproblems war schon mehrere Jahre vor dem

---

640. H. Poincaré, "La Théorie de Lorentz at le Principe de Réaction", *Archives Néerlandaises des Sciences Exactes et Naturelles*, Series 2, Volume 5, *Recueil de travaux offerts par les auteurs à H. A. Lorentz, professeur de physique à l'université de Leiden, à l'occasion du 25ᵐᵉ anniversaire de son doctorate le 11 décembre 1900*, Nijhoff, The Hague, (1900), p. 272:

"In order for the compensation to occur, the phenomena must correspond, not to the true time t, but to some determined *local time* t' defined in the following way.

I suppose that observers located at different points synchronize their watches with the aid of light signals; which they attempt to adjust to the time of the transmission of these signals, but these observers are unaware of their movement of translation and they consequently believe that the signals travel at the same speed in both directions, they restrict themselves to crossing the observations, sending a signal from A to B, then another from B to A. The local time t' is the time determined by watches synchronized in this manner.

If in such a case $1/K_0^{1/2}$ is the speed of light, and v the translation of the Earth, that I imagine to be parallel to the positive x axis, one will have: $\boldsymbol{t'} = \boldsymbol{t} - \frac{vx}{V^2}$."

"Pour que la compensation se fasse, il faut rapporter les phénomènes, non pas au temps vrai $\boldsymbol{t}$, mais à un certain *temps local* $\boldsymbol{t'}$ défini de la façon suivante.

Je suppose que des observateurs placés en différents points, règlent leurs montres à l'aide de signaux lumineux; qu'ils cherchent à corriger ces signaux du temps de la transmission, mais qu'ignorant le mouvement de translation dont ils sont animés et croyant par conséquent que les signaux se transmettent également vite dans les deux sens, ils se bornent à croiser les observations, en envoyant un signal de A en B, puis un autre de B en A. Le temps local t'est le temps marqué par les montres ainsi réglées.

Si alors $1/K_0^{1/2}$ est la vitesse de la lumière, et v la translation de la Terre que je suppose parallèle à l'axe des x positifs, on aura: $\boldsymbol{t'} = \boldsymbol{t} - \frac{vx}{V^2}$.

and *Electrité et Optique*, Gauthier-Villars, Paris, (1901), p. 530: "Allow me a couple of remarks regarding the new variable t': it is what Lorentz calls *the local time*. At a given point t and t' will not defer but by a constant, t' will, therefore, always represent the time, but the origin of the times being different for the different points serves as justification for his designation." "Disons deux mots sur la nouvelle variable t': c'est ce que Lorentz appelle *le temps local*. En un point donné t et t' ne différeront que par une constante, t' représentera donc toujours le temps mais l'origine des temps étant différente aux différents points: cela justifie sa dénomination." and from 1902, *Science and Hypothesis*, Dover, New York, (1952), p. 90: "There is no absolute time. When we say that two periods are equal, the statement has no meaning, and can only acquire a meaning by convention. Not only have we no direct intuition of the equality of two periods, but we have not even direct intuition of the simultaneity of two events occurring in two different places. I have explained this in an article entitled "Mesure du Temps."

Erscheinen von EINSTEINS grundlegender Arbeit (1905) durch H. POINCARÉ weitgehend vorbereitet worden. Dieser hatte zunächst in einem im Jahre 1898 in der Revue de Métaphysique et de Morale erscheinenen (später als Kapitel über den Begriff der Zeit in seinem Buche „Der Wert der Wissenschaft" abgedruckten) Artikel das allgemeine Zeitproblem vom physikalischen Standpunkt aus behandelt und hatte dort schon erwähnt, daß sich auf den Satz von der Konstanz der Lichtgeschwindigkeit eine Zeitdefinition gründen läßt. Er hat dann in einer Arbeit „La Théorie de LORENTZ et le principe de réaction" (Arch. Néerland. (2) Bd. 5. 1900, Lorentz-Festschrift) die LORENTZsche Ortszeit (Ziff. 23) als die Zeit definiert, die durch mit Lichtsignalen synchronisierte Uhren gemessen wird."[641]

On 7 February 1928, *The New York Times* reported on page 26,

"If [EINSTEIN] is the father of relativity, then LORENTZ is its grandfather."

In 1929, Robert P. Richardson published an extensive article on Einstein's plagiarism in *The Monist*, a publication famous for publishing the works of Mach, Hilbert, Poincaré, and others, from whom Einstein plagiarized,

"Thus, with what is known as the special theory, if we consider as paramount factor not the detail work but the guiding thoughts by which this was inspired, then the father of this special relativity theory was undoubtedly Henri Poincaré. [***] In the general theory of relativity the basic thought is that of Mach, viz. the replacement in dynamics of the law of gravitation by a law of motion. But in what Einstein built upon this basis the influence of Poincaré is again manifest. [***] And in view of all these facts one does not know at which to be most astounded: the magnanimity of Poincaré who was always over-anxious that there should be recognition of the labors of those who reaped where he himself had sown, the apathy of his friends after his death, or the peculiar attitude of Einstein and his coterie, exemplified by Born of Goettingen, who refers to Poincaré as one of those who 'collaborated' with Einstein in the development of the relativity theory!"[642]

Similar remarks are found in the writings of Haiser and Zettl.[643]

---

[641]. H. Thirring, "Elektrodynamik bewegter Körper und spezielle Relativitätstheorie", *Handbuch der Physik*, Volume 12, "Theorien der Elektrizität Elektrostatik", Springer, Berlin, (1927), p. 270, *footnote*.
[642]. R. P. Richardson, "Relativity and its Precursors", *The Monist*, Volume 39, (1929), pp. 126-152, at 136, 138.
[643]. F. Haiser, "Das Relativitätsprinzip", *Politisch-anthropoligische Revue*, Volume 19, (1920/1921), pp. 495-502. O. Zettl, "Die Idee der Relativität", *Der Weg*, Volume 1, (1924/1925), pp. 220-224, 249-254.

Accusations of plagiarism plagued Einstein throughout his career. *The New York Times* reported on 27 March 1931 on page 2 that Ira D. Edwards had attempted to sue Einstein for plagiarizing his book, which he had copyrighted in 1929. The *Times* reported that the suit was dismissed. It is difficult to prove accusations of plagiarism in a court of law, especially a specific instance of plagiarism, as opposed to a career-long pattern. This may be one reason why more individuals did not speak out against the plagiarist Einstein. They risked a defamation suit.

The *Dictionary of Scientific Biography*, in its article on Lorentz, states,

> "Einstein's 1905 special relativity paper provided Lorentz' theory with a physical reinterpretation. [***] Einstein deduced the Lorentz transformations and other results that had first been made known through Lorentz' and others' electron theories. [***] Lorentz admired, but never embraced, Einstein's 1905 reinterpretation of the equations of his electron theory. The observable consequences of his and Einstein's interpretations were the same, and he regarded the choice between them as a matter of taste. [***] Lorentz, and Einstein too, regarded the physical space of general relativity as essentially fulfilling the role of the ether of the older electron theory."[644]

This statement is very significant. It reveals that the ultimate "fiction" (Vaihinger's sense of the term in his *Die Philosophie des Als Ob*) of both Lorentz' and the Einsteins' theories is the same, with any distinctions between the two theories being *metaphysical* (truly just semantic) and not *scientific*—the theories make the same predictions; and are, therefore, *scientifically speaking*, indistinguishable. The Einsteins' theory is a quasi-positivistic mathematical analysis of Lorentz' synthetic physical theory—a "dimensional disguise" for it.[645] Albert Einstein did not grasp the distinction between Metaphysics and

---

[644]. *Dictionary of Scientific Biography*, Volume 8, Charles Scribner's Sons, New York, (1981), p. 498.

[645]. E. V. Huntington, "A New Approach to the Theory of Relativity", *Festschrift Heinrich Weber zu seinem siebzigsten Geburtstag am 5. März 1912 / gewidmet von Freunden und Schülern*, B. G. Teubner, Leipzig, (1912), pp. 147-169; reprinted "A New Approach to the Theory of Relativity", *Philosophical Magazine*, Series 6, Volume 23, Number 136, (April, 1912), pp. 494-513. *See also:* S. Mohorovičić, "Äther, Materie, Gravitation und Relativitätstheorie", *Zeitschrift für Physik*, Volume 18, Number 1, (1923), pp. 34-63, at 34. *See also:* H. Ives in, D. Turner and R. Hazelett, *The EINSTEIN Myth and the Ives Papers: A Counter-Revolution in Physics*, Devin-Adair, Old Greenwich, Connecticut, (1979). *See also:* L. Jánossy, "Über die physikalische Interpretation der Lorentz-Transformation", *Annalen der Physik*, Series 6, Volume 11, (1953), pp. 293-322; **and** *Theory of Relativity Based on Physical Reality*, Akademiai Kiadó, Budapest, (1971). *See also:* G. Builder, "Ether and Relativity", *Australian Journal of Physics*, Volume 11, (1958), pp. 279-; **and** "The Constancy of the Velocity of Light," *Australian Journal of Physics*, Volume 11, (1958), pp. 457-480; abridged form reprinted with bibliography in: *Speculations in Science and Technology*, Volume 2, (1971), p. 422.

science. He stated in 1930 that, "Science itself is metaphysics."[646] In this context, Hendrik B. G. Casimir stated,

"How[ever] brilliant Einstein's conception may have been, the quantitative treatment and the accompanying concretisation of the atomic concept [by Lorentz] proved to be a greater and as to its consequences more important occurrence."[647]

Einstein hid from the many accusations that his theory was metaphysical nonsense—an inconsistent jumble of fallacies of *Petitio Principii*—nothing but an excuse to plagiarize. Einstein conceded that he was overrated as a physicist, and that the cult of personality surrounding him was unjustified.[648] Einstein stated in 1921,

"The cult of individuals is always, in my view, unjustified. To be sure, nature distributes her gifts unevenly among her children. But there are plenty of the well-endowed, thank God, and I am firmly convinced that most of them live quiet, unobtrusive lives. It strikes me as unfair, and even in bad taste, to select a few of them for boundless admiration, attributing superhuman powers of mind and character to them. This has been my fate, and the contrast between the popular estimate of my powers and achievements and the reality is simply grotesque."[649]

A meeting was arranged to discuss Vaihinger's theory of fictions in 1920,

---

See also: S. J. Prokhovnic, *The Logic of Special Relativity*, Cambridge University Press, (1967); **and** *Light in Einstein's Universe: The Role of Energy in Cosmology and Relativity*, Dordrecht, Boston, D. Reidel Pub. Co., (1985). ***See also:*** K. Sapper, Editor, *Kritik und Fortbildung der Relativitätstheorie*, In Two Volumes, Akademische Druck- u. Verlagsanstalt, Graz, Austria, (1958/1962). ***See also:*** J. A. Winnie, "The Twin-Rod Thought Experiment," *American Journal of Physics*, Volume 40, (1972), pp. 1091-1094. M.F. Podlaha, "Length Contraction and Time Dilation in the Special Theory of Relativity—Real or Apparent Phenomena?", *Indian Journal of Theoretical Physics*, Volume 25, (1975), pp. 74-75. ***See also:*** M. Ruderfer, "Introduction to Ives' 'Derivation of the Lorentz Transformations'", *Speculations in Science and Technology*, Volume 2, (1979), p. 243. ***See also:*** D. Lorenz, "Über die Realität der FitzGerald-Lorentz Kontraktion", Zeitschrift für allgemeine Wissenschaftstheorie, Volume 13/2, (1982), pp. 308-312. ***See also:*** D. Dieks, "The 'Reality' of the Lorentz Contraction," *Zeitschrift für allgemeine Wissenschaftstheorie*, Volume 115/2, (1984), p. 341. ***See also:*** F. Winterberg, *The Planck Aether Hypothesis*, Gauss Press, Reno, Nevada, (2002), pp. 141-148.
646. A. Einstein quoted in "Einstein on Arrival Braves Limelight for Only 15 Minutes", *The New York Times*, (12 December 1930), pp. 1, 16, at 16
647. H. B. G. Casimir, "The Influence of Lorentz' Ideas on Modern Physics", in G. L. De Haas-Lorentz, Ed., *H. A. Lorentz: Impressions of His Life and Work*, North-Holland Publishing Company, Amsterdam, (1957), p. 168.
648. R. S. Shankland, "Conversations with Albert Einstein", *American Journal of Physics*, Volume 31, Number 1, (January, 1963), pp. 47-57, at 56.
649. A. Einstein, *Ideas and Opinions*, Crown, New York, (1954), p. 4.

and Einstein pledged that he would attend this meeting. Knowing that Einstein would be devoured in a debate over his mathematical fictions, which confused induction with deduction, Wertheimer and Ehrenfest helped Einstein fabricate an excuse to miss the meeting he had agreed to attend. Einstein was proven a liar.[650] He also hid from many other criticisms, and Einstein refused to answer T. J. J. See's many charges of plagiarism,[651] and refused to debate Reuterdahl or to answer his many charges of plagiarism.[652] When Robert Drill[653] criticized the theory of relativity, Einstein tried to persuade Max Born and Moritz Schlick to not respond to the critique, but if they did so, to hide from his arguments and merely ridicule Drill with insults.[654] Einstein hid from the French Academy of Sciences.[655] Einstein hid from Cardinal O'Connell.[656] Einstein hid from Dayton C. Miller's falsification of the special theory of relativity.[657] Einstein hid from Cartmel.[658] Miller hammered Einstein in the press over the course of many years. *The New York Times Index* lists several articles in which Miller's and William B. Cartmels' falsifications of the special theory of relativity are discussed.[659] Einstein and Lorentz were very worried by Miller's results and could not find

---

650. H. Goenner, "The Reaction to Relativity Theory. I: The Anti-Einstein Campaign in Germany in 1920", *Science in Context*, Volume 6, Number 1, (1993), pp. 107-133, at 111.
651. "Einstein Ignores Capt. See", *The New York Times*, (18 October 1924), p. 17.
652. "Challenges Prof. Einstein: St. Paul Professor Asserts Relativity Theory Was Advanced in 1866", *The New York Times*, (10 April 1921), p. 21. *See also:* "Einstein Charged with Plagiarism", *New York American*, (11 April 1921). *See also:* "Einstein Refuses to Debate Theory", *New York American*, (12 April 1921).
653. R. Drill, "Die Kultur der Haeckel-Zeit", *Frankfurter Zeitung*, (18 August 1919); **and** "Nachwort", *Frankfurter Zeitung*, (2 September 1919); **and** "Ordnung und Chaos. Ein Beitrag zum Gesetz von der Erhaltung der Kraft. I-II", *Frankfurter Zeitung*, (30 November 1919 / 2 December 1919).
654. *The Collected Papers of Albert Einstein*, Volume 9, Documents 198, 199 and 222, Princeton University Press, (2004).
655. *The New York Times*, (4 April 1922), p. 21.
656. "Cardinal Doubts Einstein", *The New York Times*, (8 April 1929), p. 4. *See also:* "Einstein Ignores Cardinal", *The New York Times*, (9 April 1929), p. 10. *See also:* "Cardinal Opposes Einstein", *The Chicago Daily Tribune*, (8 April 1929), p. 33. *See also:* "Cardinal Hits at Einstein Theory", *The Minneapolis Journal*, (8 April 1929). *See also:* "Cardinal Gives Further Views on Einstein", *Boston Evening American*, (12 April 1929). *See also:* "Cardinal Warns Against Destructive Theories", *The Pilot* [Roman Catholic Newspaper, Boston], (13 April 1929), pp. 1-2. *See also:* "Vatican Paper Praises Critic of Dr. Einstein", *The Minneapolis Morning Journal*, (24 May 1929).
657. M. Polanyi, *Personal Knowledge*, University of Chicago Press, (1958), p. 13. *See also:* A. Pais, *Subtle is the Lord*, Oxford University Press, (1982), pp. 113-114. *See also:* W. Broad and N. Wade, *Betrayers of the Truth: Fraud and Deceit in the Halls of Science*, Simon & Schuster, New York, (1982), p. 139.
658. *The New York Times*, (24 February 1936), p. 7. *See also:* "Calls Ether Reality; Differs with Einstein; Proof is Submitted", *The Chicago Tribune*, (23 February 1936).
659. *See also:* "Einstein Theory will be Refuted by an American", *The Chicago Tribune*, (24 October 1929), p. 18. *See also:* "Calls Ether Reality; Differs with Einstein; Proof is Submitted", *The Chicago Tribune*, (23 February 1936).

fault with them.[660] Einstein told R. S. Shankland not to perform an experiment which might falsify the special theory of relativity,

> "[Einstein] again said that more experiments were not necessary, and results such as Synge might find would be 'irrelevant.' [Einstein] told me not to do any experiments of this kind."[661]

Einstein knew he was caught at the Arbeitsgemeinschaft deutscher Naturforscher meeting in the Berlin Philharmonic, and wanted to run away from Germany. Einstein desired to hide from the Bad Nauheim debate, in which he had threatened to devour his opponents,[662] then Einstein—after being talked into appearing and after much hype promoting the event which attracted thousand of visitors—then Einstein, when losing the debate, ran away during the lunch break and again wanted to run away from Germany.[663] Einstein prospered from hype and had no legitimacy as a supposed "genius". The press rescued him again and again, while he hid. Einstein was unable to defend "his" theories in the light of strict scrutiny.

T. J. J. See wrote in *The San Francisco Journal*, on 13 May 1923, in an article entitled, "Einstein a Second Dr. Cook?":

> "THE Magazine and newspaper press for the last eight years has been so filled with systematic propaganda, undoubtedly organized and directed by Einstein and his agents, that the public has become familiar with the name of Einstein and with the phrase 'Theory of Relativity'. Not one lay person in a thousand has any idea what this all means; and as the people do not understand it, the phrases are passed on in joke, or assumed to represent something important in the higher lines of physical science. It is well known that about six years ago Einstein tried to cast a halo of glory about his head by allowing the report to go forth that not over twelve mathematicians in the world could understand his benighted theory of relativity. Of course this is preposterous, and nobody knows it better than Einstein himself. [***] In short, I have at length become convinced that Einstein is a faker, with considerable skill in deceiving the the press and public, so as to ding-dong

---

660. R. S. Shankland, "Conversations with Albert Einstein", *American Journal of Physics*, Volume 31, Number 1, (January, 1963), pp. 47-57; **and** "Conversations with Albert Einstein. II", *American Journal of Physics*, Volume 41, Number 7, (July, 1973), pp. 895-901.
661. R. S. Shankland, "Conversations with Albert Einstein", *American Journal of Physics*, Volume 31, Number 1, (January, 1963), pp. 47-57, at 54.
662. A. Einstein quoted in R. W. Clark, *Einstein: The Life and Times*, The World Publishing Company, (1971), p. 261; referencing A. Einstein to A. Sommerfeld, in A. Hermann. *Briefwechsel. 60 Briefe aus dem goldenen Zeitalter der modernen Physik*, Schwabe & Co., Basel, Stuttgart, (1968), p. 69.
663. A. Einstein, *Neues Wiener Journal*, (29 September 1920). C. Kirsten and H. J. Treder, *Albert Einstein in Berlin 1913-1933*, Akademie Verlag, Berlin, Volume 2, (1979), pp. 139, 205.

into the unthinking the idea that he is a great mathematician and philosopher, who is improving on Newton. Let us first notice the errors of Einstein, and the cunning way in which he gets away from them, owing to the layman's inability to pin him down."

T. J. J. See wrote in *The San Francisco Journal*, on 20 May 1923,

"No doubt is entertained by leading German physicists—like Professor Dr. E. Gehrcke, director of the Imperial Physical and Technical Institute of Berlin, and Dr. P. Lenard of Heidelberg, winner of the Nobel Prize in physics—that Einstein appropriated improperly the Newton-Soldner formula published 122 years before. Let the Einstein shouters explain these embarrassing coincidences if they can!

These unprofessional proceedings of Einstein have been a scandal in Europe for some time. The discussion rages all over Germany and, in fact, all over Europe. The revolt against Einstein extends from Spain to Russia, from Sweden to Italy. The learned and honored Professor Dr. Westin of Stockholm protested to the Nobel Foundation against any recognition of Einstein, accusing him of downright plagiarism, saying:

'From these facts the conclusion seems inevitable that Einstein cannot be regarded as a scientist of real note; he is not an honest investigator.'

To the present day, be it said to the honor of the Royal Swedish Academy of Sciences, they refused Einstein any recognition on the theory of relativity. Is it any wonder that the Paris Academy of Sciences (October 14, 1921) came out with conspicuous proclamations by Professors Picard and Painleve against Einsteinism, and in favor of Newtonian mechanics? It was near this time that Einstein visited Paris and sought to have the academy invite him to address the institute, though not a member of it. As this proposed proceeding was unprecedented, half a dozen leading academicians served notice on the officials of the institute that they would not have it, threatening to resign if the invitation were extended to Einstein. This put a stop to the display of Einstein planned for Paris. In fact, his reception there seems to have been quite a frost. The French are careful of the dignity of the Academy of Sciences, and in this respect they set a much better example than the Royal Society of London, which early championed Einsteinism and now is sorry for it."

T. J. J. See wrote in *The San Francisco Journal*, on 27 May 1923, in an article entitled, "Einstein a Trickster?"

"When the Lick eclipse work was reported to [Einstein], with my criticism, April 12, 1923, he admitted to the correspondent, Karl H. von Wiegand, April 14, 1923, that:

'In so far as precise measurement is concerned, Captain See may be said to be correct in denying that the tests proved the theory of relativity. But, he pointed out, under more favorable circumstances, even this might be removed.'

'Einstein said he was not worried by the attack of Captain See, but would leave it to the scientific world to settle the matter. It the fate of all scientists to arouse antagonism by revolutionary theories.' So feeble is [Einstein's] defense.

As I had recalled the charges of plagiarism made against him by Gehrcke, Leonard and Westin, it will be seen that he does not answer these charges, but adroitly evades them. Thus it looks as if he has no defense and he wishes not to discuss it. The above statement of glittering generalities show the weakness of [Einstein's] case—a tacit admission that he has no answer, and thus he prudently keeps still, hoping the public will forget the charges. So far as I can tell from the careful study of the whole business Einstein is a faker. Apparently he belongs in the company of Dr. Cook of Polar exploration noise and notoriety."

William Cardinal O'Connell gave a speech on 7 April 1929, which attracted a great deal of attention. He stated, *inter alia*:

"What does all this worked-up enthusiasm about Einstein mean? It evidently is a worked-up, fictitious enthusiasm, because I have never yet met a man who understood in the least what Einstein is driving at, and I have been so impressed by this fact I very seriously doubt that Einstein himself knows really what he means. Truth is always very clear when seen with a clear eye. The fact that any theory cannot be enunciated and only succeeds in befogging the mind, is a patent proof that it is not really truth. [***] [O]ne weakness of the American public is to run after novelties which have nothing in them but their newness. The American student body is very often misled into false channels of knowledge by the sudden appearance of these glittering meteors who from time to time shoot across the horizon. And then it seems there is some sort of organized clicque that boosts these sudden apparitions and as quickly disavows them and forgets them. [***] Now, for the moment, it is Einstein. Nobody knows what he is trying to reveal, but in a certain sense that adds mystery to his name[.] All this proves how careful the student youth must be in following this fanatical applause, which oftentimes is merely the outpouring of a sort of hero worship, but even as such can do endless harm to the impressionable mind of the young student."[664]

Cardinal O'Connell wrote in the 12 April 1929 edition of the *Boston Evening American*,

"I was rather amused the very next day to see by the Transcript that my

---

[664]. "Cardinal Warns Against Destructive Theories", *The Pilot* [Roman Catholic Newspaper, Boston], (13 April 1929), pp. 1-2

opinion of Einstein's theory and purpose had been conveyed to Einstein himself—that not he, but Frau Einstein, said that Einstein did not wish to dispute with me about his theories and that my assertions left him cold. That struck me [***] as little convincing as his general attitude to all, even the greatest scientists of Europe and America, who face him from time to time with indisputable proof of the fact that his so-called new theory of relativity is not new at all, but that whatever there is in it of scientific value is nothing but a plagiarism of Von Soldner's system explaining the deflection of light published as far back as about 1810. [***] Again and again Einstein has been faced with what appears to be clear proofs of plagiarism and absolute philosophic sophistry by the best minds in Germany, and his only answer to them is what he now answers, 'he is indifferent—it leaves him cold.'"

The Vatican newspaper *Observatore Romano* praised Cardinal O'Connell's criticism of Einstein and the theory of relativity in an editorial on 23 May 1929.[665]

Einstein's advocate, Albert von Brunn, boasted in 1931 that Einstein was not interested in "academic disputes" and presented this vice as if a virtue in order to excuse Einstein's inability to answer his critics. It was typical of the pattern of Einstein's apologists of turning Einstein's flaws into supposed virtues, his weaknesses into supposed strengths, through misguided heroic idolatry. Von Brunn wrote,

"Some reasonable critics in philosophy and physics have allowed themselves to be called in among these 'authors', with whom relativist scientists need not, and actually also do not consider it beneath their dignity to cross swords. (Although Einstein himself, by nature a pure scientist, is uninterested in such academic disputes!)"[666]

In 1931, Friedrich Jacob Kurt Geissler complained that Einstein had plagiarized his work on relativity theory, which included the relativity of time, space and simultaneity, and a relativistic analysis of mass, events and causality,

"It is completely wrong, that the expression 'theory of relativity' or even 'relativity' is inseparably tied to the name 'Einstein', as the immoderate advertising has accomplished with the lay public and some scholars. Newton has already expounded a great deal upon the relative and the absolute in Mathematics and in Physics. Modern physicists, like E. Mach, whose work Einstein knows quite well and uses, have written about generalizing the concepts of relative space, relative time and motion (long

---

[665]. "Vatican Paper Praises Critic of Dr. Einstein", *The Minneapolis Morning Journal*, (24 May 1929).

[666]. A. v. Brunn, *Die Naturwissenschaften*, Volume 19, Number 11, (13 March 1931), pp. 254-256; English translation by A. M. Hentschel appears in K. Hentschel, *Physics and National Socialism*, Birkhäuser, Basel, Boston, Berlin, (1996), p. 14.

before Einstein, 1865, 1901 'The Science of Mechanics; a Critical and Historical Account of Its Development' and later); Mansion (Paris 1863) holds that the notion of absolute motion is senseless and that the Ptolemaic and Copernican system are kinematically equally justified. Whereas Einstein first published something on relativity from 1905 on; I, myself, had already published an interdependent general 'feasible' theory of relativity in space, time, etc. in 1900; he, however, does not cite my book ('Eine mögliche Wesenserklärung... ')."

"Es ist grundverkehrt, den Ausdruck „Relativitätslehre" oder gar „Relativität" mit dem Namen „Einstein" als untrennbar zu kopulieren, wie es eine unmäßige Reklame beim Laienpublikum und einem Teil der Gelehrten fertig gebracht hat. Schon Newton spricht viel vom Relativen und Absoluten in der Mathematik und Physik. Moderne Physiker, wie E. Mach, den Einstein genau kennt und benutzt, haben über die Begriffe des relativen Raumes, der relativen Zeit und Bewegung verallgemeinernd geschrieben (längst vor Einstein, 1865, 1901 „Die Mechanik in ihrer Entwicklung" und später); Mansion (Paris 1863) hielt die absolute Bewegung für sinnlos und das Ptolemäische und Kopernikanische System für kinematisch gleichberechtigt. Eine zusammenhängende allgemeine „mögliche" Lehre der Relativität in Raum, Zeit usw. habe ich selbst schon 1900 veröffentlicht, während Einstein erst von 1905 ab einiges über Relativität veröffentlicht hat, mein Buch („Eine mögliche Wesenserklärung... ") aber nicht anführt."[667]

It is interesting to look for the source of the oft heard expression, "The Einstein Myth", which refers to the disingenuous glorification of Albert Einstein. The *Minneapolis Sunday Tribune* declared, on 10 April 1921, on page 11, that the "Einstein Theory of Relativity Is Branded Myth". Arvid Reuterdahl, a fine artist, produced a card which was distributed on the occasion of the "Albert Einstein Jubilee" at the Metropolitan Opera House in New York City on 16 April 1929 with a cartoon mockingly depicting a deified Einstein and his groveling sycophants, as well as a dignified dissenting physicist rejecting Einstein, on one side of the card, which declared on the other side,

"Einstein's message to the audience, by the Associated Press from Berlin: 'YOU MEET TO CELEBRATE A MYTH BEARING MY NAME.' Comment by Dissenting Scientist: 'THE TRUEST WORDS THAT EINSTEIN EVER SAID.'"[668]

---

[667]. I. K. Geissler, "Ringgenberg Schluss mit der Einstein-Irrung!", H. Israel, *et al*, Eds., *Hundert Autoren Gegen Einstein*, R. Voigtländer, Leipzig, (1931), pp. 10-12, at 10. Geissler refers to his book, *Eine mögliche Wesenserklärung für Raum, Zeit, das Unendliche und die Kausalität, nebst einem Grundwort zur Metaphysik der Möglichkeiten*, Gutenberg, Berlin, (1900).
[668]. Found in the Arvid Reuterdahl files in the Department of Special Collections,

On 27 November 1932, *The New York Times* published a letter by Melvin Green in section 2 on page 2 under the title, "The Einstein 'Myth.'" Melvin Green of Winchester, Virginia, wrote in his letter,

> "When I read some of Einstein's utterances, [***] and when I see all that he says taken as final absolute truth, I wonder whether we are not victims of an Einstein myth."

In 1979, Dean Turner and Richard Hazelett published a book exposing this myth, *The EINSTEIN Myth and the Ives Papers*.[669] Who first referred to the "Einstein Myth" may never be known for certain, but what is certain is that the theories are mythological and Albert Einstein was a career plagiarist.

On 23 February 1929 *The New York Times* on page 15 quoted Robert Andrews Millikan on the source of Einstein's work,

> "[Millikan] Traces Einstein's Contribution.
>
> 'Einstein in 1905 generalized [the result of the Michelson-Morley experiment] by postulating that it is in the nature of the universe impossible to find the speed of the earth with respect to the ether,' [Millikan] said. 'This postulate rests most conspicuously upon and historically grew chiefly out of the negative result of the Michelson-Morey [*sic*] experiment.[']"

Hans Reichenbach published an article "Einstein's Theory Traced to Sources" on 26 January 1929 in *The New York Times* on page 3 and stated,

> "This is the aim of Einstein's new theory, which he has now completed. [A New Field Theory]. It uses as an aid a peculiar mathematical source which, in its origin, goes back to the Zurich mathematician Weyl and the English astronomer Eddington."

The New York Times on 2 September 1936, in a story which begins on the front page, quoted Elie Joseph Cartan on page 16,

> "It is unnecessary to recall the great services which tensor analysis has rendered to geometry and to mathematical physics. Every one is aware that Einstein's general theory of relativity might not have been conceived had this admirable instrument of research not been created, under the name of 'absolute differential calculus,' by G. Ricci and T. Levi-Civita."

Sir Edmund Whittaker in his detailed survey, *A History of the Theories of*

O'Shaughnessy-Frey Library, University of St. Thomas, Minnesota.
669. Edited by, and with commentary from, D. Turner and R. Hazelett, *The EINSTEIN Myth and the Ives Papers: A Counter-Revolution in Physics*, Devin-Adair, Old Greenwich, Connecticut, (1979).

Aether and Electricity, Volume II, (1953), included a chapter entitled "The Relativity Theory of Poincaré and Lorentz". Whittaker thoroughly documented the development of the theory, documenting the authentic history, and demonstrated through reference to primary sources that Einstein held no priority for the vast majority of the theory. Einstein offered no counter-argument to Whittaker's famous book, in which the following passage appeared,

> "Einstein published a paper which set forth the relativity theory of Poincaré and Lorentz with some amplifications, and which attracted much attention. He asserted as a fundamental principle the *constancy of the velocity of light*, i.e. that the velocity of light *in vacuo* is the same in all systems of reference which are moving relatively to each other: an assertion which at the time was widely accepted, but has been severally criticized by later writers."[670]

Whittaker also wrote a biography of Einstein, in *Biographical Memoirs of Fellows of the Royal Society*, which reiterated the truth, that Einstein did not create the theory of relativity,

> "The aggregate of all the transformations so obtained, combined with the aggregate of all the rotations in ordinary space, constitutes a *group*, to which Poincaré* gave the name the group of *Lorentz Transformations*.
>   Einstein [***] adopted Poincaré's Principle of Relativity (using Poincaré's name for it) as a new basis for physics and showed that the group of Lorentz transformations provided a new analysis connecting the physics of bodies in motion relative to each other. Notable results appearing in this paper for the first time were the relativist formulae for aberration and also for the Doppler effect."[671]

Even among Einstein's admirers voices are heard which deny Einstein's priority. Max Born averred that,

> "Lorentz enunciated the laws according to which the measured quantities in various systems may be transformed into each other, and he proved that these transformations leave the field equations of the electron theory unchanged. This is the mathematical content of his discovery. Larmor (1900) and Poincaré (1905) arrived at similar results about the same time. It is interesting historically that the formula of transformation to a moving system, which we nowadays call Lorentz' transformation (see vi, 2, p. 200 formula (72)), were set up by Voigt as early as 1877 [*sic*[672]] in a dissertation

---

670. Sir Edmund Whittaker, *A History of the Theories of Aether and Electricity*, Volume II, Philosophical Library Inc., New York, (1954), p. 40.
671. *Biographical Memoirs of Fellows of the Royal Society*, Volume 1, The Royal Society, Headley Brothers LTD., (1955), pp. 37-67, at 42.
672. W. Voigt, "Ueber das Doppler'sche Princip", *Nachrichten von der Königlichen Gesellschaft der Wissenschaften und der Georg-Augusts-Universität zu Göttingen*,

which was still founded on the elastic theory of light. [***] In the new theory of Lorentz the principle of relativity holds, in conformity with the results of experiment, for all electrodynamic events."⁶⁷³

and,

"As mentioned already, Lorentz and Poincaré have succeeded in doing this by careful analysis of the properties of Maxwell's equations. They were indeed in possession of a great deal of mathematical theory. Lorentz, however, was so attached to his assumption of an ether absolutely at rest that he did not acknowledge the physical significance of the equivalence of the infinite numbers of systems of reference which he had proved. He continued to believe that one of them represented the ether at rest. Poincaré went a step further. It was quite clear to him that Lorentz's viewpoint was not tenable and that the mathematical equivalence of systems of reference meant the validity of the principle of relativity. He also was quite clear about the consequences of his theory."⁶⁷⁴

and,

"I have now to say some words about the work of these predecessors of EINSTEIN, mainly of LORENTZ and POINCARÉ. [***] H. A. LORENTZ' important papers of 1892 and 1895 on the electrodynamics of moving bodies contain much of the formalism of relativity. [***] POINCARÉ's papers [***] show that as early as 1899 he regarded it as very probable that absolute motion is indetectable in principle and that no ether exists. He formulated the same ideas in a more precise form, though without any mathematics, in a lecture given in 1904 to a Congress of Arts and Science at St. Louis, U.S.A., and he predicted the rise of a new mechanics which

---

(1887), pp. 41-51; reprinted *Physikalische Zeitschrift*, Volume 16, Number 20, (15 October 1915), pp. 381-386; English translation, as well as very useful commentary, are found in A. Ernst and Jong-Ping Hsu (W. Kern is credited with assisting in the translation), "First Proposal of the Universal Speed of Light by Voigt in 1887", *Chinese Journal of Physics* (The Physical Society of the Republic of China), Volume 39, Number 3, (June, 2001), pp. 211-230; URL's:

<http://psroc.phys.ntu.edu.tw/cjp/v39/211/211.htm>

<http://psroc.phys.ntu.edu.tw/cjp/v39/211.pdf>

See also: W. Voigt, "Theorie des Lichtes für bewegte Medien", *Nachrichten von der Königlichen Gesellschaft der Wissenschaften und der Georg-Augusts-Universität zu Göttingen*, (1887), pp. 177-238.

673. M. Born, *Einstein's Theory of Relativity*, Methuen & Co. Ltd., London, (1924), p. 188.

674. M. Born, *Einstein's Theory of Relativity*, Dover, New York, (1962), p. 224.

will be characterized above all by the rule, that no velocity can exceed the velocity of light. [***] The reasoning used by POINCARÉ was just that, which Einstein introduced in his first paper of 1905 [***] Does this mean that POINCARÉ knew all this before Einstein? It is possible [***] Many of you may have looked up his paper 'Zur Elektrodynamik bewegter Körper' in *Annalen der Physik* (4), vol. 17, p. 811, 1905, and you will have noticed some peculiarities. The striking point is that it contains not a single reference to previous literature. It gives you the impression of quite a new venture. But that is, of course, as I have tried to explain, not true."[675]

Einstein's friend, physicist Peter Gabriel Bergmann, asserted,

"The Dutch physicist, Hendrik Antoon Lorentz (1853-1928) contrived a theoretical scheme according to which absolute motion of physical objects, including measuring rods, should compress them in such a manner that differences in the speed of light remained undetectable by any conceivable apparatus. Jules Henri Poincaré (1854-1912), the French mathematician, suggested that the consistent failure to identify the frame representing absolute rest indicated that no such frame existed, and that Newton's scheme of the multiplicity of inertial frames was valid after all. In 1905, Einstein combined Lorentz' and Poincaré's ideas into a new approach to the issue of frames of reference and so was able to explain why no experiment had uncovered the absolute motion of the earth, without contradicting Maxwell's theory of electricity and magnetism."[676]

The Einsteins' 1905 paper failed to present references to the work it "combined" of Lorentz and Poincaré. That which was "new" in the "approach" is of minor significance. Poincaré's work was itself the combination of Lorentz' and Poincaré's ideas, which "combination" Mileva and Albert did not create, but simply repeated, parroting Poincaré's earlier works, virtually verbatim.
Prof. G. H. Keswani argued that,

"As far back as 1895, Poincaré the innovator had conjectured that it is impossible to detect absolute motion. In 1900 he introduced the 'The principle of relative motion' which he later called by the equivalent terms 'The law of relativity' and 'The principle of relativity' in his book *Science and Hypothesis* published in 1902. He further asserted in this book that there is no absolute time and that we have no intuition of the 'simultaneity' of two 'events' (mark the words) occurring at different places. In a lecture given in 1904, Poincaré reiterated the principle of relativity, described the method of synchronisation of clocks with light signals, urged a more satisfactory theory of the electrodynamics of moving bodies based on Lorentz's ideas

---

675. M. Born, "Physics and Relativity", *Physics in my Generation*, 2nd rev. ed., Springer, New York, (1969), pp. 101-103.
676. P. G. Bergmann, *The Riddle of Gravitation*, Scribner, New York, (1968), p. 29.

and predicted a new mechanics characterized by the rule that the velocity of light cannot be surpassed. This was followed in June 1905 by a mathematical paper entitled 'Sur la dynamique de l'électron' in which the connection between relativity (impossibility of absolute motion) and the Lorentz Transformation given by Lorentz a year earlier was recognized. In point of fact, therefore, Poincaré was not only the first to enunciate the principle, but he also discovered in Lorentz's work the necessary mathematical formulation of the principle. All this happened before Einstein's paper appeared."[677]

How do we account for the striking similarity between Lorentz' and Poincaré's writings and Einstein's words in both the "special" and "general" theories of relativity? Who published what, first? Was it mere coincidence that time after time Einstein repeated what Poincaré had earlier published? The record indicates that Poincaré held priority over Einstein, often by many years. Why is it that Albert's last name is a household word and is synonymous with "relativity", and Poincaré's name is substantially more obscure? Einstein believed,

"The secret to creativity is knowing how to hide your sources."[678]

9.3 The Æther

Many criticized Einstein's theories as metaphysical "nonsense", as purely mathematical fictions lacking physical content. As Arthur Eddington explained,

> "LET us suppose that an ichthyologist is exploring the life of the ocean. He casts a net into the water and brings up a fishy assortment. Surveying his catch, he proceeds in the usual manner of a scientist to systematise what it reveals. He arrives at two generalisations:
> (1) No sea-creature is less than two inches long.
> (2) All sea-creatures have gills.
> These are both true of his catch, and he assumes tentatively that they will remain true however often he repeats it.
> In applying this analogy, the catch stands for the body of knowledge which constitutes physical science, and the net for the sensory and intellectual equipment which we use in obtaining it. The casting of the net corresponds to observation; for knowledge which has not been or could not be obtained by observation is not admitted into physical science.

---

[677]. G. H. Keswani, "Origin and Concept of Relativity, Part I", *The British Journal for the Philosophy of Science*, Volume 15, Number 60, (February, 1965), pp.286-306, at 293-295.
[678]. This quote, which is widely attributed to Einstein, was brought to my attention in an anonymous e-mail—thank you, whoever you are!

An onlooker may object that the first generalisation is wrong. 'There are plenty of sea-creatures under two inches long, only your net is not adapted to catch them.' The icthyologist dismisses this objection contemptuously. 'Anything uncatchable by my net is *ipso facto* outside the scope of icthyological knowledge, and is not part of the kingdom of fishes which has been defined as the theme of ichtyological knowledge. In short, 'what my net can't catch isn't fish.' Or—to translate the analogy—'If you are not simply guessing, you are claiming a knowledge of the physical universe discovered in some other way than by the methods of physical science, and admittedly unverifiable by such methods. You are a metaphysician. Bah!'"[679]

The "ether", or "æther", is a hypothetical fluid, which may fill space and conduct electromagnetic waves such as light, and is perhaps an intervening medium between bodies, which causes gravity. Einstein tried to distinguish his work from Lorentz' by calling the æther "superfluous", which assertion Poincaré and countless others had long since enunciated. The existence of this "fluid" has been hotly disputed for thousands of years, but unless we deny dimension as an anthropomorphic delusion of consciousness, notional not real,[680] "space" as extension without "material" must be *something*. An empty box contains *something*, even if we evacuate the air from it. We can give this *something* any name we like, but changing its name is a matter of semantics, not discovery.

One cannot speak of "propagation" without tacitly or overtly referring to a medium, and the 1905 paper speaks of "propagation". As Sir Arthur Schuster stated,

> "Einstein, in a paper of great interest and power, has developed this idea, calling his imagined law 'The principle of relativity,' because it stipulates— *a priori*—that only the relative motion between material bodies can be detected. It is impossible for me to discuss in detail the reasoning by which this principle is justified, and an account without explanations of its consequences would lay me open to the charge that I was playing with your credulity. Suffice, therefore, it to say that strict adherers to the principle cannot admit the existence of an æther, and yet may speak of the transmission of light through space with a definite velocity. They must further accept, as a consequence of their dogma, that identical clocks placed

---

[679]. A. Eddington, *The Philosophy of Physical Science*, Ann Arbor Paperbacks, The University of Michigan Press, Second Printing, (1967), p. 16.

[680]. *Cf.* H. More, *A COLLECTION Of Several Philosophical Writings OF Dr. HENRY MORE, Fellow of* Christ's-College *in* Cambridge, Joseph Downing, London, (1712); which contains: *AN ANTIDOTE AGAINST ATHEISM: OR, An Appeal to the Natural Faculties of the Mind of Man, Whether there be not a GOD*, The Fourth Edition corrected and enlarged: WITH AN APPENDIX Thereunto annexed, "An Appendix to the foregoing Antidote," Chapter 7, pp. 199-201. F. H. Bradley, "In What Sense are Psychical States Extended?", *Mind*, New Series, Volume 4, Number 14, (April, 1895), pp. 225-235.

on two bodies moving with different velocities have different rates of going and that, even on the same body, identical clocks indicate different times, when the line joining their positions lies in the direction of motion. The motion must be determined relative to another body, which is supposed to be at rest, and a clock placed on that body must serve as the ultimate standard of time. The theory appears to have an extraordinary power of fascinating mathematicians, and it will certainly take its place in any critical examination of our scientific beliefs; but we must not let the simplicity of the assumption underlying the principle hide the very slender experimental basis on which it rests at present, and more especially not lose sight of the fact, that it goes much beyond what is proved by Michelson's experiment. In that experiment, the source of light and the mirrors which reflected the light were all connected together by rigid bodies, and their distances depended therefore on the intensity of molecular forces. Einstein's generalisation assumes that the result of the experiment would still be the same, if performed in a free space with the source of light and mirrors disconnected from each other but endowed with a common velocity. This is a considerable and, perhaps, not quite justifiable generalisation. I am well aware that Bucherer's experiments with kathode rays are taken to confirm the validity of Einstein's principle, but if we say that they are not inconsistent with it, we should probably go as far as is justifiable."[681]

The Einsteins were under the spell of the new school of positivism which was to become "Logical Positivism", and which Sir Arthur Schuster would later catagorize as a cowardly cop out to ignorance, and further which "Logical Positivism" Karl Popper would systematically discredit as solipsism.[682] The Einsteins may have believed that they could disguise their piracy of Poincaré's interpretation of Lorentz' theory, by stating it in Poincaré's quasi-positivistic form, without mentioning Poincaré. The Einsteins would have found references in Mach's work to,

"Budde's conception of space as a sort of medium."[683]

Schuster wrote against the emerging positivism, and the consequences of its cowardice,

"I have during these lectures contrasted on several occasions the former

---

681. A. Schuster, *The Progress of Physics during 33 years (1875-1908) Four Lectures delivered to the University of Calcutta during March 1908*, Cambridge University Press, (1911), pp. 109-111.
682. Among many others, Eduard Study also objected to the solipsism of positivism in a letter to Einstein of 24 May 1919, *The Collected Papers of Albert Einstein*, Volume 9, Document 45, Princeton University Press, (2004).
683. E. Mach, *Die Mechanik in ihrer Entwickelung*, 3rd Ed., F. A. Brockhaus, Leipzig, (1897), pp. 236-237.

tendency to base our technical explanations of natural phenomena on definite models which we can visualise and even constuct, with the modern spirit which is satisfied with a mathematical formula, and symbols which frequently have no strictly definable meaning. I ought to explain the distinction between the two points of view which represent two attitudes of mind, and I can do so most shortly by referring to the history of the electrodynamic theory of light, the main landmarks of which I have already pointed out in the second lecture. The undulatory theory—as it left the hands of Thomas Young, Fresnel and Stokes—was based on the idea that the æther possessed the properties of an elastic solid. Maxwell's medium being quite different in its behaviour, its author at first considered it to be necessary to justify the possibility of its existence, by showing how, by means of fly wheels and a peculiar cellular construction, we might produce a composite body having the required properties. Although later Maxwell laid no further stress on the ultimate construction of the medium, his ideas remained definite and to him the displacements which constituted the motion of light possessed a concrete reality. In estimating the importance of the support which Maxwell's views have received from experiment, we must distinguish between the fundamental assumptions on which Maxwell based his investigations and the mathematical formulæ which were the outcome of these investigations. It is clearly the mathematical formulæ only which are confirmed and the same formulæ might have been derived from quite different premises. It has always been necessary, as a second step of great discovery, to clear away the immaterial portions which are almost invariable accessories of the first pioneer work, and Heinrich Hertz, who besides being an experimental investigator was a philosopher of great perspicacity, performed this part of the work thoroughly. The mathematical formula instead of being the result embodying the concrete ideas, now became the only thing which really mattered. To use an acute and celebrated expression of Gustav Kirchhoff, it is the object of science to *describe* natural phenomena, not to *explain* them. When we have expressed by an equation the correct relationship between different natural phenomena we have gone as far as we safely can, and if we go beyond we are entering on purely speculative ground. I have nothing to say against this as a philosophic doctrine, and I shall adopt it myself when lying on my death-bed, if I have then sufficient strength to philosophise on the limitations of our intellect. But while I accept the point of view as a correct death-bed doctrine, I believe it to be fatal to a healthy development of science. Granting the impossibility of penetrating beyond the most superficial layers of observed phenomena, I would put the distinction between the two attitudes of mind in this way: One glorifies our ignorance, while the other accepts it as a regrettable necessity. The practical impediment to the progress of physics, of what may reluctantly be admitted as correct metaphysics, is both real and substantial and might be illustrated almost from any recent volume of scientific periodicals. Everyone who has ever tried to add his mite to advancing knowledge must know that vagueness of ideas is his greatest stumbling-block. But this vagueness which used to be recognised as our great enemy is now being enshrined as an idol to be worshipped. We may never know what constitutes atoms or what is the real structure of the æther, why trouble therefore, it is said, to find out more about them. Is it not safer, on the contrary, to confine ourselves to a general talk on entropy, luminiferous vectors and undefined symbols expressing vaguely certain physical relationships? What really lies at the bottom of the great fascination which these new doctrines exert on the present generation is sheer cowardice: the fear of having its errors brought home to it. As one who believes that metaphysics is a study apart from physics, not to be mixed up with it, and who considers that the main object of the physicist is to add to our knowledge, without troubling himself much as to how that knowledge may ultimately be interpreted, I must warn you against the temptation of sheltering yourself behind an illusive rampart of safety. We all prefer being right to being wrong, but it is better to be wrong than to be neither right nor wrong."[684]

James Mackaye wrote in 1931,

"Einstein's explanation is a dimensional disguise for Lorentz's. [***] Thus Einstein's theory is not a denial of, nor an alternative for, that of Lorentz. It is only a duplicate and disguise for it. [***] Einstein continually maintains that the theory of Lorentz is right, only he disagrees with his 'interpretation.' Is it not clear, therefore, that in this, as in other cases, Einstein's theory is merely a disguise for Lorentz's, the apparent disagreement about 'interpretation' being a matter of words only?"[685]

Lorentz pointed out in 1913,

"The latter is, by the way, up to a certain degree a quarrel over words: it makes no great difference, whether one speaks of the vacuum or of the æther."

"Letzteres ist übrigens bis zu einem gewissen Grade ein Streit über Worte: es macht keinen großen Untershied, ob man vom Vakuum oder vom Äther spricht."[686]

---

[684]. A. Schuster, *The Progress of Physics during 33 years (1875-1908) Four Lectures delivered to the University of Calcutta during March 1908*, Cambridge University Press, (1911), pp. 114-117. See: H. Hertz, "Author's Preface", *The Principles of Mechanics Presented in a New Form*, Macmillan, London, New York, (1899); reprinted, Dover, New York, (1956). See also: *Vorlesungen über mathematische Physik*, B. G. Teubner, Leipzig, (1876 and multiple later editions); *Gesammelte Abhandlungen*, J. A. Barth, Leipzig, (1882); *Vorlesungen über die Theorie der Wärme*, B. G. Teubner, Leipzig, (1894).
[685]. J. Mackaye, *The Dynamic Universe*, Charles Scribner's Sons, New York, (1931), pp. 42-43, 100-101.
[686]. H. A. Lorentz, *Das Relativitätsprinzip; drei Vorlesungen gehalten in Teylers Stiftung*

In 1980, Friedrich Hund wrote about the general theory of relativity and the æther,

"Man kann *Einsteins* Leistung als „Abschaffung des Äthers" bezeichnen, muß sich aber hüten, in einen Streit um Worte zu geraten. Heute, 75 Jahre später, kennen wir auch die „allgemeine Relativitätstheorie", die ein lokales „Inertialfeld" beschreibt, das was *H. Weyl* in seiner bildhaften Sprache den „Trägheitskompaß" nannte, die lorentzinvariante Einbettung des lokalen Geschehens in die weltweite Umgebung. Wir kennen weiter kosmologische Fakten, die isotrope Expansion des Systems der Galaxien und die isotrope 3K-Strahlung, die ein lokales spezielles Bezugssystem, *Weyls* „Sternenkompaß", festlegen. Diese Struktur des Universums, vielleicht nur des großen Ausschnittes aus ihm, der unserer Beobachtung zugänglich ist, sehen wir als geschichtlich geworden an. Diese Struktur hätte *H. Weyl* vielleicht Äther genannt und ihm „Kränze und Gesang geweiht."."[687]

In 1934, Albert Einstein confirmed Mackaye's assertions,
"Then came H. A. Lorentz's great discovery. All the phenomena of electromagnetism then known could be explained on the basis of two assumptions: that the ether is firmly fixed in space—that is to say, unable to move at all, and that electricity is firmly lodged in the mobile elementary particles. Today his discoveries may be expressed as follows: physical space and the ether are only different terms for the same thing; fields are physical states of space."[688]

Einstein stated in 1953,

"It was here that H. A. Lorentz' act of intellectual liberation set in. With great logic and consistency he based his investigations on the following hypotheses: The seat of the electromagnetic field is empty space. [***] The really essential step forward, indeed, was precisely Lorentz' having reduced the facts to Maxwell's equations concerning empty space, or — as it was then called — the ether. H. A. Lorentz even discovered the 'Lorentz transformation', so named after him, — though ignoring its group-like quality. For him, Maxwell's equations concerning empty space applied only to a given system of co-ordinates, which, on account of its state of rest, appeared excellent in comparison to all other existing systems of co-ordinates. This was a truly paradoxical situation, since the theory appeared to restrict the inertial system more than classical mechanics. This circumstance, proving as it did quite incompatible with the empirical

---

zu Haarlem, B. G. Teubner, Leipzig-Berlin, (1920). p. 23.
687. F. Hund, "Wer hat die Relativitätstheorie geschaffen?", *Phyiskalische Blätter*, Volume 36, Number 8, (1980), pp. 237-240, at 240.
688. A. Einstein, *Ideas and Opinions*, Crown Publishers, Inc., (1954), p. 281.

standpoint, simply *had* to lead to the special relativity theory."⁶⁸⁹

Max Abraham stated in 1908,

"The æther is empty space."

"Der Äther ist der leere Raum."⁶⁹⁰

We know that Einstein was familiar with this line from Abraham, because Gustav Mie quoted it to him in 1920 at the Bad Nauheim discussion.
Before Abraham was Horace Seal, who, in 1899, published the following,

> "All the text-books and authorities agree that the luminiferous ether fills all space and pervades all bodies, solid, gaseous, and liquid, in that space. If this is true, there is really no such thing as space as a void in which celestial objects move, but the word only remains as a term of measurement of the ether which pervades all bodies and is continuous, both in breadth, length, and depth through the whole universe. In fact, ether does not fill space, but *is* space, and the old measuring of space, which except among mathematicians excluded bodies moving in that space, with the discovery (an actual one) of the luminiferous ether, becomes obsolete. A possible objection to the above is, that loading the shoulders of what after all is only accepted as a convenient hypothesis with another one less perhaps acceptable, is unscientific. But even if the wave-theory of light, heat, &c., were not by now almost fully accepted as that of gravitation, the objection does not really apply, as this luminiferous theory is absolutely independent of hypothesis. It is not a successful guess, but an organized statement of facts, therefore its existence rests upon a solid foundation. [***] According to our theory a child gradually acquires rudimentary ideas of motion by marking the difference of quick and slow movements; but what he does not recognize until after years is, that when he is resting, this rest of his is not absolute rest, which is unknown, but only relative rest[.]"⁶⁹¹

Eugen Karl Dühring made similar arguments and even anticipated the general

---

689. A. Einstein, "H. A. Lorentz, His Creative Genius and His Personality", in G. L. de Haas-Lorentz, *H. A. Lorentz: Impressions of His Life and Work*, North-Holland Publishing Company, Amsterdam, (1957), pp. 5-9, at 6-7.
690. Mie quotes Abraham, "Allgemeine Diskussion über Relativitätstheorie", *Physikalische Zeitschrift*, Volume 21, (1920),pp. 666-668, at 667; the editors of: *The Collected Papers of Albert Einstein*, Volume 7, Document 46, Note 26, p. 359; attribute this reference to: M. Abraham, "Electromagnetische Theorie der Strahlung", *Theorie der Elektrizität*, Volume 2, Second Edition, B. G. Teubner, Leipzig, (1908).
691. H. Seal, "Space and Time", *The Westminster Review*, Volume 152, Number 6, (1899), pp. 675-679, at 676-678.

theory of relativity in 1878.[692] Bolliger also pursued this line of thought.[693]

Without an æther, there is no logical ground for assuming light speed independence from the motion of the source. Without an æther of some sort at hypothetical "absolute rest"—at rest relative to itself, anisotropic light speed in at least one of two inertial frames of reference in motion with respect to each other would *not* violate the principle of relativity, but instead would be *compelled by it*. Therefore, the Einsteins' two postulate myth of 1905 depends upon the premise of an æther, or absolute space, or "preferred frame of reference".

Obviously, Einstein's efforts to disguise his piracy through semantics and internally inconsistent Metaphysics are nonsense, for physical states compel physical substance, the æther, and Lorentz stated in 1906,

> "We shall add the hypothesis that, though the particles may move, *the ether always remains at rest*. We can reconcile ourselves with this, at first sight, somewhat startling idea, by thinking of the particles of matter as of some local modifications in the state of the ether. These modifications may of course very well travel onward while the volume-elements of the medium in which they exist remain at rest."[694]

Herbert Dingle derided Einstein's numerology, his "dimensional disguise for Lorentz's" physical theory,

> "This proposal became known as *the relativity theory of Lorentz*, and certain features of it call for attention here. [***] Like Maxwell, who realised the necessity, if he was to satisfy his mathematical desires, of postulating a 'displacement current' to justify them, so Lorentz, in order to justify his transformation equations, saw the necessity of postulating a physical effect of interaction between moving matter and ether, to give the mathematics meaning. Physics still had *de jure* authority over mathematics: it was Einstein, who had no qualms about abolishing the ether and still retaining light waves whose properties were expressed by formulae that were meaningless without it, who was the first to discard physics altogether and propose a wholly mathematical theory."[695]

As Vaihinger stated,

---

[692]. E. K. Dühring, *Neue Grundgesetze zur rationellen Physik und Chemie*, Volume 1, Chapter 1, Fues's Verlag (R. Reisland), Leipzig, (1878/1886), pp. 1-34
[693]. A. Bolliger, *Anti-Kant oder Elemente der Logik, der Physik und der Ethik*, Felix Schneider, Basel, (1882), *esp.* pp. 336-354.
[694]. H. A. Lorentz, *Theory of Electrons*, , B. G. Teubner, Leipzig, (1909), p. 11; reprinted Dover, New York, (1952). *See also:* pp. 30-31.
[695]. H. Dingle, *Science at the Crossroads*, Martin Brian & O'Keeffe, London, (1972), pp. 165-166.

"Pure mathematical space is a fiction. Its concept has the marks of a fiction: the idea of an extension without anything extended, of separation without things that are to be separated, is something unthinkable, absurd and impossible."[696]

Albert Einstein, who in 1905 had called the æther "superfluous", stated in 1920,

"To deny the ether is ultimately to assume that empty space has no physical qualities whatever. [***] Recapitulating, we may say that according to the general theory of relativity space is endowed with physical qualities; in this sense, therefore, there exists an ether. According to the general theory of relativity space without ether is unthinkable; for in such space there not only would be no propagation of light, but also no possibility of existence for standards of space and time (measuring-rods and clocks), nor therefore any space-time intervals in the physical sense."[697]

The eminent physicist Oliver Heaviside, in a hand-written letter to Prof. Vilhelm Bjerknes, discussed Einstein's compulsory shift in position from claiming that the æther was superfluous to stating directly that the æther was fundamental to "Einstein's" theories,

"I don't find Einstein's Relativity agrees with me. It is the most unnatural and difficult to understand way of representing facts that could be thought of. His distorted space is chaos [***] The Einstein enthusiasts are very patronizing about the 'classical' electromagnetics and its ether, which they have abolished. But they will come back to it by and by. [***] But you must work fairly, with the Ether, and Forces, & Momentum etc. They are the realities, without Einstein's distorted nothingness. [***] And I really think that Einstein is a practical joker, pulling the legs of his enthusiastic followers, more Einsteinisch than he. He knows the weakness of his 2nd Theory. He only does it to annoy [***] I can't get away from Einstein the Joker. [***] Did such a clever man as Einstein not see the significance of Poisson's theorem? It is said that it was by noticing some of H. A. Lorentz' formulas, and those of Minkowski, led him to the result. Well, we must believe it, if he says so, and like the silent parrot, think the more."[698]

---

[696]. H. Vaihinger, *Philosophy of the 'As if'*, Barnes & Noble, Inc., New York, (1966), p. 232. Vaihinger also pointed out that many scientists considered the hypothesis of the æther nothing more than a useful fiction, at page 40; **and**, at page 41, "In all modern science there is a tendency to depose hypotheses hitherto regarded as firmly established and to degrade them to the position of useful fictions."

[697]. A. Einstein, *Sidelights on Relativity*, translated by: G. B. Jeffery and W. Perret, Methuen & Co., London, (1922); *republished, unabridged and unaltered:* Dover, New York, (1983), pp. 16, 23.

[698]. Taken from a letter from Oliver Heaviside to Vilhelm Friman Køren Bjerknes dated

In 1938, Einstein and Infeld averred, in a statement highly reminiscent of Ernst Haeckel's *Die Welträthsel* of 1899,

"Our only way out seems to be to take for granted the fact that space has the physical property of transmitting electromagnetic waves, and not to bother too much about the meaning of this statement. We may still use the word ether, but only to express some physical property of space. This word ether has changed its meaning many times in the development of science. At the moment it no longer stands for a medium built up of particles. Its story, by no means finished, is continued by the relativity theory."[699]

Haeckel wrote,

"I. Ether fills the whole of space, in so far as it is not occupied by ponderable matter, as a *continuous substance;* it fully occupies the space between the atoms of ponderable matter.

II. Ether has probably no chemical quality, and is not composed of atoms. If it be supposed that it consists of minute homogeneous atoms (for instance, indivisible etheric particles of a uniform size), it must be further supposed that there is something else between these atoms, either 'empty space' or a third, completely unknown medium, a purely hypothetical 'interether'; the question as to the nature of this brings us back to the original difficulty, and so on *in infinitum*.

III. As the idea of an empty space and an action at a distance is scarcely possible in the present condition of our knowledge (at least it does not help to a clear monistic view), I postulate for ether a special structure which is not atomistic, like that of ponderable matter, and which may provisionally be called (without further determination) *etheric* or *dynamic* structure."[700]

Herbert Spencer addressed the root of the problem of confusing pure Mathematics with Physics,

---

"8/3/20". C. J. Bjerknes, *Albert Einstein: The Incorrigible Plagiarist*, XTX Inc., Downers Grove, Illinois, (2002), pp. 25-26.
699. A. Einstein and I. Infeld, *The Evolution of Physics*, Simon & Schuster, New York, London, Toronto, Sydney, Tokyo, Singapore, (1966), p. 153.
700. E. Haeckel, *Die Welträthsel: Gemeinverständliche Studien über Monistische Philosophie*, Emil Strauß, Bonn, (1899), pp. 243-316, *and especially* pp. 261-267, and 282-284; English translation by Joseph McCabe, *The Riddle of the Universe: At the Close of the Nineteenth Century*, Harper & Brothers, New York, (1900), p. 227. G. W. De Tunzelmann ridiculed Haeckel in *A Treatise on Electrical Theory and the Problem of the Universe. Considered from the Physical Point of View, with Mathematical Appendices*, Appendix Q, Charles Griffin, London, (1910), pp. 617-625. Haeckel's name often appears in the literature.

"To sum up this somewhat too elaborate argument:—We have seen how in the very assertion that all our knowledge, properly so called, is Relative, there is involved in the assertion that there exists a Non-relative. We have seen how, in each step of the argument by which this doctrine is established, the same assumption is made. We have seen how, from the very necessity of thinking in relations, it follows that the Relative is itself inconceivable, except as related to a real Non-relative. We have seen that unless a real Non-relative or Absolute be postulated, the Relative itself becomes absolute; and so brings the argument to a contradiction. And on contemplating the process of thought, we have equally seen how impossible it is to get rid of the consciousness of an actuality lying behind appearances; and how, from this impossibility, results our indestructible belief in that actuality."[701]

Surely, the assertion of a physical æther is a scientific hypothesis, which recognizes the need of the real behind the relative, while the abstract set of human rules which constitute "space-time" represent nothing real or imagined. Einstein failed to understand the distinction between Physics and Metaphysics. He stated,

"I believe that physics is abstract and not obvious[.]"[702]

Carlo Giannoni saw that Einstein's theory differed only philosophically from the Poincaré-Lorentz theory, and Giannoni stresses the importance of the fact that Lorentz employed the principle of relativity in his 1904 paper.[703]

9.4 The So-Called *"Lorentz* Transformation"

The mathematical transformations in relativity theory are called "Lorentz Transformations",[704] an appellation supplied by Emil Cohn[705] and Henri

---

[701]. H. Spencer, *First Principles of a New System of Philosophy*, Second American Edition, D. Appelton and Company, New York, (1874), pp. 96-97.

[702]. A. Einstein quoted in R. W. Clark, *Einstein: The Life and Times*, World Publishing, New York, (1971), p. 264.

[703]. C. Giannoni, "Einstein and the Lorentz-Poincaré Theory of Relativity", *PSA: Proceedings of the Biennial Meeting of the Philosophy of Science Association*, Volume 1970, (1970), pp. 575-589. JSTOR link:

<http://links.jstor.org/sici?sici=0270-8647%281970%291970%3C575%3AEATLTO%3E2.0.CO%3B2-Z>

Note that Giannoni undervalues the contributions of Poincaré.

[704]. M. v. Laue, *Das Relativitätsprinzip der Lorentztransformation*, Friedr. Vieweg & Sohn, Braunschweig, (1921), p. 48. A. Einstein, *Relativity, The Special and the General Theory*, Crown Publishers, Inc., New York, (1961), p. 33.

[705]. E. Cohn, "Über die Gleichungen der Electrodynamik für bewegte Körper", *Archives Néerlandaises des Sciences Exactes et Naturelles*, Series 2, Volume 5, (1900), p. 519.

Poincaré.[706] The record indicates that Woldemar Voigt,[707] Oliver Heaviside, George Francis FitzGerald, Hendrik Antoon Lorentz, Joseph Larmor, Henri Poincaré, Emil Cohn, Paul Langevin, and others, began developing the mathematical expressions of the theory of relativity some 18 years before Einstein, and completed them before Einstein published on the subject.

9.4.1 Woldemar Voigt's Space-Time Transformation

The "Lorentz Transformation" is not Lorentz' transformation, as is, and was, widely known,

> "Nor did Lorentz discover these equations. They were first used by Voight[sic]."[708]

---

[706]. J. H. Poincaré, "Sur la Dynamique de l'Électron", *Comptes rendus hebdomadaires des séances de L'Académie des sciences*, Volume 140, (1905), pp. 1504-1508; reprinted in H. Poincaré, *La Mécanique Nouvelle: Conférence, Mémoire et Note sur la Théorie de la Relativité / Introduction de Édouard Guillaume*, Gauthier-Villars, Paris, (1924), pp. 77-81 URL:

<http://gallica.bnf.fr/scripts/ConsultationTout.exe?E=0&O=N029067>

reprinted *Œuvres de Henri Poincaré*, Volume 9, Gautier-Villars, Paris, (1954), pp. 489-493; English translations appear in: G. H. Keswani and C. W. Kilmister, "Intimations of Relativity before Einstein", *The British Journal for the Philosophy of Science*, Volume 34, Number 4, (December, 1983), pp. 343-354, at pp. 350-353; **and**, translated by G. Pontecorvo with extensive commentary by A. A. Logunov, *On the Articles by Henri Poincaré ON THE DYNAMICS OF THE ELECTRON*, Publishing Department of the Joint Institute for Nuclear Research, Dubna, (1995), pp. 7-14.

[707].W. Voigt, "Ueber das Doppler'sche Princip", *Nachrichten von der Königlichen Gesellschaft der Wissenschaften und der Georg-Augusts-Universität zu Göttingen*, (1887), pp. 41-51; reprinted *Physikalische Zeitschrift*, Volume 16, Number 20, (15 October 1915), pp. 381-386; English translation, as well as very useful commentary, are found in A. Ernst and Jong-Ping Hsu (W. Kern is credited with assisting in the translation), "First Proposal of the Universal Speed of Light by Voigt in 1887", *Chinese Journal of Physics* (The Physical Society of the Republic of China), Volume 39, Number 3, (June, 2001), pp. 211-230; URL's:

<http://psroc.phys.ntu.edu.tw/cjp/v39/211/211.htm>

<http://psroc.phys.ntu.edu.tw/cjp/v39/211.pdf>

See also: W. Voigt, "Theorie des Lichtes für bewegte Medien", *Nachrichten von der Königlichen Gesellschaft der Wissenschaften und der Georg-Augusts-Universität zu Göttingen*, (1887), pp. 177-238.

[708].A. Henderson, A. W. Hobbs, J. W. Lasley, Jr., *The Theory of Relativity*, The University of North Caroline Press, Chapel Hill, North Carolina, Oxford University Press, (1924), p. 16, footnote.

The *Brockhaus Enzyklopädie* succinctly states,

"Voigt [***] presented (among the introduction of the term 'Tensor') a theory of elasticity; in the treatment of optical properties, he formulated for the first time in 1887 the formulas, which later became known through the special theory of relativity as the Lorentz-Transformation."

"Voigt [***] lieferte (unter Einführung des Begriffes >Tensor<) eine Elastizitätstheorie; bei der Behandlung der opt. Eigenschaften formulierte er 1887 erstmalig die später als Lorentz-Transformation durch die Spezielle Relativitätstheorie bekanntgewordenen Formeln."[709]

In 1887, Woldemar Voigt published the following relativistic transformation of space-time coordinates:

$$x' = x - vt, \quad y' = \frac{y}{\gamma}, \quad z' = \frac{z}{\gamma}, \quad t' = t - \frac{vx}{c^2},$$

$$\gamma = \frac{1}{\sqrt{1 - \frac{v^2}{c^2}}}.$$

Hermann Minkowski stated,

"Maxwell's and Lorentz' theory are not really opposites, but rather the rigid and the non-rigid, Zeppelin's and Parseval's electron. In the interest of history, I want yet to add, that the transformations which play the main rôle in the principle of relativity were first mathematically formulated by Voigt, in the year 1887. With the aid of these transformations, Voigt had already drawn conclusions at that time regarding the Doppler Effect."

"Nicht die Maxwellsche und die Lorentzsche Theorie sind die eigentlichen Gegensätze, sondern das starre und das unstarre, das Zeppelinsche und das Parsevalsche Elektron. Historisch will ich noch hinzufügen, daß die Transformationen, die bei dem Relativitätsprinzip die Hauptrolle spielen, zuerst mathematisch von Voigt im Jahre 1887 behandelt sind. Voigt hat damals bereits mit ihrer Hilfe Folgerungen in bezug auf das Dopplersche

---

709. *Brockhaus Enzyklopädie*, F. A. Brockhaus, Wiesbaden, Volume 19, (1974), p 697. For Voigt's use of the term 'tensor' *see:* W. Voigt, *Die fundamentalen physikalischen Eigenschaften der Krystalle in elementarer Darstellung*, Veit, Leipzig, (1898), pp. 20 ff.; **and** S. Bochner, "The Significance for Some Basic Mathematical Conceptions for Physics", *Isis*, Volume 54, (1963), pp. 179-205, at 193; **and** W. Voigt, *Elementare Mechanik als Einleitung in das Studium der theoretischen Physik*, second revised edition, Veit, Leipzig, (1901), pp. 10-26.

Prinzip gezogen."

To which Voigt responded,

> "Mr. Minkowski recalls an old work of mine. It addressed the application of the Doppler Effect to some special cases which arise due to the elastic theory of light, not the electromagnetic. It had already at that time revealed some of the consequences, which were later arrived at through the electromagnetic theory."

> "Herr Minkowski erinnert an eine alte Arbeit von mir. Es handelt sich dabei um Anwendungen des Dopplerschen Prinzips, die in speziellen Teilen auftreten, aber nicht auf Grund der elektromagnetischen, sondern auf Grund der elastischen Theorie des Lichtes. Indessen haben sich damals bereits einige derselben Folgerungen ergeben, die später aus der elektromagnetischen Theorie gewonnen sind."[710]

9.4.2 Length Contraction

In 1905, Mileva and Albert Einsteins asserted, without reference to prior authors,

> "A rigid body which, measured in a state of rest, has the form of a sphere, therefore has in a state of motion—viewed from the stationary system—the form of an ellipsoid of revolution with the axes
>
> $$R\sqrt{(1 - v^2/c^2)}, R, R.$$
>
> Thus, whereas the Y and Z dimensions of the sphere (and therefore of every rigid body of no matter what form) do not appear modified by the motion, the X dimension appears shortened in the ratio
>
> $$1 : \sqrt{(1 - v^2/c^2)}$$
>
> i.e. the greater the value of v, the greater the shortening. For $v = c$ all moving objects—viewed from the 'stationary' system—shrivel up into plane figures. [*Footnote:* That is, a body possessing spherical form when examined at rest.] For velocities greater than that of light our deliberations become meaningless; we shall, however, find in what follows, that the velocity of light in our theory plays the part, physically, of an infinitely great

---

710. *Physikalische Zeitschrift*, Volume 9, Number 22, (November 1, 1908), p. 762. While some believe that Voigt was expressing his modesty, it might also be that he was pointedly asserting the primacy of the *elastic* theory of light.

velocity."⁷¹¹

Henri Poincaré stated, in 1904,

"From all these results, if they are confirmed, would arise an entirely new mechanics, which would be, above all, characterised by this fact, that no velocity could surpass that of light, any more than any temperature could fall below the zero absolute, because bodies would oppose an increasing inertia to the causes, which would tend to accelerate their motion; and this inertia would become infinite when one approached the velocity of light."⁷¹²

Roger Joseph Boscovich argued, in 1763, in the second supplement to his *Natural Philosophy,*

"21. Again, it is to be observed first of all that from this principle of the [invariance] of those things, of which we cannot perceive the change through our senses, there comes forth the method that we use for comparing the magnitudes of intervals with one another; here, that, which is taken as a measure, is assumed to be [invariant]. Also we make use of the axiom, *things that are equal to the same thing are equal to one another*; & from this is deduced another one pertaining to the same thing, namely, *things that are equal multiples, or submultiples, of each, are also equal to one another*; & also this, *things that coincide are equal*. We take a wooden or iron ten-foot rod; & if we find that this is congruent with one given interval when applied to it either once or a hundred times, & also congruent to another interval when applied to it either once or a hundred times, then we say that these intervals are equal. Further, we consider the wooden or iron ten-foot rod to be the same standard of comparison after translation. Now, if it consisted of perfectly continuous & solid matter, we might hold it to be exactly the same standard of comparison; but in my theory of points at a distance from one another, all the points of the ten-foot rod, while they are being transferred, really change the distance continually. For the distance is constituted by those real modes of existence, & these are continually changing. But if they are changed in such a manner that the modes which follow establish real relations of equal distances, the standard of comparison will not be identically the same; & yet it will still be an equal one, & the equality of the measured intervals will be correctly determined. We can no more transfer the length of the ten-foot rod, constituted in its first position by the first real modes, to the place of the length constituted in its second

---

711. *The Principle of Relativity*, Dover, (1952), p. 48.
712. H. Poincaré's St. Louis lecture from September of 1904, *La Revue des Idées*, 80, (November 15, 1905); "L'État Actuel et l'Avenir de la Physique Mathématique", *Bulletin des Sciences Mathématique*, Series 2, Volume 28, (1904), p. 302-324; English translation, "The Principles of Mathematical Physics", *The Monist*, Volume 15, Number 1, (January, 1905), pp. 1-24.

position by the second real modes, than we are able to do so for intervals themselves, which we compare by measurement. But, because we perceive none of this change during the translation, such as may demonstrate to us a relation of length, therefore we take that length to be the same. But really in this translation it will always suffer some slight change. It might happen that it underwent even some very great change, common to it & our senses, so that we should not perceive the change; & that, when restored to its former position, it would return to a state equal & similar to that which it had at first. However, there always is some slight change, owing to the fact that the forces which connect the points of matter, will be changed to some slight extent, if its position is altered with respect to all the rest of the Universe. Indeed, the same is the case in the ordinary theory. For no body is quite without little spaces interspersed within it, altogether incapable of being compressed or dilated; & this dilatation & compression undoubtedly occurs in every case of translation, at least to a slight extent. We, however, consider the measure to be the same so long as we do not perceive any alteration, as I have already remarked.

22. The consequence of all this is that we are quite unable to obtain a direct knowledge of absolute distances; & we cannot compare them with one another by a common standard. We have to estimate magnitudes by the ideas through which we recognize them; & to take as common standards those measures which ordinary people think suffer no change. But philosophers should recognize that there is a change; but, since they know of no case in which the equality is destroyed by a perceptible change, they consider that the change is made equally.

23. Further, although the distance is really changed when, as in the case of the translation of the ten-foot rod, the position of the points of matter is altered, those real modes which constitute the distance being altered; nevertheless if the change takes place in such a way that the second distance is exactly equal to the first, we shall call it the same, & say that it is altered in no way, so that the equal distances between the same ends will be said to be the same distance & the magnitude will be said to be the same; & this is defined by means of these equal distances, just as also two parallel directions will be also included under the name of the same direction. In what follows we shall say that the distance is not changed, or the direction, unless the magnitude of the distance, or the parallelism, is altered."[713]

George Francis FitzGerald wrote, in 1889,

"I HAVE read with much interest Messrs. Michelson and Morley's wonderfully delicate experiment attempting to decide the important question as to how far the ether is carried along by the earth. Their result

---

[713]. R. J. Boscovich, *A Theory of Natural Philosophy*, M.I.T. Press, Cambridge, Massachusetts, London, (1966), pp. 204-205.

seems opposed to other experiments showing that the ether in the air can be carried along only to an inappreciable extent. I would suggest that almost the only hypothesis that can reconcile this opposition is that the length of material bodies changes, according as they are moving through the ether or across it, by an amount depending on the square of the ratio of their velocity to that of light. We know that electric forces are affected by the motion of the electrified bodies relative to the ether, and it seems a not improbable supposition that the molecular forces are affected by the motion, and that the size of a body alters consequently. It would be very important if secular experiments on electrical attractions between permanently electrified bodies, such as in a very delicate quadrant electrometer, were instituted in some of the equatorial parts of the earth to observe whether there is any diurnal and annual variation of attraction,—diurnal due to the rotation of the earth being added and subtracted from its orbital velocity; and annual similarly for its orbital velocity and the motion of the solar system."[714]

Hendrik Antoon Lorentz had averred the same in 1892,[715] and stated, in 1895,

> "The displacement would naturally bring about this disposition of the molecules of its own accord, and thus effect a shortening in the direction of motion in the proportion of 1 to $\sqrt{1 - v^2/c^2}$ in accordance with the formulæ given in the above-mentioned paragraph."[716]

In 1904, Lorentz affirmed that,

> "§ 8. Thus far we have only used the fundamental equations without any new assumptions. I shall now suppose *that the electrons, which I take to be spheres of radius R in the state of rest, have their dimensions changed by the effect of a translation, the dimensions in the direction of motion becoming kl times and those in perpendicular directions l times smaller.*

---

714. G. F. FitzGerald, "The Ether and Earth's Atmosphere (Letter to the Editor)", *Science*, Volume 13, Number 328, (1889), p. 390.
715. H. A. Lorentz, "Over de terugkaatsing van licht door lichamen die zich bewegen", *Verslagen der Zittingen de Wis- en Natuurkundige Afdeeling der Koninklijke Akademie van Wetenschappen* (Amsterdam), Volume 1, (1892), pp. 28-31; reprinted in English "On the Reflection of Light by Moving Bodies", *Collected Papers*, Volume 4, pp. 215-218; **and** "De Relatieve Beweging van de Aarde en den Aether", *Verslagen der Zittingen de Wis- en Natuurkundige Afdeeling der Koninklijke Akademie van Wetenschappen* (Amsterdam), Volume 1, (1892/1893), pp. 74-79; translated in English, "The Relative Motion of the Earth and the Ether", *Collected Papers*, Volume 4, pp. 219-223; **and** "La Théorie Électromagnétique de Maxwell et son Application aux Corps Mouvants", *Archives Néerlandaises des Sciences Exactes et Naturelles*, First Series, Volume 25, (1892), pp. 363-552; reprinted *Collected Papers*, Volume 2, pp. 164-343.
716. *The Principle of Relativity*, Dover, (1952), p. 7.

In this deformation, which may be represented by (1/*kl*, 1/*l*, 1/*l*) each element of volume is understood to preserve its charge."

9.4.2.1 Dynamic Length Contraction

In Lorentz' synthetic physical theory, length contraction is a dynamic theorem following from Maxwell's and Heaviside's[717] work on the dynamics of the æther.

9.4.2.2 Kinematic Length Contraction

In the Einsteins' fallacy of *Petitio Principii* of 1905, a change in length is merely presupposed without any physical theory to justify it, then the precise factor is arrived at through induction from the allegedly observed invariance of light speed, which is an allegedly known empirical fact, not an *a priori* postulate. No one disputes that Einstein knew Lorentz' contraction hypothesis. The Einsteins simply used the idea without crediting Lorentz, then Einstein called it a natural consequence of the "two postulates" in 1907. Since the "postulates" are empirical observations, the "natural consequences" are arrived at through induction, not deduction. In other words, the hypothesis of length contraction is more fundamental than the law of light speed invariance.

One must first propose *a priori* a change in length before one can derive the precise factor of it through induction from the supposed empirical fact of light speed invariance, and the so-called "natural consequence" is instead the inductively determined factor arrived at from the presupposed *a priori* and *ad hoc* hypothesis that length must change with velocity relative to the "resting system" (in the Einsteins' 1905 paper the "resting system" is Newton's absolute space) in order for light speed to be invariant in "moving systems" (in the Einsteins' 1905 paper "moving systems" are systems in motion relative to Newton's absolute space). This presupposed change in length is more *ad hoc* in the Einsteins' 1905 paper than it is in Lorentz' synthetic theory, which attempts a dynamic exposition on it, as physics must.

It was Poincaré, not Einstein nor Minkowski, who first recognized the group properties of the Lorentz Transformation and reciprocal length contraction and who introduced a quadri-dimensional exposition on length contraction, which renders it—in terms of a mathematical quadri-dimensional space-time—a matter of cognitive perspective. Later, many would attempt to mask Einstein's plagiarism by arguing the issue of perspective, which nowhere appeared in the

---

717. O. Heaviside, "The Electromagnetic Effects of a Moving Charge", *The Electrician*, Volume 22, (1888), pp. 147-148; **and** "On the Electromagnetic Effects due to the Motion of Electricity Through a Dielectric", *Philosophical Magazine*, Volume 27, (1889), pp. 324-339; **and** "On the Forces, Stresses and Fluxes of Energy in the Electromagnetic Field", *Philosophical Transactions of the Royal Society*, Volume 183A, (1892), p. 423; **and** "A Gravitational and Electromagnetic Analogy", *The Electrician*, Volume 31, (1893), pp. 281-282, 359; **and** *The Electrician*, Volume 45, (1900), pp. 636, 881.

Einsteins' work of 1905, where length contraction is merely presupposed without justification, then inductively demonstrated with Poincaré's operationalist thought experiment of clocks synchronized by light signals on the suppositions that light speed is invariant and that length must change to render it so.

9.4.3 Time Dilatation

Roger Joseph Boscovich argued, in 1763, in the second supplement to his *Natural Philosophy*,

> "24. What has been said with regard to the measurement of space, without difficulty can be applied to time; in this also we have no definite & constant measurement. We obtain all that is possible from motion; but we cannot get a motion that is perfectly uniform. We have remarked on many things that belong to this subject, & bear upon the nature & succession of these ideas, in our notes. I will but add here, that, in the measurement of time, not even ordinary people think that the same standard measure of time can be translated from one time to another time. They see that it is another, consider that it is an equal, on account of some assumed uniform motion. Just as with the measurement of time, so in my theory with the measurement of space it is impossible to transfer a fixed length from its place to some other, just as it is impossible to transfer a fixed interval of time, so that it can be used for the purpose of comparing two of them by means of a third. In both cases, a second length, or a second duration is substituted, which is supposed to be equal to the first; that is to say, fresh real positions of the points of the same ten-foot rod which constitute a new distance, such as a new circuit made by the same rod, or a fresh temporal distance between two beginnings & two ends. In my Theory, there is in each case exactly the same analogy between space & time. Ordinary people think that it is only for measurement of space that the standard of measurement is the same; almost all other philosophers except myself hold that it can at least be considered to be the same from the idea that the measure is perfectly solid & continuous, but that in time there is only equality. But I, for my part, only admit in either case the equality, & never the identity."[718]

Joseph Larmor agreed with Boscovich and set the scale for time dilatation thereby completing the misnamed "Lorentz Transformation", which Lorentz, Poincaré and the Einsteins later adopted.

9.4.4 The Final Form of the Transformation

---

[718]. R. J. Boscovich, *A Theory of Natural Philosophy*, M.I.T. Press, Cambridge, Massachusetts, London, (1966), pp. 204-205.

The components of the "Lorentz Transformation" evolved as follows: From the Aristotelian-Bradwardine-Galilean Transformation,[719] we have,

$$x' = x - vt, \quad y' = y, \quad z' = zt, \quad t' = t.$$

Voigt (1887) introduced the relativity of simultaneity,

$$t' = t - \frac{vx}{c^2}.$$

FitzGerald (1889) introduced the scale factor of length contraction, giving mathematical voice to Boscovich's concept,

$$x' = \frac{x - vt}{\sqrt{1 - \frac{v^2}{c^2}}}.$$

Larmor (1894-1900) introduced the scale factor of time dilatation in order to quantify the Boscovichian concept of time dilatation, and published the "Lorentz Transformation" in 1897,

$$x' = \frac{x - vt}{\sqrt{1 - \frac{v^2}{c^2}}}, \quad y' = y, \quad z' = z, \quad t' = \frac{t - \frac{vx}{c^2}}{\sqrt{1 - \frac{v^2}{c^2}}}.$$

Lorentz, himself, acknowledged Voigt's priority, and was uncomfortable with Poincaré's term "Lorentz Transformation". Lorentz wrote to Voigt,

---

719. P. Frank, "Die Stellung des Relativitätsprinzips im System der Mechanik und der Elektrodynamik", *Sitzungsberichte der mathematisch-naturwissenschaftlichen Klasse der Kaiserlichen Akademie der Wissencahften in Wien*, Volume 118, (1909), pp. 373-446; at 373, 376, 420, and 442. Frank introduced the term "Group of the Galilean Transformations" in this paper, at page 382. Peter Guthrie Tait wrote about inertial "Galilei-wise" motion in 1884, "Note on Reference Frames", *Proceedings of the Royal Society of Edinburgh*, Volume 12, (November 1883-July 1884), pp. 743-745. I do not think that Galileo was first to this concept. See: A. G. Molland, "An Examination of Bradwardine's Geometry", *Archive for History of Exact Sciences*, Volume 19, Number 2, (1978), pp. 113-175; See also: J. A. Weisheipl, *The Development of Physical Theory in the Middle Ages*, Ann Arbor Paperbacks, University of Michigan Press, (1971).

"Of course I will not miss the first opportunity to mention, that the concerned transformation and the introduction of a local time has been your idea."[720]

Lorentz kept his word:

"In a paper „Über das Doppler'sche Princip", published in 1887 (Gött. Nachrichten, p. 41) and which to my regret has escaped my notice all these years, Voigt has applied to equations of the form (6) (§3 of this book) a transformation equivalent to the formulae (287) and (288). The idea of the transformations used above (and in §44) might therefore have been borrowed from Voigt and the proof that it does not alter the form of the equations for the *free* ether is contained in his paper."[721]

and,

"It was these considerations published by me in 1904, which gave rise to the dissertation by Poincaré on the dynamics of the electron, in which he has attached my name to the transformation of which I have just spoken. I am obliged to again note the observation that the same transformation itself was previously hit upon in an article from Mr. Voigt published in 1887, and I did not remove the artifice from it to the fullest extent possible. In fact, for certain of the physical magnitudes which enter in the formulas I have not indicated the transformation which suits best. This has been done by Poincaré, and later by Einstein and Minkowski. To discover the 'transformations of relativity', as I will call them now, ... "

"Ce furent ces considérations publiées par moi en 1904 qui donnèrent lieu à POINCARÉ d'écrire son mémoire sur la Dynamique de l'électron, dans lequel il a attaché mon nom à la transformation dont je viens de parler. Je dois remarquer à ce propos que la même transformation se trouve déjà dans un article de M. Voigt publié en 1887 et que je n'ai pas tiré de cet artifice tout le parti possible. En effet, pour certaines des grandeurs physiques qui entrent dans les formules, je n'ai pas indiqué la transformation qui convient le mieux. Cela a été fait par POINCARÉ et ensuite par M. EINSTEIN et

---

720. Lorentz' letter as quoted and translated by A. Ernst and Jong-Ping Hsu (W. Kern is credited with assisting in the translation), "First Proposal of the Universal Speed of Light by Voigt in 1887", *Chinese Journal of Physics* (The Physical Society of the Republic of China), Volume 39, Number 3, (June, 2001), pp. 211-230, at 214. The authors note that the concept of general and local time was Lorentz' idea, not Voigt's. In other words, Voigt's concept was kinematic, and Einstein held no priority over Voigt for the idea of relativistic "time". The Lorentz to Voigt letter is cited as "Deutsches Museum München, Archives, HS 5549".
721. H. A. Lorentz, *Theory of Electrons*, B. G. Teubner, Leipzig, (1909), p. 198 footnote; reprinted Dover, New York, (1952).

MINKOWSKI. Pour trouver les «transformations de relativité», comme je les appellerai maintenant".[722]

Though Lorentz denied knowledge of Voigt's Transformation, it is quite likely Lorentz did know of it. Lorentz was keenly interested in theories which would explain Michelson's negative result, as did Voigt's theory which was published in the highly respected and widely read *Nachrichten von der Königlichen Gesellschaft der Wissenschaften und der Georg-Augusts-Universität zu Göttingen.*[723] Given that Voigt's Transformation differs from the "Lorentz Transformation" of modern relativity theory, some have wondered why Lorentz credited Voigt with the transformation. Prof. Wilfried Schröder published a collection of letters between Emil Wiechert and Lorentz, "Hendrik Antoon Lorentz und Emil Wiechert (Briefwechsel und Verhältnis der beiden Physiker)", *Archive for History of Exact Sciences*, Volume 30, Number 2, (1984), pp. 167-187. In addition to the fact that Lorentz again denied his friend Poincaré's legacy, Lorentz' letters are noteworthy for their elucidation of his thought process and the development of his imperfect versions of the transformation which ill-advisedly bears his name. Schröder's article should be read by all interested in the history of the "Lorentz Transformation". Among the highlights regarding Voigt's work we find: Wiechert to Lorentz 28 November 1911,

---

[722]. H. A. Lorentz, "Deux mémoires de Henri Poincaré sur la physique mathématique", *Acta Mathematica*, Volume 38, (1921), pp. 293-308; reprinted in *Œuvres de Henri Poincaré*, Volume 9, Gautier-Villars, Paris, (1954), pp. 683-695; **and** Volume 11, (1956), pp. 247-261; "for certain of the physical magnitudes which enter in the formulas I have not indicated the transformation which suits best. This has been done by Poincaré, and later by Einstein and Minkowski," taken from H. E. Ives' translation, "Revisions of the Lorentz Transformation", *Proceedings of the American Philosophical Society*, Volume 95, Number 2, (April, 1951), p. 125; reprinted R. Hazelett and D. Turner Editors, *The Einstein Myth and the Ives Papers, a Counter-Revolution in Physics*, Devin-Adair Company, Old Greenwich, Connecticut, (1979), p. 125.

[723].W. Voigt, "Ueber das Doppler'sche Princip", *Nachrichten von der Königlichen Gesellschaft der Wissenschaften und der Georg-Augusts-Universität zu Göttingen*, (1887), pp. 41-51; reprinted *Physikalische Zeitschrift*, Volume 16, Number 20, (15 October 1915), pp. 381-386; English translation, as well as very useful commentary, are found in A. Ernst and Jong-Ping Hsu (W. Kern is credited with assisting in the translation), "First Proposal of the Universal Speed of Light by Voigt in 1887", *Chinese Journal of Physics* (The Physical Society of the Republic of China), Volume 39, Number 3, (June, 2001), pp. 211-230; URL's:

<http://psroc.phys.ntu.edu.tw/cjp/v39/211/211.htm>

<http://psroc.phys.ntu.edu.tw/cjp/v39/211.pdf>

See also: W. Voigt, "Theorie des Lichtes für bewegte Medien", *Nachrichten von der Königlichen Gesellschaft der Wissenschaften und der Georg-Augusts-Universität zu Göttingen*, (1887), pp. 177-238.

"Nun kenne ich von Ihnen aus jener Zeit die Arbeit Arch. neerl. 25, 363, 1892, das in Leiden 1895 erschienene Buch, und die Arbeit Proc. Amsterdam 1904, p. 809. Giebt es wohl noch andere Arbeiten, die für die Relativitätstheorie in Betracht kommen?"

Lorentz to Wiechert 21 December 1911,

"In der Arbeit von 1899 benutze ich eine Substitution, die in der im Bornschen' Referat benutzten Bezeichnungsweise folgendermaßen lautet:

$$z' = az - bct, \quad \frac{1}{a}t' = at - \frac{b}{c}z.$$

und erst in 1904 habe ich ihre Transformation

$$z' = az - bct, \quad t' = at - \frac{b}{c}$$

eingeführt, die sich übrigens schon viel früher bei Voigt findet (Über das Dopplersche Prinzip, Gött. Nachrichten, 1887)."

Wiechert to Lorentz 15 February 1912,

"In Ihrer Arbeit von 1899 (Archives Néerlandaises) benutzen Sie die Transformation

$$t' = t - k^2 \frac{w}{c^2} x.$$

In der Arbeit 1904 (Proceedings) lautet die Gleichung 5:

$$t' = t\frac{l}{k} - kl\frac{w}{c^2}x \left(\text{statt } t' = kl\left(t - \frac{w}{c^2}x\right) \right).$$

Das ist nun doch nicht die Transformation, die man als „Lorentz-Transformation" bezeichnet. Ich vermute aber, dass es sich nur um einen Druckfehler handelt, denn die folgenden Formeln entsprechen der richtigen Formel. Dies ist doch eine richtige Ansicht?
  Sie sagen, dass Prof. W. Voigt schon 1887 die Transformation benutzt habe. Es scheint mir aber, dass dieses *nicht* der Fall ist. W. Voigt scheint mir für die Zeiten t' und t stets die gleichen Einheiten zu benutzen."

Lorentz to Wiechert 5 March 1912,

"4. Was die Formeln von Voigt betrifft, so sind diese so wenig von oben angeführten (1) verschieden, dass man, wie mir scheint, wohl sagen kann, er habe die Rel.transformation angegeben. Die von ihm zu Grunde gelegten Differentialgleichungen behalten nämlich ihre Form, wenn man x, y, z, t alle mit ein und derselben Konstante multipliziert. Man findet nun in seiner Abhandlung über das Doppler'sche Prinzip die Substitution (die Formeln 10) auf S. 45)

$$\xi = x - kt,$$
$$\eta = qy,$$
$$\zeta = qz,$$
$$\tau = t - \frac{kx}{w^2}$$

wo w die Fortpflanzungsgeschwindigkeit bedeutet, und

$$q = \sqrt{1 - \frac{k^2}{w^2}}$$

ist.
   Sie bemerken zu Recht, dass t und τ hier den gleichen Koeffizienten haben. Aber es kommt jetzt in der zweiten und dritten Gleichung der Koeffizient q vor. Setzt man

$$\xi = qx', \eta = qy', \zeta = qz', \tau = qt',$$

so verwandeln sich die Gleichungen in

$$x' = \frac{1}{q}x - \frac{k}{q}t,\ y' = y,\ z' = z,\ t' = \frac{1}{q}t - \frac{k}{qw^2}x,$$

und dies hat wirklich die Gestalt von (1), wenn man w mit c identifiziert und

$$a = \frac{1}{q},\quad b = \frac{k}{q^2}$$

setzt."

In 1900, Joseph Larmor published the following chapter in his most famous work, the award winning essay *Aether and Matter*, which was "completed at the end of the year 1898", and had Larmor already published the "Lorentz Transformation" in near modern form in 1897,[724]

"CHAPTER XI

MOVING MATERIAL SYSTEM: APPROXIMATION CARRIED TO THE SECOND ORDER

110. THE results obtained above have been derived from the correlation developed in § 106, up to the first order of the small quantity v/C, between the equations for aethereal vectors here represented by (f', g', h') and (a', b', c') referred to the axes (x', y', z') at rest in the aether and a time t'' and those for related aethereal vectors represented by (f, g, h) and (a, b, c) referred to axes (x', y', z') in uniform translatory motion and a time t' But we can proceed further, and by aid of a more complete transformation institute a correspondence which will be correct to the second order. Writing as before t'' for $t' - \frac{v}{c^2} \varepsilon x'$ the exact equations for (f, g, h) and (a, b, c) referred to the moving axes (x', y', z') and time t' are, as above shown, equivalent to

$$4\pi \frac{df'}{dt''} = \frac{dc'}{dy'} - \frac{db'}{dz'} \qquad -(4\pi C^2)^{-1}\frac{da'}{dt''} = \frac{dh'}{dy'} - \frac{dg'}{dz'}$$
$$4\pi\varepsilon \frac{dg'}{dt''} = \frac{da'}{dz'} - \frac{dc'}{dx'} \qquad -(4\pi C^2)^{-1}\varepsilon\frac{db'}{dt''} = \frac{df'}{dz'} - \frac{dh'}{dx'}$$
$$4\pi\varepsilon \frac{dh'}{dt''} = \frac{db'}{dx'} - \frac{da'}{dy'} \qquad -(4\pi C^2)^{-1}\varepsilon\frac{dc'}{dt''} = \frac{dg'}{dx'} - \frac{df'}{dy'}.$$

Now write

---

[724]. J. Larmor, *Aether and Matter*, Cambridge University Press, (1900), pp. ix, xiv, 161-179; **and** "A Dynamical Theory of the Electric and Luminiferous Medium", *Philosophical Transactions of the Royal Society of London A*, Volume 185, (1894), pp. 719-822; Volume 186, (1895), pp. 695-743; Volume 188, (1897), pp. 205-300; **and** "Dynamical Theory of the Ether I & II", *Nature*, Volume 49, (11 January 1894), pp. 260-262; and (18 January 1894), pp. 280-283; **and** *Philosophical Magazine*, Series 5, Volume 44, (1897), p. 503. **Confer:** M. N. Macrossan, "A Note on Relativity before Einstein", *The British Journal for the Philosophy of Science*, Volume 37, Number 2, (June, 1986), pp. 232-234.

(x_1, y_1, z_1) for $\left(\varepsilon^{\frac{1}{2}} x', y', z'\right)$

(a_1, b_1, c_1) for $\left(\varepsilon^{-\frac{1}{2}} a', b', c'\right)$ or $\left(\varepsilon^{-\frac{1}{2}} a, b + 4\pi vh, c - 4\pi vg\right)$

(f_1, g_1, h_1) for $\left(\varepsilon^{-\frac{1}{2}} f', g', h'\right)$ or $\left(\varepsilon^{-\frac{1}{2}} f, g - \frac{v}{4\pi C^2} c, h + \frac{v}{4\pi C^2} b\right)$

dt_1 for $\varepsilon^{-\frac{1}{2}} dt''$ or $\varepsilon^{-\frac{1}{2}} \left(dt' - \frac{v}{C^2} \varepsilon dx'\right)$,

where $\varepsilon = \left(1 - \frac{v^2}{C^2}\right)^{-1}$ and it will be seen that the factor ε is absorbed, so that the scheme of equations, referred to moving axes, which connects together the new variables with subscripts, is identical in form with the Maxwellian scheme of relations for the aethereal vectors referred to fixed axes. This transformation, from (x', y', z') to (x^1, y^1, z^1) as dependent variables, signifies an elongation of the space of the problem in the ratio $\varepsilon^{1/2}$ along the direction of the motion of the axes of coordinates. Thus if the values of (f_1, g_1, h_1) and (a_1, b_1, c_1) given as functions of x_1, y_1, z_1, t_1 express the course of spontaneous change of the aethereal vectors of a system of moving electrons referred to axes (x_1, y_1, z_1) at rest in the aether, then

$$\left(\varepsilon^{-\frac{1}{2}} f, \; g - \frac{v}{4\pi C^2} c, \; h + \frac{v}{4\pi C^2} b\right)$$

and $\left(\varepsilon^{-\frac{1}{2}} a, \; b + 4\pi vh, \; c - 4\pi vg\right)$,

expressed by the same functions of the variables

$$\varepsilon^{\frac{1}{2}} x', \; y', \; z', \; \varepsilon^{-\frac{1}{2}} t' - \frac{v}{C^2} \varepsilon^{\frac{1}{2}} x',$$

will represent the course of change of the aethereal vectors (f, g, h) and (a, b, c) of a correlated system of moving electrons referred to axes of (x', y', z') moving through the aether with uniform translatory velocity $(v, 0, 0)$. In this correlation between the courses of change of the two systems, we have

$$\frac{d\left(\varepsilon^{-\frac{1}{2}}f\right)}{d\left(\varepsilon^{\frac{1}{2}}x'\right)} \text{ equal to } \frac{df_1}{dx_1} - \frac{v}{C^2}\frac{df_1}{dt_1},$$

$$\frac{d}{dy'}\left(g - \frac{v}{4\pi C^2}c\right) \quad \text{,,} \quad \frac{dg_1}{dy_1}$$

$$\frac{d}{dz'}\left(h + \frac{v}{4\pi C^2}b\right) \quad \text{,,} \quad \frac{dh_1}{dz_1},$$

where
$$\frac{dc}{dy'} - \frac{db}{dz'} = 4\pi\left(\frac{df}{dt'} - v\frac{df}{dx'}\right)$$

and also
$$\frac{df_1}{dt_1} = \frac{df}{dt'};$$

hence $\dfrac{df}{dx'} + \dfrac{dg}{dy'} + \dfrac{dh}{dz'} - \dfrac{v}{C^2}\left(\dfrac{df}{dt'} - v\dfrac{df}{dx'}\right)$ is equal to

$$\varepsilon\frac{df_1}{dx_1} + \frac{dg_1}{dy_1} + \frac{dh_1}{dz_1} - \frac{v}{C^2}\varepsilon\frac{df}{dt'},$$

so that, up to the order of (v/C^2) inclusive,

$$\frac{df}{dx'} + \frac{dg}{dy'} + \frac{dh}{dz'} = \frac{df_1}{dx_1} + \frac{dg_1}{dy_1} + \frac{dh_1}{dz_1}.$$

Thus the conclusions as to the corresponding positions of the electrons of the two systems, which had been previously established up to the first order of v/C are true up to the second order when the dimensions of the moving system are contracted in comparison with the fixed system in the ratio $\varepsilon^{-1/2}$ or $1 - \frac{1}{2}v^2/C^2$, along the direction of its motion.

111. The ratio of the strengths of corresponding electrons in the two systems may now be deduced just as it was previously when the discussion was confined to the first order of v/C. For the case of a single electron in uniform motion the comparison is with a single electron at rest, near which (a_1, b_1, c_1) vanishes so far as it depends on that electron: now we have in the general correlation

$$g = g_1 + \frac{v}{4\pi C^2}(c_1 + 4\pi vg),$$

hence in this particular case

$$(g, h) = \varepsilon(g_1, h_1), \text{ while } f = \varepsilon^{\frac{1}{2}} f_1.$$

But the strength of the electron in the moving system is the value of the integral $\iint (f dy' dz' + g dz' dx' + h dx' dy')$ extended over any surface closely surrounding its nucleus; that is here $\varepsilon^{\frac{1}{2}} \iint (f_1 dy_1 dz_1 + g_1 dz_1 dx_1 + h_1 dx_1 dy_1)$ so that the strength of each moving electron is $\varepsilon^{1/2}$ times that of the correlative fixed electron. As before, no matter what other electrons are present, this argument still applies if the surface be taken to surround the electron under consideration very closely, because then the wholly preponderating part of each vector is that which belongs to the adjacent electron [*Footnote:* This result follows more immediately from § 110, which shows that corresponding densities of electrification are equal, while corresponding volumes are as $\varepsilon^{1/2}$ to unity.].

112. We require however to construct a correlative system devoid of the translatory motion in which the strengths of the electrons shall be equal instead of proportional, since motion of a material system containing electrons cannot alter their strengths. The principle of dynamical similarity will effect this.

We have in fact to reduce the scale of the electric charges, and therefore of $\frac{df}{dx} + \frac{dg}{dy} + \frac{dh}{dz}$ in a system at rest in the ratio $\varepsilon^{-1/2}$ Apply therefore a transformation

$$(x, y, z) = k(x_1, y_1, z_1), \quad t = l t_1,$$
$$(a, b, c) = \vartheta(a_1, b_1, c_1), (f, g, h) = \varepsilon^{-\frac{1}{2}} k(f_1, g_1, h_1);$$

and the form of the fundamental circuital aethereal relations will not be changed provided $k = 1$ and $\vartheta = \varepsilon^{-\frac{1}{2}} k$. Thus we may have k and l both unity and $\vartheta = \varepsilon^{-\frac{1}{2}} k$. so that no further change of scale in space and time is required, but only a diminution of (a, b, c) in the ratio $\varepsilon^{-1/2}$

We derive the result, correct to the second order, that if the internal forces of a material system arise wholly from electrodynamic actions between the systems of electrons which constitute the atoms, then an effect of imparting to a steady material system a uniform velocity of translation is to produce a uniform contraction of the system in the direction of the motion, of amount $\varepsilon^{-1/2}$ or $1 - \frac{1}{2} v^2 / C^2$ The electrons will occupy corresponding positions in this contracted system, but the aethereal displacements in the space around them will not correspond: if (f, g, h) and (a, b, c) are those of the moving system, then the electric and magnetic displacements at corresponding points of the fixed systems will be the values that the vectors

$$\varepsilon^{\frac{1}{2}}\left(\varepsilon^{-\frac{1}{2}}f,\ g - \frac{v}{4\pi C^2}c,\ h + \frac{v}{4\pi C^2}b\right)$$

and $\quad \varepsilon^{\frac{1}{2}}\left(\varepsilon^{-\frac{1}{2}}a,\ b + 4\pi vh,\ c - 4\pi vg\right)$

had at a time const. $+vx/C^2$ before the instant considered when the scale of time is enlarged in the ratio $\varepsilon^{-1/2}$

As both the electric and magnetic vectors of radiation lie in the wave-front, it follows that in the two correlated systems, fixed and moving, the relative wave-fronts of radiation correspond, as also do the rays which are the paths of the radiant energy relative to the systems. The change of the time variable, in the comparison of radiations in the fixed and moving systems, involves the Doppler effect on the wave-length."

In 1899, Hendrik Antoon Lorentz published his transformation in near modern form.[725] In 1904, Hendrik Antoon Lorentz published the following transformation,

"§ 4. We shall further transform these formulae by a change of variables. Putting

$$\frac{c^2}{c^2 - w^2} = k^2, \tag{3}$$

and understanding by l another numerical quantity, to be determined further on, I take as new independent variables

$$x' = klx, \quad y' = ly, \quad z' = lz, \tag{4}$$

---

725. H. A. Lorentz, "Simplified Theory of Electrical and Optical Phenomena in Moving System", *Proceedings of the Section of Sciences, Koninklijke Akademie van Wetenschappen te Amsterdam*, Volume 1, (1899), pp. 427–442; reprinted in K. F. Schaffner, *Nineteenth Century Aether Theories*, Pergamon Press, New York, Oxford, (1972), pp. 255-273; French translation, "Théorie Simplifiée des Phénomènes Électriques et Optiques dans des Corps en Mouvement", *Verslagen der Zittingen de Wis- en Natuurkundige Afdeeling der Koninklijke Akademie van Wetenschappen* (Amsterdam), Volume 7, (1899), p. 507; **and** *Archives Néerlandaises des Sciences Exactes et Naturelles*, Volume 7, (1902), pp. 64-80; reprinted *Collected Papers*, Volume 5, pp. 139-155

$$t' = \frac{l}{k}t - kl\frac{w}{c^2}x, \quad (5)"$$

In 1905, before the Einsteins, Poincaré published the following transformation and noted that it, together with all rotations of space, forms a group,

"The essential point, established by Lorentz, is that the equations of the electromagnetic field are not altered by a certain transformation (which I will call by the name of Lorentz) of the form:

$$x' = kl(x+\varepsilon t), \ y' = ly, \ z' = lz, \ t' = kl(t+\varepsilon x), \quad (1)$$

where x, y, z are the coordinates and t the time before the transformation and x, y, z' and t' after the transformation. Here ε is a constant which defines the transformation,

$$k = \frac{1}{\sqrt{1-\varepsilon^2}},$$

and l is an arbitrary function of ε One sees that in this transformation the x-axis plays an essential role, but one can evidently construct a transformation in which this role would be played by any arbitrary line passing through the origin. The ensemble of all these transformations together with all rotations of space, should form a group; but for this it is necessary that $l = 1$. One is thus forced to take $l = 1$ and this is a conclusion to which Lorentz was led by a different way."[726]

Prof. Anatoly Alexeivich Logunov has stressed the fact that Poincaré selflessly attributed to Lorentz, that which Poincaré had accomplished. Lorentz, alternately, and depending upon the audience, credited Poincaré and Einstein for the same innovations. Poincaré's priority is established by the dates of publication. Prof. Logunov has also stressed that many have failed to understand the significance of Poincaré's statements, wrongfully attributing priority to Einstein, which rightfully belongs to Poincaré. Prof. Logunov states, *inter alia*,

"Poincare writes: .«The idea of Lorentz», but Lorentz never wrote such

---

[726]. H. Poincaré, "Sur la Dynamique de l'Électron", *Comptes rendus hebdomadaires des séances de L'Académie des sciences*, Volume 140, (1905), pp. 1504-1508; reprinted *Œuvres de Henri Poincaré*, Volume 9, Gautier-Villars, Paris, (1954), pp. 489-493; English translation by G. H. Keswani and C. W. Kilmister, "Intimations of Relativity before Einstein", *The British Journal for the Philosophy of Science*, Volume 34, Number 4, (December, 1983), pp. 343-354, at 351.

words before Poincare. [***] We see that invariance of the equations of the electromagnetic field under transformations of the Lorentz group results in the relativity principle being fulfilled in electromagnetic phenomena. In other words, the relativity principle for electromagnetic phenomena follows from the Maxwell-Lorentz equations in the form of a rigorous mathematical truth. [***] It must be underlined that, by having established the group nature of the set of all purely spatial transformations together with the Lorentz transformations, that leave the equations of electrodynamics invariant, Poincare thus discovered the existence in physics of an essentially new type of symmetry related to the group of linear space-time transformations, which he called the Lorentz group. [***] Poincare thus introduces the physical concept of gravitational waves, the exchange of which generates gravitational forces, and supplies and estimation of the contribution of relativistic corrections to Newton's law of gravity. For example, he shows that the terms of first order in ϑ/c cancel out exactly and so the relativistic corrections to Newton's law are quantities of the order of $(v/c)^2$ [***] It is here that such concepts as the following first appeared: the Lorentz group, invariance of the equations of the electromagnetic field with respect to the Lorentz transformations, the transformation laws for charge and current, the addition formulae of velocities, the transformation laws of force. Here, also, Poincare extends the transformation laws to all the forces of Nature, whatever their origin might be."[727]

In 1905, without reference to prior authors, Mileva and Albert Einstein wrote,

"It follows from this relation and the one previously found that $\varphi(v) = 1$ so that the transformation equations which have been found become

$$\tau = \beta(t - vx/c^2),$$
$$\xi = \beta(x - vt),$$
$$\eta = y,$$
$$\zeta = z,$$

where $\beta = 1/\sqrt{1 - v^2/c^2}$."[728]

Given the facts that Galileo popularized the concept of the principle of relativity, Lange took from it absolute space and absolute time, Voigt introduced

---

[727]. A. A. Logunov, translated by G. Pontecorvo, *On the articles by Henri Poincare ON THE DYNAMICS OF THE ELECTRON*, Second Edition, Joint Institute for Nuclear Research, Dubna, (1995), pp. 9, 10, 14.
[728]. *The Principle of Relativity*, Dover, New York, (1952), pp. 47-48.

the relativistic transformation, and Poincaré first demonstrated relative simultaneity; why is the concept popularly referred to as "Einstein's special theory of relativity"? Einstein contributed next to nothing to the special principle of relativity. Why are the popular misconceptions of Einstein, and his supposed discoveries; which misconceptions are fed by the scientific community and the media; and the factual historic record, itself, at odds? Is exposing the truth counter-productive, if it means the downfall of a hero and the death of a religion?

Contrary to the view of some Einstein advocates that Einstein worked in near complete isolation from both the scientific literature and the physics community, many have pointed out that Einstein had easy access to the literature at the Swiss Patent Office and was heavily immersed in the most recent physics literature of the day as a prolific reviewer of that literature for the *Beiblätter zu den Annalen der Physik*. Jules Leveugle has stressed the fact that Einstein and Planck were exposed to the recent writings of Poincaré and Lorentz through many sources including the *Beiblätter zu den Annalen der Physik* and *Fortschritte der Physik*. Einstein published 21 reviews in the *Beiblätter* in 1905.[729] Jules Leveugle points out in his book *Poincaré et la Relativité : Question sur la Science*, that the *Beiblätter* published the following review of Lorentz' 1904 paper by Richard Gans, in Volume 29, Number 4, (February, 1905), pp. 168-170:

> "15. H. A. Lorentz. *Elektromagnetische Vorgänge in einem Systeme, das sich mit einer willkürlichen Geschwindigkeit (kleiner als die des Lichtes) bewegt* (Versl. K. Ak. van Wet. 12, S. 986-1009. 1904). — Durch die ursprüngliche Lorentzsche Elektronentheorie ist nicht erklärt: 1. Daß die Erdbewegung auf die Interferenz des Lichtes keinen Einfluß hat (Michelson und Morley). 2. Daß auf einen geladenen Plattenkondensator kein Drehmoment wirkt (Trouton und Noble).
>
> Die erste Tatsache ist durch eine neue Hypothese von FitzGerald und Lorentz erklärt worden, nämlich dadurch, daß die Dimensionen fester Körper in Richtung der Erdbewegung ein wenig kleiner werden.
>
> 3. Diese Hypothese verlangt eine Doppelbrechung des Lichtes in isotropen Körpern infolge der Erdbewegung; die Versuche ergaben ein negatives Resultat (Lord Rayleigh, Brace).
>
> Um diese Widersprüche zu beseitigen, stellt der Verf. folgende Betrachtungen an:
>
> Erfährt das elektromagnetische System eine konstante Geschwindigkeit w in Richtung der x —Achse, und ist die Lichtgeschwindigkeit c setzen wir ferner

---

[729]. *Cf.* J. Stachel, *et al.*, Editors, "Einstein's Reviews for the *Beiblätter zu den Annalen der Physik*", *The Collected Papers of Albert Einstein*, Volume 2, Princeton University Press, (1989), pp. 109-111.

$$\frac{c^2}{c^2 - w^2} = k^2,$$

und bilden den Raum ab durch die Transformation $x' = kx, y' = y, z' = z$ und führen anstatt der Zeit t die „Ortszeit"

$$t' = \frac{t}{k} - \frac{kwx}{c^2}$$

ein, so erhalten wir, wenn wir anstatt der elektrischen und und magnetischen Feldstärke *d* bez. *h* etwas andere Vektoren *d'* und *h'* einführen, Gleichungen im bewegten, durch die Abbildung transformierten System, welche genau so gebildet sind, wie die Lorentzschen Gleichungen im ursprünglichen ruhenden System. Es folgt daraus, daß das Feld (*d'*, *h'*) in aller Strenge dem Felde im ruhenden System an entsprechenden Punkten gleich ist, d. h. im elektrostatischen oder optischen *Felde* ist kein Einfluß irgend einer Ordnung der Bewegung zu konstatieren. Die ponderomotorischen *Kräfte* auf die Volumeinheit dagegen erleiden eine kleine Änderung entsprechend der Volumänderung, es ist

$$f'_x = f_x \quad f'_y = \frac{f_y}{k} \quad f'_z = \frac{f_z}{k},$$

wo die gestrichenen Buchstaben im bewegten System gelten.

Diese Umformung gibt die Hypothese an die Hand, daß die Dimensionen der Elektronen durch die Bewegung in derselben Weise verändert werden wie der Raum durch die oben angegebene Transformation, daß aber die Ladung entsprechender Volumelemente dieselbe bleibt.

Ferner sollen auch nicht-elektrische (z. B. elastische) Kräfte dieselbe Veränderung durch die Translation erfahren, wie oben die ponderomotorischen Kräfte *f* elektrischen Ursprungs.

Daraus folgt, daß ein Körper, der durch die Anziehungen und Abstoßungen seiner inneren Kräfte im Gleichgewicht ist, *von selbst* durch die Bewegung seine Dimensionen ändert, denn war im ruhenden System die resultierende Kraft 0 (also Gleichgewicht), so ist sie 0 im bewegten *transformierten* System (also Gleichgewicht).

So erklärt sich der Michelson und Morleysche Interferenzversuch, ferner der von Trouton und Noble über das Drehmoment eines geladenen Plattenkondensators und auch die vergeblichen Doppelbrechungsversuche von Lord Rayleigh und Brace, denn der schon früher vom Verf. (bis auf Größen zweiter Ordnung) aufgestellte Satz, daß Helligkeit, Dunkelheit, Strahl im ruhenden System Helligkeit, Dunkelheit, Strahl im bewegten transformierten entsprechen, gilt bei der jetzigen Transformation streng in

Gliedern aller Ordnungen.

Die Formeln für die elektromagnetische Masse ändern sich infolge der Abplattung der Elektronen, aber stellen trotzdem die Kaufmannschen Versuche über Becquerelstrahlen mit befriedigender Genauigkeit dar, wie eingehende Zahlenrechnungen zeigen. Gans."

Gans also published a paper, "Zur Elektrodynamik in bewegten Körpern", *Annalen der Physik*, Series 4, Volume 16, (1905), pp. 516-534.

Emil Cohn published a paper that cited Lorentz' 1904 paper containing the "Lorentz Transformation", with which Cohn paper Einstein was familiar, "Zur Elektrodynamik bewegter Systeme", *Sitzungsberichte der Königlich Preussischen Akademie der Wissenschaften zu Berlin, Sitzung der physikalisch-mathematischen Classe*, (November, 1904), pp. 1294-1303, at 1295. Einstein cited Cohn's paper in his *Jahrbuch* review article of 1907, and a copy of Cohn's 1904 paper is in his preserved collection. See: *The Collected Papers of Albert Einstein*, Volume 2, Note 128, Hardcover, p. 272. Cohn cites the Dutch version of Lorentz' work, "Electromagnetische Verschijnselen in een Stelsel dat zich met Willekeurige Snelheid, Kleiner dan die van het Licht, Beweegt." *Verslagen van de Gewone Vergaderingen der Wis- en Natuurkundige Afdeeling, Koninklijke Akademie van Wetenschappen te Amsterdam*, Volume 12, (23 April 1904), pp. 986-1009. Einstein cites Cohn in the direct context of Lorentz' 1904 paper in: A. Einstein, "Über das Relativitätsprinzip und die aus demselben gezogenen Folgerung", *Jahrbuch der Radioaktivität und Elektronik*, Volume 4, (1907), pp. 411-462, at 413.

Jules Leveugle notes that Felix Klein annotated Lorentz' article "Weiterbildung der Maxwellschen Theorie. Elektronentheorie", in Volume 2, Part 2, Chapter 14, pp. 145-280, of the *Encyklopädie der Mathematischen Wissenschaften*, with note 113:

"113) *Lorentz*, Amsterdam Zittungsverslag Akad. v. Wet 12, 1904 (Amsterdam Proceedings, 1903-1904)."

and that Max Abraham also referred his readers to Lorentz' 1904 paper, in Abraham's "Die Grundhypothesen der Elektronentheorie", *Physikalische Zeitschrift*, Volume 5, (1904), pp. 576-579:

"2) H. A. Lorentz, K. Akad. van Wetensch. te Amsterdam 1899, S. 507 und 1904, S. 809."

and that Sommerfeld cited Lorentz' 1904 paper in his paper, "Simplified Deduction of the Field and Forces of an Electron, Moving in Any Given Way" in the Koninklijke Akademie van Wetenschappen te Amsterdam Proceedings of the Meeting of Saturday, November 26, 1904, p. 346:

"1) K. Akademie van Wetenschappen te Amsterdam Mei 1904. Proceedings p. 809."

and that Grimm wrote of Lorentz' work in *Die Fortschritte der Physik*, (1905), p. 29:

"H. A. Lorentz. Electrodynamic phenomena in a system moving with any velocity smaller than that of light. Proc. Amsterdam 6. 809-831, 1904. Versl. Amsterdam 12, 986-1009, 1904.
Nachdem neuerdings eine Reihe neuer Versuche gemacht worden sind, die sämtlich das Resultat hatten, daß auch ein Einfluß zweiter Ordnung der Erdbewegung nicht zu konstatieren ist, hat Verf. es als notwendig gefunden, seiner und FITZGERALDs Hypothese, daß die Dimensionen der Körper durch ihre Bewegung geändert würden, eine allgemeinere Grundlage zu geben. Er stellt zunächst die Grundgleichungen der Elektronentheorie auf für ein sich mit einer Geschwindigkeit bewegendes System, die geringer als Lichtgeschwindigkeit ist, und dann transformiert er die Gleichungen auf ein System, das gegen das erste in der Bewegungsrichtung deformiert ist. Er erhält somit Gleichungen, die ihm gestatten, die in einem Felde gegebenen Punkte bzw. Funktionen sofort auch im anderen Felde zu finden. Hiernach führt er nun die Hypothese ein, daß die Elektronen ihre Dimensionen in der Bewegung dieser Deformation entsprechend ändern, während sie in der Ruhe Kugeln sind, und daß die Kräfte, die zwischen ungeladenen Partikeln und zwischen solchen und Elektronen bestehen, in gleicher Weise wie die elektrischen Kräfte in einem elektrostatischen System durch Translation beeinflußt werden. Es wird nun das elektromagnetische Moment eines einzelnen Elektrons berechnet und für die ARAHAMsche quasistationäre Bewegung ergibt sich dann eine rein elektromagnetische Masse des Elektrons. Dann wird der Einfluß der Bewegung auf optische Phänomene betrachtet, wobei Verf. zu dem Schlusse kommt, daß in der Deformation (l, l, kl) das $l = const$ sein muß und die Anwendung auf die übrigen neueren Versuche führt zu der allgemeinsten Hypothese, daß „die Massen aller Partikel durch die Bewegung in gleicher Weise beeinflußt werden, wie die elektromagnetischen Massen der Elektronen". Im weiteren wird die Theorie an KAUFMANNs Tabellen geprüft und gibt dabei ungefähr gleich gute Übereinstimmung, wie die KAUFMANNschen Formeln. Zum Schluß wird noch der Versuch von TROUTON diskutiert. *Grm.*"

9.4.5 Einstein's Fudge

As is well known, numerous authors have shown errors in the Einsteins' fallacy of *Petitio Principii*, including, among many others, Essen, Keswani, Miller, Planck, and Guillaume.

9.4.6 Einstein Begged the Question

Albert Einstein's arguments were almost always fallacies of *Petitio Principii*. He argued well-known experimental results as if they were *a priori* first principles. Einstein would then induce, as if deducing, the well-known hypotheses of others, and deduce from these plagiarized hypotheses the same experimental results as conclusions, which he had first stated as premises. This was Einstein's *modus operandi* for plagiarism. In the special theory of relativity, Einstein irrationally argued that light speed invariance, supposedly a well-known experimental result at the time, was an *a priori* first principle, which an empirical measurement cannot be, so that he could then induce through analysis, as if deducing in synthesis, the "Lorentz Transformation" hypotheses. Einstein then used the "Lorentz Transformation", the true set of hypotheses of the special theory of relativity, to deduce light speed invariance as a conclusion, a conclusion which Einstein had already presumed as a premise. Einstein also employed the generalized equivalence of all inertial systems he alleged was observed in the Michelson experiments, as if it were an *a priori* principle, instead of the *a posteriori* empirical observation it was, to then "deduce" from this supposed first principle, the principle itself—Michelson's result.

Einstein employed the same fallacious method in the general theory of relativity. Einstein irrationally asserted the well-known experimental gravitational-inertial mass equivalence of Newton, Bessel and Eötvös as if it were an "*a priori*" postulate, which an experimental result cannot be, only to arrive at it as an ultimate conclusion, a conclusion which was redundant to the premise. The quasi-positivistic analyses Einstein presented by turning the synthetic scientific theories of his predecessors on their heads have been applauded, ridiculed and often misrepresented as if they are synthetic, which they are not.

Albert Einstein gave a lecture at King's College in June of 1921. *The London Times* reported on 14 June 1921, on page 8,

> "PROFESSOR EINSTEIN said it gave him special pleasure to lecture in the capital of that country from which the most important and fundamental ideas of theoretical physics had spread throughout the world—the theories of motion and gravitation of Newton and the proposition of the electromagnetic field on which Faraday and Maxwell built up the theories of modern physics. It might well be said that the theory of relativity formed the finishing stone of the elaborate edifice of the ideas of Maxwell and Lorentz by endeavouring to apply physics of 'fields' to all physical phenomena, including the phenomena of gravitation.
>
> Professor Einstein pointed out that the theory of relativity was not of any speculative origin, but had its origin solely in the endeavour to adapt the theory of physics to facts observed. It must not be considered as an arbitrary act, but rather as the result of the observations of facts, that the conceptions of space, time, and motion, hitherto held as fundamental, had now been abandoned.
>
> Two main factors, continued Professor Einstein, have led modern science to regard time as a relative conception in so far as each inertial system had to be coupled with its own peculiar time: the law of constancy of the velocity of light in vacuo, sanctioned by the development of the

sciences of electro-dynamics and optics, and in connexion therewith the equivalence of all inertial systems (special principle of relativity) as clearly shown by Michelson's famous experiment. In developing this idea it appeared that hitherto the interconnexion between direct events on the one hand, and the space coordinates and time on the other, had not been thought out with the necessary accuracy.

The theory of relativity endeavours to define more concisely the relationship between general scientific conceptions and facts experienced. In the realm of the special theory of relativity the space coordinates and time are still of an absolute nature in so far as they appear to be measurable by rigid bodies, rods, and by clocks. They are, however, relative in so far as they are dependent upon the motion peculiar to the inertial system that happens to have been chosen. According to the special theory of relativity the four-dimensional *continuum*, formed by the amalgamation of time and space, retains that absolute character which, according to the previous theories, was attributed to space as well as to time, each individually. The interpretation of the spatial coordinates and of time as the result of measurements then leads to the following conclusions: motion (relative to the system of coordinates) influences the shape of bodies and the working of clocks; energy and inertial mass are equivalent.

GRAVITATIONAL FIELDS.

The general theory of relativity owes its origin, continued Professor Einstein, primarily to the experimental fact of the numerical equivalence of the inertial and gravitational mass of a body; a fundamental fact for which the classical science of mechanics offered no interpretation. Such an interpretation is arrived at by extending the application of the principle of relativity to systems of coordinates accelerated with reference to one another. The introduction of systems of co-ordinates accelerated with reference to inertial systems causes the appearance of gravitational fields relative to the systems of coordinates. That is how the general theory of relativity, based on the equality of inertia and gravity, offers a theory of the gravitational field.

Now that systems of co-ordinates, accelerated with reference to one another, have been introduced as equivalent systems of co-ordinates, based on the identity of inertia and gravity, it follows that the laws governing the position of rigid bodies in the presence of gravitational fields do not conform to the rules of Euclidean geometry. The results as regards the working of clocks is analogous. These conclusions lead to the necessity of once more generalizing the theories of space and time, because it is no longer possible directly to interpret the co-ordinates of space and time by measurements with measuring rods and clocks. This generalization of metrics, which in the sphere of pure mathematics dates back to Gauss and Riemann, is based largely on the fact that the metrics of the special theory of relativity may be considered to apply in certain cases also to the general theory of relativity. In consequence, the co-ordinate system of space and time is no longer a reality in itself. Only by connecting the space and time co-ordinates with those mathematical figures which define the gravitational field can the objects which may be measured by measuring rods and by clocks be determined.

The idea of the general theory of relativity has yet another basis. As Ernst Mach has already emphasized, the Newtonian theory of motion is unsatisfactory in the following point:—if motion is regarded not from the casual but from the purely description point of view it will be found that there exists a relative motion of bodies with reference to each other. But the conception of relative motion does not of itself suffice to formulate the factor of acceleration to be found in Newton's equations of motion. Newton was forced to introduce a fictitious physical space with reference to which an acceleration was supposed to exist. This conception of absolute space introduced by Newton *ad hoc* is unsatisfactory, although it is logically correct. Mach, therefore, endeavoured so to alter the mechanical equations that the inertia of bodies is attributed to their relative motion with reference not to absolute space but with reference to the sum total of all other measurable bodies. Mach was bound to fail considering the state of knowledge at his time. But it is quite reasonable to put the problem as he did. In view of the general theory of relativity this line of thought comes more and more to the fore, because according to the theory of relativity the physical properties of space are influenced by matter.

Professor Einstein said he was of the opinion that the general theory of relativity could only solve this problem satisfactorily by regarding the universe as spatially finite and closed. The mathematical results of the theory of relativity forced scientists to this view, if they assumed that the average density of matter within the universe was of finite, if ever so small a value."

In 1905, Mileva Einstein-Marity and Albert Einstein coauthored a paper on the "electrodynamics of moving bodies". Fallacies of begging the question emerge in the very introduction to the work. The Einsteins acknowledge in their introduction, that light speed invariance and the symmetry of electrodynamic phenomena were well-established phenomena. Well-known specific phenomena are not, by definition, "*a priori*" general concepts. However, the Einsteins asked us to abandon reason and assert specific experimental results and empirical observations, as if they were *a priori* general principles. In other words, the Einsteins engaged in an analysis of the problems of invariant light speed, and of the symmetry of electrodynamic phenomena in alleged violation of Maxwell's theory, which problems faced physicists at the end of the Nineteenth Century. The Einsteins irrationally pretended that these two problems were solutions of themselves.

Henry August Rowland stated the two main problems facing the physicists of his day, on 28 October 1899, and I have italicized that which the Einsteins would later erroneously call "two assumptions", or "postulates":

"And yet, however wonderful [the ether] may be, its laws are far more simple than those of matter. *Every wave in it, whatever its length or intensity, proceeds onwards in it according to well known laws, all with the same speed, unaltered in direction, from its source in electrified matter to the confines of the Universe, unimpaired in energy unless it is disturbed by the presence of matter.* However the waves may cross each other, each proceeds by itself without interference with the others. [***] *To detect something dependent on the relative motion of the ether and matter has been and is the great desire of physicists. But we always find that, with one possible exception, there is always some compensating feature which renders our efforts useless.* This one experiment is the aberration of light, but even here Stokes has shown that it may be explained in either of two ways: first, that the earth moves through the ether of space without disturbing it, and second, if it carries the ether with it by a kind of motion called irrotational. Even here, however, the amount of action probably depends upon relative motion of the luminous source to the recipient telescope. So the principle of Doppler depends also on this relative motion and is independent of the ether. The result of the experiments of Foucault on the passage of light through moving water can no longer be interpreted as due to the partial movement of the ether with the moving water, an inference due to imperfect theory alone. The experiment of Lodge, who attempted to set the ether in motion by a rapidly rotating disc, showed no such result. The experiment of Michelson to detect the ethereal wind, although carried to the extreme of accuracy, also failed to detect any relative motion of the matter and the ether [*Emphasis Added*]."[730]

The Einsteins turned reason on its head and called these two *a posteriori* problems, *a priori* "postulates". The Einsteins phrased their two "postulates", as follows:

1 (a). "Examples of a similar kind, as well as the failed attempts to find a motion of the earth relative to the 'light medium', lead to the supposition, that the concept of absolute rest corresponds to no characteristic properties of the phenomena not just in mechanics, but also in electrodynamics, on the contrary, for all systems of coordinates, for which the equations of mechanics are valid, the same electrodynamic and optical laws are also valid, as has already been proven for the magnitudes of the first order."

1 (b). "The laws according to which the states of physical systems change do not depend upon to which of two systems of coordinates, in uniform translatory motion relative to each other, this change of state is referred."

---

730. H. A. Rowland, *The Physical Papers of Henry August Rowland*, The Johns Hopkins Press, Baltimore, Maryland, (1902), pp. 673-674.

2 (a). "[L]ight in empty space always propagates with a determinate velocity c irrespective of the state of motion of the emitting body."

2 (b). "Every ray of light moves in the 'resting' system of coordinates with the determinate velocity $c`$, irrespective of whether this ray of light is emitted from a resting or moving body. Such that

velocity = (path of light) / (interval of time) ,

where 'interval of time' is to be construed in the sense of the definition of § 1."

Note that the first "postulate", the principle of relativity, refers only to "moving systems"; and that the second "postulate", the light "postulate", refers only to a proposed "resting system". Note further, that the light "postulate" refers only to a proposed source independence of light speed, but not to an observer independence, because this "postulate" assumes a prior privileged frame and medium in the 1905 paper, which the Einsteins identify as the "resting system". The expression "resting system" was well understood at the time to refer to "absolute space" and a system of coordinates at rest relative to the "fixed stars". The Einsteins' paper later presumes that $c' = c +/-v$ relative to the "resting system".

Many assert that the Einsteins employed only these two "*a priori* postulates" in their theorization, as opposed to FitzGerald, Larmor, Lorentz, and Poincaré, who required the additional hypotheses of length contraction, time dilatation and an æther *to arrive at the same formulation—long before the Einsteins*. Ad hoc hypotheses were frowned upon at the time, due to Newton's admonitions against them, such that the removal of hypotheses was seen as an improvement. The two postulate myth is substantially and demonstrably false. The two postulates are not postulates, but rather are the deduced conclusions of the theory— summations of the supposedly observed phenomena of the day. The "postulates" are deducible from the more fundamental hypotheses of length contraction, time dilatation, relative simultaneity, inertial motion, an æther, etc.; and these are the actual fundamental hypotheses of the special theory of relativity.

Length contraction is not deduced from invariant light speed *a priori*. It is more fundamental than light speed, which is derived from it, and is logically induced from invariant light speed *a posteriori*. Length contraction is a specific factor which deduces the broad range of all velocity comparisons, not just light speed invariance, which represents but one of these comparisons and a deduced limit. The same is true of time dilatation and relative simultaneity. A wide range of hypotheses which deduce an æther and inertial motion are far more fundamental than the deduced conclusions of light speed source independence and the covariance of the laws of nature in inertial systems. It might be true that no one has yet created a fully fundamental theory to deduce these conclusions, but that does not render empirical observations *a priori*, nor does it mean that the attempt to inductively arrive at a such a set of hypotheses *a posteriori* is futile or detrimental. In addition, the evidence taken to justify the hypotheses which are accepted in the theory of relativity has not been rationally interpreted by the "relativists".

After asserting the two "postulates", the Einsteins raised a straw man argument based a *non sequitur*. They asserted that the two "postulates" appeared irreconcilable with each other. If light speed is constant in the "resting system", then how can it also be isotropic in a "moving system" in motion relative to the "resting system"? This is a manufactured dilemma, because, in some inexplicable way, the Einsteins argue that the first postulate, the principle of relativity, compels that light speed from all sources be isotropic for all systems in uniform inertial motion with each respect to each other. However, this is clearly a *non sequitur*, because the principle of relativity no more compels light speed isotropy for all "moving systems", then the principle of relativity requires that a body resting relative to one "moving system" k also rest relative to another "moving system" K, which is in motion relative to the first.

The Einsteins also raised the opposing problem. How can light speed be isotropic in the "resting system" and also be isotropic in a "moving system"? Of course, these questions presume the conclusion before it has been proven, the conclusion being that light speed from any given signal is isotropic in the "resting system" and all "moving systems", which are in uniform translatory motion with respect to the "resting system". This conclusion is an alleged empirical observation, which much be deduced from fundamental assertions. It is not an *a priori* fundamental assertion. The Einsteins' "postulates" are in fact the very conclusions which they seek prove. The have manufactured a fallacy of *Petitio Principii*.

To knock down these straw men, the Einsteins turned the "two postulates" into one "postulate", the ultimate conclusion which is sought. The Einsteins asserted that it is the *combination* of the two postulates, not either postulate by itself, which "deduces" $c' = c$ between the moving system and the resting system, by simply asserting in their paper that $c' = c$ before it has in any way been logically proven (there is a distinction and difference between a logical proof and an empirical observation and the union of the "two postulates" does not constitute a logical proof, but rather discloses the redundancy of the "postulates" to each other—as Louis Essen has stated, they are one alleged empirical fact summarized in two redundant ways):

> "It is easy, with the help of this result, to ascertain the magnitudes ξ, η, ζ, because one expresses by means of these equations, that light (as the principle of the constancy of the velocity of light, in conjunction with the principle of relativity, requires) also propagates with the velocity c as measured in the moving system."

After irrationally presuming this conclusion that $c' = c$ before it has in any way been logically proven, the Einsteins proceeded to pretend that they had not presumed it:

> "Now, we have to prove that every ray of light propagates with the velocity c as measured in the moving system, in case this is, as we have taken for granted, the case in the resting system, because we still have not offered up the proof that the principle of the constancy of the velocity of light is reconcilable with the principle of relativity."

However, unless we presume that the "two postulates" are redundant, the combination of the two postulates results in $c' = c +/-v$, not $c' = c$ If we do not presume that the "two postulates" are redundant, then the principle of relativity applies only to "moving systems" and the principle of the constancy of the velocity of light independent of the speed of the source is an æthereal principle of the "resting system" and only of the "resting system".

In a rational approach to the problem, one must instead take the supposed empirical phenomenon of $c' = c$ as a point of departure for an *a posteriori* inductive analysis, not an *a priori* deductive synthesis, and from there induce a fundamental geometry *a posteriori*, which fundamental geometry then deduces the identity $c' = c$ and the covariance of the laws of physics *a priori*, in a synthetic scientific theory. Albert Einstein never accomplished such a theory and he politically obstructed valid criticisms of his irrationality by calling his critics "anti-Semitic" for daring to questions his fallacies of *Petitio Principii*. Albert Einstein stifled scientific progress with disingenuous "racial" politics and was himself a racist and a segregationist, and therefore a dangerous hypocrite.

The Einsteins averred, before any proof was offered:

> "It is easy, with the help of this result, to ascertain the magnitudes ξ, η, ζ, because one expresses by means of these equations, that light (as the principle of the constancy of the velocity of light, in conjunction with the principle of relativity, requires) also propagates with the velocity c as measured in the moving system. For a ray of light emitted in the direction of increasing ξ, at the time $\tau = 0$ the following equations are valid: $\xi = c\tau$"

Note the *non sequitur*, which begs the question: That allegedly if the speed of light is c in the "resting system" the principle of relativity compels that it also be measured to be c in the "moving system"; which, without the prior hypotheses of the Lorentz Transformation, clearly is not a rational conclusion, for if I rest in the resting system, the principle of relativity does not compel that I also rest in the moving system. The detection of an æther frame only violates the principle of relativity *if* we assume that the æther exists and that it is at *absolute rest*, and then only because it would provide a means to detect one's speed relative to that æther which has arbitrarily been identified as being at rest in absolute space, which is another straw man argument because rest relative to a light medium does not constitute of necessity "absolute rest"—without the metaphysical presumption of an æther at absolute rest, there is no special theory of relativity, despite its advocates assertions to the contrary. At any rate, the assertion that the detection of the æther frame would violate the principle of relativity is false and is a straw man argument made to justify the assumption that the æther rests. On the contrary, the only principle the detection of the æther frame would violate is the arbitrary principle that the æther frame cannot be detected, and the means of resolving this principle that the æther cannot be detected is the Lorentz Transformation, not the principle of relativity. It is the Lorentz Transformation which renders the laws of electrodynamics covariant, not the principle of relativity. The Einsteins simply confused their conclusion as an additional premise, which renders the two "postulates" redundant, or renders one postulate deducible from the other, and in no sense a postulate.

There is also a fallacy in the special theory of relativity of defining a violation of the principle of relativity in at least four different and distinct ways and then pretending that those different and distinct definitions are one definition. The principle of relativity is on the one hand defined as the invariance of the laws of nature in inertial frames of reference. It must be borne in mind that this principle of relativity treats of abstract idealizations and not physical reality and that inertial frames of reference do not exist in nature. This first principle is the principle of relativity of classical mechanics, which has the consequence of making it impossible to determine "absolute space" by means dynamic experiments.

Though many have averred that this principle is equivalent to, or the same as, the negative assertion that it is impossible to determine the frame of absolute space by means of the laws of mechanics, or more broadly, by any means; this consequence is not the principle itself, and it might be possible someday to determine a preferred reference frame of space (as is the case with general relativity, or the "fixed stars") without setting aside the principle of relativity. We have identified the classical principle of relativity of mechanics, and a distinct and different consequence of that principle, which is also wrongfully called the principle of relativity.

There is a third distinct and different principle of relativity introduced by Henri Poincaré, which states that the laws of electrodynamics are covariant in inertial frames of reference. This principle depends upon the presupposition of Maxwell's laws of electrodynamics and the preferred reference frame of the æther, which provides an *a priori* basis for an inertial frame of reference and for the source independence of light speed. However, this third principle is not a logical necessity, and defines the identity of the laws of physics in a different way from the classical principle of relativity by means of a different system of velocity addition. According to the classical principle of relativity, the æther ought to be detectable, and it is only rendered undetectable by the Lorentz Transformation, not the principle of relativity.

The fourth distinct and different principle of relativity is the assertion that it is impossible to detect the frame of reference of the æther itself, which is an alleged consequence of the principle of relativity of electrodynamics, not that principle itself. The æther may have properties other than electrodynamic properties which renders its position detectable, and therefore one might be able to detect the frame of the æther without violating the principle of relativity of electrodynamics, as may be the case with "tachyons" or other such proposed phenomena.

The Einsteins, following Poincaré's example, deliberately confused logical consistency between these four different definitions, an artificial consistency obtained through the *ad hoc* Lorentz Transformation; with the assertion, which is false, that logical necessity requires that if one of these principle is true, then the other three must also be true. The only binding agent between these different definitions is the tacit presumption and arbitrary definition that the detection of light speed anisotropy would constitute, of necessity, the detection of an æther at absolute rest, which would, by abstract definition alone, constitute the detection of "absolute space", which, by abstraction definition alone, is in principle not detectible in either definition.

This is a straw man argument and a *non sequitur* in that one can detect the medium of a sound wave without violating the principle of relativity, and the "relativists" have falsely and artificially confused the detection of a light medium with a violation of the "principle of relativity" and the detection of "absolute space". In addition, the "relativists" have falsely assumed that the detection of a preferred frame of reference by any means violates both the principle of relativity of mechanics and the principle of relativity of electrodynamics.

There is complete logical consistency between the detection of light speed anisotropy in a frame of reference moving with respect to the æther, and the principle of relativity of mechanics; and the entirely artificial addition to the principle of relativity of mechanics of the assertion that the principle of relativity of electrodynamics forbids the detection of an æther frame is *ad hoc* and a straw man argument, which presupposes an æther at absolute rest and which cannot exist without the supposition of an an æther at absolute rest, and which depends upon the false assumption that the detection of absolute space violates the principle of relativity of mechanics. The principle of relativity of mechanics only states that the laws of mechanics are the same in all inertial reference frames, which is different from the assertion that "absolute space" is undetectable. If "absolute space" were detected by a "resting æther" (a definition alone), this in and of itself would not be a violation of the principle of relativity of mechanics nor the principle of relativity of electrodynamics, though it would put an end to the metaphysical myth of "space-time".

Mileva and Albert have wrongly confused the fact that the *ad hoc* Lorentz Transformation renders the undetectability of the æther frame logically consist with the classical principle of relativity when it otherwise would not be, with Henri Poincaré's irrational assertion that the principle of relativity demands of logical necessity that the light medium be undetectable; as if that artificially derived logical consistency were itself a logical necessity, when it is not—quite the contrary, without the *ad hoc* Lorentz Transformation the principle of relativity demands that the æther frame be detectable, or that light speed be source and observer speed dependent. All of these tacit presumptions in the special theory of relativity presume the existence of an æther at absolute rest, and not only has the special theory of relativity not rendered an æther at absolute rest superfluous, the entire theory depends upon the tacit premise of an æther at absolute rest, which is in "principle" undetectable by means of electrodynamics, though it is theoretically detectable by means of superluminal velocities, or other means.

There is a difference between arguing that a set of circumstances renders a physical entity undetectable, and arguing that a set of circumstances renders a physical entity superfluous, and the Einsteins, following Poincaré's example, have deliberately and falsely confused undetectablity with superfluousness, just as they have deliberately and falsely confused logical consistency with logical necessity. The so-called "principle" that the æther at absolute rest is undetectable is in fact a corollary to the tacitly presumed properties of that æther and incorporates the presumption of such an æther at "absolute rest" in the very definition. The "principle" is a deducible conclusion, not a fundamental premise. The fundamental premise is the existence of an æther at "absolute rest"—though even this assertion is deducible from more fundamental elements.

There is also a difference between the assertion that the *resting frame* of an æther arbitrarily defined as at "absolute rest" is *undetecable*, and the assertion that the *æther* as a light medium is *undetecable*. In all of our human observations of physical entities we depend upon our senses and our definitions, and our consciousness of an image is not the actual entity reflected in our images of the physical world. Our knowledge of the æther exists in, among other things, the presumption of the source speed independence of light speed. The *æther* is detectable in the special theory of relativity even though its presumed *resting frame of reference* remains undetectable by means of electrodynamic experiments.

In addition, the entire structure of the Lorentz Transformation is built upon the presumption of light speed anisotropy in moving frames of references, which fact is revealed by the use of the scalar c^2. The Einsteins' assertion of the absolute velocity of light in the "resting system" as a given axiomatic fact is an acknowledgment that the "resting system" is an æther at absolute rest, and this is how the Einsteins' define it in Part 1, Section 1 of their paper. If light speed were not anisotropic in moving frames of reference, the Lorentz Transformation would not work, because light speed would not then be measured to be c in a moving frame of reference by observers relatively resting in that moving frame—moving with respect to the æther. This has been adequately proven by Guillaume, Jánossy and others.[731] Prof. Friedwardt Winterberg wrote,

---

731. E. V. Huntington, "A New Approach to the Theory of Relativity", *Festschrift Heinrich Weber zu seinem siebzigsten Geburtstag am 5. März 1912 / gewidmet von Freunden und Schülern*, B. G. Teubner, Leipzig, (1912), pp. 147-169; reprinted "A New Approach to the Theory of Relativity", *Philosophical Magazine*, Series 6, Volume 23, Number 136, (April, 1912), pp. 494-513. *See also:* S. Mohorovičić, "Äther, Materie, Gravitation und Relativitätstheorie", *Zeitschrift für Physik*, Volume 18, Number 1, (1923), pp. 34-63, at 34. *See also:* H. Ives in, D. Turner and R. Hazelett, *The EINSTEIN Myth and the Ives Papers: A Counter-Revolution in Physics*, Devin-Adair, Old Greenwich, Connecticut, (1979). *See also:* L. Jánossy, "Über die physikalische Interpretation der Lorentz-Transformation", *Annalen der Physik*, Series 6, Volume 11,

"According to Einstein, two clocks, **A** and **B**, are synchronized if

$$t_B = \frac{1}{2}\left(t_A^1 + t_A^2\right)$$

(VII.13)

where $t^1{}_A$ is the time a light signal is emitted from **A** to **B**, reflected at **B** back to **A**, arriving at **A** at the time $t^2{}_A$, and where it is <u>assumed</u> that the time t_B at which the reflection at **B** takes place is equal the arithmetic average of $t^1{}_A$ and $t^2{}_A$. Only by making this assumption does the velocity of light turn out always to be isotropic and equal to **c**. From an absolute point of view, the following is rather true: If t_R is the absolute reflection time of the light signal at clock **B**, one has for the out and return journeys of the light signal from **A** to **B** and back to **A**, if measured by an observer in an absolute system at rest in the distinguished reference system:

$$\gamma\left(t_R - t_A^1\right) = d/c_+ ,$$

$$\gamma\left(t_A^2 - t_R\right) = d/c_-$$

(VII.14)

where **d** is the distance between both clocks, and where c_+ and c_- are given by

---

(1953), pp. 293-322; **and** *Theory of Relativity Based on Physical Reality*, Akademiai Kiadó, Budapest, (1971). *See also:* G. Builder, "Ether and Relativity", *Australian Journal of Physics*, Volume 11, (1958), pp. 279-; **and** "The Constancy of the Velocity of Light," *Australian Journal of Physics*, Volume 11, (1958), pp. 457-480; abridged form reprinted with bibliography in: *Speculations in Science and Technology*, Volume 2, (1971), p. 422. *See also:* S. J. Prokhovnic, *The Logic of Special Relativity*, Cambridge University Press, (1967); **and** *Light in Einstein's Universe: The Role of Energy in Cosmology and Relativity*, Dordrecht, Boston, D. Reidel Pub. Co., (1985). *See also:* K. Sapper, Editor, *Kritik und Fortbildung der Relativitätstheorie*, In Two Volumes, Akademische Druck- u. Verlagsanstalt, Graz, Austria, (1958/1962). *See also:* J. A. Winnie, "The Twin-Rod Thought Experiment," *American Journal of Physics*, Volume 40, (1972), pp. 1091-1094. M.F. Podlaha, "Length Contraction and Time Dilation in the Special Theory of Relativity—Real or Apparent Phenomena?", *Indian Journal of Theoretical Physics*, Volume 25, (1975), pp. 74-75. *See also:* M. Ruderfer, "Introduction to Ives' 'Derivation of the Lorentz Transformations'", *Speculations in Science and Technology*, Volume 2, (1979), p. 243. *See also:* D. Lorenz, "Über die Realität der FitzGerald-Lorentz Kontraction", *Zeitschrift für allgemeine Wissenschaftstheorie*, Volume 13/2, (1982), pp. 308-312. *See also:* D. Dieks, "The 'Reality' of the Lorentz Contraction," *Zeitschrift für allgemeine Wissenschaftstheorie*, Volume 115/2, (1984), p. 341. *See also:* F. Winterberg, *The Planck Aether Hypothesis*, Gauss Press, Reno, Nevada, (2002), pp. 141-148.

$$c_+ = \sqrt{c^2 - v^2 \sin^2 \psi} - v \cos \psi$$

$$c_- = \sqrt{c^2 - v^2 \sin^2 \psi} + v \cos \psi$$

Adding the equations (VII.14) one obtains

$$c\left(t_A^2 - t_A^1\right) = 2\gamma d \sqrt{1 - (v^2/c^2) \sin^2 \psi} \qquad \text{(VII.15)}$$

If an observer at rest with the clock wants to measure the distance from **A** to **B**, he can measure the time it takes a light signal to go from **A** to **B** and back to **A**. If he assumes that the velocity of light is constant and isotropic in all inertial reference systems, including the one he is in, moving together with **A** and **B** with the absolute velocity v, this distance is

$$d' = (c/2)\left(t_A^2 - t_A^1\right) \qquad \text{(VII.16)}$$

and because of (VII.15)

$$d' = \gamma d \sqrt{1 - (v^2/c^2) \sin^2 \psi} \qquad \text{(VII.17)}$$

Comparing this result with,

$$l' = l\sqrt{1 - (v^2/c^2) \cos^2 \varphi} = \frac{l}{\gamma \sqrt{1 - (v^2/c^2) \sin^2 \psi}}$$

one sees that he would obtain the same distance d', if he uses a contracted rod as a measuring stick, of Einstein's constant light velocity postulate. The velocity of light between **A** and **B** by using a rod to measure the distance and the time it takes a light signal in going from **A** to **B** and back to **A**, of course, will turn out to be equal to c, because according to (VII.16)

$$\frac{2d'}{t_A^2 - t_A^1} = c \qquad \text{(VII.18)}$$

Rather than using a reflected light signal to measure the distance d', the observer at **A** may try to measure the one-way velocity of light by first synchronizing the clock **B** with **A** and then measure the time for a light signal to go from **A** to **B**. However, since this synchronization procedure also uses reflected light signals, the result is the same. For the velocity he finds

$$\frac{d}{t_B - t_A^1} = \frac{d'}{(1/2)\left(t_A^1 + t_A^2\right) - t_A^1} = \frac{2d'}{t_A^2 - t_A^1} = c \qquad \text{(VII.19)}$$

By subtracting the equations (VII.14) one finds that

$$t_R = t_B + \left(\gamma/c^2\right) vd \cos \psi \qquad \text{(VII.20)}$$

which shows that from an absolute point of view the 'true' reflection time t_R at clock B is only then equal to t_B if $v = 0$. From an absolute point of view the propagation of light is isotropic only in the distinguished reference system, but anisotropic in a reference system in absolute motion against the distinguished reference system. This anisotropy remains hidden due to the impossibility to measure the one way velocity of light. This impossibility is expressed in the Lorentz transformations themselves, containing the scalar c^2 rather than the vector $\underline{c}$, through which an anisotropic light propagation would have to be expressed."[732]

The expected anisotropy from which the transformation evolved exhibits itself in the predictions the theory makes for an interferometer constructed and calibrated in an inertial reference system k_0 without rigid attachments, but instead assembled with rockets or automobiles at each of the relevant surfaces, which after being adjusted are then simultaneously and uniformly accelerated with respect to k_0 then allowed to travel in inertial motion in inertial reference system k_1, but which do not suffer a Lorentz contraction due to the lack of rigid attachments between them and the uniform manner in which they are accelerated. The special theory of relativity predicts a shift in the interference fringe pattern on the interferometer, which matches the exact result for which Michelson and Morley originally sought but did not find, and which prediction results from light speed anisotropy in at least one of the two inertial reference systems employed in the experiment.

Lajos Jánossy proved this argument,

"§7. Im vorigen Abschnitt haben wir gezeigt, wie man ein materialles Bezugssystem K_1 konstruieren kann, das eine vollkommene Galileische Transformation des Systems K_0 ist. Das System K_1 ist jedoch ein sehr unbequemes Bezugssystem. Wir finden nämlich, daß 1. das Licht sich in K_1 nicht isotrop ausbreitet, und 2. daß bewegte Uhren Phasenverschiebungen erleiden, auch wenn sie sehr langsam in K_1 bewegt werden; die Phasenverschiebung verschwindet auch im Grenzfall der verschwindenden Verschiebungsgeschwindigkeit nicht.

---

[732]. F. Winterberg, *The Planck Aether Hypothesis: An Attempt for a Finitistic Non-Archmedean Theory of Elementary Particles*, Carl Friedrich Gauss Academy of Science Press, Reno, Nevada, (2002), pp. 144-145.

Wir zeigen zunächst, daß diese erwähnte, unbequeme Eigenschaft in K_1 tatsächlich auftritt.

1. Daß Licht sich in K_0 isotrop ausbreitet, kann durch den Michelson-Morley-Versuch gezeigt werden. Betrachten wir nun ein Interferometer in K_0, das aus vier unzusammenhängenden Teilen besteht (s. Abb. 2 [*Figure deleted*]): Eine halbversilberte Platte P, zwei Spiegel M_1 and M_2 und ein Fernohr T. Wenn wir das System drehen, so daß die relativen Entfernungen von M_1, M_2 , P und T unverändert bleiben, dann wird auch das Streifensystem in T unverändert bleiben. Wenn wir nun die vier Teile des Systems unabhängig, aber gleichzeitig beschleunigen, dann bringen wir das Interferometer in des System K_1 . Diese Beschleunigung wird aber das Streifensystem, das man in T sieht, beeinflussen. Diese Beschleunigung würde in der Tat eine Streifenverschiebung hervorrufen, die in Lichtzeit ausgedrückt folgenden Wert besitzt.

$$\Delta T = l\left(\frac{4}{c} - \frac{2}{\sqrt{c^2 - v^2}} - \frac{1}{c-v} - \frac{1}{c+v} \right) = -\frac{lv^2}{c^2} + \cdots \qquad (13)$$

Der obige wert der Verschiebung ist nämlich genau der, den seinerzeit Michelson und Morley erwartet hatten, aber nicht fanden. Der Unterschied zwischen dem hier beschriebenen Experiment und dem wirklichen Michelson-Morley-Experiment ist nämlich der, daß das wirkliche Interferometer nicht aus unabhängigen Bestandteilen „zusammengesetzt" ist, sondern ein festes System bildete. Wenn die Teile unseres gedachten Interferometers durch materielle Stäbe verbunden wären, dann würden die einzelnen Teile nach Vollzug der Beschleunigung durch die in den Stäben auftretenden, elastischen Kräfte verschoben werden. Wenn wir also den elastischen Kräften freies Spiel gewähren würden, dann würden sie das Interferometer im Vergleich zum System K_1 in einer solchen Weise verzerren, daß die Verzerrung die Phasenverschiebung (13) genau kompensieren würde.

Um dies ganz klar zu machen, betrachten wir schematisch ein Interferometer, dessen vier Bestandteile auf vier Autos montiert sind. Setzen wir nun voraus, daß diese Autos gleichzeitig in der in §6 beschriebenen Weise losfahren. (Wir setzen voraus, daß die Autos so glatt fahren, daß die Interferenzstreifen während der Fahrt bestehen bleiben.) Das Interferometer, das auf diese Weise in Bewegung gesetzt worden ist, wird sicher eine Phasenverschiebung zeigen. Wir haben in §6/1 darauf hingewiesen, daß elastische Bänder, die zwischen Autos gespannt sind, in Spannung geraten, wenn die Autos sich in Bewegung setzen, weil nämlich diese Bänder sich zusammenzuziehen versuchen, aber daran verhindert werden durch die Autos. Wenn wir jetzt die Autos sich einander soweit nähern lassen, daß die elastische Spannung aufhört, dann verschieben wir damit die Spiegel genau in der richtigen Weise, um die nach der Beschleunigung aufgetretene Phasenverschiebung rückgängig zu machen. Zusammenfassend sehen wir, daß die Lichtfortpflanzung in K_1 nicht der isotrop erfolgt. Dieses Resultat setzt natürlich voraus, daß wir mit der Methode der Konstruktion von K_1, wie sie in §6 beschreiben wurde, einverstanden sind."[733]

Metaphysical four-dimensional expositions, which would obfuscate these facts with the obvious fiction of a false *ad hoc* fourth dimension, are not science and depend upon an imaginary dimension to perform the mutations of physical bodies which must have a physical basis if they in fact occur.

As Einstein, himself, avowed, "the real basis of the special relativity theory" is not the deduced conclusion of light speed invariance and the covariance of the laws of electrodynamics in Ludwig Lange's "inertial systems". As Albert Einstein later admitted, the real set of *a priori* postulates is the *ad hoc* "Lorentz Transformation", replete with its dreaded *ad hoc* hypotheses of length contraction and time dilatation. The Lorentz Transformation deduces all velocity comparisons, not just invariant light speed, which is a specific speed, and a derived unit,[734] not a general and fundamental geometry. Therefore, the Lorentz Transformation is more fundamental than light speed invariance and the principle of relativity.

In the modern metaphysical theory of special relativity first developed by Henri Poincaré through the use of his pseudo-Euclidean geometry, it is space-time which is fundamental, and which provides the basis to deduce the quadri-dimensionality of numerous non-physical quantities.[735] Space-time is not the principle of relativity, nor is it the principle of light speed invariance. Space-time is more fundamental than either and both are deducible from space-time. But it must be borne in mind that when speaking of space-time one is dealing in metaphysical quantities and qualities, not physical and measurable ones. In other words, one is pretending in lieu of a formulating a rational physical theory.

Later formulations of the special theory of relativity change the 1905 light postulate, from the Einsteins' constant speed of light exclusively in the "resting system", into the invariance of light speed in all of Lange's inertial systems. But this renders the principle of relativity redundant to, or deducible from, the light "postulate", and, therefore, not a "postulate", *per se*, because the light "postulate" then asserts the identity of Lange's inertial systems as light speed invariance, and the principle of relativity is already proven in the light "postulate". On the other hand, if we pretend that the principle of relativity is the

---

[733]. L. Jánossy, "Über die physikalische Interpretation der Lorentz-Transformation", *Annalen der Physik*, Series 6, Volume 11, (1953), pp. 293-322, at 306-307. *See also:* L. Jánossy, *Theory of Relativity Based on Physical Reality*, Akadémiai Kiadó, Budapest, (1971).
[734]. J. C. Maxwell, *A Treatise on Electricity and Magnetism*, Volume I, Third Edition, Clarendon, Oxford, (1892), pp. 5-6. W. S. Jevons, *The Principles of Science*, Second Revised Edition, Macmillan, London, New York,(1877), p. 321. E. Mach, *The Science of Mechanics*, Open Court, La Salle, Illinois, (1960), p. 367.
[735]. A. A. Logunov, *Henri Poincare i TEORIA OTNOSITELNOSTI*, Nauka, Moscow, (2004), pp. 45-49.

covariance of the laws of physics embracing Maxwell's theory of the æther, given the "Lorentz Transformation" as a premise, then the second "postulate" is already incorporated in the first "postulate".

If we are to assume that the Einsteins, in their 1905 paper, deduced, not induced, the Lorentz Transformation from invariant light speed; we would further have to fallaciously assume that empirically observed Lorentz Transformation metrics provoked the Einsteins to induce an unobserved invariant light speed and the unobserved symmetry of electrodynamic phenomena, as self-evident general truths induced *a posteriori* from empirically observed and reciprocally measured: length contraction, time dilatation, relative simultaneity and inertial relative motion between two systems devoid of any net force. Such is obviously not what happened, and such is not what is argued in the 1905 paper.

On the contrary, supposedly observed invariant light speed and the supposedly observed symmetry of electrodynamic phenomena led Voigt, FitzGerald and Larmor to scientifically induce, *a posteriori*, the general geometry of the (misnamed) "Lorentz Transformation", which general set of hypotheses supposedly deduced all "known" phenomena in non-existent hypothetical "inertial systems". The Einsteins pseudo-Metaphysics, their ontology of redundancy, simply disguised the more scientific, though likewise irrational, work of their predecessors, in a way which attempted to make it appear that the Einsteins had deduced that which must be induced, and had avoided hypotheses, which they had not avoided, but rather attempted to induce, through fallacies of *Petitio Principii*.

Most of the post-1905 statements of the special theory of relativity substitute a completely different proposition for the "two postulates". Einstein, himself, substituted one light theorem in 1907 for the "two postulates" of 1905:

> "the 'principle of the constancy of the velocity of light' [***] for a system of coordinates in a definite state of motion [as opposed to solely in the 'resting system' as in 1905.]"[736]

which presumes the Lorentz Transformation from which this supposed "postulate" is deduced, and which presumes the tacit hypotheses of an isotropic and homogenous absolute space[737] and "a definite state of motion" relative to

---

[736]. A. Einstein, "Über das Relativitätsprinzip und die aus demselben gezogenen Folgerung", *Jahrbuch der Radioaktivität und Elektronik*, Volume 4, (1907), pp. 411-462, at 416.

[737]. *See:* A. Pais, *Subtle is the Lord*, Oxford University Press, (1982), p. 142; where Pais refers to Einstein's so-called "Morgan manuscript" of 1921, which is reproduced in *The Collected Papers of Albert Einstein*, Volume 7, Document 50, Princeton University Press, (2002), pp. 372-378. **Einstein plagiarized this from:** N. R. Campbell, "The Common Sense of Relativity", *Philosophical Magazine*, Series 6, Volume 21, Number 124, (April, 1911), pp. 502-517, at 505. *See also:* R. D. Carmichael, "On the Theory of Relativity: Analysis of the Postulates", *The Physical Review*, First Series, Volume 35, (September, 1912), pp. 153-176; and "On the Theory of Relativity: Mass, Force and Energy", *The*

that absolute space. This new light "postulate" represents, therefore, not a postulate, but a deduction, a theorem, and a phenomenon.

Einstein admitted in 1907 that this "postulate" could not be *a priori*, but must instead be *a posteriori*:

> "That the supposition made here, which we want to call the 'principle of the constancy of the velocity of light', is actually met in Nature, is by no means self-evident, nevertheless, it is—at least for a system of coordinates in a definite state of motion—rendered probable through its verification, which Lorentz' theory based upon an absolutely resting aether has ascertained through experiment."[738]

The so-called "postulates" are simply a restatement of supposed experimental facts, and are not postulates, but empirical facts generalized as "laws" and "theorems". As Robert Daniel Carmichael stated:

> "The experiments which we have described (and others related to them) are fundamental in the theory of relativity. The postulates in the next chapter are based on them. These postulates are in the nature of generalizations of the facts established by experiment. [***] In the next chapter we shall begin the systematic development of the theory of relativity. It will be seen that its fundamental postulates, or laws, are based on the experiments of which we have given a brief account and on others related to them. [***] The postulates, as we shall see, are simply generalizations of experimental facts; and, unless an experiment can be devised to show that these generalizations are not legitimate, it is natural and in accordance with the usual procedure in science to accept them as 'laws of nature.'"[739]

There is an obnoxious pun in Carmichael's argument related to the use of the word "generalization". The generalization expressed is that: what happens in experiment *A* must happen in experiment *B*, given like conditions; and *not* that the like results of experiments *A* and *B* are general principles, *per se*. The "laws of nature" incorporate general principles to deduce the generalized experimental results, and there is an absolute distinction between the general principles and the generalization of experimental results, which the general principles must deduce. Carmichael blurs the distinction with a pun.

Hendrik Antoon Lorentz questioned Albert Einstein's "method" of pretending that induction is deduction:

Physical Review, Series 2, Volume 1, (February, 1913), pp. 161-197.
738. A. Einstein, "Über das Relativitätsprinzip und die aus demselben gezogenen Folgerung", *Jahrbuch der Radioaktivität und Elektronik*, Volume 4, (1907), pp. 411-462, at 416.
739. R. D. Carmichael, *The Theory of Relativity*, Mathematical Monographs No. 12, John Wiley & Sons, Inc., New York, Chapman & Hall, Limited, London, (1920), pp. 13-14.

"Einstein simply postulates what we have deduced, with some difficulty and not altogether satisfactorily, from the fundamental equations of the electromagnetic field. [***] I have not availed myself of his substitutions, only because the formulae are rather complicated and look somewhat artificial".[740]

We soon discover in the introduction of the Einsteins' 1905 paper a clear statement of the fallacious objective of their entire paper:

"These two assumptions are sufficient in order to arrive at a simple and consistent electrodynamics of moving bodies, taking as a basis Maxwell's theory for resting bodies."

Is Maxwell's theory for resting bodies a third postulate? One of the "two assumptions", the first "postulate", is that the laws electrodynamics of moving bodies be consistent among systems of reference in uniform translatory motion with respect to the "resting system". Of course, the reasoning presented is circular, first assuming via the first "postulate" that the laws of electrodynamics are consistent, then arguing that this mandated consistency, as a premise, deduces consistency as a conclusion. It is the first of many circular arguments found in the Einsteins' 1905 paper. How are we to determine that which constitutes an "inertial system", other than circularly, as in: An inertial system is one in which there is no net force acting on the system; *i. e.* there is no net force acting on a system, when it is in inertial motion?

Maxwell's theory for resting bodies is Maxwell's theory of the medium, a privileged frame, the æther. However, the Einsteins alleged that the aether was "superfluous" to their theory. The Einsteins irrationally wrote with the same pen that the æther was superfluous, while assuming it and its laws and properties as a basis for "their" theory.

In the introduction to the 1905 paper, we are being primed to venture forth from Maxwell's theorems for bodies resting in the æther, so that we can return to them, *Petitio Principii*, as the covariant laws of moving bodies, while being asked to pretend that the æther is superfluous, so that we aren't too shocked when simultaneity is claimed to be relative, again, *Petitio Principii*, via an impossible light signal clock synchronization operation which is itself based on the unproven assumption of light speed invariance, or $c' = c$, which premise of light speed invariance is also the conclusion of the theory. The unproven conclusion is redundant to the unproven premise. The Lorentz Transformations are then plagiarized as if from nowhere to save the day and provide the proof which otherwise does not exist, and which begins from the true postulates of length contraction, time dilatation, relative simultaneity, inertial motion, the æther, etc. For example, Albert Einstein stated in 1949:

---

740. H. A. Lorentz, *The Theory of Electrons*, Dover, New York, (1952), p. 230.

"[T]he following postulate is [***] sufficient for a solution [***] *L[ight]-principle holds for all inertial systems* (application of the special principle of relativity to the L[ight]-principle) [***] With the help of the Lorentz transformations the special principle of relativity can be expressed thus: The laws of nature are invariant with respect to Lorentz-transformations".[741]

Compare Albert Einstein's later statement to Willem de Sitter's statement of 1911:

"The principle of relativity can be enunciated as the postulate that the transformations, with respect to which the laws of nature shall be invariant, are 'Lorentz-transformations.'*"[742]

Einstein, ever the plagiarist, stated in 1952:

"The whole content of the special theory of relativity is included in the postulate: The laws of Nature are invariant with respect to the Lorentz transformations."[743]

Einstein disclosed his *modus operandi* for manipulating credit for the synthetic scientific theories of others, when he stated in 1936:

"There is no inductive method which could lead to the fundamental concepts of physics. Failure to understand this fact constituted the basic philosophical error of so many investigators of the nineteenth century. [***] Logical thinking is necessarily deductive; it is based upon hypothetical concepts and axioms. How can we expect to choose the latter so that we might hope for a confirmation of the consequences derived from them? The most satisfactory situation is evidently to be found in cases where the new fundamental hypotheses are suggested by the world of experience itself."[744]

This is a clear statement by Einstein that he would have science deduce a thing from itself, taking the world of experience as a hypothesis, only to deduce the world of experience as an effect, of itself. Albert Einstein avowed that,

"[A]ll knowledge of reality starts from experience and ends in it. [***]

---

741. A. Einstein, *The Theory of Relativity and other Essays*, Carol Publishing Group, (1996), pp. 6-8.
742. W. de Sitter, "On the Bearing of the Principle of Relativity on Gravitational Astronomy", *Monthly Notices of the Royal Astronomical Society*, Volume 71, (March, 1911), pp. 388-415, at 388-389.
743. A. Einstein, *Relativity, the Special and the General Theory*, Crown Publishing, Inc., New York, (1961), p. 148.
744. A. Einstein, *Ideas and Opinions*, Crown Publishers, Inc., New York, (1954), p. 307.

[E]xperience is the alpha and omega of all our knowledge of reality."[745]

Of course, Mileva and Albert were forced to present the real hypotheses, which they stuck in the middle of their arguments by way of induction, or an attempt at induction, which analyses they attempted to disguise as deductions from *a priori* principles, but which "*a priori* principles" were well-known summations of physical phenomena.

Einstein wanted people to believe that it is irrelevant that his predecessors induced the theories he later copied, because Einstein just invented them, *sua sponte*, irrationally, after he had read them, and therefore deserved credit for them. Einstein stated,

> "Invention is not the product of logical thought, even though the final product is tied to a logical structure."[746]

Einstein stated, together with Infeld:

> "Physical concepts are free creations of the human mind, and are not, however it may seem, uniquely determined by the external world."[747]

This was a philosophy they took over from Henri Poincaré.[748]

Certainly, the two "postulates" of the theory of relativity were not, "free creations of the human mind," but were, instead, summations of the empirical observations of the well-known phenomena of the day framed with the familiar concepts of the day. What Infeld and Einstein meant by "free" is difficult to fathom, and it is simply repetitive to say that creations of the mind are creations of the mind. Einstein's vague notions are perhaps the result of his plagiarizing Newton, Mach, Pearson, and others, on the principle of logical economy and watering down what they had written with Einstein's simplistic and naïve talk. If "free" is to mean unrestricted in any sense, no human mind is "free". We are limited in our concepts, experience, and scope, and we are socialized,

---

745. A. Einstein, "On the Method of Theoretical Physics", *Ideas and Opinions*, Crown, New York, (1954), p. 271.

746. A. Einstein, quoted in A. Pais, *Subtle is the Lord*, Oxford University Press, Oxford, New York, Toronto, Melbourne, (1982), p. 131.

747. A. Einstein and I. Infeld, *The Evolution of Physics*, Simon & Schuster, New York, London, Toronto, Sydney, Tokyo, Singapore, (1966), p. 31. Compare to the more lucid, prior statements of: W. K. Clifford, *The Common Sense of the Exact Sciences*, Dover, New York, (1955), pp. 193-194. E. Mach, "The Economy of Science", *The Science of Mechanics*, Open Court, LaSalle, Illinois, (1960), pp. 577-595. K. Pearson, *The Grammar of Science*, Second Revised and Enlarged Edition, Adam and Charles Black, London, (1900), pp. 30-37. H. Poincaré, *Dernières Pensées*, Flammarion, Paris, (1913); English translation, *Mathematics and Science: Last Essays*, Dover, New York, (1963), pp. 22-23. Einstein often plagiarized these works.

748. P. Frank, "Einstein's Philosophy of Science", *Reviews of Modern Physics*, Volume 21, Number 3, (July, 1949), 349-355.

indoctrinated and inculcated into certain beliefs.

Despite Einstein's assertions to the contrary, there is no mutual exclusion between being creative and being logical. A true scientist can create logical hypotheses through creative induction, even though Albert Einstein lacked the talent needed to do it for himself.

It is the Lorentz Transformation which is the product of creative inductive logic, with its hypotheses of length contraction, time dilatation and relative simultaneity, and which is the fundamental postulation of the special theory of relativity. Invariant light speed and the covariance of the laws of physics, were observed, not induced, and are deducible from the Lorentz Transformation, the laws of physics, and the definition of inertial motion, which are more fundamental in the special theory of relativity than invariant light speed. Speed must be composed of the more fundamental elements of distance and duration. Speed is a derived unit. Therefore, the synthesis of the special theory of relativity comes in deducing invariant light speed from the hypotheses of an isotropic and homogenous space, Maxwell's theory of the medium, the theory of inertial motion, and the hypotheses of length contraction, time dilation and relative simultaneity. This is precisely the conclusion Einstein was obliged to admit in 1935:

> "The special theory of relativity grew out of the Maxwell electromagnetic equations. So it came about that even in the derivation of the mechanical concepts and their relations the consideration of those of the electromagnetic field has played an essential role. The question as to the independence of those relations is a natural one because the Lorentz transformation, the real basis of the special relativity theory[...]"[749]

To argue, as the Einsteins did argue in 1905, that invariant light speed and the mandated identity of Lange's inertial systems deduces invariant light speed and the mandated identity of Lange's inertial systems, is to argue in fallacies of *Petitio Principii*, which is precisely what the Einsteins did do, in an attempt to hide their plagiarism of the induced hypotheses of Boscovich, Voigt, FitzGerald and Larmor.

9.5 The "Two Postulates"

The two postulates, are not in fact postulates, but are instead summations of well-known empirical facts; which are deducible from more fundamental principles, and even from each other. Henry August Rowland stated the two "postulates" on October 28th, 1899,

---

[749]. A. Einstein, "Elementary Derivation of the Equivalence of Mass and Energy", *Bulletin of the American Mathematical Society*, Series 2, Volume 41, (1935), pp. 223-230, at 223.

"And yet, however wonderful [the ether] may be, its laws are far more simple than those of matter. *Every wave in it, whatever its length or intensity, proceeds onwards in it according to well known laws, all with the same speed, unaltered in direction, from its source in electrified matter to the confines of the Universe, unimpaired in energy unless it is disturbed by the presence of matter.* However the waves may cross each other, each proceeds by itself without interference with the others. [***] *To detect something dependent on the relative motion of the ether and matter has been and is the great desire of physicists.* But we always find that, with one possible exception, there is always some compensating feature which renders our efforts useless. This one experiment is the aberration of light, but even here Stokes has shown that it may be explained in either of two ways: first, that the earth moves through the ether of space without disturbing it, and second, if it carries the ether with it by a kind of motion called irrotational. Even here, however, the amount of action probably depends upon relative motion of the luminous source to the recipient telescope. So the principle of Doppler depends also on this relative motion and is independent of the ether. The result of the experiments of Foucault on the passage of light through moving water can no longer be interpreted as due to the partial movement of the ether with the moving water, an inference due to imperfect theory alone. The experiment of Lodge, who attempted to set the ether in motion by a rapidly rotating disc, showed no such result. The experiment of Michelson to detect the ethereal wind, although carried to the extreme of accuracy, also failed to detect any relative motion of the matter and the ether [*Emphasis Added*]."[750]

9.5.1 The "Principle of Relativity"

Boscovich wrote of length contraction, time dilatation, relative simultaneity, and the "Principle of Invariance" resulting from these long ago in the 1700's.[751] Stallo, Streintz, Everett and Lange stressed the principle of relativity. The term "principle of relativity" was not original to the Einsteins. It was, in fact, a common term long before they entered the scene. It was found in German in:

---

[750]. H. A. Rowland, *The Physical Papers of Henry August Rowland*, The Johns Hopkins Press, Baltimore, Maryland, (1902), pp. 673-674.
[751]. R. J. Boscovich, *A Theory of Natural Philosophy*, M.I.T. Press, Cambridge, Massachusetts, London, (1966), pp. 197-205. **Confer:** H. V. Gill, *Roger Boscovich, S. J. (1711-1787) Forerunner of Modern Physical Theories*, M. H. Gill and Son, LTD., Dublin, (1941).

Lange,[752] Stallo,[753] Violle,[754] Poincaré,[755] and the German translation, with notes by Felix Hausdorff,[756] of Huyghens' Seventeenth Century seminal paper on relativity theory, "Über die Bewegung der Körper durch den Stoss / Über die Centrifugalkraft"; all before 1905. The term also appeared in many other languages, and was used by many other authors prior to 1905. Poincaré frequently iterated his electrodynamics-based "principle of relativity" long before the Einsteins repeated the same principle. Rowland had expressed it by 1900 and Maxwell in 1872.

Though it was an ancient notion, Galileo Galilei made the principle of relativity of mechanics famous,

> "When you have observed all these things carefully (though there is no doubt that when the ship is standing still everything must happen in this way), have the ship proceed with any speed you like, so long as the motion is uniform and not fluctuating this way and that. You will discover not the least change in all the effects named, nor could you tell from any of them whether the ship was moving or standing still."[757]

Boscovich argued in 1763 in the second supplement to his *Natural Philosophy,*

"§ II
Of Space & Time, as we know them
18. We have spoken, in the preceding Supplement, of Space & Time, as they are in themselves; it remains for us to say a few words on matters that pertain to them, in so far as they come within our knowledge. We can in no direct way obtain a knowledge through the senses of those real modes of existence, nor can we discern one of them from another. We do indeed perceive, by a difference of ideas excited in the mind by means of the senses, a determinate relation of distance & position, such as arises from any two local modes of existence; but the same idea may be produced by

---

[752]. L. Lange, "Das Inertialsystem vor dem Forum der Naturforschung: Kritisches und Antikritisches", *Philosophische Studien*, Volume 20, (1902), p. 18.

[753]. J. B. Stallo, *Die Begriffe und Theorieen der modernen Physik*, Johann Ambrosius Barth, Leipzig, (1901), pp. 205, 331.

[754]. J. Violle, *Lehrbuch der Physik*, Julius Springer, Berlin, (1892), p. 90; *cited in* J. Stachel, Ed., *The Collected Papers of Albert Einstein*, Volume 2, Princeton University Press, (1989), p. 255, Note 13.

[755]. H. Poincaré, *Wissenschaft und Hypothese*, B. G. Teubner, Leipzig, (1904), pp. 113-114, 119, *especially* 243, 340.

[756]. F. Hausdorff, translator, annotator, *Über die Bewegung der Körper durch den Stoss / Über die Centrifugalkraft*, Ostwald's Klassiker der exakten Wissenschaften, Nr. 138, Wilhelm Engelmann, Leipzig, (1903), pp. 64, 73.

[757]. Galileo Galilei, translated by S. Drake, *Dialogue Concerning the Two Chief World Systems—Ptolemaic & Copernican*, University of California Press, Berkeley, Los Angeles, London, (1967), p. 187.

innumerable pairs of modes or real points of position; these induce the relations of equal distances & like positions, both amongst themselves & with regard to our organs, & to the rest of the circumjacent bodies. For, two points of matter, which anywhere have a given distance & position induced by some two modes of existence, may somewhere else on account of two other modes of existence have a relation of equal distance & like position, for instance if the distances exist parallel to one another. If those points, we, & all the circumjacent bodies change their real positions, & yet do so in such a manner that all the distances remain equal & parallel to what they were at the start, we shall get exactly the same ideas. Nay, we shall get the same ideas, if, while the magnitudes of the distances remain the same, all their directions are turned through any the same angle, & thus make the same angles with one another as before. Even if all these distances were diminished, while the angles remained constant, & the ratio of the distances to one another also remained constant, but the forces did not change owing to that change of distance; then if the scale of forces is correctly altered, that is to say, that curved line, whose ordinates express the forces; then there would be no change in our ideas.

19. Hence it follows that, if the whole Universe within our sight were moved by a parallel motion in any direction, & at the same time rotated through any angle, we could never be aware of the motion or the rotation. Similarly, if the whole region containing the room in which we are, the plains & the hills, were simultaneously turned round by some approximately common motion of the Earth, we should not be aware of such a motion; for practically the same ideas would be excited in the mind. Moreover, it might be the case that the whole Universe within our sight should daily contract or expand, while the scale of forces contracted or expanded in the same ratio; if such a thing did happen, there would be no change of ideas in our mind, & so we should have no feeling that such a change was taking place.

20. When either objects external to us, or our organs change their modes of existence in such a way that that first equality or similitude does not remain constant, then indeed the ideas are altered, & there is a feeling of change; but the ideas are the same exactly, whether the external objects suffer the change, or our organs, or both of them unequally. In every case our ideas refer to the difference between the new state & the old, & not to the absolute change, which does not come within the scope of our senses. Thus, whether the stars move round the Earth, or the Earth & ourselves move in the opposite direction round them, the ideas are the same, & there is the same sensation. We can never perceive absolute changes; we can only perceive the difference from the former configuration that has arisen. Further, when there is nothing at hand to warn us as to the change of our organs, then indeed we shall count ourselves to have been unmoved, owing to a general prejudice for counting as nothing those things that are nothing in our mind; for we cannot know of this change, & we attribute the whole of the change to objects situated outside of ourselves. In such manner any one would be mistaken in thinking, when on board ship, that he himself was

motionless, while the shore, the hills & even the sea were in motion."[758]

Newton stated, in the fifth corollary to his *Principia*,

> "Corollary V.
> "*The motions of bodies included in a given space are the same among themselves, whether that space is at rest, or moves uniformly forwards in a right line without any circular motion.*
> For the differences of the motions tending towards the same parts, and the sums of those that tend towards contrary parts, are at first (by supposition) in both cases the same; and it is from those sums and differences that the collisions and impulses do arise with which the bodies mutually impinge one upon another. Wherefore (by Law 2.) the effects of those collisions will be equal in both cases; and therefore the mutual motions of the bodies among themselves in the one case will remain equal to the mutual motions of the bodies among themselves in the other. A clear proof of which we have from the experiment of a ship: where all motions happen after the same manner, whether the ship is at rest, or is carried uniformly forwards in a right line."[759]

J. D. Everett expressly stated the principle of relativity at least as early as 1883, in anticipation of Lange,

> "We cannot even assert that there is any such thing as absolute rest, or that there is any difference between absolute rest and uniform straight movement of translation."[760]

and, in 1895, Everett asserted the principle of relativity as a negative assertion,

> "[T]here is no test by which we can distinguish between absolute rest and uniform velocity of translation".[761]

As Joseph Larmor noted in 1898, and as G. H. Keswani and C. W. Kilmister clarified,[762] James Clerk Maxwell stated the principle of relativity of

---

[758]. R. J. Boscovich, *A Theory of Natural Philosophy*, M.I.T. Press, Cambridge, Massachusetts, London, (1966), p. 203.
[759]. I. Newton, *The Mathematical Principles of Natural Philosophy*, Volume 1, Benjamin Motte, London, (1729), p. 30.
[760]. J. D. Everett, *Elementary Treatise on Natural Philosophy by A. Privat Deschanel*, 6th Ed., D. Appelton and Company, New York, (1883), p. 43. ***See also:*** H. Streintz, *Die physikalischen Grundlagen der Mechanik*, B.G. Teubner, Leipzig, (1883).
[761]. J. D. Everett, "On Absolute and Relative Motion", *Report of the Sixty-Fifth Meeting of the British Association for the Advancement of Science*, Volume 65, (1895), p. 620.
[762]. J. Larmor, *Aether and Matter*, Cambridge University Press, (1900), p. 18. G. H. Keswani and C. W. Kilmister, "Intimations of Relativity: Relativity before Einstein",

electromagnetism in 1873 in his *Treatise on Electricity and Magnetism* §§ 600, 601,

> "On the Modification of the Equations of Electromotive Intensity when the Axes to which they are referred are moving in Space.
>
> 600.] Let x', y', z' be the coordinates of a point referred to a system of rectangular axes moving in space, and let x, y, z be the coordinates of the same point referred to fixed axes.
>
> Let the components of the velocity of the origin of the moving system be u, v, w and those of its angular velocity $\omega_1, \omega_2, \omega_3$ referred to the fixed system of axes, and let us choose the fixed axes so as to coincide at the given instant with the moving ones, then the only quantities which will be different for the two systems of axes will be those differentiated with respect to the time. If $\frac{\delta x}{\delta t}$ denotes a component velocity at a point moving in rigid connexion with the moving axes, and $\frac{dx}{dt}$ and $\frac{dx'}{dt}$ those of any moving point, having the same instantaneous position, referred to the fixed and the moving axes respectively, then
>
> $$\frac{dx}{dt} = \frac{\delta x}{\delta t} + \frac{dx'}{dt}, \tag{1}$$
>
> with similar equations for the other components.
>
> By the theory of the motion of a body of invariable form,
>
> $$\left. \begin{aligned} \frac{\delta x}{\delta t} &= u + \omega_2 z - \omega_3 y, \\ \frac{\delta y}{\delta t} &= v + \omega_3 x - \omega_1 z, \\ \frac{\delta z}{\delta t} &= w + \omega_1 y - \omega_2 x. \end{aligned} \right\} \tag{2}$$
>
> Since F is a component of a directed quantity parallel to x if $\frac{dF\prime}{dt}$ be the value of $\frac{dF}{dt}$ referred to the moving axes, it may be shewn that

---

British Journal for the Philosophy of Science, Volume 34, Number 4, (1983), pp. 343-354.

$$\frac{dF'}{dt} = \frac{dF}{dx}\frac{\delta x}{\delta t} + \frac{dF}{dy}\frac{\delta y}{\delta t} + \frac{dF}{dz}\frac{\delta z}{\delta t} + G\omega_3 - H\omega_2 + \frac{dF}{dt}.$$
(3)

Substituting for $\dfrac{dF}{dy}$ and $\dfrac{dF}{dz}$ their values as deduced from the equations (A) of magnetic induction, and remembering that, by (2),

$$\frac{d}{dx}\frac{\delta x}{\delta t} = 0, \quad \frac{d}{dx}\frac{\delta y}{\delta t} = \omega_3, \quad \frac{d}{dx}\frac{\delta z}{\delta t} = -\omega_2,$$
(4)

we find

$$\frac{dF'}{dt} = \frac{dF}{dx}\frac{\delta x}{\delta t} + F\frac{d}{dx}\frac{\delta x}{\delta t} + \frac{dG}{dx}\frac{\delta y}{\delta t} + G\frac{d}{dx}\frac{\delta y}{\delta t} + \frac{dH}{dx}\frac{\delta z}{\delta t} + H\frac{d}{dx}\frac{\delta z}{\delta t}$$
$$- c\frac{\delta y}{\delta t} + b\frac{\delta z}{\delta t} + \frac{dF}{dt}.$$
(5)

If we now put

$$-\Psi' = F\frac{\delta x}{\delta t} + G\frac{\delta y}{\delta t} + H\frac{\delta z}{\delta t},$$
(6)

$$\frac{dF'}{dt} = -\frac{d\Psi'}{dx} - c\frac{\delta y}{\delta t} + b\frac{\delta z}{\delta t} + \frac{dF}{dt}.$$
(7)

The equation for P the component of the electromotive intensity parallel to x is, by (B),

$$P = c\frac{dy}{dt} - b\frac{dz}{dt} - \frac{dF}{dt} - \frac{d\Psi}{dx},$$
(8)

referred to the fixed axes. Substituting the values of the quantities as referred to the moving axes, we have

$$P' = c\frac{dy'}{dt} - b\frac{dz'}{dt} - \frac{dF'}{dt} - \frac{d(\Psi + \Psi')}{dx},$$
(9)

for the value of P referred to the moving axes.

601.] It appears from this that the electromotive intensity is expressed by a formula of the same type, whether the motions of the conductors be referred to fixed axes or to axes moving in space, the only difference between the formulæ being that in the case of moving axes the electric potential Ψ must be changed into $\Psi + \Psi'$

In all cases in which a current is produced in a conducting circuit, the electromotive force is the line-integral

$$E = \int \left(P\frac{dx}{ds} + Q\frac{dy}{ds} + R\frac{dz}{ds} \right) ds, \qquad (10)$$

taken round the curve. The value of Ψ disappears from this integral, so that the introduction of Ψ' has no influence on its value. In all phenomena, therefore, relating to closed circuits and the currents in them, it is indifferent whether the axes to which we refer the system be at rest or in motion. See Art. 668."

Maxwell wrote in his *Matter and Motion*,

"18. ABSOLUTE SPACE

Absolute space is conceived as remaining always similar to itself and immovable. The arrangement of the parts of space can no more be altered than the order of the portions of time. To conceive them to move from their places is to conceive a place to move away from itself.

But as there is nothing to distinguish one portion of time from another except the different events which occur in them, so there is nothing to distinguish one part of space from another except its relation to the place of material bodies. We cannot describe the time of an event except by reference to some other event, or the place of a body except by reference to some other body. All our knowledge, both of time and place, is essentially relative. When a man has acquired the habit of putting words together, without troubling himself to form the thoughts which ought to correspond to them, it is easy for him to frame an antithesis between this relative knowledge and a so-called absolute knowledge, and to point out our ignorance of the absolute position of a point as an instance of the limitation of our faculties. Anyone, however, who will try to imagine the state of a mind conscious of knowing the absolute position of a point will ever after be content with our relative knowledge.

[***]

102. RELATIVITY OF DYNAMICAL KNOWLEDGE

Our whole progress up to this point may be described as a gradual development of the doctrine of relativity of all physical phenomena. Position we must evidently acknowledge to be relative, for we cannot describe the position of a body in any terms which do not express relation. The ordinary language about motion and rest does not so completely exclude the notion of their being measured absolutely, but the reason of this is, that in our ordinary language we tacitly assume that the earth is at rest.

As our ideas of space and motion become clearer, we come to see how the whole body of dynamical doctrine hangs together in one consistent system.

Our primitive notion may have been that to know absolutely where we are, and in what direction we are going, are essential elements of our knowledge as conscious beings.

But this notion, though undoubtedly held by many wise men in ancient times, has been gradually dispelled from the minds of students of physics.

There are no landmarks in space; one portion of space is exactly like every other portion, so that we cannot tell where we are. We are, as it were, on an unruffled sea, without stars, compass, soundings, wind, or tide, and we cannot tell in what direction we are going. We have no log which we can cast out to take a dead reckoning by; we may compute our rate of motion with respect to the neighbouring bodies, but we do not know how these bodies may be moving in space."

Poincaré stated the principle of relativity of electrodynamics in 1895,

"Experience reveals an abundance of facts, which can be summed up in the following formula: it is impossible to make manifest the absolute motion of matter, or, more correctly, the relative motion of ponderable matter with reference to the æther; the only thing which can be observed is the motion of ponderable matter with reference to ponderable matter."

"L'expérience a révélé une foule de faits qui peuvent se résumer dans la formule suivante: il est impossible de rendre manifeste le mouvement absolu de la matière, ou mieux le mouvement relatif de la matière pondérable par rapport à l'éther; tout ce qu'on peut mettre en évidence, c'est le mouvement de la matière pondérable par rapport à la matière pondérable."[763]

In 1899, Poincaré declared the principle of relativity to be rigorously valid,

"This strange property would appear to be a veritable '*fudging factor*' given by nature to prevent the detection of the absolute movement of the Earth by optical phenomena. I find that unsatisfactory, and I feel a duty to express my feelings: I look upon it as very probable that the optical phenomena depend only on the relative movements of the material source of light, related bodies or optical apparatus; and *then not only with the quantities close to the order of the square or the cube of aberration, but rigorously.* As the experiments become more exact, this principle will be checked with greater precision. [***] a well made theory should enable us to demonstrate the principle in one fell swoop in all its rigor."

---

[763]. H. Poincaré, *Œuvres de Henri Poincaré*, Volume 9, Gautier-Villars, Paris, (1954), p. 412; *reprinted from:* "A Propos de la Théorie de M. Larmor", *L'Éclairage électrique*, Volume 5, (October 5th, 1895) pp. 5-14.

"Cette étrange propriété semblerait un véritable «*coup de pouce*» donné par la nature pour éviter que le mouvement absolu de la terre puisse être révélé par les phénomènes optiques. Cela ne saurait me satisfaire et je crois devoir dire ici mon sentiment: je regarde comme très probable que les phénomènes optiques ne dépendent que des mouvements relatifs des corps matériels en présence, sources lumineuses ou appareils optiques et *cela non pas aux quantités près de l'ordre du carré ou du cube de l'aberration, mais rigoureusement*. A mesure que les expériences deviendront plus exactes, ce principe sera vérifie avec plus de précision. [***] une théorie bien faite devrait permette de démontrer le principe d'un seul coup dans toute sa rigueur."[764]

In 1900, Poincaré declared,

"I do not believe, in spite of Lorentz, that more exact observations will ever make evident anything else but the relative displacements of material bodies. [***] No; the same explanation must be found for the two cases, and everything tends to show that this explanation would serve equally well for the terms of the higher order, and that the mutual destruction of these terms will be rigorous and absolute."[765]

Poincaré reiterated the principle of relativity in 1902 in his book *La Science et l'Hypothèse*, E. Flammarion, Paris, (1902); and we know from Solovine's accounts[766] that Einstein had read Poincaré's book,

"*The Principle of Relative Motion.*—Sometimes endeavours have been made to connect the law of acceleration with a more general principle. The movement of any system whatever ought to obey the same laws, whether it is referred to fixed axes or to the movable axes which are implied in uniform motion in a straight line. This is the principle of relative motion; it is imposed upon us for two reasons: the commonest experiment confirms it; the consideration of the contrary hypothesis is singularly repugnant to the mind."[767]

---

[764]. H. Poincaré, *Electrité et Optique*, Gauthier-Villars, Paris, (1901), p. 536.
[765]. H. Poincaré, "RELATIONS ENTRE LA PHYSIQUE EXPÉRIMENTALE ET LA PHYSIQUE MATHÉMATIQUE", *RAPPORTS PRÉSENTÉS AU CONGRÈS INTERNATIONAL DE PHYSIQUE DE 1900*, Volume I, Gauthier-Villars, Paris, (1900), pp. 1-29; translated into German "Über die Beziehungen zwischen der experimentellen und mathematischen Physik", *Physikalische Zeitschrift*, Volume 2, (1900-1901), pp. 166-171, 182-186, 196-201; English translation in *Science and Hypothesis*, Chapters 9 and 10, Dopver, New York, (1952), p. 172.
[766]. J. Stachel, Editor, *The Collected Papers of Albert Einstein*, Volume 2, Princeton University Press, (1989), pp. xxiv-xxv.
[767]. H. Poincaré, *La Science et l'Hypothèse*, E. Flammarion, Paris, (1902); translated into English *Science and Hypothesis*, Dover, New York, (1952), at p. 111, which work also appears in *The Foundations of Science*.

Poincaré's 1904 principle of relativity states, and we know from Solovine's accounts[768] that Einstein had read this lecture, which was reprinted as Chapters 7 and 8 of Poincaré's book *La Valeur de la Science*, E. Flammarion, Paris, (1904),

> "The principle of relativity, according to which the laws of physical phenomena should be the same, whether for an observer fixed, or for an observer carried along in a uniform movement of translation; so that we have not and could not have any means of discerning whether or not we are carried along in such a motion."[769]

Poincaré stated, in 1905, before the Einsteins,

> "It appears at first sight that the aberration of light and other related optical phenomena would furnish us a means of determining the absolute motion of the earth, that is, its motion relative to ether rather than relative to the stars; there are no such phenomena. The experiments in which one takes account only of the first power of aberration have been unsuccessful, and one knows the reasons for that. But Michelson, having thought of an experiment in which one could measure effects depending on the second power of aberration, was equally unsuccessful. It appears that this impossibility of demonstrating the absolute motion of the earth is a general law of nature."[770]

In 1908, Poincaré reaffirmed the principle of relativity,

> "*The Principle of Relativity* [***] Whatever be the means employed there will never be disclosed anything but relative velocities; I mean the velocities of certain material bodies with reference to other material bodies. [***] We have seen above the reasons which impel us to regard the principle of

---

768. J. Stachel, Editor, *The Collected Papers of Albert Einstein*, Volume 2, Princeton University Press, (1989), pp. xxiv-xxv.

769. H. Poincaré's St. Louis lecture from September of 1904, *La Revue des Idées*, Volume 80, (15 November 1905); "L'État Actuel et l'Avenir de la Physique Mathématique", *Bulletin des Sciences Mathématique*, Series 2, Volume 28, (1904), pp. 302-324; **reprinted:** *La Valeur de la Science*, Chapters 7 and 8, E. Flammarion, Paris, (1904). **English translation:** "The Principles of Mathematical Physics", *The Monist*, Volume 15, Number 1, (January, 1905), pp. 1-24; **alternative English translation:** *The Value of Science*, Chapters 7 and 8, The Science Press, New York, (1907), pp. 91-105, at 94.

770. H. Poincaré, "Sur la Dynamique de l'Électron", *Comptes rendus hebdomadaires des séances de L'Académie des sciences*, Volume 140, (1905), pp. 1504-1508; reprinted *Œuvres de Henri Poincaré*, Volume 9, Gautier-Villars, Paris, (1954), pp. 489-493; English translation by G. H. Keswani and C. W. Kilmister, "Intimations of Relativity before Einstein", *The British Journal for the Philosophy of Science*, Volume 34, Number 4, (December, 1983), pp. 343-354, at 350.

relativity as a general law of nature."[771]

It was Lorentz, who properly phrased the *corollary of relativity* in 1904,

> "It would be more satisfactory, if it were possible to show, by means of certain fundamental assumptions, and without neglecting terms of one order of magnitude or another, that many electromagnetic actions are entirely independent of the motion of the system."

The Einsteins wrote, in 1905, without reference to previous authors,

> "Examples of a similar kind, as well as the failed attempts to find a motion of the earth relative to the 'light medium', lead to the supposition, that the concept of absolute rest corresponds to no characteristic properties of the phenomena not just in mechanics, but also in electrodynamics, on the contrary, for all systems of coordinates, for which the equations of mechanics are valid, the same electrodynamic and optical laws are also valid, as has already been proven for the magnitudes of the first order."

and,

> "The laws according to which the states of physical systems change do not depend upon to which of two systems of coordinates, in uniform translatory motion relative to each other, this change of state is referred."

9.5.2 The "Light Postulate"

The Einsteins asserted the "light postulate", in 1905, without reference to previous authors,

> "[L]ight in empty space always propagates with a determinate velocity c irrespective of the state of motion of the emitting body."

> "Every ray of light moves in the 'resting' system of coordinates with the determinate velocity c, irrespective of whether this ray of light is emitted from a resting or moving body. Such that
>
> velocity = (path of light) / (interval of time) ,
>
> where 'interval of time' is to be construed in the sense of the definition of § 1."

---

[771]. H. Poincaré, *Science and Method*, reprinted in *The Foundations of Science*, The Science Press, Lancaster, Pennsylvania, (1946), pp. 498-499, 505.

The references in Lorentz' and Poincaré's works to this velocity are too numerous to repeat. In the Einsteins' 1905 paper, this velocity is the absolute velocity of light in its medium, absolute space. Einstein stated in 1912,

"To fill this gap, I introduced the principle of the constancy of the velocity of light, which I borrowed from H. A. Lorentz's theory of the stationary luminiferous ether, and which, like the principle of relativity, contains a physical assumption that seemed to be justified only by the relevant experiments (experiments by Fizeau, Rowland, etc.)."[772]

We know that Einstein believed in absolute space, the "reference frame of the vacuum", the "resting system",

"Then I tried to discuss the Fizeau experiment on the assumption that the Lorentz equations for electrons should hold in the frame of reference of the moving body as well as in the frame of reference of the vacuum as originally discussed by Lorentz."[773]

---

[772]. A. Einstein, translated by A. Beck, "Relativity and Gravitation: Reply to a Comment by M. Abraham", *The Collected Papers of Albert Einstein*, Volume 4, Document 8, (1996), p. 131.

[773]. A. Einstein, *Physics Today*, **35**, 8, (August, 1982), p. 46. Note the reference is to the frame of the moving body, not the frame of a moving medium, and to electrons, not light. As usual, Einstein is nonspecific in his reference. Fizeau performed numerous experiments, some to measure light's speed on Earth: H. Fizeau, *Comptes rendus hebdomadaires des séances de L'Académie des sciences*, 29, (1849), p. 90; which could be compared to Bradley's interpolation of light speed in the reference frame of the vacuum, as Einstein did in 1952, "But the result of Fizeau's experiment and the phenomenon of aberration also guided me."—J. Stachel, Ed., *The Collected Papers of Albert Einstein*, Volume II, Princeton University Press, (1989), p. 262; to infer a Michelson-like test, as a positive or a negative result, depending, *inter alia*, on the accuracy attributed to the results; and Fizeau conducted some experiments to test Fresnel's coefficient of drag in moving media: H. Fizeau, *Comptes rendus hebdomadaires des séances de L'Académie des sciences*, 33, (1851), pp. 349-355; **and** *Annales de Chimie et de Physique*, Series 2, Volume 9, p. 57-66; and Einstein would have interpreted these as taking place with reference to the resting æther, "In Aarau ist mir eine gute Idee gekommen zur Untersuchung, welchen Einfluß die Relativbewegung der Körper gegen den Lichtäther auf die Fortpflanzungsgeschwindigkeit des Lichtes in durchsichtigen Körpern hat."—J. Stachel, Ed., *The Collected Papers of Albert Einstein*, Volume I, Princeton University Press, (1987), p. 230; had he ever considered them.

Though Lorentz averred that the Fizeau result compelled a resting æther, in 1895, it was Jakob Laub, in 1907, who first sought to arrive at the Lorentz Transformation by means of the Fizeau experiment of moving medium with respect to Fresnel's coefficient of drag, and it was Max von Laue who corrected Laub's formulation: J. Laub, *Annalen der Physik*, 23, (1907), pp. 738-744; **and**, 25, (1908), pp. 175-184. M. v. Laue, *Annalen der Physik*, 23, (1907), pp.989-990.

For an older interpretation, see also: P. Drude, *The Theory of Optics*, Longmans, Green and Co., London, New York, Toronto, (1902), pp. 457-482. ***See also:***

Lorentz pointed out in 1913,

"The latter is, by the way, up to a certain degree a quarrel over words: it makes no great difference, whether one speaks of the vacuum or of the æther."

"Letzteres ist übrigens bis zu einem gewissen Grade ein Streit über Worte: es macht keinen großen Untershied, ob man vom Vakuum oder vom Äther spricht."[774]

Lorentz, who knew the Einsteins' theory well, would not have alleged that it made no difference to speak of vacuum as opposed to æther, if Einstein had discounted absolute space, a "resting system" in which light propagates independently of the speed of the source. Both Sommerfeld and Pauli also recognized that the "resting system" of the Einsteins' 1905 paper was simply another appellation for Lorentz' æther, with absolute *celeritas* being an æther concept. Einstein described the light postulate as an æthereal idea to Peter A. Bucky.[775] Pauli stated, regarding *celeritas* in absolute space, that,

"There is no question of a *universal* constancy of the velocity of light in vacuo, if only because it has the constant value *c* only in Galilean systems of reference. On the other hand its independence of the state of motion of the light source obtains equally in the general theory of relativity. It proves to be the true essence of the old aether point of view."[776]

J. Larmor, *Aether and Matter*, Cambridge University Press, (1900), p. 177-179. *See also:* H. Poincaré, "La Théorie de Lorentz at le Principe de Réaction", *Archives Néerlandaises des Sciences Exactes et Naturelles*, Series 2, Volume 5, *Recueil de travaux offerts par les auteurs à H. A. Lorentz, professeur de physique à l'université de Leiden, à l'occasion du 25me anniversaire de son doctorate le 11 décembre 1900*, Nijhoff, The Hague, (1900), pp. 252-278; reprinted *Œuvres*, Volume IX, p. 488. *See also:* H. A. Lorentz, *Versuch einer Theorie der electrischen und optischen Erscheinungen in bewegten Körpern*, §68, E. J. Brill, Leiden, (1895), p. 101.
    Einstein made no mention of Fresnel's drag coefficient until after Laub published on the subject, and then he criticized Laub in Einstein's *Jahrbuch* review of 1907: A. Einstein, "Über das Relativitätsprinzip und die aus demselben gezogenen Folgerung", *Jahrbuch der Radioaktivität und Elektronik*, 4, (1907), p. 414. Einstein later failed to cite Laub or Laue, when he parroted their work in 1916: A. Einstein, *Relativity, the Special and the General Theory*, Crown, New York, (1961), pp. 38-41.
[774]. H. A. Lorentz, *Das Relativitätsprinzip; drei Vorlesungen gehalten in Teylers Stiftung zu Haarlem*, B. G. Teubner, Leipzig-Berlin, (1920). p. 23.
[775]. P. A. Bucky, Einstein, and A. G. Weakland, *The Private Albert Einstein*, Andrews and McMeel, Kansas City, (1992), p. 93.
[776]. W. Pauli, *Theory of Relativity*, Pergamon Press, London, Edinburgh, New York, Toronto, Sydney, Paris, Braunschweig, (1958), p. 5.

And Sommerfeld held it up as,

"The only valid remnant of the ether concept"[777]

We discover in "Part I" of the Einsteins' 1905 paper, that the "resting system" of the light postulate signifies absolute space, the "reference frame of the vacuum" a. k. a. the "æther", as Albrecht Fölsing has noted,

"To that end he proceeds from a 'system at rest,' the customary three-dimensional Euclidean space with Cartesian coordinates, in which the movement of a body is described by its coordinates as a function of time. This is so conventional that many readers must have asked themselves why it was even mentioned. [***] For the 'system at rest' for which these observations were initially made, it may be stated 'in accordance with experience'—i. e., in line with Maxwell-Lorentz theory—that the velocity of light in a vacuum is a universal constant. [***] To be sure, Einstein is using almost 'prerelativist' terminology by referring, throughout this section, to a system 'at rest' in which the rod, either at rest or in motion, is observed. While this formulation lets the background of Lorentzian theory—a motionless ether—shine through, it also leads to complications in which even an attentive reader can lose the thread."[778]

Philipp Frank makes clear that Einstein effectively adopted Lorentz' æther, and certainly adopted Lorentz' light postulate of the "resting system",

"This law [***] may be called the *relativity principle of mechanistic physics*. It is a deduction from the Newtonian laws of motion and deals only with relative motions and not, as Newton's laws proper, with absolute motion. In this form it is a positive assertion, but it can also be formulated in a negative way, thus: It is impossible by means of experiments such as those described above to differentiate one inertial system from another. [***] Besides this 'principle of relativity,' Einstein needed a second principle dealing with the interaction of light and motion. He investigated the influence of the motion of the source of light on the velocity of light emitted by it. From the standpoint of the ether theory, it is self-evident that it makes no difference whether or not the source of light moves; light considered as mechanical vibration in the ether is propagated with a constant velocity with respect to the ether. [***] Dropping the ether theory of light, Einstein had to reformulate this law into a statement about observable facts. There is one system of reference, F (the fundamental system), with respect to which light is propagated with a specific speed, c. No matter with what velocity the light source moves with respect to the fundamental system (F),

---

777. A. Sommerfeld, *Electrodynamics*, Academic Press, New York, (1952), p. 235.
778. A. Fölsing, *Albert Einstein, A Biography*, Viking, New York, (1997), pp. 184-185.

the light emitted is propagated with the same specific velocity (c) relative to F. This statement is usually called the 'principal [sic] of the of the constancy of the speed of light.'"[779]

Immanuel Kant and Carl Neumann reawoke an interest in the Newtonian concept of absolute space, and Hobbes had suggested that the æther far from major bodies is quiescent—a belief that held sway among many at least as late as Lorentz, Larmor and Volkmann. Thomas Young argued that the aether rests.[780] Neumann argued that absolute space is definable through a body, which is taken to be at absolute rest, the so-called "body Alpha". Fresnel[781] proposed that the æther only participates in the motion of bodies to a limited degree and rests outside of ponderable bodies. Many like Larmor, Lorentz, Volkmann, Maxwell, Heaviside, Hertz, Volterra and Drude believed that Young and Fresnel's resting æther signified Neumann's "body Alpha", an absolute space endowed with special properties, as opposed to an absolute space of true vacuum, and they used the same nomenclature of "resting system" and "moving system" which the Einsteins used without distinctions and to mean absolute space and motion relative to it.[782] Michelson set out to find the relative motion

---

[779]. P. Frank, *Einstein, His Life and Times*, Alfred A. Knopf, New York, (1967), pp. 32, 54-55. Note that Frank uses Budde's terminology of the 'Fundamental System' 'F '. Compare to Streintz' use of the term: H. Streintz, *Physikalische Grundlagen der Mechanik*, B. G. Teubner, Leipzig, (1883).
[780]. T. Young, *Philosophical Transactions of the Royal Society A*, Volume 94, (1804), p. 1; *Works*, Volume I, p. 188.
[781]. F. Fresnel, *Annales de Chimie et de Physique*, Series 2, Volume 9, pp. 57-66.
[782]. J. C. Maxwell, *Treatise on Electricity and Magnetism*, Multiple Editions, §§ 600, 601. H. R. Hertz, "Ueber die Grundgleichungen der Elektrodynamik für ruhende Körper", *Nachrichten von der Königlichen Gesellschaft der Wissenschaften und der Georg-Augusts-Universität zu Göttingen*, (1890), pp. 106-149; reprinted *Annalen der Physik und Chemie*, Volume 40, (1890), pp. 577-624; reprinted *Untersuchung über die Ausbreitung der Elektrischen Kraft*, Johann Ambrosius Barth, Leipzig, (1892), pp. 208-255; translated into English by D. E. Jones, as: "On the Fundamental Equations of Electromagnetics for Bodies at Rest", *Electric Waves*, Macmillan, London, (1894, 1900), pp. 195-239; and "Ueber die Grundgleichungen der Elektrodynamik für bewegte Körper", *Annalen der Physik und Chemie*, Volume 41, (1890), pp. 369-399; reprinted *Untersuchung über die Ausbreitung der Elektrischen Kraft*, Johann Ambrosius Barth, Leipzig, (1892), pp. 256-285; translated into English by D. E. Jones, as: "On the Fundamental Equations of Electromagnetics for Bodies in Motion", *Electric Waves*, Macmillan, London, (1894, 1900), pp. 241-268. P. Volkmann, *Einführung in das Studium der theoretischen Physik*, B. G. Teubner, Leipzig, (1900), pp. 53-54. J. Larmor, *Aether and Matter*, Cambridge University Press, (1900). p. 273. V. Volterra, "Sulle Funzioni Coniugate", *Atti della Reale Accademia dei Lincei. Rendiconti. Classe di scienze fisiche, mathematiche e naturali*, Volume 5, (1889), pp. 599-611. P. Drude, *The Theory of Optics*, Longmans, Green and Co., London, New York, Toronto, (1902), p. 457. C. Neumann, *Ueber die Principien der Galilei-Newton'schen Theorie*, B. G. Teubner, Leipzig, (1870); English translation, "The Principles of the Galilean-Newtonian Theory", *Science in Context*, Volume 6, (1993), pp. 355-368.

of the Earth in the supposedly still sea of æther, but wrecked on the static shores of his interferometer.

The Einsteins again and again refer to a "Resting System" with "resting" rods, clocks and observers and an empirically observed absolute speed of light and an absolute time in the "resting system"; and they asserted $c \pm v$ in the "moving system". The nomenclature of the day, which stemmed from Newton, Maxwell, Larmor and Lorentz, among many others, was clearly that the "resting system" was a system of coordinates at rest with respect to the fixed stars, and *not* any and all inertial systems. Einstein wrote to Mach on 25 June 1913, "relative to the fixed stars ('Restsystem')",[783] which confirms Frank's analysis of Einstein's thought process.

In 1911, Albert Einstein again confirmed that it was his essential belief that the "resting system" is Lorentz' æther at rest with respect to itself and with respect to the "fixed stars", as expressed ontologically as "absolute space",

"[W]e will extract from Lorentz's theory of the stationary luminiferous ether the following aspects most essential to us. What is the physical meaning of the statement that there exists a stationary luminiferous ether? The most important content of this hypothesis can be expressed as follows: There exists a reference system (called in Lorentz's theory 'a system at rest relative to the ether') with respect to which every light ray propagates in a vacuum with the universal velocity c. This ought to hold independently of whether the light-emitting body is in motion or at rest."[784]

The detection of an æther frame in no sense violates the principle of relativity unless the æther is defined to be at absolute rest—whatever that "absolute rest" should ultimately be interpreted to mean.

Max Abraham wrote in 1904,

"The electromagnetic theory addresses the absolute motion of light, which light issues forth in every direction with the same velocity (c)"

"Die elektromagnetische Theorie spricht von einer absoluten Bewegung des Lichtes, die nach jeder Richtung hin mit derselben Geschwindigkeit (c) erfolgt"[785]

The absolute velocity of light was stated numerous times in history, for

---

[783]. A. Einstein to E. Mach, translated by A. Beck, *The Collected Papers of Albert Einstein*, Volume 5, Document 448, Princeton University Press, (1995), p. 340.
[784]. A. Einstein, "Relativitäts-Theorie", *Vierteljahrsschrift der Naturforschenden Gesellschaft in Zürich*, Volume 56, (1911), pp. 1-14; quoted from the English translation by Anna Beck, *The Collected Papers of Albert Einstein*, Volume 3, Princeton University Press, (1993), pp. 343-345.
[785]. M. Abraham, "Zur Theorie der Strahlung und des Strahlungsdruckes", *Annalen der Physik*, Series 4, Volume 14, (1904), p. 238.

example, as an observed empirical result, by Cassini and Roemer (ca. 1676) and Bradley (ca. 1729).

Maxwell created his theorem of the velocity of light as a dynamic process in its medium. W. Stanley Jevons wrote in the 1870's,

> "In a first subclass we may place the velocity of light or heat undulations, the numbers expressing the relation between the lengths of undulations, and the rapidity of the undulations, these numbers depending only on the properties of the ethereal medium, and being probably the same in all parts of the universe."[786]

Willem de Sitter stated in his famous paper of 1911,

> "The principle of relativity can be enunciated as the postulate that the transformations, with respect to which the laws of nature shall be invariant, are 'Lorentz-transformations.'*"[787]

Einstein, ever the plagiarist, stated in 1952:

> "The whole content of the special theory of relativity is included in the postulate: The laws of Nature are invariant with respect to the Lorentz transformations."[788]

The Einsteins argued, in 1905, that the æther is "superfluous", without reference to prior authors,

> "The introduction of a 'luminiferous ether' will prove to be superfluous inasmuch as the view here to be developed will not require an 'absolutely stationary space' provided with special properties".

Johann Heinrich Ziegler gave widely-discussed lectures in Switzerland, in which he sought to abolish the concept of the æther. Ziegler directly accused Einstein of plagiarism. Ziegler wrote, in 1902,

> "Und doch ist diese Annahme nichts anderes als ein greifbarer Unsinn. Der den Raum oder die Stofflosigkeit überall erfüllende stofflose Stoff, genannt Weltäther, ist ein unbegreiflicher Begriff, und alle Lehren, welche auf ihm beruhen, sind genau ebenso unvollkommen und trügerisch, wie die Grundlage. Keine der Wellenbewegungen, die man jenem wesenlosen Ding

---

[786]. W. S. Jevons, *The Principles of Science*, 2nd Ed., Macmillan, London, (1877), p. 331.
[787]. W. de Sitter, "On the Bearing of the Principle of Relativity on Gravitational Astronomy", *Monthly Notices of the Royal Astronomical Society*, Volume 71, (March, 1911), pp. 388-415, at 388-389.
[788]. A. Einstein, *Relativity, the Special and the General Theory*, Crown Publishing, Inc., New York, (1961), p. 148.

andichtet, um die Fortpflanzung des Lichtes zu erklären, ist wirklich vorhanden. Es sind dies bloß mathematische Fiktionen, die ausschließlich in der Einbildung der Physiker vorhanden sind, gerade wie jener phantomhafte Stoff selbst, der bald dem bewegten Wasser, bald einem geschlagenen, gespannten Seil ähnliche Schwingungen ausführen soll."[789]

Lorentz stated in 1895,

"It does not suit my purpose to examine more thoroughly such speculations, or to express presumptions about the nature of the æther. I merely wish, as far as possible, to free myself of all preconceived notions regarding this substance and not to ascribe to it, for example, any of the qualities of ordinary liquids and gasses. Should it be shown, that a description of the phenomena is best arrived at through the assumption of absolute permeability, then one must surely in the meantime adopt this sort of hypothesis, and leave it to further research, if possible, to open up a deeper understanding to us.

It stands to reason, that there can be no question of the *absolute* rest of the æther; the phrase would not even have made sense. When I concisely state, the æther rests, it is only meant that one part of this medium does not displace the other, and that all perceptible motions of the heavenly bodies are relative motions in reference to the æther."

"Es liegt nicht in meiner Absicht, auf derartige Speculationen näher einzugehen oder Vermuthungen über die Natur des Aethers auszusprechen. Ich wünsche nur, mich von vorgefassten Meinungen über diesen Stoff möglichst frei zu halten und demselben z. B. keine von den Eigenschaften der gewöhnlichen Flüssigkeiten und Gase zuzuschreiben. Sollte es sich ergeben, dass eine Darstellung der Erscheinungen am besten unter der Voraussetzung absoluter Durchdringlichkeit gelänge, dann müsste man sich zu einer solchen Annahme einstweilen schon verstehen und es der weiteren Forschung überlassen, uns, womöglich, ein tieferes Verständniss zu erschliessen.

Dass von *absoluter* Ruhe des Aethers nicht die Rede sein kann, versteht sich wohl von selbst; der Ausdruck würde sogar nicht einmal Sinn haben. Wenn ich der Kürze wegen sage, der Aether ruhe, so ist damit nur gemeint, dass sich der eine Theil dieses Mediums nicht gegen den anderen verschiebe und dass alle wahrnehmbaren Bewegungen der Himmelskörper relative Bewegungen in Bezug auf den Aether seien."[790]

---

[789]. J. H. Ziegler, *Die universelle Weltformel und ihre Bedeutung für die wahre Erkenntnis aller Dinge*, 1 Vortrag, Kommissionsverlag Art. Institut Orell Füßli, Zürich, (1902), p. 9.
[790]. H. A. Lorentz, *Collected Papers*, Volume 5, Martinus Nijhoff, (1937), pp. 3-4; reprint of *Versuch einer Theorie der Electrischen und optischen Erscheinungen in bewegten Körpern*, E. J. Brill, Leiden, (1895); unaltered reprint by B. G. Teubner,

Joseph Larmor wrote, in 1900,

"At the same time all that is known (or perhaps need be known) of the aether itself may be formulated as a scheme of differential equations defining the properties of a *continuum* in space, which it would be gratuitous to further explain by any complication of structure; though we can with great advantage employ our stock of ordinary dynamical concepts in describing the succession of different states thereby defined."[791]

In 1900, Paul Drude stated,

"*The velocity of light in space* [***] independent of what is understood by a light vector. [***] The conception of an ether absolutely at rest is the most simple and the most natural,—at least if the ether is conceived to be not a substance but merely space endowed with certain physical properties."[792]

Poincaré asserted in 1900,

"Does our ether actually exist? We know the origin of our belief in the ether. If light takes several years to reach us from a distant star, it is no longer on the star, nor is it on the earth. It must be somewhere, and supported, so to speak, by some material agency.
   The same idea may be expressed in a more mathematical and more abstract form."[793]

Maxwell stated,

"These are some of the already discovered properties of that which has often been called vacuum, or nothing at all. They enable us to resolve several kinds of action at a distance into actions between contiguous parts of a contiguous substance. Whether this resolution is of the nature of explication or complication, I must leave to the metaphysicians."[794]

Poincaré also asserted in 1889 that,

Leipzig, (1906).
791. J. Larmor, *Aether and Matter*, Cambridge University Press, (1900), p. 78.
792. P. Drude, *The Theory of Optics*, Longmans, Green and Co., London, New York, Toronto, (1902), pp. 261, 457.
793. H. Poincaré, *Science and Hypothesis*, Dover, New York, (1952). p. 169; quoting, "RELATIONS ENTRE LA PHYSIQUE EXPÉRIMENTALE ET LA PHYSIQUE MATHÉMATIQUE", *RAPPORTS PRÉSENTÉS AU CONGRÈS INTERNATIONAL DE PHYSIQUE DE 1900*, Gauthier-Villars, Paris, (1900), Volume I, p. 21.
794. J. C. Maxwell, "Ether", *Scientific Papers*, Dover, (1952), p. ???

"Whether the ether exists or not matters little—let us leave that to the metaphysicians; what is essential for us is, that everything happens as if it existed, and that this hypothesis is found to be suitable for the explanation of phenomena. After all, have we any other reason for believing in the existence of material objects? That, too, is only a convenient hypothesis; only, it will never cease to be so, while some day, no doubt, the ether will be thrown aside as useless."[795]

Poincaré likened the æther to "Shinola",

"What is meant by the ether? In France or in Germany, it is little more than a system of differential equations; provided that these equations are internally consistent and account for the observed facts, one won't worry if the picture which they suggest is more or less strange or unprecedented. On the other hand, W. Thomson immediately tries to carve out the figure of a familiar substance which has a greater likeness to the æther, it appears that it is *scotch shoe wax*, which is to say, a very tough species of shoemaker's wax."

"Que dire de l'éther? En France ou en Allemagne, ce n'est guère qu'un système d'équations différentielles; pourvu que ces équations n'impliquent pas contradiction et rendent compte des faits observés, on ne s'inquiétera pas si l'image qu'elles suggèrent est plus ou moins étrange ou insolite. W. Thomson, au contraire, cherche tout de suite quelle est la matière connue qui ressemble le plus à l'éther; il paraît que c'est le *scotch shoe wax*, c'est-à-dire une espèce de poix très dure."[796]

Poincaré stated,

"[If the ether] is able to explain everything, this is because it does not enable us to foresee anything; it does not enable us to decide between the different possible hypotheses, since it explains everything beforehand. It therefore becomes useless."[797]

In 1901, Cohn averred,

"Like Maxwell and Hertz we address a chemically and physically homogenous medium as an entity, which is also completely characterized at all points electromagnetically by the same value of some constants. This

---

[795]. H. Poincaré, *Sceince and Hypothesis*, Dover, New York, (1952). pp. 211-212.
[796]. H. Poincaré, *Savants et Écrivains*, Flammarion, Paris, (1910), p. 235.
[797]. H. Poincaré, "The Value of Science", *Popular Science Monthly*, Volume 70, Number 4, (1907), p. 348; as quoted in J. E. Boodin, "Energy and Reality II. The Definition of Energy", *The Journal of Philosophy, Psychology and Scientific Methods*, Volume 5, Number 15, (16 July 1908), pp. 393-406, at 404-405.

type of medium fills each element of our space; it is perhaps a certain ponderable substance, or it may also be the vacuum. In light of this, we will avoid continuing to speak of an 'æther'."

"Wie Maxwell und Hertz behandeln wir ein chemisch und physikalisch homogenes Medium als ein Gebilde, welches auch elektromagnetisch in allen Punkten durch die gleichen Werte einiger Constanten vollständig charakterisiert ist. *Ein solches Medium erfüllt jedes Element unseres Raumes; es kann eine bestimmte ponderable Substanz oder auch das Vacuum sein. Daneben noch von einem „Aether" zu sprechen, werden wir vermeiden.*"[798]

Faraday argued, in April of 1846,

"The point intended to be set forth for consideration of the hearers was, whether it was not possible that the vibrations which in a certain theory are assumed to account for radiation and radiant phænomena may not occur in the lines of force which connect particles, and consequently masses of matter together; a notion which as far as it is admitted, will dispense with the æther, which, in another view, is supposed to be the medium in which these vibrations take place.

You are aware of the speculation[2] which I some time since uttered respecting that view of the nature of matter which considers its ultimate atoms as centres of force, and not as so many little bodies surrounded by forces, the bodies being considered in the abstract as independent of the forces and capable of existing without them. In the latter view, these little particles have a definite form and a certain limited size; in the former view such is not the case, for that which represents size may be considered as extending to any distance to which the lines of force of the particle extend: the particle indeed is supposed to exist only by these forces, and where they are it is. The consideration of matter under this view gradually led me to look at the lines of force as being perhaps the seat of the vibrations of radiant phænomena.

[***]

The view which I am so bold as to put forth considers, therefore, radiation as a high species of vibration in the lines of force which are known to connect particles and also masses of matter together. It endeavours to dismiss the æther, but not the vibration. The kind of vibration which, I believe, can alone account for the wonderful, varied, and beautiful phænomena of polarization, is not the same as that which occurs on the

---

798. Emil Cohn, *Nachrichten von der Königlichen Gesellschaft der Wissenschaften zu Göttingen. Mathematisch-physikalische Klasse*, (1901), p. 74; *Annalen der Physik*, 7, (1902), p. 30.

surface of disturbed water, or the waves of sound in gases or liquids, for the vibrations in these cases are direct, or to and from the centre of action, whereas the former are lateral. It seems to me, that the resultant of two or more lines of force is in an apt condition for that action which may be considered as equivalent to a *lateral* vibration; whereas a uniform medium, like the æther, does not appear apt, or more apt than air or water.

The occurrence of a change at one end of a line of force easily suggests a consequent change at the other. The propagation of light, and therefore probably of all radiant action, occupies *time*; and, that a vibration of the line of force should account for the phænomena of radiation, it is necessary that such vibration should occupy time also. I am not aware whether there are any data by which it has been, or could be ascertained whether such a power as gravitation acts without occupying time, or whether lines of force being already in existence, such a lateral disturbance of them at one end as I have suggested above, would require time, or must of necessity be felt instantly at the other end.

As to that condition of the lines of force which represents the assumed high elasticity of the æther, it cannot in this respect be deficient: the question here seems rather to be, whether the lines are sluggish enough in their action to render them equivalent to the æther in respect of the time known experimentally to be occupied in the transmission of radiant force.

The æther is assumed as pervading all bodies as well as space: in the view now set forth, it is the forces of the atomic centres which pervade (and make) all bodies, and also penetrate all space. As regards space, the difference is, that the æther presents successive parts or centres of action, and the present supposition only lines of action; as regards matter, the difference is, that the æther lies between the particles and so carries on the vibrations, whilst as respects the supposition, it is by the lines of force between the centres of the particles that the vibration is continued."[799]

Faraday's ideas were very influential. William Kingdon Clifford speculated in the year of his death and of Einstein's birth, 1879, that light may be naught but flickering "space",

"In order to explain the phenomena of light, it is not necessary to assume anything more than a periodical oscillation between two states at any given point of space."[800]

---

[799]. M. Faraday, "Thoughts on Ray-vibrations", *Philosophical Magazine*, Series 3, Volume 28, Number 188, (May, 1846), pp. 345-350; reprinted in *Experimental Researches in Electricity*, Three Volumes Bound as Two, Volume 3, Dover, New York, (1965), pp. 447-452.

[800]. W. K. Clifford, *Lectures and Essays*, Volume I, Macmillan, London, (1879), p. 85; See also: *The Common Sense of the Exact Sciences*, Edited by K. Pearson, D. Appleton, New York, Macmillan, London, (1885), *especially* Chapter 4, "Position", Section 1, "All Position is Relative", pp. 134-135; **and** Section 19, "On the Bending of Space", p. 201,

Karl Pearson noted, as second editor and annotator of Clifford's *The Common Sense of the Exact Sciences* in 1884-1885,

> "The most notable physical quantities which vary with position and time are heat, light, and electro-magnetism. It is these that we ought peculiarly to consider when seeking for any physical changes, which may be due to changes in the curvature of space. If we suppose the boundary of any arbitrary figure in space to be distorted by the variation of space-curvature, there would, by analogy from one and two dimensions, be no change in the volume of the figure arising from such distortion. Further, if we *assume* as an axiom that space resists curvature with a resistance proportional to the change, we find that waves of 'space-displacement' are precisely similar to those of the elastic medium which we suppose to propagate light and heat. We also find that 'space-twist' is a quantity exactly corresponding to magnetic induction, and satisfying relations similar to those which hold for the magnetic field. It is a question whether physicists might not find it simpler to assume that space is capable of a varying curvature, and of a resistance to that variation, than to suppose the existence of a subtle medium pervading an invariable homaloidal space."[801]

In 1934, Einstein repeated Clifford's idea without an attribution, which idea appeared before Lorentz' theory appeared,

> "Then came H. A. Lorentz's great discovery. All the phenomena of electromagnetism then known could be explained on the basis of two assumptions: that the ether is firmly fixed in space—that is to say, unable to move at all, and that electricity is firmly lodged in the mobile elementary particles. Today his discoveries may be expressed as follows: physical space and the ether are only different terms for the same thing; fields are physical states of space."[802]

Note 1.
801. K. Pearson in W. K. Clifford, *Common Sense of the Exact Sciences*, Dover, New York, (1955), p. 203.
802. A. Einstein, *Ideas and Opinions*, Crown Publishers, Inc., (1954), p. 281.

OTHER TITLES

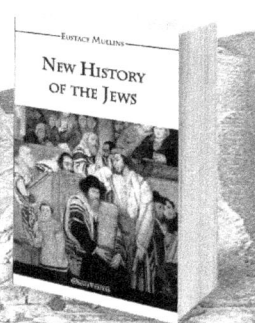

 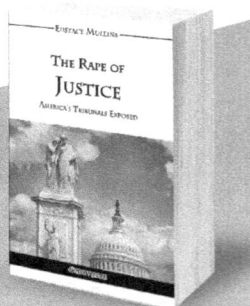

The Priority Myth

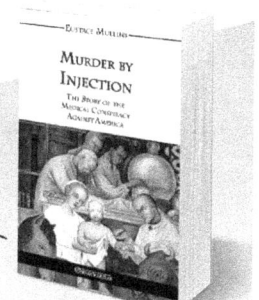

Omnia Veritas Ltd presents:

MURDER BY INJECTION

by

EUSTACE MULLINS

THE STORY OF THE MEDICAL CONSPIRACY AGAINST AMERICA

The cynicism and malice of these conspirators is something beyond the imagination of most Americans.

Omnia Veritas Ltd presents:

EZRA POUND
THIS DIFFICULT INDIVIDUAL

by

EUSTACE MULLINS

Ezra's interest in money as a phenomenon, in contrast to the usual attitude toward money as something to get, is a legitimate one.

An illustration for his own monetary theories...

OMNIA VERITAS LTD PRESENTS:

MY LIFE IN CHRIST

BY

EUSTACE MULLINS

Christ did not wish to be followed by robots and sleepwalkers, He desired man to awaken, and to attain the full use of his earthly powers.

THIS is the story of my life in Christ

The Manufacture and Sale of Saint Einstein

The Priority Myth

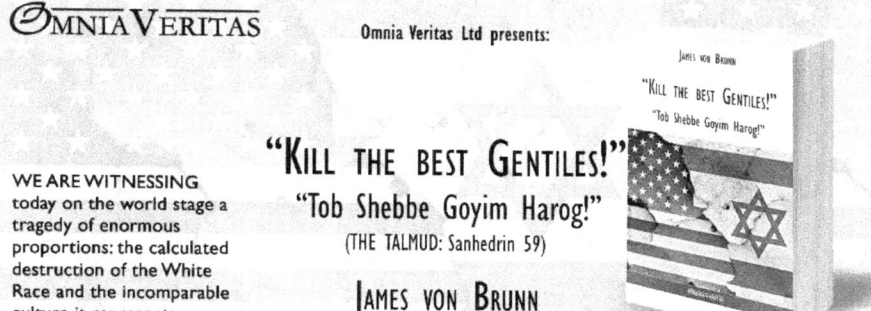

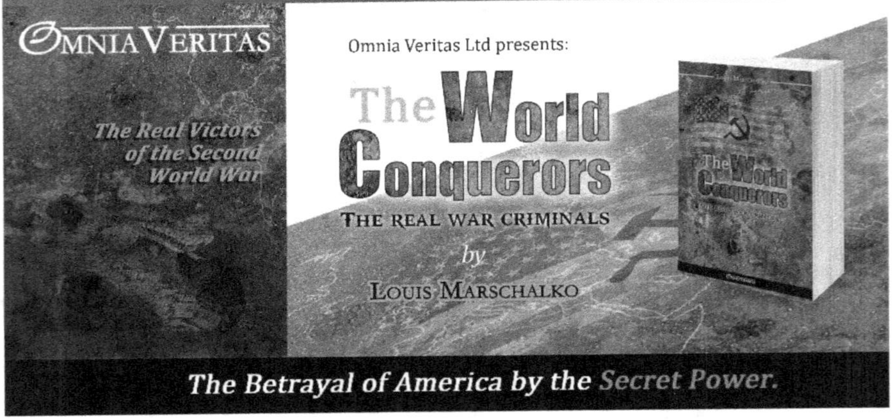

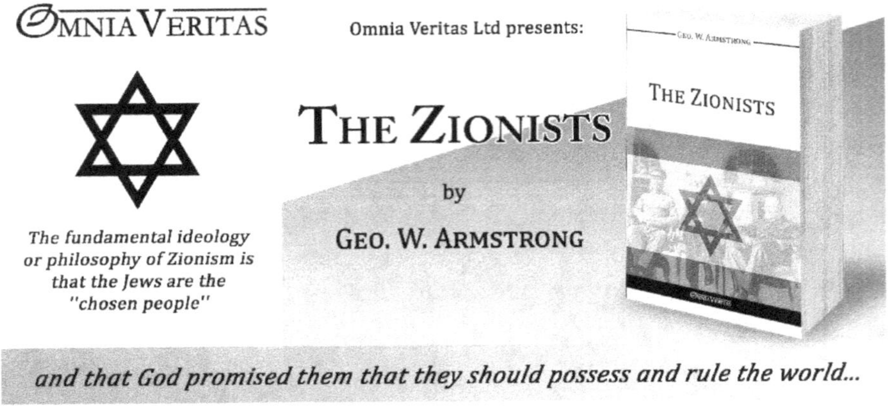

www.omnia-veritas.com

https://www.instagram.com/omnia.veritas/

https://twitter.com/OmniaVeritasLtd

www.ingramcontent.com/pod-product-compliance
Lightning Source LLC
Chambersburg PA
CBHW071359230426
43669CB00010B/1389